Review of Medical Embryology

Text and Illustrations by

BEN PANSKY, Ph.D., M.D.
Professor of Anatomy,
Medical College of Ohio at Toledo,
Toledo, Ohio

Review of Medical Embryology

Macmillan Publishing Co., Inc.
New York

Collier Macmillan Canada, Inc.
Toronto

Collier Macmillan Publishers
London

Copyright © 1982, Ben Pansky

Printed in the United States of America

All rights reserved. No part of this book may be reproduced or transmitted in any form or by any means, electronic or mechanical, including photocopying, recording, or any information storage and retrieval system, without permission in writing from the Publisher.

Macmillan Publishing Co., Inc.
866 Third Avenue, New York, New York 10022

Collier Macmillan Canada, Inc.

Collier Macmillan Publishers · London

Library of Congress Cataloging in Publication Data

Pansky, Ben.
   Review of medical embryology.

   Includes bibliographical references and index.
   1. Embryology, Human. I. Title. [DNLM: 1. Embryology. QS 604 P196r]
QM601.P17   1982   612′.64   82-12748
ISBN 0-02-390620-0

Printing: 1 2 3 4 5 6 7 8   Year: 2 3 4 5 6 7 8 9 0

This book is dedicated to my mother and father, **Leah** and **Abraham,** who supplied the initial gametes that made me human, and who nurtured the embryo, fetus, and offspring into manhood; to my wife, **Julie,** who helped the man mature and grow, and who kept the spark of life alive with her love and understanding; and to my son, **Jon,** who carries the genes, old and new, into the next generation.

Preface

Review of Medical Embryology has been written for medical students, physicians, and others who require a visual comprehension and understanding of the fundamentals of human development. The material is presented in a simplified, concise, and outline form, which emphasizes applications to medicine and biology, rather than theoretical aspects of the subject. For example, the very early stages of embryo formation (embryogenesis), the mechanisms of uterine implantation, and the correlated changes of the mother's reproductive organs are important in obstetrics, gynecology, and pediatrics. The advanced stages in the formation of organs (organogenesis) are helpful in understanding the "body plan" and find their correlation in anatomy, medicine, and even surgery, where repair mechanisms may serve to alter embryonic and fetal malformations. And knowledge about histogenesis of major organs provides a basis for understanding and dealing with organ architecture, as seen in microscopic anatomy and pathology.

Throughout the text, every effort has been made to present and illustrate developmental processes as a sequence of dynamic events. Sections interpreting functional significance of relations are interspersed with descriptions of developmental morphology, in order to highlight correlations and to enhance learning and understanding.

In a subject such as embryology, in which the interrelations of a number of growing parts are of critical significance, well-planned illustrations can provide a three-dimensional visualization of what is taking place. Thus, with this fact in mind, the book has been written around its pictures rather than having its pictures planned around the text. Over 1000 original line-cut illustrations, most of which appear on right-hand pages, give a clear, concise view of what is taking place amid the intricacies of human development. To facilitate the rapid location of each of the major units within the text, gray tabs and black-stripe tabs are printed in the margins of left-hand pages.

Appendixes at the end of the book, which summarize some of the more general concepts of embryology, include sections on correlated human development, germ layer derivatives, critical periods of human development, teratogens known to cause human malformations, physiologic development of the central nervous system, and prematurity. In addition, a list of general references appears at the end of the book.

The author is indebted to the Medical College of Ohio for providing him with the intellectual atmosphere and facility that permitted the book to be completed; to Dr. Liberato J. A. DiDio, Professor and Chairman of Anatomy, for his encouragement and support; and to Ms. Joan C. Zulch, Editor-in-Chief, Medical Books Department, Macmillan Publishing Co., Inc., for her enthusiasm, guidance, advice, and encouragement, without which the book could not have been prepared.

B. P.

Contents

PART I. EMBRYOGENESIS

1. Terms of Description   2–3
2. The Male Reproductive System   4–5
   Microscopy of the Male Testis (Figure 1)   6
   Electron Micrographs of Sperm (Figure 2)   7
3. Gamete (Germ Cell) Formation, or Gametogenesis: Spermatogenesis   8–9
   Mitosis and Meiosis (Figure 3)   10
   Human Spermatozoa and Spermiogenesis (Figure 4)   11
   Electron Micrographs of Spermatocytes and Spermatogonium (Figure 5)   12
   Electron Micrographs of Spermatids, Sertoli Cells, and Leydig Cells (Figure 6)   13
4. Gametogenesis: Oogenesis   14–15
   Micrographs of Pronuclear Stage, First Cleavage Division, and Two-Cell Embryo (Figure 7)   16
   Micrographs of Four-Cell Embryo, Morula, and Blastocyst Stages (Figure 8)   17
5. Anomalies (Abnormalities) of Gametogenesis   18–19
6. The Adult Female Uterus   20–21
7. The Adult Uterine Tubes (Oviduct, Fallopian Tube, or Salpinx)   22–23
8. Reproductive Cycles: The Ovarian Cycle and Ovulation   24–25
9. Reproductive Cycles: The Ovarian Cycle and the Corpus Luteum   26–27
10. Reproductive Cycles: The Menstrual (Uterine) Cycle   28–29
11. Germ Cell Viability and Movement and Abnormal Implantation Sites   30–31
12. Fertilization   32–33
13. Implantation and Its Preparation: General Concepts   34–35
14. Week 1 of Embryonic Development: Ovulation to Implantation   36–37
15. Week 2 of Development: Bilaminar Germ Disk Embryo   38–39
16. Week 2 of Development: Days 10 to 14   40–41
17. Review of Week 2 and Abnormal Development   42–43
18. Week 3 of Development: Trilaminar Germ Disk Embryo Formation and Gastrulation   44–45
19. Week 3 of Development: The Notochord, Neural Tube, and Allantois   46–47
    Transverse, Horizontal, and Sagittal Sections of Early Embryogenesis (Figure 9)   48
    Neural Tube Development (Figure 10)   49
20. Week 3 of Development: Intraembryonic Mesoderm, Somite Development, and the Intraembryonic Coelom   50–51
    Development of the Yolk Sac and Coelomic Cavities (Figure 11)   52
    Somite Differentiation (Figure 12)   53

21. Week 3 of Development: Cardiovascular System Development   54–55
22. Week 3 of Development: Trophoblast and Villus Development   56–57
23. Weeks 4 to 6 of Development: The Embryonic Period   58–59
24. Embryonic Period: Weeks 7 and 8 and External Embryo Appearance   60–61
25. Germ Layers and Their Derivatives   62–63
26. Embryonic Folding and Flexion of the Embryo   64–65
    Embryonic Folding and Flexion: Transverse and Sagittal Sections (Figure 13)   66
    Sections of Flexed Embryo Caudal and Cephalic to the Umbilicus (Figure 14)   67
27. General Mechanisms of Normal Development   68–69
28. The Fetal Period: Weeks 9 to 20 of Development   70–71
29. The Fetal Period: Weeks 21 to Term   72–73
30. Multiple Pregnancies   74–75
    Monozygotic, Genetically Identical Twin Pregnancies (Figure 15)   76
    Conjoined Twins (Figure 16)   77
31. Congenital Malformation and Their Causes   78–79
    Turner's and Cri-du-chat Syndrome Karyotypes (Figure 17)   80
    Down's and Klinefelter's Syndrome Karyotypes (Figure 18)   81
32. Congenital Malformations and Their Causes: Human Malformations   82–83
33. The Fetal Membranes   84–85
34. The Amnion, Allantois, and Yolk Sac   86–87
35. The Umbilical Cord   88–89
36. Uterine Growth During Pregnancy and Parturition   90–91
37. Fetal-Maternal Incompatibility   92–93
38. The Placenta: General Discussion   94–95
39. The Placental Villi   96–97
40. The Placenta: Decidual Formation   98–99
41. Placental Physiology   100–101
42. Placental Circulation   102–103
43. Hormonal Balance and Tests for Pregnancy   104–105

PART II. ORGANOGENESIS

Unit One. BODY CAVITIES AND MESENTERIES

44. Body Cavities: Coelomic Divisions   110–111
    The Mesenteries in Cross-Section (Figure 19)   112
    Continuity of Pericardial Portion of Coelom with Paired Coelomic Chamber of Midbody Area (Figure 20)   113
45. Body Cavities: Pleuropericardial and Pleuroperitoneal Membranes   114–115
46. Body Cavities: The Lesser Peritoneal Sac (Omental Bursa) and Dorsal Mesogastrium   116–117

47. Development of the Mesenteries 118–119
    Transverse Sections of Mesentery Development (Figure 21) 120
    Cross-Sections of Mesentery Development (Figure 22) 121
48. Development of the Diaphragm 122–123
    Development of the Diaphragm (Figure 23) 124
    Congenital Diaphragmatic Defects and Normal Diaphragmatic Parts and Relations (Figure 24) 125

Unit Two. THE BRANCHIAL APPARATUS: The Face, Pharynx, and Related Branchial Derivatives

49. The Branchial Apparatus: The Branchial (Pharyngeal) Arches 128–129
    The Branchial (Pharyngeal) Arches and Associated Nerves (Figure 25) 130
    The Oral Cavity and Pharynx (Figure 26) 131
50. The Pharyngeal Clefts and Pouches 132–133
51. Malformations Related to the Branchial Apparatus 134–135
52. Branchial Arch Derivatives: The Thyroid Gland 136–137
53. The Branchial Apparatus: The Floor of the Pharynx—Tongue and Associated Structures 138–139
54. The Face 140–141
55. Development of the Palate 142–143
56. Congenital Malformations of the Lip and Palate 144–145

Unit Three. THE RESPIRATORY SYSTEM: Nasal Cavities, Larynx, Trachea, Bronchi, and Lungs

57. Development of the Nasal Cavities 148–149
58. Development of the Lower Respiratory System: Larynx and Trachea 150–151
59. Development of the Lower Respiratory System: The Bronchi and Surrounding Structures 152–153
60. Development of the Lower Respiratory System: The Lungs and Terminal Respiratory Tubes 154–155
61. Development of the Lower Respiratory System: Surfactant and Respiratory Movements 156–157
62. Malformations of the Lower Respiratory Tract 158–159

Unit Four. THE MUSCULAR SYSTEM

63. Development of the Muscular System 162–163
    General Body Muscle Groups (Figure 27) 164
    Branchiomeric and Adjacent Myotomic Muscles (Figure 28) 165

Unit Five. THE SKELETAL AND ARTICULAR SYSTEMS

64. Development of the Skeletal and Articular Systems: Cartilage and Bone Histogenesis   168–169
    Endochondral Bone Growth (Figure 29)   170
    Transformation of Cancellous to Compact Bone (Figure 30)   171
65. Bone Histogenesis: Secondary Ossification Centers and Joint Development   172–173
66. Development of the Axial Skeleton   174–175
    Vertebral Development (Figure 31)   176
    Development of the Sternum and Ribs (Figure 32)   177
67. Appendicular Skeleton and Skull Development   178–179
    Viscerocranial Cartilage Derivatives and Epiphyses and Diaphyses of the Extremities (Figure 33)   180
    Skull Development (Figure 34)   181
68. Congenital Malformations of the Skeletal System   182–183
69. Development of the Limbs   184–185
70. Dermatone and Cutaneous Innervation of the Limbs   186–187
71. Malformations of the Appendicular Skeleton (the Limbs)   188–189

Unit Six. THE INTEGUMENTARY SYSTEM: The Skin, Cutaneous Appendages, and Teeth

72. Development of the Integumentary System: Ectodermal Derivatives   192–193
73. Congenital Malformations of the Integumentary System   194–195
74. Development of the Hair and Associated Structures   196–197
75. Development of the Nails   198–199
76. Development of the Mammary Glands   200–201
77. Development of the Teeth   202–203
78. Tooth Eruption and Malformations of the Teeth   204–205

Unit Seven. THE DIGESTIVE SYSTEM

79. The Digestive System: General Introduction   208–209
80. The Foregut: Esophagus and Stomach   210–211
81. The Foregut: The Omental Bursa and Duodenum   212–213
82. The Foregut: The Liver and Biliary Apparatus (Gallbladder and Ducts)   214–215
83. The Foregut: The Pancreas and Spleen   216–217
84. Development of the Midgut: General Introduction   218–219
85. The Midgut: Fixation, the Cecum and Appendix   220–221
86. Development of the Hindgut   222–223

87. Congenital Malformations of the Digestive System: Foregut Malformations   224–225
    Congenital Anomalies and Malformations of the Biliary Apparatus (Figure 35)   226
    Annular Pancreas and Heterotopic Pancreatic Tissue (Figure 36)   227
88. Congenital Malformations of the Digestive System: Midgut Malformations   228–229
    Anomalies of the Small Intestine and Colon (Figure 37)   230
    Atresia, Stenosis, and Meckel's Diverticulum (Figure 38)   231
89. Midgut Malformations   232–233
90. Congenital Malformations of the Digestive System: Hindgut Malformations   234–235

Unit Eight. THE URINARY SYSTEM

91. The Urinary or Excretory System: Intermediate Plate, Nephrogenic Cord, and Pronephros   238–239
92. The Urinary or Excretory System: The Mesonephros   240–241
    Developing Mesonephros (Figure 39)   242
    Relationship of Mesonephros to Surrounding Developing Structures (Figure 40)   243
93. The Urinary or Excretory System: The Metanephros   244–245
94. The Urinary or Excretory System: The Definitive Kidney   246–247
95. The Urinary or Excretory System: The Urinary Bladder and Urethra   248–249
96. Malformations of the Urinary System   250–251
97. Malformations of the Urinary System   252–253

Unit Nine. THE GENITAL OR REPRODUCTIVE SYSTEM

98. The Genital or Reproductive System: The Primitive Genital System   256–257
99. Development of the Testis   258–259
100. The Genital or Reproductive System: Primitive Genital Tracts and Sex Discrimination   260–261
    Stages in Development of the Definitive Male Genitourinary System (Figure 41)   262
    Differentiation and Development of the Male Excretory System (Figure 42)   263
101. Differentiation of the Male Genital Tracts and Auxiliary Glands   264–265
102. Development of the Male External Genital Organs   266–267
103. Inguinal Canal Development and Testicular Migration   268–269
104. Development of the Female Genital System: Ovarian Differentiation   270–271

105. Differentiation of the Female Genital Tracts: Uterus, Vagina, Auxiliary Glands, Mesenteries   272–273
   Development of Uterus and Vagina (Figure 43)   274
   Migration of Ovaries and Ligaments of the Uterus, Ovaries, and Uterine Tubes (Figure 44)   275
106. Development of the Upper and Lower Portions of the Female Genital Tract   276–277
107. Development of the Female External Genital Organs   278–279
108. Sexual Anomalies of Genetic and Hormonal Origin   280–281
109. Genital Malformations in the Male   282–283
   Hydrocele, Inguinal Hernia, and Cryptorchid Testes (Figure 45)   284
   Anomalies of the Male and Female Reproductive Tracts (Figure 46)   285
110. Uterovaginal Malformations of the Female   286–287
111. Adult Derivatives of Embryonic Urogenital Structures   288
   Differential Diagnosis of Patients with Ambiguous External Genitalia   289

Unit Ten. THE CIRCULATORY SYSTEM: CARDIOVASCULAR AND LYMPHATIC SYSTEMS

112. Hematopoiesis and General Development of the Circulatory System   292–293
113. Cardiovascular Circulatory and Lymphatic Systems: Early Development   294–295
114. Development of the Heart: Cardiac Tube Development   296–297
   Longitudinal Sections: Changes in Position of the Heart and Pericardial Cavity Due to Head Folding (Figure 47)   298
   Coronal Sections: Changes in Position of the Heart and Pericardial Cavity Due to Head Folding (Figure 48)   299
115. Development of the Heart: Formation of the Heart Loop   300–301
116. Pericardial Cavity Development and Primitive Heart Circulation   302–303
117. Atrioventricular and Interatrial Septation and Development   304–305
   Partitioning of the Primitive Atrium (Figure 49)   306
   Formation of the Septum in the Atrioventricular Canal (Figure 50)   307
118. Development of the Sinus Venosus and Associated Veins   308–309
119. The Right and Left Atrial Walls and the Venous Valves   310–311
120. Septation of Ventricles, Truncus Arteriosus, and Conus Cordis   312–313
121. The Cardiac Valves and Conducting System   314–315
   Conduction System and Fibrous Skeleton of the Heart (Figure 51)   316
   Development of the Atrioventricular Valves, Chordae Tendineae, and Papillary Muscles (Figure 52)   317
122. The Primitive Circulatory Network   318–319
123. Development of the Arterial System   320–321
   Development of the Arteries of the Upper Extremity (Figure 53)   322
   Development of the Arteries of the Lower Extremity (Figure 54)   323
124. The Aortic Arches   324–325
125. Development of the Venous System: Primitive Venous Network and Superior Vena Cava   326–327

126. Development of the Venous System: The Inferior Vena Cava 328–329
127. Development of the Venous System: The Portal System and Pulmonary Veins 330–331
128. Development of the Lymphatic System 332–333
    Development of a Lymph Node (Figure 55) 334
    Development of the Tonsils, the Thymus, and the Spleen (Figure 56) 335
129. The Circulatory System Before and After Birth 336–337
130. Adult Derivatives of Fetal Structures 338–339
131. Malformations of the Cardiovascular System 340–341
    Aortic Arch Abnormalities (Figure 57) 342
    Abnormal Development of the Superior and Inferior Venae Cavae and Pulmonary Venous Drainage (Figure 58) 343
132. Malformations of the Heart and Great Vessels 344–345
    Interatrial Septal Abnormalities (Figure 59) 346
    Interatrial Septal Abnormalities (Figure 60) 347
133. Malformations of the Heart and Great Vessels 348–349
    Failure of Separation of Aortic and Pulmonary Trunks (Figure 61) 350
    Transposition of the Great Vessels and Pulmonary Stenosis (Figure 62) 351
134. Complex Cardiac Malformations 352–353
    Valvular Atresia and Stenosis (Figure 63) 354
    Abnormalities in the Development of the Lymphatic System (Figure 64) 355

Unit Eleven. THE NERVOUS SYSTEM

135. Early Development of the Nervous System 358–359
136. Early Nervous System Development: The Neural Tube and Neural Crest 360–361
137. General Development of the Central Nervous System 362–363
    Primary Brain Vesicles and Neural Tube Flexures: Days 23–30 (Figure 65) 364
    Major Components of the Developing Brain: Weeks 5–7 (Figure 66) 365
138. Phylogenesis of the Nervous System 366–367
    Annelid and Worm Nervous Systems (Phyla Annelida and Chordata) (Figure 67) 368
    Comparative Anatomy of the Chordate Brain (Figure 68) 369
139. Metameric Organization of the Nervous System 370–371
    Metameric Organization of the Central Nervous System (Figure 69) 372
    Dermatomes and Cutaneous Nerve Distribution in the Adult (Figure 70) 373
140. General Considerations Related to the Anatomy of the Spinal Cord 374–375
141. The Spinal Cord: Normal Development 376–377
142. The Spinal Cord: Differentiation of Nerve and Glial Cells 378–379
143. The Spinal Cord: Neural Crest Cells and Myelination 380–381

144. Spinal Cord Length and Spinal Meninges   382–383
145. Malformations of the Spinal Cord   384–385
146. Introduction to Brainstem Development   386–387
147. The Brainstem: Myelencephalon (Fifth Vesicle)—Basal Motor Plate   388–389
    Brainstem Nuclei and Nerves (Dorsal View) (Figure 71)   390
    Basal and Alar Plate Development of Myelencephalon and Transverse Sections of Medulla at Various Levels (Figure 72)   391
148. The Brainstem: Myelencephalon (Fifth Vesicle)—Alar Sensory and Roof Plates   392–393
149. The Brainstem: Metencephalon (Fourth Vesicle)   394–395
150. The Brainstem: Metencephalon (Fourth Vesicle)—The Cerebellum   396–397
    Cerebellar Development and Histogenesis (Figure 73)   398
    Adult Cerebellum—Components and Terminology of Description (Figure 74)   399
151. The Brainstem: Mesencephalon (Third Vesicle)   400–401
152. The Peripheral Nervous System and Cranial Nerves   402–403
    Cranial Nerves (4- to 5-Week Embryo) (Figure 75)   404
    Cranial Nerves (Fetal and Adult) (Figure 76)   405
153. The Diencephalon (Second Vesicle)   406–407
154. The Telencephalon (First Vesicle): Phylogenesis   408–409
155. The Brain: Telencephalon (First Vesicle)   410–411
156. The Brain: Telencephalon (First Vesicle)—Lobes and Pallial Development   412–413
157. The Telencephalon (First Vesicle): Development of the Rhinencephalon   414–415
158. Histogenesis of the Cerebral Cortex   416–417
159. Commissures of the Telencephalon   418–419
160. The Coverings and Vascularization of the Brain   420–421
161. Malformations of the Brain   422–423
162. Malformations of the Brain   424–425
163. Malformations of the Brain: Hydrocephalus   426–427
164. The Autonomic Nervous System: The Sympathetic System   428–429
165. The Autonomic Nervous System: The Parasympathetic System   430–431
    The Sympathetic Nervous System and Urinary Retention Mechanism (Figure 77)   432
    The Parasympathetic Nervous System and Urinary Expulsion Mechanism (Figure 78)   433
166. The Olfactory System   434–435
167. The Eye: Optic Cup and Lens Vesicle, Retina, Iris, and Ciliary Body   436–437
    Anatomy of the Mature Eye (Figure 79)   438
    Retinal Histogenesis (Developmental and Adult) (Figure 80)   439
168. The Eye: Lens, Choroid, Sclera, Cornea, and Optic Nerve   440–441
169. Congenital Malformations of the Eye   442–443
170. The Vestibulocochlear System: The External Ear and the Eardrum (Tympanic Membrane)   444–445

171. The Vestibulocochlear System: The Internal Ear—Membranous Labyrinth   446–447
172. The Vestibulocochlear System: The Internal Ear—Bony Labyrinth   448–449
173. The Vestibulocochlear System: Histogenesis of the Internal Ear   450–451
174. The Vestibulocochlear System: The Middle Ear   452–453
175. Congenital Malformations of the Vestibulocochlear System   454–455
176. The Hypophysis (Pituitary Gland): Glandular Primordium   456–457
177. The Hypophysis (Pituitary Gland): Neural Primordium and Portal System   458–459
178. Role of the Hypophysis (Pituitary Gland): Physiology and Pathology   460–461
179. The Paraganglionic System: The Paraganglia   462–463
180. Development of the Adrenal (Suprarenal) Gland   464–465
181. Pathology Associated with the Adrenal Gland   466–467

APPENDIXES

I. Correlated Human Development   471
II. Germ Layer Derivatives   478
III. Critical Periods of Human Development (Sensitivity to Teratogens)   480
IV. Teratogens Known to Cause Human Malformations   481
V. Physiologic Development of the Central Nervous System   483
VI. Prematurity   486
VII. References   487

Index   491

PART I
Embryogenesis

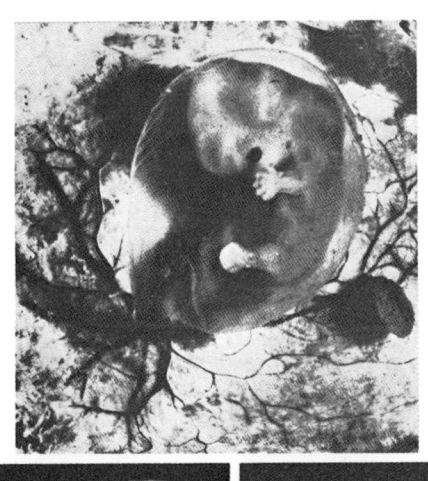
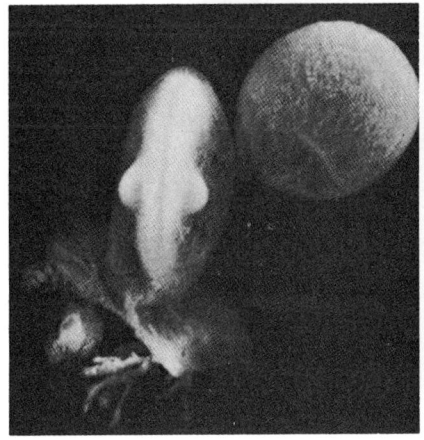
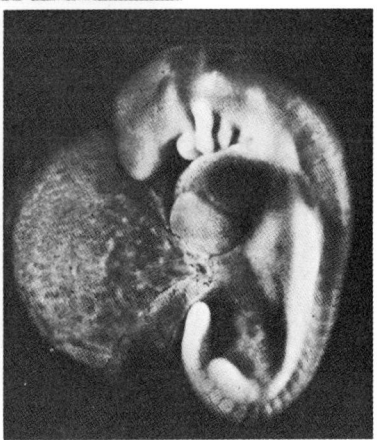

1. TERMS OF DESCRIPTION

I. Special features of the embryo
  A. THE ADULT: it is assumed that the body is erect, with the arms by the sides and the palms directed forward—*the anatomic position*. The terms *anterior* or *ventral* and *posterior* or *dorsal* describe the front or back of the body or limbs, as well as the relations of structures inside the body to one another. *Superior* and *inferior* indicate the relative levels of different structures
  B. THE EMBRYO is curved (or flexed), therefore a reference position is not as easily defined as in the adult. In the embryo, the terms dorsal and ventral are nearly always used, and *cranial* (*cephalic*) and *caudal* commonly denote relationships to the head and tail ends, respectively
   1. The term *rostral* indicates the relationships of structures to the nose
   2. The terms *proximal* or *distal* are described as distances from the source of attachment of a structure; e.g., in the upper limb, the elbow is proximal to the wrist and the wrist is distal to the elbow

II. Planes of section: the classic planes of space seen in 3 dimensions in the embryo and fetus are the same as described in the adult, only the terminology is special, with the subject described as being in the upright position facing the observer. However, due to embryonic flexion, the reciprocal relationships of the frontal and transverse planes are modified; thus, a section which may be frontal at the level of the head can be transverse at the level of the trunk
  A. THE MEDIAN PLANE is a vertical plane passing through the center of the body. Median sections divide the body into right and left halves. *Lateral* and *medial* refer to structures that are, respectively, farther from or nearer to the median plane of the body
  B. THE SAGITTAL PLANE is any vertical plane perpendicular to the forehead and parallel with the long axis of the nose and the median plane that divides the subject into right and left parts. "Parasagittal" and "median sagittal" are redundant terms
  C. THE TRANSVERSE OR HORIZONTAL PLANE is any plane at right angles to both the median and frontal planes. It is parallel with an imaginary line joining the eyes and divides the subject into superior and inferior or top and bottom parts
  D. A FRONTAL OR CORONAL PLANE is any vertical plane that intersects the median plane at right angles. It is parallel with the forehead and divides the subject into front (anterior or ventral) and back (posterior or dorsal) parts

III. Embryonic sections usually used
  A. A MEDIAN (MIDSAGITTAL) SECTION is cut through the median plane. Longitudinal sections parallel to the median plane, but not through it, are called sagittal sections
  B. A FRONTAL OR CORONAL SECTION is a vertical section through the frontal or coronal plane
  C. TRANSVERSE OR HORIZONTAL SECTIONS are sections through the transverse plane; they also are called *cross-sections*
  D. OBLIQUE SECTIONS are neither perpendicular nor horizontal but are inclined or slanted

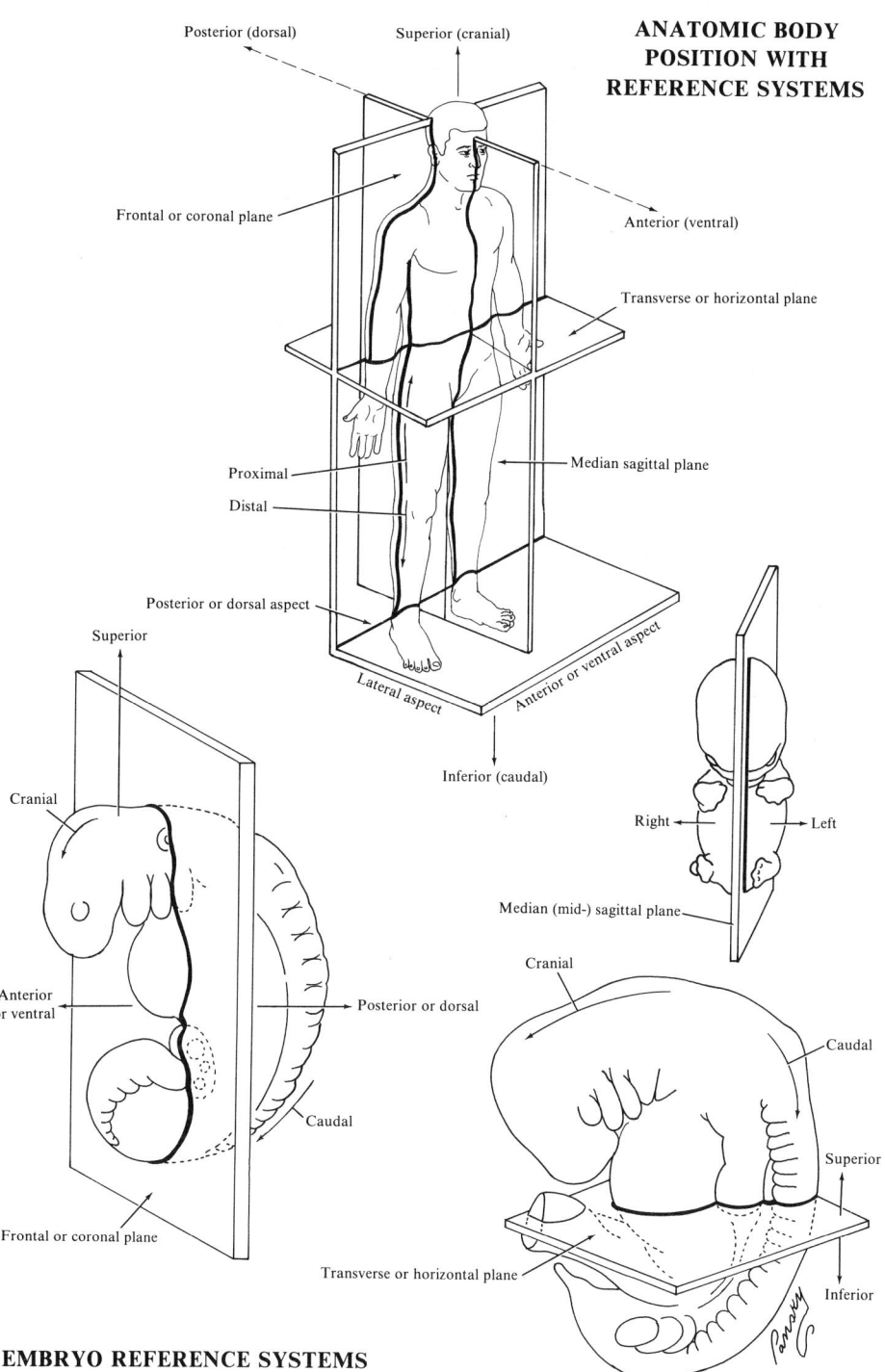

2. THE MALE REPRODUCTIVE SYSTEM

I. **The testes** are the male gonads, normally found outside the body proper in a sac called the *scrotum*
   A. THE TESTES ARE EGG-SHAPED ORGANS about 3.7–5.0 cm (1.5–2.0 in.) in length and about 2.5 cm (1 in.) in each of its other dimensions
   B. THE MASS OF THE SPECIALIZED TESTICULAR TISSUE is arranged in tubules, in the walls of which the spermatozoa are produced
   C. THE INTERSTITIAL CELLS between the tubules are small groups of cells that secrete the hormone *testosterone*

II. **The male tubes** are tubes for carrying the spermatozoa. They begin with the *seminiferous tubules* inside the testes themselves. From these tubules, the spermatozoa are collected and transported via the following tubes (in order)
   A. THE STRAIGHT TUBULES to the *rete testis* to the *efferent ductules of the epididymis*
   B. THE EPIDIDYMIS is a coiled, C-shaped, long tube (600 cm or 20 ft) found inside the scortal sac. It has a *head,* at the upper pole of the testis; a *body,* found on the back of the testis; and a *tail,* attached to the lower pole of the testis
      1. While temporarily stored here, the spermatozoa mature and become mobile
      2. The coiled epididymal ducts all empty into a single duct, the *ductus epididymis* which itself becomes more and more convoluted as it forms the body and tail of the epididymis. The ductus is 0.4 mm in diameter and about 6.5 m (21 ft.) long
   C. THE DUCTUS (VAS) DEFERENS OF THE SPERMATIC CORD is a continuation of the ductus epididymis and is 2–3 mm in diameter and only 45 cm (18 in.) long. It runs in the spermatic cord, through the inguinal canal, accompanied by fascias, the cremaster muscle of the cord, the testicular artery, the pampiniform plexus of veins, and the lymph and nerve supply of the testis and epididymis
      1. The ductus curves behind the urinary bladder, over the ureter, and just medial to the seminal vesicle. Each ductus dilates into an *ampulla* and joins with the seminal vesicle (one on each side) to form the *ejaculatory duct*
   D. THE EJACULATORY DUCTS (2.0 cm or 0.8 in.) pierce the glandular tissue of the prostate to open into the prostatic urethra at the urethral crest

III. **Auxiliary male genital glands**
   A. THE SEMINAL VESICLES are tortuous muscular tubes with small outpouchings and are about 7.5 cm (3 in) long and are at back of bladder. They produce a thick, yellow secretion that forms most of the semen volume and helps nourish the spermatozoa
   B. THE PROSTATE GLAND is a pyramid-shaped fibromuscular gland about 3.7 cm (1.5 in) in diameter and is wrapped around the urethra at the base of the urinary bladder
      1. Tubules from the gland enter the prostatic urethra and add prostatic secretion to the sex cells as they pass through. This helps maintain motility
   C. THE BULBOURETHRAL OR COWPER'S GLANDS are the largest of the mucus-secreting glands in the male reproductive system and are a pair of pea-shaped organs in the pelvic floor tissue just below the prostate gland. Their ducts empty into the urethra.
      1. Other small urethral glands also secrete mucus into the penile urethra

IV. **The urethra and penis**
   A. THE URETHRA serves a dual role: conveying urine from the bladder and carrying reproductive cells and their accompanying secretions to the outside
   B. THE EJECTION OF SEMEN is made possible by the erection of the penis. The latter consists of a spongelike tissue containing many blood spaces that are relatively empty during organ flaccidity, but fill with blood and distend on erection
   C. THE PENIS AND SCROTUM are the male external genitalia
   D. THE SEMEN passes from the prostatic urethra to the membranous urethra to the penile urethra before it reaches the outside

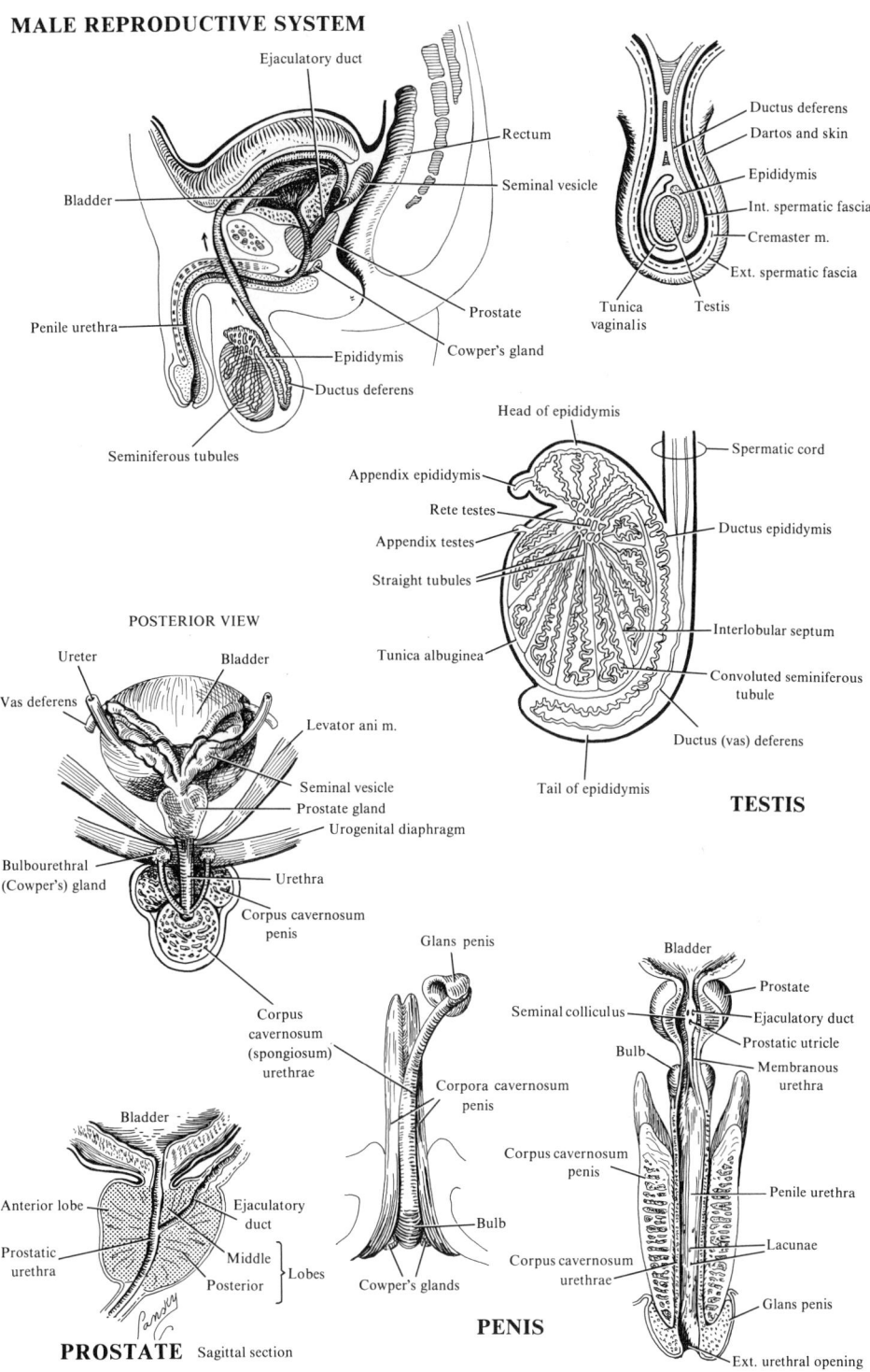

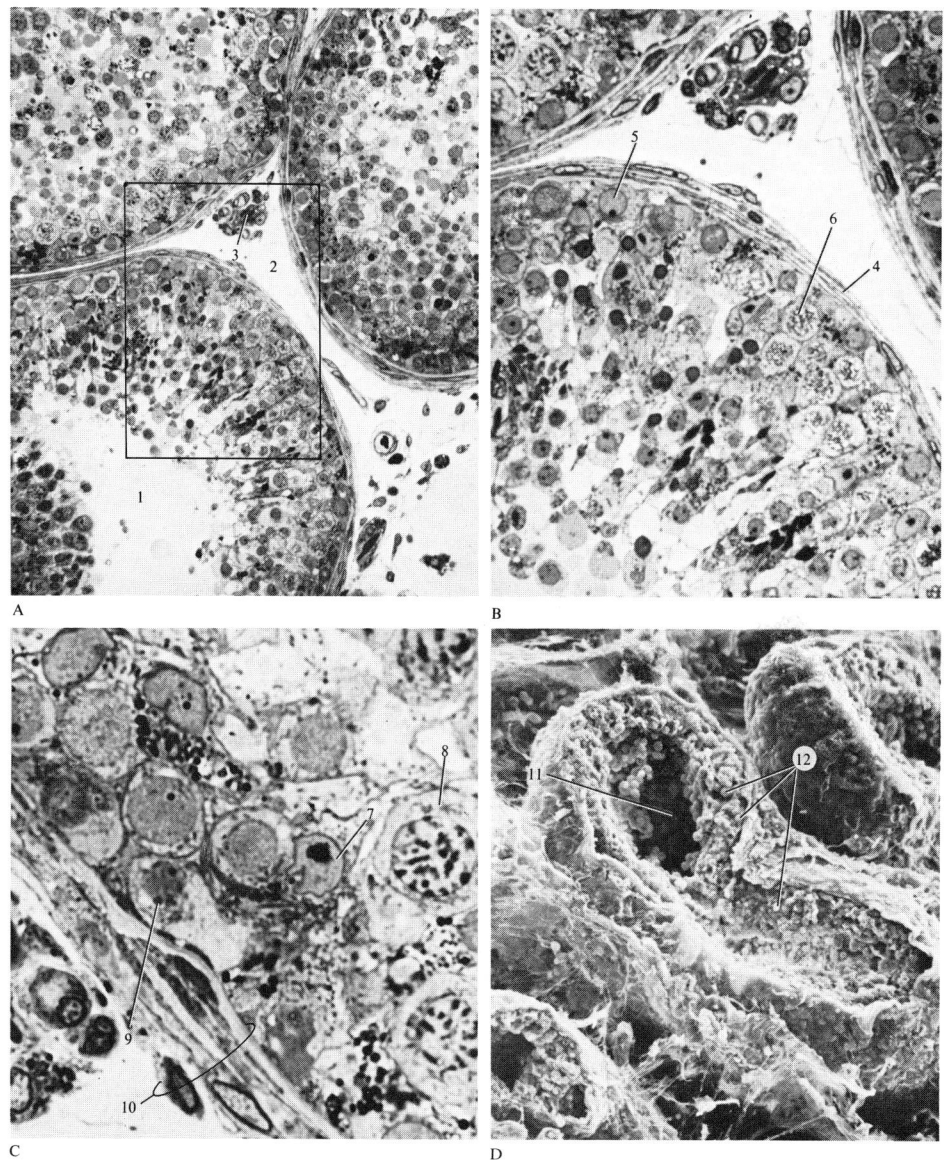

FIGURE 1. **Microscopy of the male testis.** **A.** Light micrograph of portions of 3 seminiferous tubules ($\times$ 270). **B.** High magnification of rectangular area seen in **A** ($\times$ 600). **C.** High magnification of upper left portion of **B** ($\times$ 1,600). **D.** Scanning electron micrograph of the testis, showing portions of 3 seminiferous tubules ($\times$ 350). *1*, Lumen of tubule; *2*, peritubular space; *3*, peritubular and Leydig cells; *4*, basement membrane; *5*, spermatogonium; *6*, primary spermatocyte; *7*, Sertoli cell; *8*, primary spermatocyte; *9*, spermatogonium; *10*, basement membrane; *11*, lumen of tubule; and *12*, cells of the tubule.

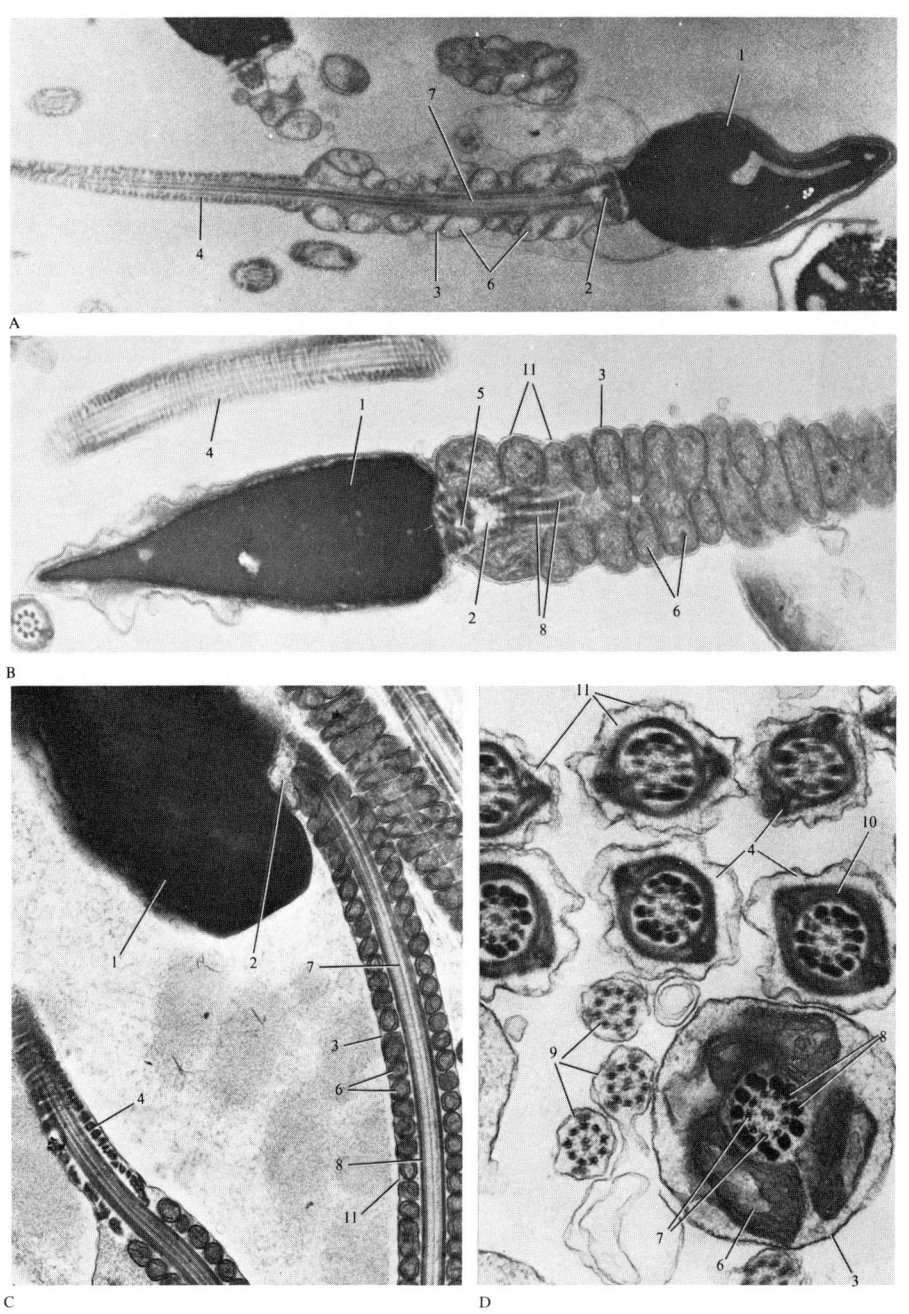

FIGURE 2. **Electron micrographs of sperm.** **A.** Human sperm (× 22,000). **B.** Monkey sperm (× 32,832). **C.** Mouse sperm (× 25,380). **D.** Mouse sperm (cross-section) (× 36,708). *1*, Sperm head; *2*, neck of sperm; *3*, middle piece of tail; *4*, principal piece of tail; *5*, proximal centriole; *6*, mitochondria; *7*, axonemal complex (2 + 9); *8*, dense fibers; *9*, end pieces of tail; *10*, fibrous sheath; and *11*, sperm plasma membrane.

3. GAMETE (GERM CELL) FORMATION, OR GAMETOGENESIS: SPERMATOGENESIS

I. **The gametes** are formed in the gonads, which also have hormonal functions. The *sperm* and *oocyte* (the male and femal germ cells or gametes) are specialized sex cells containing one-half the regular number of chromosomes

II. **Gametogenesis** is a delicate, specialized maturation process called *spermatogenesis* in males and *oogenesis* in females and has 2 major functions
   A. REDUCTION TO HALF THE NUMBER OF CHROMOSOMES AS WELL AS REDISTRIBUTION OF THE HEREDITARY MATERIAL: accomplished by *meiosis* (the combination of 2 divisions involving a single synthesis of DNA and an exchange of chromosome segments)
      1. Each daughter cell formed by meiosis (i.e., secondary spermatocyte) has only half the number of chromosomes of the parent cell
      2. There are 2 successive meiotic divisions
         a. In the first meiotic division, homologous chromosomes pair during prophase and segregate at anaphase
         b. At the second meiotic division, the centromere of each chromosome divides, and chromatids are drawn to opposite poles
      3. Significance of meiosis
         a. Constancy of chromosome number from generation to generation by producing haploid sex cells
         b. Allows independent assortment of maternal and paternal chromosomes
            i. Crossing over (relocating maternal and paternal chromosomes) allows genetic mixing and recombination of genetic material
   B. ACQUISITION OF SPECIAL FORM AND FUNCTION BY THE REPRODUCTIVE CELLS prepares them and makes them suitable for fertilization

III. **Spermatogenesis (sperm maturation)**
   A. SPERMATOZOA are formed in the seminiferous tubules of the testis, from basic cells called spermatogonia which have been dormant in the tubules since the fetal period. Production of spermatozoa is continuous from puberty until death. Transformation of an immature germ cell or spermatogonium into a mature sperm takes about 64 days
      1. Spermatogonia begin to increase in number at puberty (13–16 years of age in male)
      2. After several mitotic divisions, spermatogonia grow and undergo changes to form *primary spermatocytes* (largest germ cells in the tubules)
      3. Each primary spermatocyte undergoes a reduction division called *the first meiotic division* to form 2 haploid *secondary spermatocytes* (half the size of primary)
      4. Secondary spermatocytes undergo a second meiotic division to form 4 haploid *spermatids* (half the size of secondary spermatocytes)
      5. The spermatids gradually form 4 mature sperm or spermatozoa via *spermiogenesis*
   B. BIOLOGIC CHARACTERISTICS OF NORMAL HUMAN SPERMATOZOA
      1. Length: 65 μm
      2. Number: 100 million per ml of semen
      3. Motile at emission: more than 80%
      4. Rate of movement in the genital tract: 1.5 mm per minute
      5. Survival in the genital tract: 3 to 4 days

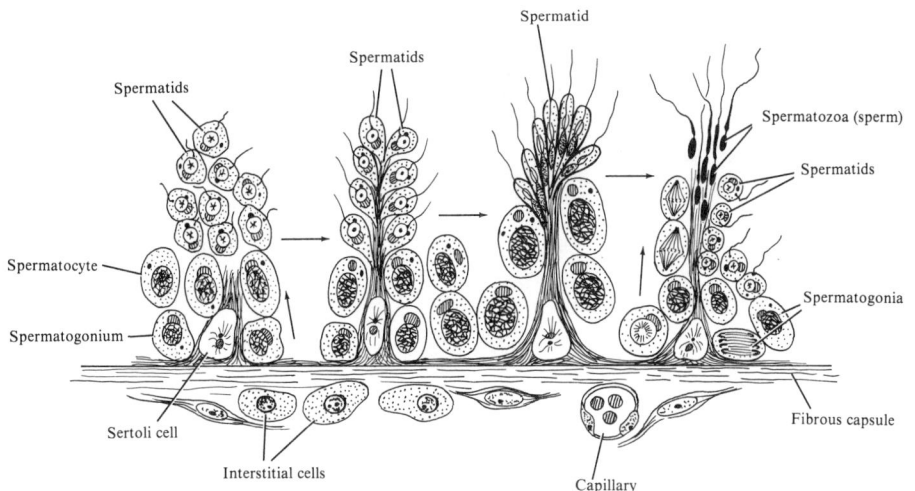

WALL OF SEMINIFEROUS TUBULE: STAGES IN SPERMATOGENESIS

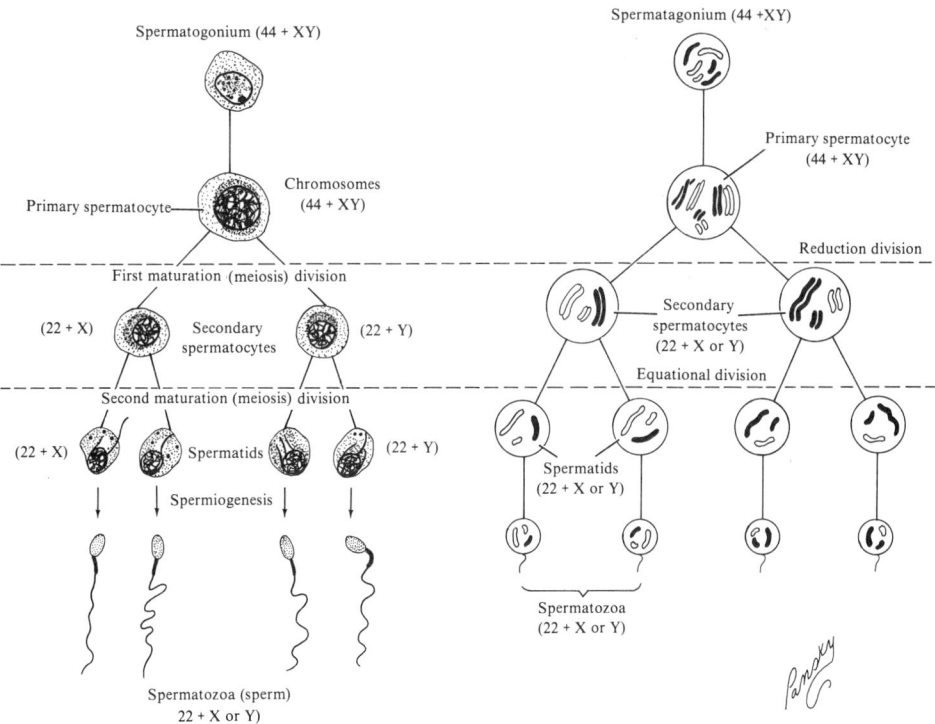

GAMETOGENESIS: SPERMATOGENESIS

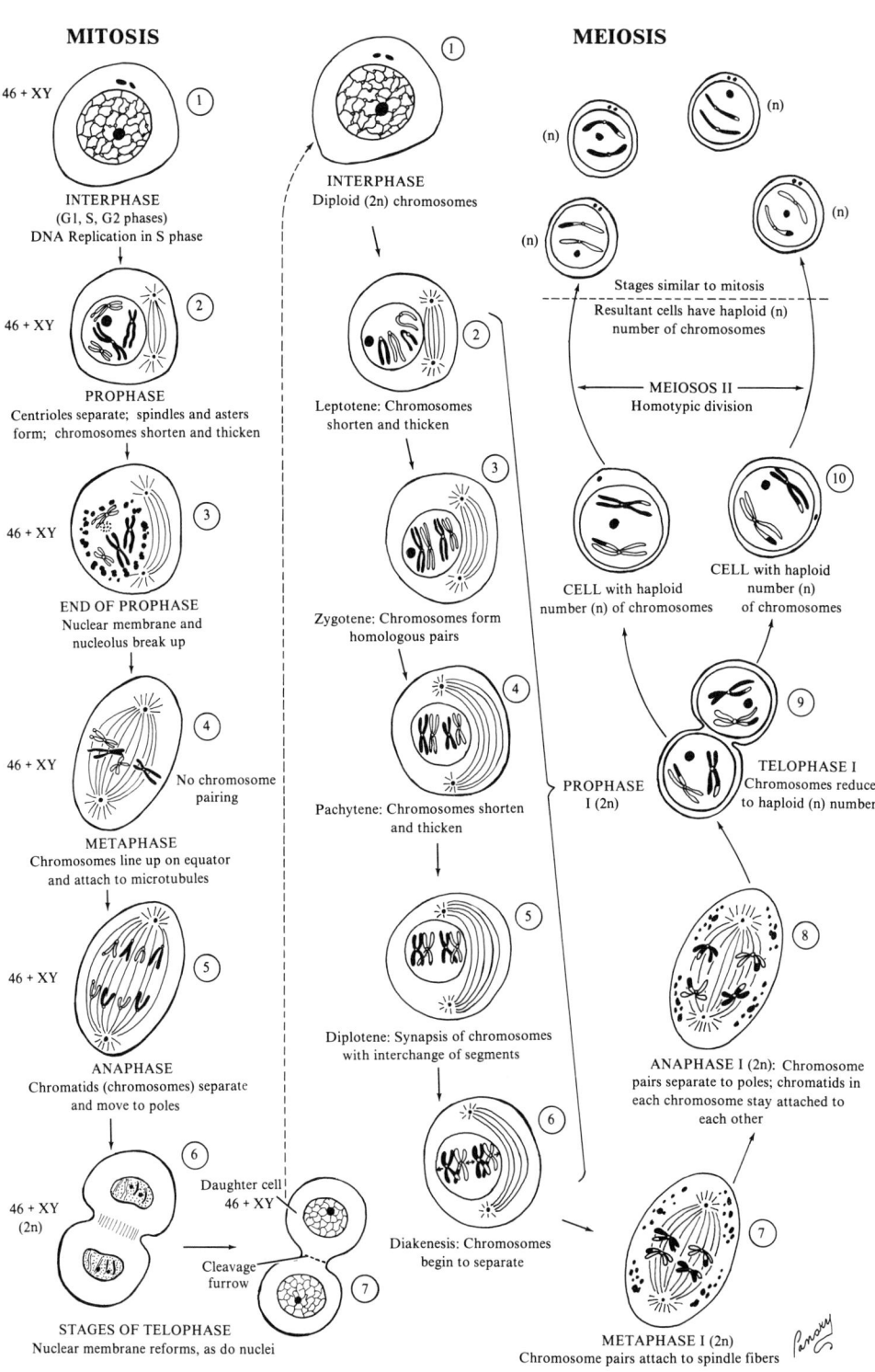

FIGURE 3. **Mitosis and meiosis.**

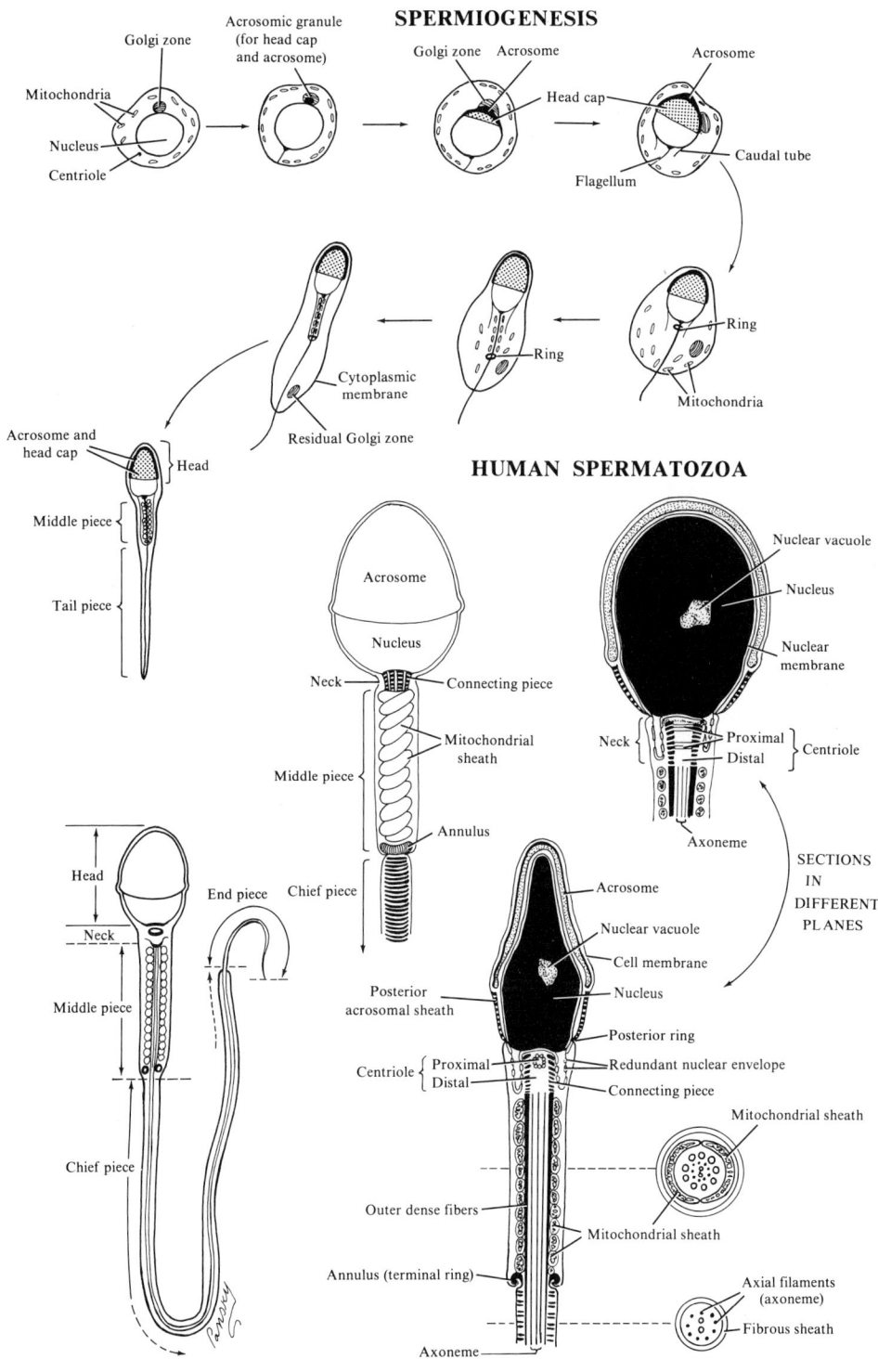

FIGURE 4. **Human spermatozoa and spermiogenesis.**

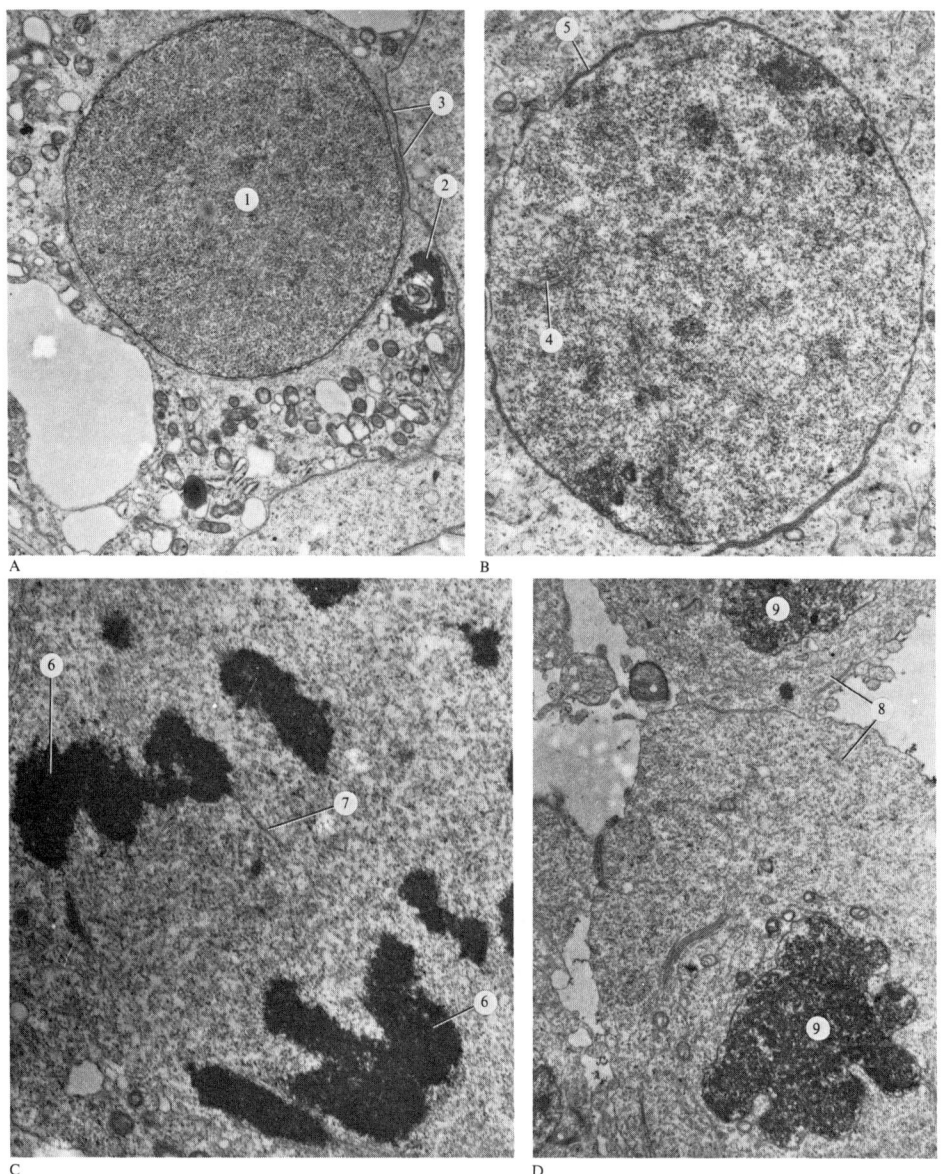

FIGURE 5. **Electron micrographs of spermatocytes and spermatogonium.** **A.** Spermatogonium, showing nucleus and some cytoplasmic organelles (× 9,700). **B.** Primary spermatocyte showing the synaptonemal complexes in the nucleus. This cell is at the first meiotic prophase (× 11,600). **C.** Primary spermatocyte at the anaphase stage. Note section of chromosomes and microtubules (× 18,468). **D.** Primary spermatocyte at the late telophase stage. Early formation of 2 daughter secondary spermatocytes (× 10,070). *1*, Nucleus; *2*, typical crystalloid; *3*, plasma (cell) membrane; *4*, synaptonemal complex; *5*, nuclear membrane; *6*, chromosomes; *7*, microtubule; *8*, daughter secondary spermatocytes; and *9*, nucleus.

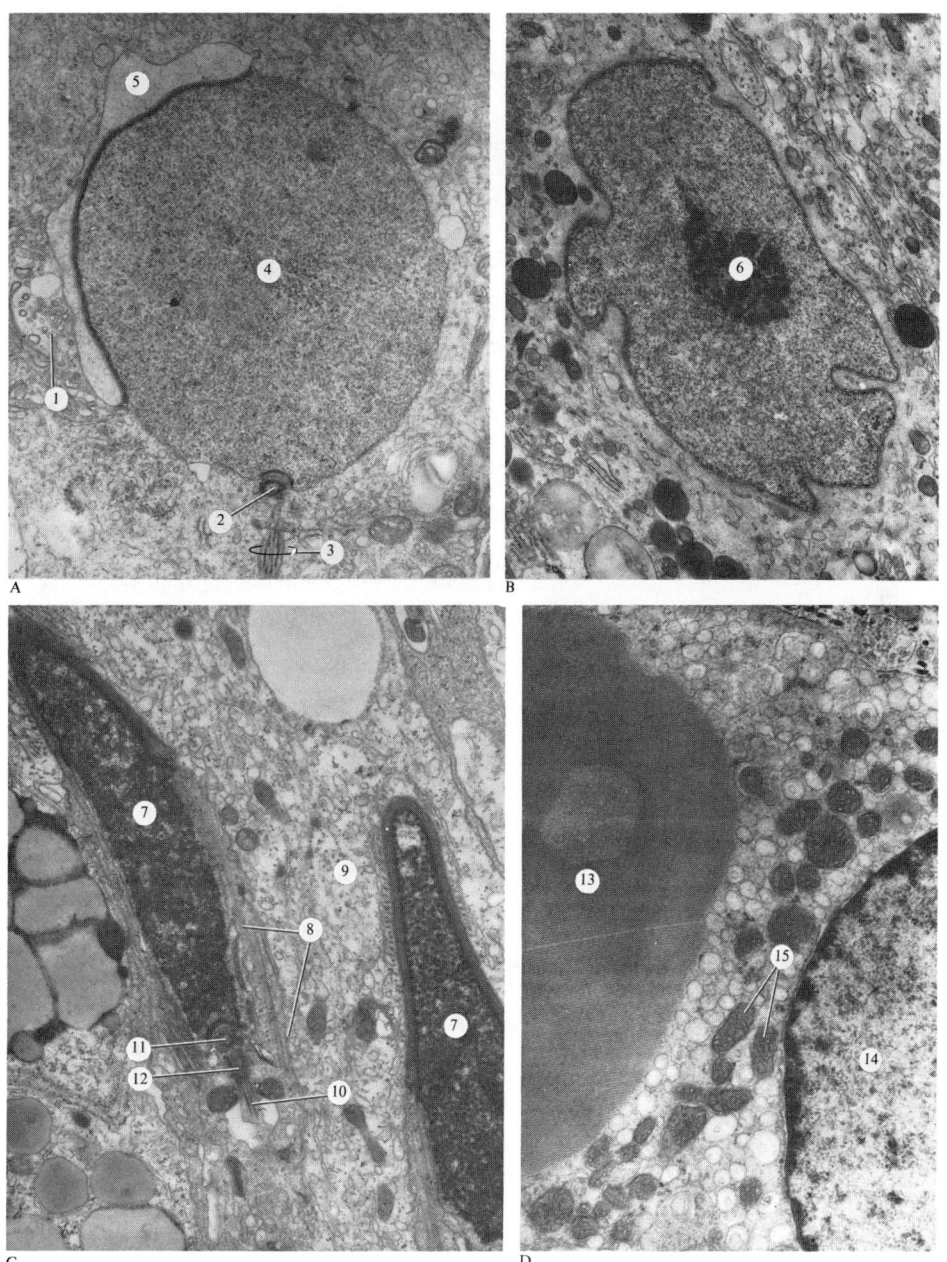

FIGURE 6. **Electron micrographs of spermatids, Sertoli cells, and Leydig cells.** A. Early spermatid showing head and tail formation ($\times$ 17,780). B. Sertoli cell, showing atypical indented nucleus ($\times$ 10,465). C. Late spermatids. Advanced stage of head and tail formation ($\times$ 18,800). D. Leydig cells, showing portion of nucleus ($\times$ 23,598). *1*, Golgi apparatus; *2*, centriole; *3*, portion of tail; *4*, spermatid nucleus; *5*, acrosome head cap; *6*, nucleolus; *7*, spermatid heads; *8*, manchette microtubules; *9*, portion of Sertoli cell cytoplasm; *10*, portion of spermatid tail; *11*, proximal centriole; *12*, distal centriole; *13*, Reinke's crystal; *14*, nucleus; and *15*, mitochondria.

4. GAMETOGENESIS: OOGENESIS

I. Oogenesis (oocyte maturation)
A. PRENATAL MATURATION: the ova are formed in the ovary from cells called *oogonia* which proliferate by mitotic division. All of the oogonia enlarge to form primary oocytes, of which about 2 million are present at birth. No primary oocytes form after birth (in contrast to the continuous production of primary spermatocytes in the male after puberty)
  1. Ovarian stromal cells surround the developing primary oocyte to form a single layer of flattened follicular cells. The primary oocyte and its follicular cells constitute the *primordial follicle*
     a. The follicular cell layer becomes cuboidal and then columnar as the primary oocyte enlarges at puberty and a *primary follicle* is formed
        i. A primary follicle with more than one layer of cuboidal follicular cells is called a *growing follicle*
  2. Primary ooctyes begin the first meiotic division before birth but do not complete prophase until after puberty (arrested in the dictyotene stage until before ovulation)
     a. Long duration of the meiotic division may account for the high frequency of meiotic errors such as *nondisjunction*
B. POSTNATAL MATURATION: the primary oocytes stay dormant in the ovaries until puberty
  1. The primary oocyte increases in size and a membrane, the *zona pellucida,* forms around it as the follicle matures
  2. Just before ovulation, the primary oocyte completes the *first meiotic division,* but unlike its male counterpart, the division of cytoplasm is unequal
     a. The secondary ooctye gets almost all the cytoplasm
     b. The first polar body receives little cytoplasm and is a small, nonfunctional cell that degenerates
  3. At ovulation, the nucleus of the secondary ooctye begins the second meiotic division progressing only to metaphase, then division arrests
     a. If fertilization occurs, the second meiotic division is completed, and the mature ooctye retains most of the cytoplasm, whereas the second polar body is small and degenerates
  4. The secondary oocyte released at ovulation is surrounded by the zona pellucida and a follicular cell layer, the *corona radiata*. It is a large cell
  5. About 2 million primary oocytes are found in the ovaries of a newborn female. Many regress during childhood so that at puberty about 30-40 thousand remain. Only about 200-400 of these ever reach full maturity after puberty and are expelled at ovulation during the female's reproductive life
C. EVERY MENSTRUAL CYCLE corresponds to the maturation of an oocyte, which becomes an ovum through division, yielding cells of unequal size (oocytes and polar bodies). This unequal division produces
  1. The ovum, which measures about 120-150 μm and which alone is fertilizable
  2. The polar bodies, which are no larger than 10 μm and are not fertilizable

II. Sperm versus oocyte
  1. Oocyte is immobile and massive when compared to the highly motile sperm
  2. Oocyte contains much cytoplasm with yolk granules for nutrition during early development, whereas the sperm has sparse cytoplasm and is specialized for motility
  3. There are 2 kinds of normal sperm, with respect to sex chromosomes: 23,X and 23,Y; there is but 1 kind of normal oocyte: 23,X (refers to 23 chromosomes in the complement, made up of 22 autosomes and 1 sex chromosome, X or Y)

GAMETOGENESIS: OOGENESIS

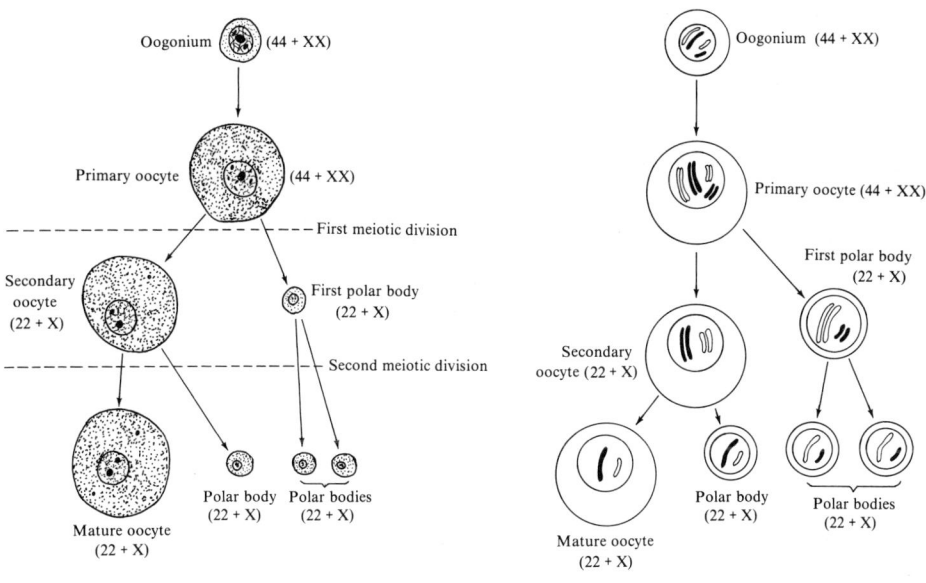

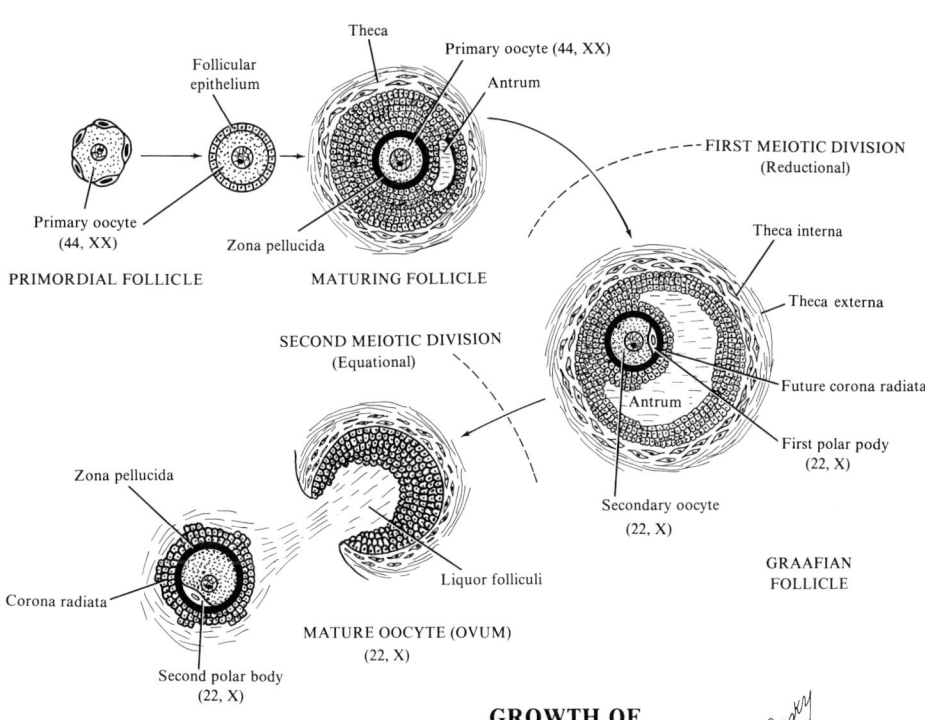

GROWTH OF OVARIAN FOLLICLE

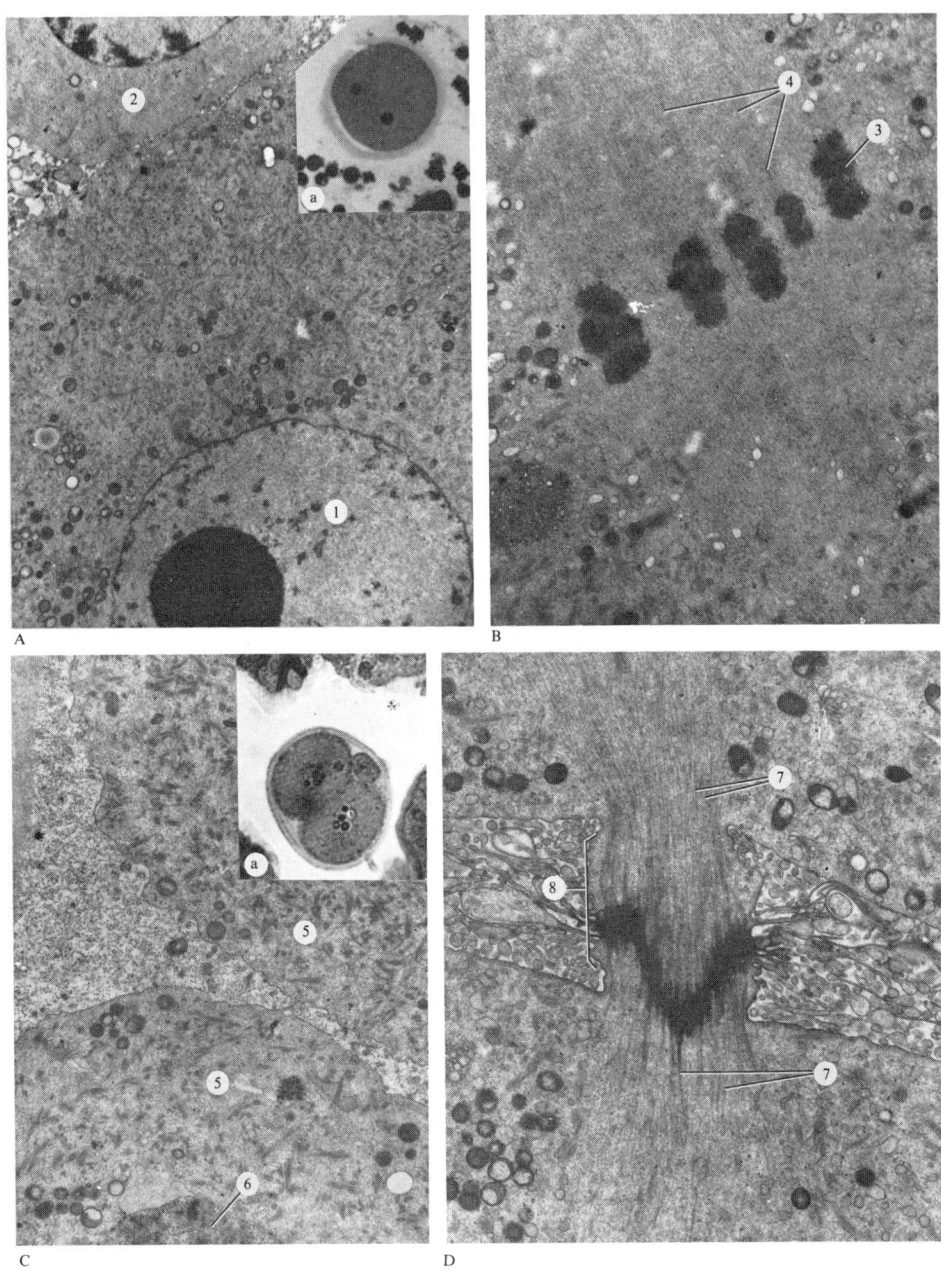

FIGURE 7. **Micrographs of pronuclear stage, first cleavage division, and two-cell embryo.**
A. Pronuclear stage. One pronucleus and the polar body are seen ($\times$ 11,400). **Aa.** Light micrograph of male and female pronuclei. Dispersed corona radiata cells also are seen ($\times$ 340). **B.** First cleavage division. See metaphase chromosomes and spindle microtubules ($\times$ 11,400). **C.** Two-cell mouse embryo. Portions of 2 daughter cells with a part of a nucleus in one cell are seen ($\times$ 10,000). **D.** Midbody of a 2-cell mouse embryo showing microtubules (electron micrograph) ($\times$ 18,800). *1*, Pronucleus; *2*, polar body; *3*, chromosome; *4*, spindle microtubules; *5*, daughter cells; *6*, nucleus; *7*, microtubules; and *8*, midbody.

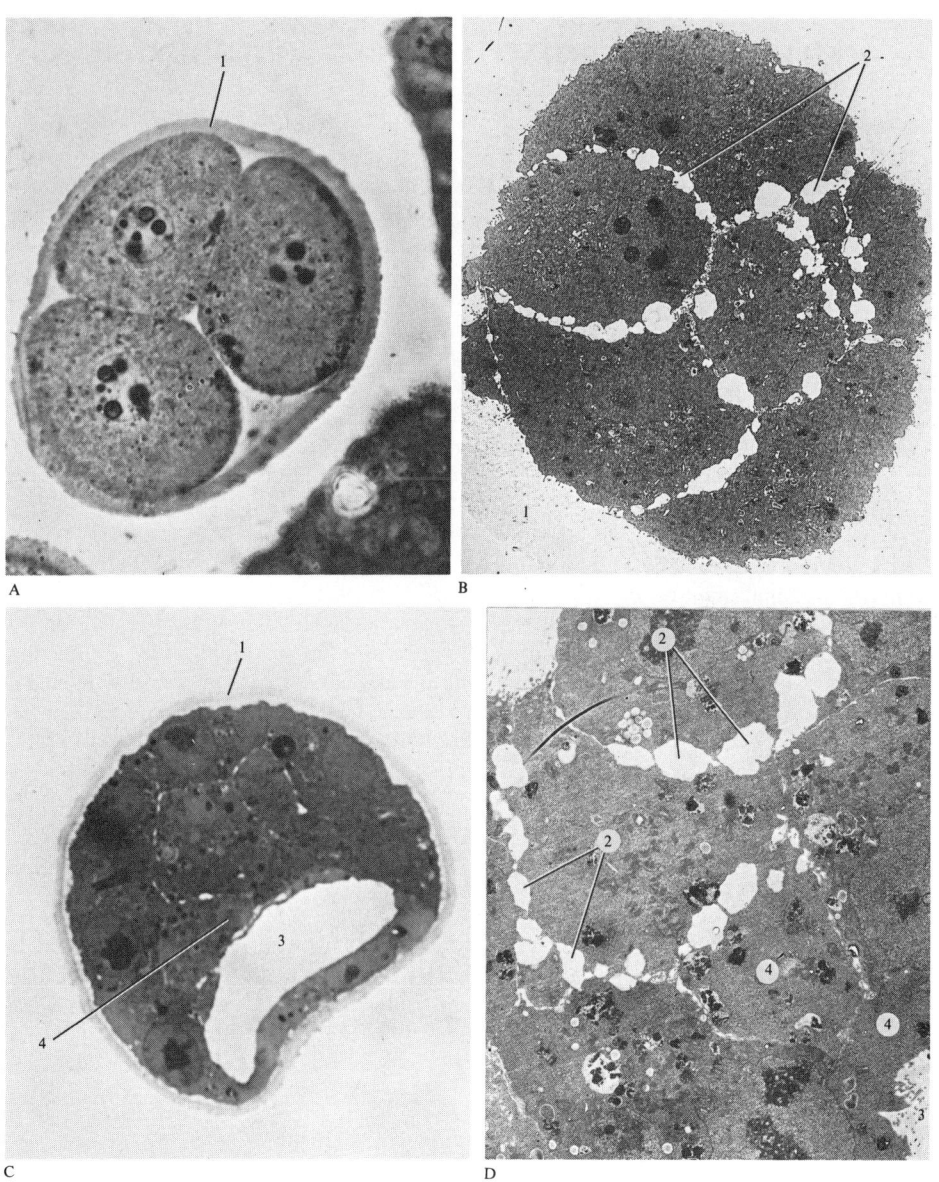

FIGURE 8. **Micrographs of four-cell embryo, morula, and blastocyst stages.** **A.** Light micrograph of a 4-cell mouse embryo within the fallopian tube (× 1,000). Note that in this plane of section, only 3 of the 4 cells are visible. **B.** Electron micrograph of a mouse embryo at the morula stage (× 2,625). **C.** Light micrograph of a mouse blastocyst (× 1,400). **D.** Electron micrograph of a portion of a mouse blastocyst showing the outer trophoblast and inner cell mass. Portion of the blastocele cavity also is seen. (× 4,750). *1*, Zona pellucida; *2*, intercellular spaces; *3*, blastocele cavity; and *4*, inner cell mass.

5. ANOMALIES (ABNORMALITIES) OF GAMETOGENESIS

I. Morphologic anomalies
  A. ABNORMAL SPERMATOZOA can be seen even in normal semen. Up to 10% of sperm in an ejaculate may be grossly abnormal. They generally do not fertilize ooctyes due to their poor motility and fertilizing capabilities. They generally do not affect fertility unless their numbers exceed 20%
   1. Double sperm forms may be the result of failure of disjunction during spermatogenesis
   2. The number of abnormal sperm may be increased by x-rays, severe allergic reactions, antispermatogenic agents, and other factors
  B. UNUSUAL CELL TYPES IN A FETAL OVARY
   1. Oocyte with 2 nuclei—usually fail to mature
   2. Two ooctyes in the same follicle—infrequent in the human female. Most of these never mature

II. Chromosomal anomalies: abnormalities occur during meiosis when distribution of the chromosomal material between the gametes takes place
  A. ANOMALIES INVOLVING THE AUTOSOMES (SOMATIC CHROMOSOMES): meiotic division includes a stage of chromosome pairing. This provides the possibility of nondisjunction where homologous chromosomes may fail to separate and pass to opposite poles, resulting in some germ cells with 24 chromosomes, while others have only 22
   1. If a germ cell with 24 autosomal chromosomes unites with one that has a normal complement (23), a zygote with 47 chromosomes is formed —*trisomy* (presence of 3 representatives of a certain chromosome rather than 2)
   2. If a germ cell with 22 autosomal chromosomes fuses with a normal one, the zygote ends up with 45 chromosomes—*monosomy* (presence of only 1 representative of a particular chromosome instead of the usual pair)
  B. ANOMALIES INVOLVING THE SEX CHROMOSOMES
   1. The same type of abnormalities as described under IIA are seen here. Certain cells have no sex chromosomes, while others have 2 (or sometimes even more)
   2. Generally speaking, chromosome abnormalities more often affect the female gametes. The greater vulnerability of the female as compared with the male gametes is due to sex differences in the chronology of maturation, despite almost identical fundamental mechanisms

III. Chronology of gametogenesis
  A. IN THE MALE: the fundamental difference apparent between the male and female is that there is an unequal duration of meiosis in the 2 sexes. In the male, meiosis comes about within several days
  B. IN THE FEMALE: the process begun during fetal life is suspended for a "considerable" time, almost a dozen years. It is this very delay which may result in the number of chromosomal abnormalities seen in the female

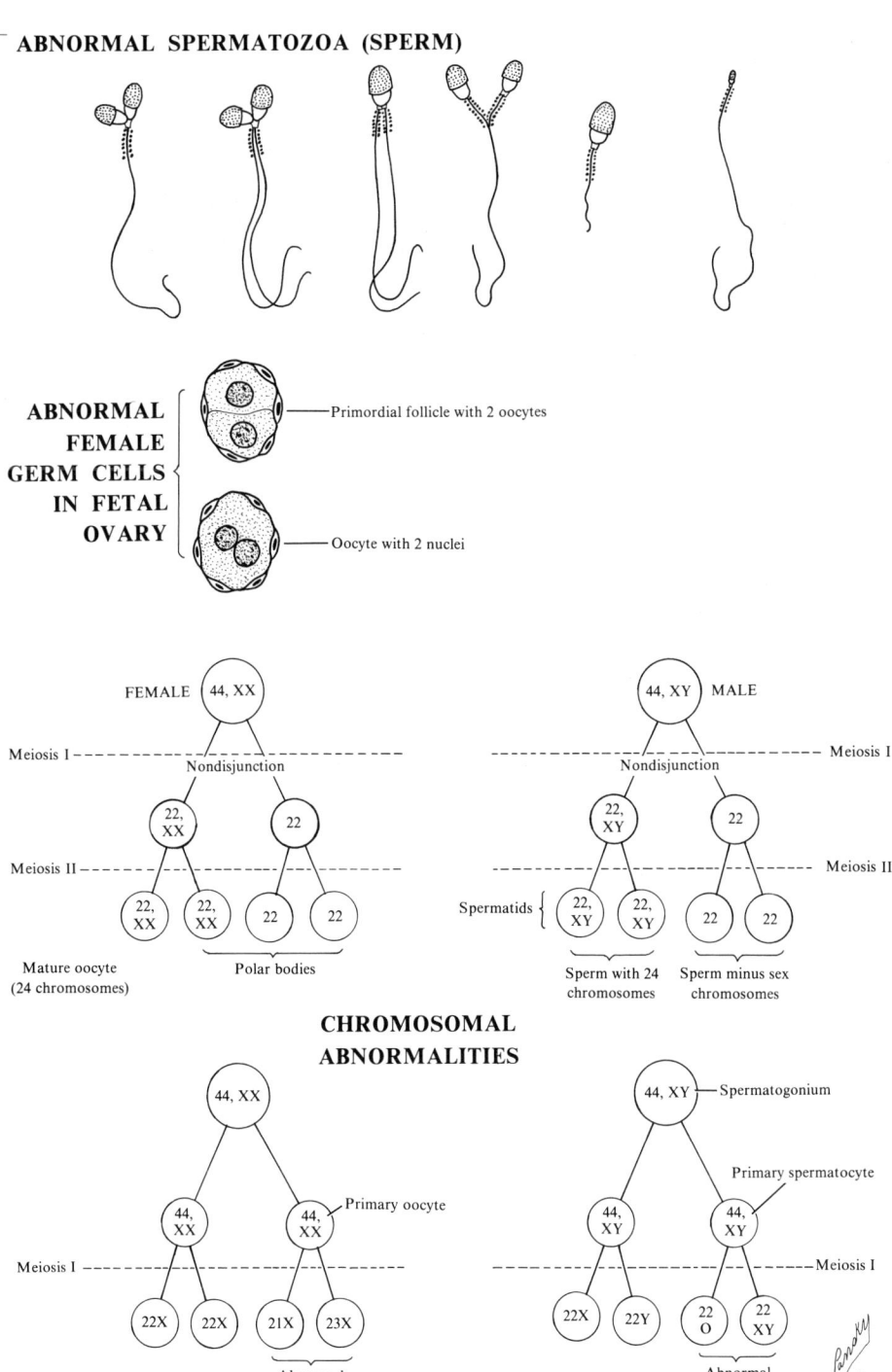

6. THE ADULT FEMALE UTERUS

I. Uterine anatomy: the uterus is a thick-walled, pear-shaped organ, 7.0 × 5.0 × 2.5 cm, consisting of an upper body (corpus) with its rounded, dome-shaped top or *fundus;* and a narrow, cylindrical *neck or cervix* whose terminal portion projects into the vagina as the *portio vaginalis*. The part between body and cervix is the *isthmus*. The walls consist of 3 layers

A. PERIMETRIUM: a very thin outer serosa which is the peritoneal layer of the broad ligament and is firmly attached to the underlying muscularis

B. MYOMETRIUM (MUSCLE): a thick middle smooth muscle layer about 15 mm thick having 3 layers of muscle
   1. Stratum subvasculare: inner longitudinal muscle layer
   2. Stratum vasculare: middle circular or spiral layer forming the bulk of the muscularis with many large blood vessels, especially veins
   3. Stratum supravasculare: outer, thin layer with circular and longitudinal fibers

C. ENDOMETRIUM: thin inner layer lined by simple columnar epithelium with many tubular glands. There is no submucosa, and the mucosa is closely attached to the myometrium
   1. During secretory phase of menstrual cycle, 3 layers of endometrium can be seen
      a. Compact layer (compacta): thin, narrow superficial layer of densely packed stromal cells around the straight necks of glands. Little edema here
      b. Spongy layer (spongiosa) makes up the bulk of the endometrium composed of edematous stroma with dilated, tortuous bodies of glands
      c. Basal layer (basalis): deepest layer, relatively thin and narrow, containing the blind ends of glands. The latter undergo little or no change. This layer has its own blood supply and is not lost at menstruation or at parturition
      d. Functionalis layer consists of layers a and b. Disintegrates and is shed at menstruation and parturition

D. BLOOD SUPPLY: via the uterine artery (usually a branch of the internal iliac but may arise as a common trunk with the vaginal or with the middle rectal artery). Ends as the ovarian branch which anastomoses with the ovarian artery. In addition, a variable number of branches go to the cervix, upper vagina, medial part of uterine tube (tubal branch), to the round ligament of the uterus and to the ligament of the ovary. Blood is returned via a venous plexus that follows the uterine artery

E. NERVE SUPPLY: receives autonomic and sensory fibers via the uterovaginal plexuses which run along the uterine artery
   1. The uterus is painless to most stimuli, but pain may be felt when the cervix is grasped with a forceps or is dilated. These nerve fibers may ascend and enter the spinal cord via the lumbar splanchnic nerves

F. LYMPH DRAINAGE
   1. From fundus and upper body drain into lumbar (or aortic) nodes
   2. From lower body drain into external iliac nodes
   3. From cervix drain into external iliac, internal iliac, and sacral nodes
   4. From area near the uterine tubes, drainage follows the round ligament and may drain into superficial inguinal nodes

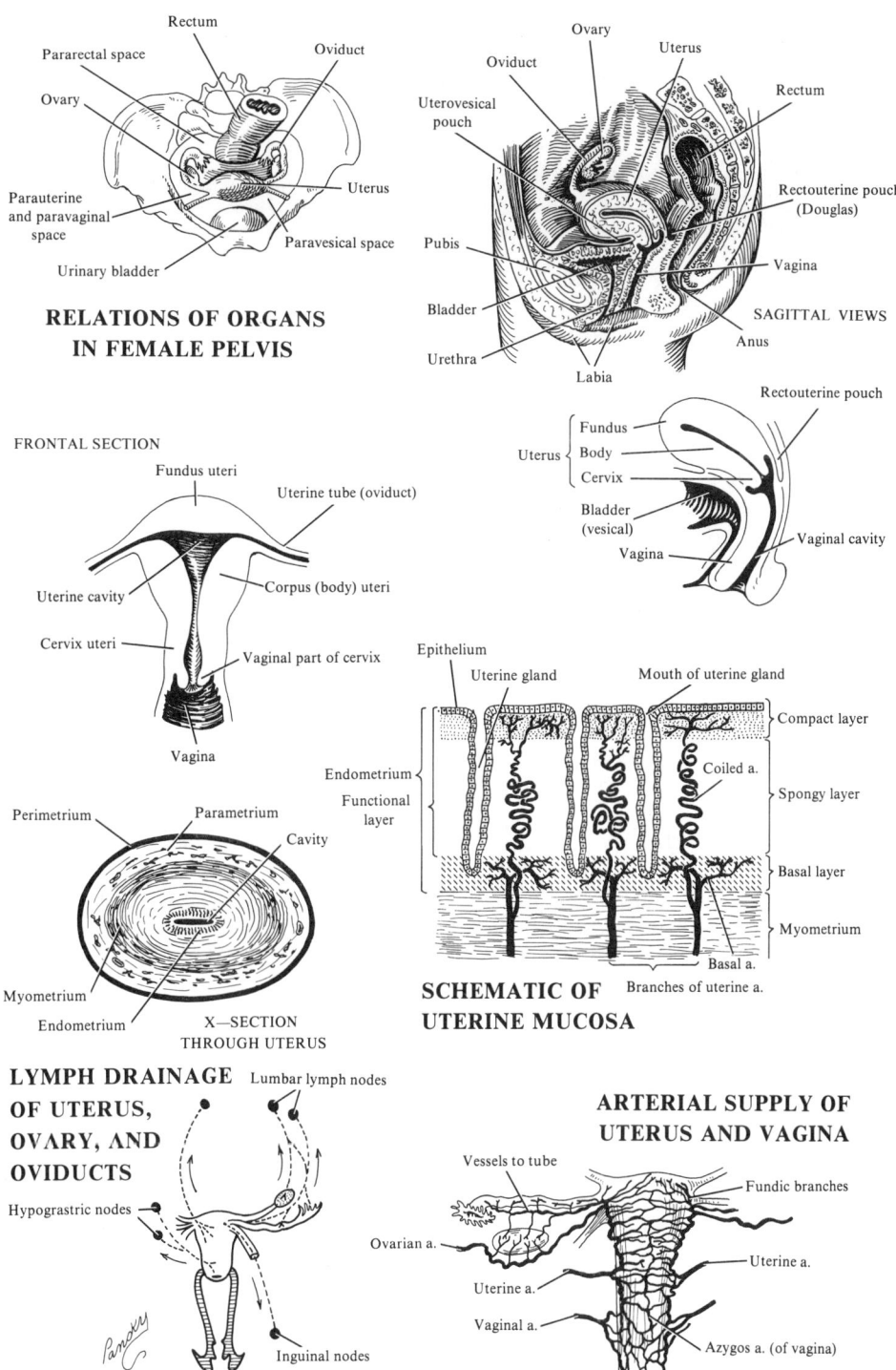

7. THE ADULT UTERINE TUBES (OVIDUCT, FALLOPIAN TUBE, OR SALPINX)

I. **Introduction:** the uterine tubes are paired structures about 10–15 cm long and 6–8 mm in diameter. One end opens into the peritoneal cavity near the ovary; the other end opens into the superior lateral part of the uterine cavity. The tube conducts the ova, discharged at ovulation, to the uterine cavity

II. **Regions of the tube**
 A. INFUNDIBULUM: funnel-shaped, formed of a number of processes or fimbriae
 B. AMPULLA: largest segment, thin-walled like the infundibulum
 C. ISTHMUS: short segment, smaller in diameter and thicker walled than the ampulla
 D. UTERINE OR INTERSTITIAL: segment embedded in uterine wall. Of small diameter (1 mm)

III. **Histology of uterine tube:** wall consists of a series of layers
 A. MUCOSA: simple columnar type epithelium with some ciliated cells and others being narrow, peg-shaped, and nonciliated
  1. Secretes mucus and other substances to maintain ovum's journey through tube
  2. Epithelial height and proportion of ciliated to nonciliated secretory cells vary and correlate with menstrual cycle changes; e.g., epithelium is taller in the first half of the follicular phase than second half, and the relative number of nonciliated cells increases in the corpus luteum phase. In addition, the epithelium is low in pregnancy and the number of "peg" cells increases
  3. Cilia of the epithelium beat toward the uterus
  4. There are no glands in the tube
  5. The mucosa of the infundibulum and ampulla have many tall folds with corresponding deep grooves. The lumen is irregular. The folds decrease in height toward the uterus and are low in the isthmus. The uterine portion of the tube has slight folds
 B. MUSCULARIS: thickest in the isthmus and thins toward the fimbriated end
  1. Has a well-developed inner circular layer and a thin outer longitudinal layer (the latter is complete only in the isthmus)
  2. The longitudinal muscle bundles are discontinuous in ampulla and may be absent in the fimbria
 C. SEROSA has the usual structure of peritoneum

IV. **Blood supply:** tubal branch of the uterine artery and small branches from the ovarian artery. The arteries run in the stroma along the bases of the folds, giving rise to a dense capillary stromal network
  1. The veins course similar to the arteries

V. **Nerve supply:** via the ovarian plexus and fibers from the inferior hypogastric plexus. Some fibers are sensory, others are autonomic to muscle coats, and still others are vasomotor to the blood vessels

VI. **Lymphatic drainage:** lymphatics follow the blood vessels and drain into the lumbar (or aortic) nodes

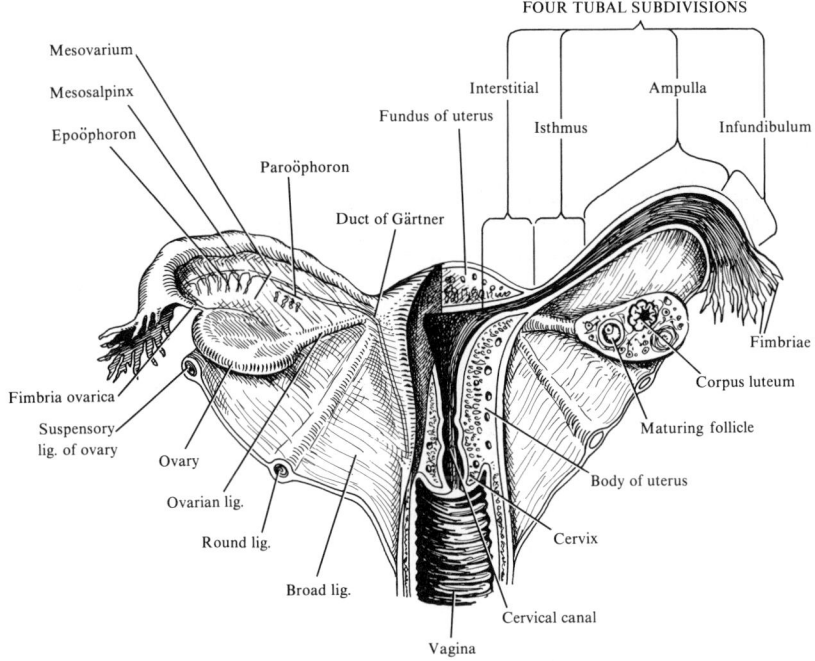

UTERUS, OVARIES, AND FALLOPIAN (UTERINE) TUBES

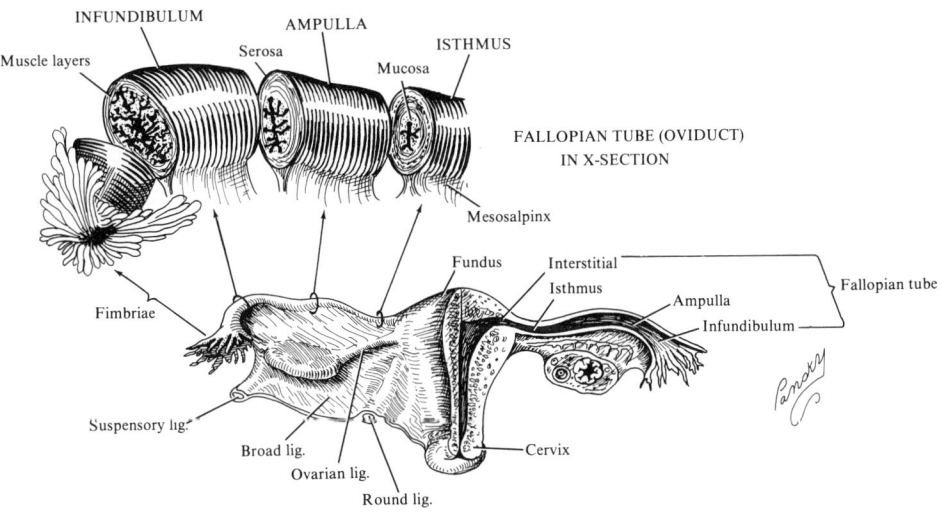

8. REPRODUCTIVE CYCLES: THE OVARIAN CYCLE AND OVULATION

I. Introduction: the recurring periods of sexual excitement in adult females, other than primates, are called *estrus* or "heat." In mammals other than man we speak of the *estrous cycle* (series of physiologic uterine, ovarian, and other changes that occur which consist of proestrus, estrus, postestrus, and anestrus or diestrus). In humans there is the *menstrual cycle* (period in which the ovum matures, is ovulated, and enters the uterine lumen via the uterine tubes)
  A. MENSTRUATION begins at puberty, about 12 to 15 years of age, and continues throughout the reproductive years in the human female
  B. REPRODUCTIVE OR SEXUAL CYCLES occur monthly and involve activities of the hypothalamus, hypophysis, ovaries, uterus, uterine tubes, vagina, and mammary glands
  C. CYCLES prepare the reproductive system for pregnancy. A hormone-releasing factor synthesized in the hypothalamus and carried via the hypophyseal portal system of vessels to the anterior lobe of the hypophysis, causes the cyclic release of the gonadotropic hormones, follicle-stimulating hormone (FSH), and luteinizing hormone (LH)

II. Ovarian cycle
  A. THE GONADOTROPINS (FSH and LH) produce cyclic changes in the ovaries such as development of follicles, ovulation, and corpus luteum formation—ovarian cycle
    1. FSH promotes growth of several ovarian follicles, but usually only one forms a mature follicle which finally expels its oocyte
  B. FOLLICULAR DEVELOPMENT is characterized by
    1. Growth and differentiation of the primary oocyte
    2. Proliferation of the follicular cells
    3. Formation of the zona pellucida
    4. Development of the theca folliculi (connective tissue capsule) from ovarian stroma
  C. THE FOLLICULAR CELLS actively divide, producing a stratified layer around the ovum
    1. The follicle becomes oval and the oocyte eccentric in position because the follicular cell proliferation is greater on one side
    2. Fluid-filled spaces then appear around the cells, coalesce, and form a single large cavity, the *follicular antrum*, and the ovarian follicle is now called a *secondary* or *vesicular follicle*
    3. The primary oocyte is pushed to one side of the follicle where it is surrounded by a follicular cell mound, the *cumulus oophorus* which projects into the antrum
  D. EARLY FOLLICLE DEVELOPMENT is induced by FSH. Final stages of maturation require LH as well
    1. Growing follicles produce estrogen (female sex hormone) that regulates development and functions of reproductive organs
      a. Estrogens are predominantly formed by the *theca interna*
  E. OVULATION
    1. Under influence of FSH and LH, around midcycle or 14 days ± 1 day, the follicle grows rapidly producing a bulge or cystic swelling on the ovarian surface, and a small oval avascular spot, the *stigma*, is seen on the swelling
    2. Before ovulation, the oocyte and some cells of the cumulus oophorus detach from the inside of the distended follicle
    3. At ovulation, there is a "surge" of LH release, the stigma balloons out, forming a surface vesicle, then it ruptures, expelling the oocyte with follicular fluid
    4. The oocyte is covered by the zona pellucida and one or more layers of follicular cells which radially arrange themselves as the *corona radiata*
    5. Signs of ovulation include *mittelschmerz* or intermenstrual pain and basal body temperature rise (slightly). Although the time between ovulation and succeeding menstrual bleeding is constant, the time between ovulation and the preceding menstruation is highly variable and depends on how long the follicle needs to mature. One cycle of maturation may need more time than another

DEVELOPING OVUM AND OVULATION

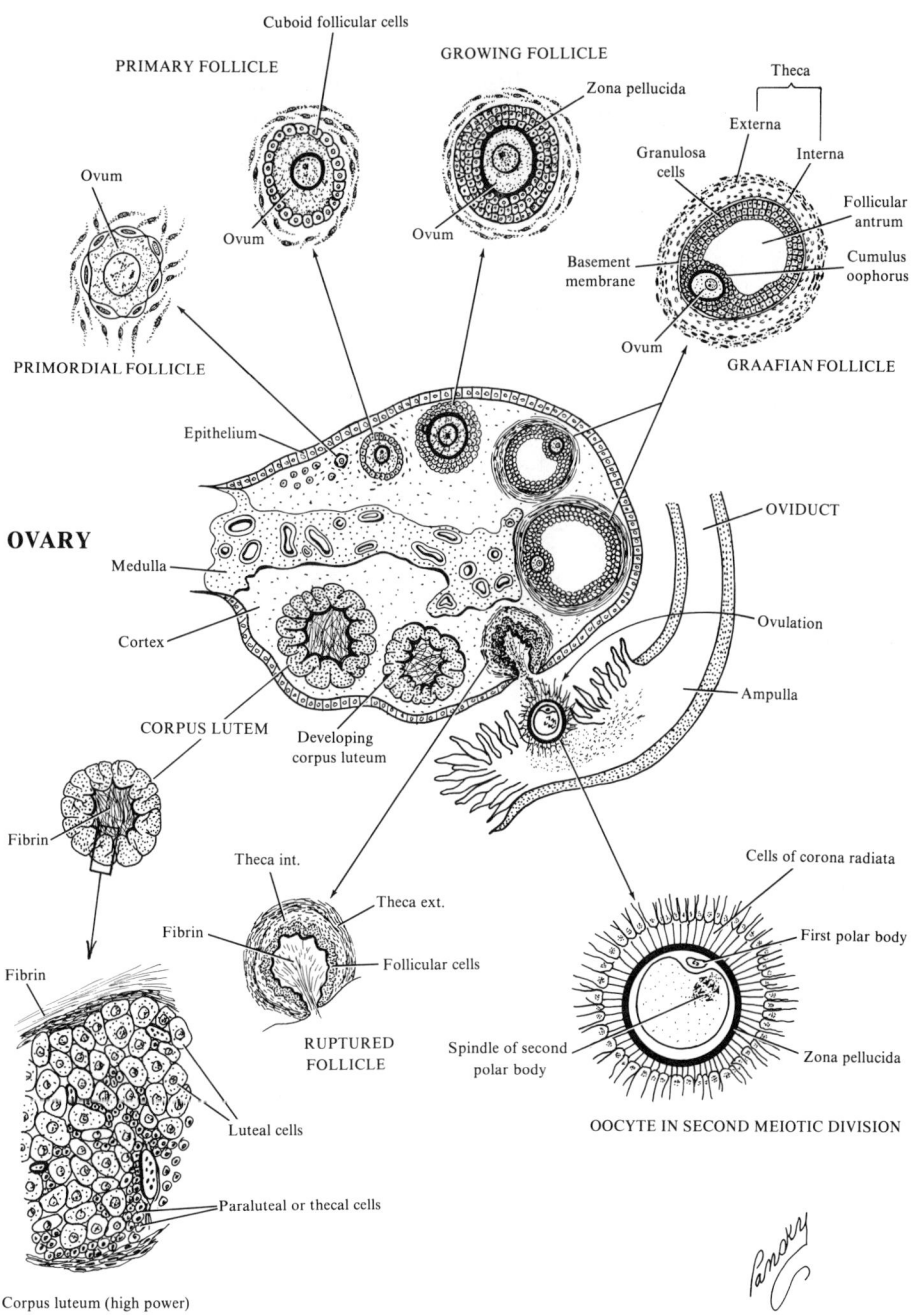

9. REPRODUCTIVE CYCLES: THE OVARIAN CYCLE AND THE CORPUS LUTEUM

I. **Introduction:** at the onset of each ovarian cycle, a number of primary follicles begin to grow and mature. Usually, only 1 follicle reaches full maturity and only 1 oocyte is discharged. The others degenerate and become atretic. In the next cycle, another group of primary follicles matures and again only 1 develops. When a follicle becomes atretic, the oocyte and follicular cells degenerate and are gradually replaced by connective tissue, forming the *corpora atretica*

II. **Corpus luteum (CL) formation**
A. AFTER OVULATION: the follicular walls and theca folliculi collapse and become folded and under LH influence develop into the corpus luteum, a glandular structure which secretes mostly *progesterone* (also some estrogen)
   1. Progesterone causes the endometrial glands to secrete, preparing the endometrium for blastocyst implantation
   2. If the oocyte is fertilized, the corpus luteum enlarges to form a *corpus luteum of pregnancy* and increases hormone production
      a. With pregnancy, chorionic gonadotropin, secreted by the trophoblast of the chorion, prevents degeneration of the corpus luteum
   3. If the oocyte is not fertilized, the corpus luteum degenerates in 10–12 days after ovulation and forms the *corpus luteum of menstruation* and is eventually transformed into a white scar, the *corpus albicans*
B. AFTER OVULATION: when the wall of the ruptured follicle collapses, the rest of the follicular cells gradually become vascularized by vessels growing in from the periphery. The follicular cells begin to hypertrophy, become polyhedral, and develop a yellowish pigment. The modified yellowish cells are called *luteal cells*
   1. The corpus luteum of pregnancy, by the end of the third month of pregnancy, may constitute as much as one-third to one-half the total size of the ovary
   2. The luteal cells continue to secrete progesterone until the end of the fourth month, but thereafter regress slowly
   3. Whether during this period new luteal cells are added to the periphery by differentiation of the surrounding stroma cells, or by active division of the existing luteal cells, is unknown
   4. Removal of the corpus luteum of pregnancy before the fourth month usually results in abortion

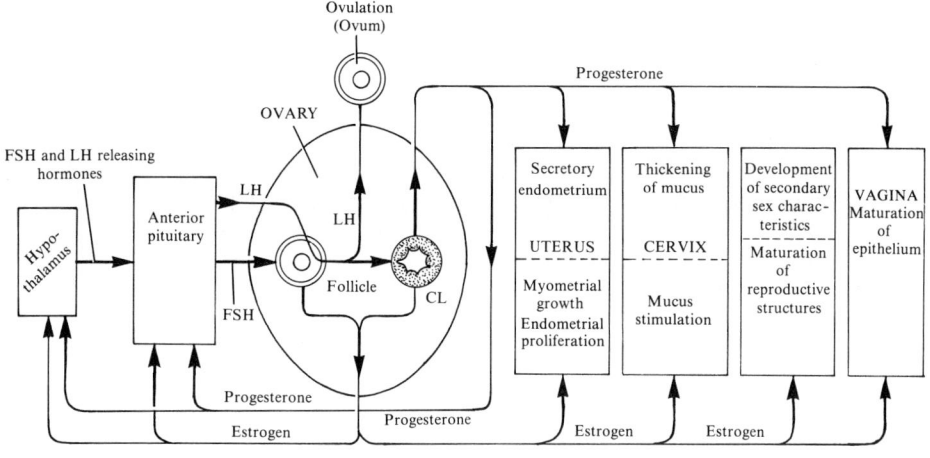

HORMONE RELATIONSHIPS IN THE FEMALE

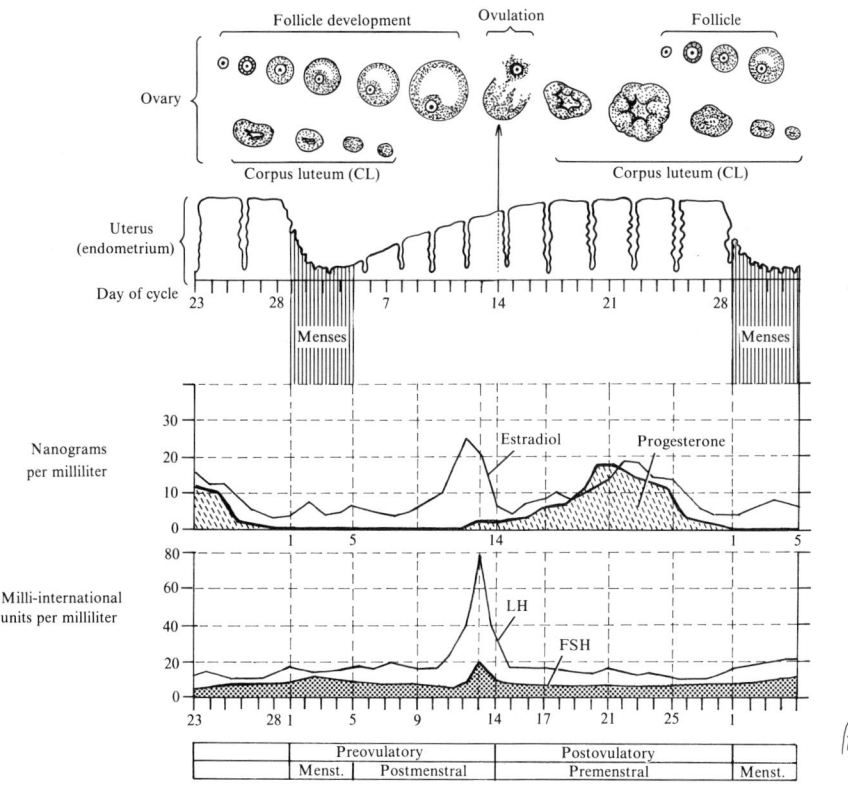

10. REPRODUCTIVE CYCLES: THE MENSTRUAL (UTERINE) CYCLE

I. **Introduction:** the menstrual cycle is caused by changes in the uterine endometrium. Menstruation is a discharging of blood secretion and tissue debris from the uterus that recurs in nonpregnant, breed-age, primate females at about monthly intervals and represents a readjustment of the uterus to the nonpregnant state
   A. ANOVULATORY CYCLE: the ovary fails to produce a mature follicle, endometrial changes are minimal, and the proliferative endometrium develops as usual
      1. There is no ovulation, and there is no corpus luteum formation
      2. The endometrium does not progress to a secretory phase but continues to be of proliferative type until the beginning of menstruation
      3. This state also can be produced by giving steroid hormones which act on the hypothalamus and hypophysis to inhibit secretion of hypothalamic releasing factors and pituitary gonadotropins needed for ovulation (like taking birth control pills)
      4. Birth control pills: estrogen inhibits ovulation; progesterone induces the secretory phase of the menstrual cycle
   B. OVARIAN HORMONES cause the cyclic changes in the endometrium of the uterus. There are 3 cyclic phases, each passing into the next in a continuous process
      1. The *menstrual phase:* day 1 of menstruation is the beginning of this cycle
         a. The functional layer of the uterine wall is sloughed off and discarded
         b. Typically occurs at about 28-day intervals and lasts about 3–6 days
      2. The *proliferative or follicular phase* occurs between days 6 to 14 and coincides with the growth of the ovarian follicles
         a. Controlled by estrogens secreted by the theca interna (around follicle)
         b. Two- to threefold increase in endometrial thickness (repair and proliferation)
         c. Continuous surface epithelium covers the endometrium; the glands increase in length and number, and spiral arteries elongate but do not reach the surface
      3. The *secretory or luteal phase* occurs between days 16 to 28 and coincides with the formation and growth of the corpus luteum (CL)
         a. Progesterone secreted by the CL stimulates the glandular epithelium to produce a glycogen-rich material. Endometrium thickens due to fluid in stroma
         b. The glands become wide, tortuous, and saccular
         c. Spiral arteries grow into the superficial compact layer and become coiled
         d. If oocyte, released at ovulation, is fertilized, the blastocyst begins to implant in the endometrium at about day 20 of this phase
         e. If fertilization does not occur, the secretory endometrium goes into an *ischemic* or *premenstrual phase* during the last 2 days of the menstrual cycle
            i. *Ischemic phase:* last part of the secretory phase with ischemia due to blood deficiency, endometrium pales, and spiral arteries constrict intermittently due to decreased hormone secretion by the degenerating corpus luteum
            ii. Hormonal decrease results in stoppage of glandular secretion, loss of interstitial fluid, and a shrinking of the endometrium
            iii. As the ischemic period nears its end, the spiral arteries constrict longer, blood seeps through the arterial ruptured walls into the surrounding stroma, and pooled blood breaks through the endometrial surface, resulting in bleeding into the uterine lumen and the beginning of a new menstrual phase
   C. ENDOMETRIAL DETACHMENT AND ARTERIAL BLEEDING INTO THE UTERINE CAVITY lead to the loss of about 35 ml of blood and, over the 3-to-6-day period, the entire compact and most of the spongy uterine layers are lost in the menstrual flow. The remaining spongy and basal layers undergo regeneration during the next proliferative phase
   D. MENSTRUAL CYCLES normally continue until the end of the reproductive life, approximately to the ages of 47 to 52 years
   E. IF PREGNANCY OCCURS, menstruation ceases, and the endometrium passes into a *pregnancy phase*. After pregnancy, ovarian and menstrual cycles resume after a variable period of time, about 6 to 10 weeks later, if there is no breast feeding

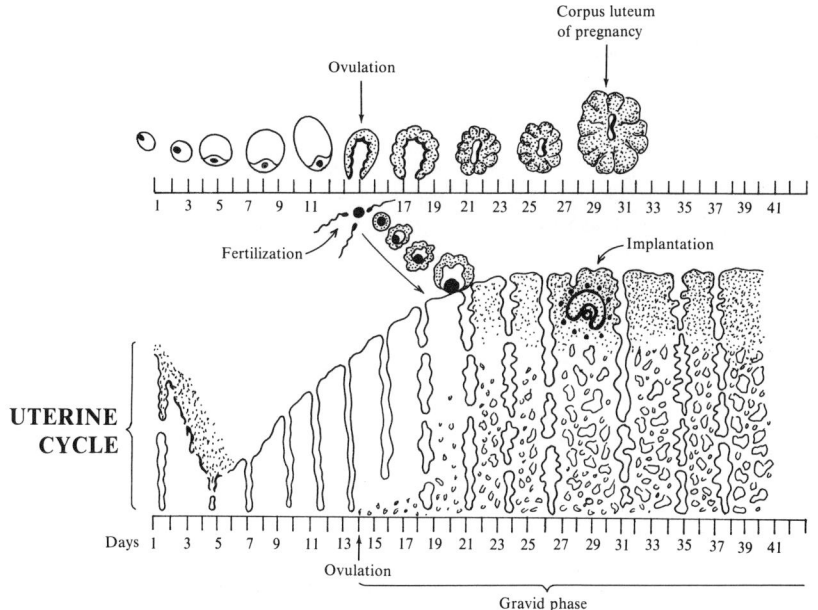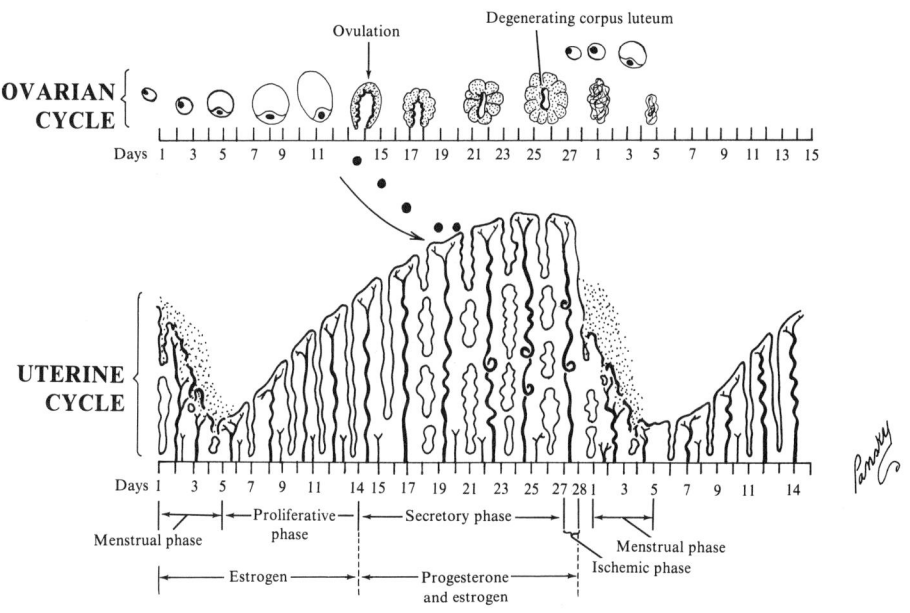

11. GERM CELL VIABILITY AND MOVEMENT AND ABNORMAL IMPLANTATION SITES

I. Sperm transport
A. THREE TO FIVE-HUNDRED MILLION SPERM are placed in the posterior fornix of the vagina during intercourse, near the external os of the cervical canal
  1. The sperm pass through the cervical canal by movement of their tails, whereas passage through the uterus and uterine tubes is facilitated by the muscular contractions of the walls of these organs
  2. Transport time to the fertilization site is short and takes about an hour
  3. About 300 to 500 sperm reach the fertilization site

II. Oocyte transport
A. THE OOCYTE, at ovulation, is carried in a peritoneal fluid stream, produced by the movements of the fimbriae of the uterine tube, into the infundibulum of the tube
  1. The oocyte passes into the ampulla of the uterine tube due to action of the cilia of the epithelial cells and by muscular contraction of the tubal wall

III. Fertilization site is in the tubal ampulla, its widest and longest portion
A. UNFERTILIZED OOCYTES undergo dissociation in the uterus

IV. Abnormal fertilization
A. PARTHENOGENESIS: oocyte is activated without sperm penetration and development may begin. No record of viable birth via this method
  1. Cleaving oocytes in ovary may develop into an *ovarian teratoma*
B. SUPERFECUNDATION may follow polyovulation. An oocyte is fertilized by spermatozoa from one male and another oocyte is fertilized by a second male. Seen in various mammals, not usual in man.
C. SUPERFETATION: ovulation and fertilization occur during an established pregnancy

V. Viability of the germ cell
A. SPERM remain alive *in vivo* for about a day or so
  1. Semen can be preserved *in vitro* for about 4 days and thus may actually survive that long in the female reproductive tract
  2. After freezing ($-79°$ C to $-196°$ C), semen may be kept for about 10 years
B. OOCYTES are usually fertilized within 12 hours after ovulation
  1. Unfertilized oocytes, *in vitro,* die within 12-24 hours

VI. Abnormal implantation sites:
A. THE HUMAN BLASTOCYST normally implants in the endometrium along the posterior wall of the body of the uterus, where it becomes attached between the openings of the endometrial glands or occasionally in the mouth of a glandular duct
B. NOT INFREQUENTLY, THE BLASTOCYST IMPLANTS IN ABNORMAL LOCATIONS outside the uterine body. This usually leads to the death of the embryo and severe hemorrhage of the mother during the second month of pregnancy. Such an implantation is called an *extrauterine* or *ectopic pregnancy* and may occur in the abdominal cavity, the ovary, the uterine tube or pelvis. Rarely does an extrauterine embryo come to full term
  1. Tubal pregnancy is the most frequent ectopic site. The tube usually ruptures during the second month of pregnancy, resulting in severe internal hemorrhaging
  2. Abdominal pregnancy: the peritoneal lining of the rectouterine cavity is the most frequent implantation site. Also on peritoneum of the intestinal tract or omentum
C. OCCASIONALLY, IMPLANTATION IN THE UTERUS ITSELF may lead to serious complications, particularly if implantation occurs near the internal os (low uterus). The placenta then bridges the os and we have what is called *placenta previa* which results in severe bleeding in the latter or second part of pregnancy and during delivery
D. FERTILIZED OVUM MAY ABNORMALLY MOVE to contralateral tube

FERTILIZATION SITE

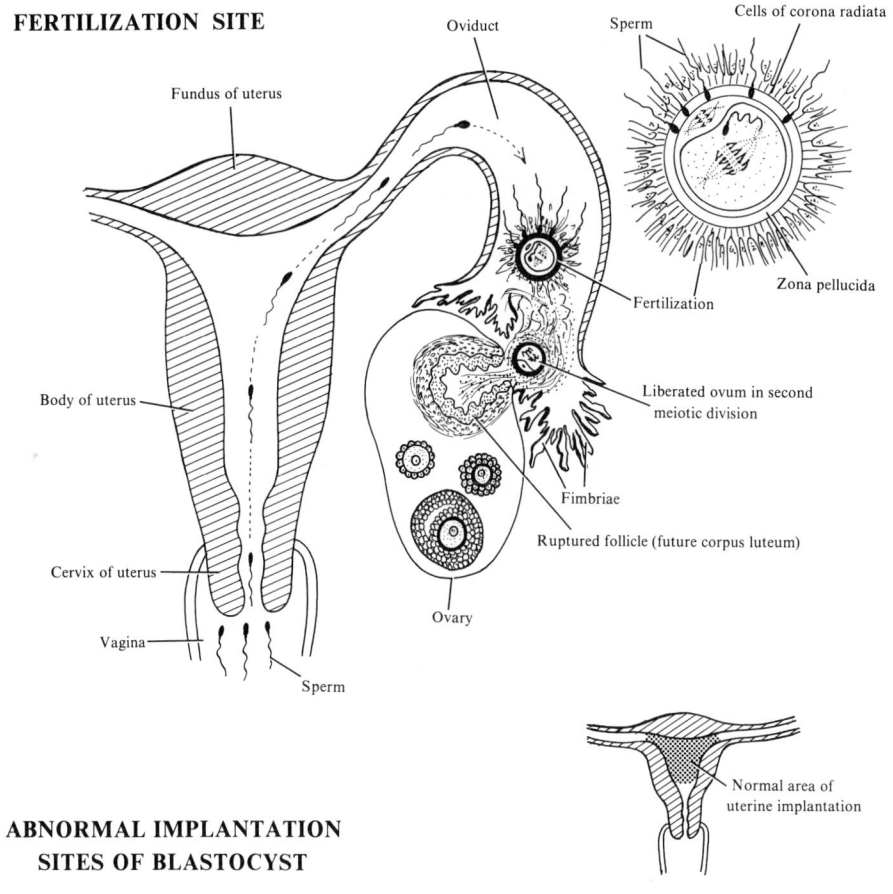

ABNORMAL IMPLANTATION SITES OF BLASTOCYST

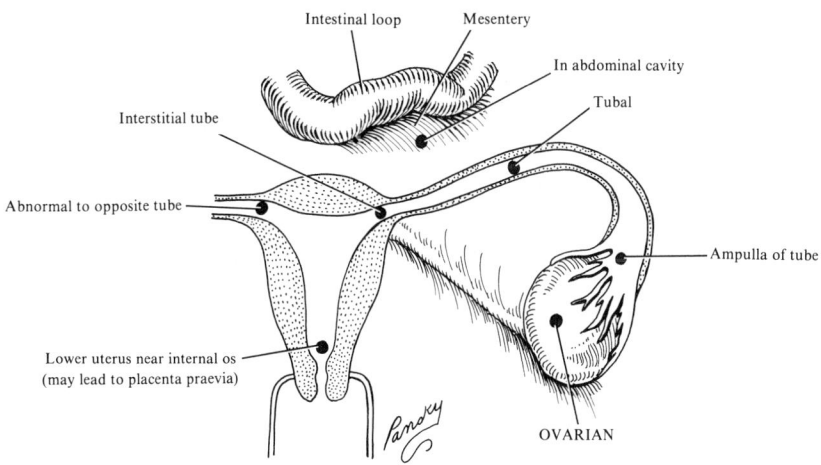

12. FERTILIZATION

I. **Definition:** fertilization is the union of the male (sperm) and female (oocyte) gametes to form a zygote and marks the beginning of pregnancy. Embryonic life begins with fertilization. Fertilization process requires about 24 hours

II. **Maturation of a follicle** takes place in the ovary from an oocyte to a graafian follicle

III. **Ovulation** coincides with the first maturation division and with the elimination of the first polar body. The ovum is "captured" by the ampulla of the uterine tube whose fimbriae sweep over the ovary

IV. **Fertilization** takes place in the distal third of the uterine tube. Spermatozoa arrive about 10 hours after coitus. The ovum must be fertilized within 24 hours after ovulation
   A. MORPHOLOGIC CHANGES IN FERTILIZATION
   1. Sperm passes through the corona radiata
      a. Dispersal of cells *in vitro* is the result of enzymatic action of tubal mucosa and semen. Sperm tail movements also help penetration of corona and zona pellucida
   2. Sperm penetrates the zona pellucida: digests a path by action of enzymes released from its acrosome
      a. Only 1 sperm enters the oocyte and fertilizes it, even though several may penetrate the zona pellucida
      b. Two sperm may take part in fertilization during an abnormal process called *dispermy* resulting in a triploid embryo (69 chromosomes), but it nearly always aborts or dies shortly after birth
      c. If 2 female pronuclei take part in fertilization, it is called *polygyny*
   3. Sperm head attaches to surface of the ooctye, plasma membranes of oocyte and sperm fuse, and then break at contact point
      a. Head and tail of sperm enter ooctye cytoplasm with sperm's plasma membrane being attached to oocyte's plasma membrane. Once inside the cytoplasm of the oocyte, the sperm tail degenerates
   4. Ooctye responds by
      a. Zonal reaction: change in zona pellucida inhibits entry of more sperm, due to substance of oocyte cytoplasm
      b. Secondary oocyte completes second meiotic division and its chromosomes (22 plus X) arrange themselves in a vesicular nucleus called the *female pronucleus*. The second polar body is extruded
   5. Sperm head enlarges and forms the *male pronucleus*
   6. The male and female pronuclei approach each other in the oocyte center, meet, and lose their nuclear membranes. They resolve their chromatin into a complete single haploid set of chromosomes which become organized on a spindle
   7. After the maternal and paternal chromosomes intermingle, metaphase of the first cleavage mitosis takes place, and the normal chromosome number is reconstituted
   8. Anaphase of the first cleavage mitosis then occurs
   9. The first 2 *blastomeres* are next seen, following cell division, and they are surrounded by the zona pellucida

V. **Consequences of fertilization**
   A. ACTIVATION OF THE OVUM
   B. MODIFICATION OF THE CYTOPLASM and of the membranes
   C. MODIFICATION OF THE NUCLEUS (Species variations occur)
      1. Reconstruction or restoration of the diploid number of chromosomes
      2. Determination of sex by the X and Y chromosomes of the sperm gamete
      3. Initiation of cleavage, stimulating the zygote to undergo rapid cell division

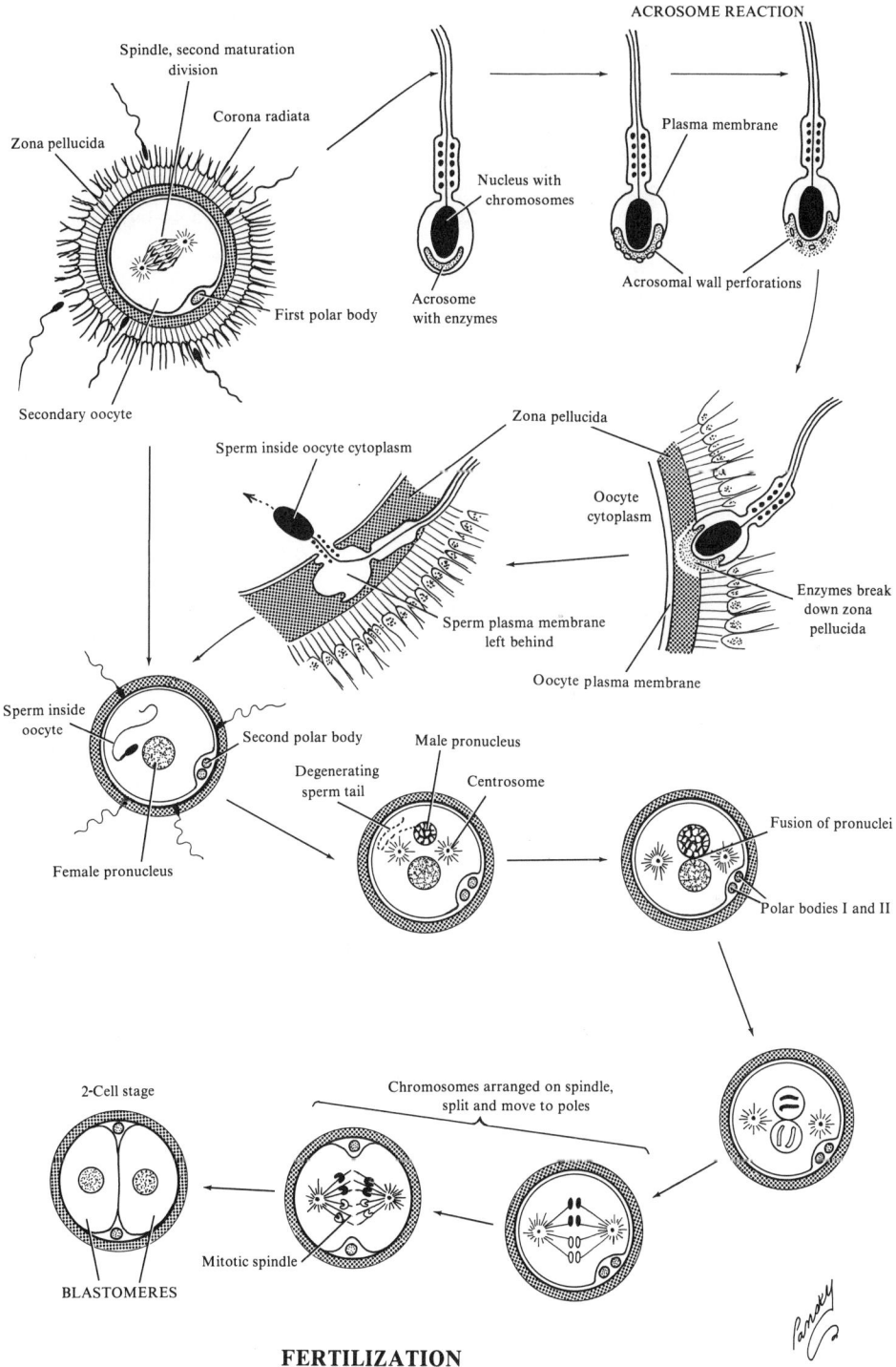

FERTILIZATION

13. IMPLANTATION AND ITS PREPARATION: GENERAL CONCEPTS

I. Changes in the uterine mucosa
A. IMPLANTATION generally takes place on the 21st day of the menstrual cycle during the progestational phase
   1. At this time, the mucosa is thick, highly vascularized, and contains a large amount of glyogen
   2. There are proliferation and predominance of secretion, congestion, and edema of the uterine wall
   3. The blastocyst finds conditions in the uterus very favorable for its implantation, especially for its nutrition

II. Hormonal aspects of implantation
A. ACTION OF OVARIAN HORMONES ON THE ENDOMETRIUM
   1. During each menstrual cycle, the uterine mucosa undergoes preparation for implantation which is directly conditioned by the ovarian hormones estrogen and progesterone
      a. The theca interna of the graafian follicle is the major source of estrogen, and the corpus luteum of pregnancy is the principal source of progesterone
   2. Morphologic changes in the uterine mucosa during the menstrual cycle result in proliferation of the endometrium involving not only the epithelium, the glands, and the stroma, but also, in a very essential way, the blood vessels

III. Hypophyseal-ovarian relationship
A. HYPOPHYSIS: endocrine activity of the ovary is under the control of the anterior lobe of the pituitary gland. The latter, in the human, secretes 2 gonad-stimulating hormones called *gonadotropins* or *gonadotropes*
   1. Follicle-stimulating hormone (FSH) is secreted from the very beginning of the menstrual cycle and determines the growth of the ovarian follicle
   2. Luteinizing hormone (LH) is secreted in the middle of the cycle and acts synergistically with FSH to provoke ovulation. LH stimulates the development of the corpus luteum
B. THE OVARY: the endocrine activity of the ovary, under the influence of the hypophyseal gonadotropins, is diphasic
   1. During phase 1, it secretes estrogen
   2. During phase 2, both estrogen and progesterone are secreted
      a. The secretion of progesterone is detectable even before the formation of the corpus luteum

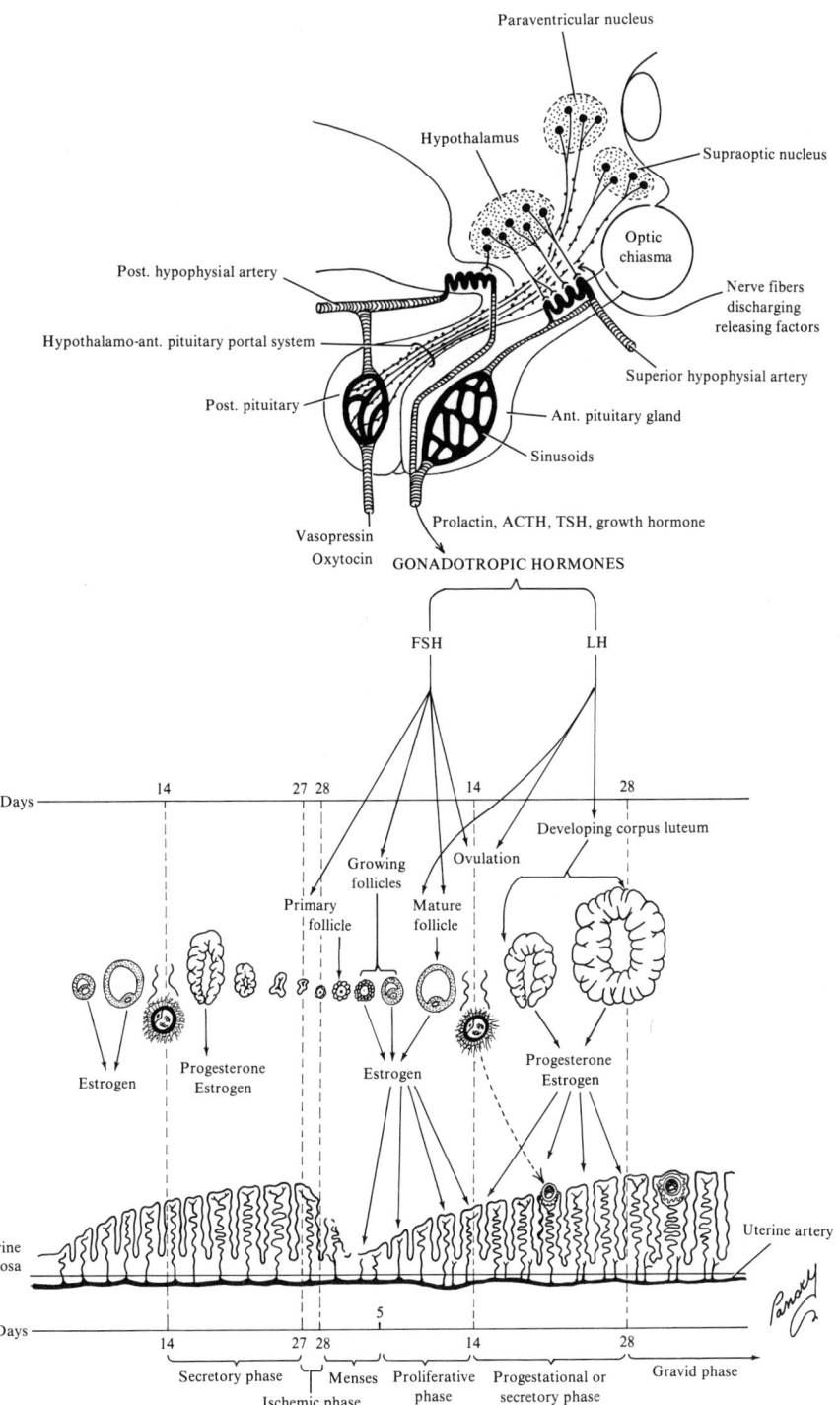

14. WEEK 1 OF EMBRYONIC DEVELOPMENT: OVULATION TO IMPLANTATION

I. **Introduction:** the unfertilized ovum reaches the ampulla of the uterine tube and is fertilized, in the distal third of the tube by 12-24 hours, to form a *zygote*. During its passage through the uterine tube, until the end of its morula stage, the egg undergoes almost no change in volume and is about 150 μm. It remains surrounded by the zona pellucida which it loses upon entering the uterine cavity. The zygote has made its way under the influence of peristaltic movements of the uterine tube and of ciliary movements of the tube epithelium. During its passage, the egg maintains itself on its own reserves, which are reduced since it is an alecithal* egg, and on tubular secretions. Survival of the egg and its transport down the tube, as well as implantation of blastocyst, depends on hormonal secretions of the ovary and anterior pituitary gland

II. **Cleavage** is the rapid mitotic cell division that the zygote, a single cell, undergoes as it passes down the uterine tube. It may occur without fertilization as a part of parthenogenesis (naturally occurring or artificially induced)
   A. DIVISION OF THE ZYGOTE into 2 daughter cells, the *blastomeres*, takes place by 30 hours. Further divisions follow rapidly upon one another, forming progressively smaller and smaller blastomeres: 4 are seen in 40-50 hours, 8 by 60 hours, and 12-16 by day 3 or 4
      1. The 12-16 blastomere stage, arrived at by cleavage of the fertilized ovum, is a solid ball resembling a mulberry and is called a *morula* (morula stage). As it forms, the morula enters the uterine cavity from the tube
   B. ABOUT DAY 4, fluid enters the morula from the uterine cavity and occupies the intercellular spaces. The fluid-filled spaces fuse to form a single, large cavity, the *blastocele*, and the morula is now called a *blastocyst* (blastocyst stage)
      1. As fluid increases, the cells separate into 2 major areas
         a. An outer cell layer, the *trophoblast*, which gives rise to the placenta
         b. A group of centrally located cells, the *inner cell mass* or *embryoblast*, which gives rise to the embryo proper
      2. The free blastocyst is seen in the uterine cavity on day 4 or 5. The zona pellucida disappears rapidly
   C. IMPLANTATION: normal area is the upper posterior wall of the uterine mucosa
      1. In the human, the trophoblast cells over the embryonic pole of the blastocyst penetrate the epithelial cells of the uterine mucosa at about day 6 or 7 or about 20 days after the beginning of the last menstruation
         a. Penetration and erosion of the epithelial cells of the mucosa result from proteolytic enzymes produced by the trophoblast
         b. The uterine mucosa also promotes proteolytic action of the blastocyst so implantation is a mutual action of the endometrium and trophoblast
      2. As invasion of uterus proceeds, the trophoblast differentiates into 2 layers
         a. An inner *cytotrophoblast* or *cellular trophoblast*
         b. An outer *syncytiotrophoblast* or *syncytial trophoblast* consisting of a multinucleated protoplasmic mass in which intercellular boundaries are absent
      3. The fingerlike processes of the syncytial trophoblast grow into the endometrial epithelium and invade the endometrial stroma
      4. By the end of week 1, the blastocyst is superficially implanted in the compact layer
      5. At the time of implantation, the uterine mucosa is at day 21 of the menstrual cycle and is richly vascularized, edematous, and secreting mucus and glycogen—all favoring implantation of the blastocyst
      6. As the blastocyst is implanting, early differentiation of the inner cells mass occurs
         a. The embryonic endoderm (entoderm), a flattened layer of cells, appears on the surface of the inner cell mass facing the blastocyst cavity, at about day 7. It is the first of 3 germ layers of the embryo that forms during the first 3 weeks

*Alecithal: without yolk; ova with little or no deutoplasm.

FIRST WEEK OF DEVELOPMENT

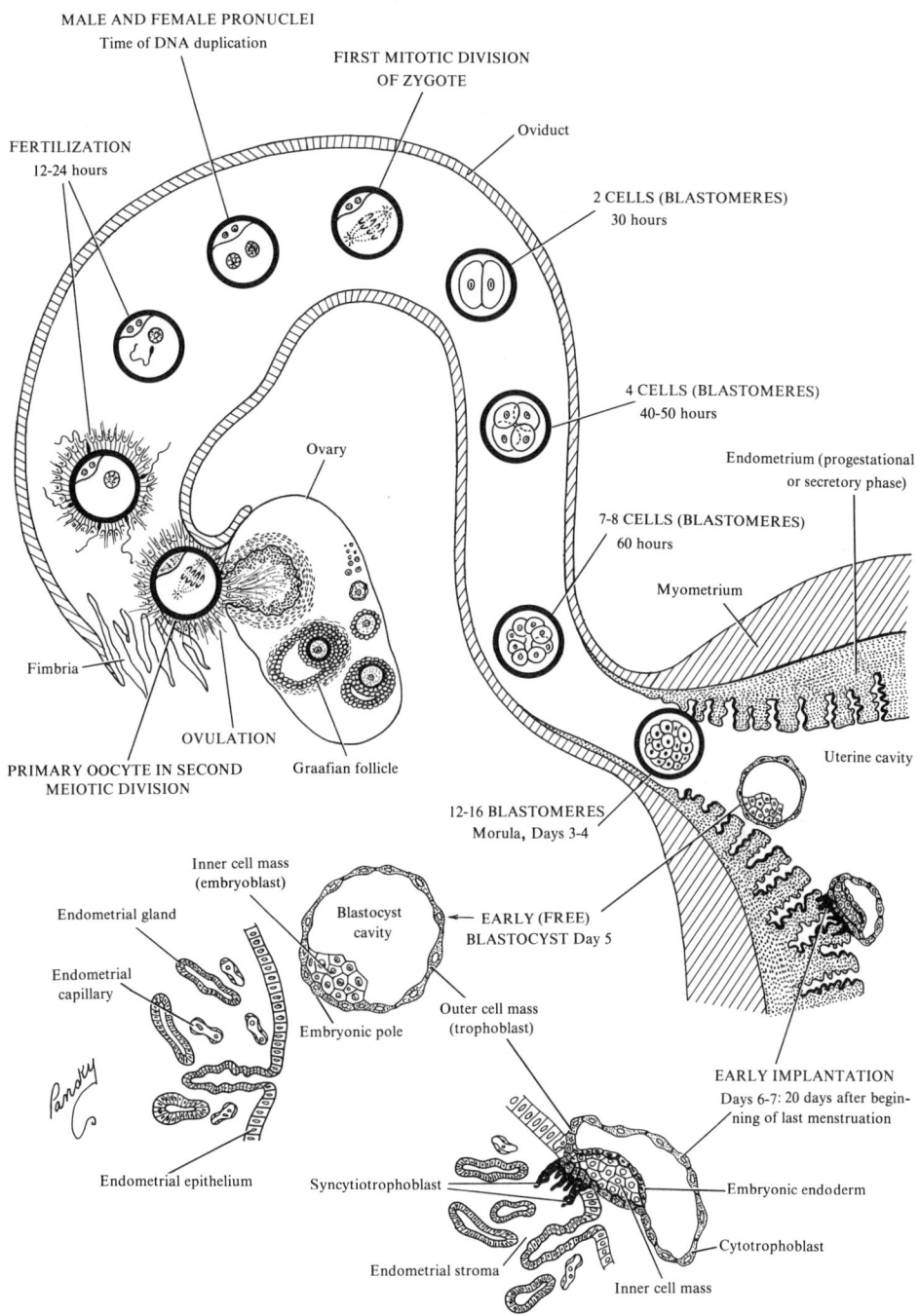

15. WEEK 2 OF DEVELOPMENT: BILAMINAR GERM DISK EMBRYO*

I. Introduction: on day 7, endometrial invasion has begun, and the trophoblast differentiates into 2 layers: a cytotrophoblast and a syncytiotrophoblast or syncytium. The active erosive trophoblast invades the endometrial stroma containing capillaries and glands, and the blastocyst slowly sinks into the endometrium. The syncytium at the embryonic pole, near the developing embryo, rapidly becomes a thick, multinucleated protoplasmic mass with no cell boundaries. It is invasive, ingestive, and digestive, and the conceptus derives its initial nourishment from endometrial tissues. Later it receives nutrients directly from maternal blood. The cytotrophoblast is mitotically active and forms new cells that migrate into the increasing mass of the syncytiotrophoblast

In week 2 of implantation, the trophoblast penetrates deeper into the endometrium, and the blastocyst changes morphologically. The inner cell mass produces a bilaminar embryonic disk composed of epiblast (future embryonic ectoderm and mesoderm) and embryonic endoderm. Concomitantly, the amniotic cavity, yolk sac, connecting stalk, and chorion develop

II. Normal stages of week 2 development

A. DAY 8
 1. Cells of the inner cell mass differentiate into 2 distinct layers
    a. Endodermal (entodermal) germ layer: layer of small, cuboidal cells
    b. Ectodermal germ layer: layer of high columnar cells
 2. The cells of each germ layer form a flat disk and together are known as the *bilaminar germ disk*
 3. Cells of the ectodermal layers, initially firmly attached to the cytotrophoblast, develop small clefts between their layers as development proceeds
    a. The clefts coalesce and form a cavity, the *amniotic cavity*
    b. Amnioblasts, large, flattened cells, are seen along the trophoblastic border of the newly formed amniotic cavity (probably derived from trophoblast)
       i. The cells are continuous with the ectoderm and together line the amniotic cavity
    c. Endometrial stroma adjacent to the implantation site is edematous, highly vascular, with large tortuous glands that secrete glycogen and mucus

B. DAY 9: blastocyst embeds deeper into endometrium, and a fibrin coagulum "plug" (blood clot and cellular debris) closes the penetration defect in uterine epithelial surface—*interstitial implantation*
 1. Trophoblast progresses in development, especially at the embryonic pole, and vacuoles appear in the syncytium. The vacuoles fuse to form large lacunae (lakes), and we have the lacunar stage of trophoblast development
 2. Endometrial stroma around the trophoblast has vascular congestion, and the cells are rich in glycogen
 3. Flattened cells delaminate from the inner surface of the cytotrophoblast, at the abembryonic pole, and form a thin membrane called *Heuser's exocoelomic membrane* which is continuous with the edges of the entodermal layer. Together, they form the lining of the *exocoelomic cavity* or the *primitive yolk sac*

---

*Embryos of the same fertilization size do not necessarily develop at the same rate. Considerable differences are seen in early developmental stages.

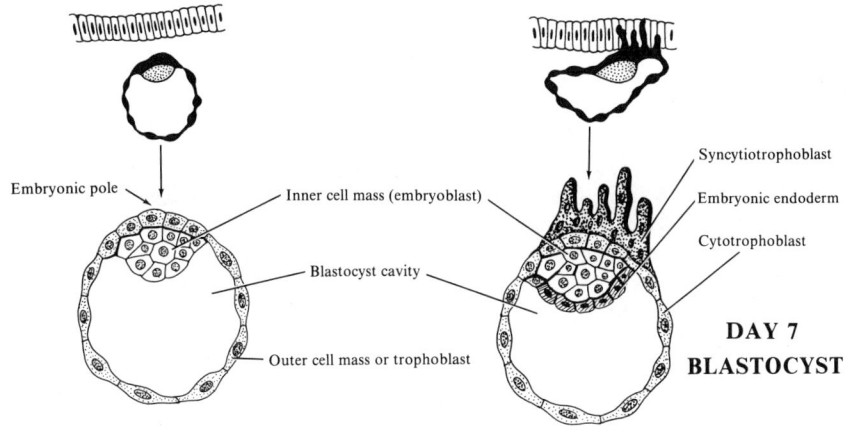

DAY 6 BLASTOCYST

DAY 7 BLASTOCYST

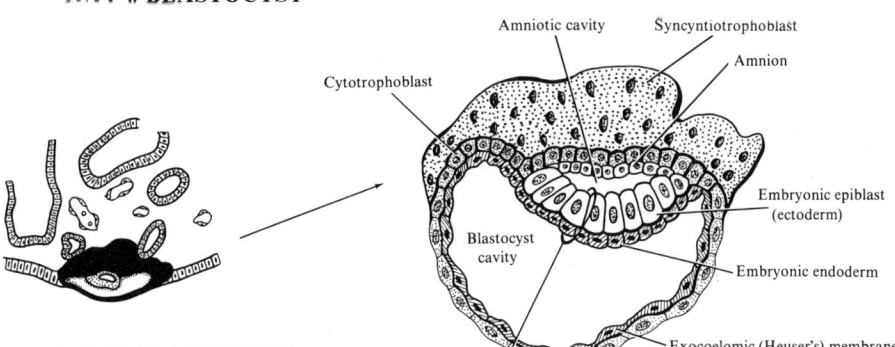

DAY 8 BLASTOCYST

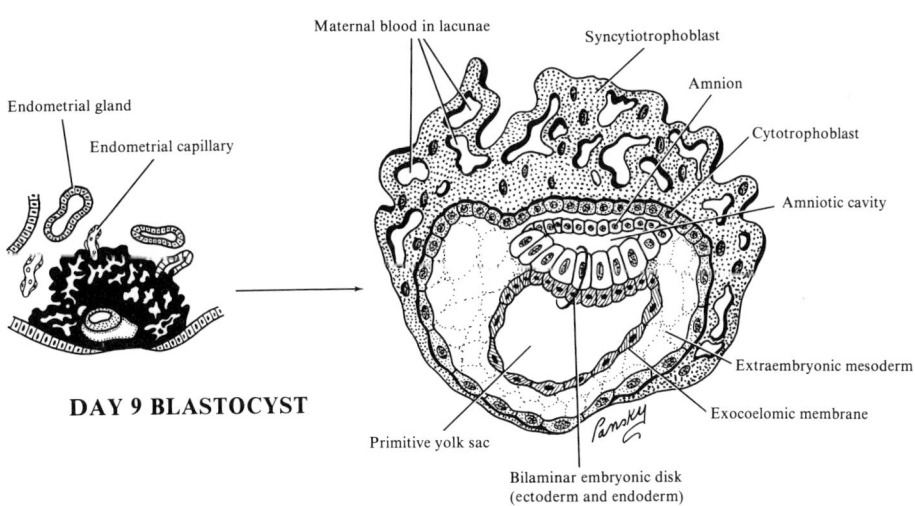

DAY 9 BLASTOCYST

16. WEEK 2 OF DEVELOPMENT: DAYS 10 TO 14

I. Normal stages of week 2 development (cont.)
C. DAYS 10, 11, AND 12
  1. The blastocyst becomes completely embedded in the endometrial stroma, and the uterine surface epithelium almost entirely covers the original epithelial lining defect of the mucosa. Only a slight protrusion is seen in the uterine lumen
  2. The trophoblast is characterized by lacunar spaces in the syncytium, and they form an interconnecting network, particularly at the embryonic pole
  3. At the abembryonic pole, the trophoblast consists of cytotrophoblastic cells and only a few lacunar spaces
  4. The syncytial cells penetrate deep into the stroma and erode the endothelial lining of maternal congested and dilated capillaries called *sinusoids*
     a. The syncytium becomes continuous with the endothelial cells of the vessels, and maternal blood enters the lacunar system
     b. With more and more sinusoid invasion by the trophoblast, the lacunae eventually become continuous with the arterial and venous systems. Pressure differences between arterial and venous capillaries result in maternal blood flowing through the trophoblastic lacunar system to form the *uteroplacental circulation*
  5. Cytotrophoblast also differentiates. On its inner surface, cells delaminate to form a fine, loose tissue, the *extraembryonic mesoderm,* which fills the space between external trophoblast and amnion and internal yolk sac
     a. Large cavities develop in this extraembryonic mesoderm, become confluent, and form the *extraembryonic coelom,* which surrounds the primitive yolk sac and amniotic cavity, except where the extraembryonic mesoderm forms the future connection between the germ disk and the trophoblast
        i. The extraembryonic mesoderm lining the cytotrophoblast and amnion is called the *extraembryonic somatopleuric mesoderm;* that covering the yolk sac is called the *extraembryonic splanchnopleuric mesoderm*
  6. The bilaminar germ disk grows slowly compared to the trophoblast, but by the end of day 12, entodermal cells begin to spread over the inside of Heuser's membrane
  7. Endometrial cells become polyhedral, are loaded with glycogen and lipids, and the intercellular spaces fill with extravasate; the tissue is edematous—all a process of the *decidual reaction,* initially at implantation site but then throughout endometrium
D. DAYS 13 AND 14: the endometrial surface defect is usually healed, but there may be occasional bleeding at the site of implantation due to the increased blood flow in the lacunar spaces at the abembryonic pole. Can be confused with menstrual bleeding
  1. The trophoblast shows more organization at the embryonic pole
  2. The cytotrophoblast cells proliferate, penetrate the syncytium, and form cellular columns surrounded by syncytium—together forming the *primary stem villi*
  3. The entodermal germ layer proliferates and newly formed cells gradually line a new cavity known as the *secondary* or *definitive yolk sac* (smaller than original)
  4. The extraembryonic coelom expands to form the *chorionic cavity*
  5. The extraembryonic mesoderm then lines the cytotrophoblast and is called the *chorionic plate.* It also forms a covering layer for the secondary yolk sac and amnion
     a. The extraembryonic mesoderm only traverses the chorionic cavity in the connecting stalk (connecting the embryo with the trophoblast)
     b. With development of blood vessels, the stalk becomes the *umbilical cord*
  6. By the end of week 2, the germ disk consists of two apposed cell disks: the ectodermal germ layer, forming the floor of the expanding amniotic cavity, and the entodermal germ layer, forming the roof of the secondary yolk sac
     a. In its cephalic region, the entodermal disk is thickened to form the *prochordal plate,* an area of columnar cells attached to the overlying ectodermal disk
  7. The primitive streak appears and indicates the onset of gastrulation

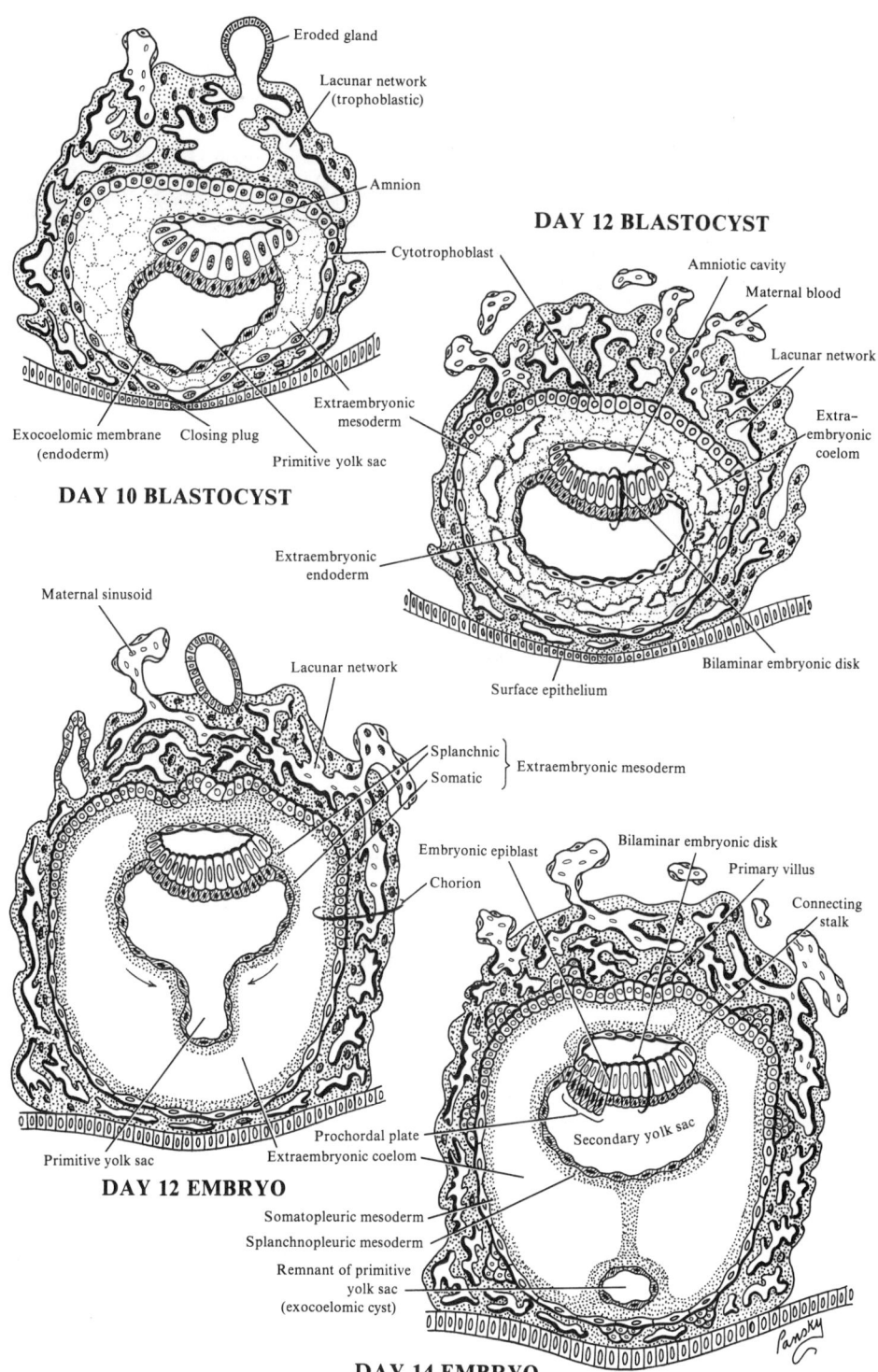

17. REVIEW OF WEEK 2 AND ABNORMAL DEVELOPMENT

I. Review of week 2
A. BLASTOCYST IMPLANTATION begins at the end of week 1 and ends during week 2
   1. The zona pellucida disappears (days 4 to 5) probably due to enzymatic lysis released by the acrosomes of sperm that penetrate it
   2. The blastocyst attaches to the endometrial wall (days 5 to 6)
   3. The trophoblast erodes the uterine epithelium as it differentiates (days 7 to 8) into an inner cytotrophoblast and an outer syncytiotrophoblast
   4. The lacunae are seen in the syncytiotrophoblast (day 9)
   5. The lacunae fuse to form lacunar networks (days 10 to 11)
   6. The trophoblast invades the endometrial sinusoids permitting the maternal blood to flow into the lacunar networks, thus, a uteroplacental circulation is formed (days 11 to 12)
   7. The endometrial surface defect disappears (days 12 to 13)
   8. One sees a marked decidual reaction in the endometrium around the implanting embryo (days 13 to 14)

II. Abnormal development
A. ABOUT 15% OF OOCYTES fail to become fertilized
B. ABOUT 10–15% OF OOCYTES begin cleavage, but fail to implant
C. OF THE 70–75% OF OOCYTES that do implant, only about 58% survive until week 2
   1. Of the above, about 16% are abnormal
D. WHEN THE FIRST EXPECTED MENSTRUATION IS MISSED, only about 42% of the eggs exposed to sperm usually survive and, of these, a number will abort in subsequent weeks, and some will be abnormal at birth
E. ABOUT ONE-THIRD TO ONE-HALF OF ALL ZYGOTES never survive to implant
   1. May be the result of a poorly developed endometrium
   2. Most likely caused by chromosomal abnormalities in the zygote itself
F. EXPOSURE OF THE EMBRYOS TO TERATOGENS during the first 2 weeks usually does not result in congenital malformations, but some do kill the blastocysts or cause early abortions
G. ABOUT 25% OF EARLY ABORTED EMBRYOS (days 7 to 17) have abnormal chromosomes and are lost during the menstrual flow, which occurs later than usual and the woman may not even be aware of the fact that she had been pregnant
H. IF THE TROPHOBLAST BECOMES HIGHLY PROLIFERATIVE (or even after embryonic death), we may find a noninvasive *hydatidiform mole* or a highly malignant tumor, a *chorioepithelioma* (from chorionic epithelium) which may produce high levels of human chorionic gonadotropin

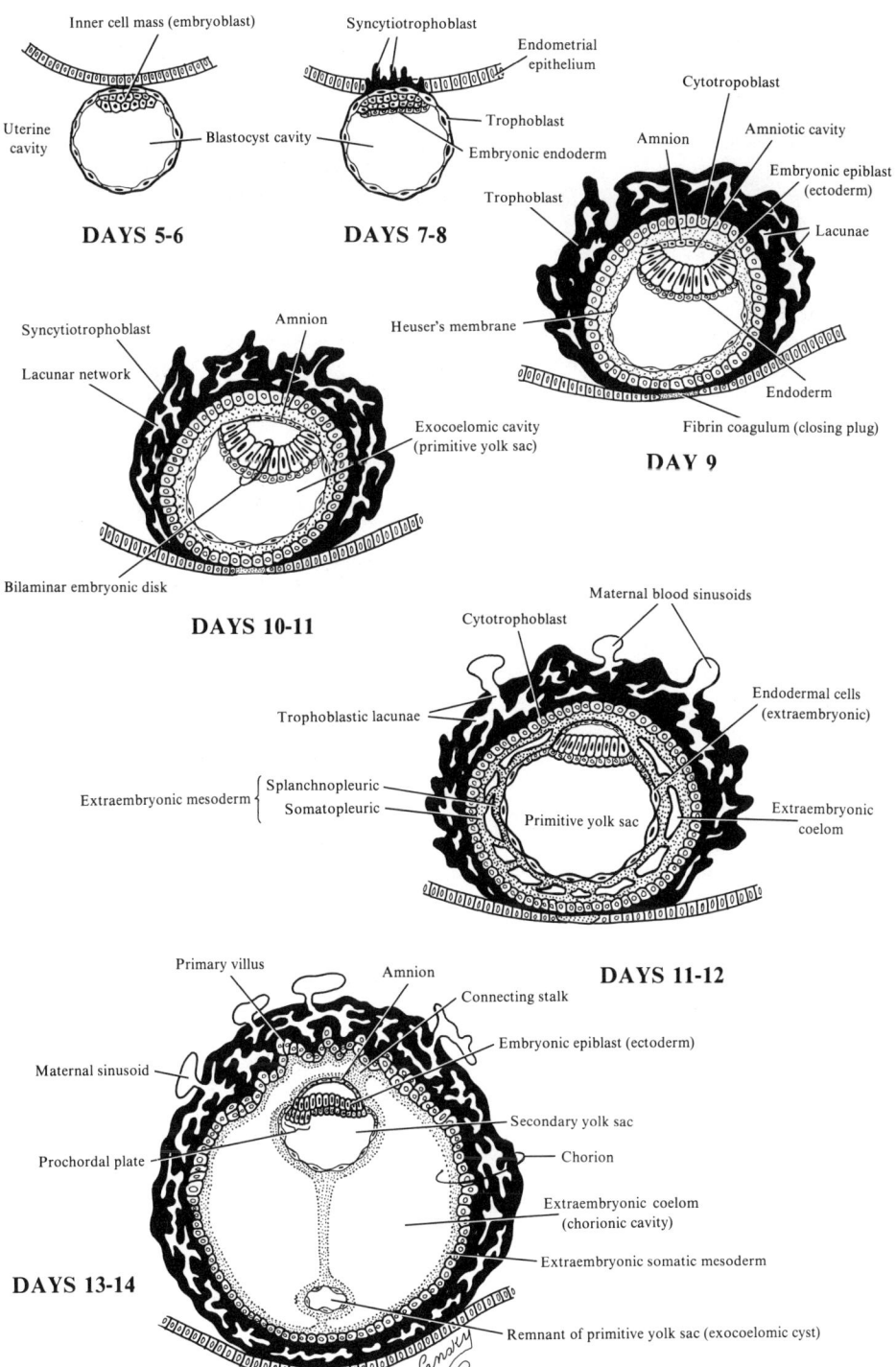

18. WEEK 3 OF DEVELOPMENT: TRILAMINAR GERM DISK EMBRYO FORMATION AND GASTRULATION

I. **Introduction:** week 3 is a period of rapid development of the conceptus coinciding with the first missed menstrual period. By days 15–16, the embryo is 1.5 mm long, and one clearly sees the primitive streak, Hensen's node, and the notochordal process—all morphologic indications characteristic of gastrulation. The latter is the formation of the third embryonic layer, the mesoderm

II. **Trilaminar germ disk**
A. PRIMITIVE STREAK appears near the end of week 2 and is clearly seen about day 15 as a narrow midline, thickened linear band of embryonic epiblast in the caudal end of the dorsal aspect of the embryonic disk. Its appearance enables identification of embryonic axes, cranial and caudal ends, top and bottom surfaces, and sides of the embryo
  1. As the streak elongates by addition of cells at its caudal end, its cranial end thickens to form the *primitive knot or node* (*Hensen's*); simultaneously a narrow *primitive groove* develops in it which continues into a depression in the knot called *the primitive pit*
  2. About day 16, cells of the epiblast migrate medially toward the streak, enter the primitive groove, then leave the basal layer of the groove and migrate laterally between the embryonic ectoderm and entoderm to organize into a layer, the *intraembryonic mesoderm*
     a. Cells migrate from the intraembryonic mesoderm to become mesenchymal cells and form mesenchyme which has a variety of differentiation: become fibroblasts, osteoblasts, and chrondroblasts. The mesenchyme functions as a supporting tissue
     b. The epiblast layer is called the *embryonic ectoderm* (second germ layer) after the streak begins to form cells destined to become mesoderm
  3. About day 16, cells migrate cranially from the primitive knot and form a midline cord, the *notochordal process* which grows between ectoderm and entoderm to reach the *prochordal plate* (small circular area of columnar entodermal cells). The latter is firmly attached to overlying ectoderm forming the *oropharyngeal* or *buccopharyngeal membrane*
     a. Other cells of the streak and notochordal process migrate laterally and cranially to reach margins of the embryonic disk and there join the extraembryonic mesoderm covering the yolk sac and amnion. Thus, most extraembryonic mesoderm comes from trophoblast, but some may originate from the primitive streak
     b. Some cells of the primitive streak migrate cranially, pass on each side of the notochordal process, around the prochordal plate, and meet cranially in the cardiogenic area where the heart will be formed
  4. The *cloacal membrane* is a circular area caudal to the primitive streak. The embryonic disk remains bilaminar here and at the oropharyngeal membrane because embryonic ectoderm and entoderm fuse at these sites and prevent mesenchymal cells from migrating between them
  5. By middle of week 3, intraembryonic mesoderm separates the ectoderm and entoderm everywhere except at the oropharyngeal membrane cranially, the cloacal membrane caudally, and in the midline, cranial to the primitive knot, where the notochordal process extends
  6. The embryonic disk is at first flat and circular but eventually elongates and becomes pear-shaped as the notochordal process grows. The disk expands mostly in the cranial region; the caudal region remaining mostly unchanged
     a. The growth and elongation of the disk are caused by the continuous migration of cells from the primitive streak
  7. The primitive streak actively forms the intraembryonic mesoderm until the end of week 4 and then slows down as it diminishes in relative size. It usually undergoes degeneration and disappears but may in some cases persist and give rise to a teratoma

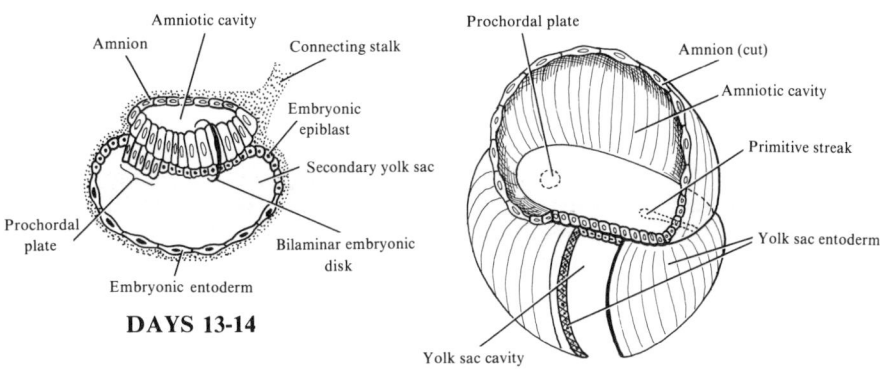

DAYS 13-14

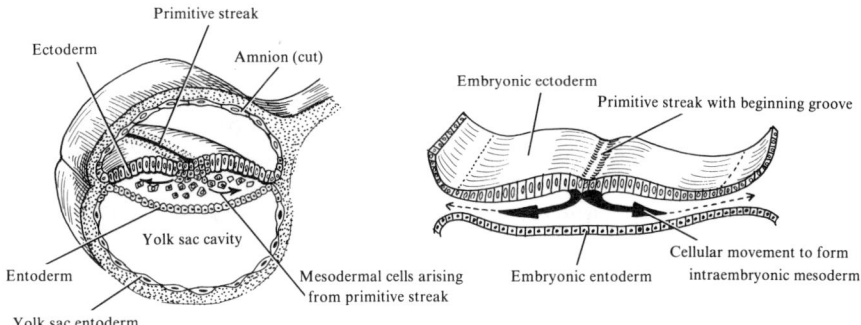

DAYS 15-16

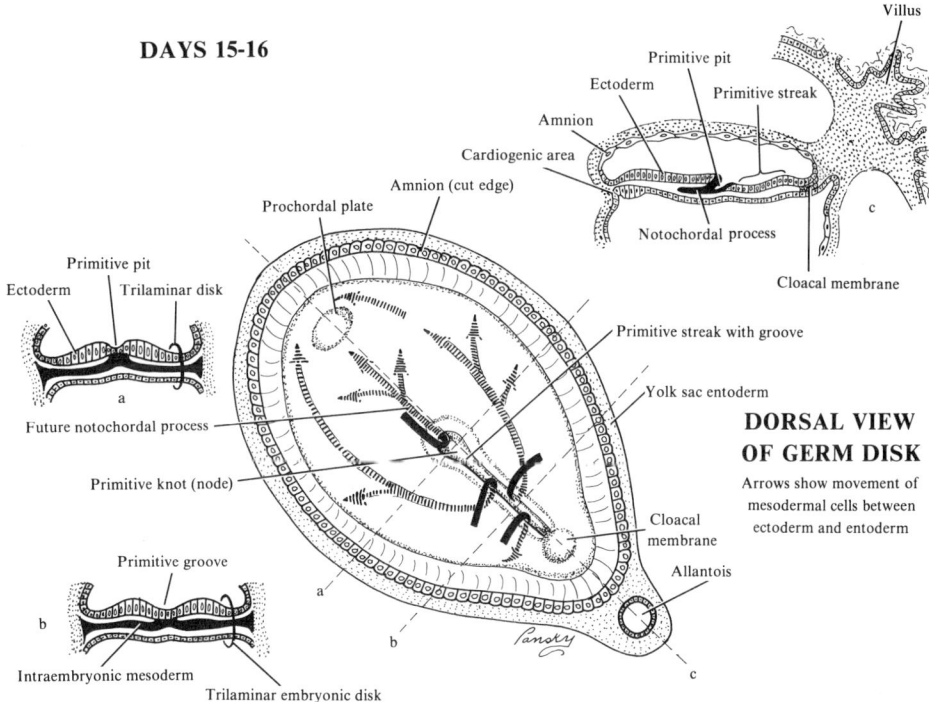

DORSAL VIEW OF GERM DISK

Arrows show movement of mesodermal cells between ectoderm and entoderm

19. WEEK 3 OF DEVELOPMENT: THE NOTOCHORD, NEURAL TUBE, AND ALLANTOIS

I. **Notochord development:** in the human, the notochord is a cellular rod that develops from the prochordal process and forms the first longitudinal midline axis around which the vertebral bodies are organized and is the basis for the axial skeleton. It will later regress. By day 12 or 13, the notochord is visible throughout the length of the embryo and around it are layered concentrations of cells, representing the primordia of the future vertebral bodies
  A. STAGE OF THE NOTOCHORDAL PROCESS (entire area cephalic to the primitive streak): about day 17
     1. The floor of the notochordal process fuses with the underlying entoderm as it undergoes preferential growth. Hensen's node seems to recede toward the caudal end
  B. PROCHORDAL STAGE: about day 19
     1. Degeneration of the fused region takes place, and openings appear in the floor of the notochordal process (resorption of the floor), opening a communication between the yolk sac and the *notochordal canal,* a lumen which is formed as the primitive pit extends into the notochordal process during its development
     2. The openings become confluent, and the floor of the notochordal canal disappears. A small passage, the *neurenteric canal,* temporarily connects the yolk sac and the amniotic cavity
  C. NOTOCHORD STAGE: about day 20
     1. The notochord process remains and forms a grooved, flattened plate, the *notochordal plate,* which, beginning at its cranial end, infolds to form the *notochord.* The embryonic entoderm again forms a continuous layer below the notochord. The latter is thus the primary skeleton of the 3-layer embryo

II. **Neural tube development (neurulation)**
  A. THE EMBRYONIC ECTODERM over the developing notochord thickens to form a *neural plate* (about day 18) which apparently is enduced by the developing notochord and paraxial mesoderm on either side
     1. The plate first appears cranial to the primitive knot and dorsal to the notochordal process with mesoderm adjacent to it
     2. With the elongation of the notochordal process, the neural plate broadens and extends cranially to the oropharyngeal membrane
     3. The ectoderm of the plate is called *neuroectoderm* and eventually gives rise to the central nervous system (brain and spinal cord)
  B. THE NEURAL PLATE, on about day 20, invaginates along its central axis to form the *neural groove* with *neural folds* created on each side of the groove
     1. By the end of week 3, the neural folds move together, fuse, and convert the neural plate into the *neural tube*. Closure begins in the middle of the embryo and progresses toward both cephalic and caudal ends. It begins on day 21
        a. Closure of the neural groove is more rapid toward the cephalic (anterior) end than toward the caudal (posterior) end
           i. The anterior or cranial neuropore closes in week 4 (day 26), whereas the posterior or caudal neuropore closes near day 28
     2. Neuroectodermal cells at the lateral edge of the neural plate do not become part of the tube but form a *neural crest* over the neural tube and give rise to the neural crest cells

III. **The allantois:** appears on day 16 as a small, fingerlike outpouching or diverticulum from the caudal wall of the yolk sac. It remains small in the human embryo, is involved with early blood formation, and is related to the development of the urinary bladder

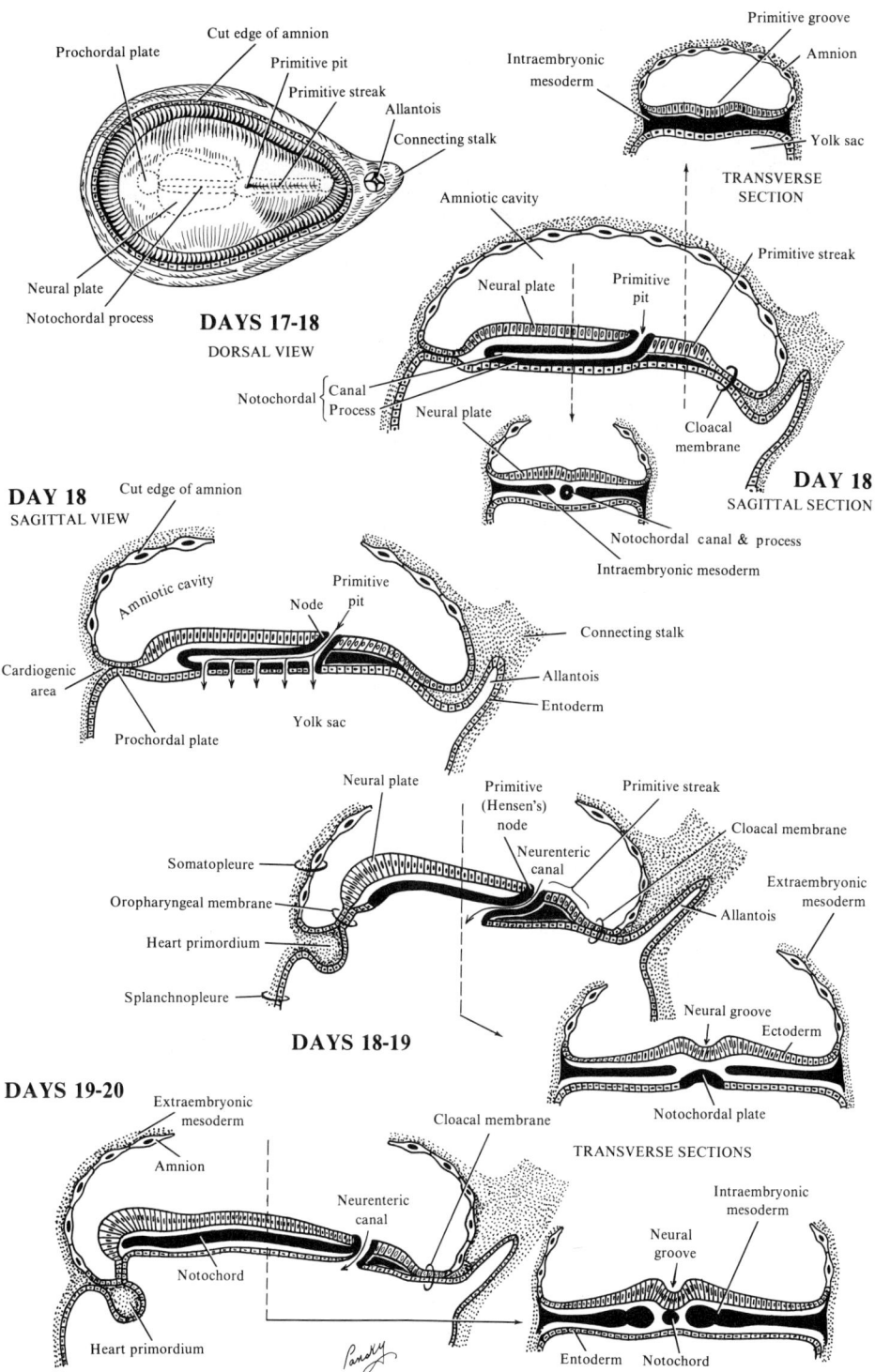

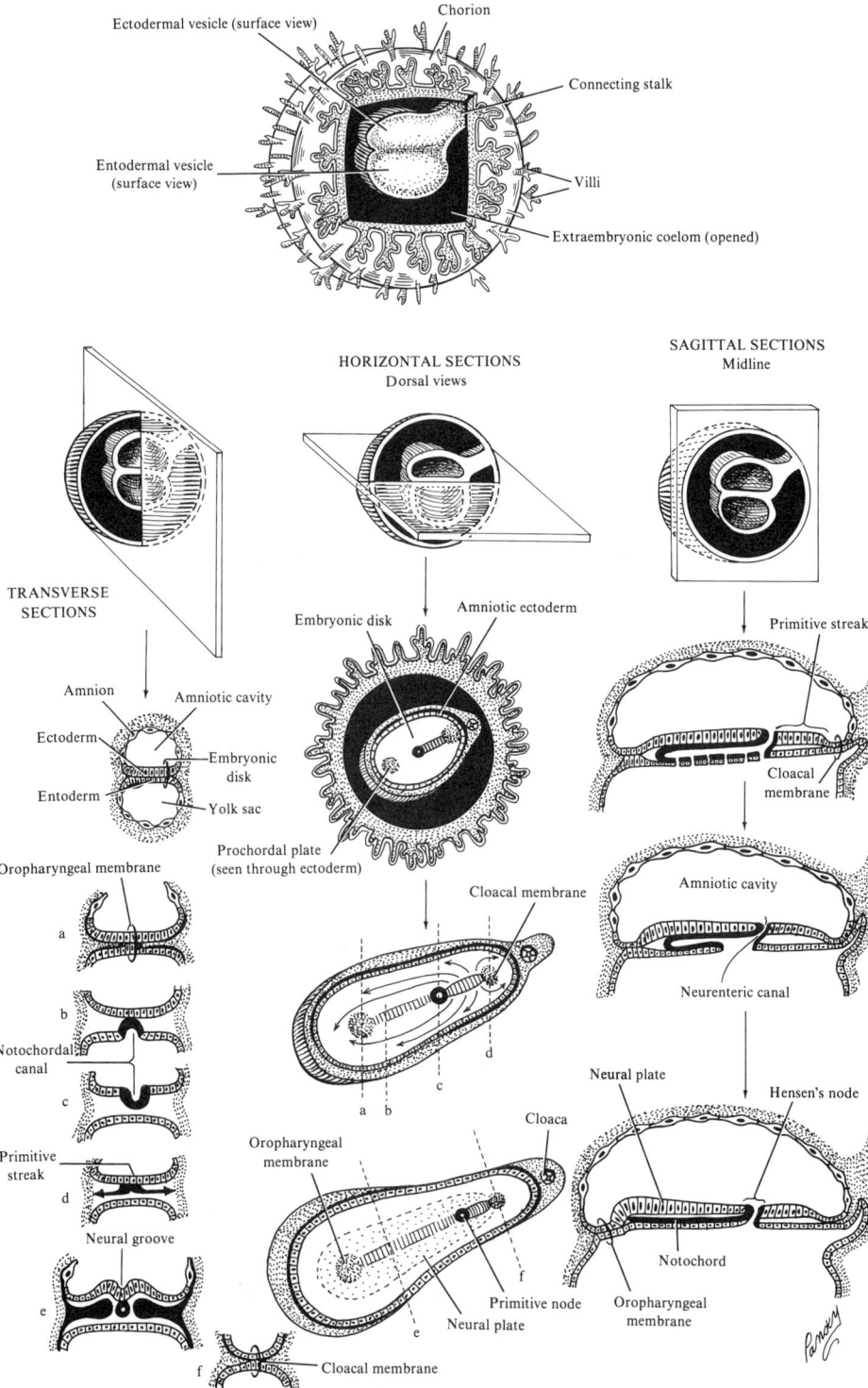

FIGURE 9. **Transverse, horizontal, and sagittal sections of early embryogenesis.**

10. WEEK 3 OF DEVELOPMENT: INTRAEMBRYONIC MESODERM, SOMITE DEVELOPMENT, AND THE INTRAEMBRYONIC COELOM

I. **The intraembryonic mesoderm**
   A. AS THE NOTOCHORD AND NEURAL TUBE FORM, the intraembryonic mesoderm on each side forms longitudinal columns, the *paraxial mesoderm*, each in turn being continuous laterally with the *intermediate mesoderm*, and the latter gradually thinning out further laterally into the *lateral mesoderm*
   B. PARAXIAL MESODERM AND SOMITE FORMATION: somite development begins about day 20 and is the result of segmentation of the paraxial mesoderm
      1. The paraxial mesoderm thickens and fragments metamerically, dividing into paired cuboid bodies called somites which give rise to most of the axial skeleton and associated musculature as well as much of the dermis of the skin
      2. The first pair of somites develops just caudal to the cranial end of the notochord (future occipital area), and subsequent pairs form in a craniocaudal sequence after the appearance of the first somites
      3. About 38 somite pairs form during days 20–30, the so-called *somite period*. Eventually about 42–44 somite pairs develop by the end of week 5
         a. The somites form distinct surface elevations and are triangular in shape when seen in a transverse section
         b. Each somite develops a slitlike cavity, the *myocele*, which eventually is occluded
         c. Somites give origin to the *sclerotome*, whose cells condense around the notochord and give rise to the vertebral primordia and the *myotome*, which gives rise to the vertebral muscles
            i. The myotome with the somatopleure gives origin to the muscles of the limbs and the anterior lateral body wall
   C. INTERMEDIATE AND LATERAL PLATE MESODERM
      1. Intermediate mesoderm gives rise to the nephrogenic cord from which the excretory apparatus originates
         a. This mesodermal cell aggregation undergoes metameric segmentation parallel to the somites and forms nephrotomes. Segmentation, however, is never completed and never reaches the extreme caudal end of the nephrogenic cord
      2. Lateral plate mesoderm splits into 2 layers
         a. The intraembryonic splanchnopleure, which gives rise to muscle and connective tissue layers of the trunk, lines the entoderm and continues in the extraembryonic splanchnopleure
         b. The intraembryonic somatopleure or outer layer which lines the ectoderm and helps form the lateral and ventral trunk walls

II. **The intraembryonic coelom** first appears as many small isolated coelomic spaces in the lateral mesoderm and cardiogenic mesoderm (between the 2 layers of the lateral plate mesoderm) which coalesce to form a horseshoe-shaped cavity, the intraembryonic coelom, which is lined by flattened epithelial (mesothelial) cells. It will become the pleuro-pericardial-peritoneal cavity
   A. THE COELOM DIVIDES the lateral mesoderm into 2 layers: a somatic (parietal) layer continuous with the extraembryonic mesoderm over the amnion and a splanchnic (visceral) layer, which is continuous with the extraembryonic mesoderm over the yolk sac
      1. Somatic mesoderm plus overlying embryonic ectoderm forms the body wall or *somatopleure*
      2. Splanchnic mesoderm plus embryonic entoderm forms the wall of the primitive gut and is called the *splanchnopleure*
   B. DURING THE SECOND MONTH, the intraembryonic coelom is divided into the body cavities, namely, the pericardial cavity, the pleural cavities, and the peritoneal cavity

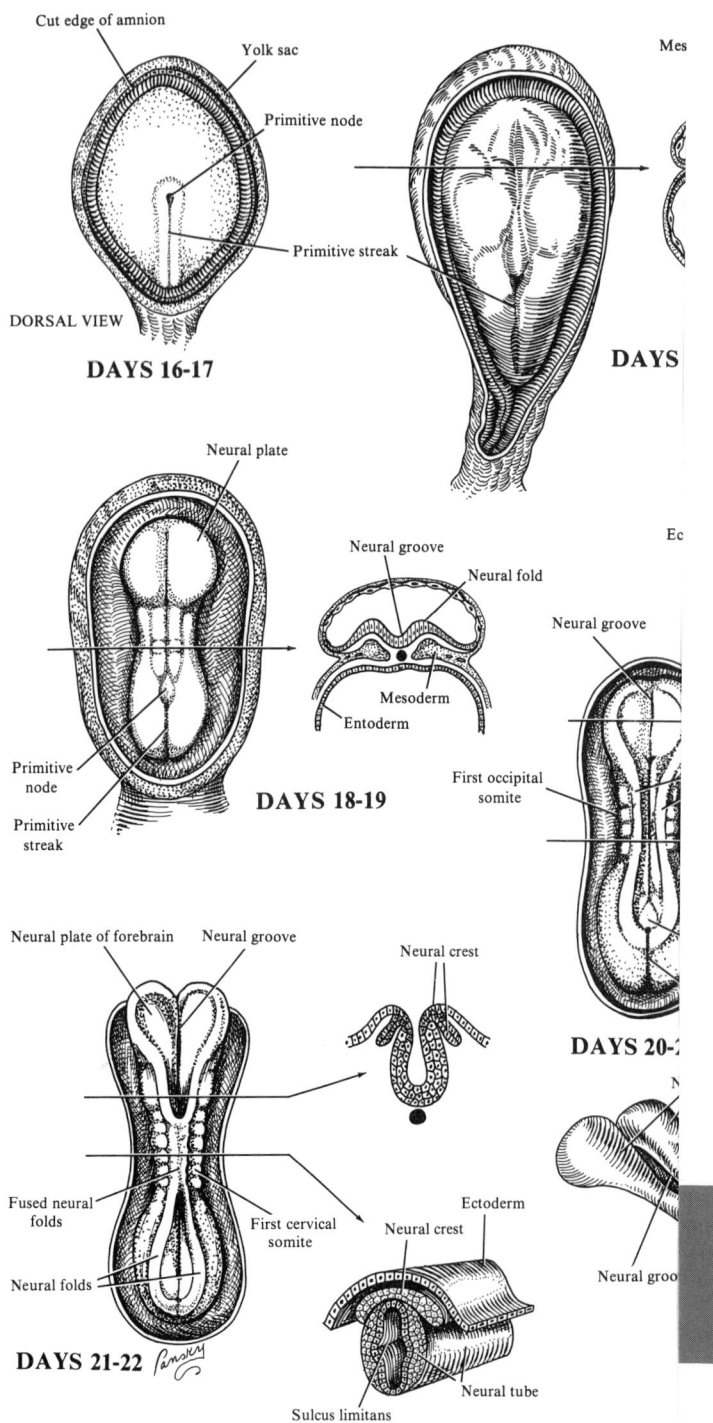

FIGURE 10. **Neural tube development.**

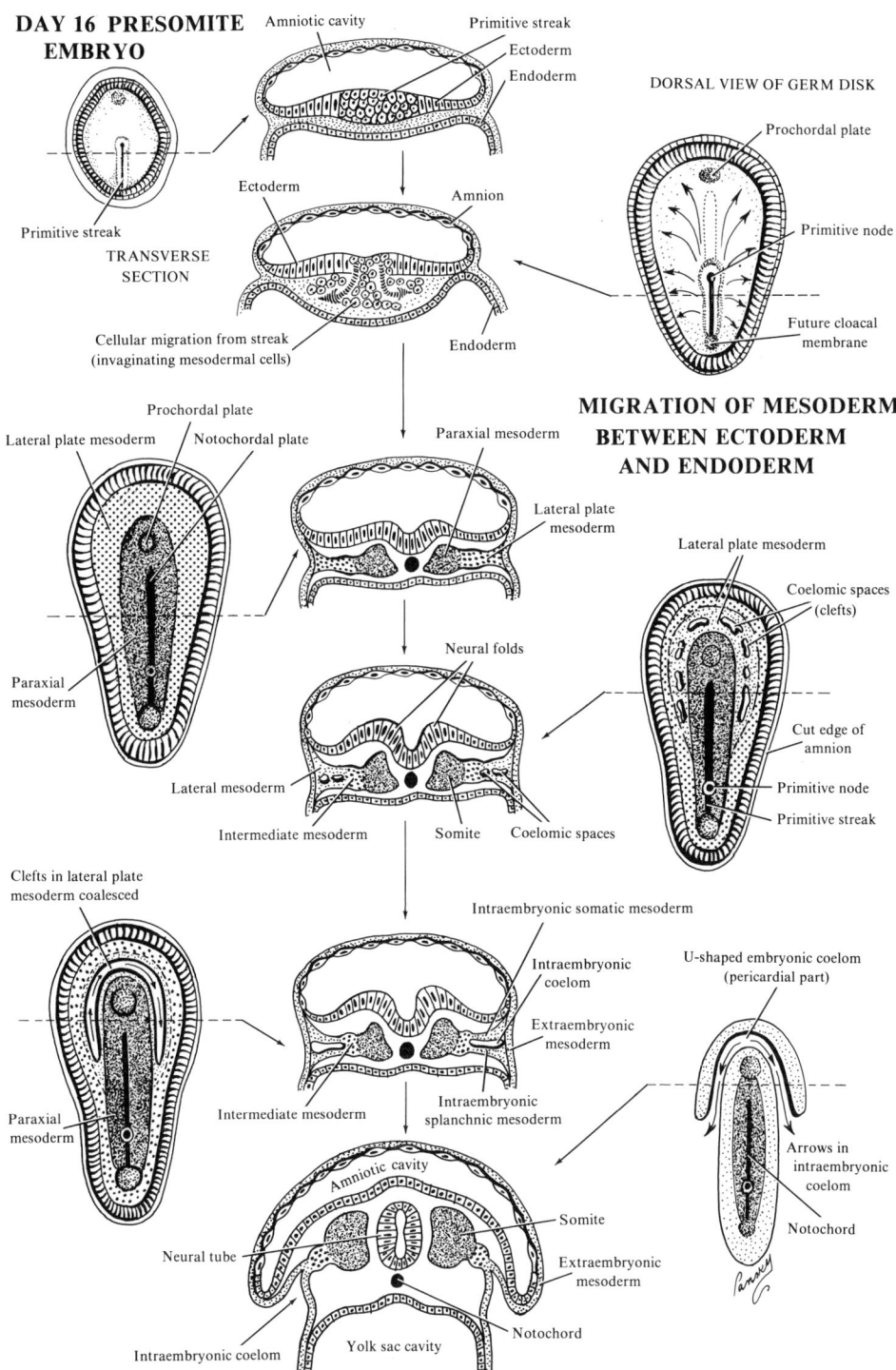

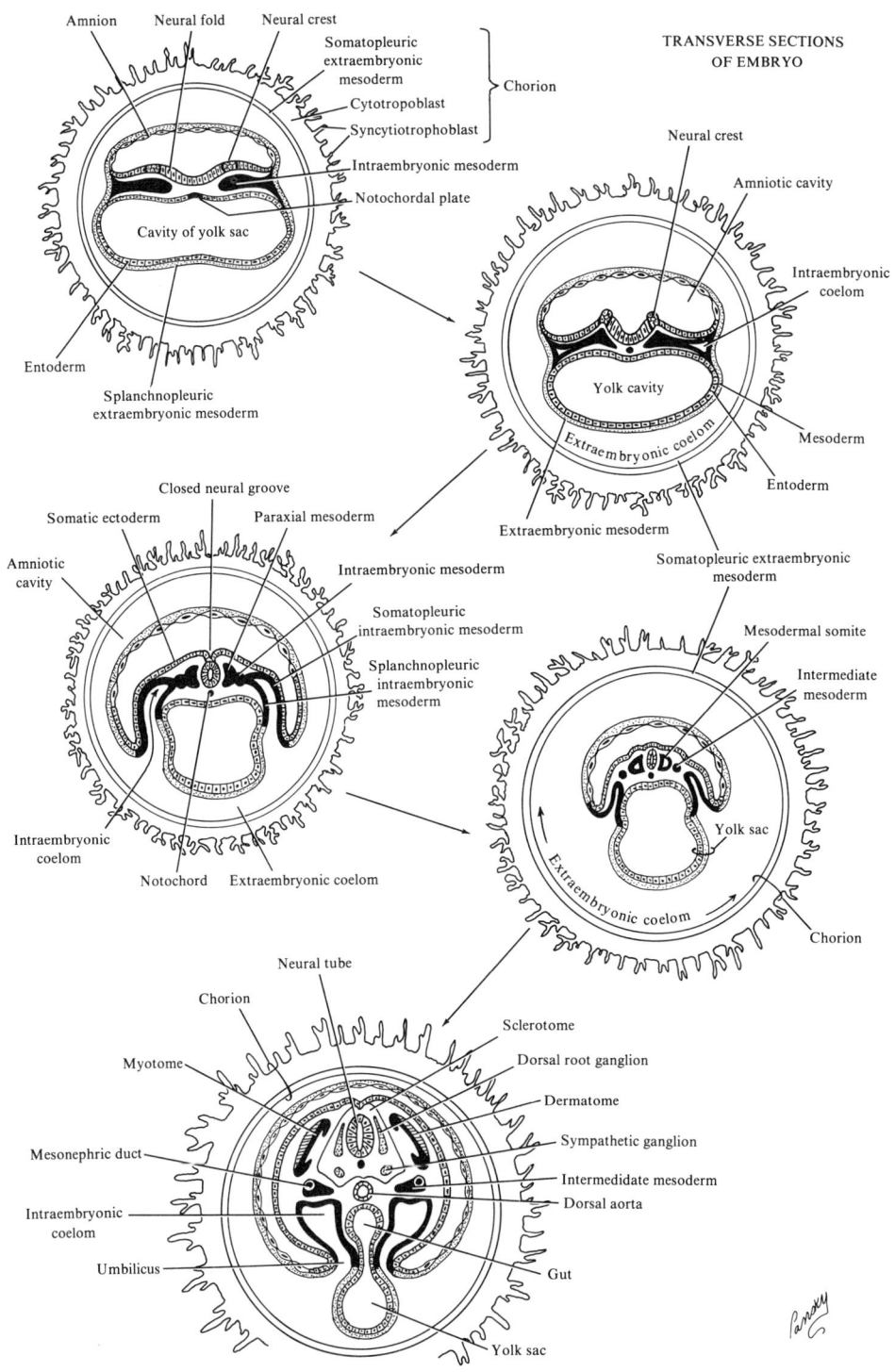

FIGURE 11. **Development of the yolk sac and coelomic cavities.**

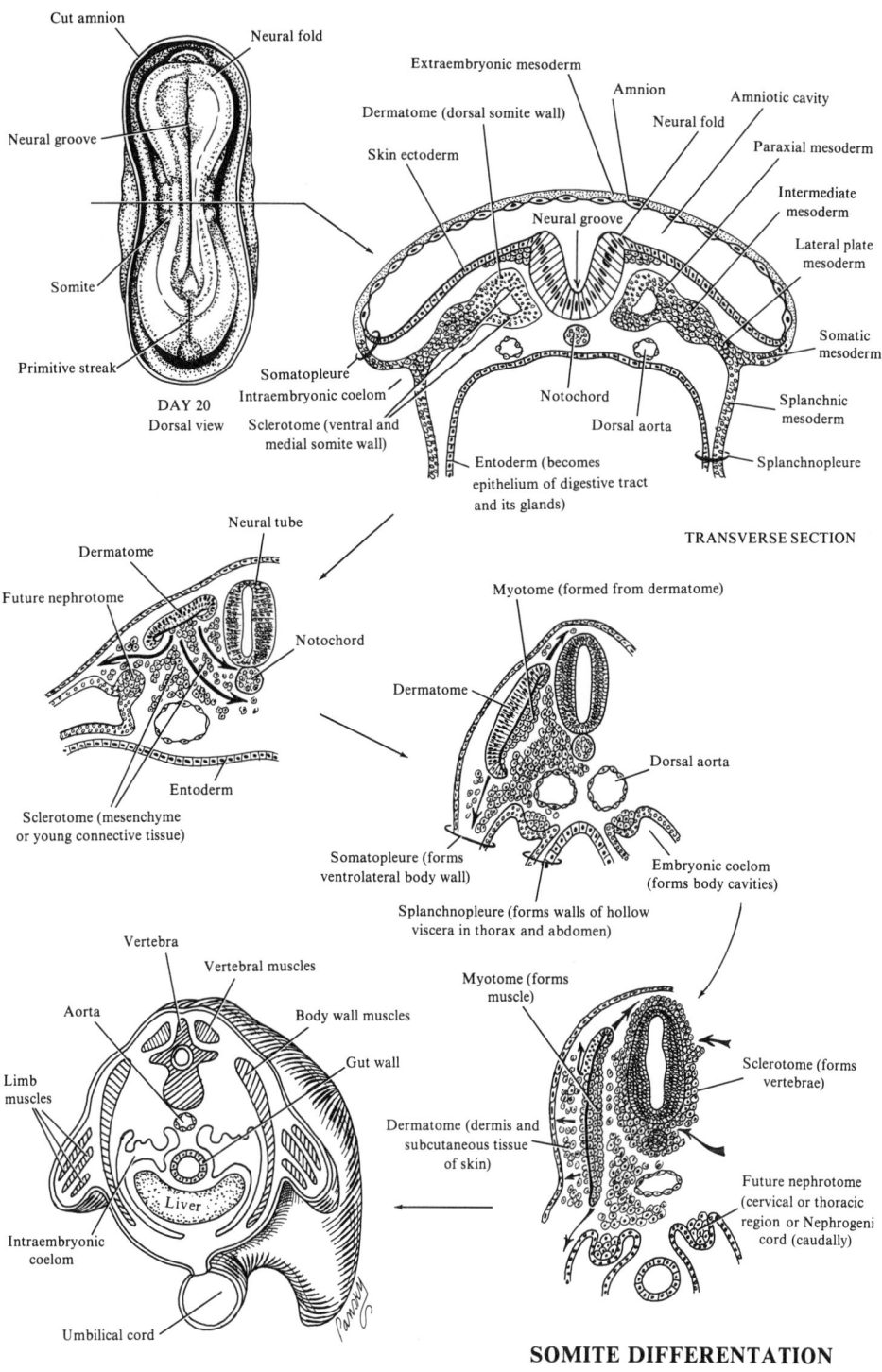

FIGURE 12. **Somite differentiation.**

21. WEEK 3 OF DEVELOPMENT: CARDIOVASCULAR SYSTEM DEVELOPMENT

   I. **Introduction:** early cardiovascular system formation can be related to the absence of a significant amount of yolk material in the ovum and yolk sac with the need for vessels to carry both nutrient materials and oxygen

  II. **Angiogenesis or blood formation** begins in the extraembryonic mesoderm of the yolk sac, connecting stalk, and chorion during days 13–15. The embryonic vessels begin to develop approximately 2 days later
   A. MESENCHYMAL CELLS or *angioblasts* aggregate to form isolated masses and cords called *blood islands*
      1. Spaces accumulate in the islands, angioblasts arrange themselves around the cavities to form the primitive endothelium, then isolated vessels fuse to form networks of endothelial channels
      2. Vessels continue to extend into adjacent areas by endothelial budding and fusion with other vessels being formed independently

 III. **Blood cells and primitive plasma** develop from the endothelial cells as the vessels develop on the allantois and yolk sac
   A. BLOOD FORMATION begins in week 5, occurring in various portions of the embryonic mesenchyme, particularly in the liver, then later in the spleen, bone marrow, and lymph nodes
   B. MESENCHYMAL CELLS around the primitive endothelial vessels differentiate into the muscles and connective tissue of the blood vessels
   C. THE PRIMITIVE ENDOTHELIAL CARDIAC TUBES form from mesenchymal cells in the cardiogenic area
      1. Longitudinally paired endothelial channels, the heart tubes, develop before the end of week 3 and begin to fuse into the primitive heart tube
      2. By day 21, the paired tubes link up with blood vessels in the embryo, connecting stalk, chorion, and yolk sac to form a primitive cardiovascular system. The cardiovascular system is the first organ system to reach a functional state
      3. Blood circulation usually is started by the end of week 3

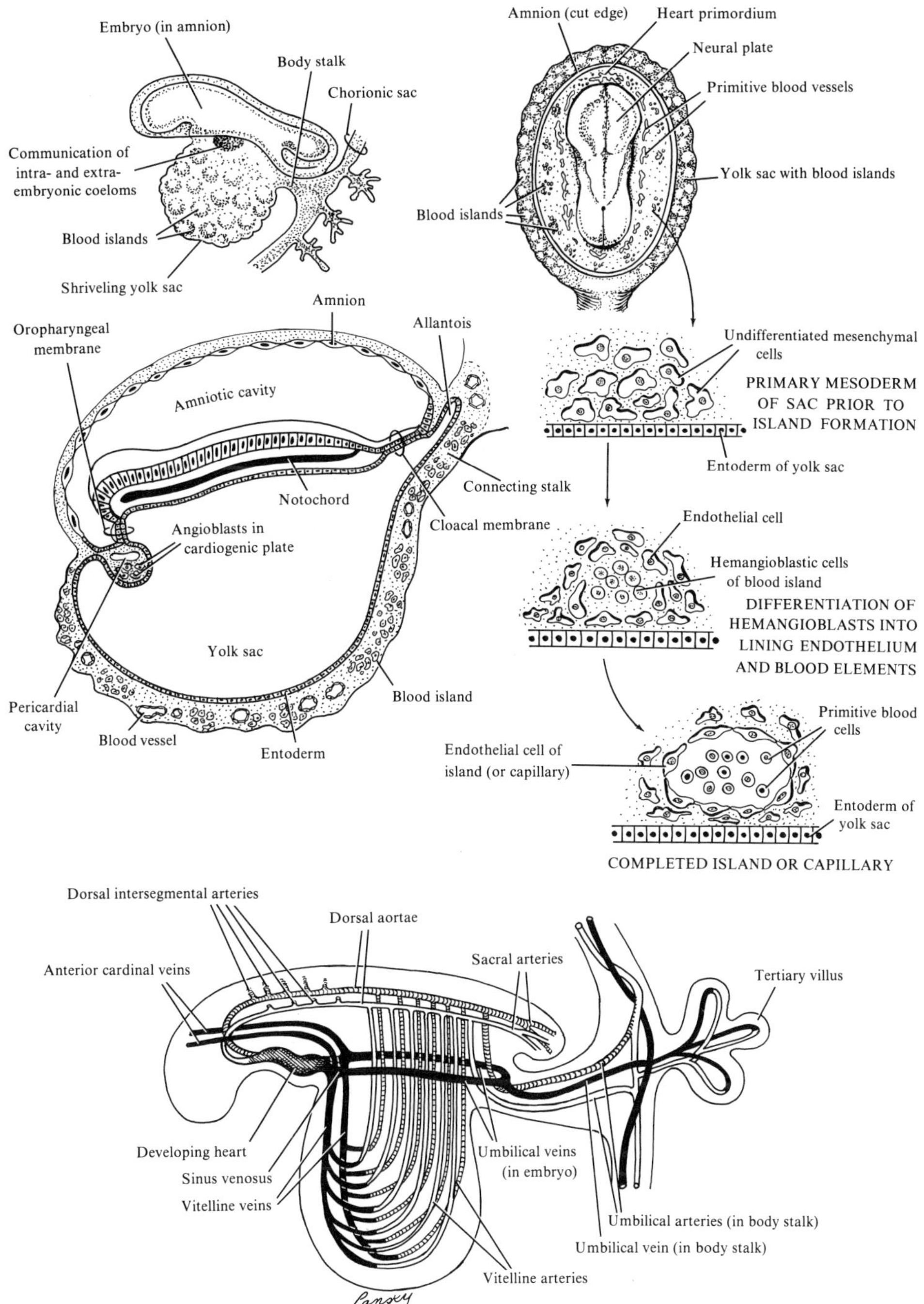

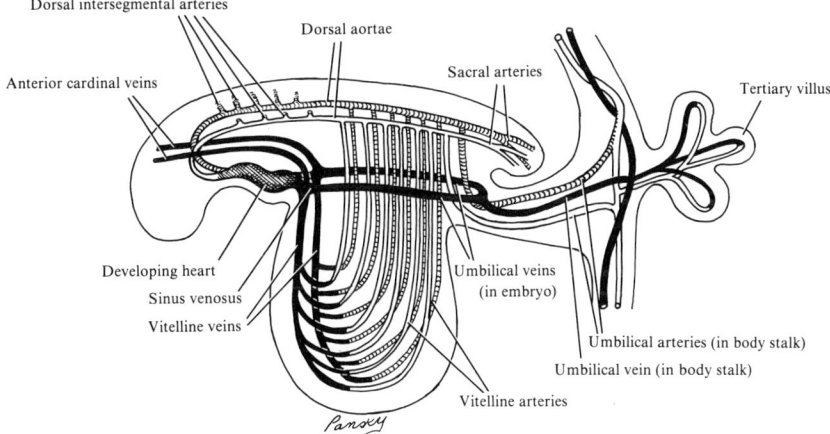

SCHEME OF EXTRA - AND INTRAEMBRYONIC VASCULARIZATION IN AN EMBRYO (ABOUT THREE WEEKS)

22. WEEK 3 OF DEVELOPMENT: TROPHOBLAST AND VILLUS DEVELOPMENT

I. Trophoblast and villus development
A. THE TROPHOBLAST is characterized by many primary stem villi, consisting of a cytotrophoblast core covered by a syncytial layer, at the beginning of week 3
   1. With development, mesodermal cells from the extraembryonic somatopleuric mesoderm or cytotrophoblast penetrate the core of the primary villi and grow in the direction of the decidua to form the *secondary stem villi* which consist of a loose connective tissue core covered by a cytotrophoblastic layer which, in turn, is covered by a thin syncytial layer
B. BY THE END OF WEEK 3, mesodermal cells in the villus core differentiate into blood cells and small blood vessels, forming the villous capillary system, and thus create the *tertiary villi*
   1. By week 4, the tertiary villi are seen over the entire surface of the chorion
   2. The capillaries in the tertiary villi contact capillaries developing in the mesoderm of the chorionic plate and in the connecting stalk, eventually contact the intraembryonic circulatory system, and connect the placenta and the embryo. Thus, in week 4, when the heart begins to beat, the villous system is able to supply the embryo with oxygen and nutrients, whereas prior to that time it was all done by diffusion
C. CYTOTROPHOBLAST CELLS in the villi penetrate the overlying syncytium to reach the maternal endometrium
   1. They establish contact with similar extensions of neighboring villous stems to form a thin outer *cytotrophoblast shell*
      a. The cytotrophoblast shell is seen on the embryonic pole initially and then expands toward the abembryonic pole until it covers the entire trophoblast, thus attaching the chorionic sac firmly to the maternal endometrial tissue
   2. Villi attached to the maternal tissues via the trophoblastic shell are called *stem* or *anchoring villi*
   3. Villi that grow from the sides of the stem villi are called *branch villi,* and it is through these that the major exchange of materials between the mother and the embryo takes place
D. BY DAYS 19 AND 20, the extraembryonic coelom or chorionic cavity enlarges, and the embryo is attached to its trophoblast shell only by a narrow connecting stalk
   1. The stalk is composed of extraembryonic mesoderm which is continuous with the chorionic plate and is attached to the embryo at its caudal end
   2. The connecting stalk or *body stalk* later develops into the *umbilical cord* to connect the placenta and the embryo

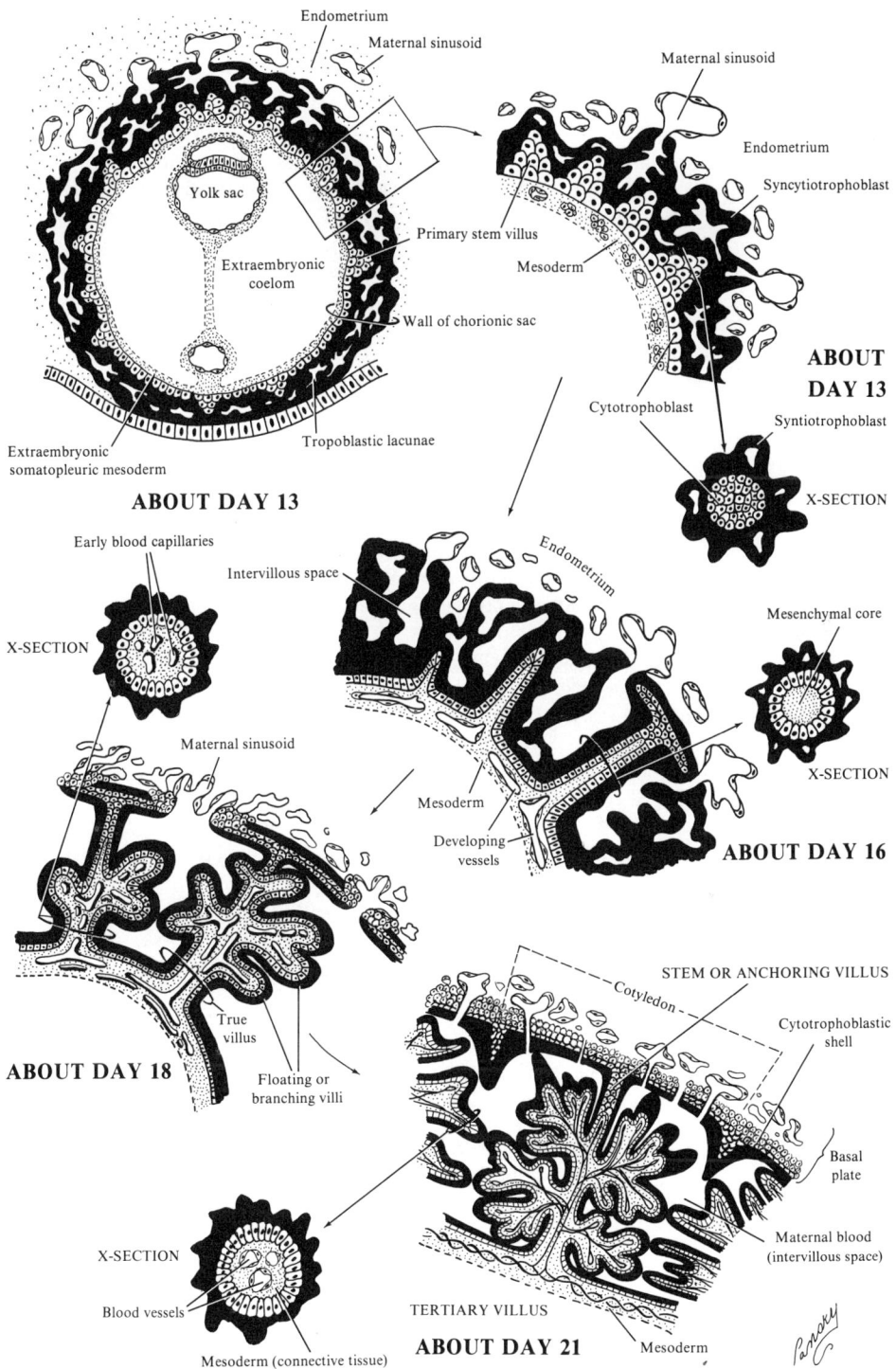

23. WEEKS 4 TO 6 OF DEVELOPMENT: THE EMBRYONIC PERIOD

I. **Introduction:** during this relatively short embryonic period (weeks 4 to 8), one sees the beginnings of all major internal and external structural (organ and organ systems) develop during which time the 3 germ layers give rise to specific tissues and organs—the period of *organogenesis*. The shape of the embryo changes, and major features of the external body form (morphogenesis) become recognizable by the end of month 2. In addition, major congenital malformations can occur due to exposure of the embryo to teratogens during this developmental period

II. **Week 4 of development**
   A. ABOUT DAYS 22 TO 23: embryo is almost straight or slightly curved, and somites create conspicuous surface elevations. The neural tube is closed opposite the somites but is open at its caudal and rostral neuropores
   B. ABOUT DAY 24: the first (mandibular) and second (hyoid) branchial arches become distinct
      1. Most of the mandibular process of the first arch gives rise to the lower jaw and a rostral extension of it; the maxillary process helps form the upper jaw
      2. Head- and tailfolds cause a slight curvature of the embryo
      3. The heart produces a large ventral prominence
   C. ABOUT DAY 26: 3 pairs of branchial arches are seen and the rostral neuropore closes
      1. The forebrain creates a distinct elevation on the head and, with longitudinal folding, the embryo now has a distinct C-shaped curvature to it
      2. Transverse folding causes a narrowing of the connection between yolk sac and embryo
      3. The arm buds are now recognizable as small swellings on the body wall's ventral surface
      4. The otic pits, the primordia for the inner ears, are clearly seen
   D. ABOUT DAY 28 (END OF WEEK 4): the fourth pair of branchial arches and the leg buds are seen
      1. The lens placodes (ectodermal thickenings) represent the future lenses on the side of the head

III. **Week 5 of development:** there are fewer body form changes this week
   A. HEAD GROWTH is accelerated as a result of rapid brain development
      1. The face contacts the heart prominence
      2. The second (hyoid) branchial arch overgrows arches 3 and 4 to form an ectodermal depression, the *cervical sinus*
      3. The forelimbs show some regional differentiation as the *hand plates* develop

IV. **Week 6 of development**
   A. LIMB BUDS, especially the forelimbs, show regional differentiation. Hind limbs develop later
      1. Elbow and wrist areas are identifiable
      2. Paddle-shaped hand plates develop digital ridges, the *finger rays,* for future fingers
   B. SOME SMALL SWELLINGS appear around the groove between the first branchial arches. The groove becomes the external auditory meatus, the swellings the ear auricle
   C. THE EYE becomes obvious due to appearance of retinal pigment
   D. THE HEAD appears larger relative to the trunk and bends even farther over the heart prominence
      1. Bending is due to the cervical flexure as a result of bending of the brain in the cervical region
      2. The trunk and neck begin to straighten out
   E. SOMITES are visible in the lumbosacral region by the middle of this week

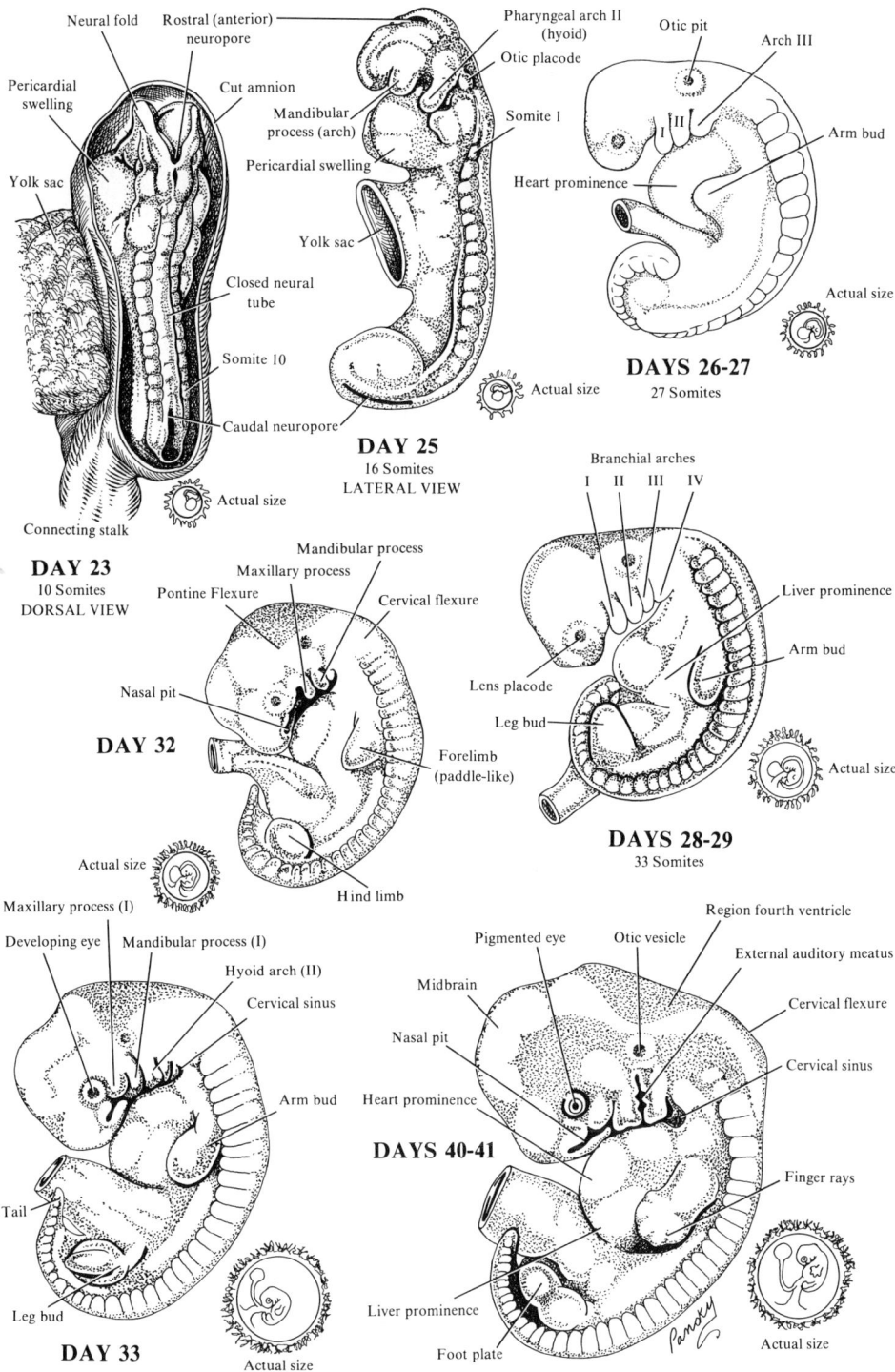

24. EMBRYONIC PERIOD: WEEKS 7 AND 8 AND EXTERNAL EMBRYO APPEARANCE

I. Week 7 of development
A. CONNECTION BETWEEN THE GUT AND YOLK SAC consists of only the small yolk stalk
   1. Umbilical herniation occurs; intestines during rotation of the gut enter the extraembryonic coelom in the proximal portion of the umbilical cord
B. THE LIMBS change markedly during this week
   1. The forelimbs project over the heart
   2. Notches are seen between the rays in the hand plates indicating future fingers

II. Week 8 of development: this is the final week of the embryonic period
A. THE FINGERS are noticeably webbed and short
B. NOTCHES now are seen between the toe rays, and the tail bud is still visible
C. THE LIMB REGIONS are clear, fingers lengthen, toes are distinct, and the tail bud disappears by the end of this week
D. THE EMBRYO has human characteristics, but the head is distinctly large (about one-half of embryo)
   1. The neck region is established, and the eyelids are obvious
   2. The abdomen is less bulging, and the umbilical cord is reduced in size
   3. The intestine is still within the proximal portion of the umbilical cord
E. THE EYES usually open, but near the end of week 8 the eyelids begin to meet and fuse
F. THE EXTERNAL EARS (AURICLES) assume their final shape but are still low set
G. THE EXTERNAL GENITALIA are not distinct enough for accurate sexual identification

III. External embryo appearance
A. BY THE END OF WEEK 4 the embryo has about 28 somites, the ventral body wall has closed, and the major external features are somites and pharyngeal arches. The age of the embryo now is often expressed in somites or as the crown-rump (CR) length (sitting height) in millimeters. The standing height or crown-heel (CH) length is sometimes used for 8-week and older specimens, but these are hard to make accurately
   1. The CR length is measured from the skull vertex to the midpoint between the apices of the buttocks. Variations in flexion result only in approximate real age measurements
      a. 5 weeks: 5–8 mm; 6 weeks: 10–14 mm; 7 weeks: 17–22 mm; 8 weeks: 28–30 mm
B. THE EXTERNAL APPEARANCE during month 2 changes due to the great size of the head and formation of the limbs, ears, nose, and eyes. By the beginning of week 5, the fore- and hind-limbs appear as paddle-shaped buds
C. AGE ESTIMATION usually relies on 2 commonly used references, namely, the onset of the last menstrual period (LMP) and the time of fertilization
   1. Since the zygote does not form until the second week after the onset of the last normal menstural period, 14 ± 2 days are deducted from the so-called menstrual age to get the actual or fertilization age of the embryo
   2. Foot-length correlates with CR length and is used in aging an incomplete or macerated fetus
   3. Fetal weight is not too accurate for use in age determination in light of any maternal metabolic disturbance

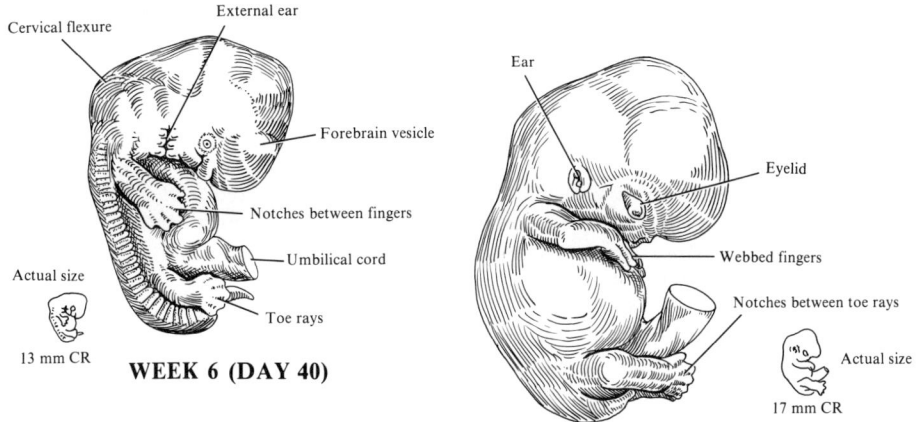

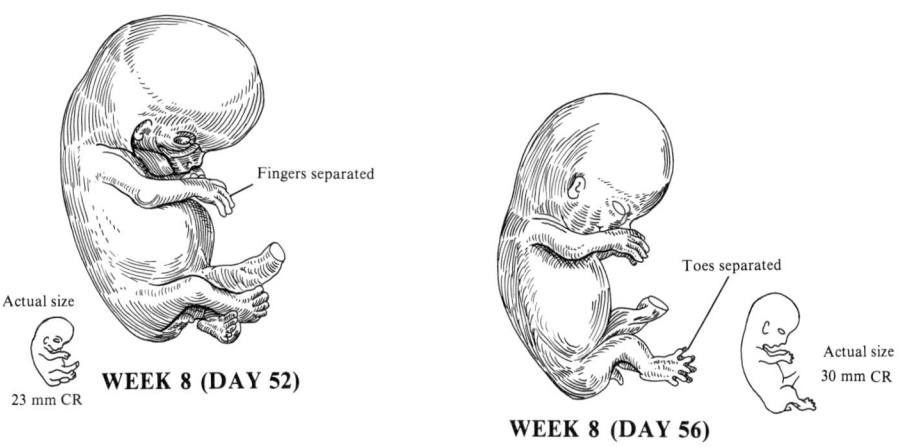

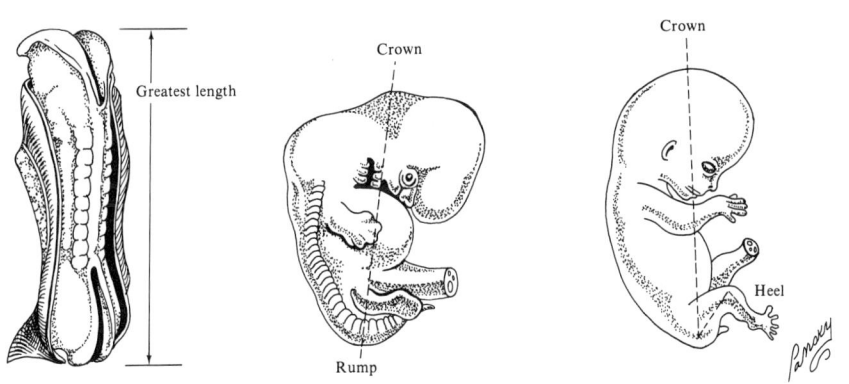

EMBRYONIC MEASUREMENTS

25. GERM LAYERS AND THEIR DERIVATIVES

I. **The 3 germ layers—the ectoderm, the mesoderm, and the entoderm (endoderm):** are in place at the end of gastrulation
   A. THE ECTODERM gives rise to the central nervous system (the brain and spinal cord); the peripheral nervous system; the sensory epithelia of the eye, ear, and nose; the epidermis and its appendages (the nails and hair); the mammary glands; the hypophysis; the subcutaneous glands; and the enamel of the teeth
      1. Ectodermal development is called neurulation in regard to nervous tissue
   B. THE MESODERM gives rise to connective tissue, cartilage, and bone; striated and smooth muscles; the heart walls, blood and lymph vessels and cells; the kidneys; the gonads (ovaries and testes) and genital ducts; the serous membranes lining the body cavities; the spleen; and the suprarenal (adrenal) cortices
   C. THE ENTODERM gives rise to the epithelial lining of the gastrointestinal and respiratory tracts; the parenchyma of the tonsils, the liver, the thymus, the thyroid, the parathyroids, and the pancreas; the epithelial lining of the urinary bladder and urethra; and the epithelial lining of the tympanic cavity, tympanic antrum, and auditory tube
      1. The entoderm development is simpler than that of either mesoderm or ectoderm. It is a monocellular layer lining the yolk sac until cephalocaudal flexion of the embryo takes place
         a. Flexion takes the embryo from a flat disk to its basic embryonic body form. The *primitive gut* originates from entoderm at the time of its flexion
         b. The yolk sac constricts, thus the intraembryonic entoderm (future digestive tube) and the extraembryonic entoderm (forms the inner lining of the yolk sac) are delineated
      2. Three major parts of the primitive gut are the foregut, the midgut, and the hindgut (including the cloaca)
      3. The oropharyngeal (buccopharyngeal) and cloacal membranes temporarily close the 2 ends of the primitive gut
         a. In humans, the buccopharyngeal membrane disappears at the beginning of week 4
         b. The cloacal membrane lasts longer and at week 7, like the cloaca, it divides into an anterior *urogenital membrane* and posterior *anal membrane,* the latter being absorbed by week 9

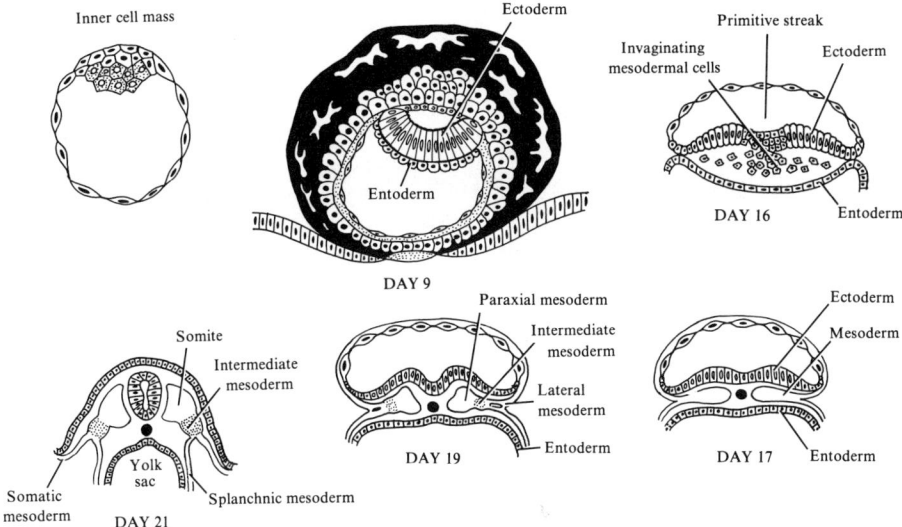

GERM LAYER DERIVATIVES

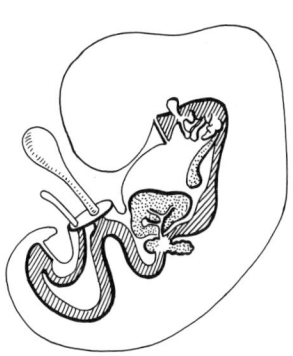
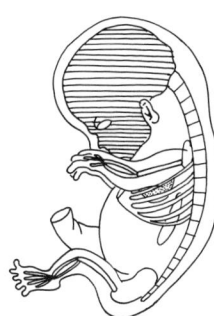
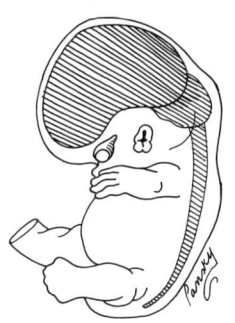

ENTODERM

Epithelium of GI tract
Liver
Pancreas
Urachus
Urinary bladder

Epithelial portions
  Pharynx
  Thyroid
  Trachea, bronchi, lungs
  Tympanic cavity
  Pharyngotympanic tube
  Tonsils
  Parathyroids

MESODERM

Skeleton (head and body)
Muscle
Connective tissue
Circulatory system
  Cardiovascular
  Lymphatic
Urinary system
Spleen
Adrenal cortex
Genital system:
  gonads, ducts, accessory glands
Dermis
Dentine of teeth

ECTODERM

NERVOUS TISSUE

Neural tube	Neural crest
CNS	Pigment cells
Retina	Adrenal medulla
Post. pituitary	Cranial and sensory nn.
Pineal gland	Cranial and sensory ganglia

EPIDERMIS

Hair
Nails
Mammary glands
Cutaneous glands
Ant. pituitary
Teeth enamel
Inner ear
Eye lens

26. EMBRYONIC FOLDING AND FLEXION OF THE EMBRYO

I. **Embryonic folding:** the flat trilaminar embryonic disk becomes a more cylindric embryo due to the longitudinal and transverse folding that occurs as a result of embryonic growth, especially of the neural tube. The foldings occur simultaneously and are not separate sequential events. Flexion, a process of curving, transforms the embryo into a sort of "tube" and isolates it from the embryonic membranes, to which it is eventually attached only by a thin stalk, the umbilical cord. The embryo increases rapidly in its long axis due to central growth being greater than peripheral growth, and the dorsal region of the embryo grows more rapidly than its ventral region, resulting in the embryo curving itself around the umbilical region. The dorsal region also thickens, especially in the midline, and the edges of the disk swing ventrally carrying the amnion with them. Thus, the embryo is surrounded by its amniotic cavity
   A. LONGITUDINAL FOLDING produces both head- and tailfolds, or flexion, and creates a cranial and caudal region to the embryo
      1. Headfold: neural folds (end of week 3) begin to develop into the brain and project dorsally into the amniotic cavity
         a. The forebrain grows cranially beyond the oropharyngeal membrane and overhangs the primitive heart. At the same time, the *septum transversum* (a mass of mesoderm cranial to the pericardial coelom), the heart, the pericardial coelom, and the oropharyngeal membrane turn under onto the ventral surface
            i. During folding, part of the yolk sac is incorporated as the foregut (between brain and heart, ending blindly at the oropharyngeal membrane). The membrane separates the foregut from the stomodeum or primitive mouth cavity
         b. After folding, the septum transversum lies caudal to the heart and develops into a major portion of the diaphragm
         c. Before folding, the intraembryonic coelom is a flattened horseshoe-shaped cavity. After folding, the pericardial coelom lies ventrally and the pericardioperitoneal canals run dorsally over the septum transversum to join the peritoneal coelom which, on each side, communicates with the extraembryonic coelom
      2. The tailfold (caudal end) takes place later than the headfold and results from the dorsal and caudal growth of the neural tube
         a. As the embryo grows, the tail region projects over the cloacal membrane which eventually comes to lie ventrally
         b. During folding, part of the yolk sac is incorporated into the embryo as the *hindgut*, the terminal portion of which soon dilates and forms the *cloaca*, separated from the amniotic cavity by the *cloacal membrane*
         c. Before folding, the primitive streak lies cranial to the cloacal membrane, but, after folding, lies caudal to it
         d. The connecting stalk now attaches to the ventral embryonic surface, and the allantois is partly incorporated into the embryo
   B. TRANSVERSE FOLDING (FLEXION) produces right and left lateral folds
      1. Each lateral body wall (somatopleure) folds toward the midline, rolling the edges of the embryonic disk ventrally to form a cylindric embryo
      2. As lateral and ventral body walls form, part of the yolk sac is incorporated into the embryo as the *midgut;* simultaneously, the connection of the midgut with the yolk sac is reduced to a *yolk stalk* or *vitelline duct*
      3. After folding, the area of the amnion attachment to the embryo is reduced to a narrow *umbilicus* on its ventral surface
      4. As the midgut is separated from the yolk sac, it attaches to the dorsal abdominal wall via a thin *dorsal mesentery*
      5. As the umbilical cord forms, the ventral fusion of the lateral folds reduces the area of communication between the intra- and extraembryonic coelom
      6. As the amniotic cavity enlarges and obliterates the extraembryonic coelom, the amnion forms an outer covering for the umbilical cord

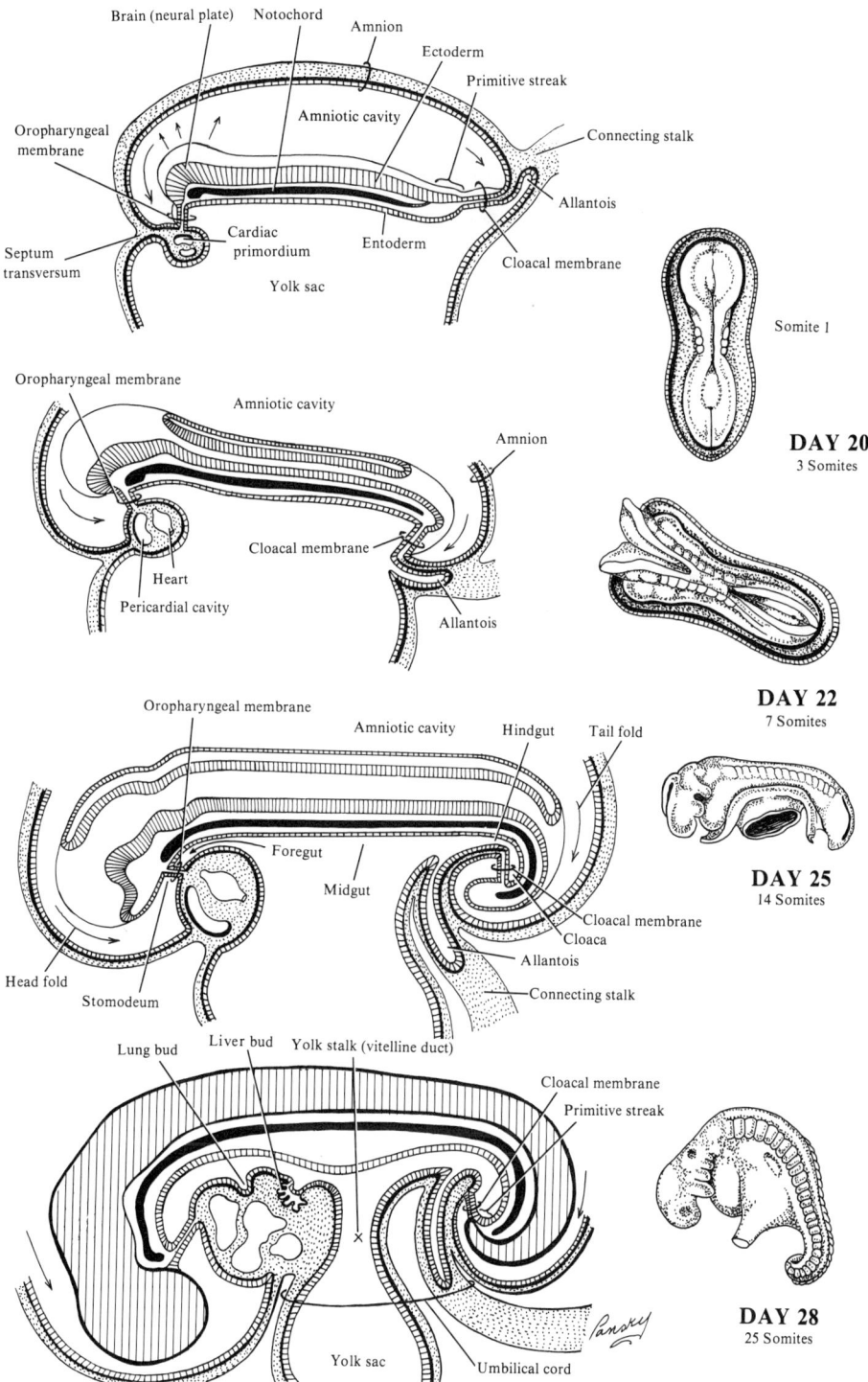

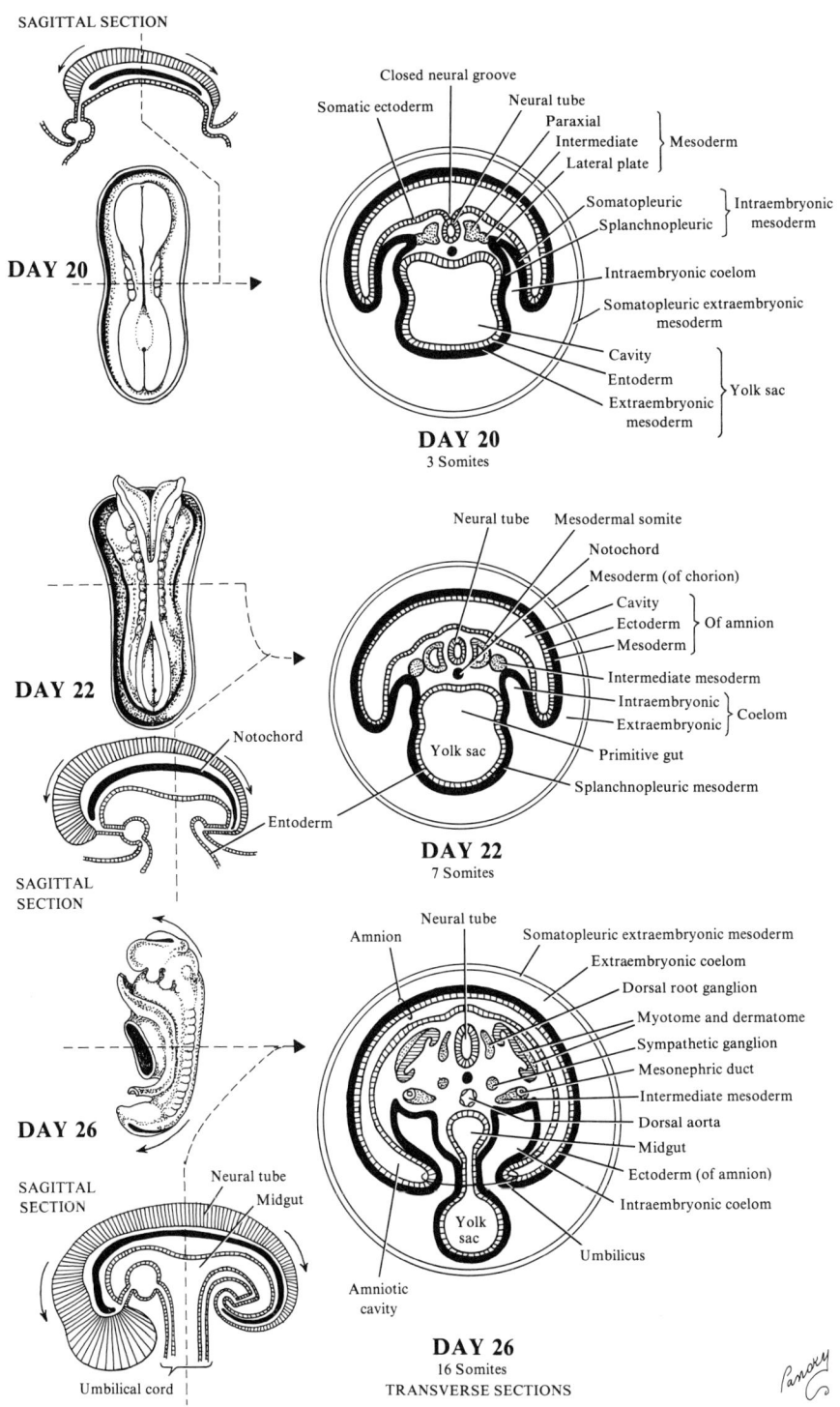

FIGURE 13. **Embryonic folding and flexion: transverse and sagittal sections.**

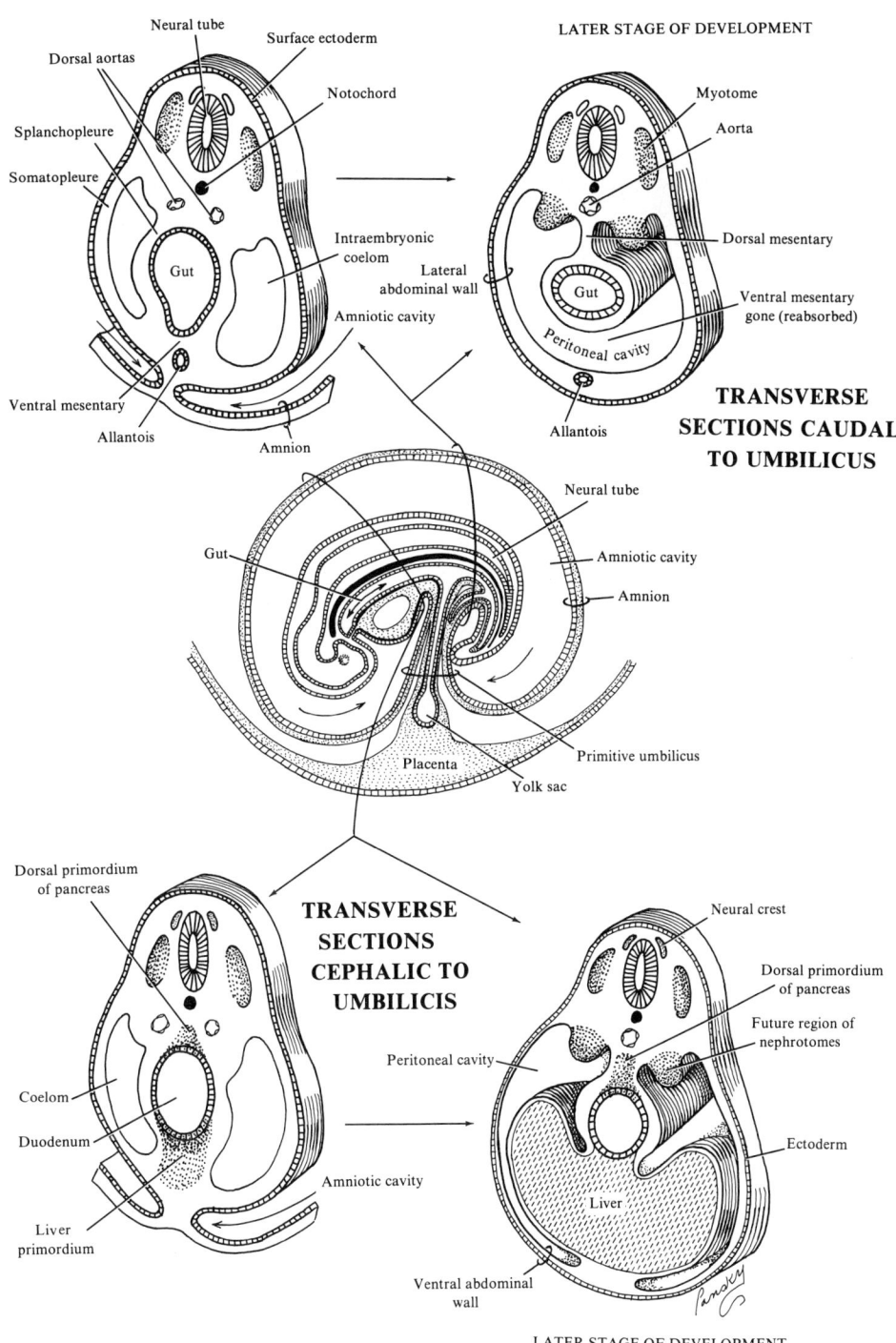

FIGURE 14. Sections of flexed embryo caudal and cephalic to the umbilicus.

27. GENERAL MECHANISMS OF NORMAL DEVELOPMENT

I. **Normal development** involves *growth* and *differentiation,* both under very strict coordination and rigorous and precise organization. The processes that accomplish this development depend on 2 major overall controls
   A. GENETIC CONTROL determines the inherent potential of the organism
   B. EPIGENETIC CONTROL assures progressive formation of the primordia, then of the definitive organs, and includes a series of complex mechanisms all acting simultaneously or successively,
      1. Cellular movements are based entirely on *migration* and *invagination* of primary ectoderm cells: gastrulation is a good example; another is migration of sclerotome cells toward the notochord region to furnish material for the future spinal axis
      2. Induction represents reciprocal influences between cellular groups. It is the essential determinant of embryonic development and is the process by which a cellular group, the *inductor* or organizer, influences or induces differentiation of another cellular group, the *competent* or *induced tissue*
         a. The primary inductor or organizer: for a period of time in early development, certain embryonic tissue is capable of inducing the development of adjacent tissue. The best-known inductive tissues are the primitive streak, the notochord, and paraxial mesoderm which act as primary organizers of the CNS; e.g., development of the forebrain; and the notochord as an inductor of sclerotome which will form the vertebral primordia
         b. Secondary induction: once the basic embryonic plan has been established by the primary organizers, a chain of secondary indutions takes place. Development of the eye and ear is an example of the forebrain reacting to the secondary induction of its adjacent mesenchyme
         c. Nature of the inductor: induction has its effect by provoking cellular differentiation (at cellular level) or synthesis of a new type of protein (at the molecular level). Some substance, probably protein in nature, passes from the inducing tissue to the tissue being induced
            i. When the sclerotome begins to form the cartilaginous matrix of the vertebral primordium, this cellular morphologic change is preceded by synthesis in the mesenchymal cells of the constituent proteins of cartilage
      3. Regression: the notochord again serves as a prime example since it disappears or regresses almost entirely after having contributed to the formation of the vertebral column. It persists only at the level of the intervertebral disk
      4. Regulation is shown by an egg reconstituting a harmonious whole when part of its substance is removed. Monozygotic twins represent an apparent demonstration of the existence of the phenomenon of regulation in the human species

CELLULAR MOVEMENTS

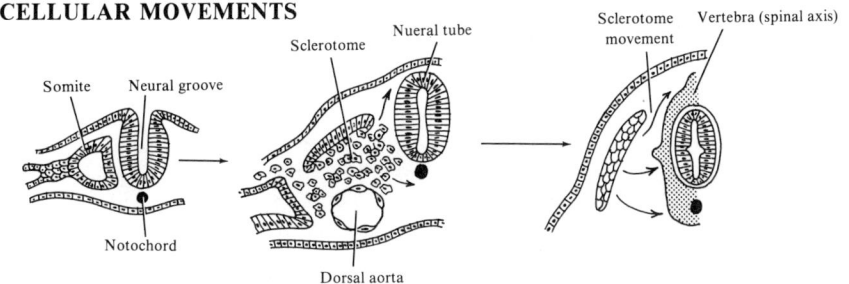

INDUCTION

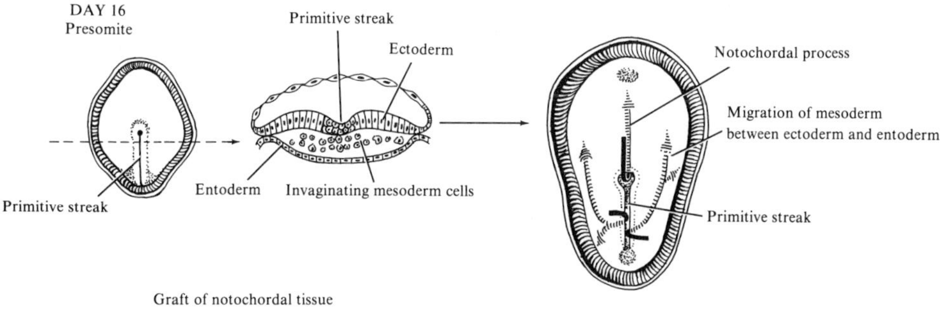

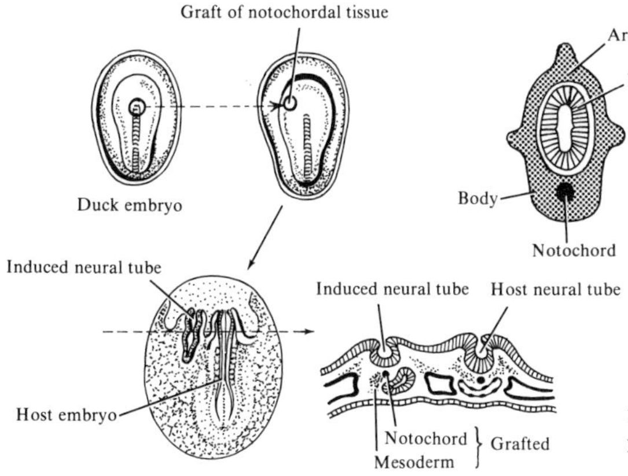

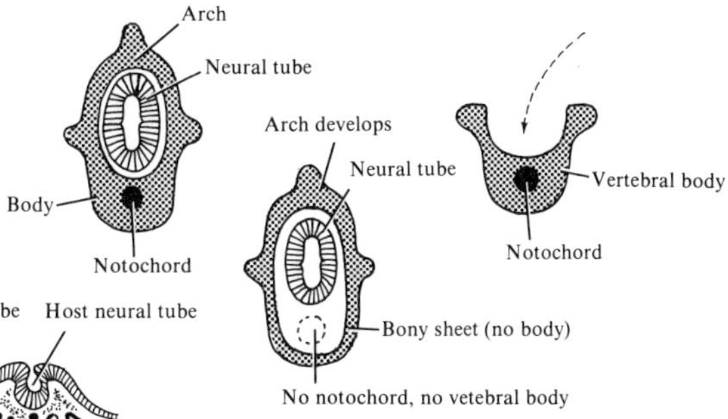

NOTOCHORD AND NEURAL TUBE INDUCE FORMATION OF VERTEBRAL PARTS FROM SCLEROTOME OF SOMITE

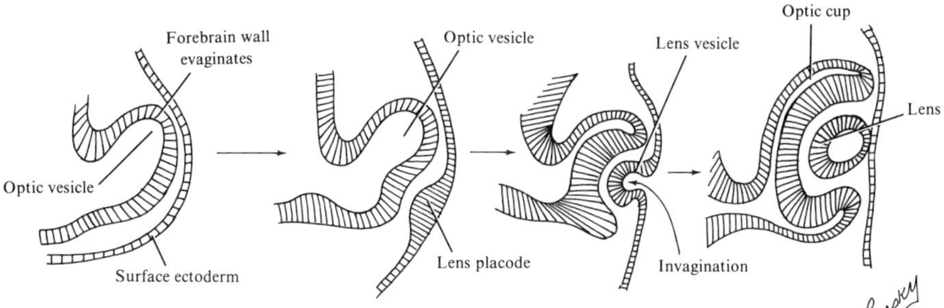

(After Waddington and Schmidt, 1933)

SECONDARY INDUCTIONS: EYE DEVELOPMENT

28. THE FETAL PERIOD: WEEKS 9 TO 20 OF DEVELOPMENT

I. **Introduction:** development from embryo to fetus is not abrupt, but the embryo changes to a recognizable human being and develops all the basic outlines of its organs and is then called a fetus. This long (7-month) fetal period is concerned with growth and differentiation of tissues and organs that began to develop in the embryonic period, maturation of the primordia, reorganization of spatial relationships of primordia, and the embryo begins to make functional use of its organs for part of its needs. Its volume and weight increase proportionally, and it grows considerably, from about 30 mm (CR) to about 330 mm (CR). Fetal growth is complex and is really a phenomenon which results from the sum of very asynchronous growth of different organs and parts even at the histologic level (histogenesis). Body proportions at term are very different from the fetus of 2 or 3 months

II. **Fetal age** is best expressed in weeks. Most uncertainty comes from the use of both calendar months (28–31 days) or lunar months (28 days). Most use calendar months
   A. GESTATIONAL PERIOD is divided into 3 trimesters or parts of 3 calendar months each
      1. By the end of the first trimester, all major systems are developed, and CR length is about as wide as one's palm
      2. At the end of the second trimester (26 weeks according to LMP but only 24 weeks related to time of fertilization), the fetus is too immature to live independently and is about the size of a handspan

III. **Fetal period changes**
   A. WEEKS 9 TO 12 (STAGE OF INITIAL FETAL ACTIVITY)
      1. Beginning of week 9: the head is about half the fetal size, but growth of the body in CR length more than doubles by the end of week 12
      2. One sees a broad face, eyes widely separated, ears low set, and eyelids fused so that the conjunctival sacs are closed
      3. Beginning of week 9, the legs are short and thighs relatively small, but by the end of week 12, the upper limbs reach relatively normal lengths, even though the lower limbs are not well developed and are relatively shorter than normal
      4. The external genitalia of male and female appear similar until end of week 9
      5. Intestinal loops are clearly seen in the proximal end of the umbilical cord until the middle of week 10 when they return to the abdomen
      6. By the end of week 12, the fetus will react to stimuli (seen in aborted fetuses)
   B. WEEKS 13 TO 16 (PERIOD OF RAPID FETAL GROWTH)
      1. By end of week 16, the head is relatively small compared to a 12-week fetus
      2. Legs have become longer
      3. Skeletal ossification progresses rapidly and is seen on x-ray by week 16
      4. Scalp hair pattern gives some clue to early brain development
   C. WEEKS 17 TO 20
      1. Growth slows, but fetal CR length increases by 50 mm and lower limbs reach their final relative proportions
      2. Fetal movements (quickening) are generally felt by mother
      3. Skin is covered by the *vernix caseosa* (greasy, cheeselike material) by week 20 due to fetal sebaceous gland secretion and dead epithelial cells
         a. Protects fetal skin from chapping, abrasions, and hardening due to amniotic fluid around it
      4. By week 20, the fetus is covered by *lanugo,* a fine downy hair
      5. The eyebrows and head hair are visible
      6. Brown fat forms and is the site of heat production (in the newborn)
         a. Brown fat is specialized adipose tissue that produces heat by oxidizing fatty acids and is chiefly found in the floor of the anterior triangle of the neck surrounded by the subclavian and carotid vessels, behind the sternum, and in the perirenal areas

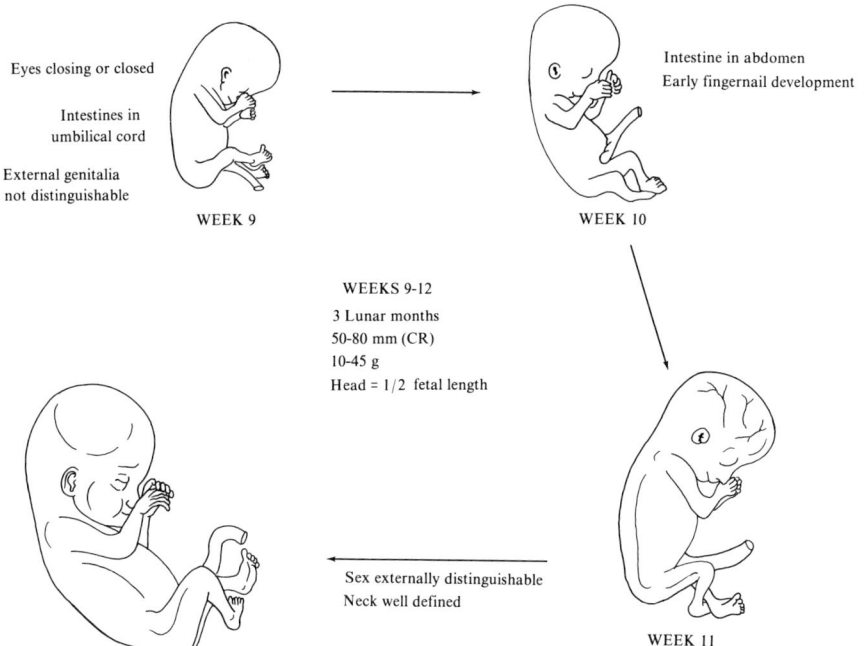

FETAL GROWTH: ACTUAL SIZE
WEEKS 9-20

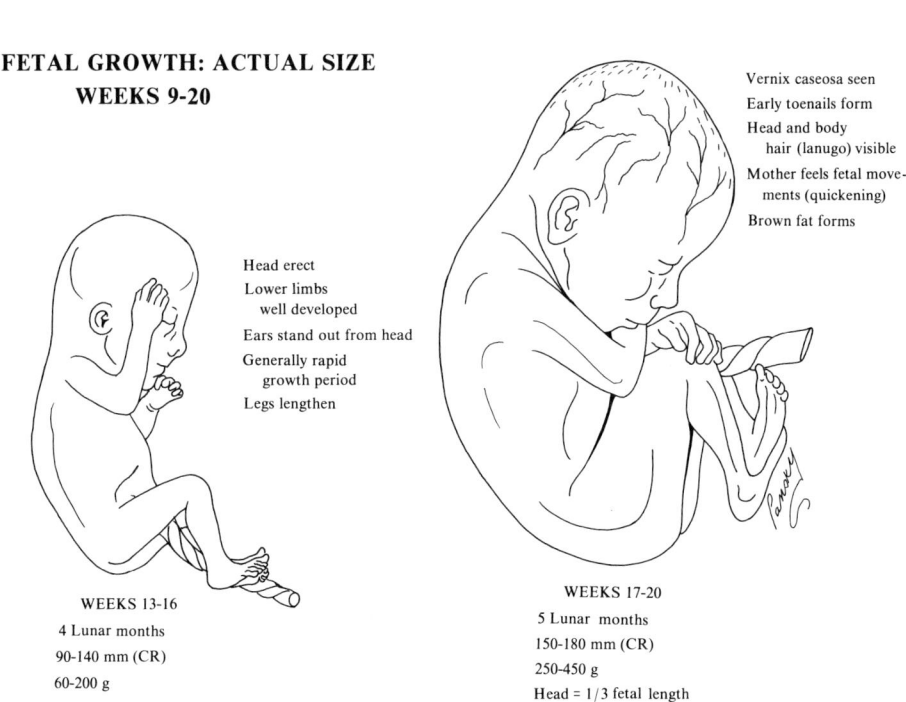

29. THE FETAL PERIOD: WEEKS 21 TO TERM

I. Fetal period changes
- A. WEEKS 21 TO 25
  1. There is much weight gain; the body is proportioned; the skin is wrinkled but is very translucent and is pink to red in color, with blood in the capillaries visible
  2. If born at this time, the fetus usually dies due to an immature respiratory system
- B. WEEKS 26 TO 29
  1. The fetal eyes reopen, and the head and lanugo hair are well developed
  2. There is much subcutaneous fat formed, and the wrinkles of skin smooth out
  3. White fat increases to 3.5%
  4. The fetus is viable and can survive if born prematurely, but the mortality is high as a result of respiratory problems, even though the lungs and pulmonary vasculature are developed for gas exchange
  5. The CNS is mature and can control breathing and body temperature
- C. WEEKS 30 TO 34
  1. The skin is smooth and pink, and the arms and legs are round and full
  2. The body white fat is now about 7 to 8%
- D. WEEKS 35 TO 38
  1. The fetus is now "plump," and growth slows down with approaching parturition
  2. The CR length is about 360 mm and the weight about 3400 g
  3. By term, the white fat is about 16% with the fetus producing about 14 g of fat per day during the last week of gestation
  4. The female fetus grows slower than the male and weighs less
  5. By full term, the skin is white or bluish-pink, the chest is prominent, and the breasts protrude in male and female
     - a. The testes are in the scrotum at full term since descent begins weeks 28–32
     - b. The head is smaller in relation to the body but still has a large circumference
- E. AT BIRTH (TERM)
  1. Usually takes place about 266 days or 38 weeks after fertilization, or 280 days (about 40 weeks) from the onset of the last menstrual period. Variations are due to irregular menstrual periods and difficulty of counting and may be from 2 to 3 weeks
  2. Most fetuses are born within 10 to 15 days of the above times but may go to 276 to 286 days after fertilization
  3. Postmature infants are thin and have dry, parchmentlike skin
  4. For calculations of birth date: in the typical 28-day menstrual cycle, you count back 3 calendar months from the first day of the last menstrual period and then add 1 year and 1 week
- F. FACTORS INFLUENCING GROWTH AND METABOLISM
  1. Amino acids and glucose (primary source of energy): derived from mother via placenta
  2. Insulin: for glucose metabolism and stimulation of fetal growth. Secreted by the fetal pancreas; little of the mother's insulin reaches the fetus normally
     - a. Infants of diabetic mothers tend to be larger than normal
  3. Maternal malnutrition can cause reduced fetal growth
  4. Smoking can reduce fetal growth rate if in excess
  5. Multiple pregnancies: each fetus tends to be smaller than normal single births
  6. Impaired uteroplacental blood flow can cause fetal starvation and even fetal growth may decrease
  7. Placental dysfunction or defects may result in a growth decrease
  8. Genetic factors and chromosomal aberrations may lead to retarded fetal growth

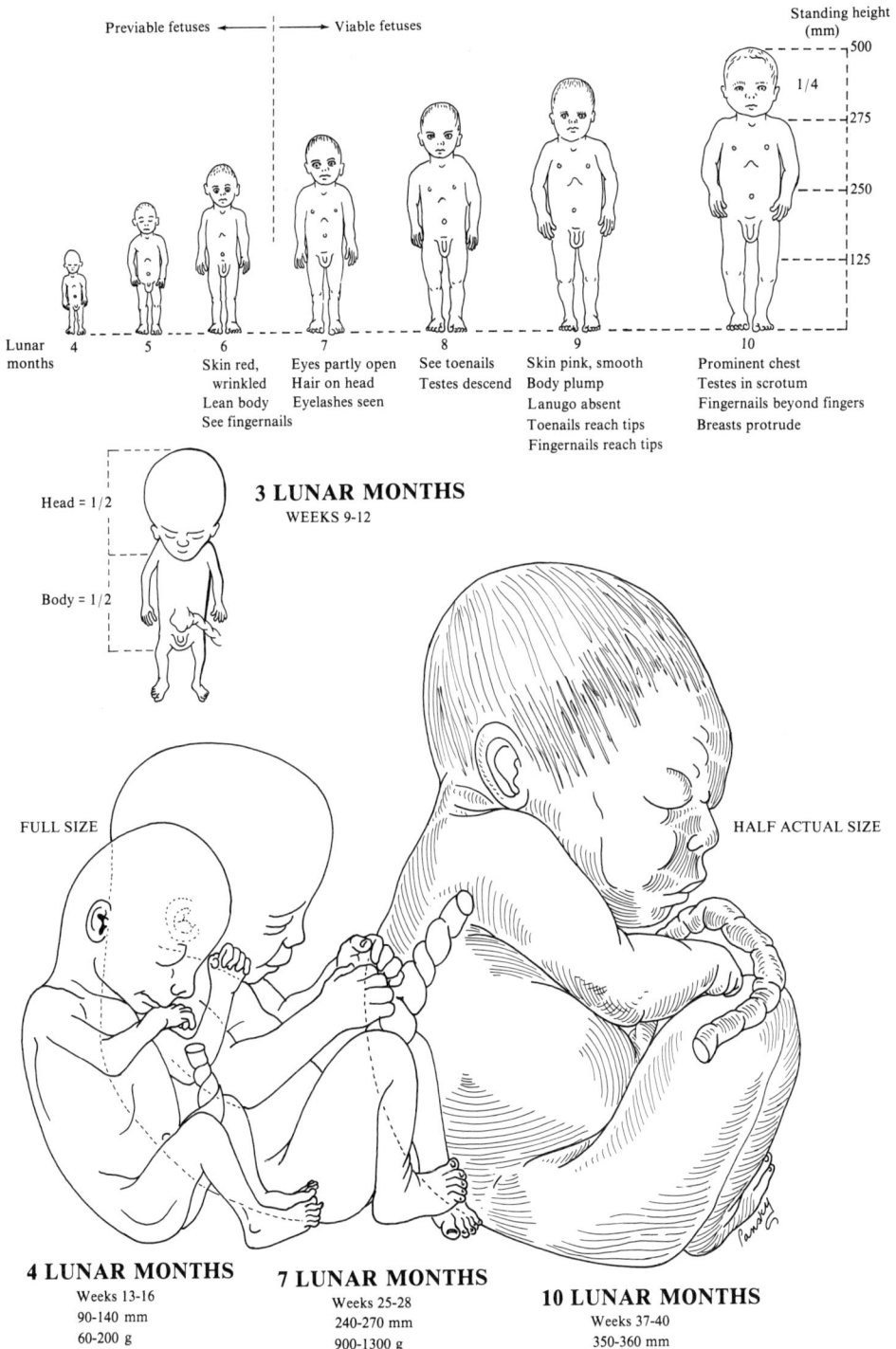

30. MULTIPLE PREGNANCIES

I. **Twinning** may originate from 2 zygotes (dizygotic, nonidentical, or fraternal) or from 1 zygote (monozygotic, identical). Fetal membranes and placenta(s) vary with the derivation of the twins or the time when twinning occurs; e.g., should the embryo duplication occur after the amniotic cavity forms (about day 8), the embryos will be in the same chorionic and amniotic sacs
   A. TWINS occur about once in 80–90 pregnancies (about 1%) and two-thirds or 70% are dizygotic
      1. Dizygotic twinning shows racial differences; the incidence of monozygotic twinning, however, is the same in all populations
      2. The rate of monozygotic twinning shows little variation with the mother's age; dizygotic twinning increases with maternal age
      3. The anastomoses between blood vessels of fused placentas of human dizygotic twins may occur and result in erythrocyte mosaicism (red cells of 2 different types)
         a. In cattle, the above results in a *freemartin*
         b. In man, if one fetus is male and the other female, masculinization of the female fetus does not occur, but a *chimera* is created (has blood cells of 2 genotypes)
   B. DIZYGOTIC (DIOVULAR AND DIVITELLINE) TWINS result from fertilization of 2 oocytes by different sperms
      1. The twins may be of the same or different sexes
      2. The offspring are no more alike genetically than brothers and sisters born at different times (physically dissimilar)
      3. Always have 2 amnions and 2 chorions, but the chorions and placentas may be fused
      4. They are different in blood characteristics and have no tolerance for transplants
   C. MONOZYGOTIC (MONOVULAR AND MONOVITELLINE) TWINS result from fertilization of 1 oocyte. Occur in about 30% of twins and in 3 of 1000 births. They share fetal membranes almost completely
      1. They are morphologically and physiologically identical and of the same sex
      2. They are genetically identical and have identical blood characteristics
      3. They have a tolerance for transplants
      4. Any physical differences are due to environmental factors
      5. Twinning usually begins at the end of week 1 and is the result of division of the inner cell mass into 2 embryonic primordia
         a. There are 2 embryos, each in its own amniotic sac which develop in one chorionic sac and have a common placenta
         b. Early division of the blastomeres results in monozygotic twins with 2 amnions, 2 chorions, and 2 placentas. They cannot be typed by their membranes
         c. Later division of the embryonic cells may result in monozygotic twins with 1 amnion and 1 chorion sac. Such twins are rarely delivered alive because the cords are entangled and circulation stops, and 1 or both embryos die
   D. CONJOINED TWINS: result of incomplete division of embryonic disk and are named by regions of attachment, e.g., thoracopagus (fusion of anterior thoracic region); pygopagus (fusion at back); or craniopagus or cephalopagus (fusion at head)
      1. One in 400 monozygotic twins are conjoined or Siamese
      2. Some may be successfully separated surgically
   E. OTHER MULTIPLE BIRTHS
      1. Triplets are seen in 1 of about 8000 pregnancies: may be identical (1 zygote); identical twins (2 zygotes) and a single infant; or 3 zygotes. May be of the same or different sexes
      2. May see above in quadruplets (4); quintuplets (5); etc. These are rare but seen now as a result of treatment with gonadotropins for ovulatory failure
      3. Superfetation: implantation of 1 or more blastocysts in a uterus that already has a developing embryo
      4. Superfecundation: fertilization of 2 oocytes at the same time by sperm from different males (not proved in the human)

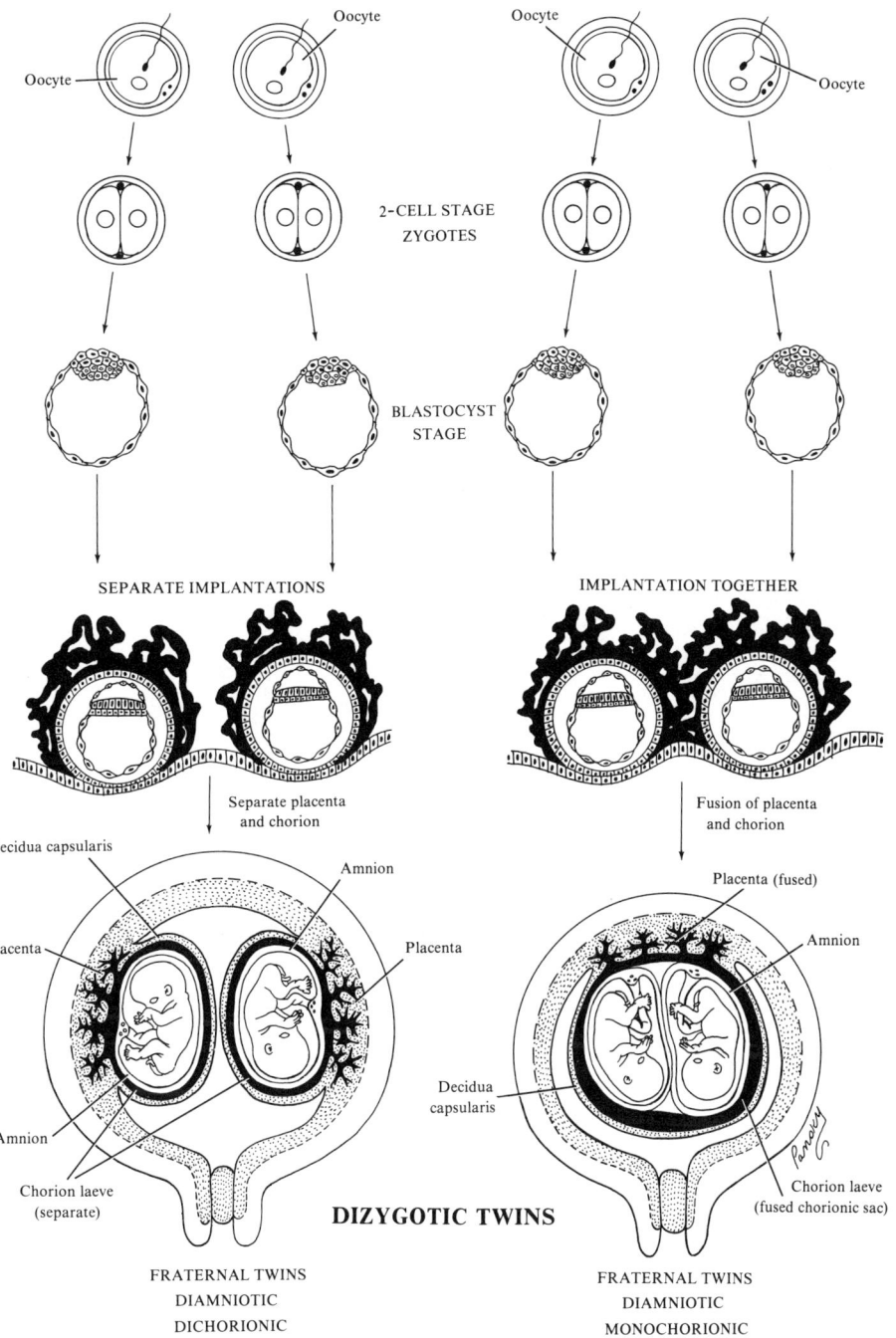

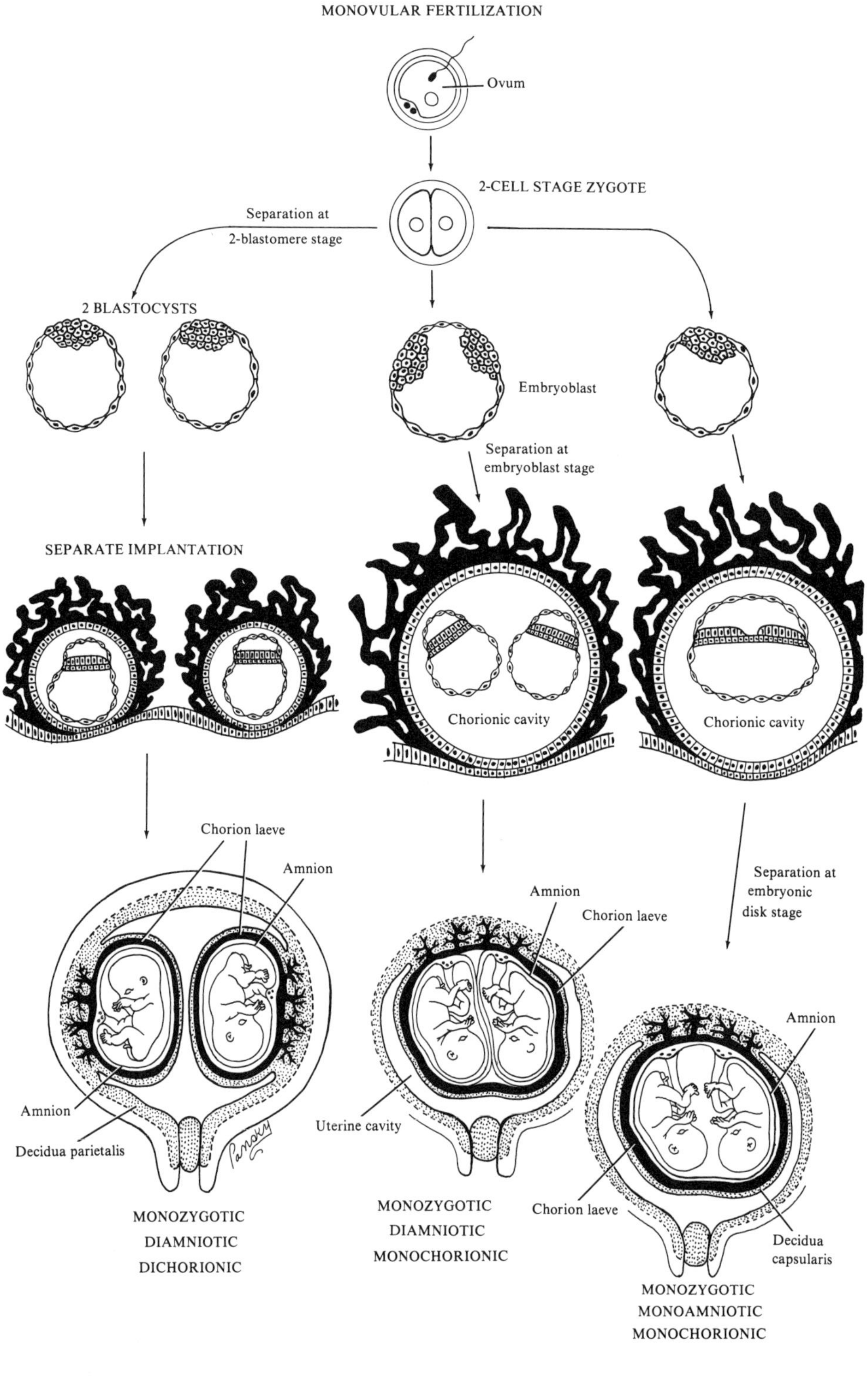

FIGURE 15. **Monozygotic, genetically identical twin pregnancies.**

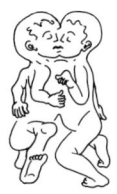
CEPHALOTHORACOPAGUS
Thorax and head fusion

THORACOPAGUS
Chest-to-chest fusion

CEPHALOPAGUS
Head to-head fusion

PYOPAGUS
Rump-to-rump fusion

Upper parts separate

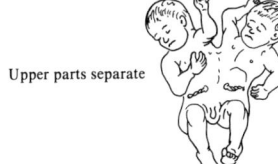
DUPLICITAS ANTERIOR
Fusion of lower parts

Reduced by thoracic fusion
(note only four extremities)

Parasitic twin attached
to rump of autosite
(normal)

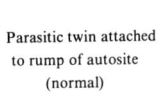
PYOPAGUS PARASITICUS

UNEQUAL TWINS

Parasitic twin attached to
thorax of autosite (normal)

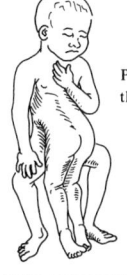
THORACOPAGUS PARASITICUS
(Conjoined duplicitas posterior)

FIGURE 16. **Conjoined twins.**

31. CONGENITAL MALFORMATIONS AND THEIR CAUSES

I. Anomalies of development
A. ETIOLOGIC FACTORS: malformations develop as a result of 3 mechanisms
1. Noxious influence of external factors during the first phase of development. This type of teratogenic agent has been studied experimentally
2. Transmission of a genetic abnormality by the parents: is inscribed in the genetic code of 1 or both parents and transmitted according to the laws of heredity
   a. The fetus may be the carrier of a purely molecular abnormality, such as *hemophilia,* due to the absence of a globulin necessary for coagulation
      i. This molecular abnormality also may be translated on the cellular scale: an *abnormal hemoglobin* can bring about a visible RBC deformity
   b. More rarely, certain metabolic anomalies bring about a morphologically visible malformation, such as the dwarfism and body deformities of *achondroplasia*
3. Chromosome aberration existing in one of the gametes or appearing during the first division
   a. The mechanism can result in an abnormality of chromosome constitution at the time of gametogenesis
   b. At fertilization, the zygote resulting from the above gamete will have an abnormal karyotype which will be transmitted to all cells of the embryo. Thus, the abnormality can involve either an autosome or a sex chromosome

B. AUTOSOMAL ABERRATIONS
1. The most frequent is *mongolism* or *Down's syndrome* which is caused by trisomy 21; that is, chromosome pair 21 consists of 3 chromosomes instead of 2. One gamete, the ovum (most frequently abnormal) carries two 21 chromosomes, whereas the other, the spermatozoon, normally carries only 1. The zygote thus contains three 21 chromosomes and 47 chromosomes in total
   a. Characteristics: mental deficiency, brachycephaly, flat nasal bridge; slant to eyelid fissures; protruding tongue, simian crease, congenital heart defects
2. Trisomy 17–18 results in mental retardation, congenital heart defects, low-set ears, and flexion of the fingers and hands
   a. Frequently see micrognathia, renal anomalies, syndactyly, and skeletal malformations
   b. Seen in about 0.3 of 1000 births, and infants usually die by the age of 2 months
3. Trisomy 13–15 results in mental retardation, congenital heart defects, deafness, and cleft lip and palate. Also show microphthalmia, anophthalmia, and coloboma
   a. Incidence is about 0.2 per 10,000 newborns. Most die by 3 months

C. SEX CHROMOSOME ABERRATIONS
1. Appear to be more frequent than autosomal abnormalities, but this may be due to the fact that the latter usually lead to precocious death of the embryo
2. Initiated by a chromosomal anomaly of the ovum; 2 major sex chromosome aberrations are described
   a. *Turner's syndrome:* the ovum does not carry any sex chromosome; there is a female phenotype (appearance), and one sees dwarfism, malformation, and gonadal aplasia or dysgenesis
      i. All cells are sex-chromatin negative, and cells have only 45 chromosomes with an XO chromosomal complement
      ii. Usually due to nondisjunction in male gamete during meiosis
   b. *Klinefelter's syndrome:* the ovum carries 2 sex chromosomes; the phenotype is male, and one notes sterility, testicular atrophy, hyalinization of the seminiferous tubules, and usually gynecomastia
      i. Cells have 47 chromosomes with a sex chromosomal complement of XXY type, and a sex chromatin body is seen in 80% of cases
      ii. Seen in 1 of 500 male births; caused by nondisjunction of the XX homologs
   c. *Triple-X syndrome:* from fertilization of an XX oocyte by an X sperm
      i. Patients are infantile with scanty menses and a degree of mental retardation
      ii. Have 2 sex chromatin bodies in their cells and are called "superfemale"
      iii. Some are of proven fertility, and offspring may be normal

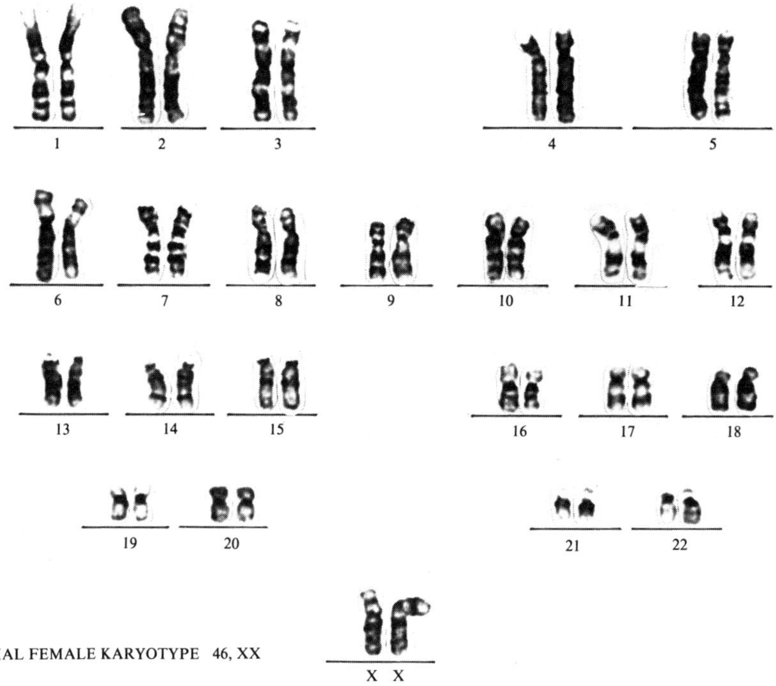

NORMAL FEMALE KARYOTYPE 46, XX

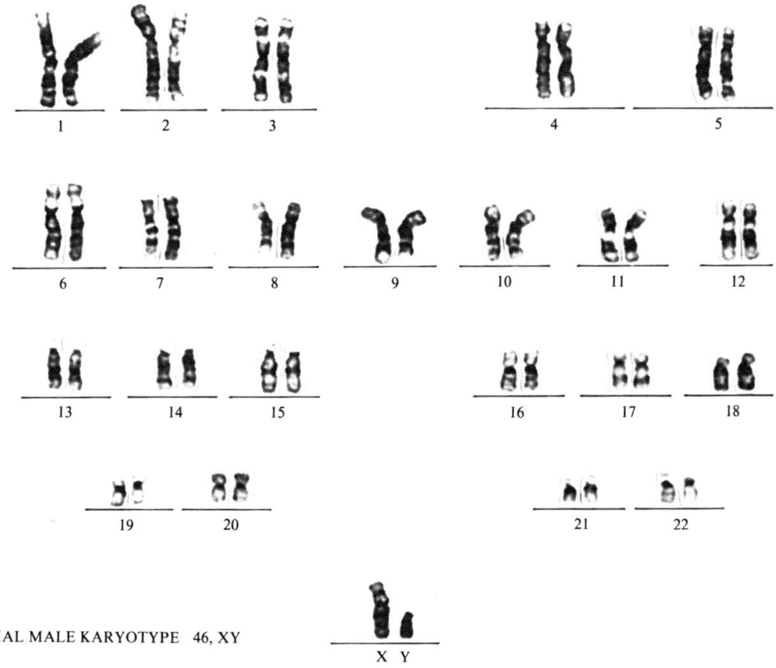

NORMAL MALE KARYOTYPE 46, XY

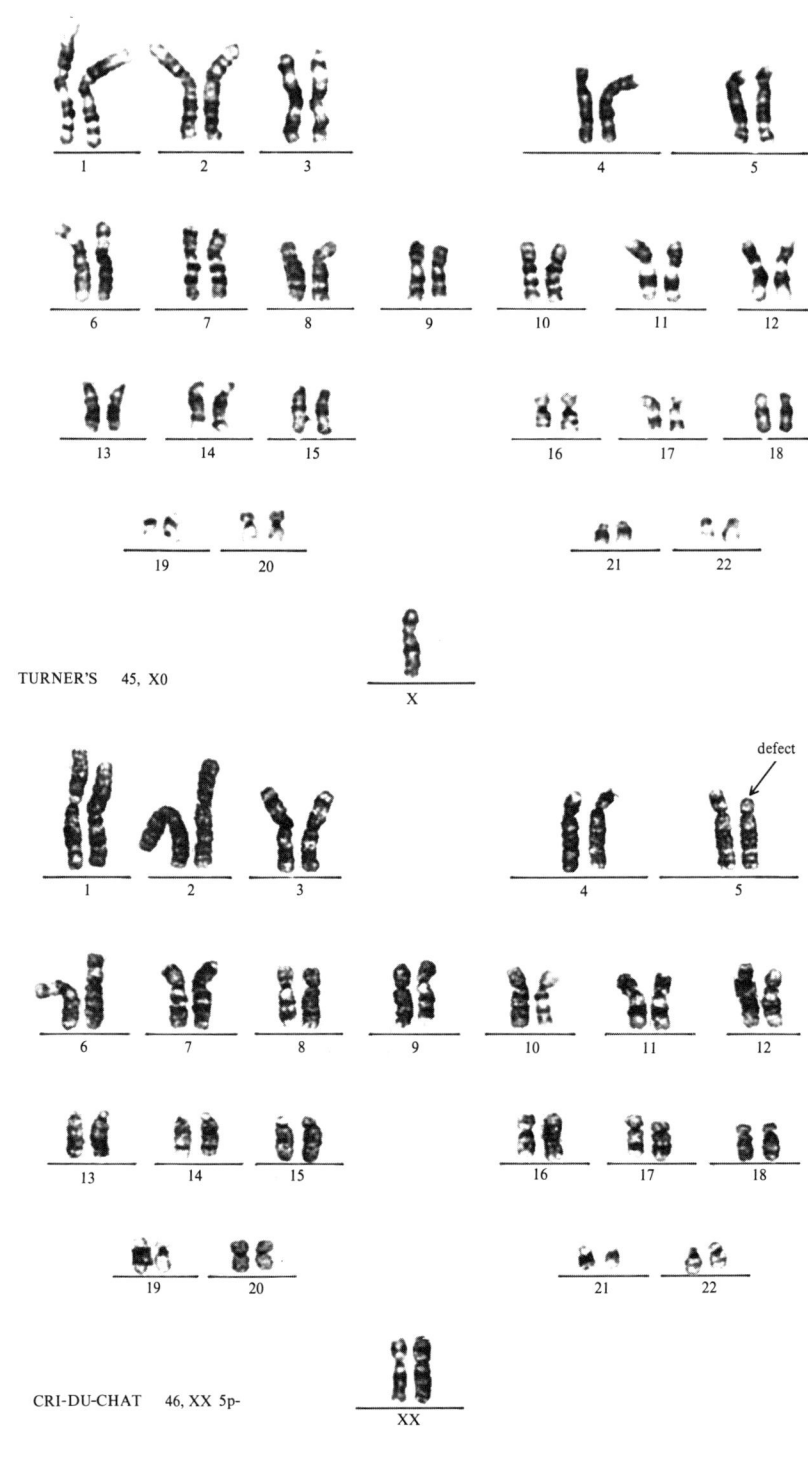

FIGURE 17. **Turner's and Cri-du-chat syndrome karyotypes.**

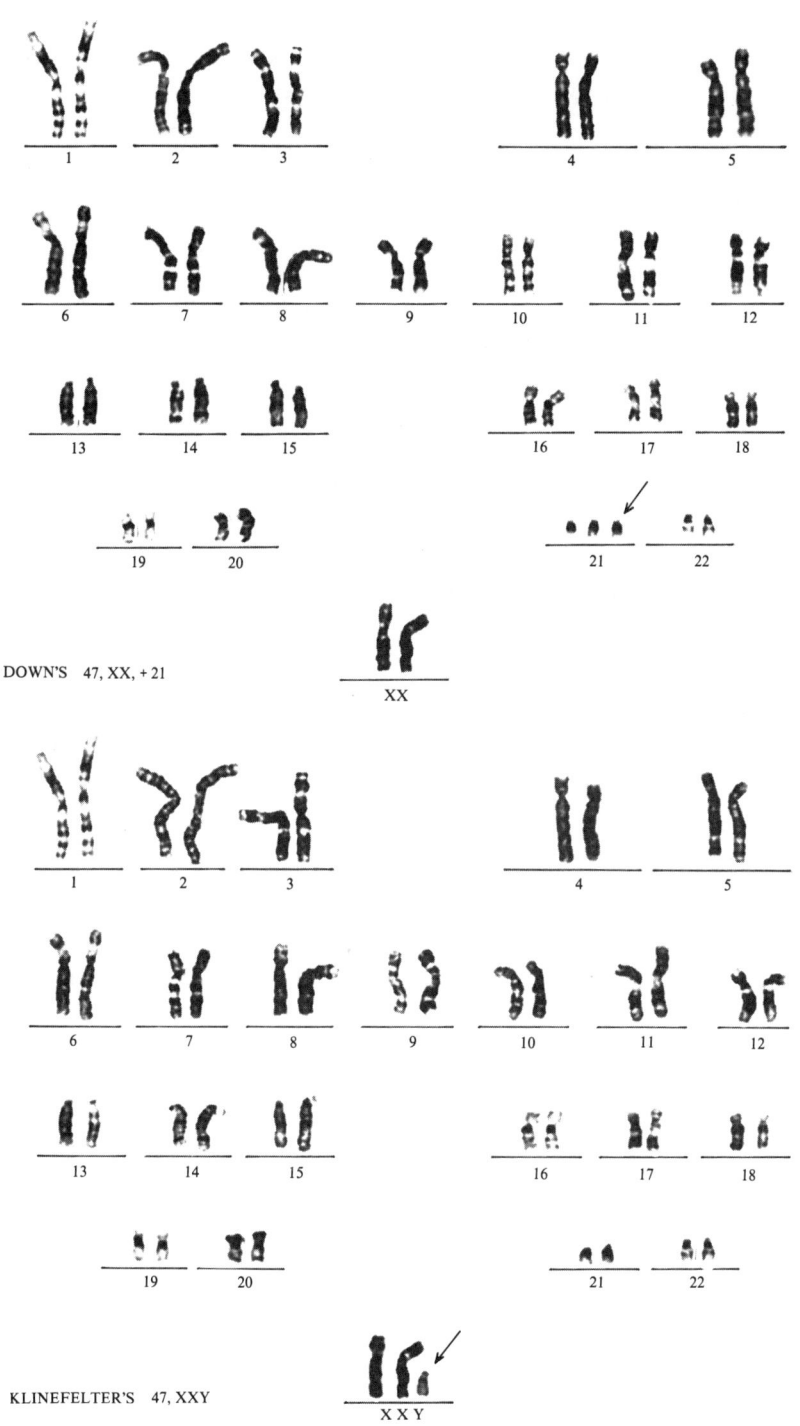

FIGURE 18. **Down's and Klinefelter's syndrome karyotypes.**

32. CONGENITAL MALFORMATIONS AND THEIR CAUSES: HUMAN MALFORMATIONS

I. **Human malformations** occur in 2–3% of births. Some examples of these are
A. SINGLE MONSTERS
   1. Phocomelia: limb anomaly spontaneously created. Seen in 1/100,000 births
      a. A typical lesion of thalidomide: 10% of the women who took the drug during the critical period had babies with this anomaly
   2. Coelosomy: a defect of closure of the abdominal wall whereby the normally developed abdominal viscera are found in an extra-abdominal position
   3. Craniorrhachischisis: complete failure of the neural tube to close
      a. There are angiomatous degeneration of nervous tissue, absence of the cranial vault, and absence of the posterior arches of the vertebrae
B. DOUBLE MONSTERS OR DOUBLE-TYPE MALFORMATIONS can be considered as nonseparated twins with the degree and type of fusion being variable
   1. Janus-type (janiceps) cephalothoracopagus (pagus, meaning something fastened)
   2. Asymmetric thoracopagus
   3. Acardia is one of a pair of monozygotic twins which has degenerated after a failure of vascularization. Structures already present regress, ending in the formation of what looks like an amorphous mass with no organization
C. EXPERIMENTAL MALFORMATIONS: the mammalian embryo, despite its apparent protection, is very sensitive to the influence of various external teratogenic agents. Most of the malformations, seen in humans clinically, have been reproduced experimentally
   1. Classification of teratogenic factors: usually described in 5 major groups
      a. Physical factors: x-rays, radiation, etc.
      b. Chemical factors: hypoglycemia, antitumor drugs, neuroepileptics, etc.
      c. Nutritional factors: hyper- or hypovitaminosis, mineral excess or deficiencies, vitamin imbalance, etc.
      d. Hormonal factors: use of androgens, synthetic progesterones, cortisone, etc.
      e. Infectious factors: toxoplasmosis, rickettsioses, Asian flu, and viruses
         i. Viruses: especially rubella or German measles, cytomegalovirus, herpes simplex virus, measles, mumps, hepatitis, poliomyelitis, chickenpox, syphilis, ECHO virus, and Coxsackie virus. A number have been implicated but not all cause malformations
   2. Mode of action of teratogenic factors: the effect depends predominantly on the stage of intervention of the agent (chronologic factor) and the genetic constitution (constitutional factor). Several types of sensitivity are listed
      a. Time or stage of sensitivity
         i. Before implantation: external agents, according to their intensity, provoke either completely reversible lesions or definitive mortal lesions
         ii. After implantation and during the entire period of active morphogenesis: this is the principle teratogenic period because a primordium is most sensitive to teratogenic actions at the time of its appearance
            a) The same substance can produce different malformations if given at different stages of morphogenesis
            b) When more than 1 primordium develop simultaneously, the same agent can result in multiple malformations
      b. Species sensitivity: an agent teratogenic for 1 species may not be so for another
      c. Strain sensitivity: even in the same species, the percentage of malformations seen with any substance can vary according to strain and even the line
      d. Individual sensitivity: even in the same animal litter subjected to a teratogenic influence, certain individuals react differently and may be free of any malformations. Even those malformed are not so to the same degree necessarily. Different metabolic peculiarities may explain these individual variations and also those seen between strains

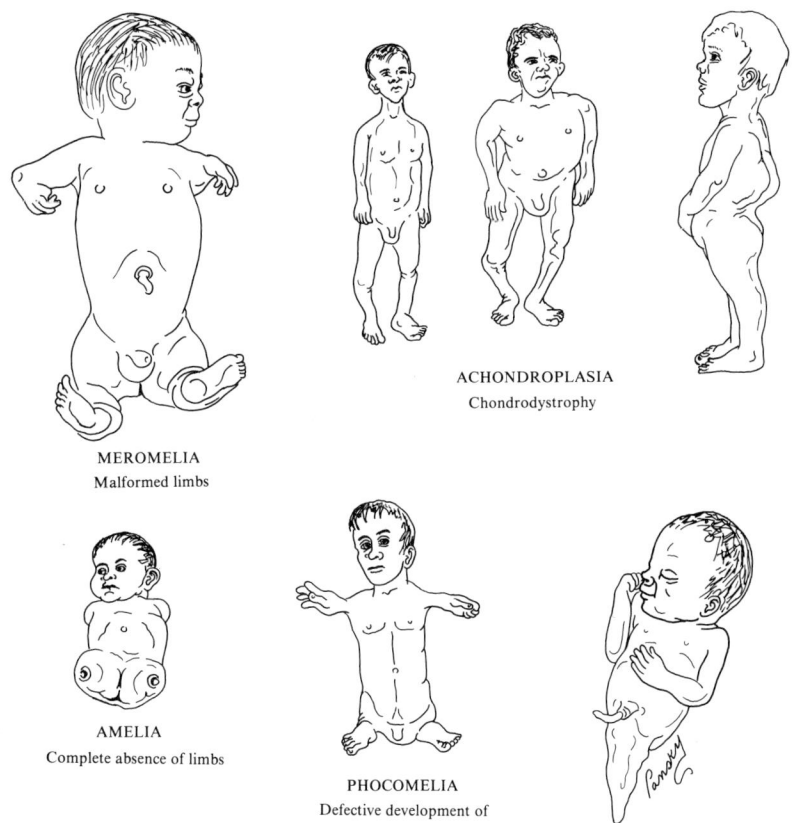

MEROMELIA
Malformed limbs

ACHONDROPLASIA
Chondrodystrophy

AMELIA
Complete absence of limbs

PHOCOMELIA
Defective development of arms or legs or both

SIRENOMELIA
Union of legs with total fusion; may see partial fusion

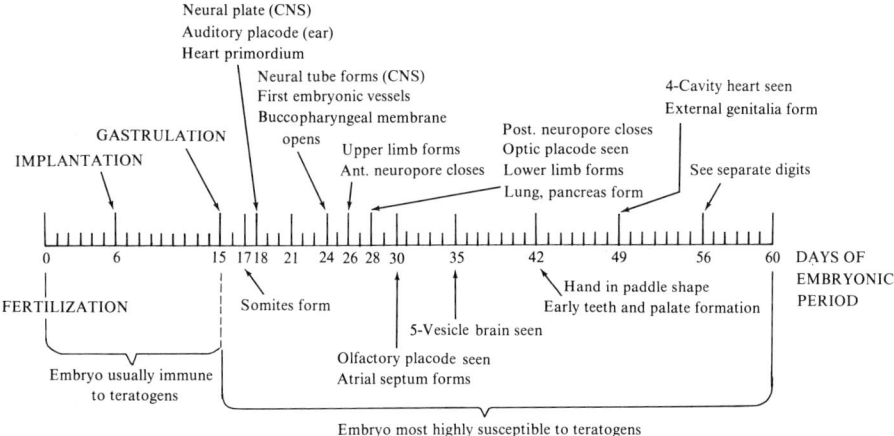

33. THE FETAL MEMBRANES

I. **Week 1 of development:** the following structures appear in succession
   A. THE TROPHOBLAST, during the morula stage, begins to differentiate from the superficial cellular layer of the fertilized egg
   B. BY THE BLASTOCYST STAGE, at about day 5, the trophoblast is differentiated (described with the placenta since it gives rise to it)
   C. IMPLANTATION begins about day 6
   D. THE AMNIOTIC CAVITY, which is hollowed out from the middle of the inner cell mass (embryoblast) makes its appearance with the beginning of the primitive yolk sac at about day 7 and is bounded by entoderm and Heuser's membrane. By day 9, the primitive yolk sac is clearly seen

II. **Weeks 2 and 3**
   A. THE EXTRAEMBRYONIC MESENCHYME CONDENSES TO FORM 2 LAYERS
      1. The external layer is fused to the trophoblast to form the *chorion*
      2. The internal layer is attached to the amnion and with it forms the *somatopleure* and with the yolk sac forms the *splanchnopleure*
      3. Between the 2 layers appears the *extraembryonic coelom,* except for the region of the connecting stalk where the embryo is connected to the wall of the egg
   B. THE AMNIOTIC CAVITY enlarges and is carried along by the edges of the embryonic disk during cephalocaudal flexion
   C. THE PRIMITIVE YOLK SAC is entirely bordered by entoderm which develops along Heuser's membrane
   D. THE ALLANTOIS appears at the union of the caudal portions of the disk and the primitive yolk sac

III. **Week 4**
   A. THE EXTRAEMBRYONIC COELOM gets smaller, and the amniotic cavity enlarges at the expense of the coelom
   B. THE PRIMITIVE YOLK SAC constricts and the yolk sac appears, attached to the area of the primitive gut by the future *vitelline duct*
   C. THE ALLANTOIS progresses in the connecting stalk, along with the umbilical-allantoic vessels

IV. **Week 8**
   A. THE EXTRAEMBRYONIC COELOM disappears as it is affected by the development of the amniotic cavity
   B. THE YOLK SAC is found up against the placental area, at the end of the long vitelline duct which later regresses
   C. THE ALLANTOIS, after being extended over almost the entire length of the umbilical cord, disappears distally. The umbilical vessels, however, continue to develop

V. **From the third month**
   A. THE YOLK SAC has disappeared almost completely
   B. THE UMBILICAL CORD now contains only the umbilical vessels and remnants of the allantois and vitelline duct
   C. THE AMNIOTIC CAVITY continues to grow until term, at which time it contains almost a liter (1000 ml) of liquid—the so-called "bag of waters"
   D. AT TERM, the membranes surround the placenta with the cord and fetal vessels in its center. The membranes are torn during labor in order to permit delivery

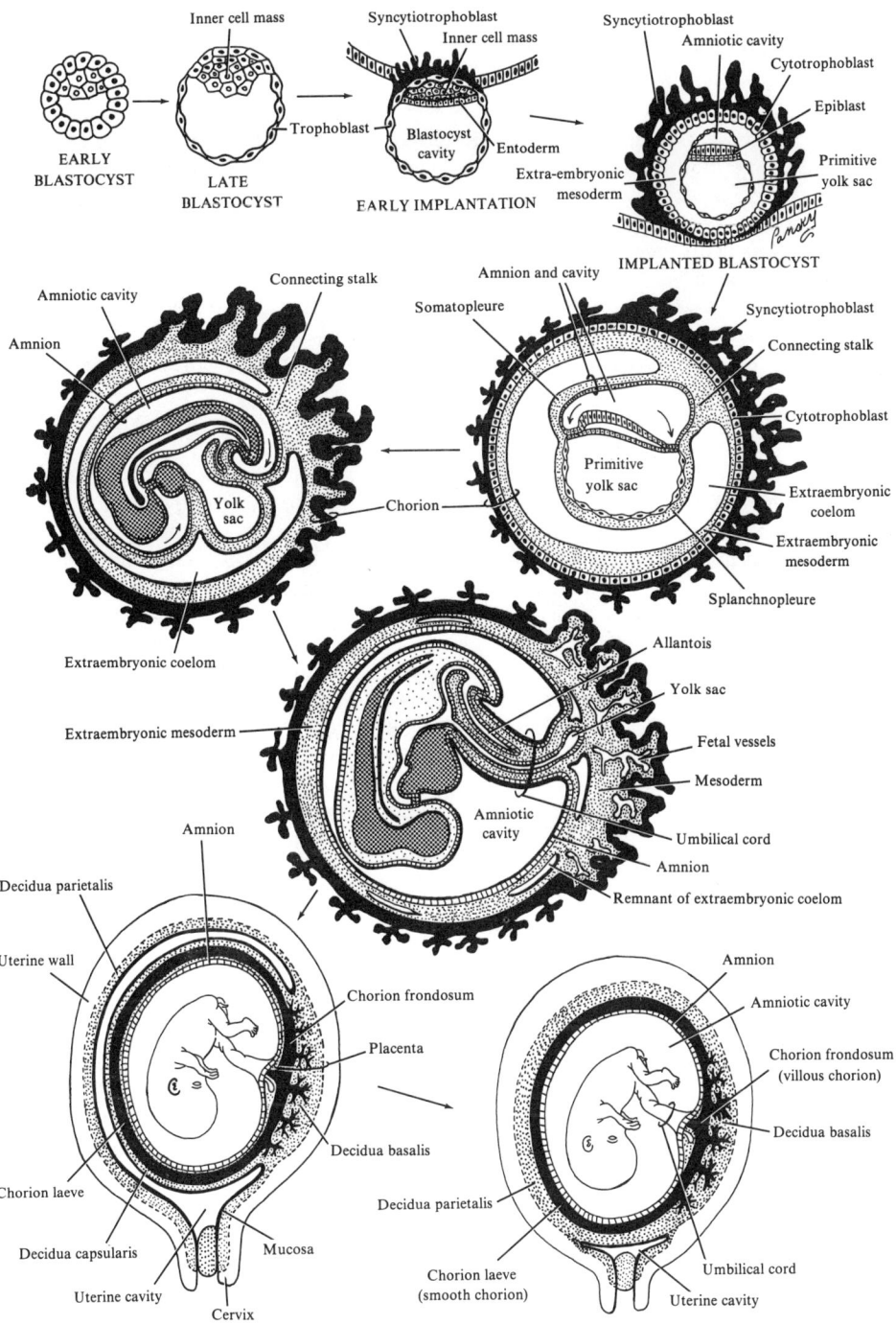

34. AMNION, ALLANTOIS, AND YOLK SAC

I. **The amnion** is the membrane around the fetus. *Amniocentesis* is the transabdominal aspiration of fluid from the amniotic sac (the innermost of the membranes enveloping the embryo *in utero*). The amnion's epithelial cells have microvilli which may play a role in fluid transfer
   A. ORIGIN OF FLUID: may initially be secreted by amniotic cells but most comes from the maternal blood
      1. The fetus also contributes by excreting urine into the amniotic fluid
      2. By late pregnancy, one-half a liter of fetal urine is added daily
   B. VOLUME OF FLUID increases slowly to about 30 ml at 10 weeks, 350 ml at 20 weeks, and 1000 ml by 37 weeks. It then decreases very sharply
      1. Low volume (about 400 ml) results in *oligohydramnios,* a consequence of placental insufficiency and decreased blood flow
      2. *Polyhydramnios* (excess fluid) may occur when the fetus does not drink its usual amount of fluid and is often associated with malformations of the CNS
   C. FLUID EXCHANGE: water in the amniotic fluid changes every 3 hours
      1. Fluid is normally swallowed by the fetus, up to 400 ml/day near term, and absorbed back into the fetal circulation via the fetal gastrointestinal tract
   D. COMPOSITION OF FLUID: it is really a suspension consisting of desquamated fetal epithelial cells and equal parts of organic and inorganic salts in 98–99% water
      1. Changes occur as fetal excreta is added with pregnancy development
   E. SIGNIFICANCE OF FLUID: embryo floats freely in the fluid
      1. Permits symmetric external growth of the embryo
      2. Prevents adherence of the amnion to the embryo
      3. Cushions the embryo against some trauma
      4. Helps control embryonic body temperature
      5. Helps fetus move freely, aiding in skeletomuscular development

II. **The allantois**
   A. SIGNIFICANCE: blood formation occurs in its walls during weeks 3–5, and its blood vessels become the umbilical arteries and vein
   B. FATE: its intraembryonic portion runs from the umbilicus to the urinary bladder with which it is continuous
      1. As the bladder enlarges, it involutes to form the *urachus*
      2. After birth, the urachus becomes a fibrous cord, the *median umbilical ligament*

III. **The yolk sac:** nonfunctional as yolk storage site in the human
   A. SIGNIFICANCE
      1. Role in transfer of nutrients in the embryo during weeks 2–3
      2. Blood development occurs in its walls beginning in week 3 and continues to form there until the hematopoietic activity begins in the liver at about week 5
      3. During week 4, its dorsal part is incorporated into the embryo as an entodermal tube, the *primitive gut,* and gives rise to the epithelium of the trachea, bronchi, lungs, and digestive tract
      4. Primordial germ cells appear in the yolk sac wall in week 3 and migrate to the area of developing sex glands (gonads) where they become germ cells (oogonia and spermatogonia)
   B. FATE: by week 12, small yolk sac lies in chorionic cavity between amnion and chorionic sac; it shrinks as pregnancy proceeds and eventually gets smaller and solid and may even persist through pregnancy but is of no significance
      1. The yolk stalk usually detaches from the gut by the end of week 5, but in 2% of cases, the proximal intra-abdominal part of the yolk stalk persists as a diverticulum of the ileum, *Meckel's diverticulum*

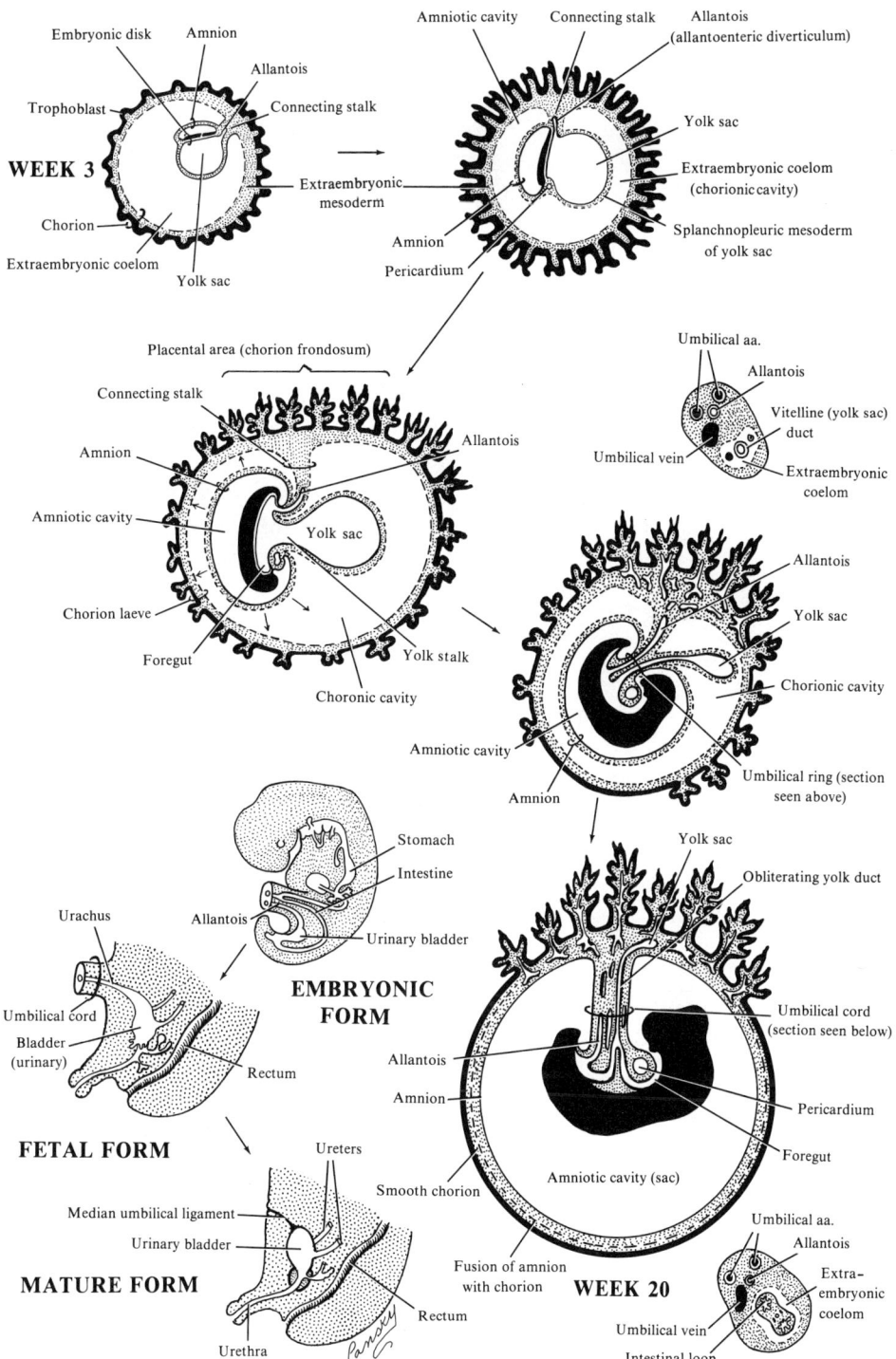

35. THE UMBILICAL CORD

A. Two stalks are seen at the beginning of development
   1. A ventral one, the *yolk sac stalk,* containing the vitelline duct and the vitelline vessels
   2. A caudal one, the *connecting stalk,* containing the allantois and the umbilical vessels
   3. The connecting stalk progresses ventrally as a result of the cephalocaudal flexion of the embryo and fuses with the yolk sac stalk to form the umbilical cord
B. The umbilical cord brings together, in the same mesenchymal core, the components of the connecting stalk (the allantois and umbilical vessels) and the vitelline duct and vessels
C. The umbilical cord is covered by amnion which is continuous with the outer epithelial layer of the embryo at the attachment of the umbilicus
D. The umbilical cord is short and thick in the young embryo and is inserted in the lower portion of the ventral region of the embryo
   1. With the development of the anterior abdominal wall, the region of umbilical implantation contracts, the cord elongates, and also becomes slender
   2. At term, the cord contains only the umbilical vessels surrounded by *Wharton's jelly* (a smooth mesenchymous material, a mucous differentiation product of mesoderm) and is about 50–60 cm long and about 2 cm in diameter
      a. It is tortuous, which may result in the so-called *false-knots* (not significant)
      b. An extremely long cord may encircle the neck of the fetus, creating some problems at the time of delivery; or a very short cord may cause difficulties during delivery by pulling the placenta from its attachment
      c. In about 1% of cases there are so-called *true knots* in the cord, which in most cases form during labor as a result of the fetus passing through a loop. Since the knots are usually loose, they have no clinical significance.
         i. If the true knot forms early in pregnancy and tightens during active fetal movements, it may interfere with the fetal circulation and cause death and abortion of the embryo or fetus
      d. Looping of the cord around the fetus occasionally occurs
         i. In one-fifth of all deliveries it loops once around the neck and does not create any fetal risk
   3. In about 1/200 newborns, only one umbilical artery is present and may be associated with cardiovascular abnormalities in 15–20% of cases

UMBILICAL CORD

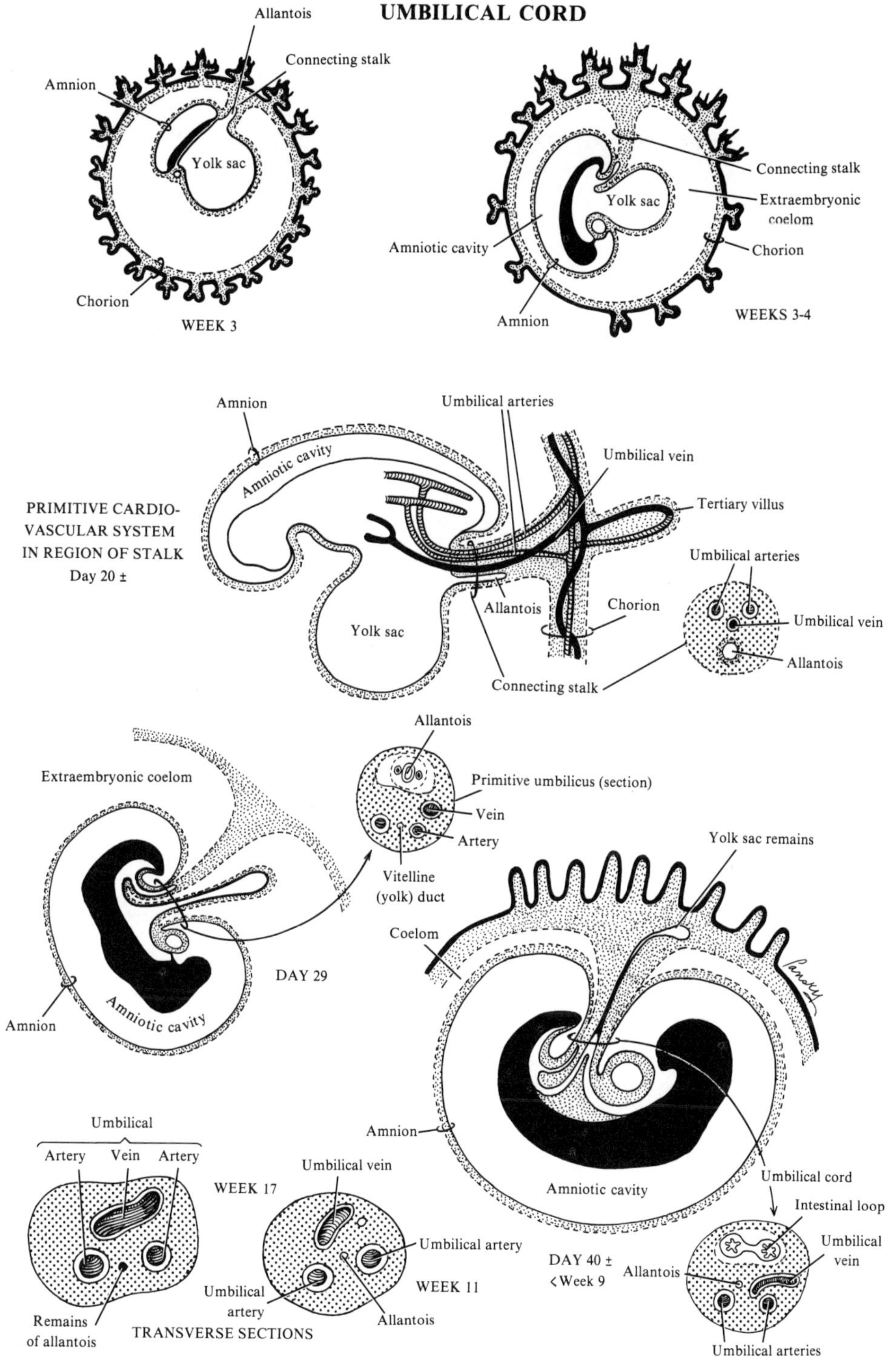

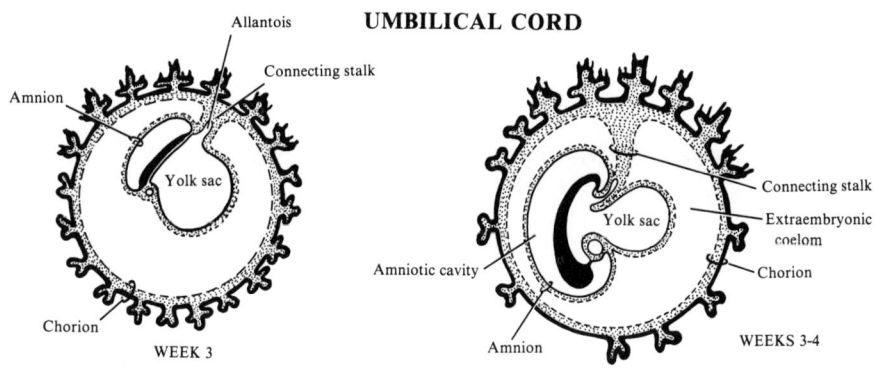

36. UTERINE GROWTH DURING PREGNANCY AND PARTURITION

I. **The nonpregnant uterus** lies within the pelvis

II. **With pregnancy**
   A. THE UTERUS grows in size, increases in weight, and its walls thin out
   B. DURING THE FIRST TRIMESTER, THE UTERUS rises out of the pelvic cavity and reaches the level of the umbilicus by week 20
      1. By weeks 28-30, the uterus reaches the epigastric region
      2. The mother's abdominal viscera are displaced, and the skin and muscles of the anterior abdominal wall are greatly stretched
   C. UTERINE SIZE INCREASE is due to hypertrophy of preexisting muscle fibers for the most part and partly to development of some new fibers

III. **Parturition or labor:** the process of expelling the fetus, placenta, and fetal membranes. The factors causing parturition are unclear, but opinion favors the intermittent release of oxytocin from the maternal neurohypophysis as being important in determining the strength and duration of uterine contraction once labor is established and may well initiate labor as well. Steroid quantity produced by the fetal adrenal cortex may trigger some event in the placenta or myometrium of the mother that effects the onset of labor
   A. STAGES OF LABOR are usually described as being 3
      1. Stage I: when there is evidence of progessive dilatation of the cervix
         a. Occurs with the onset of regular uterine contractions that are less than 10 minutes apart and are painful
         b. Ends with complete cervix dilatation
         c. Average duration of labor is about 12 hours for the first pregnancy and about 7 hours for the multiparous mother
      2. Stage II begins when the cervix is fully dilated and ends with delivery
         a. Average duration for the prima gravida is 50 minutes; for the multipara is 20 minutes
      3. Stage III begins when the baby is born and ends when the placental membranes are delivered
         a. Average duration of this stage is less than 30 minutes
   B. AFTER DELIVERY OF THE BABY, THE UTERUS continues to contract, and a hematoma is formed behind the placenta which separates it from the decidua
   C. AFTER PLACENTAL DELIVERY, MYOMETRIAL CONTRACTIONS constrict the spiral arteries that supplied the intervillous space
      1. Contractions are tonic and prevent excessive bleeding from the placental site
      2. A relaxed uterus is the most common cause of postpartum hemorrhage

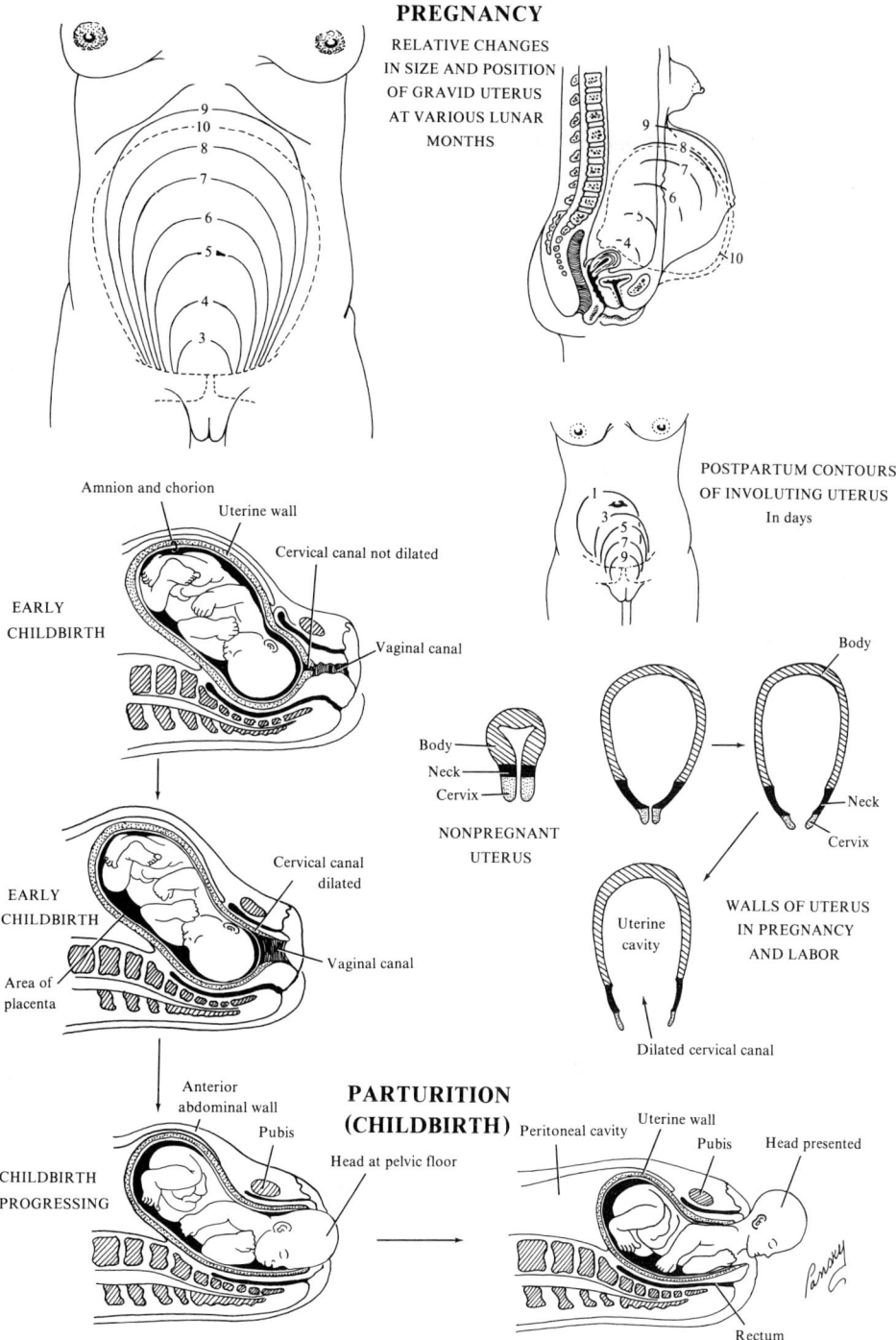

37. FETAL-MATERNAL INCOMPATIBILITY

I. **Introduction:** the disease caused by fetal-maternal incompatibility begins in the fetus and manifests itself after birth. It results from incompatibility of the blood between the fetus and mother and involves immunization of the mother against a blood group antigen carried by the fetus. It occurs in 1 in 150 births and is the most frequent fetal disease. The most common cause is *Rhesus* or *Rh immunization*

II. **Mechanism in Rh immunization**
A. R<small>H</small> is a blood group carried by the red blood cells in about 85% of individuals and is referred to as Rh+. An Rh− subject can be immunized against the Rh+ antigen if Rh+ red blood cells are introduced into his or her body (e.g., by transfusion). Immunization is manifested by the appearance of anti-Rh antibodies
   1. Immunization of the Rh− woman who has an Rh+ husband can occur during pregnancy. Fetal red blood cells may cross placenta and enter the maternal circulation. If the fetus is Rh+, Rh+ antigen is introduced into the Rh− maternal blood.
      a. The mother is immunized against the Rh+ antigen, and in her blood one sees anti-Rh+ antibodies. The latter, like most antibodies, easily cross the placental barrier and pass into the fetal blood
      b. The maternal antibodies attach themselves to the red blood cells of the fetus, resulting in their destruction in the spleen, causing *hemolytic perinatal disease*
B. SUMMARY
   1. If the mother is Rh− and the father Rh+ when an immunization takes place:
      a. If the father is homozygous, all the children will be Rh+. *The first pregnancy will be normal.* Maternal immunization can appear from the second pregnancy on and will become progressively greater with each pregnancy
      b. If the father is heterozygous, some children will be Rh− and be born completely normal, since the maternal antibody will have no effect on their red cells. In contrast, the Rh+ children will be affected as previously described
   2. Blood incompatibility affects only couples in which the mother is Rh− and the father is Rh+. Even in these cases, however, only 5% are affected

III. **Consequences of fetal-maternal incompatibility for the child**
A. BEFORE BIRTH
   1. Maternal antibodies are fixed on the fetal red blood cells and cause their destruction in the spleen which, as a result, hypertrophies (enlarges), resulting in hemolytic anemia and the release of hemoglobin pigment (from destroyed RBCs). The latter is immediately denatured to yellow bilirubin
   2. The pigment crosses the placenta and is eliminated by the maternal organism
   3. The fetus reacts to the anemia by the formation of new red blood cells in the liver, resulting in an erythropoiesis and liver hypertrophy
      a. As a result of the increased red cell production, some of the red blood cells in the blood are young, nucleated, immature forms or erythroblasts, resulting in an intense erythroblastosis at birth
B. AT BIRTH severe anemia with erythroblastosis and an enlarged liver and spleen are seen
   1. Icterus (yellow skin color) occurs very rapidly since the pigments are no longer being eliminated by the mother in her circulation
C. AFTER BIRTH: if the infant is not treated, it is threatened with death as a result of the anemia or by specific complications, such as *nuclear icterus,* caused by the accumulation and toxic action of bilirubin in the gray nuclei of the brain
D. IN CERTAIN VERY SERIOUS FORMS, in addition to the symptoms described above, generalized fluid infiltration results in a *fetal-placental hydrops,* which can produce fetal death in the last months of pregnancy. The placenta becomes exceedingly large and can be more than twice the size of a normal placenta

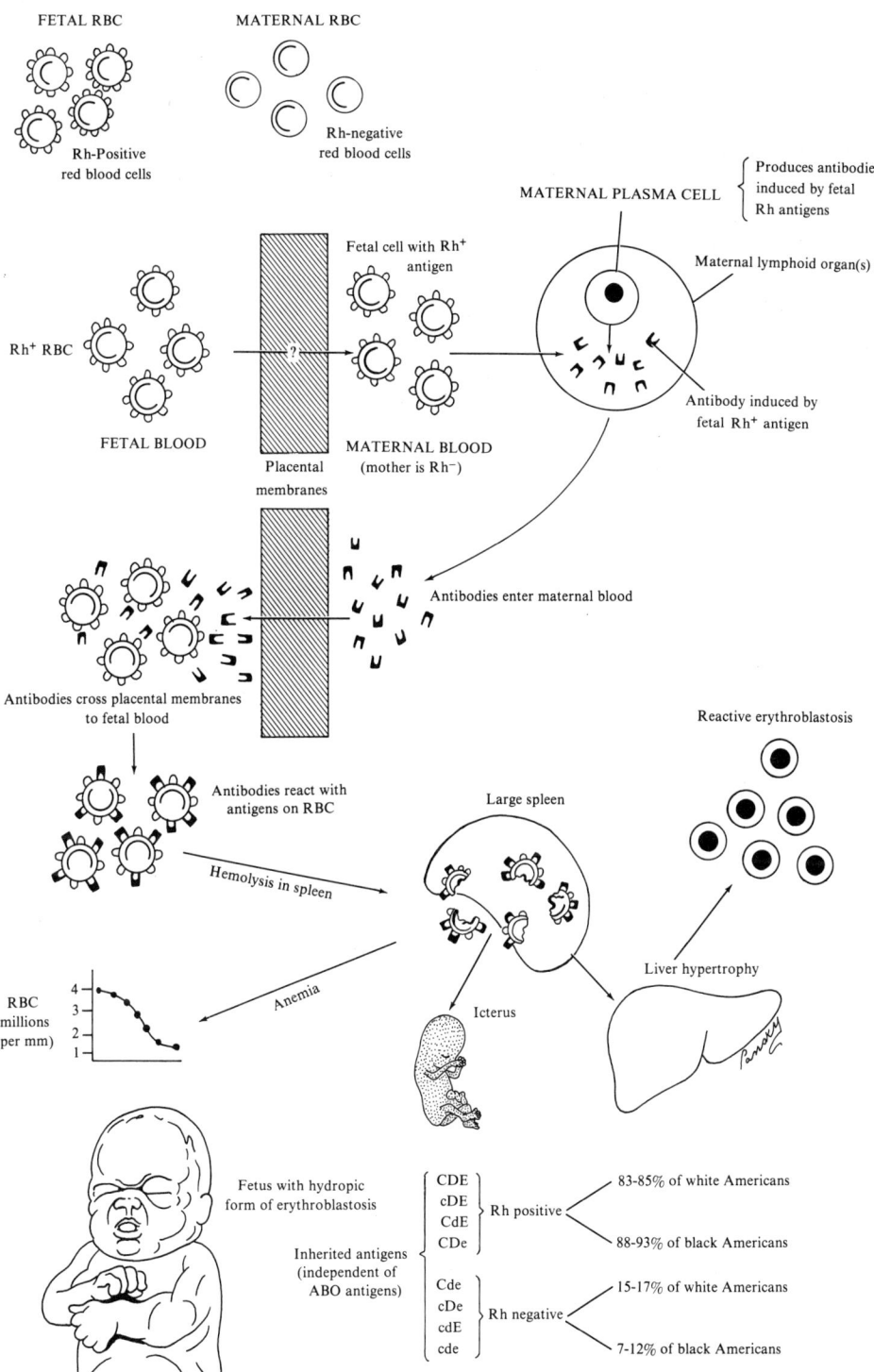

38. THE PLACENTA: GENERAL DISCUSSION

I. Morphology: the placenta is the most important acessory fetal structure and brings the fetal and maternal circulations into close relationship. Morphologically, it is partly of fetal origin (the trophoblast) and partly of maternal origin (arising from the transformation of the uterine mucosa)

A. EXTERNAL APPEARANCE OF THE PLACENTA
1. The trophoblast is seen at about day 5. By days 6 to 7, it ensures implantation of the egg in the uterine mucosa due to its proteolytic activity, and at that time consists of an inner cellular layer, the *cytotrophoblast,* and an outer syncytial layer, the *syncytiotrophoblast*
2. By week 7, the trophoblast has proliferated into *villi* which are visible over the entire surface of the chorion (placenta)
3. At the end of month 2, the villi begin to group together, forming the villous chorion or the *chorion frondosum* (bushy). The umbilical cord is very thick at this stage and is moving toward this region
4. By $2\frac{1}{2}$ months, rarefaction of the villi at 1 of the poles is seen
5. A human 3-month-old egg shows differentiation of a clearly defined placenta. The villi are grouped at 1 pole of the egg and form the placenta. The rest of the embryonic vesicle is devoid of villi and is smooth (*chorion laeve*), and through it the outline of the fetus can be seen indistinctly
6. After 3 months, the placenta grows, thickens, and spreads out, developing along with the uterus
7. The placenta at term is a disk about 20 cm in diameter, 3 cm thick, and weight about 500 g (about one-sixth of the fetal weight). A ratio markedly different than this indicates a pathologic condition
    a. From the maternal side, one sees that deep furrows divide the placental mass into a number of lobes or *cotyledons*
    b. Examination of the placenta provides information about placental dysfunction, fetal growth, retardation, neonatal illness, and infant death
    c. Retention of a cotyledon in the uterus may lead to late puerperal hemorrhage, but, more often, retained placental tissue is the cause

B. FULL-TERM PLACENTA SUMMARY
1. Maternal surface: cobblestone appearance caused by 10 to 38 cotyledons separated by grooves formerly occupied by the placental septa
    a. The cotyledon surface is covered by shreds of the decidua basalis. Most of the latter, however, is temporarily retained in the uterus and shed with subsequent uterine bleeding
2. Fetal surface: the umbilical cord attaches to this surface
    a. The amniotic covering of the cord is continuous, with the amnion adherent to this surface of the placenta
    b. The vessels radiating from the umbilical cord are clearly seen through the transparent amnion

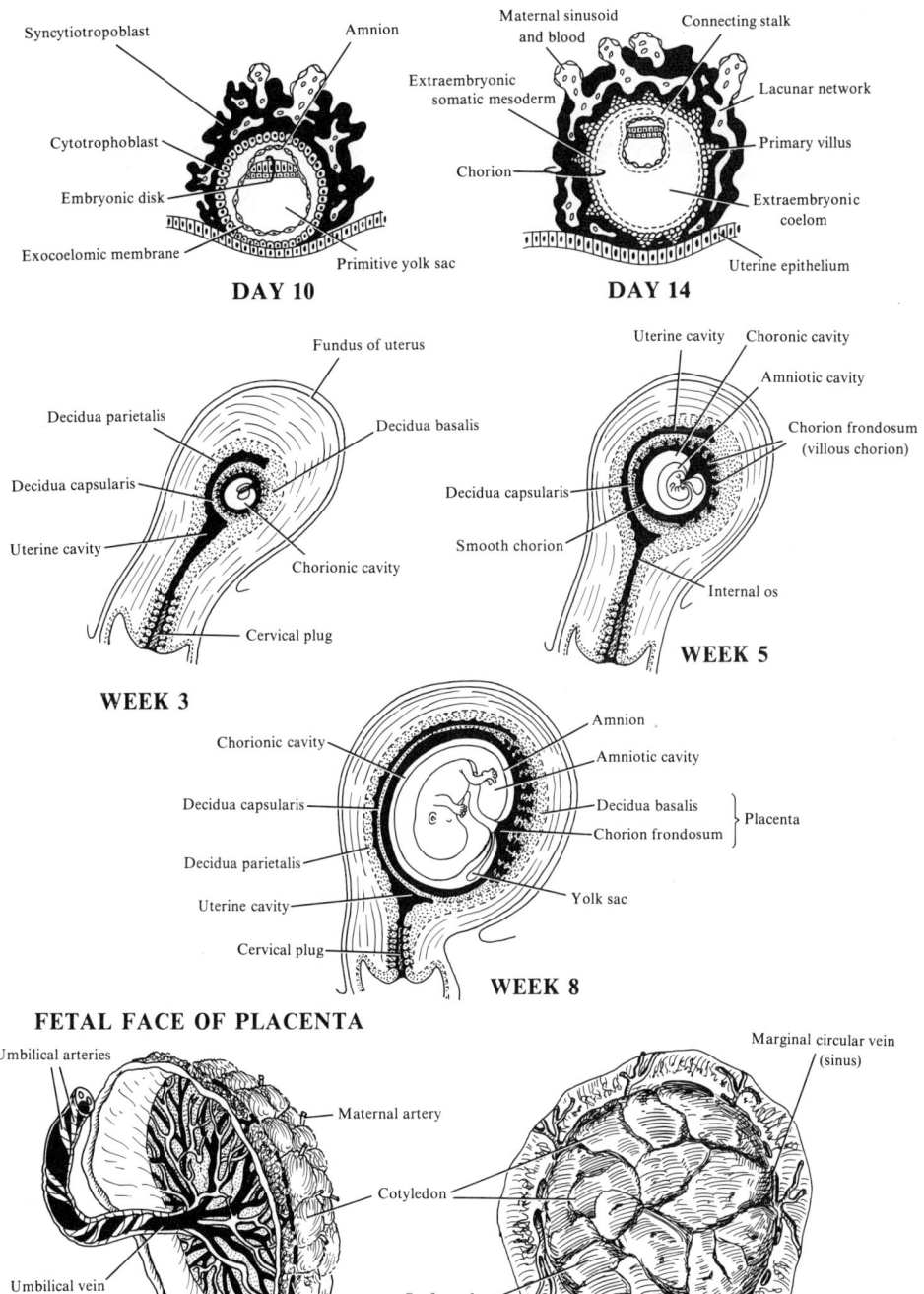

39. THE PLACENTAL VILLI

I. **Structure and development of the placental villi:** the human placenta is classified as *villous, hemochorial,* and *chorioallantoic,* since the placental villi are bathed directly by maternal blood and are transversed by vessels coming from the allantoic circulation of the fetus
   A. THE VILLUS PRIOR TO MONTH 2
      1. About day 13: the villi begin to appear in the form of *syncytial branches* separated by lacunae or *intervillous spaces*
      2. About day 15: a *cytotrophoblast core* is seen inside each column with its outer covering of *syncytiotrophoblast.* As the cell columns develop and progress, they open maternal vessels whose blood spreads out in the intervillous spaces—the beginnings of the maternal-placental circulation
      3. About day 18: the villus consists of a mesenchymal core surrounded by a double layer of cyto- and syncytiotrophoblast. In the center of the mesenchyme, *vascular islets* appear which are the beginnings of the future fetal circulation. In contrast, the lacunae, which are now intervillous spaces, are already sites of intense maternal circulation
         a. The intervillous space is divided into compartments by the *placental septa,* but, because the septa do not reach the chorionic plate, there is communication between the intervillous spaces of different compartments
         b. The intervillous spaces are drained by veins found over the entire surface of the decidua basalis
      4. About day 21: the intravillous vascular network connects with the umbilical-allantoic vessels establishing the fetal placental circulation (chorioallantoic)
   B. THE VILLUS FROM THE SECOND TO THE FOURTH MONTH
      1. The villus develops a treelike shape and is bordered by a double trophoblast layer, the superficial syncytiotrophoblast, and the deeper or inner cytotrophoblast or the cells of Langerhans
      2. The villi that make contact with the maternal tissue are called *stem* or *anchoring villi.* The others remain free in the intervillous spaces and are referred to as *branch* or *floating villi.* In sections, the villus of 4 months is far more dense than the villus of 2 months
      3. As the villi invade the decidua basalis, they leave several wedge-shaped areas of decidua tissued called *placental septa,* which appear in month 4 starting out from the maternal (decidua) plate but do not reach the chorionic plate. These septa divide the fetal part of the placenta into 10–38 convex areas composed of lobes and lobules, the so-called *cotyledons*
         a. *Each cotyledon is made up of 2 or more mainstem villi and their many branches*
   C. DEVELOPMENT OF VESSELS IN THE VILLOUS TRUNK
      1. The placenta of 1 month already reveals some single vessels in the villus. By the second month, the vascular core has developed. At term, large vessels follow the villous core up to the basal plate, and these vascular trunks in the anchoring villi give rise to the capillary networks which are involved in all branches of the villi
   D. THE VILLUS AFTER MONTH 4
      1. After 4 months, the villus is richly vascular and has a thin coat as a result of the disappearance of the cytotrophoblast (contrast with the villus of 2 months)
         a. The placenta at term may show a few large, persistent cytotrophoblastic areas (particularly on the maternal plate), but, as term nears, the cytotrophoblast vanishes in this region as well and is replaced by a fibrinoid layer
      2. As a result of numerous branchings, the villus has become comparable to a "bushy tree." Its branches form a tangled mass in whose meshes the maternal blood circulates

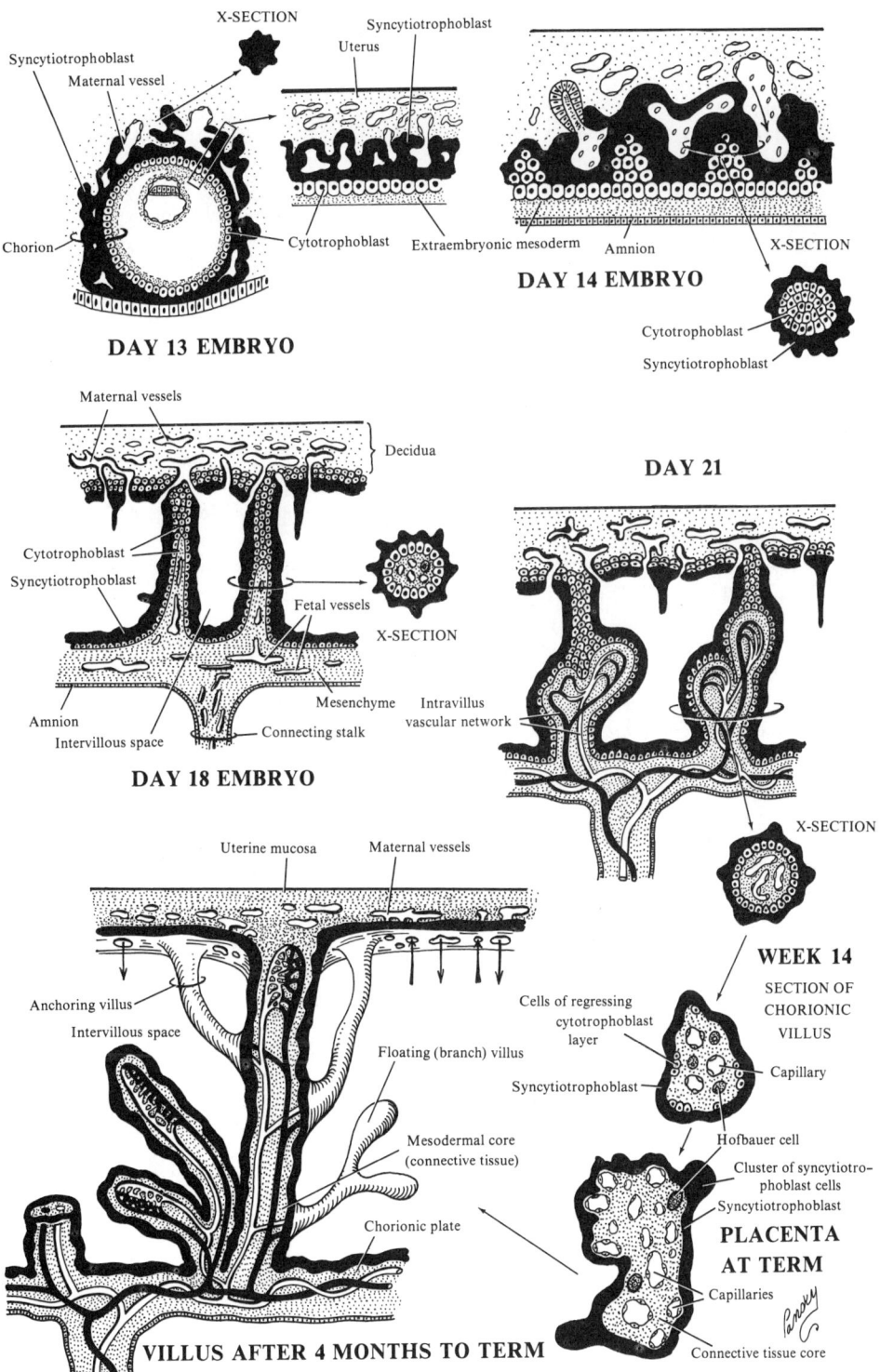

40. THE PLACENTA: DECIDUAL FORMATION

I. **Formation of the decidua** results from the changes in the uterine mucosa that accompany pregnancy
  A. THE 3 REGIONS OF THE DECIDUA that are described according to the implantation site are
    1. The decidua basalis: the portion underlying the conceptus and forming the maternal component of the placenta. Its compact layer is called the *decidual plate*
    2. The decidua capsularis is the superficial portion overlying the conceptus
    3. The decidua parietalis: all the remaining uterine mucosa
  B. AT THE TIME OF IMPLANTATION, THE CONNECTIVE TISSUE CELLS OF THE MATERNAL MUCOSA undergo epithelioid transformation, the so-called *decidual reaction,* which forms the compact zone of the mucosa (here no trace of the surface portion of the uterine glands is seen)
    1. The stromal cells increase in size and number, and blood vessels and glands undergo changes, all dependent on progesterone
    2. In the layer just beneath this compact zone, the cul-de-sacs of the glands persist to form the so-called *spongy zone* through which the plane of cleavage will pass at the time of parturition
  C. FROM MONTH 4 ON, THE DECIDUA CAPSULARIS bulges into the uterine cavity and becomes attenuated. The decidua parietalis and capsularis come in contact, fuse, and obliterate the uterine cavity
    1. By week 22, reduced blood supply causes the decidua capsularis to degenerate and disappear
    2. The amniotic sac enlarges faster than the chorionic sac, and their walls fuse to form the *amniochorionic membrane.* The latter fuses with the decidua capsularis and, following the disappearance of the capsularis, it fuses with the parietalis
  D. SUMMARY OF FETAL AND MATERNAL CONSTITUTENTS
    1. There are up to 38 large villous trunks corresponding to the cotyledons (described earlier) and seen as lobes on the maternal side of the placenta at term. Each trunk with its branchings lies in a space partitioned laterally by the decidual septa which appear month 4 of development, starting out from the maternal (decidual) plate but not reaching the chorionic plate
    2. The chorionic plate consists of the amnion, connective tissue, the syncytiotrophoblast, and the cytotrophoblast
    3. The basal plate consists of the syncytiotrophoblast, cytotrophoblast, the compact zone (decidua), and the spongy layer or zone
      a. At the time of delivery, the placenta separates from the uterus at the spongy zone

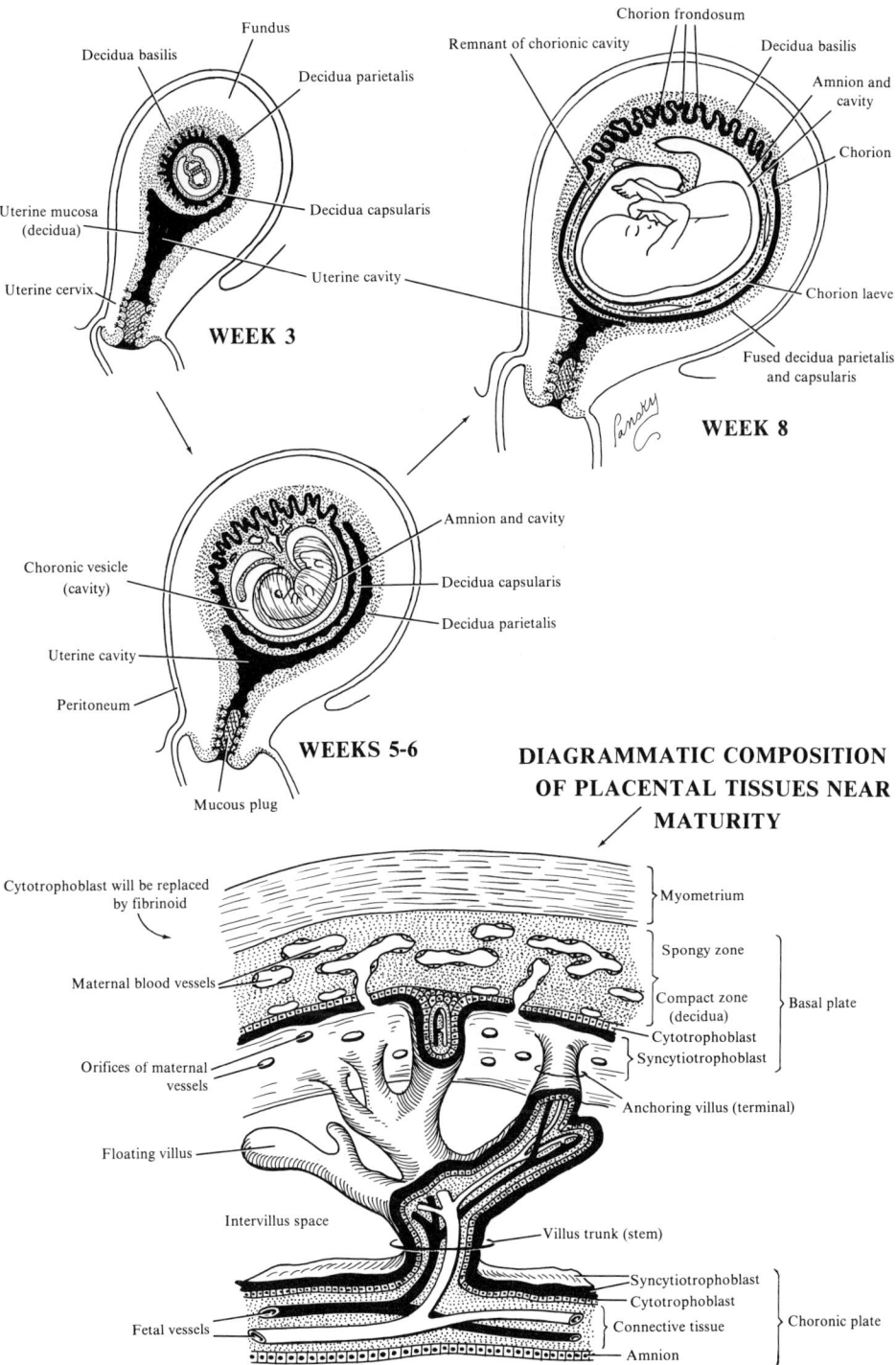

41. PLACENTAL PHYSIOLOGY

I. **Three major activities of the placenta:** metabolism, transfer, and endocrine secretion
   A. DURING EARLY PREGNANCY, in particular, the placenta synthesizes glycogen, cholesterol, and fatty acids
      1. Almost all materials are transported across placental membranes by simple diffusion, facilitated diffusion, active transport, and pinocytosis

II. **Types of placentas:** the human placenta is of the hemochorial type: the fetal tissue or chorion is directly in contact with the maternal blood. The membrane consists of only 3 layers: the syncytiotrophoblast, connective tissue, and the vascular fetal endothelium
   A. OTHER TYPES OF PLACENTAS INCLUDE
      1. Endotheliochorial type: seen in the cat and dog, consisting of the fetal vessels, connective tissue of villus, trophoblast, and the maternal vessel endothelium
      2. Syndesmochorial type: seen in the sheep and ruminants, consisting of the same layers as in 1 above, plus the connective tissue of the maternal mucosa
      3. Epitheliochorial type: seen in the pig and horse, consisting of the same 5 layers as in the syndesmochorial type plus the epithelium of the maternal mucosa which persists.

III. **Permeability according to placental type**
   A. STUDIES OF THE PASSAGE OF CERTAIN SUBSTANCES ACROSS THE PLACENTA have shown that the intensity of exchange is inversely proportional to the thickness of the placental membranes and increases regularly during gestation
      1. In most types of placentas, it reaches a maximum just before normal term
      2. The decrease at the end of gestation can be attributed, in the hemochorial type, to the deposit of fibrinoid on the exchanging surface

IV. **Placental membrane and fetal-maternal exchange**
   A. DURING PREGNANCY, the placental membrane becomes progressively thinner, and by month 4, exchange is favorable. It consists of 3 layers: the syncytiotrophoblast, the fetal vascular endothelium, and a thin sheet of connective tissue between the 2
   B. AT THE END OF PREGNANCY, the placental membrane has a thickness of 2 to 6 μm
      1. Exchange occurs both by passive diffusion and, especially, by selective and active transport resulting from activity of the membrane itself
      2. The exchanging surface is further increased by the presence of microvilli. The "brush borders" seen in light microscopy correspond to these microvilli
         a. In addition, one sees many mitochondria, ribosomes, pinocytotic vacuoles and lipid enclosures, indicating functional activity of synthesis and exchange
      3. Exchange may involve not only physiologic necessities, but also elements or substances which could create a pathologic risk for the developing fetus
         a. *Gases:* oxygen, carbon dioxide, carbon monoxide cross by simple diffusion
            i. Near term, uterus extracts 20–35 ml of oxygen per minute from maternal blood
         b. *Nutrients:* water is freely exchanged; vitamins (water soluble cross faster than fat soluble); glucose; small amounts of free fatty acids; little to no transfer of maternal cholesterol, triglycerides, or phospholipids
         c. *Hormones:* unconjugated steroid hormones pass freely; testosterone and synthetic progestins cross; protein hormones do not reach fetus in large amounts
         d. *Electrolytes:* freely exchanged
         e. *Antibodies* give fetus some passive immunity; gamma globulin (7S, IgG) reaches fetus readily
         f. *Wastes:* CO_2, urea, uric acid, bilirubin, etc., clear fetus
         g. *Drugs:* most (and their metabolites) cross placenta freely by diffusion
         h. *Infectious agents:* rubella and Coxsackie viruses and those associated with variola, varicella, measles, encephalitis, and polio pass across the placenta

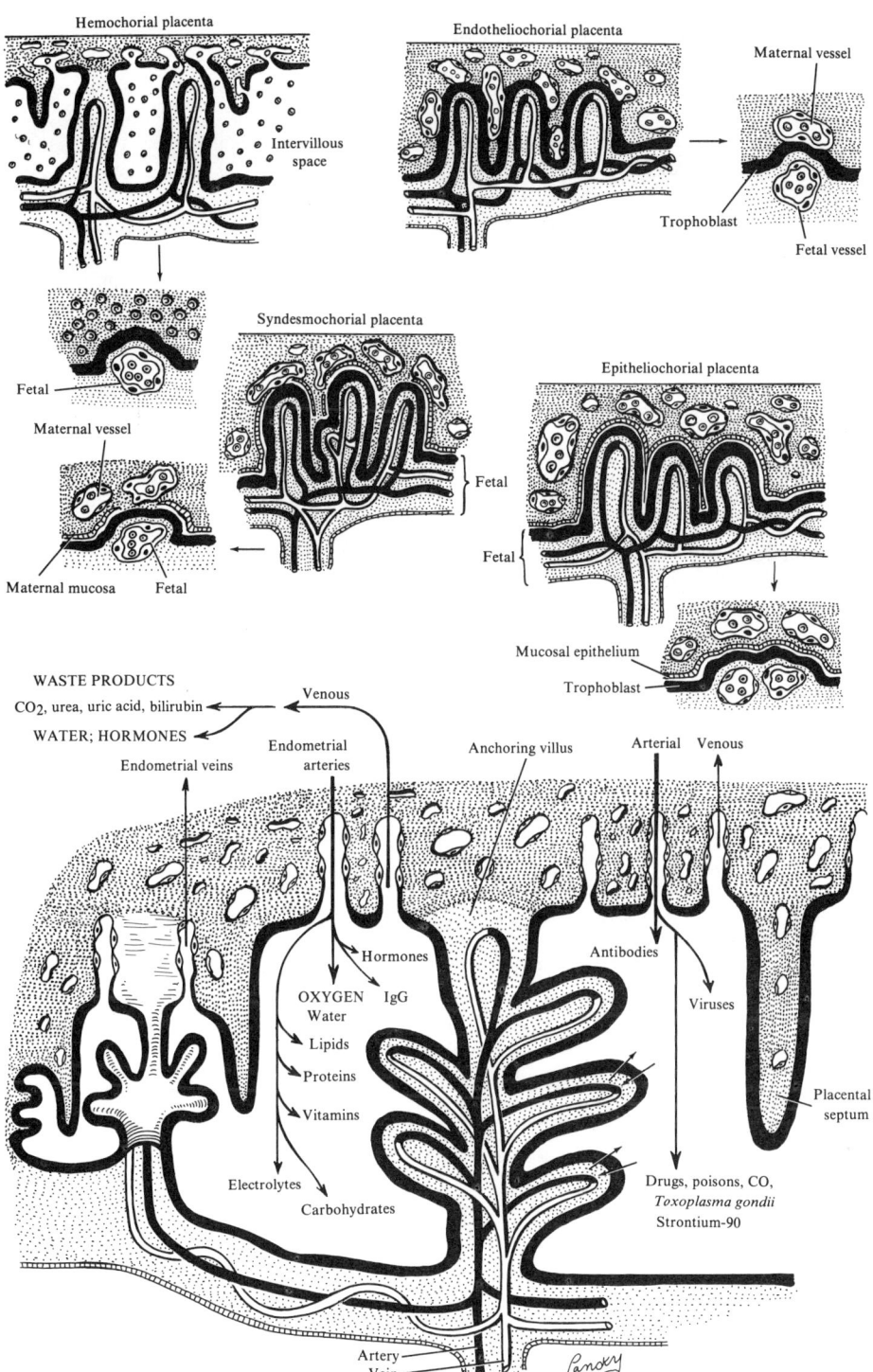

42. PLACENTAL CIRCULATION

I. **Introduction:** the intervillous space is limited on the maternal side by the *basal plate* and on the fetal side by the *chorionic plate*. It is incompletely limited laterally by the *decidual septa*. The complex division of the villi provides a great deal of placental surface and is a very important factor affecting the rate of fetal-maternal exchange. At term, the placental surface consists of more than 10 m². The anchoring or stem villi are attached to the basal plate and define a general circular area. The villus tree, in its entirety, forms a very complex system consisting of a major anatomic unit, the *cotyledon*

II. **The fetal circulation** can be compared to the pulmonary circulation of the adult in that desaturated blood enters throgh the fetal arteries and oxygenated blood returns by way of the veins
  A. BLOOD ARRIVES via the 2 umbilical arteries which are branches of the iliac arteries of the fetus. It is dispersed in a highly dense network whch penetrates even the smallest villous division
  B. BLOOD IS RETURNED via the umbilical vein and finally reaches the inferior vena cava system of the fetus
  C. FETAL CIRCULATION is carried out in a closed vascular system where the average pressure is about 30 mm Hg, which is much higher than that seen in the intervillous space where it is about 10 mm Hg. The difference in pressure prevents the collapse of the villous vessels

III. **The maternal circulation**
  A. BLOOD ARRIVES at the uterus by the branches of the uterine artery, spreads out in the intervillous spaces, and circulates between the branches of the villous trees. It is returned by branches of the uterine veins. The flow in these two circulations is very high, about 500 ml/min, which favors fetal-maternal exchange
  B. MATERNAL CIRCULATION results from a difference in pressure which is very high in the artery (about 70 mm Hg) and relatively low in the intervillous space (about 10 mm Hg). Blood spurts up to the chorionic plate, then comes toward the basal plate and is taken up by the uterine veins where the pressure is even lower than that found in the intervillous space

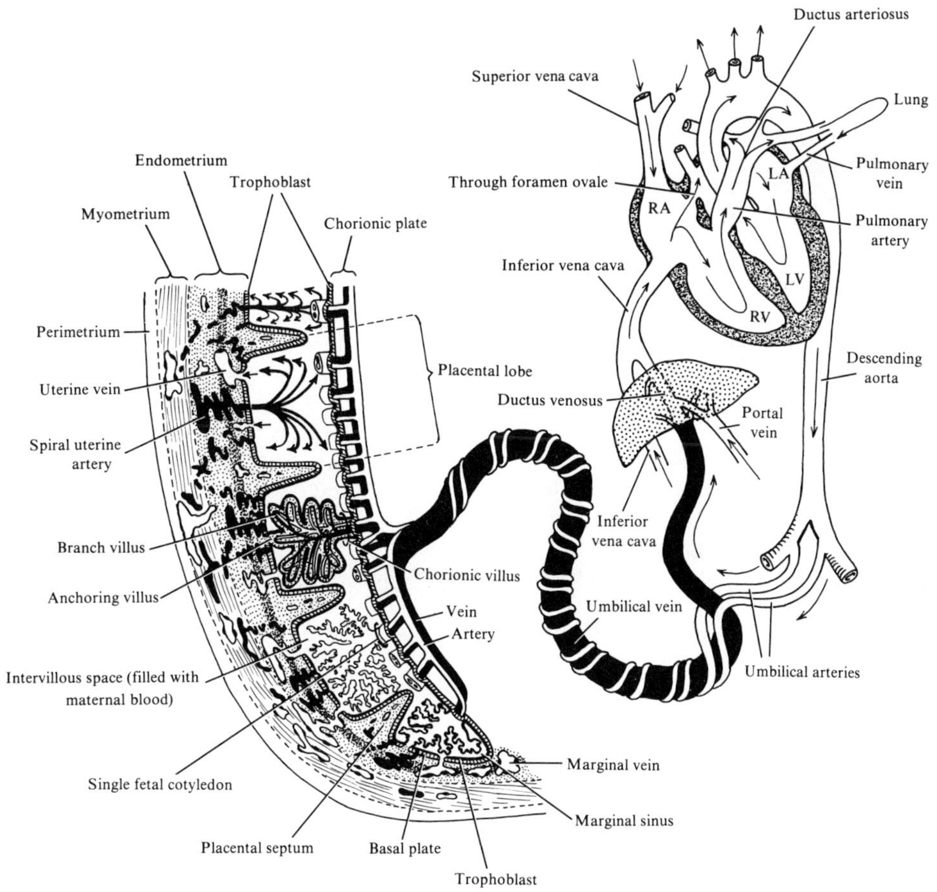

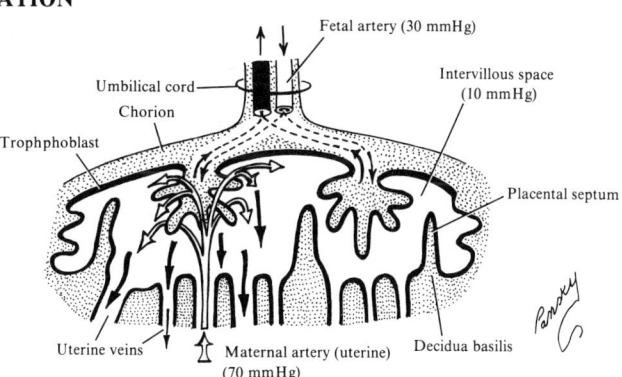

PLACENTAL CIRCULATION

43. HORMONAL BALANCE AND TESTS FOR PREGNANCY

I. Hormonal balance in pregnancy
A. BEFORE IMPLANTATION, MAINTENANCE OF PREGNANCY is assured by the ovarian and pituitary hormones
B. AFTER PREGNANCY, HORMONAL CONTROL OF PREGNANCY is assured by the combined action of the pituitary, ovarian, and placental hormones
  1. Only four hormones are known to be produced by the placenta
     a. Protein hormones are human chorionic gonadotropic hormone (hCG), human chorionic somatomammotropic hormone (hCS), or placental lactogen (hPL)
     b. Steroid hormones: progesterone (from maternal cholesterol or pregnenolone) and estrogen (from 19-carbon precursors)
  2. Chorionic gonadotropins are discernible very early, several days after nidation
  3. Although the levels of estrogen and progesterone increase regularly during pregnancy until term, the gonadotropins, after reaching their peak on about day 60, decrease and are maintained at a relatively low level until the end of pregnancy

II. Test for pregnancy: biologic diagnosis of pregnancy is based on the detection of chorionic gonadotropins
A. BIOLOGIC TEST IN ANIMALS can be made in a variety of animal species, namely, the rat, the mouse, the rabbit, the frog, and even the toad. Injected gonadotropins provoke "characteristic" changes in the genital tracts of the animals. As seen in the genital tract of the injected virgin female rabbit
  1. A urine or blood sample from a woman suspected of being pregnant is injected into the rabbit, and its genital tract is examined about 36 hours later
     a. A positive reaction reveals congestion and hyperemia of the uterine horns and one or more hemorrhagic follicles are seen in the rabbit ovaries
B. IMMUNOLOGIC TEST: the existence of chorionic gonadotropins in the urine of a pregnant female can be demonstrated by anitgonadotropic serum, obtained by immunization of an animal against human gonadotropin
  1. The reaction system contains 2 components
     a. Antigonadotropic serum or reactive agent "a"
     b. Red blood cells artificially covered with gonadotropin or reactive agent "b"
  2. Negative or control reaction: if agent "a" and agent "b" are mixed, antibodies in the serum agglutinate the red blood cells through the gonadotropins which are attached, and this reaction is directly visible on a slide or in a test tube
  3. Positive or no agglutination reaction: if several milliliters of urine containing free gonadotropins from a pregnant woman are mixed with reactive agent "a," these gonadotropins block the serum "a" antibodies and prevent them from agglutinating the red blood cells with the agent "b" on it
  4. Currently, the red blood cells of reactive agent "b" have been replaced by the use of inert particles such as latex, which serve the same role and have the added advantage of not altering or degrading with time
C. HYDATIDIFORM MOLE
  1. Hormonal equilibrium gives one an idea of the progress of the pregnancy, but also allows the diagnosis of anomalies, particularly of placental degeneration
  2. In hydatidiform mole and its malignant transformation, the chorioepithelioma, chorionic gonadotropins are abnormally high

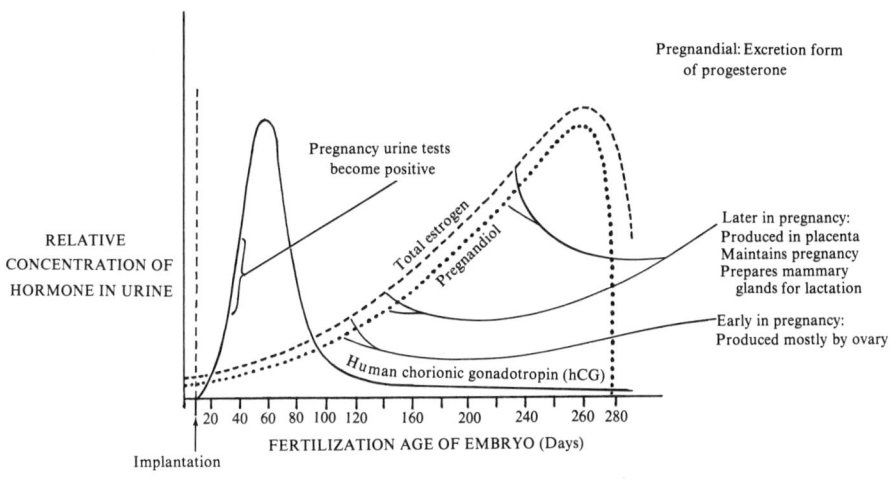

CHANGE IN EXCRETION RATE OF CERTAIN HORMONES DURING PREGNANCY

UTERUS WITH A HYDATIDIFORM MOLE

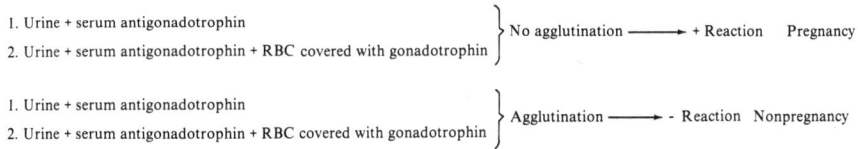

1. Urine + serum antigonadotrophin
2. Urine + serum antigonadotrophin + RBC covered with gonadotrophin
} No agglutination ⟶ + Reaction   Pregnancy

1. Urine + serum antigonadotrophin
2. Urine + serum antigonadotrophin + RBC covered with gonadotrophin
} Agglutination ⟶ − Reaction   Nonpregnancy

IMMUNOLOGIC TEST FOR PREGNANCY

PART II
Organogenesis

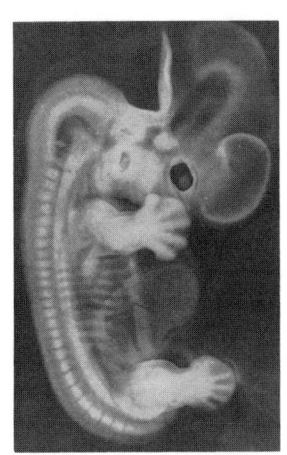
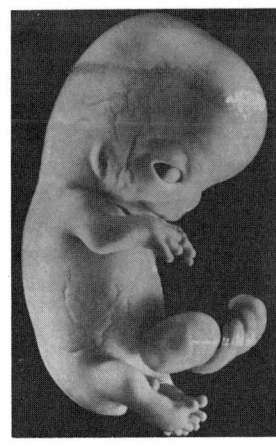
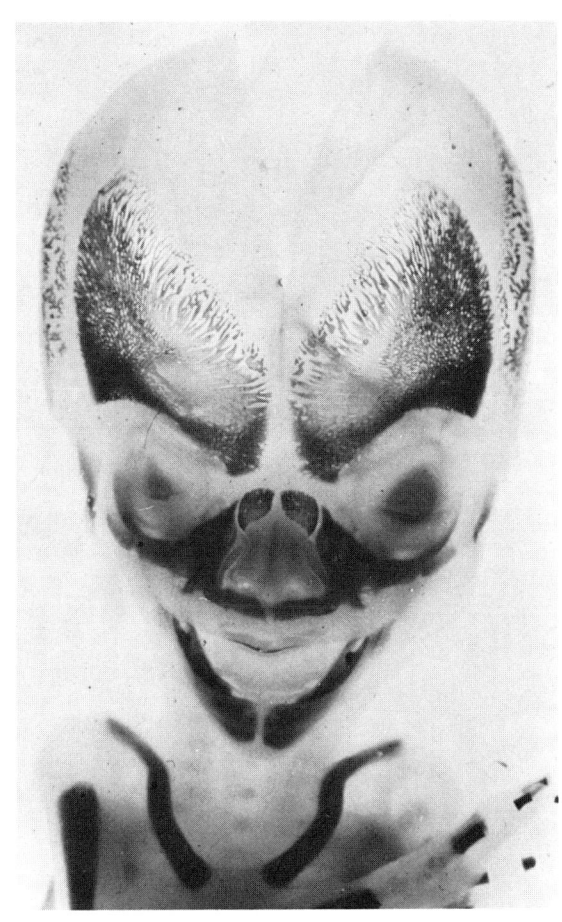

UNIT ONE

BODY CAVITIES AND MESENTERIES

44. BODY CAVITIES: COELOMIC DIVISIONS

I. **Introduction:** by week 4, the intraembryonic mesoderm on each side of the midline forms a paraxial portion, an intermediate portion, and a lateral plate. With the appearance and coalescence of many intercellular clefts, the lateral plates divide into 2 layers: a *somatic mesoderm layer* which continues with the extraembryonic mesoderm covering the wall of the amniotic cavity and a *splanchnic mesoderm layer* which is continuous with the mesoderm of the yolk sac wall. The spaces between these layers are the *intraembryonic coelomic cavities**

   A. THE RIGHT AND LEFT INTRAEMBRYONIC COELOMIC CAVITIES, at first, are widely connected with the extraembryonic coelom, but with development and embryonic folding, they lose this connection. The 2 intraembryonic coelomic cavities are then separated by a double-layer membranous partition formed by fusion of the right and left splanchnic mesoderm layers, the so-called ventral mesentery, as the lateral folds move below the embryo and join each other

   B. THIS SEPTUM OR MESENTERY persists indefinitely in some portions of the body, particularly in the upper abdominal region, but in others, e.g., the thorax, it partly disappears to unite the right and left coelomic cavities

   C. IN WEEK 5, AS THE HEADFOLD FORMS, the heart and pericardial cavity move ventrally below the foregut, but open dorsally into the pericardioperitoneal canals which pass above the septum transversum on each side of the foregut, and the intraembryonic coelom consists of a thoracic and abdominal portion connected by the 2 canals, the *pericardioperitoneal canals*

   D. AFTER FOLDING OF THE EMBRYO, the caudal foregut, midgut, and hindgut are suspended in the peritoneal cavity by the *dorsal mesentery*. Thus, temporarily, the dorsal and ventral mesenteries divide the peritoneal cavity into 2 separate halves. However, the ventral one soon disappears except at its attachment to the caudal portion of the foregut, and the peritoneal cavity once again is a large continuous space

   E. IN THE ADULT, THE INTRAEMBRYONIC COELOM is divided into 3 well-defined compartments
      1. A pericardial cavity containing the heart
      2. The pleural cavities containing the lungs
      3. The peritoneal cavity with the viscera, caudal to the diaphragm

   F. THE SEPTUM between the thoracic and abdominal cavities is formed by the *diaphragm*, and between the pericardial and pleural cavities is found the *pleuropericardial membranes*

---

*Provides short-term storage for excretory products and room for organ development and movement. May aid in transfer of fluid and nutrients to early embryo where intra- and extraembryonic coeloms communicate. The latter are occluded with folding of the embryonic disk.

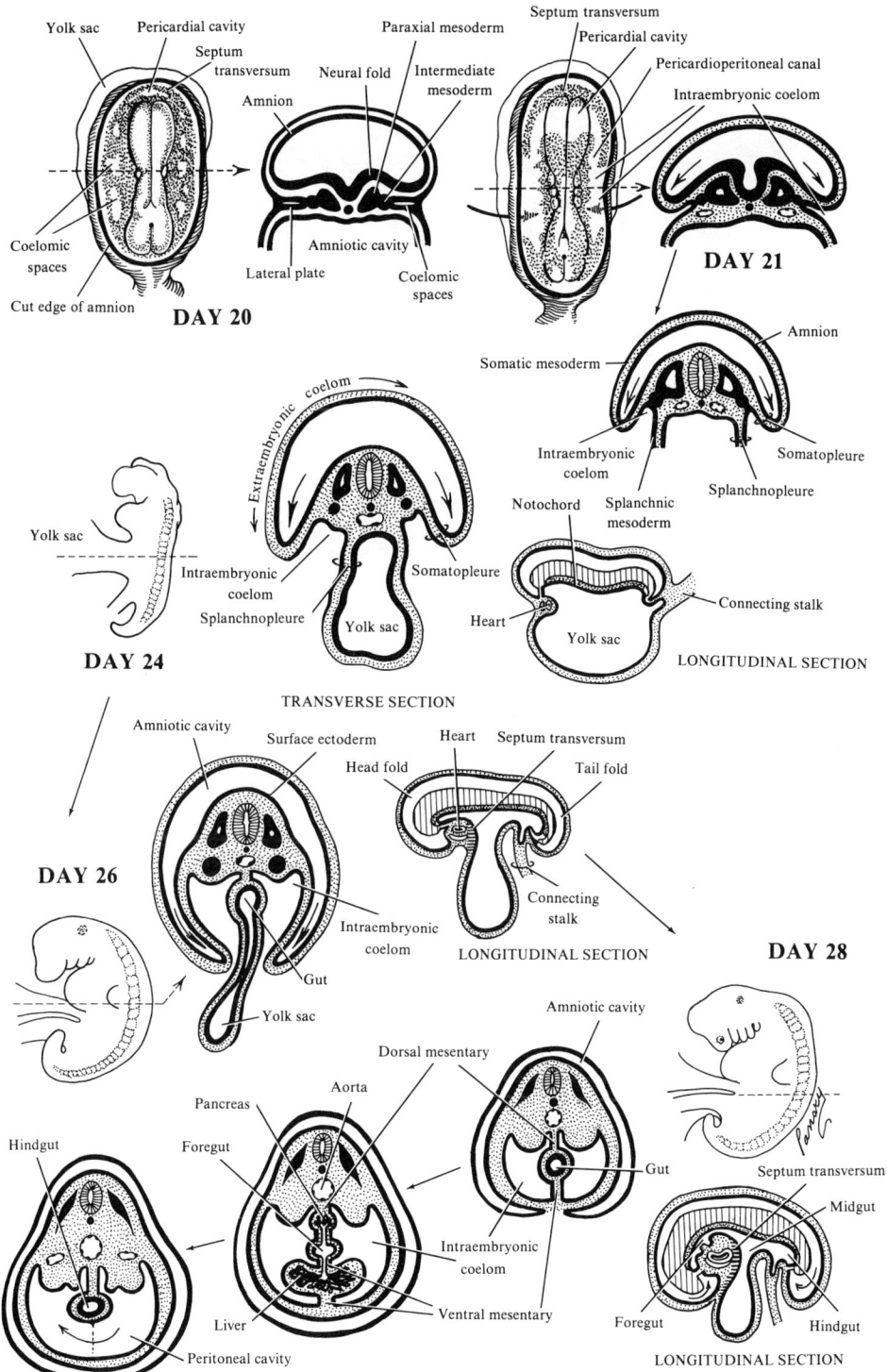

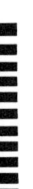

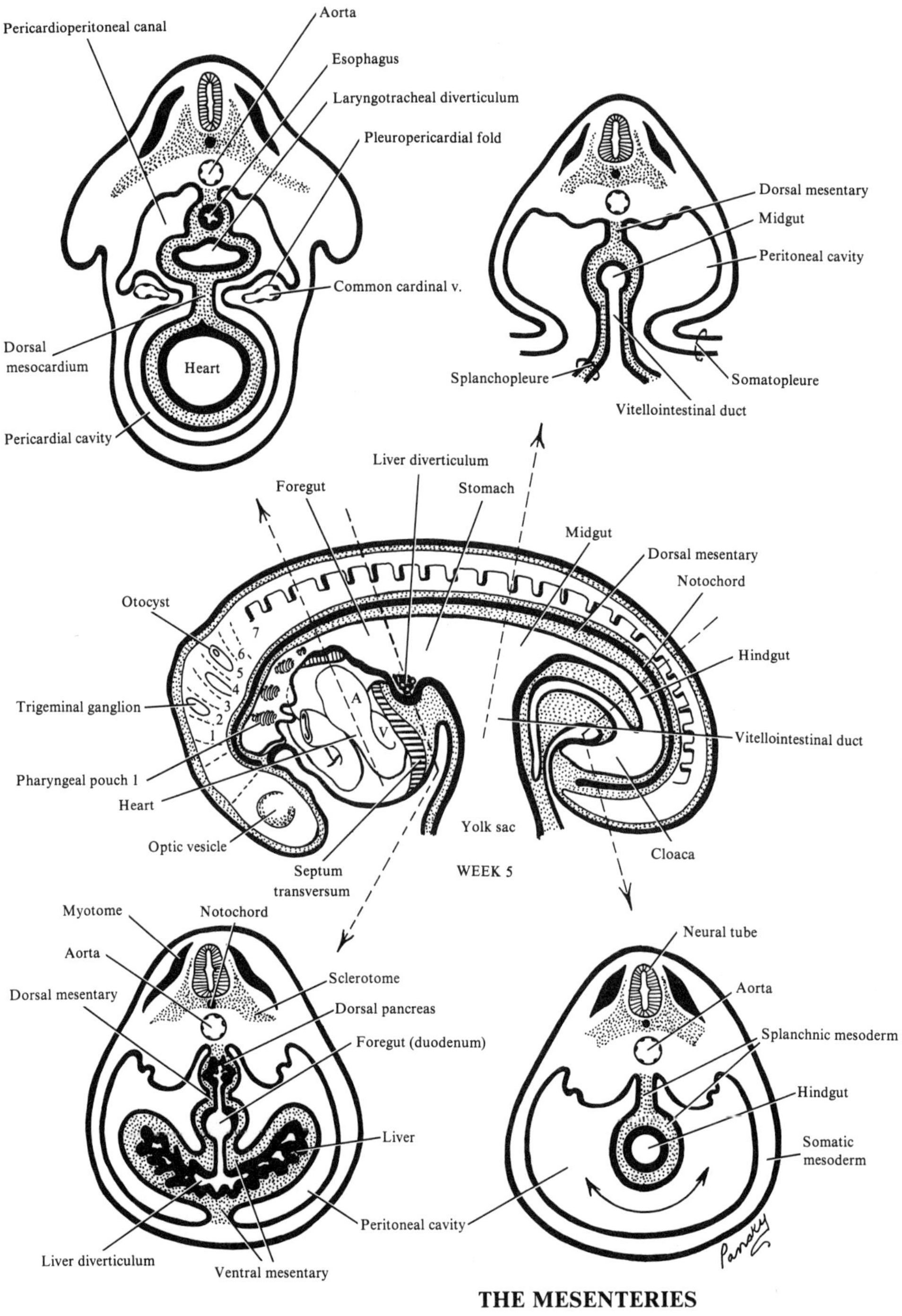

THE MESENTERIES

FIGURE 19. **The mesenteries in cross-section.**

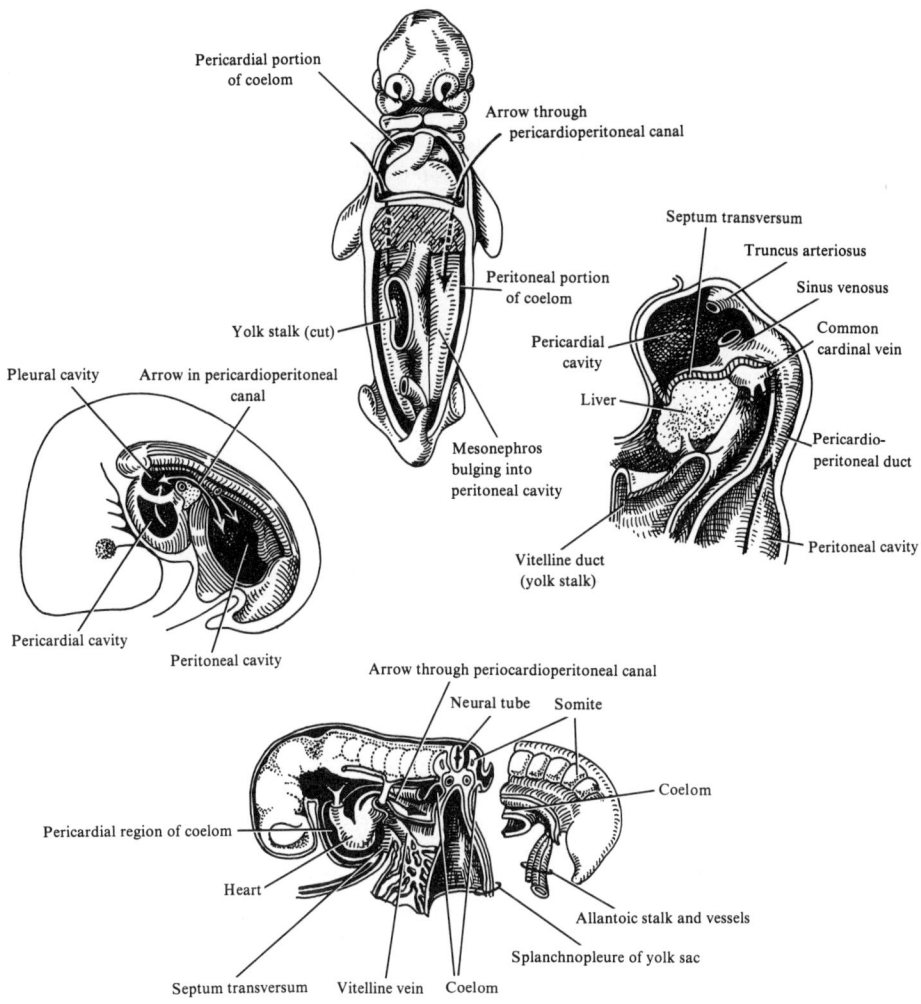

CONTINUITY OF PERICARDIAL PORTION OF COELOM WITH PAIRED COELOMIC CHAMBERS OF MIDBODY AREA
(PERITONEAL PORTION OF COELOM)

FIGURE 20. Continuity of pericardial portion of coelom with paired coelomic chamber of midbody area.

45. BODY CAVITIES: PLEUROPERICARDIAL AND PLEUROPERITONEAL MEMBRANES

I. **Pleuroperitoneal membranes** are a pair of membranes which gradually separate the pleural and peritoneal cavities, produced as the pleural cavities expand by invading the body wall
   A. THE MEMBRANES ARE ATTACHED dorsolaterally to the body wall, and their free edges project into the caudal end of the pericardioperitoneal canals
      1. During week 6, they grow medially and ventrally, and by the end of the week, their free edges fuse with the dorsal mesentery of the esophagus and with the septum transversum to separate the pleural and pericardial cavities
      2. Closure of the openings is further enhanced by the growth of the liver and muscle tissue extension into the membranes. The right-side opening closes before that of the left

II. **Pleuropericardial membranes** initially appear as small folds or ridges of mesenchyme projecting into the primitive undivided thoracic cavity. The folds contain the common cardinal veins which drain the primitive venous system into the sinus venosus of the primitive heart
   A. AS A RESULT OF SUBSEQUENT GROWTH of the common cardinal veins, descent of the heart, and expansion of the pleural cavities, the membranes are drawn out in a mesentery like fold that extends from the lateral wall
   B. BY WEEK 7, the membranes fuse with the mesoderm ventral to the esophagus or primitive mediastinum (the dorsal mesocardium) and divide the thoracic cavity into a single pericardial cavity and 2 pleural cavities
      1. The mediastinum is filled with a mass of mesenchyme and separates the developing lungs as it extends from the sternum to the vertebral column
   C. THE RIGHT PLEUROPERICARDIAL OPENING closes before the left one since the right common cardinal vein is larger and produces a larger membrane
   D. SUBSEQUENTLY, THE LUNG BUDS grow into the medial walls of the pericardiopleural canals (primitive pleural cavities); and the pleural cavities expand around the heart into the body wall and split the mesenchyme into an outer layer that becomes the chest wall and an inner layer that forms the fibrous pericardium

III. **Congenital malformations**
   A. CONGENITAL PERICARDIAL DEFECTS: defective formation and/or fusion of the pleuropericardial membrane(s) which normally separate(s) the pericardial from the pleural cavities may result in congenital defects of the pericardium, usualy on the left side. It is a rare abnormality.
      1. The pericardial cavity then communicates with the pleural cavity, and in rare instances, a part of the atrium may herniate into the pleural cavity with each heartbeat

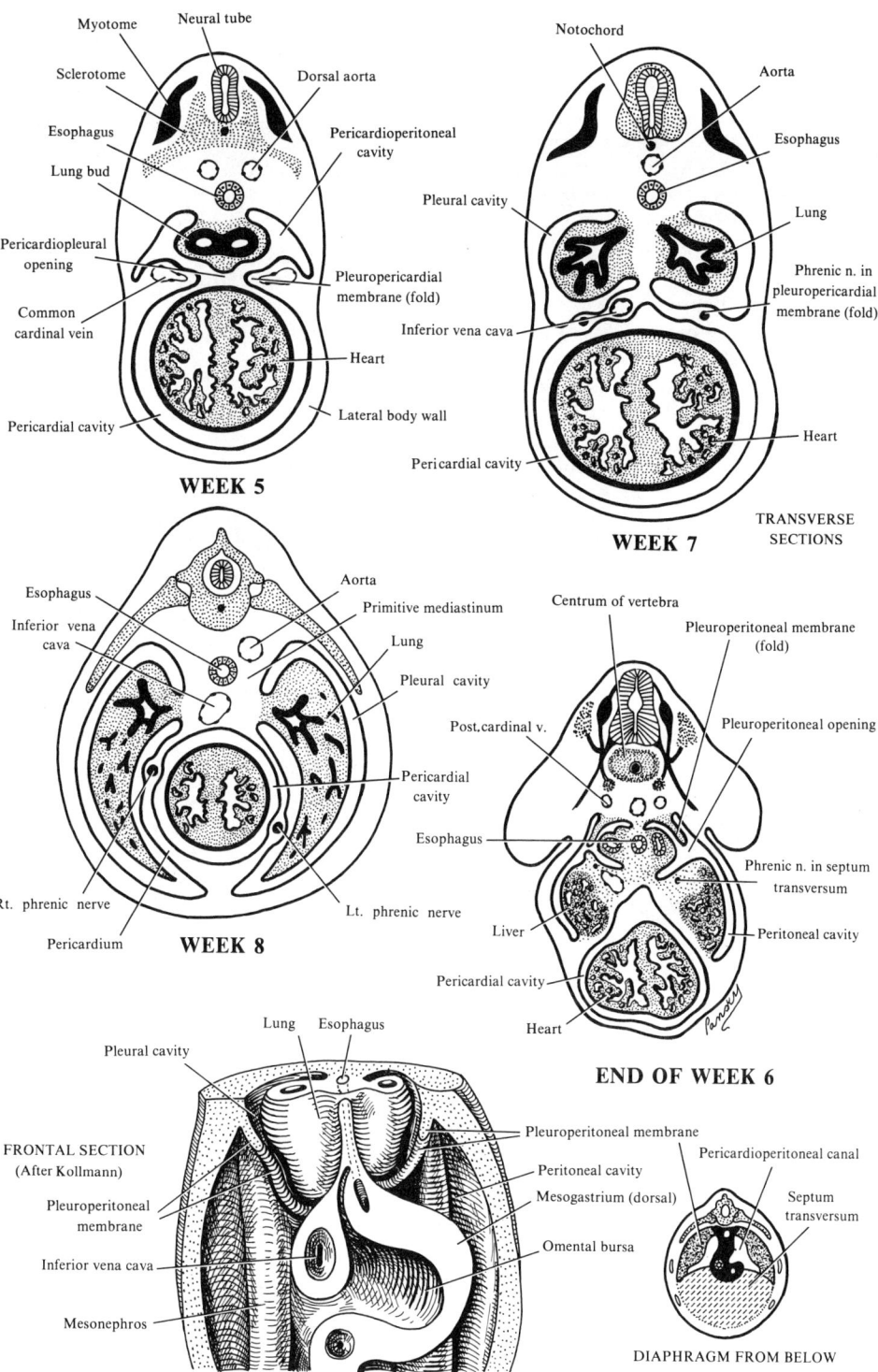

46. BODY CAVITIES: THE LESSER PERITONEAL SAC (OMENTAL BURSA) AND DORSAL MESOGASTRIUM

I. **Small intercellular clefts** are seen first in the mesoderm lateral to the foregut in week 4. The clefts fuse and form the *pneumatoenteric recess or cavity*
   A. THE CAVITY is initially bilateral, but its left side is soon obliterated, and on the right side it comes to extend cranially between the esophagus and the right lung bud
   B. WITH PLEUROPERITONEAL MEMBRANE DEVELOPMENT, the cranial portion of the recess is isolated and froms a small supradiaphragmatic bursa, the *infracardiac bursa*
   C. SIMULTANEOUS WITH THE APPEARANCE OF THE PNEUMATOENTERIC RECESS is the formation of another recess on the right side of the mesogastrium which expands toward the left side of the body, resulting in the formation of the *omental bursa* (lesser peritoneal sac) (an extension of the right half of the peritoneal cavity)
   D. AFTER ROTATION OF THE STOMACH is complete, the dorsal mesogastrium continues to grow and forms a double-leaved apron which extends in front of the transverse colon and small loops of the intestine as the *greater omentum*
      1. Its leaves fuse to form a single sheet that hangs over the greater curvature of the stomach, with the upper portion of the posterior leaf fusing to the mesentery of the transverse colon
   E. THE FORMATION OF THE SPLEEN complicates the dorsal mesogastrium development
      1. Its primordium appears in week 5 as a mesenchymal condensation between the 2 leaves of the mesogastrium, but soom bulges into the left peritoneal cavity
      2. With formation of the omental bursa, a part of the dorsal mesogastrium, between the spleen and the dorsal midline, fuses with the posterior adbominal wall, while the remainder remains and connects the spleen to the kidney as the *lienorenal ligament*
         a. The connection of the spleen to the stomach forms the *gastrolienal ligament*
         b. Thus, the spleen always keeps an intraperitoneal position
   F. THE FORMATION OF THE OMENTAL BURSA also influences the position of the pancreas which initially grows into the dorsal mesoduodenum, but in time, its tail portion expands into the dorsal mesogastrium
      1. Since this portion of the mesogastrium (the left leaf) fuses with the peritoneum of the body wall during rotation, the pancreatic tail comes to lie retroperitoneally

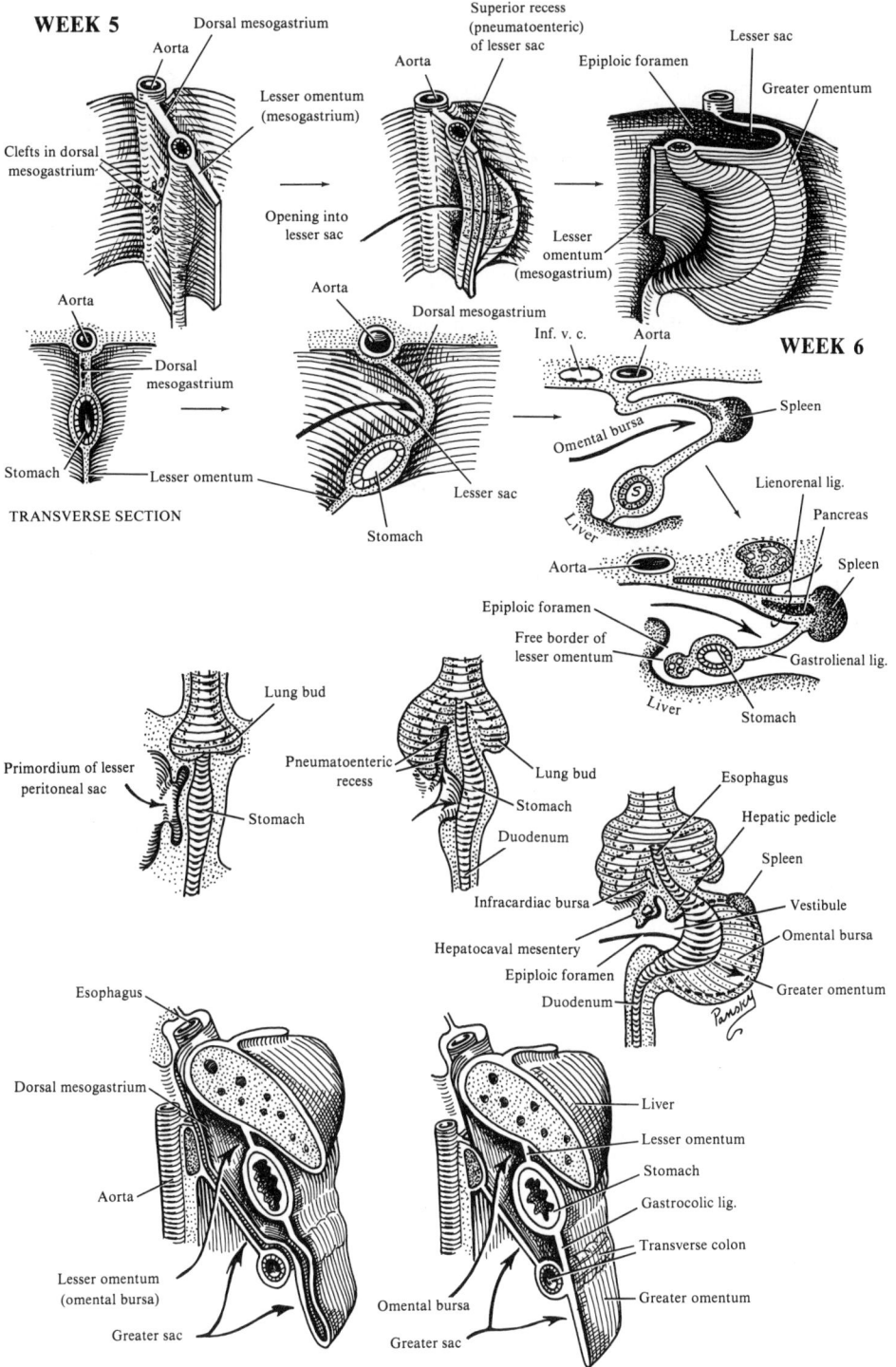

47. DEVELOPMENT OF THE MESENTERIES

I. **Introduction:** the 2 apposing splanchnic mesoderm layers fuse to form a double-layer membrane, the *primitive mesentery,* with the closing of the abdominal wall and the formation of the peritoneal cavity
   A. THE VENTRAL MESENTERY
      1. With formation and closure of the abdominal wall, the stomach and upper duodenum, which are initially in contact with the septum transversum, draw away from the septum, except for their ventral borders which stay connected to the underside of the septum and the anterior abdominal wall by the ventral mesentery
      2. During formation of the ventral mesentery, the liver grows so rapidly that the septum transversum can no longer accommodate it, so it protrudes between the 2 leaves of the mesentery, dividing the latter into an anterior part, the *falciform ligament* (from liver to anterior abdominal wall) and a posterior part, the *lesser omentum* (between the liver and ventral borders of the stomach and duodenum)
         a. The free edge of the falciform ligament contains the umbilical vein which is obliterated after birth to from the *ligamentum teres hepatis* (round ligament)
         b. The free edge of the lesser omentum contains the common bile duct, the portal vein, and the hepatic artery
      3. Initially, the lesser omentum is oriented in a sagittal plane, but with rotation of the stomach and growth of the liver, it acquires a frontal position so that its lower edge forms the upper margin of the epiploic foramen of Winslow—the entrance into the lesser peritoneal sac behind the stomach
      4. The liver is completely surrounded by ventral mesentery except on its upper surface where it adjoins the diaphragm—so-called *bare area of the liver*
         a. The lines of reflection, where peritoneum over the liver is continuous with that on the underside of the diaphragm, are called the *coronary ligaments*
   B. THE DORSAL MESENTERY extends from the lower esophagus to rectum and throughout its length serves as a pathway to the gut for blood vessels, nerves, and lymphatics
      1. In the area of the duodenum, it is called the *dorsal mesoduodenum*
         a. Rotation of the stomach and duodenum with growth of the head of the pancreas, causes the duodenum to swing from midline to the right side of the peritoneal cavity, pushing the duodenum and pancreatic head against the dorsal body wall and fusing the right surface of the dorsal mesoduodenum with wall peritoneum
            i. Both layers disappear completely, except in the area of the pylorus of the stomach, where a small part of duodenum remains intraperitoneal, and the duodenum and pancreatic head become fixed in a retroperitoneal position
      2. In the area of the colon, it is called the *dorsal mesocolon*
         a. When the ascending and descending colons acquire their definitive positions, their mesenteries are pressed against the abdominal wall peritoneum, and the colons are permanently anchored or fused as retroperitoneal. The appendix and lower end of the cecum, however, retain a free mesentery
         b. The transverse mesocolon, at first, covers the duodenum with an additional peritoneal layer, but later fuses with the posterior wall of the omental bursa. Its final lines of attachment extend from the hepatic to the splenic flexures
      3. In the area of the stomach, it is called the *dorsal mesogastrium* or *greater omentum*
      4. The dorsal mesentery of the jejunum and ileum is the *mesentery proper* which changes with rotation and coiling of the intestinal loops
         a. When the lower limb of the primitive intestinal loop moves to the right side of the abdominal cavity, the dorsal mesentery twists around the origin of the superior mesenteric artery
         b. The mesentery of the jejunoileal loops is initially continuous with that of the ascending colon, but when the ascending mesocolon fuses with the posterior abdominal wall, the mesentery of the loops obtains a new line of attachment, extending from where the duodenum is intraperitoneal to the ileocecal junction

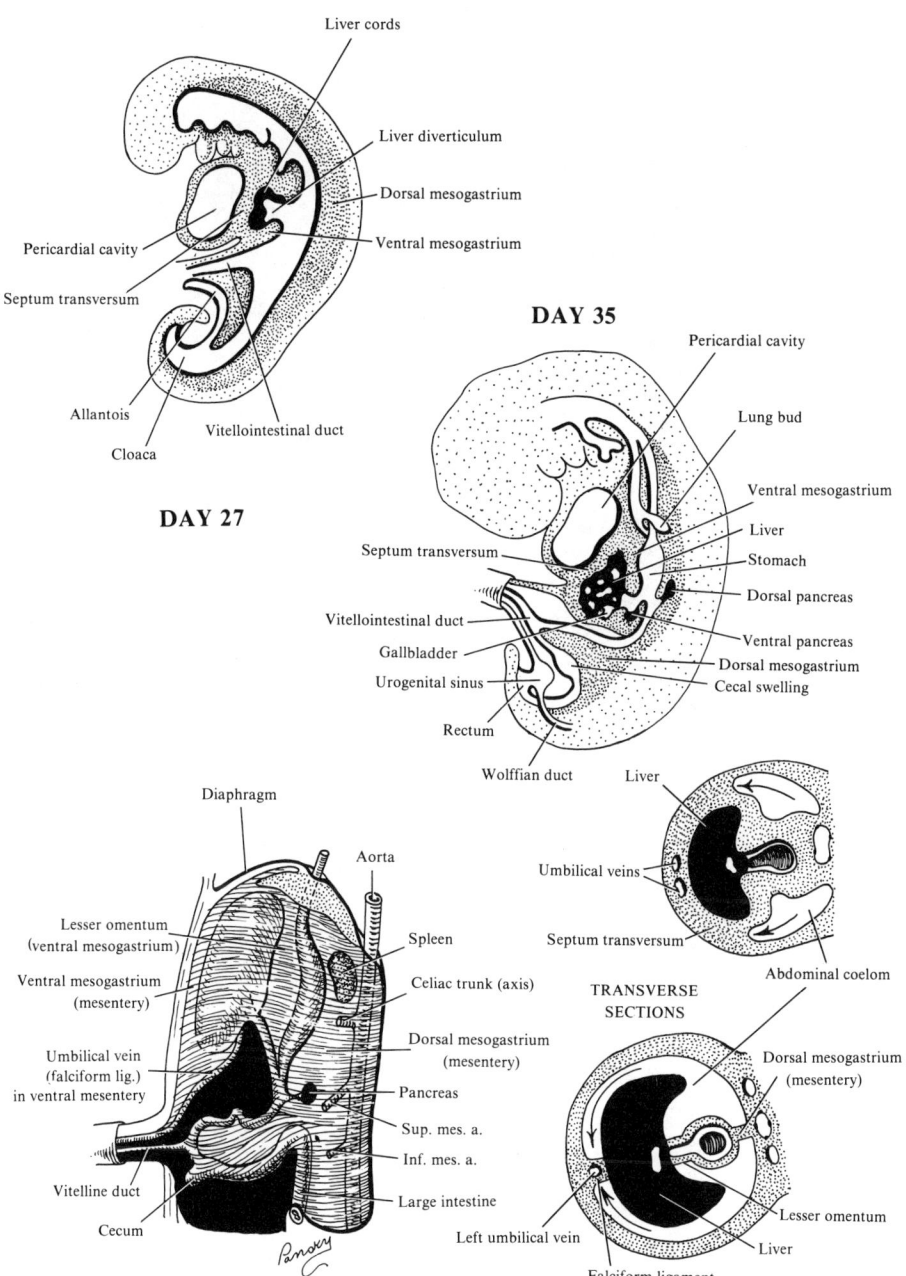

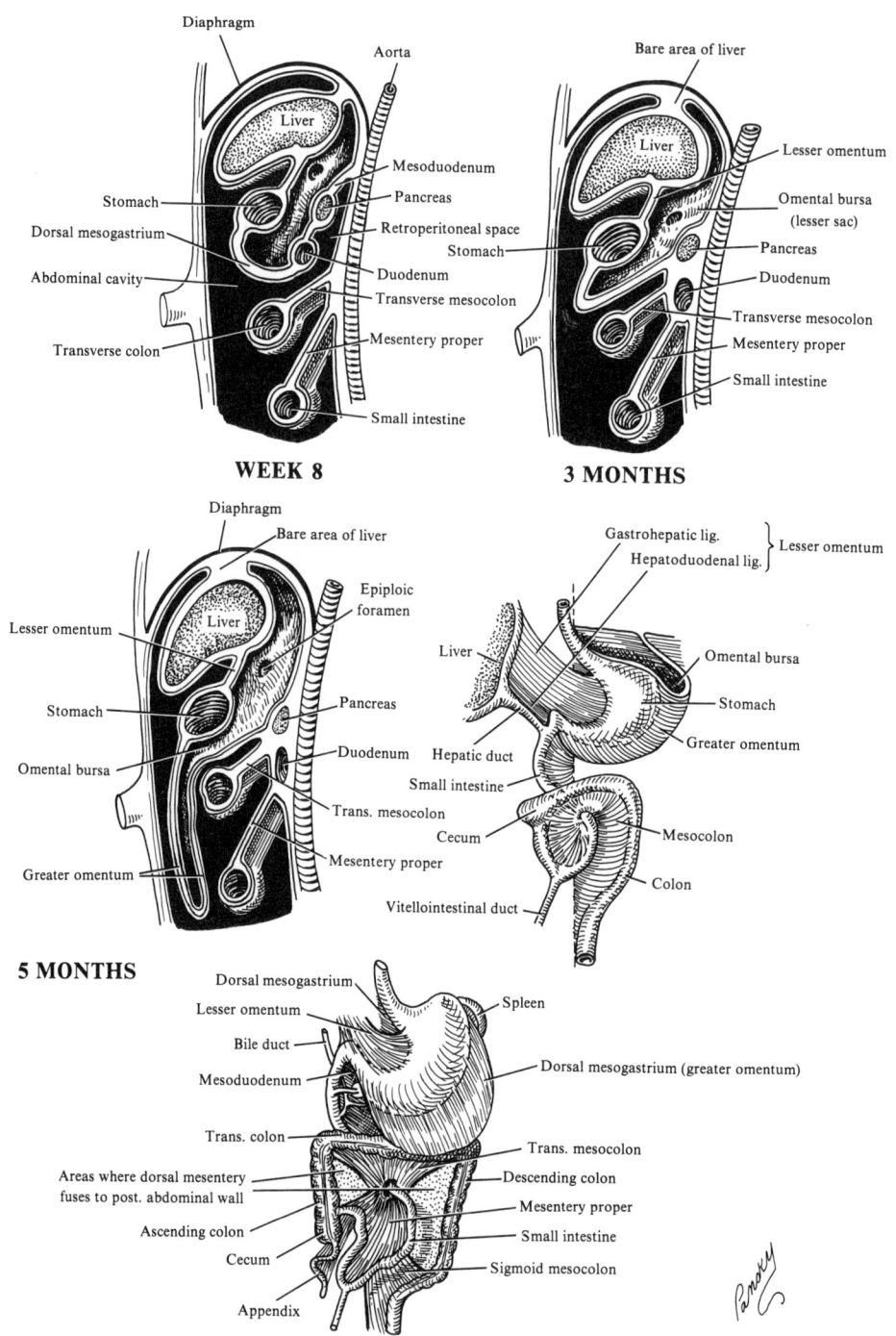

FIGURE 21. **Transverse sections of mesentery development.**

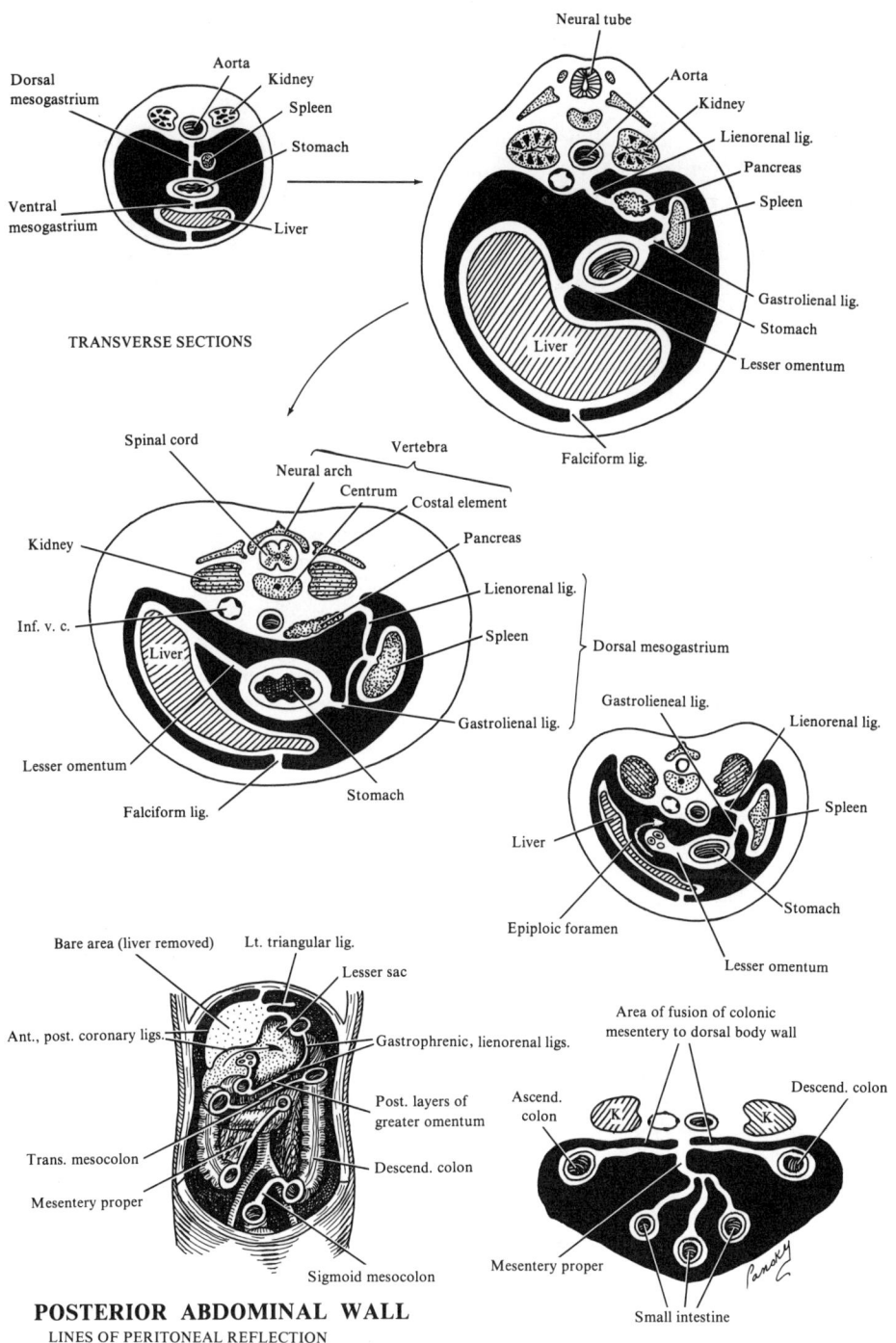

FIGURE 22. **Cross-sections of mesentery development.**

48. DEVELOPMENT OF THE DIAPHRAGM

I. **The diaphragm** is a musculotendinous, dome-shaped partition between the thoracic and abdominal cavities and develops from 4 major structures
   A. THE SEPTUM TRANSVERSUM (most important component) forms the *central tendon* and is first seen as a thick mesodermal plate cranial to the pericardial cavity between the base of the thoracic cavity and the stalk of the yolk sac
      1. The septum does not separate the thoracic and abdominal cavities entirely, but after the headfold forms (week 4), it becomes a thick incomplete partition between the cavities with an opening on each side of the gut, the *pleural canals*
      2. The septum fuses dorsally with the primitive mediastinal mesenchyme below the esophagus and later with the pleuroperitoneal membranes
   B. PLEUROPERITONEAL MEMBRANES fuse wth the dorsal mesentery of the esophagus and with the dorsal part of the septum transversum to complete the partition between the thoracic and abdominopelvic cavities to form the *primitive diaphragm*. They represent only a small portion of the final adult structure
   C. THE DORSAL ESOPHAGEAL MESENTERY (mesoesophagus) fuses with both A and B. This mesentery forms the median portion of the diaphragm. The *crura* of the diaphragm develop from muscle fibers which grow into the esophageal mesentery
   D. THE BODY WALL: during weeks 9 to 12, the pleural cavities enlarge and invade the lateral body walls. Body wall tissue, at this time, splits off medially to form the peripheral parts of the diaphragm outside that formed by the membranes (B)
      1. Extensions of the pleural cavities into the body walls form the *costodiaphragmatic recesses*

II. **Innervation and position of the diaphragm**
   A. DURING WEEK 4, THE SEPTUM TRANSVERSUM lies opposite the upper cervical somites, and during week 5, nerves from the cervical spinal segments, C3, C4, and C5 grow into the septum and form the *phrenic nerve*. These nerves pass to the septum via the pleuropericardial membrane, thus, the nerves lie in the fibrous pericardium
   B. RAPID GROWTH OF THE DORSAL EMBRYO BODY compared to its ventral part results in an apparent descent of both diaphragm and nerves, by week 6, to thoracic somite level
   C. BY WEEK 8, the dorsal part of the diaphragm lies at the level of the first lumbar vertebrae, thus, its nerve has been carried down with it from the cervical region

III. **Congenital malformations**
   A. CONGENITAL DIAPHRAGMATIC HERNIA: a common malformation in the newborn seen in 1/2200 births and usually as a posterolateral defect of the diaphragm
      1. Usually results as a defective formation and/or fusion of the pleuroperitoneal membrane(s) which normally separate(s) the pleural and peritoneal cavities
      2. Defect is usually unilateral with a large opening (foramen of Bochdalek) in the posterolateral part of diaphragm. It is seen more often on the left as a result of an earlier closure of the right pleuroperitoneal opening
      3. If the pleuroperitoneal membrane is not fused when the intestines return to the abdomen from the umbilical cord (week 10), the intestines may pass into chest
      4. Occasionally see stomach, spleen, cecum, appendix, and parts of colon in the chest cavity. If present at birth, may interfere with respiration
      5. Heart and mediastinum are often displaced. Lungs are small and hypoplastic
   B. CONGENITAL HIATAL HERNIA: rare; abdominal viscera herniate through a large esophageal hiatus or opening. Usually an acquired lesion seen in adult life
   C. ESOPHAGEAL HERNIA: if esophagus is shorter than normal, part of stomach may appear in the thorax and be constricted as it passes through the enlarged esophageal hernia
   D. RETROSTERNAL OR PARASTERNAL HERNIA (of Morgagni): a rare defect between sternum and sternocostal parts of diaphragm. A small peritoneal sac with intestinal loops often seen in chest
   E. CONGENITAL EVENTRATION OF DIAPHRAGM: rare; half of diaphragm has defective muscles and balloons up into chest cavity. Upward displacement of abdominal contents

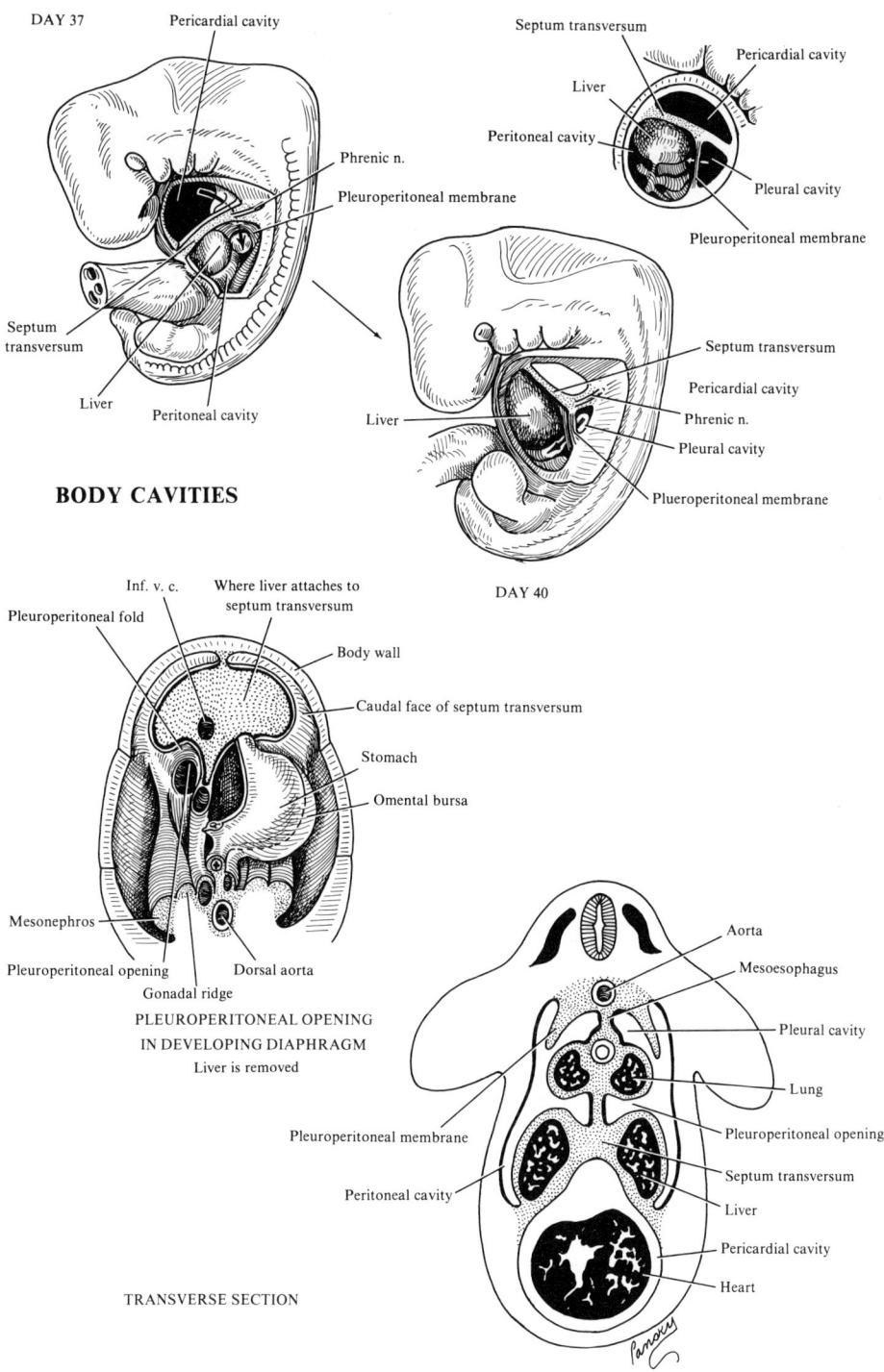

BODY CAVITIES

PLEUROPERITONEAL OPENING
IN DEVELOPING DIAPHRAGM
Liver is removed

TRANSVERSE SECTION

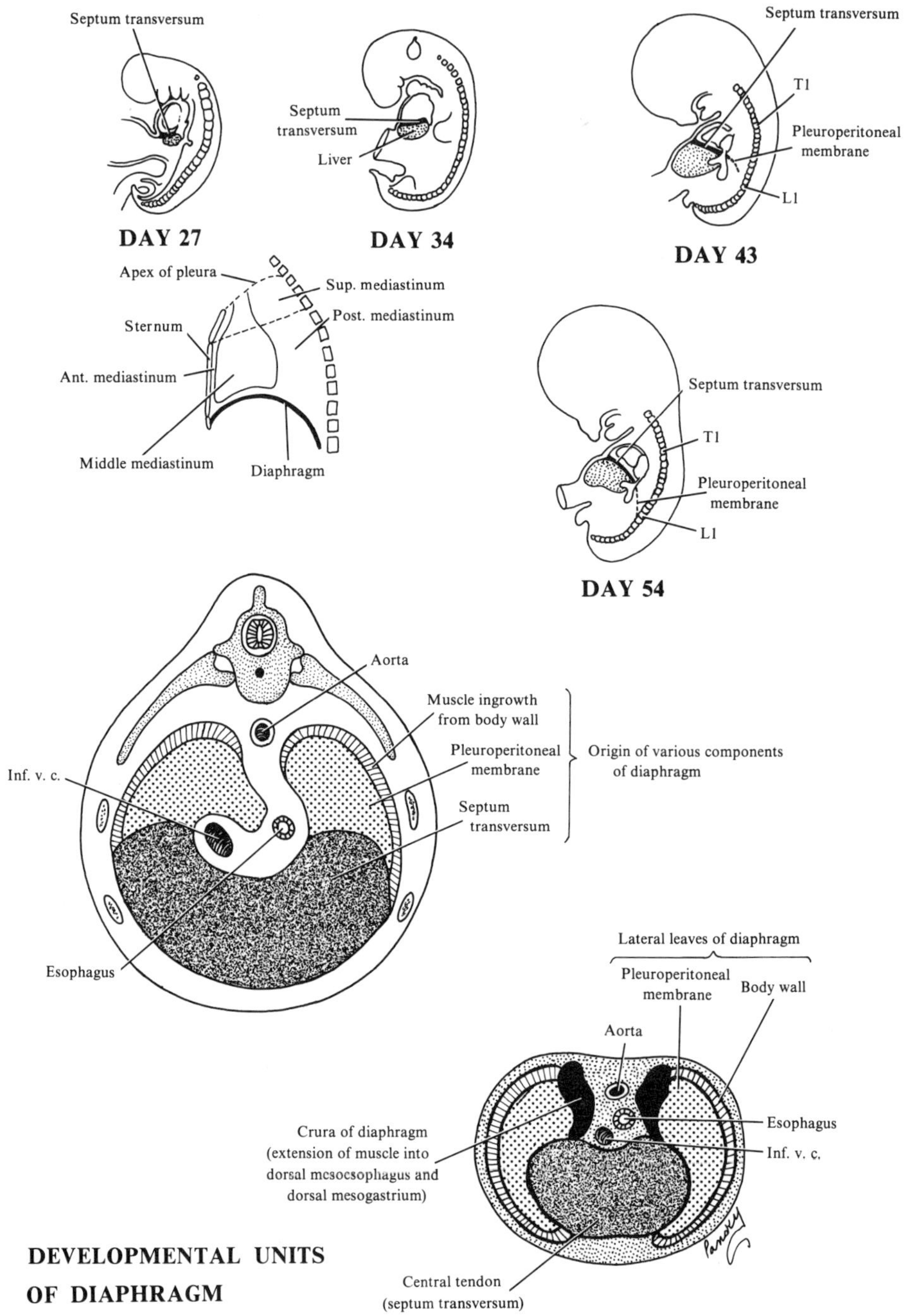

FIGURE 23. **Development of the diaphragm.**

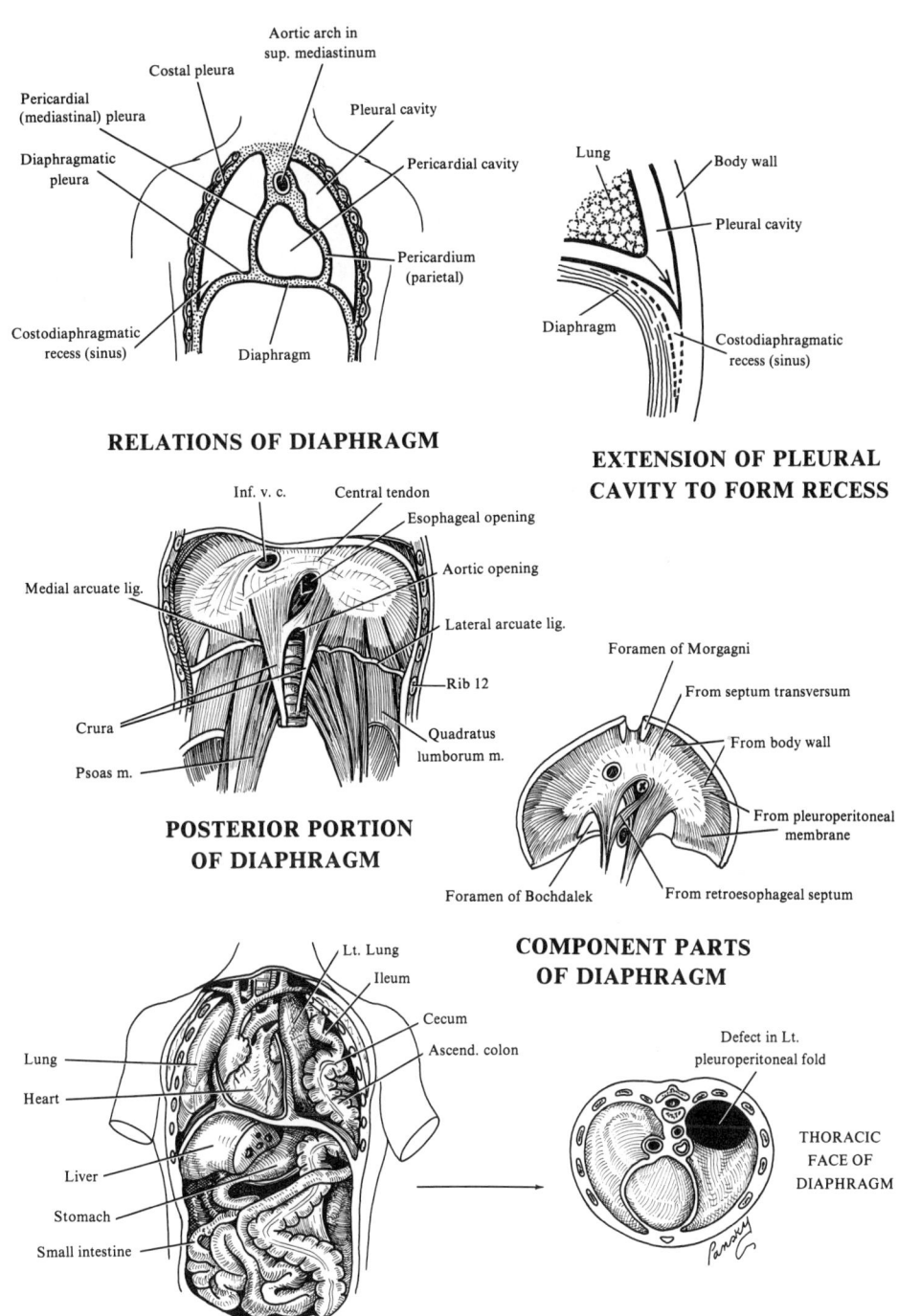

FIGURE 24. Congenital diaphragmatic defects and normal diaphragmatic parts and relations.

UNIT TWO

THE BRANCHIAL APPARATUS:

The Face, Pharynx, and Related Branchial Derivatives

49. THE BRANCHIAL APPARATUS: THE BRANCHIAL (PHARYNGEAL) ARCHES

I. **The branchial apparatus (branchial, meaning gills)** consists of the *branchial arches*, the *pharyngeal pouches*, the *branchial grooves*, and the *branchial membranes*. Modeling and development of the cephalic end and face of the embryo result in a complex form for the pharyngeal segment of the digestive tract. The lateral walls of the developing pharyngeal gut form the branchial system or apparatus

II. **Branchial or pharyngeal arches** are masses of mesoderm covered by ectoderm and lined by entoderm. Within these masses, muscular and skeletal components develop, as well as aortic arches and nerve networks. The arches are separated by grooves, visible on the surface of the embryo as *pharyngeal clefts* and in the interior as the *pharyngeal pouches*

   A. THE BRANCHIAL SYSTEM is only transitory. Continuous modification of the cephalic end of the embryo, during brain development, also affects the basic architecture. The arches give rise to skeletal structures from predominantly cartilaginous precursors; the pouches and clefts either are effaced or persist as ducts or canals. Some give rise to important glandular structures. The branchial system is formed at the 5-mm embryo stage, in week 4, with its 4 arches, 4 pharyngeal clefts, and 5 pharyngeal pouches

   1. The mandibular arch (arch I) is centered on *Meckel's cartilage*. The *malleus* and *incus*, ossicles of the middle ear, develop from its posterior portion. The *mandible* and *muscles of mastication* form from tissues (mesenchyme and invading neural crest cells) surrounding Meckel's cartilage
      a. It is supplied by aortic arch I (facial artery) and the mandibular division of the trigeminal nerve (V3)
   2. The hyoid arch (arch II) is centered on *Reichert's cartilage*. It gives rise to the *stapes* of the middle ear, the *styloid process* of the temporal bone, the *stylohyoid ligament*, the *lesser horns* and *upper part of the body of the hyoid bone*, the *platysma muscle*, and the *muscles of facial expression*
      a. It is supplied by aortic arch II (external carotid artery) and the facial nerve (VII)
   3. The thyrohyoid arch (arch III) produces the *body and greater horns of the hyoid bone* and the *stylopharyngeus muscle*
      a. It is supplied by aortic arch III (internal carotid artery) and the glossopharyngeal nerve (IX)
   4. The (unnamed) arch IV is much less clearly differentiated. It gives rise to the *cartilages of the larynx* and the *cricothyroid muscle*
      a. It is supplied by aortic arch IV and the external branch of the superior laryngeal nerve of the vagus nerve (X)
   5. Pharyngeal arches V and VI (unnamed) are never seen in humans. Their corresponding aortic arches, however, do occur. The muscle mass of arch V forms some of the intrinsic muscles of the larynx, which are supplied by the recurrent or inferior laryngeal branch of the vagus nerve (X)
      a. The cartilages of arches IV and V become the framework of the larynx
      b. The thyroid cartilage originates from both arches IV and V
      c. The cricoid, arytenoids, the rings of the trachea, and the bronchi are formed from arch VI
      d. Arch VI is distinctly differentiated, whereas V is transitory

   B. AT 8 MM, arch II develops more rapidly than the others and overlaps the other arches caudally. This process is further accentuated by the flexion of the head at this stage of development

   C. AT ABOUT 13 MM (about 34 days), arch II has entirely overlapped arches III and IV. It also has closed the second, third, and fourth pharyngeal clefts to form the *cervical sinus*. Only the first pharyngeal cleft persists, and by day 42, it is seen as the external auditory meatus, the only exterior evidence of the pharyngeal system

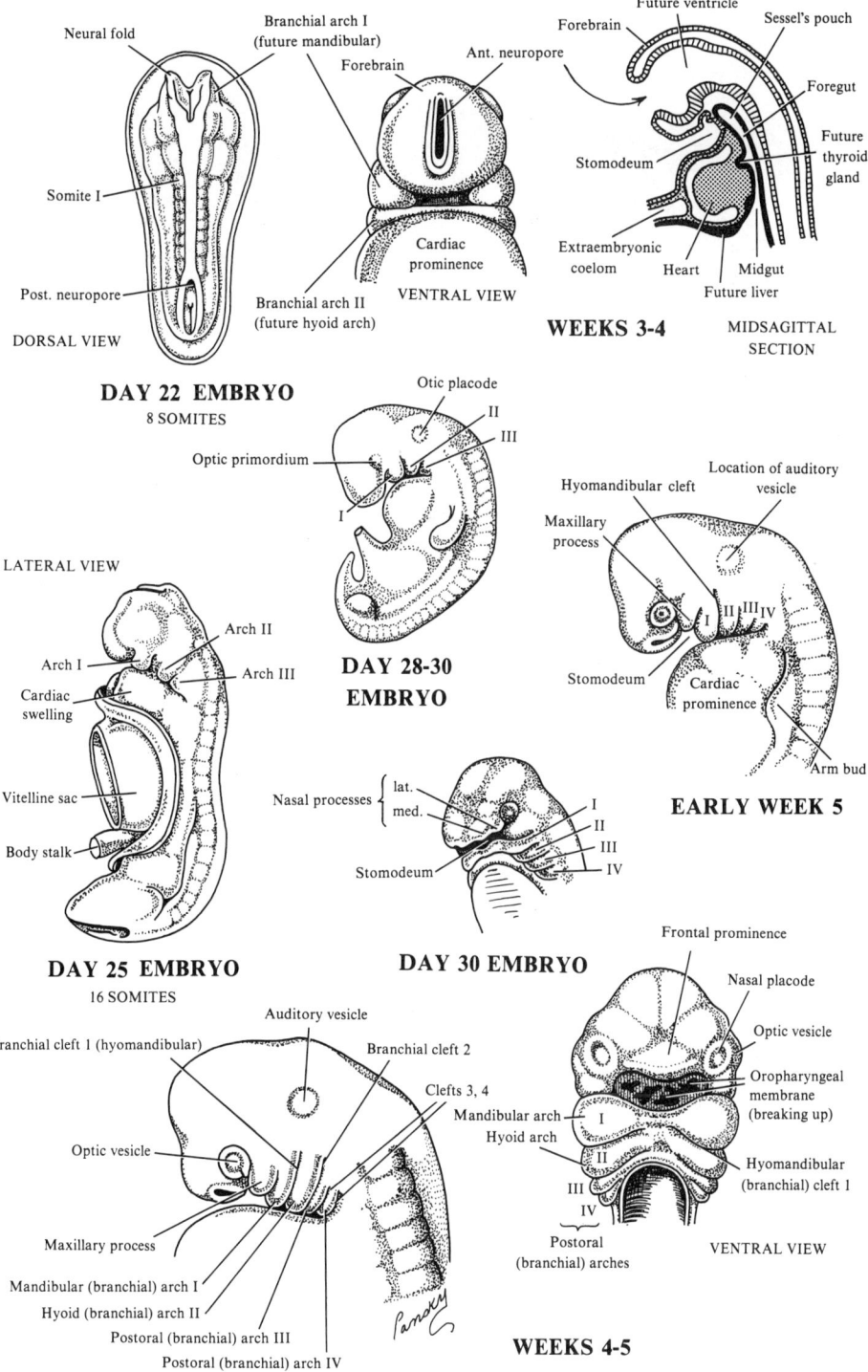

- 129 -

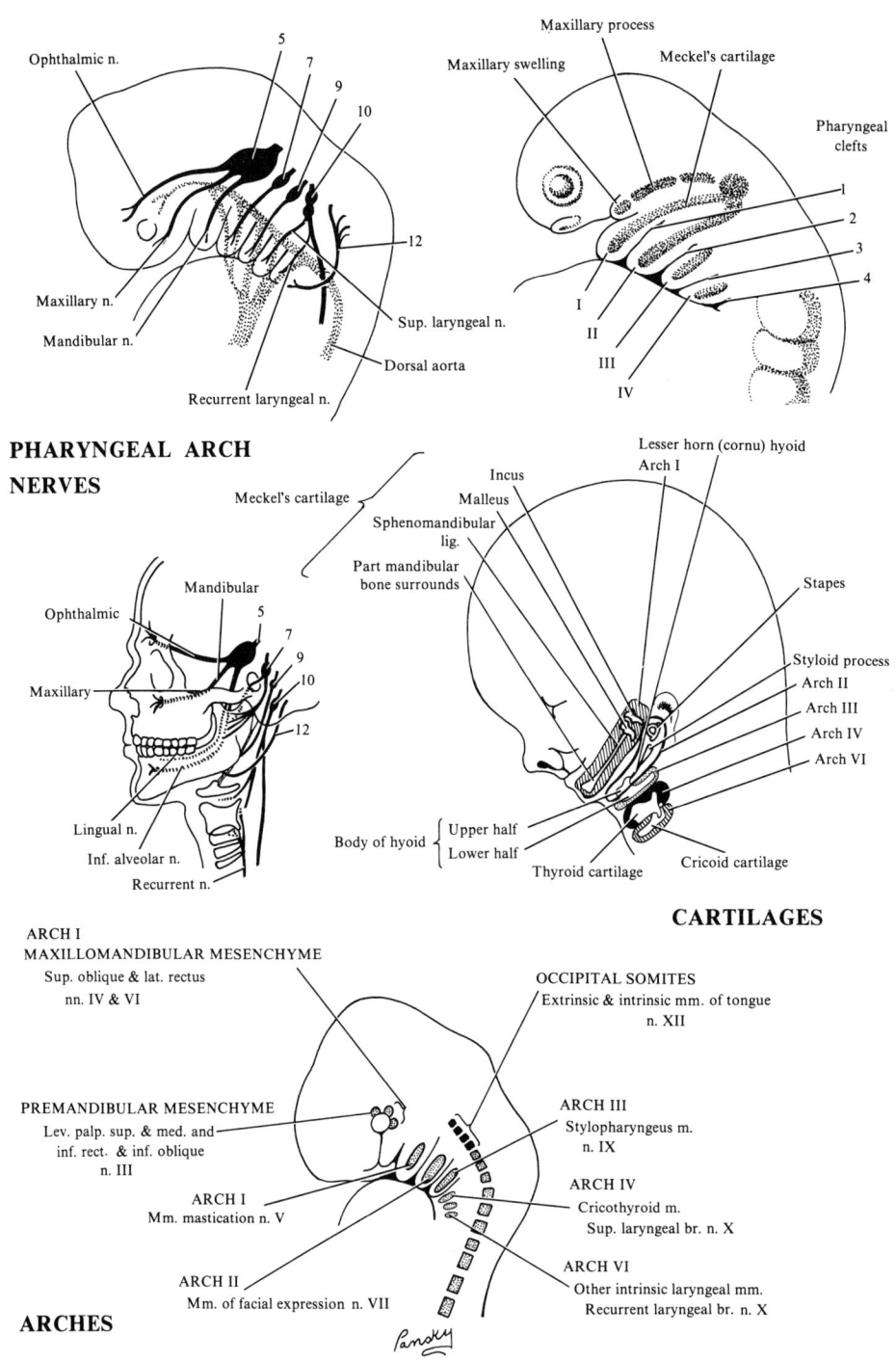

FIGURE 25. The branchial (pharyngeal) arches and associated nerves.

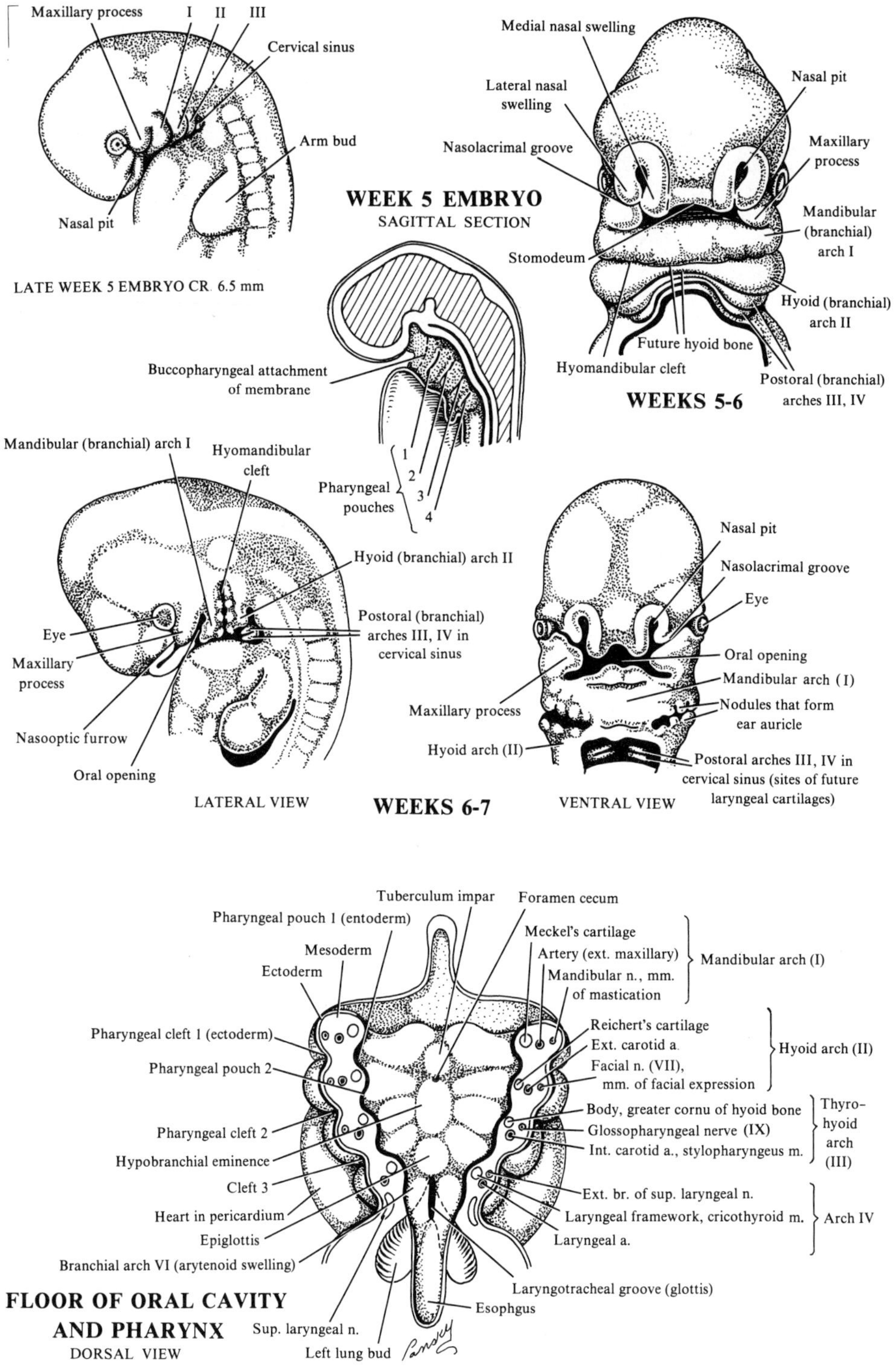

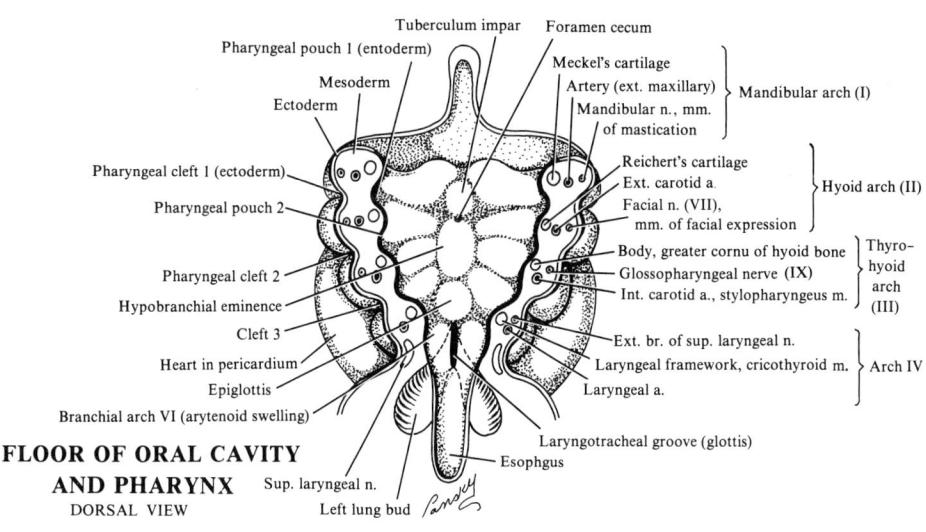

FIGURE 26. **The oral cavity and pharynx.**

50. THE PHARYNGEAL CLEFTS AND POUCHES

I. **The pharyngeal clefts** are seen in the 5-week embryo and in development almost completely disappear. The second, third, and fourth clefts are overlapped by the development of arch II and form a space lined by squamous epithelium, the so-called *cervical sinus*. This, too, disappears during extension of the cervical flexure
   A. THE FIRST PHARYNGEAL CLEFT is the only one to persist, but only partly, to form the epithelium of the *external auditory meatus* and *part of the tympanic membrane*
   B. AT THE EXTERNAL ORIFICE OF THE AUDITORY MEATUS, swellings arising from the mandibular (I) and hyoid (II) arches participate in the formation of the external ear
   C. THE CERVICAL SINUS sometimes persists in vestigial form and forms a *branchial cyst*. If it communicates only with the outside, it forms a *pharyngeal fistula*, which is harmless. If it opens to both the interior and exterior, it forms a *pharyngocutaneous fistula*, which allows saliva to run out during mastication

II. **The pharyngeal pouches** are balloonlike diverticulae of the pharyngeal entoderm that line the inside of the branchial arches. They develop in a craniocaudal sequence between the arches, e.g., pouch 1 lies between arches I and II. There are 4 well-developed pairs of pouches; the fifth pair is rudimentary or absent. The pouch entoderm reaches the branchial groove ectoderm to form the double-layer *branchial membranes* that separate them. The arches enclose the primitive pharynx within which develop the important T structures: the tongue, tonsils, tube (eustachian), thyroid, thymus, and parathyroids
   A. THE FIRST PHARYNGEAL POUCH elongates as the tubotympanic recess and appears between the external and internal ear, enveloping the middle ear bones. The distal portion reaches the first branchial groove and forms the *tympanic cavity* and *mastoid antrum*. The remainder forms the *eustachian tube*, which opens into the pharynx
      1. Fusion of the ectoderm and the entodermal layers forms the *tympanic membrane* or *eardrum*
   B. THE SECOND PHARYNGEAL POUCH does not elongate as much as the first. It forms the *tonsillar fossa*. At its extremity, the entodermal epithelium swells and invades the surrounding mesenchyme to form the *palatine tonsil*, which develops *in situ*. During months 3 to 5, the tonsil is gradually infiltrated by lymphatic tissue and eventually forms lymph nodules
   C. THE THIRD PHARYNGEAL POUCH is the area from which the primordia for the *thymus gland* is formed. These paired symmetric primordia, originally from a long ventral wing which proliferates and obliterates its cavities, migrate to form a single median gland found in the anterior portion of the upper thoracic region
      1. The *inferior parathyroid glands* arise from the dorsal border or wing of this pouch and later migrate toward the posterior inferior end of the lateral lobe of the thyroid gland
      2. Both parathyroids and thymus primordia lose pharyngeal connections and migrate caudally and then separate from each other. Growth and development of the thymus are not completed by birth, but the gland is relatively large during the perinatal period and grows until puberty, at which time it begins to diminish progressively and finally atrophies
   D. THE FOURTH PHARYNGEAL POUCH also gives rise to a thyroid primordium, but in humans, this regresses. The *superior parathyroid glands* arise from the dorsal border or wing of this pouch and later migrate to reach the superior end of the lateral thyroid lobe on its posterior side. Its pharyngeal connection degenerates
   E. THE FIFTH PHARYNGEAL POUCH forms the *ultimobranchial body,* which is thought to participate in formation of the thyroid gland. Cells of this body are thought to disseminate and give rise to *parafollicular* or *C cells* of the thyroid gland (store and secrete calcitonin) involved in normal calcium level regulation in the body fluids. These cells are apparently of neural crest origin which have migrated here

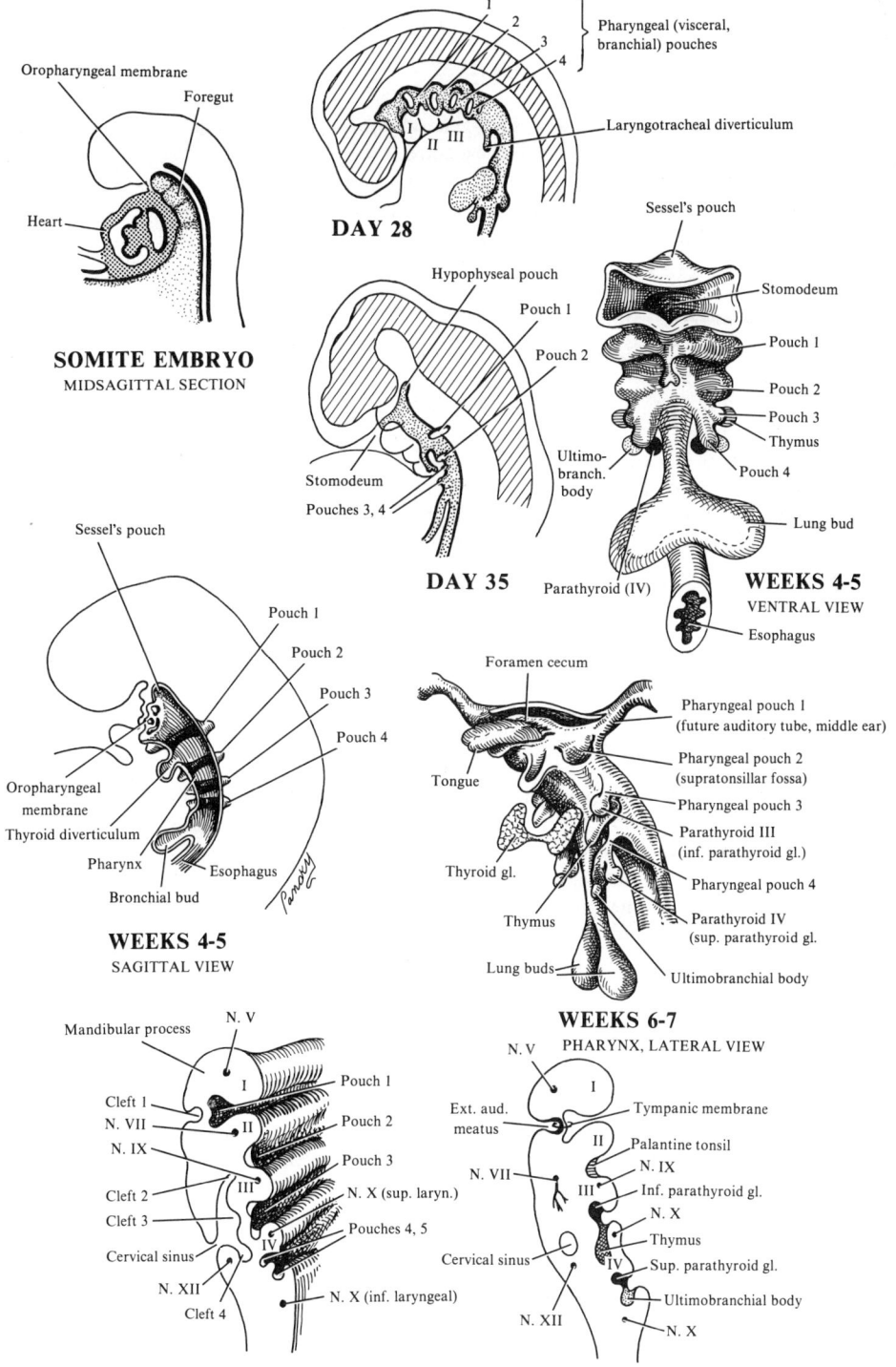

51. MALFORMATIONS RELATED TO THE BRANCHIAL APPARATUS

I. **Malformations of the head and neck** generally originate during the transformation of the branchial apparatus into adult structures
   A. CONGENITAL AURICULAR PITS AND CYSTS: blind pits or cysts in the skin are commonly found in a small area anterior to the ear and are probably ectodermal folds sequestered during the formation of the external ear
   B. BRANCHIAL OR LATERAL CERVICAL SINUSES are relatively rare and almost always open externally on the side of the neck. They are due to a failure of the second branchial cleft or groove to obliterate
      1. A blind pit or channel opens on the lower third of the neck near the sternocleidomastoid muscle
         a. There may be a periodic mucus discharge
      2. A branchial sinus opening into the pharynx is rare, but one may open into the tonsillar fossa as a result of the persistence of part of the second pharyngeal pouch
   C. BRANCHIAL FISTULA is an abnormal tract opening both on the side of the neck and in the pharynx (tonsillar fossa)
      1. It is the result of the persistence of parts of the second branchial groove and second pharyngeal pouch
      2. It provides drainage for a lateral cervical cyst (remnants of the cervical sinus) and is found most often just below the angle of the jaw
   D. BRANCHIAL VESTIGE is rare; seen as cartilaginous or bony remnants of branchial arch cartilages, which normally disappear. It is seen on the side of the neck in front of the lower third of the sternocleidomastoid muscle
   E. THE FIRST ARCH SYNDROME: a number of malformations result from the disappearance or abnormal development of various parts of pharyngeal arch I. They are probably due to insufficient migration of arch I neural crest cells and are associated with anomalies of the mandibular swelling and ear. Two rare symptom complexes are described
      1. Treacher Collins syndrome (mandibulofacial dysostosis): due to a dominant gene. One sees abnormal external ears, anomalies of the middle and inner ears, hypoplasia of the malar bone and mandible, defects of the lower eyelid, and downslanting of the palpebral fissures
      2. Pierre Robin syndrome: hypoplasia of the mandible, cleft palate, and defects of the eye and ear
   F. DIGEORGE'S SYNDROME: congenital thymic aplasia and absence of the parathyroid glands. Characterized by congenital hypoparathyroidism, susceptibility to infection, mouth malformations, low-set notched external ears, nasal clefts, thyroid hypoplasia, and cardiac anomalies
      1. Due to a failure of the third and fourth pharyngeal pouches to differentiate into a thymus and parathyroid glands. Some abnormal development of arch I components also are seen
   G. CERVICAL THYMUS AND ACCESSORY THYMUS: very rare; cords of thymus are seen in the course of its descent from the third pharyngeal pouches

II. **Doubling of the cephalic extremity at the time of gastrulation (15-20 days)** leads to
   A. APODYMY: lateral doubling leading to a single or double median eye
   B. RHINODYMY: doubling seen only at the nose
   C. STOMODYMY: doubling seen only at the mouth

III. **Abnormality during facial swelling formation (weeks 3 to 4)**
   A. AGENESIS OF THE FRONTAL PROMINENCE
      1. Major type (cyclopy): incompatible with life
      2. Median type (archiencephaly): agenesis of the corpus callosum
      3. Minor type (agenesis of the nasal septum): a median nasal fissure with a median cleft lip

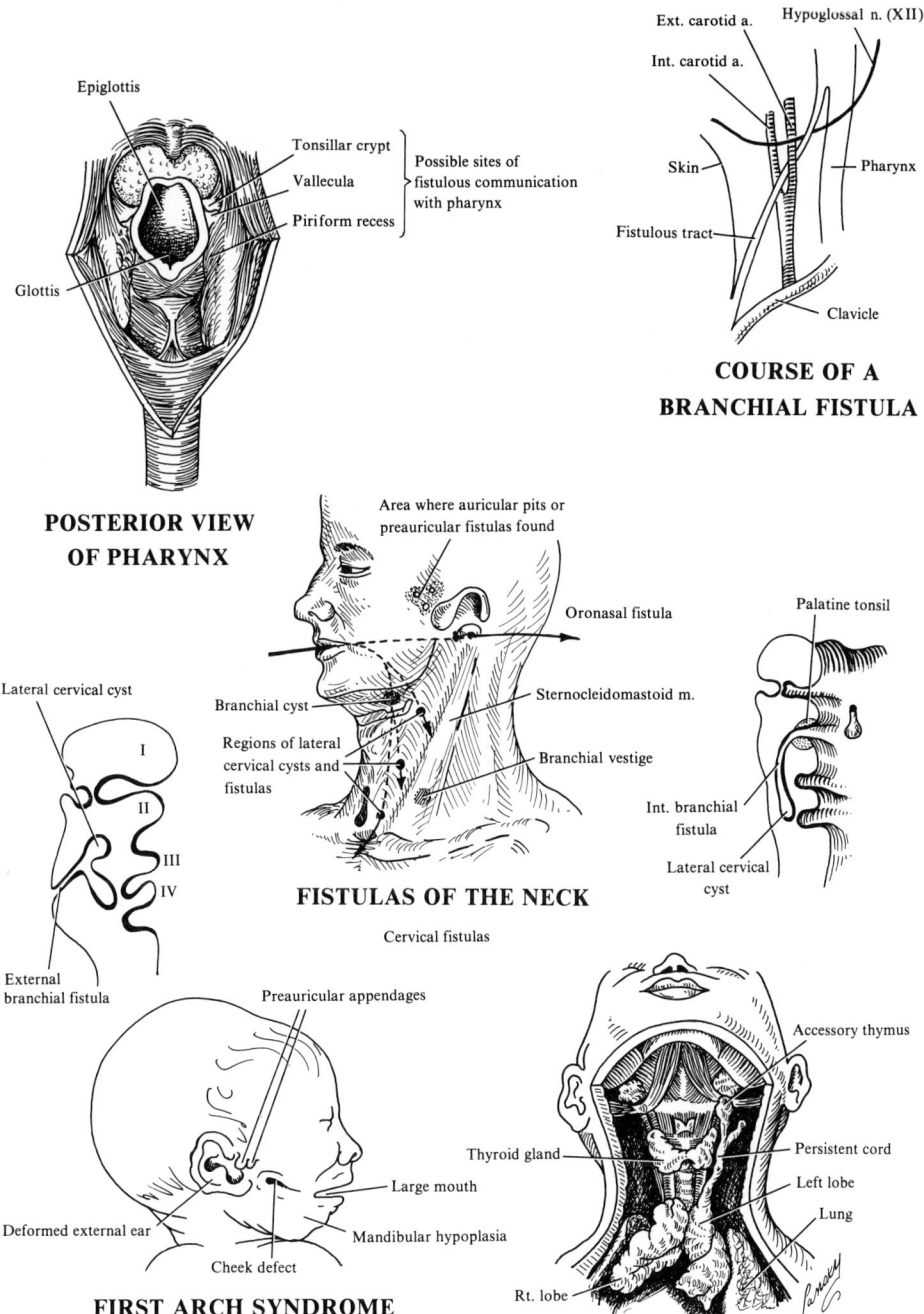

52. BRANCHIAL ARCH DERIVATIVES: THE THYROID GLAND

I. **Introduction:** the thyroid gland begins its development by about day 24 (in week 4) from a median entodermal thickening in the floor of the primitive pharynx just caudal to the future site of the tuberculum impar
  A. THE THICKENING forms a downgrowth, the *thyroid diverticulum,* which grows into the underlying mesoderm, and as the embryo elongates and the tongue grows, the diverticulum descends in front of the neck and pharyngeal gut
   1. The diverticulum is connected to the tongue by a narrow canal, the *thyroglossal duct,* which opens in the tongue via the *foramen cecum,* which persists as a vestigial pit on the tongue
  B. THE DIVERTICULUM grows rapidly and forms 2 lobes and by week 7 of embryonic development, it reaches its final position anterior to the trachea, having acquired a small median isthmus and 2 lateral lobes. By then, the thyroglossal duct usually has disappeared
   1. A pyramidal lobe, extending from the isthmus, is seen in about 50% of thyroid glands and is derived from the thyroglossal duct
  C. THE THYROID GLAND begins to function at about the end of month 3, at which time, the first follicles containing colloid can be seen
  D. AT FIRST, THE THYROID PRIMORDIUM is made up of a solid mass of entodermal cells
   1. It later breaks up into a network of epithelial cords or plates by invasion of the surrounding mesenchyme
   2. By week 10, the cords have divided into small cellular groups, and a lumen forms in each cellular cluster. The cells then arrange themselves in a single layer around the lumen
   3. During week 11, colloid is seen in these *follicle* structures, and even thyroxine can be demonstrated

II. **Congenital malformations**
  A. REMNANTS OF THE THYROGLOSSAL DUCT
   1. The normal remains of the thyroglossal duct are the vestigial foramen cecum (of the tongue) and the functional pyramidal lobe of the thyroid gland
  B. THYROGLOSSAL DUCT CYSTS AND SINUSES
   1. Cysts can form anywhere along the course of the developing thyroglossal duct during descent of the developing thyroid gland from the tongue
   2. Remnants of the duct may persist and give rise to cysts in the tongue or in the midline of the neck, usually below the hyoid bone
   3. In about 30% of cases, an opening through the skin is found as a result of perforation following infection of a cyst, and this forms the so-called *thyroglossal duct sinus,* which usually opens in the midline of the neck in front of the laryngeal cartilages
  C. ETOPIC THYROID GLAND AND ACCESSORY THYROID TISSUE
   1. Very rarely the thyroid fails to descend from the tongue area resulting in a *lingual thyroid*
   2. Incomplete descent, which is rare, may result in a *cervical thyroid* that is seen in the neck at or just below the hyoid bone
   3. Accessory thyroid tissue often is fully functional, originates from remnants of the thyroglossal duct, thus can be found anywhere from the level of the tongue to where the thyroid gland comes to rest in the neck

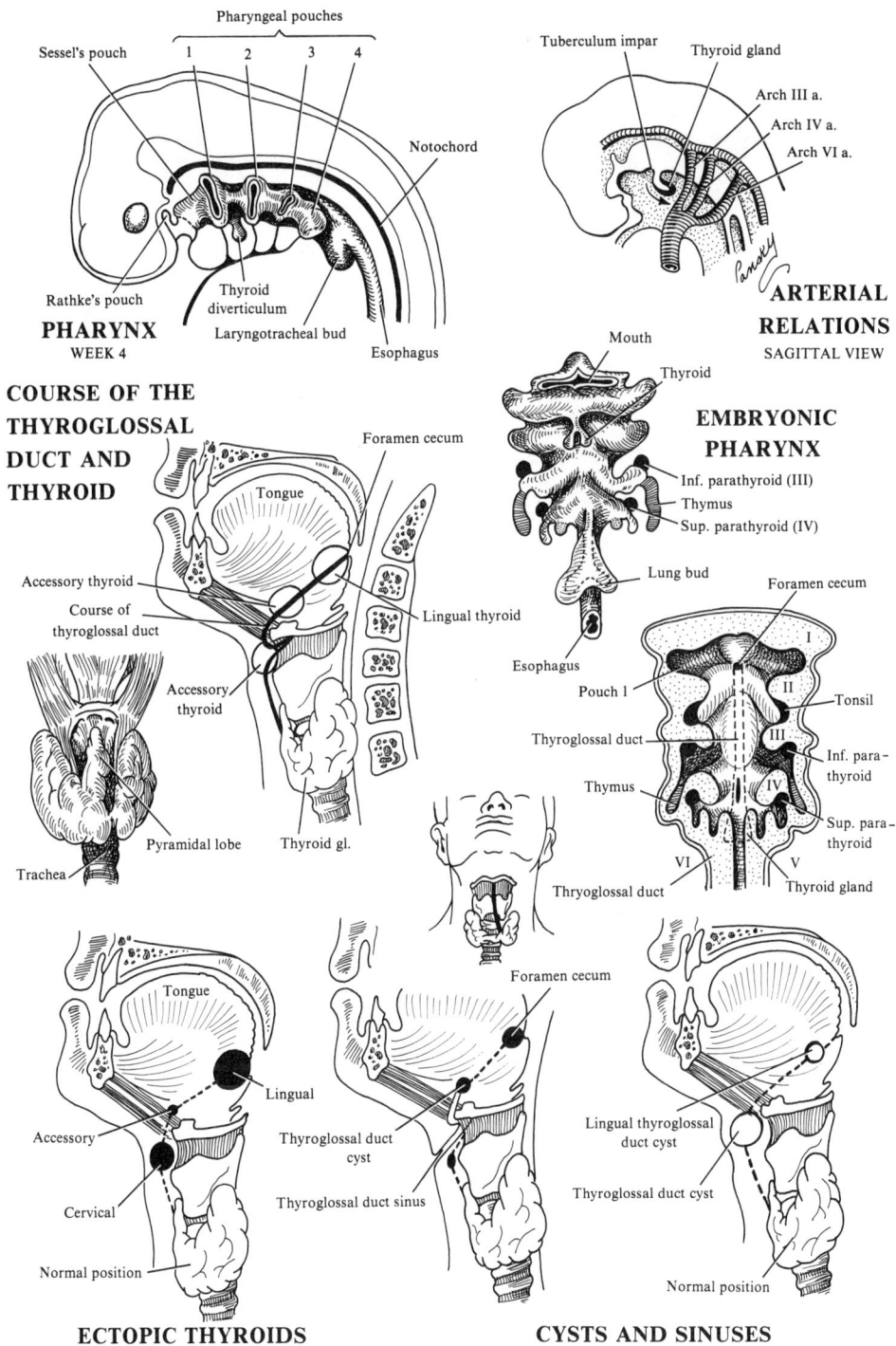

53. THE BRANCHIAL APPARATUS: THE FLOOR OF THE PHARYNX—TONGUE AND ASSOCIATED STRUCTURES

I. **The tongue** appears in the floor of the pharynx, in embryos of about 4 weeks of age, in the form of 2 oval lateral *lingual swellings* (distal tongue buds) and 1 median triangular swelling, the *tuberculum impar*. They are the result of proliferation of mesenchyme in the ventromedial part of the first pair of branchial (mandibular) arches
  A. THE LATERAL LINGUAL SWELLINGS increase in size, merge together, and overgrow the tuberculum impar and form the anterior two-thirds or body of the tongue
    1. Their fusion is superficially marked by the *median sulcus* of the tongue and internally by the fibrous *median septum*. The tuberculum impar forms no significant portion of the adult tongue
  B. THE POSTERIOR THIRD OR ROOT of the tongue first is seen as 2 elevations caudal to the foramen cecum
    1. A median, *copula (connector)*, is formed by fusion of the ventromedial parts of the second branchial arches, and a large *hypobranchial eminence*, caudal to the copula is formed from mesoderm of arches III and IV
      a. The copula is overgrown by the hypobranchial eminence and disappears. Thus, the posterior third of the tongue comes from the cranial portion of the hypobranchial eminence
  C. THE LINE OF FUSION of the anterior two-thirds and the posterior third of the tongue is marked by the V-shaped *terminal sulcus*
  D. BRANCHIAL ARCH MESODERM forms the connective tissue, lymphatics, and blood vessels of the tongue and some of its muscle fibers. However, most tongue muscle is derived from myoblasts that migrate from the *myotomes of the occipital somites*
  E. PAPILLAE OF THE TONGUE appear about day 54 with the vallate and foliate seen first in relation to the terminal branches of cranial nerve IX and the fungiform seen later, induced by the chorda tympani nerve of cranial VII
    1. All develop taste buds
    2. Reflex pathways between taste buds and facial muscles are present by weeks 26 to 28
  F. NERVE SUPPLY OF TONGUE is explained by the branchial arch development. The sensory supply to the mucosa of the entire anterior two-thirds of the tongue is via the lingual branch of the mandibular division of cranial V
    1. The chorda tympani branch of VII supplies the taste buds in the anterior two-thirds of the tongue except for the vallate papillae. Cranial VII does not supply any tongue mucosa since arch III overgrows arch II
    2. The vallate papillae in the anterior two-thirds of the tongue and the posterior third of the tongue are innervated by cranial IX, with the superior laryngeal of X (arch IV) supplying a small area of the tongue anterior to the epiglottis
    3. Nerve XII supplies all the extrinsic and intrinsic muscles of the tongue since it follows the occipital myotomes

II. **Salivary glands** develop as solid proliferations of cells from the epithelium of the primitive mouth during weeks 6 and 7
  A. THE PAROTID GLAND: from buds of ectodermal lining of the stomodeum. Duct is *Stensen's*
  B. THE SUBMANDIBULAR GLANDS: from entoderm in floor of mouth. Duct is *Wharton's*
  C. THE SUBLINGUAL GLANDS: from multiple buds of entoderm in the paralingual sulcus

III. **Congenital malformations of the tongue**
  A. ANKYLOGLOSSIA (tongue-tie) occurs in 1/300 North American infants. There is a shortening of the lingual frenulum so tip of tongue is tied to floor of mouth
  B. MACROGLOSSIA: an excessively large tongue resulting from generalized hypertrophy
  C. MICROGLOSSIA: an abnormally small tongue associated with micrognathia
  D. CLEFT TONGUE: incomplete fusion of the lateral lingual swellings posteriorly
  E. BIFID TONGUE: complete failure of fusion of lateral lingual swellings

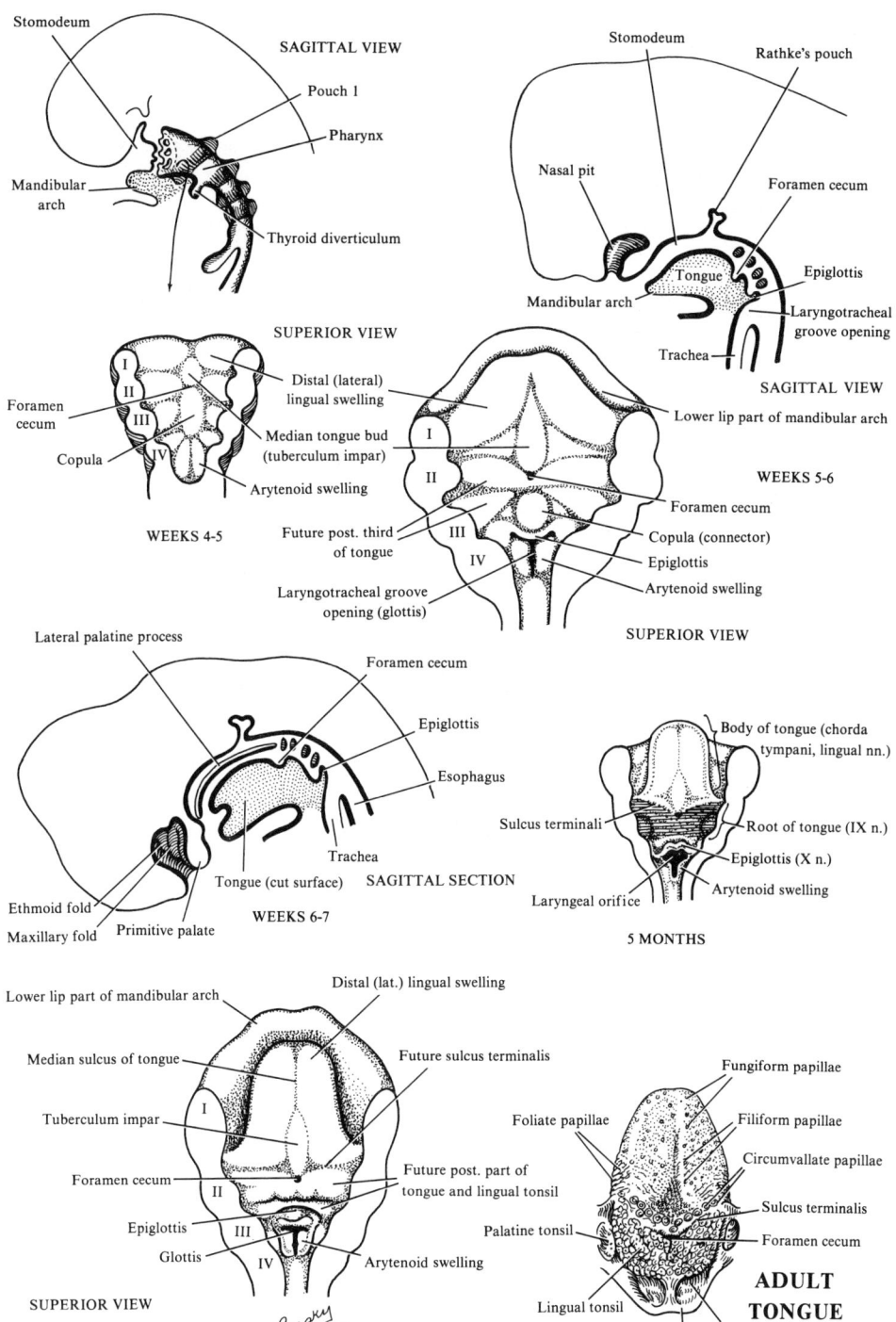

FLOOR OF ORAL CAVITY AND PHARYNX

54. THE FACE

I. **The face** is built up from "facial swellings" as a result of mesodermal masses lifting the surface ectoderm. At the end of week 4, the center of the developing facial structures is formed by an ectodermal depression, the *stomodeum*, or primitive mouth surrounded by the first pair of pharyngeal arches. The 5 facial primordia, formed by proliferation of mesenchyme, appear around the stomodeum early in week 4
   A. THE UNPAIRED MEDIAN FRONTONASAL (FRONTAL) PROMINENCE (SWELLING) constitutes the upper boundary of the stomodeum resulting from proliferation ventral to the brain
      1. By weeks 3 to 4, the anterior neuropore can be seen in the center of this swelling
   B. THE PAIRED MAXILLARY PROMINENCES (SWELLINGS) of branchial arch I form the lateral boundaries or sides of the stomodeum
   C. THE PAIRED MANDIBULAR PROMINENCES (SWELLINGS) of arch I make up the lower stomodeal boundary or floor

II. **The nasal placodes:** bilateral oval-shaped thickenings of surface ectoderm that develop on each side of the lower part of the frontonasal prominence and just above the stomodeum by the end of week 4
   A. THE HORSESHOE-SHAPED MEDIAL AND LATERAL NASAL PROMINENCES (RIDGES) are produced by mesenchymal proliferation at the placode margins during week 5, and the nasal placodes now lie in depressions called *nasal pits*
      1. The lateral swellings form the alae of the nose; the medial swellings form the middle part of the nose, the middle upper lip, the middle part of the maxilla, and the entire primary palate. By weeks 6 to 7, the nasal cavities are well formed

III. **The maxillary prominences** approach each other and the medial nasal prominences
   A. THE NASOLACRIMAL GROOVE marks the cleft or furrow separating the lateral nasal and maxillary prominences
      1. The nasolacrimal ducts form from a linear ectodermal thickening that forms in the floor of the nasolacrimal groove. This epithelial cord sinks into the mesenchyme, canalizes and forms the duct, while its upper part expands to form the *lacrimal sac*. The duct drains into the inferior meatus of the nose
   B. DURING WEEKS 6 TO 7, the medial nasal and maxillary prominences merge
      1. Merging of the former forms the intermaxillary segment of the upper jaw, giving rise to middle part of philtrum of upper lip, middle part of upper jaw and associated gingiva (gums), and primary palate
      2. The maxillary prominences form the lateral parts of the upper lip, the upper jaw, and the secondary palate. They merge laterally with the mandibular
   C. THE PRIMITIVE LIPS AND CHEEKS are invaded by branchial arch II (nerve VII) mesenchyme, giving rise to the facial muscles (innervated by cranial nerve VII)

IV. **The frontonasal prominence** forms the forehead and dorsum and apex of the nose

V. **The mandibular prominences** merge in week 4, and the groove between them vanishes by the end of week 5. They give rise to the lower jaw, lower lip, and lower part of face

VI. **Primitive jaws until end of week 6** are solid tissue masses
   A. LIPS AND GINGIVAE (GUMS) develop when the ectodermal linear thickening, the *labiogingival lamina,* grows into the mesenchyme
      1. It gradually degenerates leaving a labiogingival groove or lip sulcus between the gingivae and lips. Only the frenulum remains between the lips and gingiva

VII. **Slowness of facial development** is result of proportional and position change of components. Face smallness at birth is due to: unerupted teeth, small maxillary air sinuses and nasal cavities, and rudimentary upper and lower jaws. The fetal face is almost definitively formed by about weeks 9 to 10

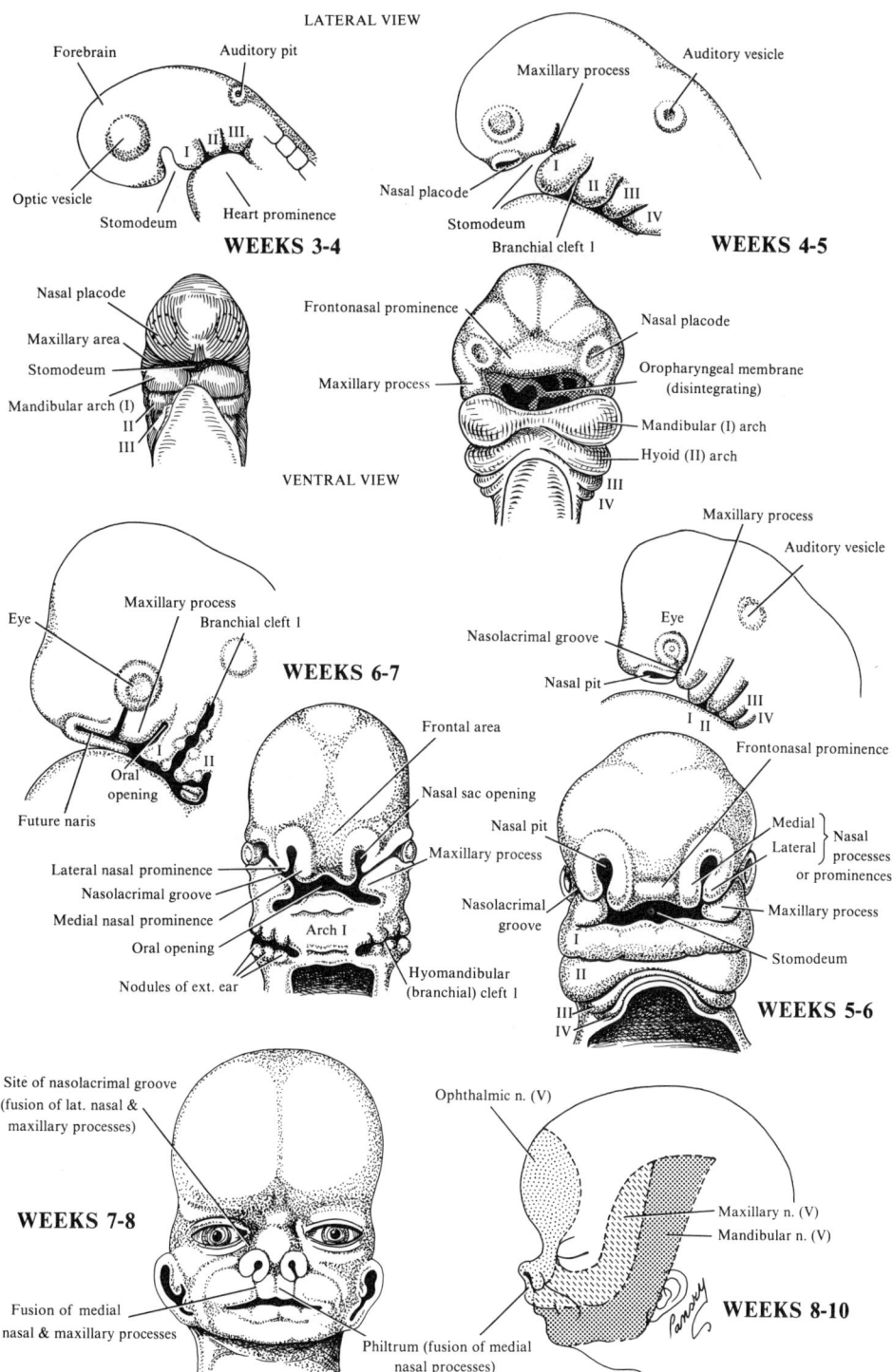

55. DEVELOPMENT OF THE PALATE

I. **The palate:** development begins during week 5, but fusion of its component parts is not complete until week 12. The palate forms from 2 major parts: the *primary* and *secondary palates*
   A. THE PRIMARY PALATE OR MEDIAN PALATINE PROCESS develops from the innermost or ventral portion of the intermaxillary segment of the upper jaw at the end of week 5. The segment is covered with ectoderm
      1. The intermaxillary segment is formed by merging of the medial nasal prominences
      2. The segment forms a wedge-shaped mesodermal mass between the maxillary prominences of the developing upper jaw
   B. THE SECONDARY PALATE develops from 2 horizontal mesodermal projections called the *lateral palatine processes* or *palatine shelves,* formed on the inner surfaces of the maxillary prominences which appear in week 6
      1. They project obliquely downward on each side of the tongue, but as the jaw develops, the tongue moves down and the lateral palatine processes grow toward each other and fuse
         a. They also fuse with the primary palate and the nasal septum, the latter developing as a downward growth from the merged medial nasal prominences
      2. Fusion begins anteriorly in week 9 and is completed posteriorly by week 12
   C. MEMBRANE BONE develops in the primary palate, forming the premaxillary part of the upper jaw and carrying the incisor teeth
      1. At the same time, bone extends into the lateral palatine processes to form the *hard palate*
      2. The posterior portions of the lateral palatine processes do not become ossified, but extend past the nasal septum and fuse to form the *soft palate* and *uvula*. This is the last portion of the palate to form
         a. The palatine raphé indicates the line of fusion of the lateral palatine processes
      3. The nasopalatine canal persists in the palatine midline between the premaxillary portion of the maxilla and the palatine processes of the maxillae. Most of it is obliterated except for the *incisive foramen*

ROOF OF MOUTH

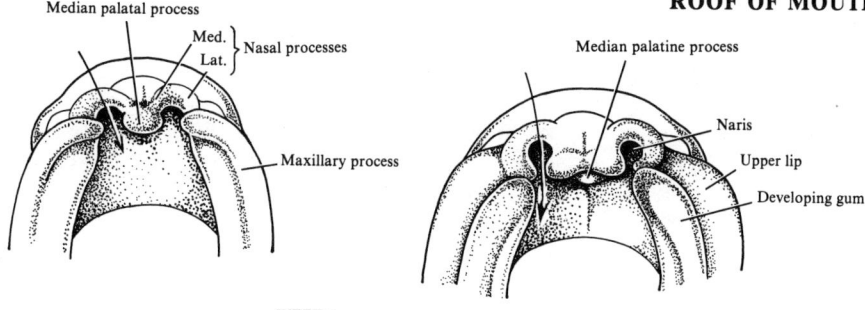
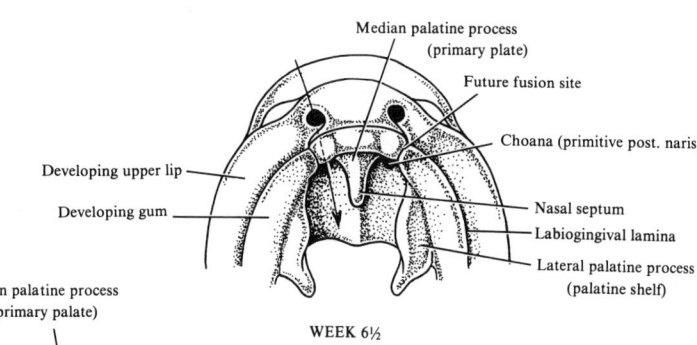
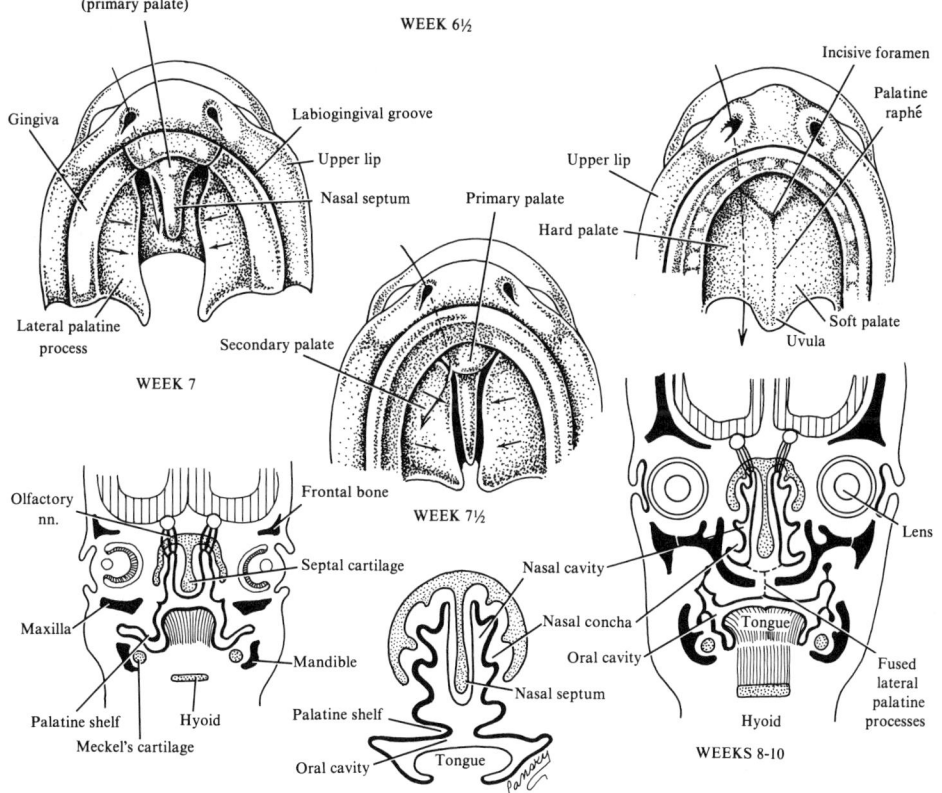

56. CONGENITAL MALFORMATIONS OF THE LIP AND PALATE

I. **Cleft lip and palate** are common malformations of the face and palate. Although often associated, cleft lip and palate are embryologically and etiologically separate malformations. They originate at different times in development and relate to different developmental processes
   A. CLEFT PALATE AND CLEFT LIP have mixed environmental and genetic causes. The genetic causes are of greater importance in cleft lip with or without cleft palate than they are in cleft palate
      1. Clefts of the lip and alveolar processes that continue through the palate are usually transmitted by a sex-linked male gene. Bilateral cleft lip and palate are common in trisomy 13 syndrome and occur at the time of coalescence of the facial swellings (weeks 5 to 8) which results in an abnormal persistence of a fissure
   B. CLEFT lip is usually an upper lip malformation with or without cleft palate and is seen in 1/900 births and more frequently in males. Clefts vary from small notching of the lip's red border to extending into the floor of the nostril and through the alveolar bridge. It may be unilateral or bilateral
      1. Unilateral cleft lip: failure of the maxillary prominence on the affected side to join with the merged medial nasal prominences, resulting in a *persistent labial groove*
      2. Bilateral cleft lip: failure of the mesenchymal masses of the maxillary prominences to meet and merge with the merged medial nasal prominences. Defects may or may not be similar, with varying degree of defects on each side
      3. Median cleft lip: very rare; caused by a mesodermal deficiency. Partial or complete failure of the medial nasal prominences to merge and form the intermaxillary segment
         a. Characteristic feature of Mohr's syndrome (transmitted as an autosomal recessive trait)
         b. Is really the only type of a true harelip
   C. CLEFT PALATE: with or without cleft lip is seen in about 1/2500 births and may involve only the uvula or extend through the soft and hard palates. In severe cases, with cleft lip, it may extend through the alveolar process and lips of both sides
      1. Its embryologic basis is failure of the mesenchymal masses of the lateral palatine processes to meet and fuse with each other, with the nasal septum, and/or with the posterior margin of the median palatine process or primary palate
      2. It may be either unilateral or bilateral
      3. Clefts of the anterior or primary palate occur anterior to the incisive foramen and are caused by a failure of the lateral palatine processes to meet and fuse with the primary palate
      4. Clefts of the anterior and posterior palate involve both the primary and secondary palate and are caused by failure of the lateral palatine processes to meet and fuse with each other, the primary palate, and the nasal septum
      5. Clefts of the posterior or secondary palate: these clefts are posterior to the incisive foramen and are caused by a failure of the lateral palatine processes to meet and fuse with each other and the nasal septum

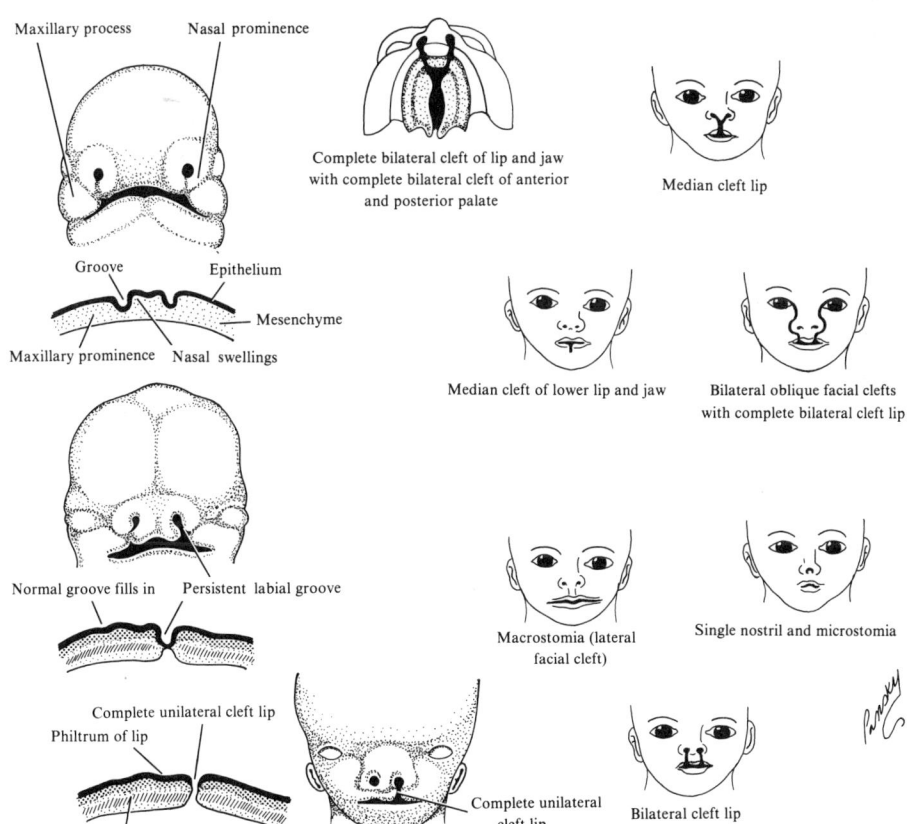

UNIT THREE

THE RESPIRATORY SYSTEM:

Nasal Cavities, Larynx, Trachea, Bronchi, and Lungs

57. DEVELOPMENT OF THE NASAL CAVITIES

I. **The nasal pits** deepen during week 6 due to the growth of the surrounding nasal swellings, also their penetration into the underlying mesenchyme. Thus, the primitive nasal cavities or nasal sacs (pits) each grow dorsocaudally in a position which is ventral to the developing brain
  A. EACH SAC (PIT), at first, is separated from the primitive oral cavity by the so-called *oronasal membrane* which soon breaks down and allows the nasal and oral cavities to communicate with each other via the *primitive choanae,* which lie posterior to the primary palate
    1. After the secondary palate develops, the choanae are at the junction of the nasal cavities and the pharynx

II. **Lateral palatine processes:** when the lateral palatine processes fuse with each other and the nasal septum, the oral and nasal cavities are again separated. This results in a separation of the nasal cavities from each other

III. **The superior, middle, and inferior conchae or turbinates** develop as elevations on the lateral nasal wall of each nasal cavity
  A. THE ECTODERMAL EPITHELIUM in the roof of the nasal cavities becomes specialized for *olfaction*

IV. **The paranasal sinuses** develop during late fetal life and in infancy as diverticula of the lateral nasal walls
  A. THE SINUSES extend into the maxilla, the ethmoid, and the frontal and the sphenoid bones during childhood and reach their mature size in the early twenties, whereupon they enlarge very slowly until death

V. **Developmental malformations of nasal cavities and nose**
  A. ABSENCE OF NOSE: no nasal placodes form
  B. A SINGLE NOSTRIL: only one nasal placode forms
  C. BIFID NOSE: the medial nasal prominences do not merge completely. The nostrils are widely separated and the nasal bridge is bifid

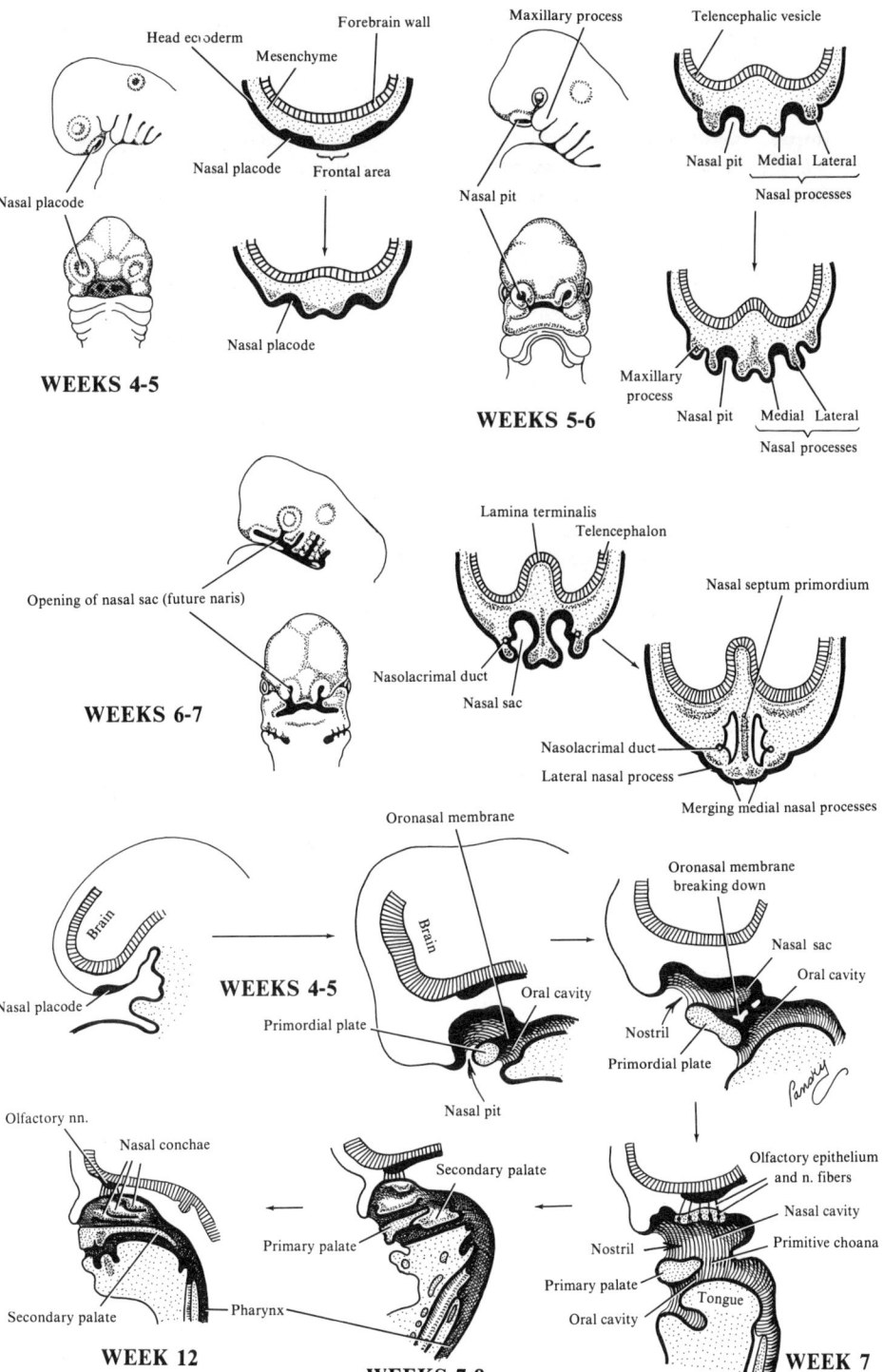

58. DEVELOPMENT OF THE LOWER RESPIRATORY SYSTEM: LARYNX AND TRACHEA

I. **Introduction:** the first indication of the future lower respiratory system appears in the 4-mm embryo, in the primitive pharyngeal floor just behind the pharyngeal pouches, early in week 4 of embryonic life, as a longitudinal groove, the *laryngotracheal groove*. Externally it is seen as a ridge. The entodermal lining of the laryngotracheal groove forms the epithelium and glands of the larynx, trachea, bronchi, and pulmonary lining epithelium. The splanchnic mesenchyme ventral to the foregut will give rise to the connective tissue, the cartilage, and smooth muscle accompanying these structures. When the groove appears, optic vesicle and auditory placode are already present, pharyngeal pouches are forming, oropharyngeal membrane (between stomodeum and foregut) is disintegrating, and the vessels that form heart have fused into a single tube
   A. THE LARYNGOTRACHEAL GROOVE deepens and, with development, the external ridge grows caudally below the pharynx to become a diverticulum, the *tubular lung bud*
      1. As the diverticulum grows from the pharyngeal floor, it is invested by splanchnic mesenchyme. The cranial part of the tube becomes the laryngeal epithelium; the caudal part forms the epithelium of the lower respiratory system
      2. As the diverticulum grows, it becomes separated from the pharynx by a partition, the *tracheoesophageal septum,* which divides the foregut into the laryngotracheal tube and the esophagus
         a. The laryngotracheal tube and surrounding splanchnic mesenchyme give origin to the larynx, the trachea, the bronchi, and the lungs
      3. When the tubular lung bud forms, it develops 2 knoblike enlargements at its distal end, the so-called *bronchial buds*

II. **The larynx** develops from the entodermal lining of the cranial end of the laryngotracheal tube and surrounding mesenchyme (from branchial arches IV to VI)
   A. THE MESENCHYME proliferates to produce paired *arytenoid swellings,* giving the primitive glottis a T-shaped appearance and reducing the laryngeal lumen to a slit
      1. The *laryngeal cartilages* develop within the arytenoid swellings from the cartilage bars of the branchial arches
      2. The *epiglottis* develops from the caudal half of the *hyopbranchial eminence,* a derivative of branchial arches III and IV
   B. THE ENTRANCE TO THE LARYNX ends blindly, between weeks 7 to 10, because of the fusion of epithelium, but as the epithelium breaks down, the *laryngeal aditus* enlarges and recanalizes. A pair of lateral recesses, the *laryngeal ventricles,* form which are bound cranially and caudally by anteroposterior folds of mucous membrane, the future *vestibular (false)* and *vocal (true) folds,* respectively
      1. The laryngeal muscles develop from muscle elements in branchial arches IV to VI and are innervated by laryngeal branches of the vagus nerve

III. **The trachea:** the entodermal lining of the middle segment of the laryngotracheal tube forms the epithelium and glands of the trachea. Mesenchymal cells (from splanchnic mesenchyme) surround the tracheal tube and ultimately form the cartilage, connective tissue, and smooth muscles of its walls
   A. BY WEEK 8 (28-30-MM EMBRYO): mesenchymal rudiments of the 16-20 tracheal cartilages are seen and, in the following 2 weeks, the masses form cartilage beginning cranially and extending caudally. Simultaneously, fibroelastic tissue of the tracheal wall arises from mesenchyme between the cartilage and, posteriorly, between the ends of the embryonic rings smooth muscle (the trachealis) arises
      1. Cilia appear at 10 weeks (51-53-mm embryo)
      2. By week 12, mucosal glands are seen and develop in a craniocaudal direction
      3. By the end of week 20, all major microscopic features of the trachea are visible, but it is short and narrow while the larynx is relatively long. This relationship remains until after birth when the trachea outgrows the larynx to reach its final form

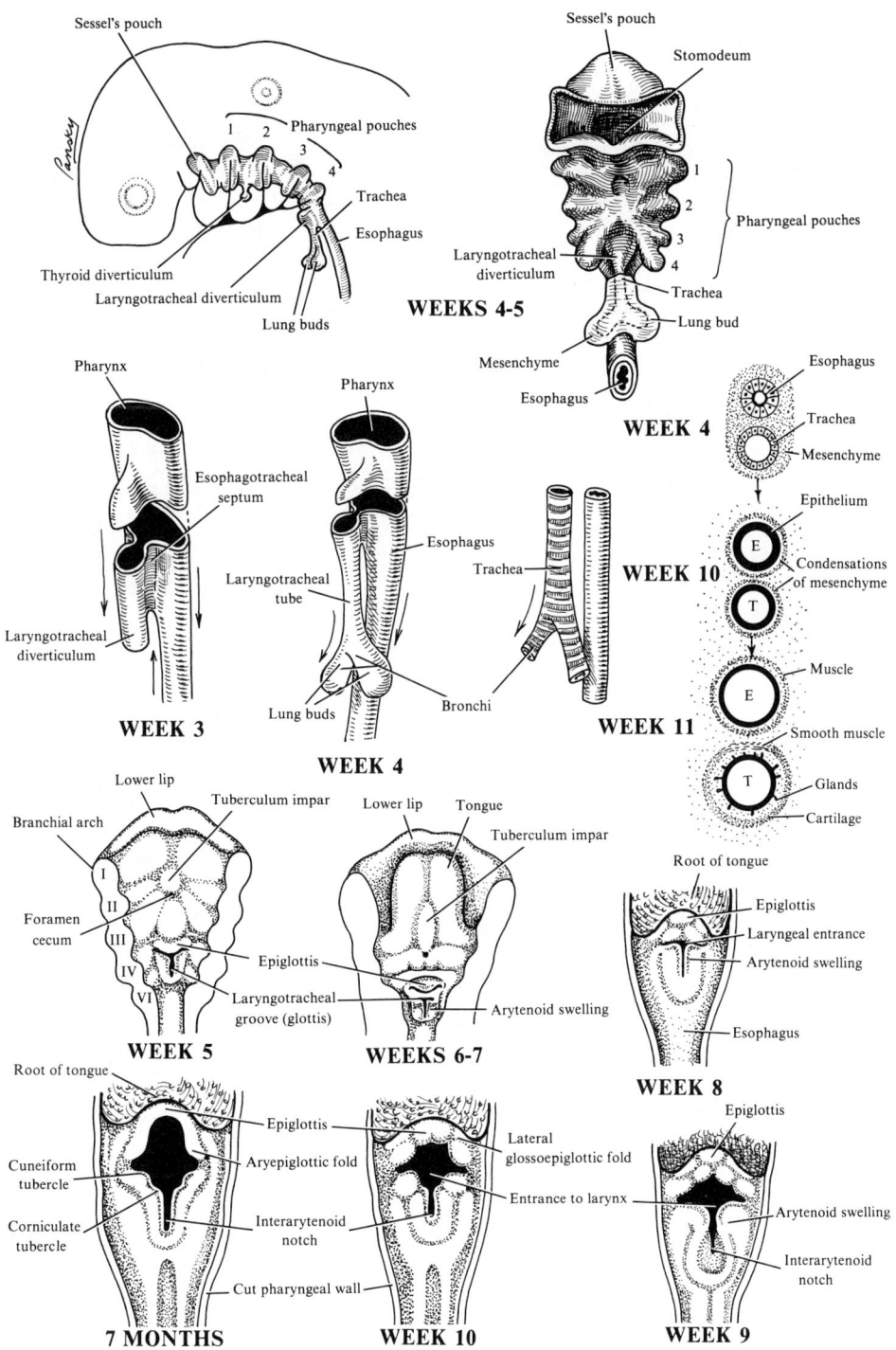

59. DEVELOPMENT OF THE LOWER RESPIRATORY SYSTEM: THE BRONCHI AND SURROUNDING STRUCTURES

I. **The bronchi:** by the early part of week 5, the entodermal lung bud, surrounded by splanchnic mesenchyme, has developed 2 *bronchial buds* that will differentiate into the major bronchi as well as their ramifications in the lungs. Each bud enlarges to form a *primitive primary bronchus* with the right being slightly larger and more vertical than the left
   A. DURING WEEK 5, EACH PRIMARY BRONCHUS grows laterally into the medial walls of the pericardioperitoneal canals or primitive pleural cavities. Simultaneously, the right primary bud gives rise to 2 secondary buds (bronchi), whereas the left only gives rise to 1. Thus, in the adult, there are 3 secondary bronchi and lobes (superior, middle, and inferior) on the right, but only 2 secondary bronchi and lobes on the left (superior and inferior)
      1. At the same time, the surrounding mesenchyme forms definitive masses which become the lung lobes, and all grow caudally into the coelomic cavity of the embryo
   B. EACH SECONDARY LUNG BUD then undergoes progressive dichotomous branching to form *tertiary (segmental) bronchi:* 10 in the right lung and 8 or 9 in the left, which begin to appear by week 7
      1. Each tertiary bronchus with its surrounding mass of mesenchyme will eventually form a *bronchopulmonary segment*
   C. BY WEEK 24, ABOUT 17 ORDERS OF AIRWAYS are formed, and the respiratory bronchioles are present. From then on to term, alveolar ducts and primitive alveoli develop
   D. AT BIRTH, THE LUNGS are not completely developed, and additional airways form until the age of 8 years when the full complement of about 24 orders of airways is established. Once complete, no further new airways replace those lost to trauma or disease

II. **Cartilage, smooth muscle, and connective tissue** develop from the surrounding mesenchyme as the bronchial tree is developing
   A. CARTILAGE seen in primary bronchi by week 10, in segmental bronchi by week 12
   B. SMOOTH MUSCLE CELLS begin differentiating at the end of week 7 and by week 12 help form the posterior wall of the larger bronchi where cartilage is absent
      1. In the smaller bronchi and bronchioles, muscle forms an interlacing network in a double spiral pattern which extends into the alveolar ducts
   C. FIBROELASTIC STRUCTURES differentiate about week 7 with fibrils in the mesenchyme. By week 12, fibrils form bundles and look like mature, fibrous, collagenous tissue. In week 24, the elastic fibers begin to differentiate

III. **Epithelial structures:** initially, the lining of the bronchi is nonciliated columnar cells, but by week 12 the cells become cuboidal and develop cilia
   A. BY WEEK 13 one sees cilia in the segmental bronchi, and at the end of week 20 the cells are cuboidal; at birth, cilia are present in the terminal bronchioles
   B. SUBMUCOSAL, MUCUS-SECRETING GLANDS of the bronchi and bronchioles arise from the epithelial cells which migrate into the submucosa
      1. By week 13, the cells produce mucous glands which secrete 1 week later
      2. At the end of week 28, over 85% of adult mucous glands are present in the system

IV. **Pulmonary vessels**
   A. EARLY IN WEEK 4, a primitive pulmonary artery arises from the ventral aspect of both right and left sixth (pulmonary) aortic arches and extends caudally toward the developing tubular lung bud, eventually being incorporated into the mesenchymal tissue around the primitive trachea and bronchial buds
      1. The major branch of each pulmonary arch is the pulmonary artery which accompanies each developing primary bronchus (*see* Circulatory system for details)
   B. THE VENOUS RETURN develops from the mesenchyme and the heart wall and, instead of following the bronchial tree, runs between bronchopulmonary segments

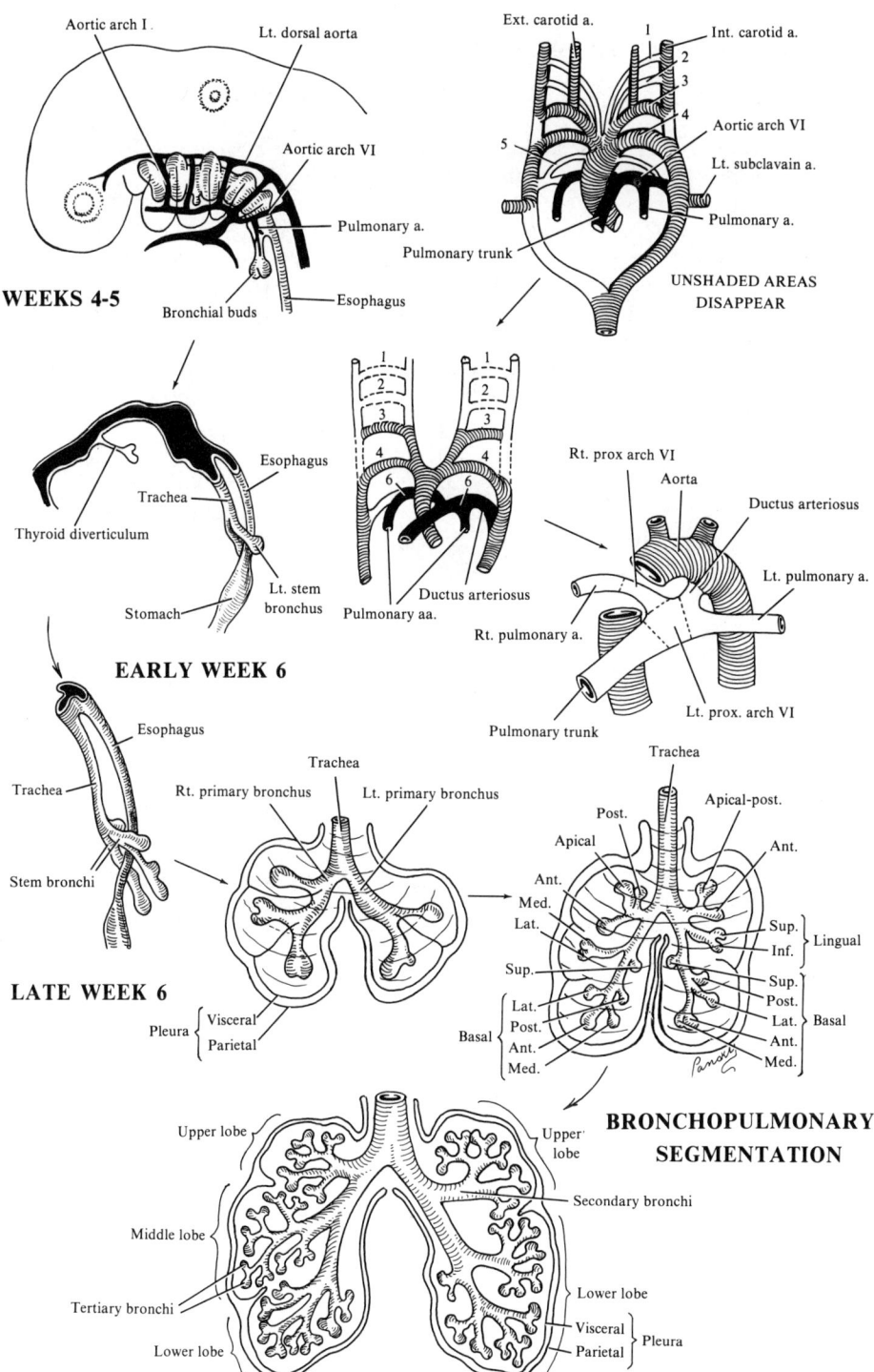

60. DEVELOPMENT OF THE LOWER RESPIRATORY SYSTEM: THE LUNGS AND TERMINAL RESPIRATORY TUBES

I. **Introduction:** as the lungs develop, they acquire a layer of *visceral pleura* from the splanchnic mesenchyme, and with expansion the lungs and pleural cavities grow caudally into the mesenchyme of the body wall and come to lie near the heart. The thoracic cage becomes lined by *parietal pleura* derived from somatic mesoderm

A. DURING WEEKS 4 AND 5, the respiratory system roughly mimics that of amphibians (lungs have 2 air sacs); during weeks 7 and 8, when segmental bronchi and bronchopulmonary segments are seen, the lungs resemble those of reptiles. From here, 4 stages of development are described

1. Pseudoglandular period (weeks 5-17): the developing lung resembles a gland, and during this period, the bronchial divisions are differentiated and the air-conducting system is established.
    a. By week 17, all elements of the lung are formed except those related to gas exchange. Respiration is not feasible because airways are blind-ending tubules
2. Canalicular period (weeks 13-25): a time overlap exists because cranial lung segments mature earlier than caudal ones
    a. The lumina of the bronchi and bronchioles get larger, and lung tissue becomes more highly vascularized
    b. By week 24, each terminal bronchiole has formed 2 or more *respiratory bronchioles,* and respiration is possible by the end of this period since some terminal sacs or primitive alveoli are present at the ends of respiratory bronchioles and vascularization is very good
3. Terminal sac period (week 24 to birth): many terminal sacs develop, lungs have lost their canalicular appearance, and the sac epithelium becomes thin with capillaries bulging into them
    a. The terminal sacs are lined by a continuous flattened epithelium of entodermal origin and are called *type I alveolar epithelial cells*
    b. Capillary networks proliferate greatly in the surrounding mesenchyme, and there is a concurrent development of lymphatic capillaries
    c. By weeks 25-28 (1000-g fetus), sufficient terminal sacs are present so that survival of a premature birth is possible. The adequate pulmonary vasculature and the alveolar surface area are the critical considerations
    d. In addition to type I cells, *type II alveolar cells* (cuboidal cells with a brush border of the terminal sacs) or *secretory cells* differentiate (weeks 23-24). They secrete *surfactant* (see later discussion, p. 156)
4. Alveolar period (late fetal to 8 years): terminal sac epithelial lining attenuates to a very thin squamous type
    a. Type I cells become so thin that underlying capillaries bulge into space of each terminal sac
    b. By late fetal period (week 28), the lungs are capable of respiration because alveolar-capillary (respiratory) membrane (blood-gas barrier) is thin enough to allow gas exchange and an adequate amount of surfactant is being produced. The lungs begin their vital function at birth
    c. At beginning of this period, respiratory bronchioles terminate in a cluster of thin-walled terminal sacs (future alveolar ducts) separated by loose connective tissue. Thus, *alveolar ducts* probably do not exist before birth
    d. Mature alveoli do not form for some time after birth
    e. At birth, air expands the primitive alveoli slightly and the lungs expand. Increase in lung size after birth is really due to an increase in the number of primary alveoli rather than an increase in alveolar size. One-eight to one-sixth of adult number are seen in the newborn.
        i. From years 3 to 8, the number and size of immature alveoli increase as does the potential to form additional primitive alveoli. As the primitive alveoli increase in size, they also mature

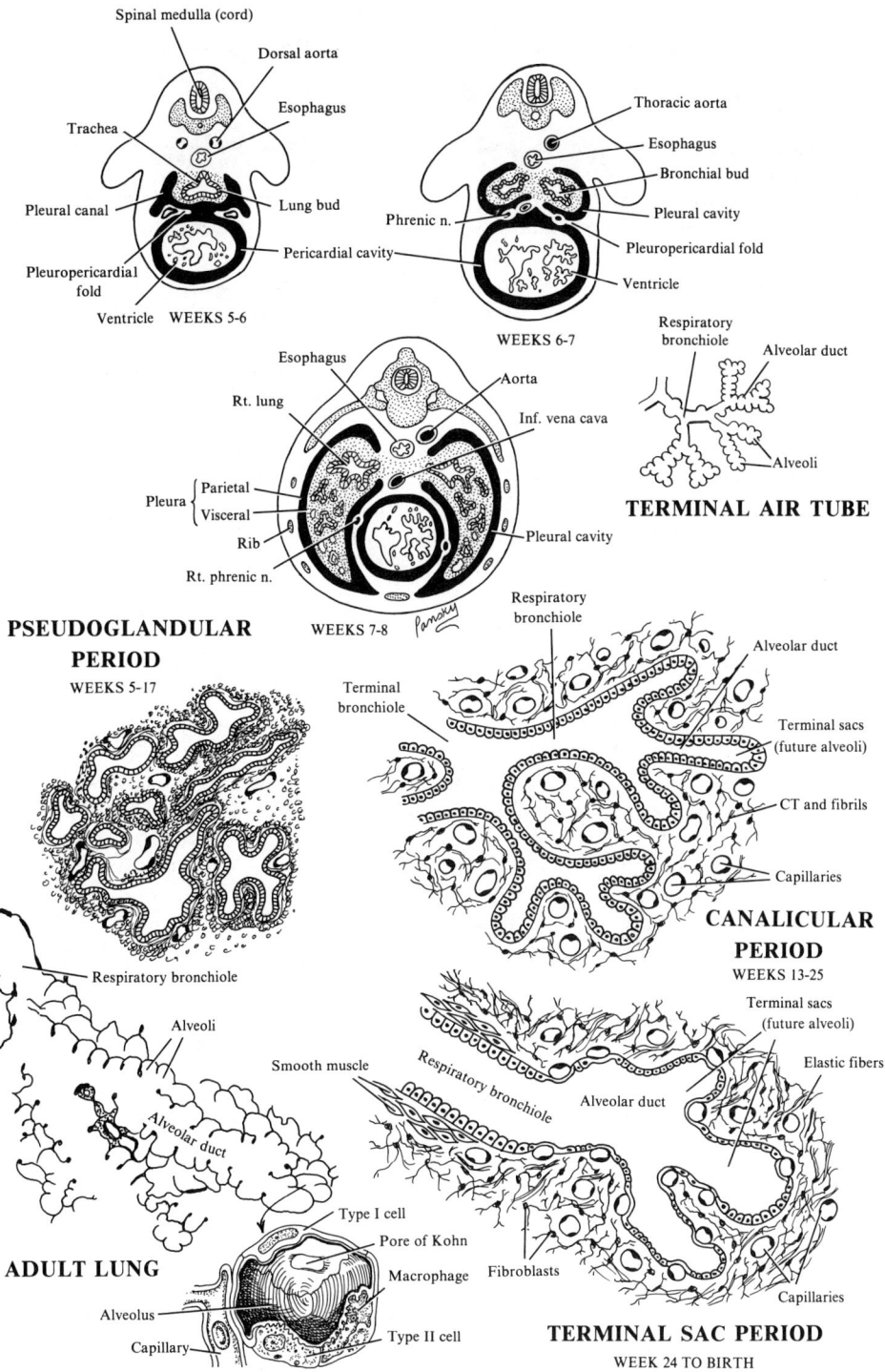

61. DEVELOPMENT OF THE LOWER RESPIRATORY SYSTEM: SURFACTANT AND RESPIRATORY MOVEMENTS

I. **Respiratory movements** occur before birth, causing aspiration of amniotic fluid into the lungs
  A. At birth, the developing lungs are half-inflated with fluid from the lungs themselves, the amniotic cavity, and tracheal glands. Therefore, aeration is the replacement of intra-alveolar fluid for air
   1. During and after birth: one-third of the fluid expelled by the lungs is via the mouth and nose; one-third enters the pulmonary capillaries; and one-third passes into the lymphatic system
   2. In the adult, the average surface area of the alveolar-capillary membrane is between 70 and 80 m², favoring the daily loss of about 800 ml of water in the expired air
   3. The alveolar-capillary membrane, in the adult, consists of a thinned-out cell wall plus the cytoplasm of a type I cell with its basement membrane and the thinned-out cell wall and cytoplasm of a capillary endothelial cell with its basement membrane. Where they meet, the two basement membranes fuse
      a. In the mature, fully distended alveolus, the thickness of the alveolar-capillary membrane varies from 0.2 μm to 2.5 μm. In the newborn, this membrane is 0.4 μm in thickness
      b. The extreme thinness of the alveolar-capillary membrane favors diffusion of oxygen and carbon dioxide

II. **Pulmonary surfactant**
  A. Type II alveolar epithelial cells (cuboidal cells of the terminal sacs) secrete surfactant by weeks 23 to 24
   1. The cells contain osmophilic inclusions or granules, said to be either surfactant or a precursor of surfactant
   2. Surfactant is capable of lowering surface tension at the air-alveolar surface, forming a monomolecular layer at the interface between the air in the alveoli and the fluid layer covering the alveolar cells. This helps maintain the patency of the alveoli and prevent atelectasis (lung collapse)
      a. Surfactant is a mixture of lipoproteins rich in phospholipids, especially dipalmityl lecithin, and has a half-life of 14–24 hours. It is present in lungs of all air-breathing vertebrates
  B. By weeks 25-28, the amount of surfactant is enough to prevent alveolar collapse when breathing begins
  C. The absence or deficiency of surfactant is a major cause of *hyaline membrane disease* and is one major cause of respiratory stress syndrome in newborns
   1. The lungs are underinflated, and the alveoli contain fluid high in protein content
   2. The membrane is derived from a combination of substances in the circulation and injured pulmonary epithelium
  D. Thyroxine is known to be a potent stimulator of surfactant production
  E. Suprarenal cortical hormone, cortisol, controls both maturation of type II cells during uterine life and the production of surfactant
  F. Prolonged intrauterine asphyxia may produce irreversible changes in the alveolar cells, making them incapable of producing surfactant. However, there are undoubtedly many causes for the absence or deficiency of surfactant in both premature and full-term infants

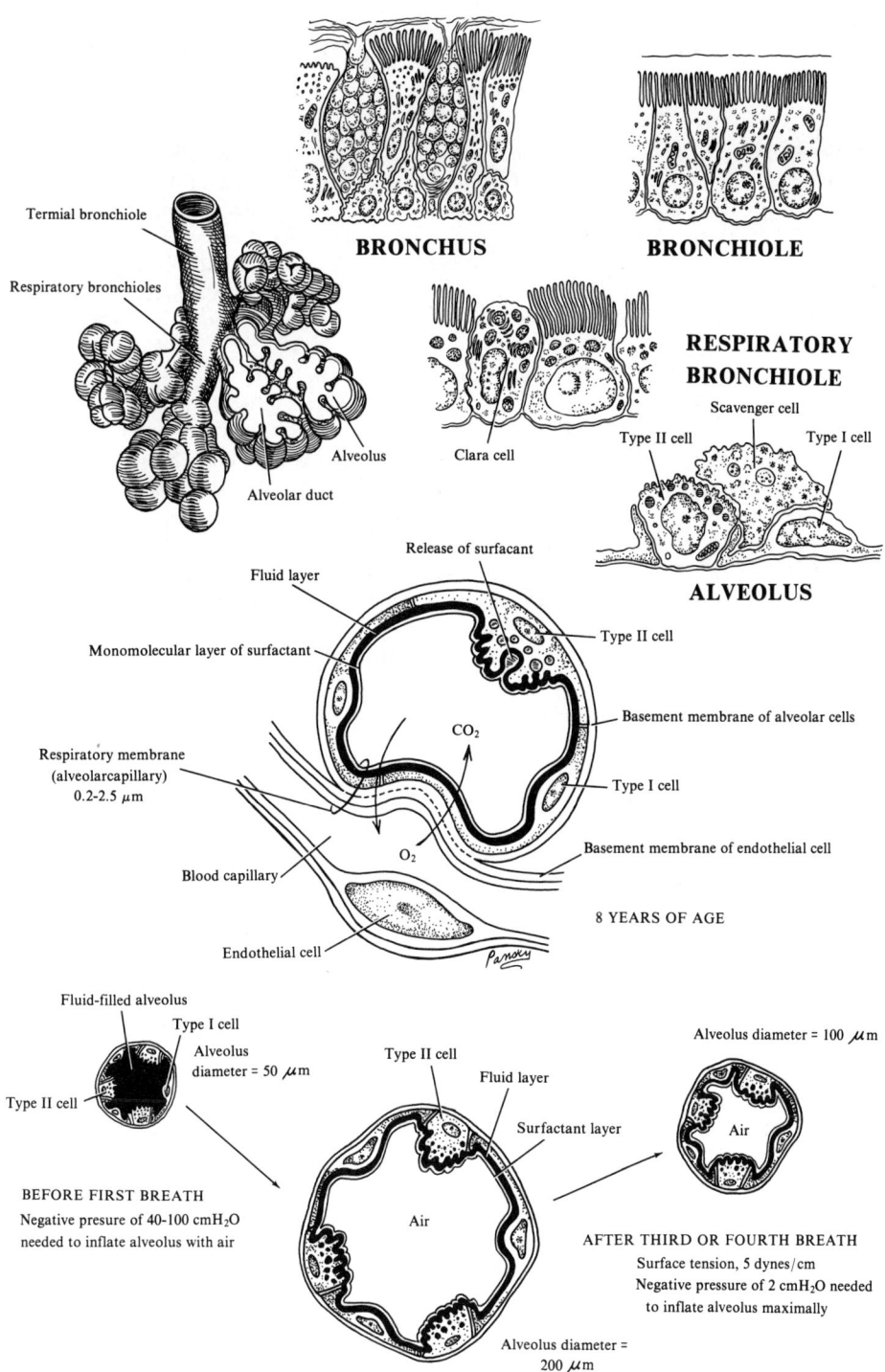

62. MALFORMATIONS OF LOWER RESPIRATORY TRACT

I. Bronchial anomalies as a result of failure of the lung bud to develop
   A. AGENESIS: resulting in a total absence of the bronchial tree, alveolar tissue, and vascular system. It is unusual to find bilateral agenesis with the trachea ending in a blind end. In unilateral agenesis, there is no lung bud on the affected side
   B. APLASIA is generally unilateral. There is a stump of the lung bud present
   C. HYPOPLASIA is characterized by insufficient development. It may be *partial*, involving only a restricted segment, or *total*, involving an entire lung

II. Bronchial anomalies due to malposition
   A. BRONCHIAL VARIATIONS: where the abnormality involves only the division of the bronchopulmonary tree. These anomalies are frequent
   B. ABNORMALITIES OF BRONCHIAL DIVISION resulting in supernumerary bronchi and lungs
   C. ABNORMALITIES OF SYMMETRY leading to situs inversus and mirror-image lungs

III. Anomalies due to bronchial detachment or sequestration
   A. THESE ABNORMALITIES of bronchopulmonary tissue most often may show clear predominance of bronchial structures with cysts and yet retain a normal appearance even though the alveoli are not expanded
   B. THE SEQUESTRATIONS are most often intralobar (sometimes extralobar)
      1. Their bronchi are not part of the tracheobronchial tree
      2. Their vascularization is systemic, coming directly off the aorta or one of its collateral branches. There is no blood supply by way of the pulmonary artery
   C. SUCH CYSTIC STRUCTURES, regardless of origin, result in those parts of a lung being poorly ventilated or not ventilated at all. As a result they cannot oxygenate blood

IV. Failure of cleavage of the tracheal groove
   A. TRACHEOESOPHAGEAL FISTULAS AND ESOPHAGEAL ATRESIA: about 1 in 2500 births
      1. Tracheoesophageal fistulas are caused by incomplete separation of the esophagus and trachea at the time of cleavage of the tracheoesophageal groove, during week 4 of development. Fistulas almost always accompany esophageal atresia
      2. There are 5 anatomic types, according to Ladd
         a. Types 1, *simple atresia without fistula*, and 2, *atresia with fistula*, are extremely rare
         b. Types 3 and 4 are by far the most frequent and occur in 95% of all cases. They are *tracheoesophageal* and *bronchoesophageal fistula* on the lower portion of the esophagus, below the atresia
         c. Type 5 is rare and consists of a *double fistula, surrounding the atresia*

V. Tracheal stenosis or atresia: narrowing (stenosis) and closure (atresia) of trachea are rare malformations and associated with one of the varieties of tracheoesophageal fistulas. Probably due to unequal partitioning of foregut into trachea and esophagus

VI. Laryngeal web: rare malformation due to incomplete recanalization of larynx in week 10. A membranous web forms at vocal cord level and partly obstructs passage

VII. Tracheal diverticulum: rare deformity consisting of a blind bronchuslike projection from trachea. May end in normal-appearing lung tissue to form a *tracheal lobe*

VIII. Azygos lobe: abnormal fissures and lobes are common and usually insignificant. This one is due to azygos vein not moving medially and forming a fissure in the developing lung in the area of the main upper right lobe

IX. Congenital bronchial cysts: the terminal bronchioles rarely may form abnormal saccular enlargements which may give rise to cysts

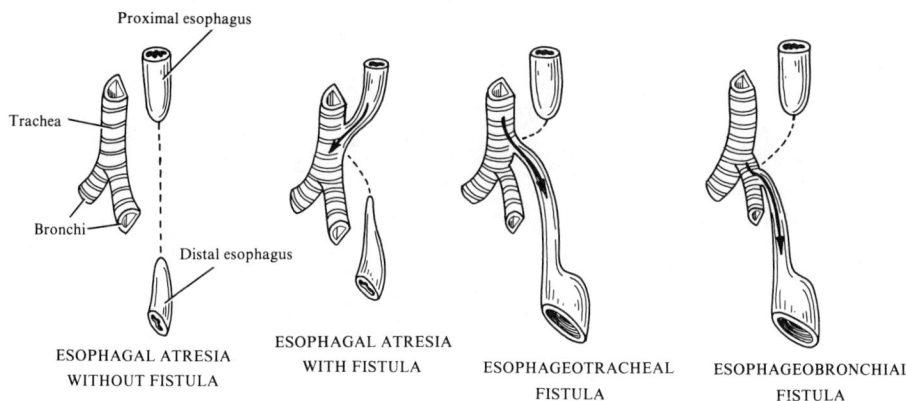
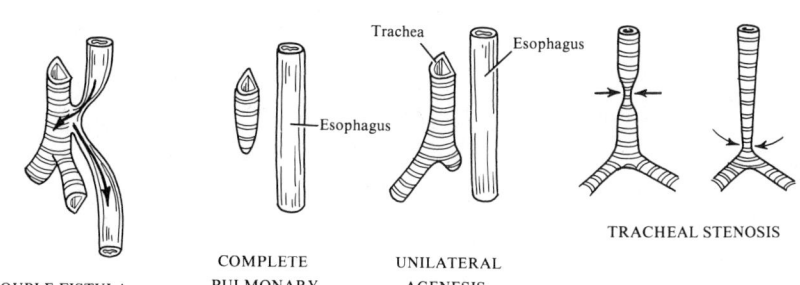
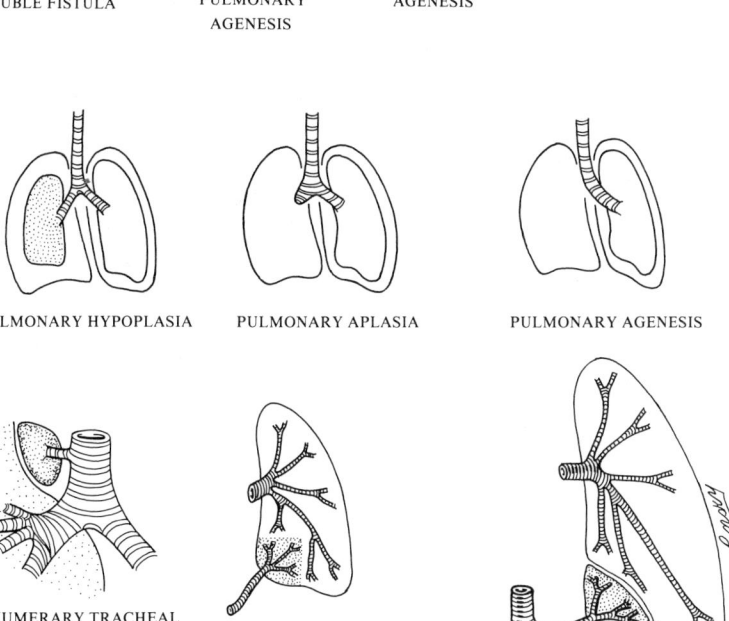

UNIT FOUR
THE MUSCULAR SYSTEM

63. DEVELOPMENT OF THE MUSCULAR SYSTEM

I. **Introduction:** at day 17, the 3 germ layers are seen (ectoderm above, entoderm below, and mesoderm between). At day 19, the lateral mesodermal plate cleaves and the intraembryonic coelom appears, and differentiation of a *somite plate* is seen on the side of the neural tube. Metamerization begins at days 20 to 21 with embryonic flexion. Segmentation proceeds caudally, resulting in 42 to 44 pairs of somites by the end of week 5. Each somite develops a *myocele*. The *sclerotome* forms on its ventromedial portion and migrates to the notochord where it gives rise to fibroblasts, chondroblasts, and osteoblasts according to location. The rest of the somite, its dorsolateral part, is called the *dermomyotome*, which forms a *dermatome* (spreads under the surface ectoderm to form the subcutaneous tissue) and a *myotome* which forms the skeletal muscles

II. **Skeletal muscles** are derived from mesenchymal myoblasts which originate in the myotome portion of the dermomyotome. Skeletal muscles may also arise from mesenchyme in the branchial arches and somatic mesoderm
   A. THE MYOBLASTS elongate, combine to form parallel bundles, and fuse to form multinucleated cells. The central nuclei move to the periphery, and during fetal life myofibrils are seen in the cytoplasm. By month 3, cross-striations are also visible
   B. THE MYOTOME: most myotome development occurs in the thoracic region by week 5. Each myotome divides into a small dorsal *epaxial division* (*epimere*) and a larger ventral *hypaxial division* (*hypomere*). Each spinal nerve also divides, sending branches to each division, a dorsal primary ramus to the epimere and a ventral primary ramus to the hypomere. Most myotomes migrate to form nonsegmented muscles; some remain segmentally arranged like the somites (e.g., the intercostals of the thorax)
      1. Epaxial derivatives: these myoblasts form the extensor muscles of the neck, vertebral column, and lumbar region. Extensors from the caudal sacral and coccygeal myotomes degenerate and become the adult dorsal sacrococcygeal ligament
      2. Hypaxial derivatives: myoblasts of cervical myotomes form the scalene, prevertebral, infrahyoid, and geniohyoid muscles. Thoracic myotomes become the lateral and ventral flexors of the vertebral column. Lumbar myotomes become the quadratus lumborum muscle. The sacrococcygeal myotomes form the muscles of the pelvic diaphragm, anus, and sex organs
   C. BRANCHIAL ARCH MUSCLES: myoblasts from the arches migrate to form the muscles of mastication, of facial expression, and muscles of the pharynx and larynx. They are innervated by branchial arch nerves V, VII, IX, and X, respectively
   D. OCULAR MUSCLES are probably derived from mesenchymal cells around the prochordal plate which gives rise to 3 preoptic myotomes. Groups of myoblasts with cranial nerves III, IV, and VI form the extrinsic muscles of the eyeball
   E. TONGUE MUSCLES: 4 occipital myotomes are seen first, but the first pair disappears. The last 3 pairs form the tongue muscles, innervated by cranial nerve XII
   F. LIMB MUSCLES develop *in situ* from mesenchyme around the developing limb bones. The mesenchyme comes from the somatic layer of the lateral plate mesoderm

III. **Visceral (smooth and cardiac) muscle**
   A. SMOOTH MUSCLE forms from splanchnic mesenchyme around the primitive gut and its derivatives. Elsewhere, it forms from local mesenchyme
      1. Muscles of the iris (sphincter and dilator pupillae) and the myoepithelial cells of the breast and sweat glands come from mesenchymal cells of ectodermal origin
   B. CARDIAC (HEART) MUSCLE: from splanchnic mesenchyme around the embryonic heart
      1. Special bundles of muscle cells develop with few myofibrils. These atypical cells form the *Purkinje fibers* of the conduction system of the heart

IV. **Congenital malformations of muscle** may be due to muscle failure to develop or a pathologic process affecting the muscle or nerve during embryonic development

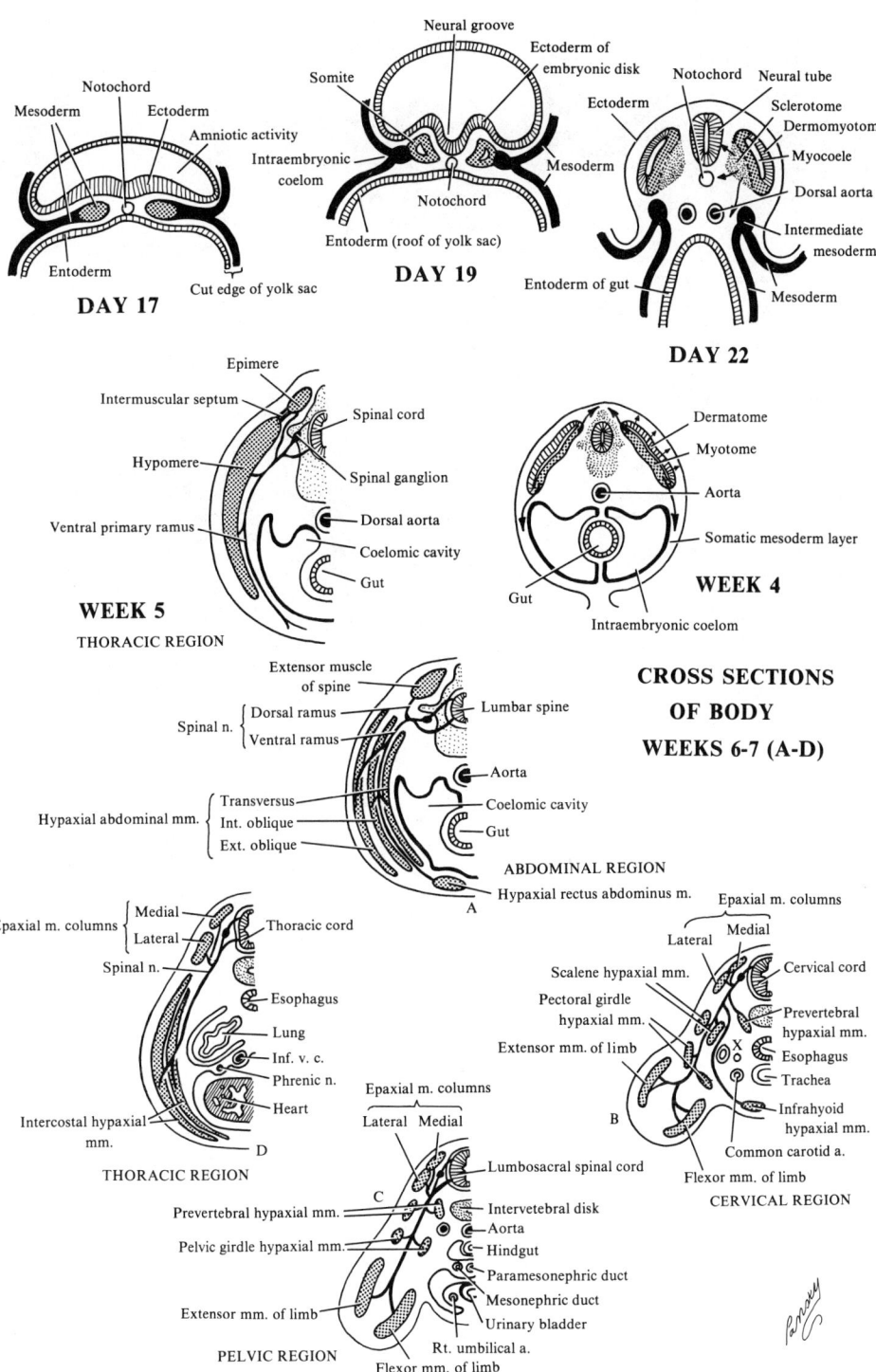

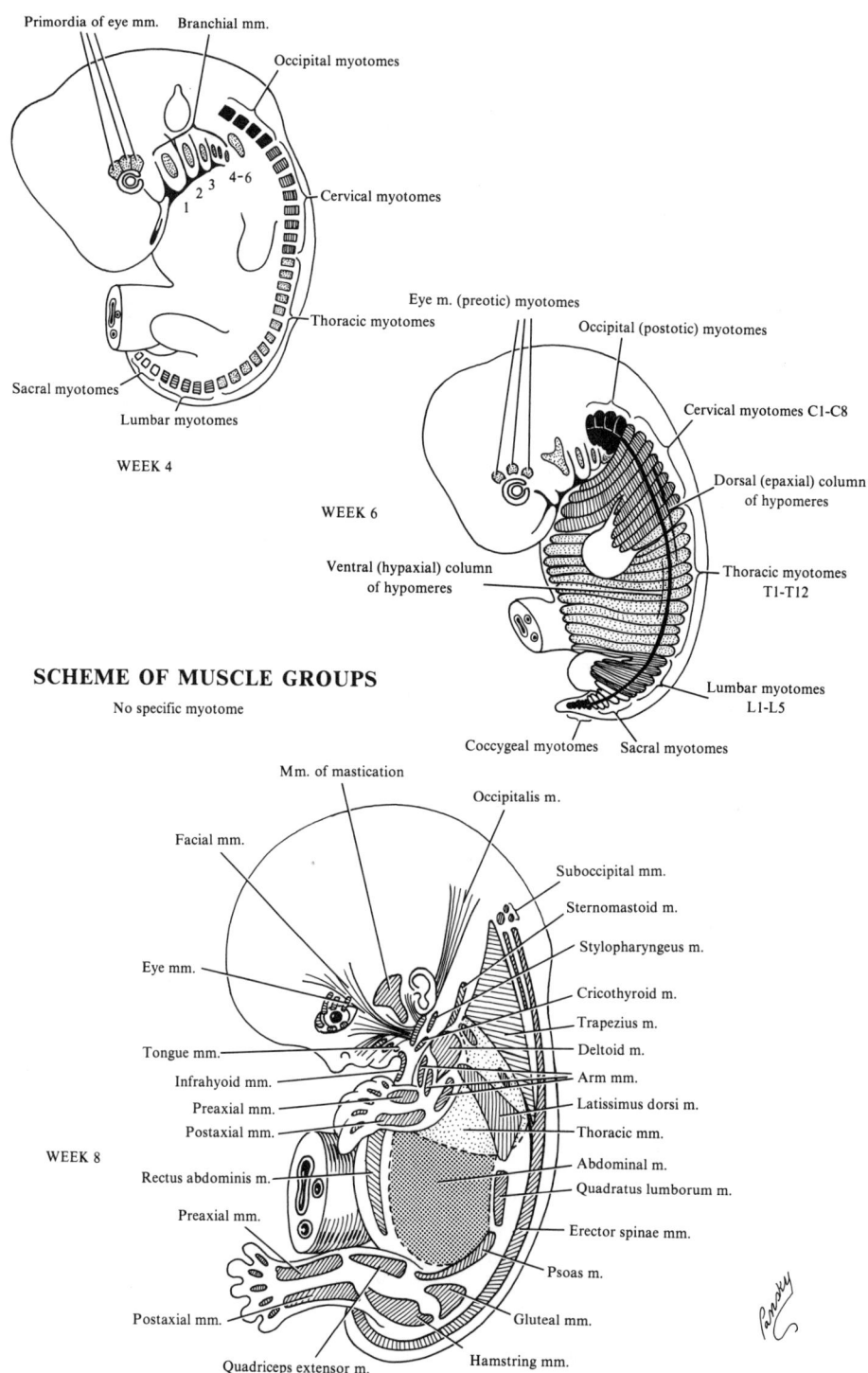

FIGURE 27. **General body muscle groups.**

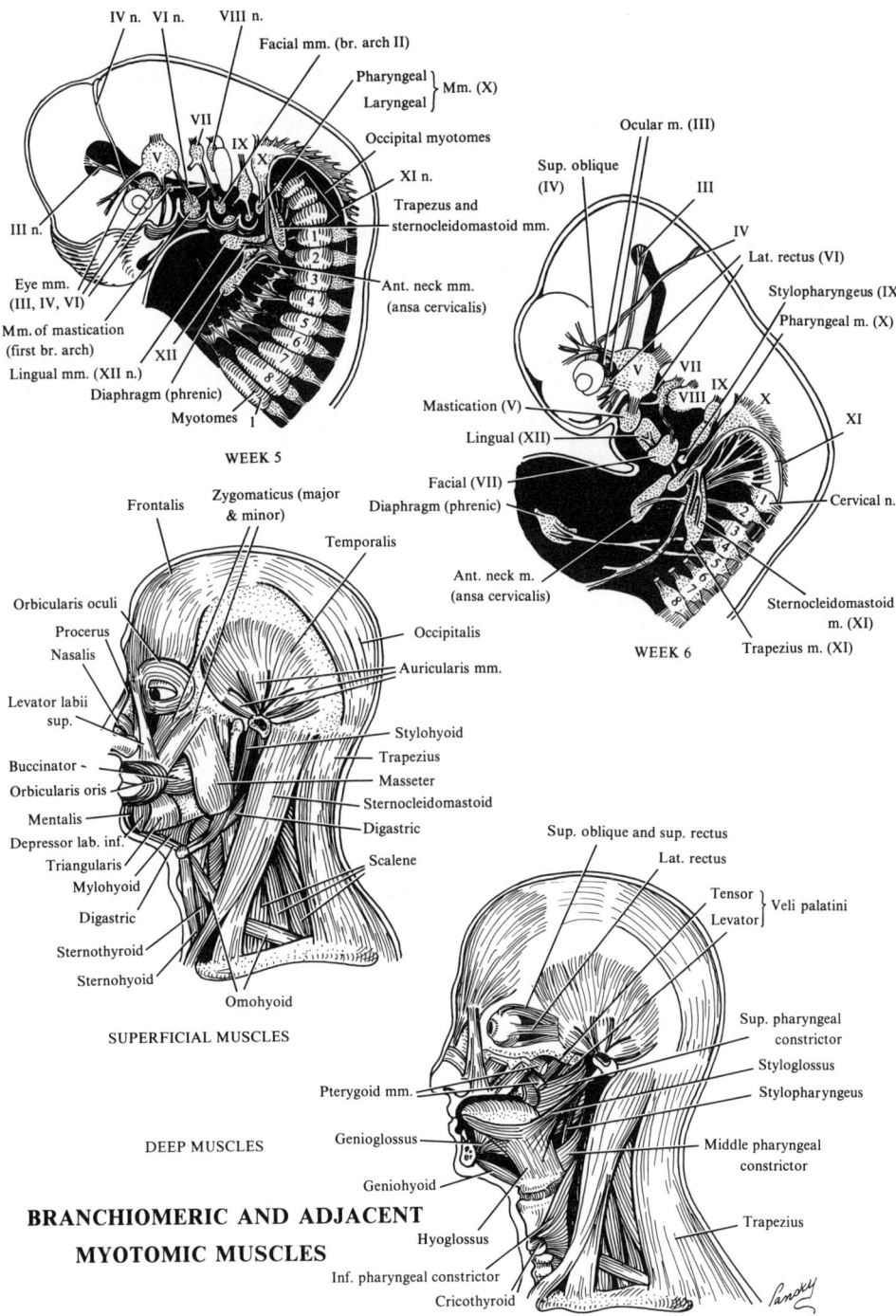

FIGURE 28. **Branchiomeric and adjacent myotomic muscles.**

UNIT FIVE
THE SKELETAL AND ARTICULAR SYSTEMS

64. DEVELOPMENT OF THE SKELETAL AND ARTICULAR SYSTEMS: CARTILAGE AND BONE HISTOGENESIS

I. **Introduction:** these systems develop from mesoderm. The somites differentiate into 2 parts: cells of the *sclerotome* give rise to bone, cartilage, and ligaments; those of the *dermomyotome* give rise to skeletal muscle. The mesodermal cells form mesenchyme (embryonic connective tissue) which can differentiate into fibroblasts, chondroblasts, and osteoblasts. In addition to somite mesenchyme, the splanchnic and somatic mesoderm also can form mesenchyme (some head mesenchyme even arises from neuroectoderm). Most bones first appear as condensations of mesenchymal cells which give rise to hyaline cartilage models that ossify via *endochondral ossification*. Others develop in mesenchyme by *intramembranous bone formation*

II. **Cartilage histogenesis:** cartilage is seen as mesenchymal condensations, at about week 5, where it is to develop. The cells proliferate, round up, and elastic or cartilaginous fibers are deposited in the intercellular substance (matrix). Three types are described: hyaline, fibrocartilage, and elastic cartilage, depending on matrix

III. **Bone histogenesis:** bone develops in 2 types of preexisting connective tissue, namely, in mesenchyme or cartilage
   A. INTRAMEMBRANOUS FORMATION develops in mesenchyme
      1. The mesenchyme condenses, becomes very vascular, and the cells differentiate into *osteoblasts* (bone-forming) which deposit an intercellular matrix
      2. The matrix is calcified to form spicules of spongy bone
      3. Some osteoblasts are trapped in the matrix to become *osteocytes* (bone cells) as successive layers of lamellae are deposited by other osteoblasts
      4. The spicules thicken, fuse, and form plates of compact bone. With internal reorganization, *haversian systems* develop
      5. Between the plates of bone, the intervening bone stays spongy and the mesenchyme forms *bone marrow*
      6. Both osteoblasts and osteoclasts continue to remodel the bone
      7. Ossification begins at the end of the embryonic period
   B. ENDOCHONDRAL OR INTRACARTILAGINOUS OSSIFICATION takes place in a preexisting cartilage model
      1. In a long bone (e.g., the femur), the *primary ossification center* is seen in the diaphysis or shaft (between ends of the bone). Here the cartilage increases in size (hypertrophies), the matrix is calcified, and the cells die
         a. Concurrently, a thin layer of bone is laid down under the perichondrium around the diaphysis and will become the *periosteum*
         b. Vascular connective tissue invades from the periosteum and breaks up the cartilage. Some of the mesenchymal cells form hematopoietic cells of the bone marrow and others form osteoblasts which deposit bone matrix on the spicules of calcified cartilage. The spicules are remodeled by osteoblasts and osteoclasts, and the process continues toward both ends of the bones (*epiphysis*)
         c. The bone grows in length at the *diaphyseoepiphyseal junction* where the cartilage cells proliferate by mitosis
         d. The cartilage cells facing the diaphysis hypertrophy, the matrix is broken up into spicules by the vascular tissue from the marrow, and bone is deposited on the spicules
            i. Resorption of bone keeps the bone mass relatively constant in length and enlarges the marrow cavity
            ii. At birth, the diaphyses are largely ossified, but most epiphyses are still cartilaginous

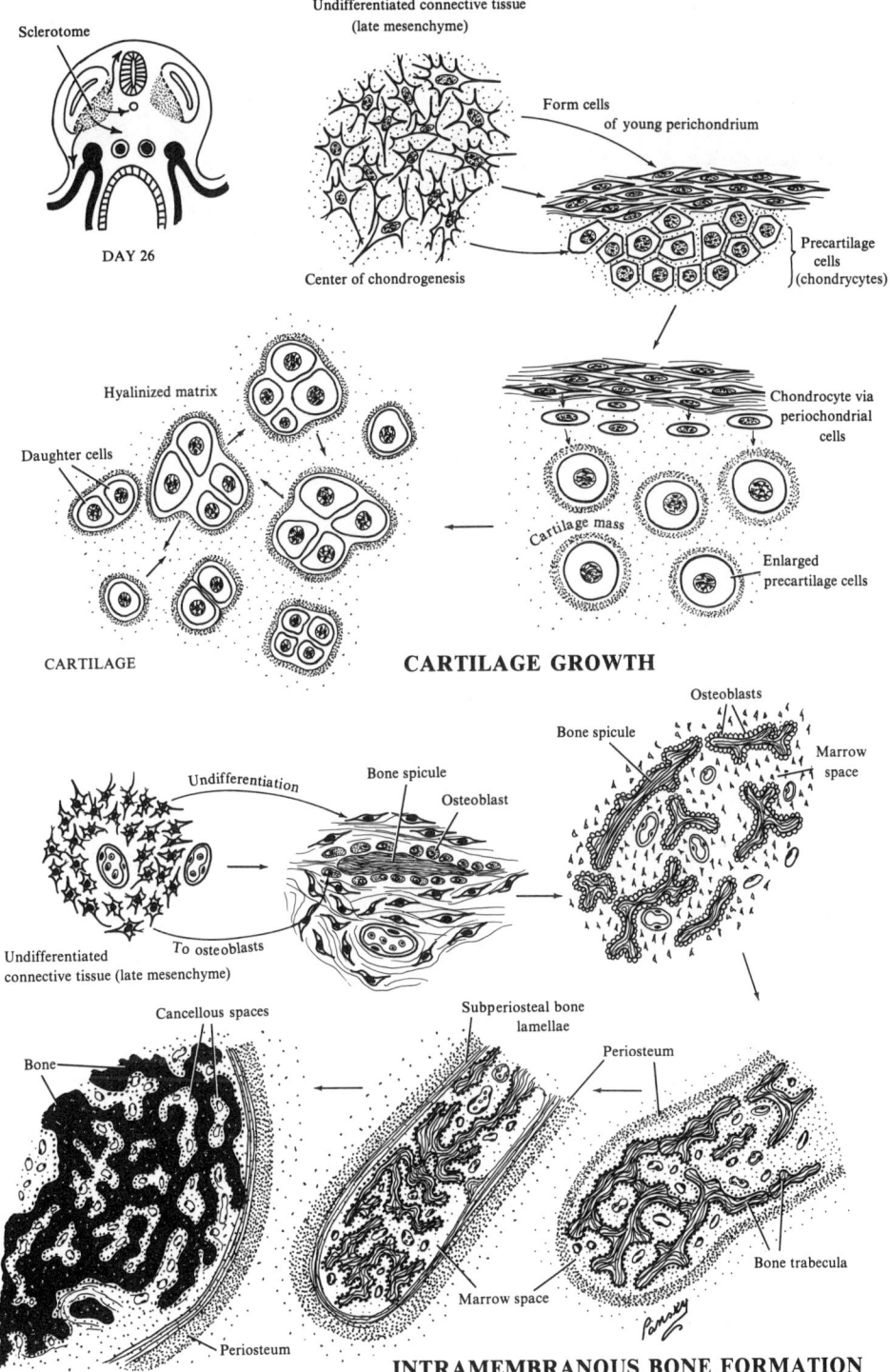

CARTILAGE GROWTH

INTRAMEMBRANOUS BONE FORMATION

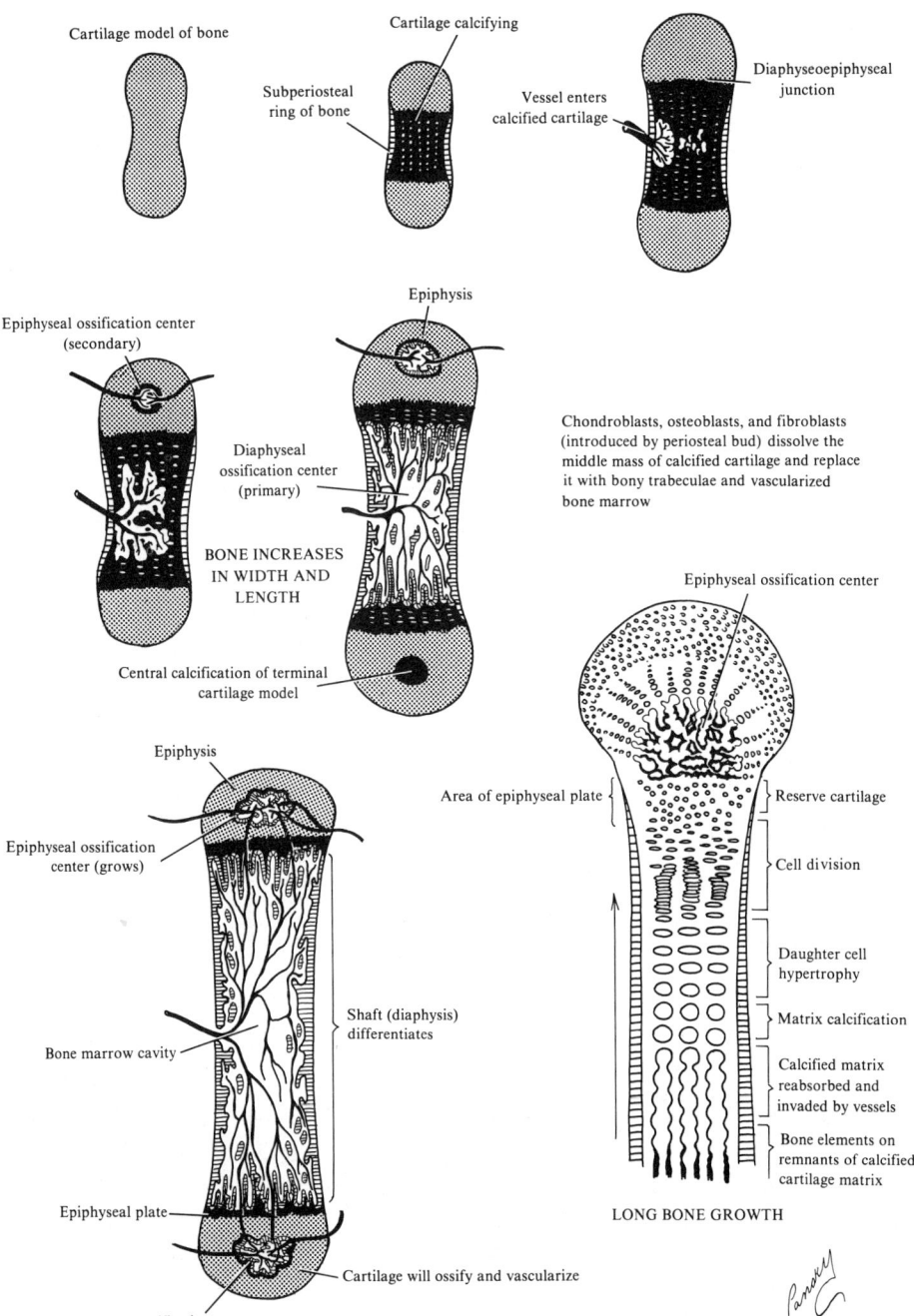

FIGURE 29. **Endochondral bone growth.**

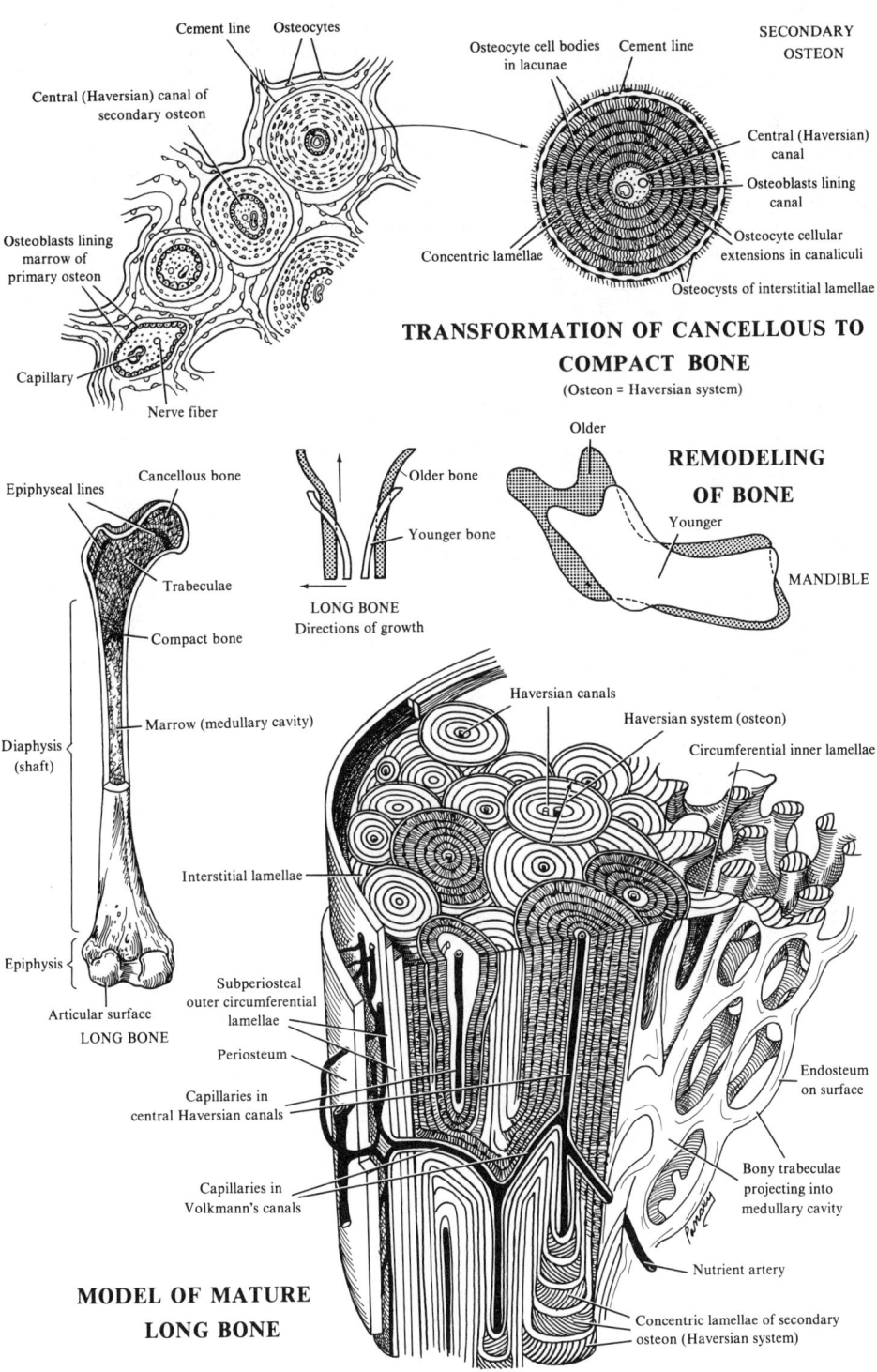

FIGURE 30. **Transformation of cancellous to compact bone.**

65. BONE HISTOGENESIS: SECONDARY OSSIFICATION CENTERS AND JOINT DEVELOPMENT

I. Secondary (epiphyseal) ossification centers
A. THE SECONDARY CENTERS APPEAR, during the first years of postnatal life, in the epiphyses of the bone (at the ends of the bone)
  1. Epiphyseal cartilage cells hypertrophy, and vascular connective tissue invades the epiphysis
    a. The articular cartilage and the epiphyseal plate remain cartilaginous as ossification spreads in all directions
    b. After the epiphyses become ossified, growth takes place on the diaphyseal side of the epiphyseal plate only
    c. When growth is finally complete, the epiphyseal plate is replaced by spongy bone, the epiphyses and diaphyses fuse, and no further growth occurs
B. THE EPIPHYSES AND DIAPHYSES usually fuse by the 20th year
C. DIAMETER GROWTH is due to bone deposition at the periosteum and resorption on the medullary surface. Balanced deposition and resorption regulate the size of the bone and its marrow cavity
D. IRREGULAR BONES develop by a similar process with ossification beginning in the center of the bones and spreading outward

II. Joint development
A. SYNOVIAL JOINTS (e.g., shoulder, elbow, knee): the mesenchyme between the developing bones differentiates
  1. The mesenchyme gives rise to capsular and other joint ligaments peripherally
  2. The mesenchyme disappears in the middle, between the bones, to form the joint cavity proper
  3. The mesenchyme lining the capsule and the articular surfaces forms the *synovial membrane*
B. CARTILAGINOUS JOINTS (e.g., neurocentral and symphysis pubis): the mesenchyme between the developing bones forms hyaline cartilage or fibrocartilage. Hyaline cartilage covers the bones of the joints at their articular surfaces
C. FIBROUS JOINTS (e.g., the skull sutures): the mesenchyme between the developing bones forms a dense fibrous tissue

III. Skeletal malformations in general
A. HYPERPITUITARISM causes the infant to grow at a very abnormally fast rate. This can result in *gigantism* (increased height and body proportion) or *acromegaly* (great enlargement of the hands, face, and feet). The condition can be caused by a pituitary gland tumor. Congenital infantile hyperpituitarism is rare
B. HYPOTHYROIDISM AND CRETINISM: due to a deficiency of fetal thyroid hormone leading to mental deficiency, skeletal abnormalities, as well as auditory and neurologic deficiencies
  1. Cretinism is generally very rare these days
  2. Agenesis of the thyroid gland is one possible cause of cretinism
C. ACHONDROPLASIA (hypoplastic chondrodystrophy): a common cause of dwarfism. The extremities are short due to a disturbance of endochondral ossification at the epiphyseal plates of long bones during fetal existence
  1. The trunk is usually of normal length
  2. The head may be slightly smaller
  3. It is transmitted as a mendelian dominant character

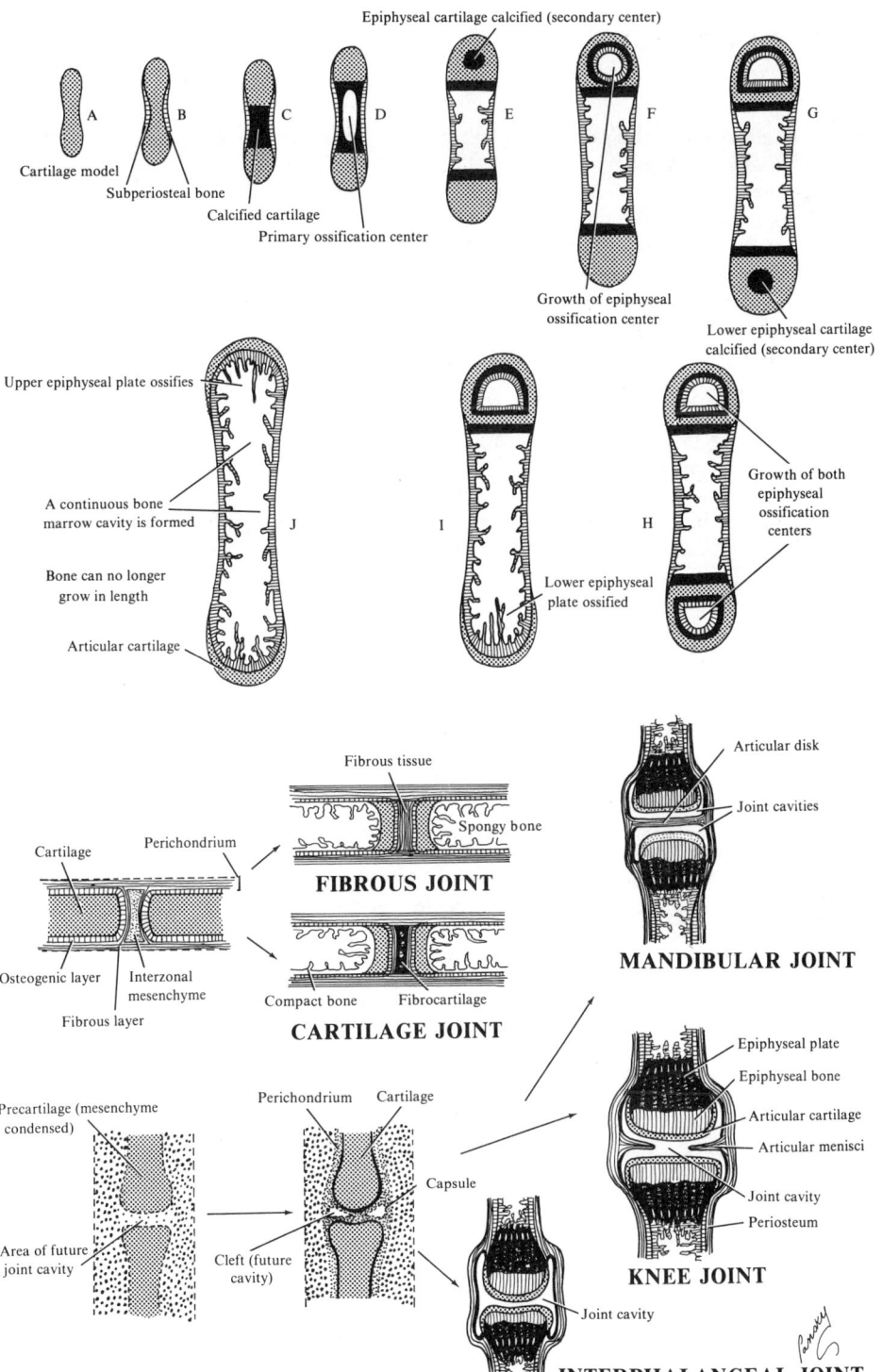

66. DEVELOPMENT OF THE AXIAL SKELETON

I. **Development of the vertebral column:** during week 4, cells of the sclerotome migrate medially to surround the spinal cord and notochord to form a long mesenchymal column. The development takes place in 3 essential stages
   A. THE PRECARTILAGE STAGE: the cells of the sclerotome migrate in 3 directions
      1. Ventromedially to surround the notochord. The caudal part of each sclerotome condenses into densely packed cells; the cranial part forms loosely packed cells
         a. Some of the densely packed cells migrate opposite the center of the myotome and give origin to the *intervertebral disk*
         b. The rest fuse with the loosely arranged cells of the underlying sclerotome to form the mesenchymal *centrum* (central mass of the body) of the vertebra
            i. Thus, each vertebra develops from 2 adjacent sclerotomes and is referred to as an intersegmental structure
         c. The spinal nerves lie near the intervertebral disks, thus leave the column via the intervertebral foramina
         d. The intersegmental (intercostal) arteries, which at first lie between sclerotomes, pass midway and come to lie on each side of the vertebral bodies
         e. The notochord regresses entirely in the region of the vertebral bodies, persists and enlarges in the region of the intervertebral disks, and undergoes mucoid degeneration to form the gelatinous disk center, the *nucleus pulposus*
            i. The nucleus is later surrounded by circular fibers, the *anulus fibrosus*. Thus, the intervertebral disk consists of the nucleus and anulus
      2. Dorsally the mesenchyme covers the neural tube and forms the *vertebral arch*
      3. Ventrolaterally the mesenchyme forms costal processes (in the thorax form *ribs*)
   B. CHONDRIFICATION STAGE: during week 6, centers of chondrification are seen in the mesenchymal vertebrae
      1. Two centers in each centrum fuse at the end of the embryonic period to form the *cartilaginous centrum*. In addition, centers are seen in the vertebral arches, which fuse with each other as well as with the centrum
         a. Extensions of chondrification centers in the vertebral arch give form to the *spinous and transverse processes*
   C. OSSIFICATION STAGE begins in the embryonic period and ends at about 25 years of age
      1. Prenatal period: 3 primary centers of ossification are seen at the end of the embryonic period: 1 in the centrum and 1 in each half of the vertebral arch. Thus, at birth, each vertebra has 3 bony parts connected by cartilage
      2. Postnatal period: the vertebral arch halves fuse by the first year
         a. The vertebral arch articulates with the centrum at the neurocentral joints, the latter disappearing when the arches finally fuse with the centrum at 3 to 6 years
         b. After puberty, 5 secondary centers are seen: 1 for the tip of each transverse process, 1 for the tip of the spinous process, and 2 annular epiphyses on the upper and lower surfaces of the body of the vertebra
         c. The vertebral body consists of the centrum, parts of the neural arch, and facets for the rib heads. Thus, the centrum and body are not synonymous
         d. All secondary centers unite with the rest of the vertebra by age 25 years

II. **Rib development:** from the mesenchymal costal processes of the thoracic vertebrae. They became cartilaginous in embryonic life and ossify later. The vertebral and costal junctions are synovial joints

III. **Development of the sternum:** a pair of mesenchymal *sternal bands* develops ventrolaterally in the body wall, independent of the ribs. Chondrification occurs in the bands and the costal cartilages attach to them. The bands fuse craniocaudally, in the median plane, to form a model for the manubrium, body segments, and xiphoid. Ossification centers appear before birth except for the xiphoid (occurs in childhood)

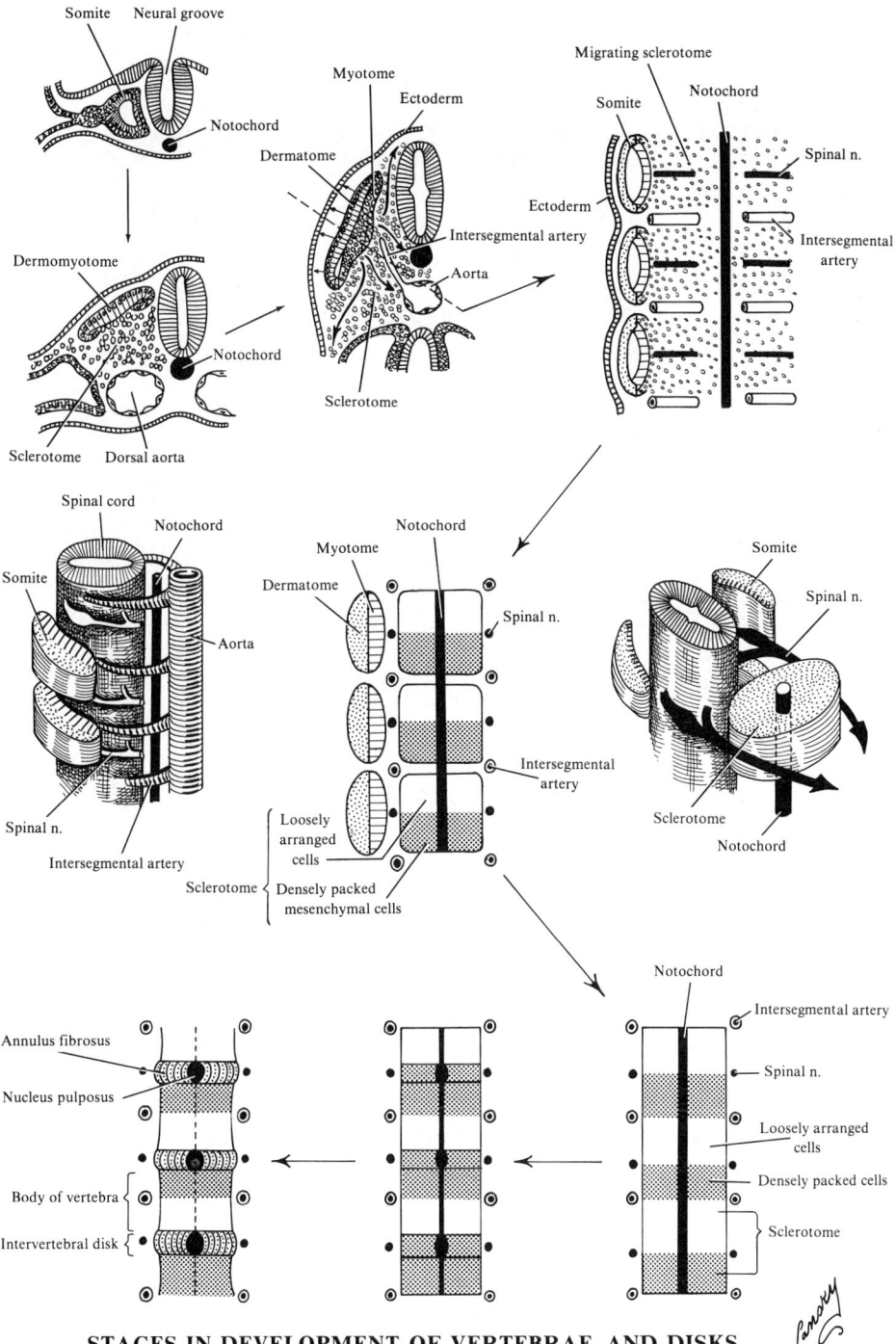

STAGES IN DEVELOPMENT OF VERTEBRAE AND DISKS

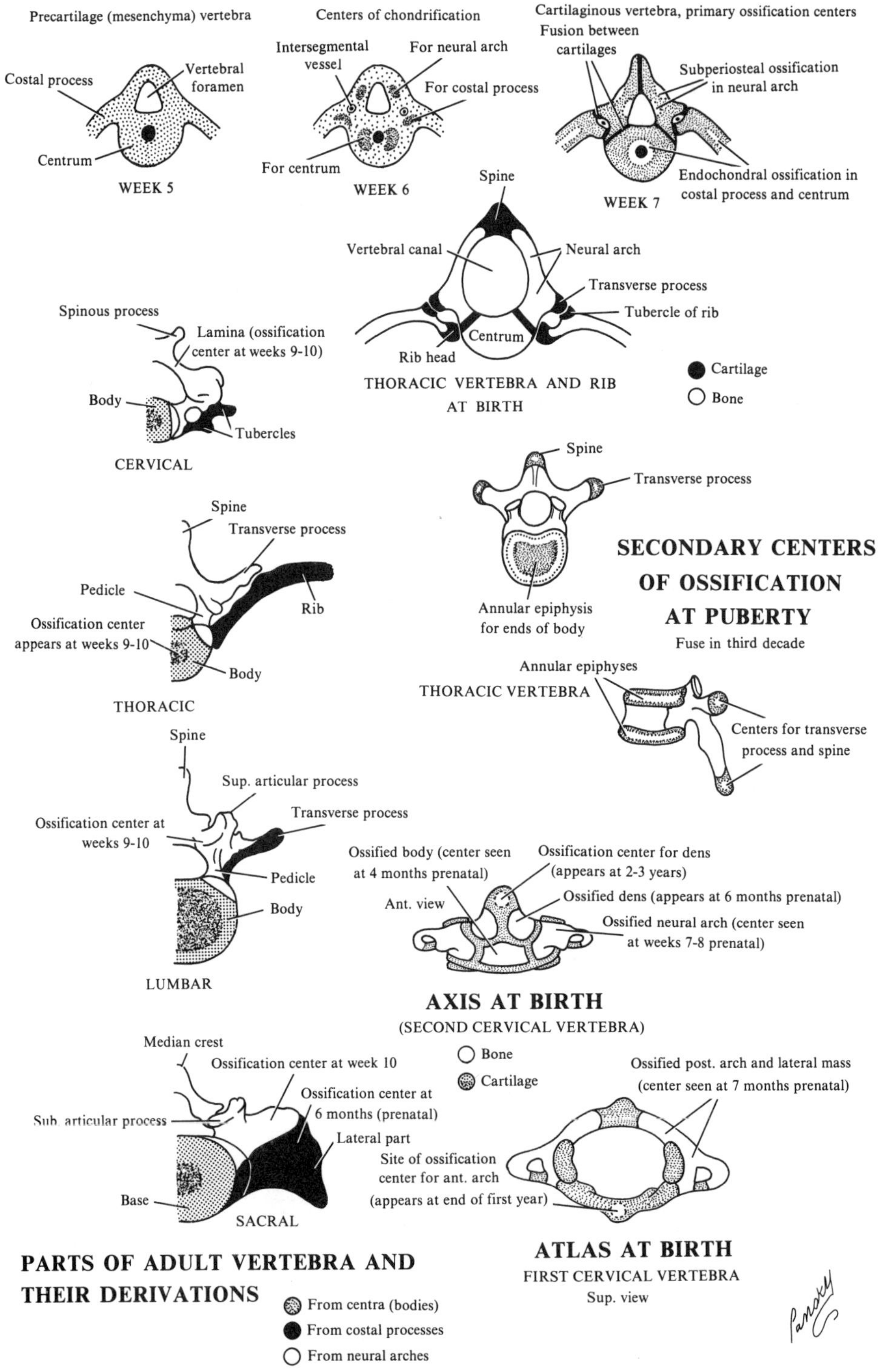

FIGURE 31. **Vertebral development.**

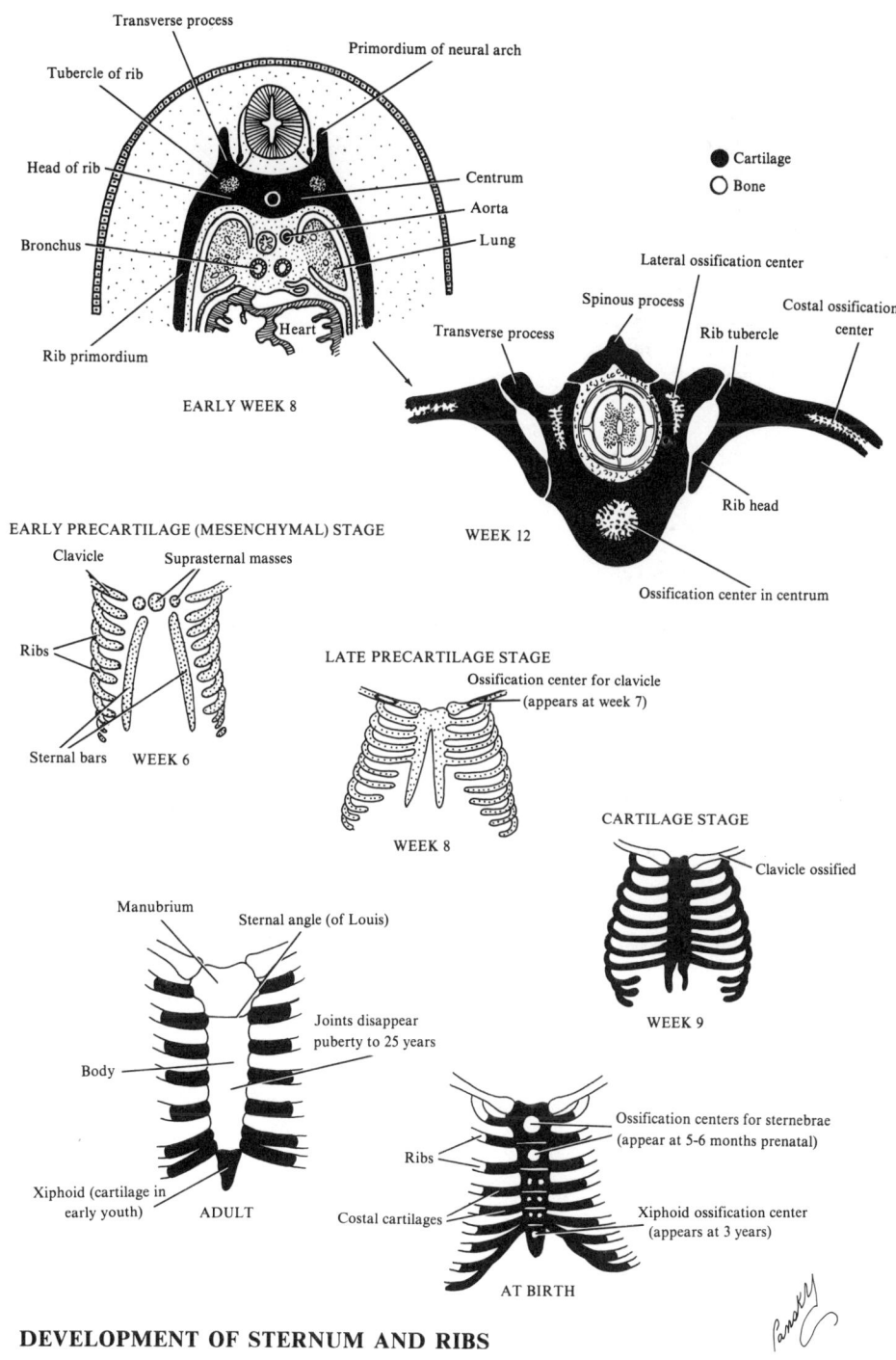

FIGURE 32. Development of the sternum and ribs.

67. APPENDICULAR SKELETON AND SKULL DEVELOPMENT

I. **The appendicular skeleton** consists of the shoulder (pectoral) and pelvic girdles as well as the bones of the limbs
   A. DURING WEEK 6, the mesenchymal primordia of bones in the limb buds undergo chondrification to form hyaline cartilage models of the future bones
      1. The clavicle initially develops by intramembranous ossification but does form growth cartilages at both ends
      2. The pectoral girdle and upper limb cartilages appear before those of the pelvic girdle and lower limbs. The cartilages appear in a proximodistal sequence
   B. OSSIFICATION in long bones begins by the end of the embryonic period; the primary centers are seen by week 12 in almost all bones of the extremities; and secondary ossification centers are seen after birth

II. **The skull** develops from mesenchyme around the developing brain and consists of a *neurocranium*, which forms a protective case around the brain, and a *viscerocranium*, which forms the skeleton of the face
   A. THE NEUROCRANIUM consists of 2 parts: the *cartilaginous base* of the skull or *chondrocranium* and a *membranous portion* which forms the flat bones around the brain
      1. Cartilaginous or chondrocranial part makes up the cartilaginous base of the skull formed by fusion of cartilages. Endochondral ossification finally forms the bones of this portion of the skull: base of the occipital bones around the foramen magnum (with contribution from the occipital somites), body of the sphenoid bone around the pituitary gland, body of the ethmoid, greater and lesser wings of the sphenoid, and petrous and mastoid portions of the temporal bone
      2. Membranous part: intramembranous ossification in mesenchyme over brain forms the cranial vault (frontal, parietal, squamous temporal, and part of occipital bones)
         a. At birth, the flat bones of the skull are separated from each other by narrow seams of dense connective tissue, the sutures or fibrous joints
            i. At points where more than 2 bones meet, the sutures are wide and called *fontanelles*, of which there are 6. The most prominent of these is the anterior fontanelle, at the point where the 2 parietals and 2 frontals meet
            ii. The softness of the bones and the loose connections at the sutures allow for shaping and molding as well as enlarging of the skull in development
   B. THE VISCEROCRANIUM also consists of both cartilaginous and intramembranous anlage
      1. Cartilaginous part consists of the cartilaginous skeleton of the first 2 pairs of branchial arches and involves endochondral ossification
         a. Dorsal end of arch I cartilage (Meckel's) forms malleus and incus of middle ear
         b. The dorsal end of arch II cartilage (Reichert's) forms the stapes of middle ear and styloid process of the temporal bone; its ventral end forms the lesser cornu and upper part of the body of the hyoid bone
         c. Cartilages of arches III, IV, and VI are found only in ventral parts of the arches
            i. Arch III gives rise to the greater cornu and lower part of the body of the hyoid
            ii. Arches IV and VI fuse to form the laryngeal cartilages except for the epiglottis
      2. Membranous part: intramembranous ossification is seen in the maxillary process of arch I to form the maxilla, the zygomatic, and squamous temporal bones
         a. Ossification around arch I cartilage gives rise to the mandible. (There is endochondral ossification at the chin center and at the mandibular condyles with disappearance of Meckel's cartilage below the sphenomandibular ligament. Thus, Meckel's cartilage does not actually form the adult mandible)

III. **Size of the skull** is large in proportion to the rest of the skeleton, and the face is small when compared to the cranium in the newborn. Postnatally, the cranial vault grows rapidly to 7 years as a result of brain growth. The face and jaws also develop rapidly with the development of the teeth and paranasal sinuses

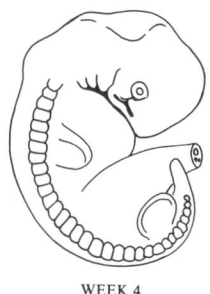
WEEK 4

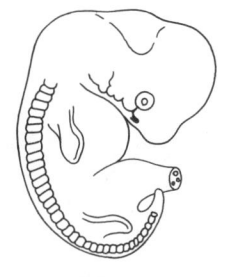
WEEK 5

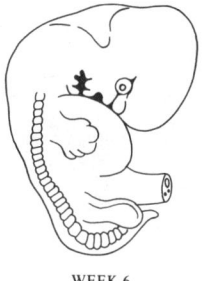
WEEK 6

PRIMARY OSSIFICATION CENTERS (Weeks)

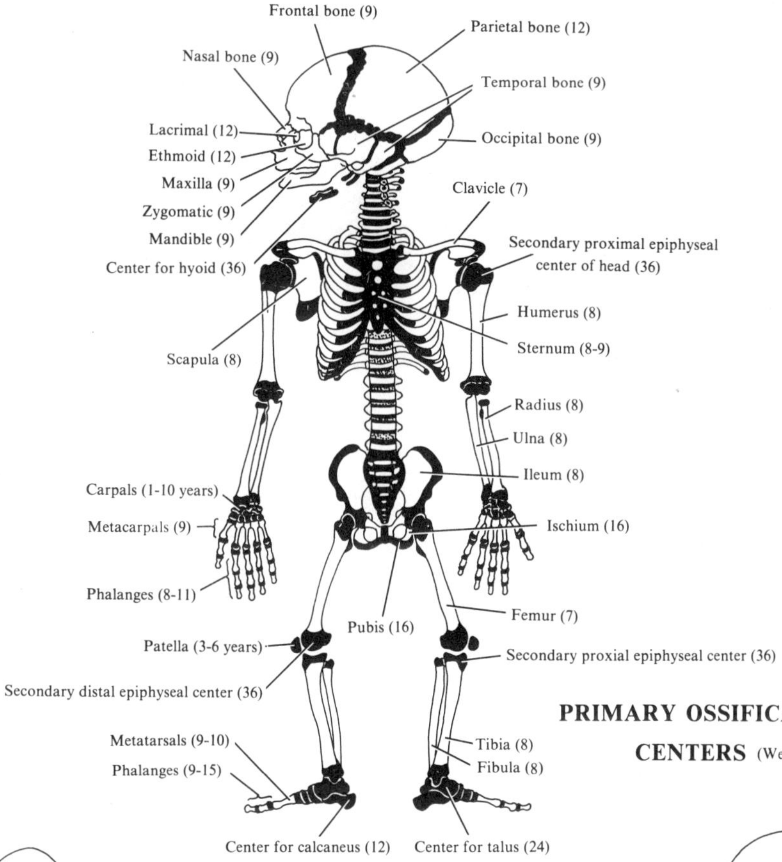

FULL TERM NEWBORN

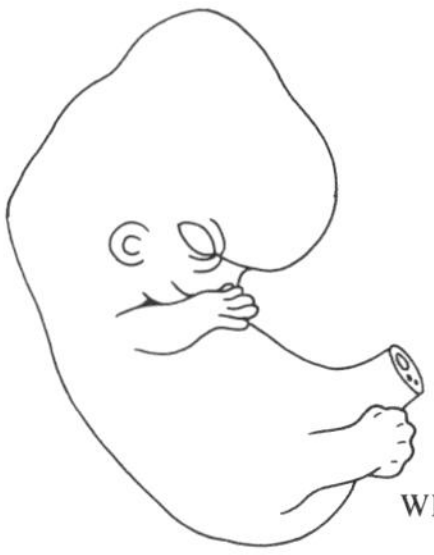

- Frontal bone (9)
- Parietal bone (12)
- Nasal bone (9)
- Temporal bone (9)
- Lacrimal (12)
- Occipital bone (9)
- Ethmoid (12)
- Maxilla (9)
- Zygomatic (9)
- Mandible (9)
- Clavicle (7)
- Center for hyoid (36)
- Secondary proximal epiphyseal center of head (36)
- Humerus (8)
- Scapula (8)
- Sternum (8-9)
- Radius (8)
- Ulna (8)
- Ileum (8)
- Carpals (1-10 years)
- Ischium (16)
- Metacarpals (9)
- Phalanges (8-11)
- Pubis (16)
- Femur (7)
- Patella (3-6 years)
- Secondary proxial epiphyseal center (36)
- Secondary distal epiphyseal center (36)
- Metatarsals (9-10)
- Tibia (8)
- Phalanges (9-15)
- Fibula (8)
- Center for calcaneus (12)
- Center for talus (24)

HAND DEVELOPMENT (Weeks)
5, 5.5, 6, 7-8

WEEK 7

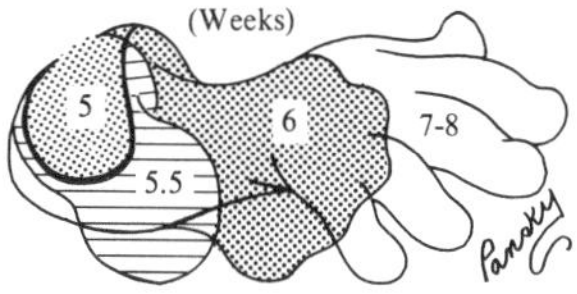

WEEK 8

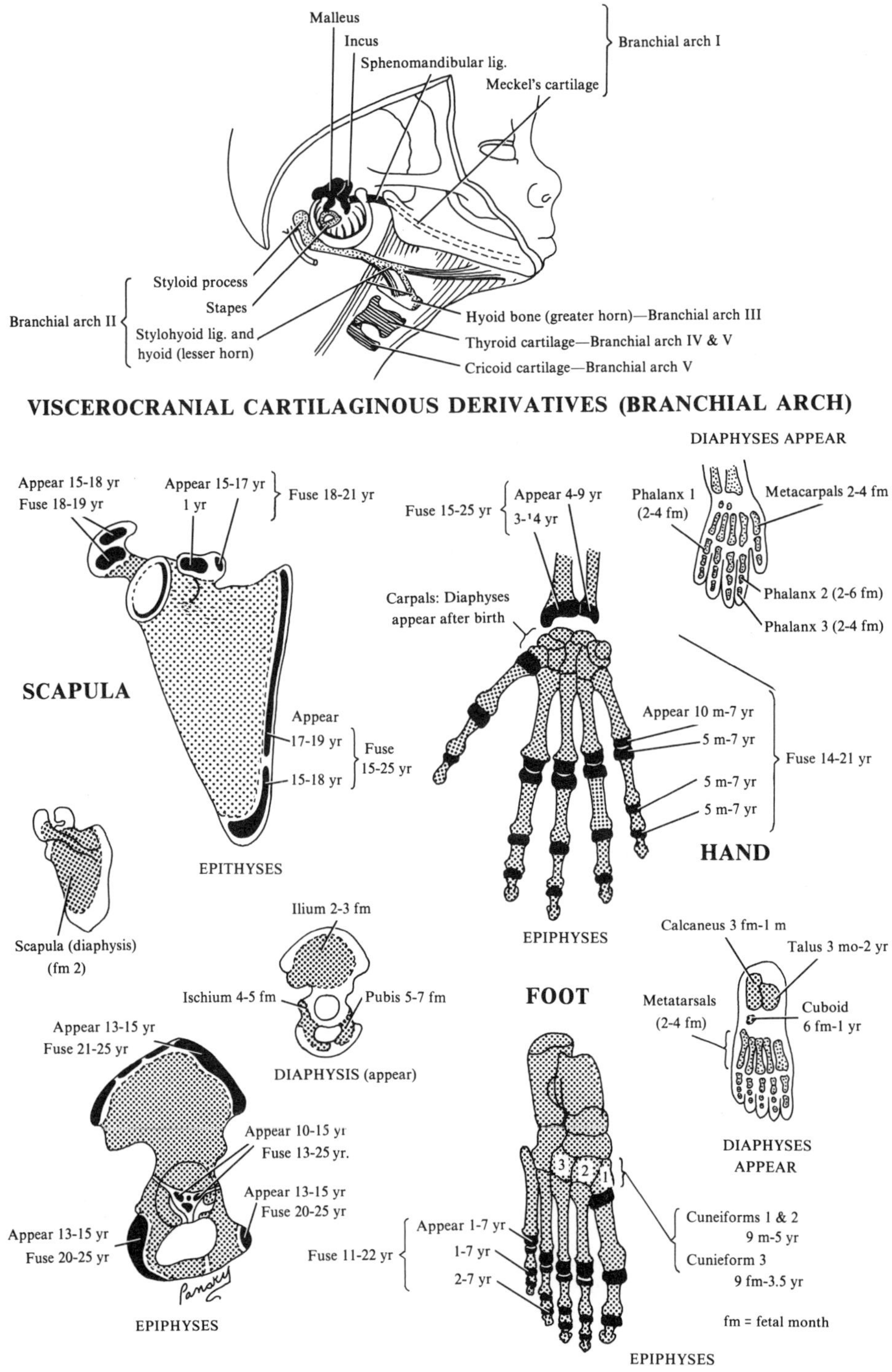

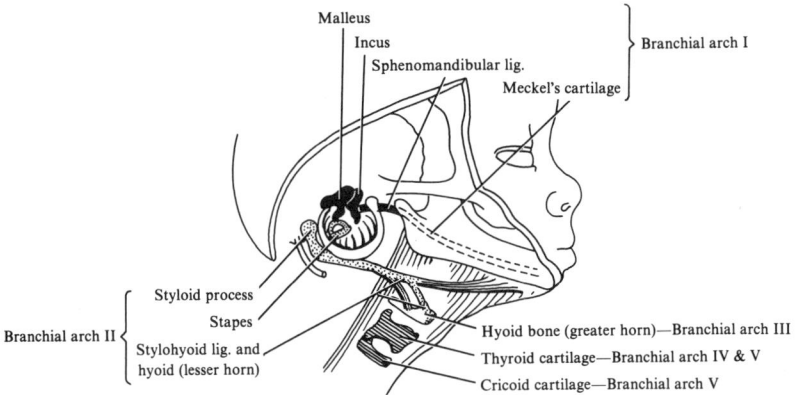

FIGURE 33. Viscerocranial cartilage derivatives and epiphyses and diaphyses of the extremities.

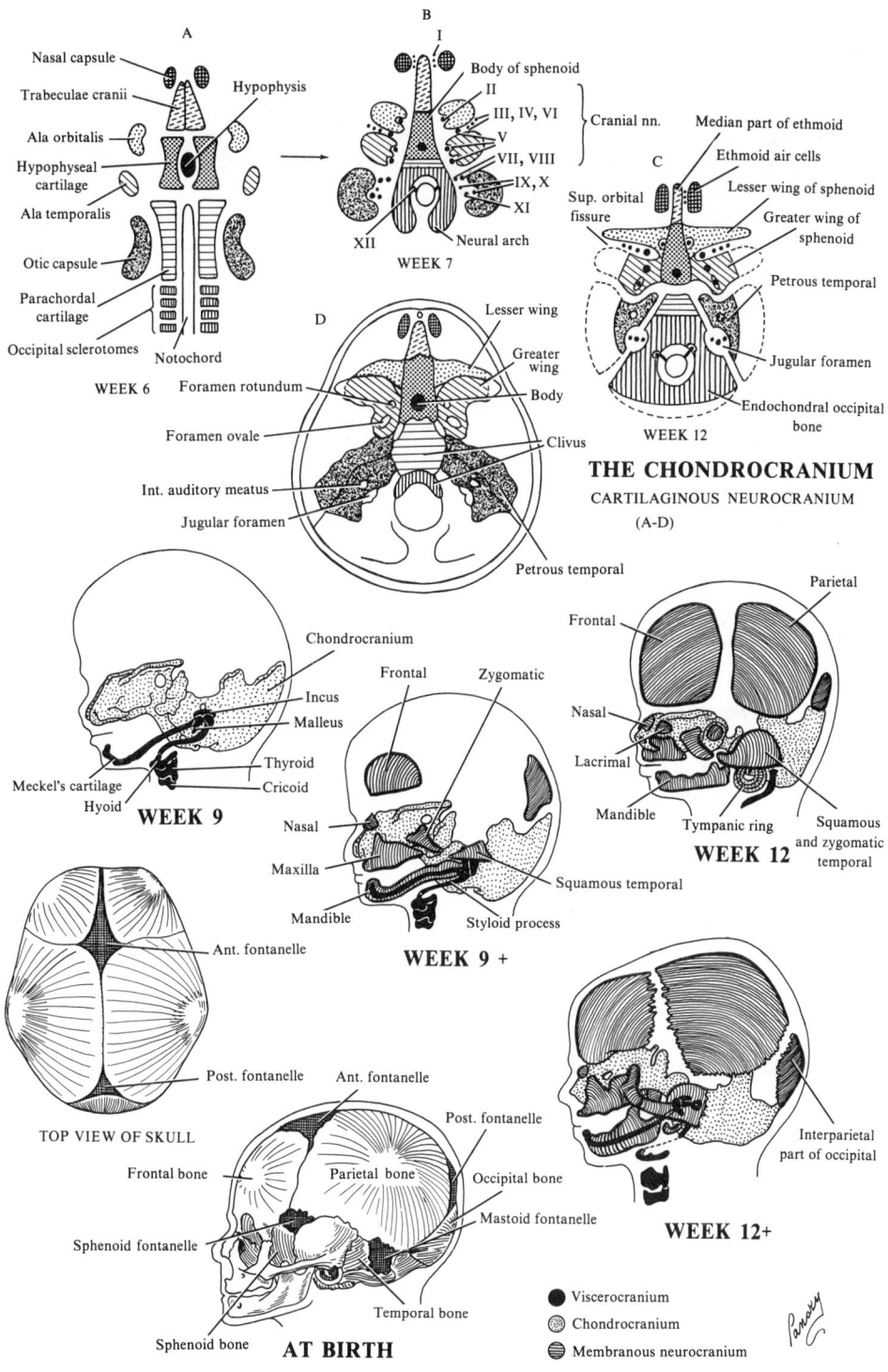

FIGURE 34. Skull development.

68. CONGENITAL MALFORMATIONS OF THE SKELETAL SYSTEM

I. Vertebral column
A. SPINA BIFIDA OCCULTA results from failure of fusion of the halves of the vertebral arch, most often in the lumber and sacral regions
   1. It is a common defect but usually of no significance
   2. The skin over the bifid spine is usually intact, and there may be no visible signs of the defect except for a "dimple" or tuft of hair
   3. Severe types do exist and are described under Nervous system development
B. KLIPPEL-FEIL SYNDROME (brevicollis): very rare; extreme shortening of the neck due to a reduced number of cervical vertebrae. The rest of the cervical vertebrae are usually abnormal in shape and may be fused. Associated with other abnormalities
C. ASYMMETRICALLY FUSED VERTEBRAE or parts of vertebrae missing; an increase or decrease in vertebral number is not uncommon due to the complicated process of formation and rearrangement of the segmental sclerotomes in development

II. Ribs: defects are mostly secondary to malformations of the vertebral column
A. IF PART OR ALL OF A VERTEBRA is missing, the corresponding ribs are generally gone
B. IN SEVERE CONGENITAL SCOLIOSIS, the ribs on the concave side of the chest are often fused or branched
C. ACCESSORY RIBS: usually the cervical rib (lumbar ribs are less common)
   1. Attached to the seventh cervical vertebra; may be unilateral or bilateral
   2. Pressure effects on the brachial plexus or subclavian vessels may produce symptoms
   3. From retention and development of costal processes of cervical or lumbar vertebrae
D. FUSED RIBS: this may occur posteriorly when 2 or more ribs arise from a single vertebra
   1. Often associated with a hemivertebra which may produce scoliosis

III. Sternum
A. CLEFT STERNUM
   1. Minor clefts or notches are common and are seen as isolated anomalies
   2. Major clefts are usually associated with severe malformations of the chest
   3. Large clefts are rare (e.g. heart); associated with herniation of thoracic viscera

IV. The clavicles
A. CLEIDODYSOSTOSIS: absence of all or part of the clavicle
   1. Usually bilateral and the shoulders are drawn forward to meet under the chin
   2. Often associated with skull defects (cleidocranial dysostosis)

V. Skull malformations range from major defects incompatible with life to those that are minor and relatively unimportant. The abnormalities are manifold, and either all or part of the skull may be involved. They are frequently associated with brain defects
A. CRANIOSCHISIS OR ACRANIA: the cranial vault is almost absent and a large spinal defect is often present. Also associated with *anencephaly*
   1. Due to a failure of the cranial end of the neural tube to close during week 4, thus the cranial vault does not form
B. CRANIOSYNOSTOSIS OR CRANIOSTENOSIS: due to premature closure of skull sutures
   1. More common in male than female; associated with other skeletal abnormalities
   2. Type of deformed skull depends on which sutures close prematurely
      a. If sagittal suture: a long, narrow, wedge-shaped skull (scaphocephaly)
      b. If the coronal suture: a high, towerlike skull (oxycephaly or acrocephaly)
      c. If coronal or lambdoid suture closes on one side: twisted and asymmetric skull (plagiocephaly)
C. MICROCEPHALY: cranium is normal size or slightly small, but there is no abnormal closure of the sutures. It is primarily an abnormality of the CNS in which the brain and skull both fail to grow

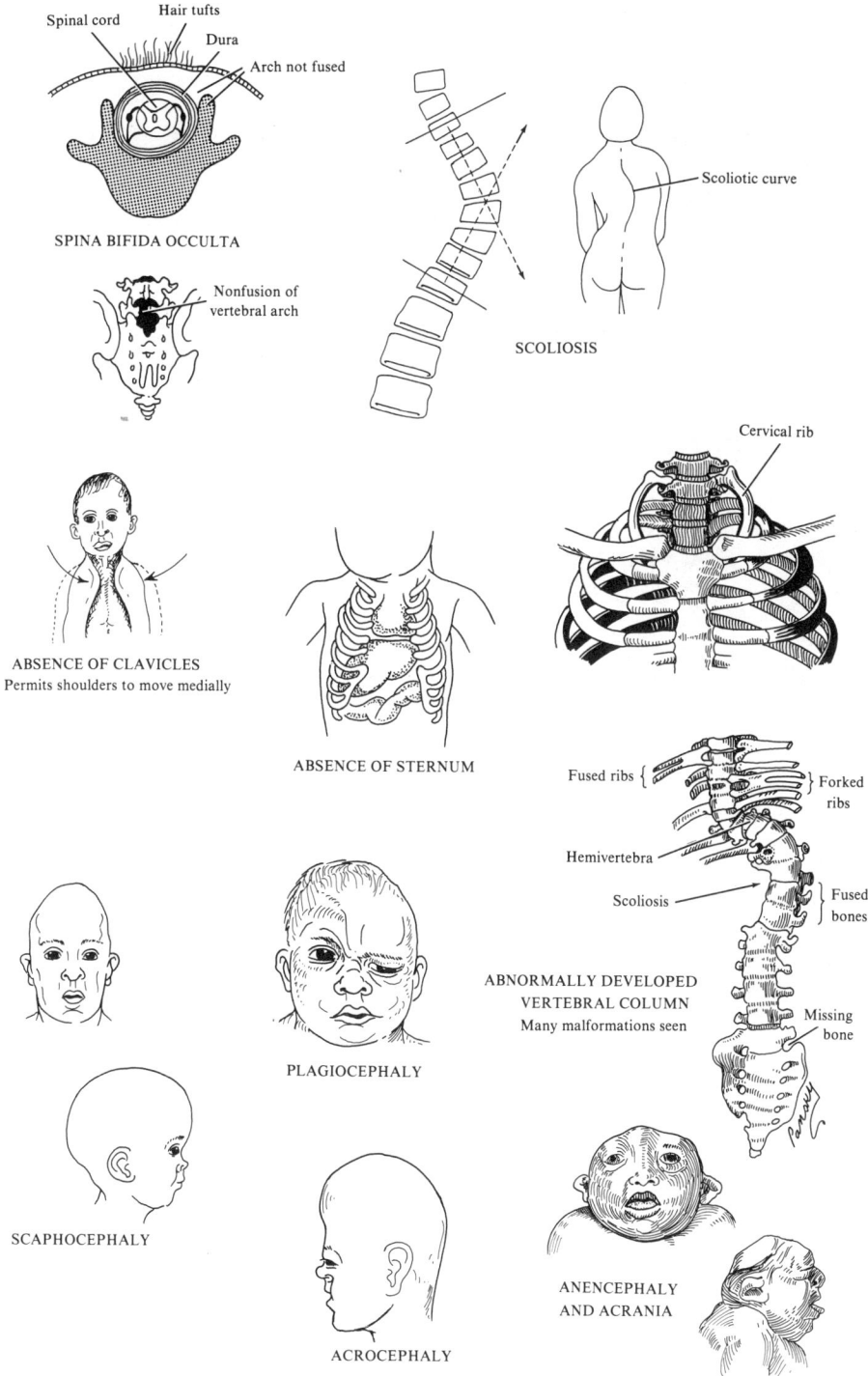

69. DEVELOPMENT OF THE LIMBS

I. Limb primordia
A. THE LIMB BUD PRIMORDIA appear at the end of week 4 as small elevations of the ventrolateral body wall, but most development occurs in week 6. Early developmental stages for both the upper and lower limbs are identical, except that the arm buds precede those of the leg buds by several days
B. THE BUDS are formed by a series of reciprocal inductions of mesoderm and ectoderm
   1. The lateral mesoderm of the somatopleure induces a transitory longitudinal thickening of the surface ectoderm, the wolffian crest, a fold which can be seen in front of the somite column. The middle portion of the column disappears rapidly, leaving only 2 nodes at the extremities of the crest which are at the level of the future osseous girdles
      a. The arm buds develop opposite the caudal cervical segments; the leg buds develop opposite the lumbar and upper sacral segments
   2. The ectodermal nodule or apical ectoderm ridge, located on the proximal side of the column, induces the mesenchyme to grow and develop the limbs, in successive waves. There is no apparent contribution from the myotome regions of the somite
      a. Each bud is thus initially a mass of mesenchyme, of somatic mesodermal origin, covered by ectoderm
   3. Innervation of the limbs is an early phenomenon. The upper limb is supplied from the last 6 cervical and first 2 thoracic metameres (brachial plexus); the lower limb from the last 4 lumbar and first 3 sacral metameres (lumbosacral plexus)

II. Limb development
1. The first primordium of the upper limb appears about the 24th day and that of the lower limb at about day 26. The essential basic constituents of the limbs are distinguishable at day 34.
2. The distal ends of the limb buds flatten into paddle-shaped hand or foot plates, and the respective digits form at the margins of these plates
3. The limb acquires its distal segment in week 7. Shortly after this, a groove divides the proximal segment, and the limb now consists of its 3 definitive segments. Development of the upper limb is more advanced than that of the lower
4. Chondroblasts appear in the precartilaginous matrix which fragments to form the various skeletal parts. Between them, the first joint structures make their appearance toward week 8.
5. As the bones form and limbs elongate, myoblasts aggregate and form the large muscle masses in each limb
   a. The muscle masses separate into dorsal (extensor) and ventral (flexor) components
6. Early in week 7, the limbs move ventrally, and the developing arms and legs rotate to different degrees and in opposite directions
   a. Initially, the flexor surface of the limbs is ventral and the extensor surface is dorsal, with the preaxial and postaxial borders being cranial and caudal, respectively
   b. With rotation, the upper limbs rotate laterally through 90° on their long axes, the elbows come to face posteriorly, and the extensor muscles come to lie on the outer or dorsal aspect of the arm
   c. With rotation, the lower limbs rotate medially through 90° on their long axes, the knees face forward or ventrolaterally, and the extensor muscles come to lie on the ventral aspect of the legs

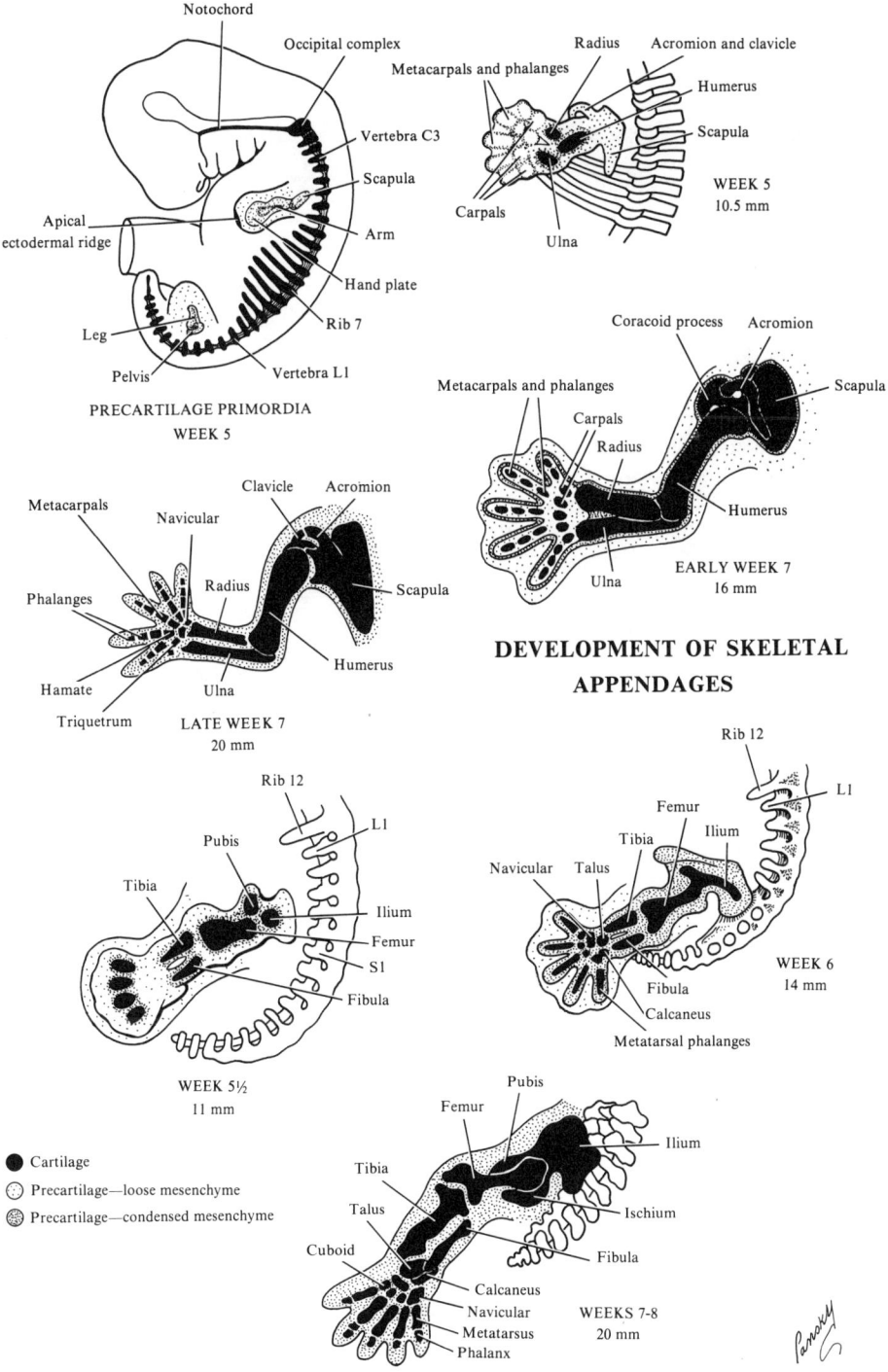

70. DERMATOME AND CUTANEOUS INNERVATION OF THE LIMBS

I. **Dermatome distribution and cutaneous innervation** are related to the growth and rotation of the limbs
   A. A DERMATOME is an area of skin (sensory) supplied by a single spinal nerve and its dorsal root ganglion
      1. Peripheral spinal nerves grow from the brachial plexi into the upper limb buds, and lumbosacral plexi grow into the lower limb buds, during week 5
      2. The spinal nerves (dermatome nerves) are distributed in segmental bands to supply the dorsal and ventral surfaces of the limb buds
         a. As the bud elongates, the cutaneous nerves migrate out along the limbs, and although the original dermatomal patterns change with growth, there is still an orderly sequence of nerve distribution
         b. The limb dermatomes can be traced down the lateral side of the arm and up its medial side, while those of the lower limb can be traced down the ventral side and up the dorsal side
         c. Autonomic nerves that supply the blood vessels grow out into the limb buds with the spinal nerves
   B. A CUTANEOUS NERVE supplies an area of the skin that is related to a peripheral nerve. Note that any cutaneous nerve may contain fibers from several individual spinal nerves. Therefore, cutaneous nerve areas and dermatome areas show much overlapping. A cutaneous nerve area is generally broader and wider than an area supplied by only a single spinal (dermatome) nerve

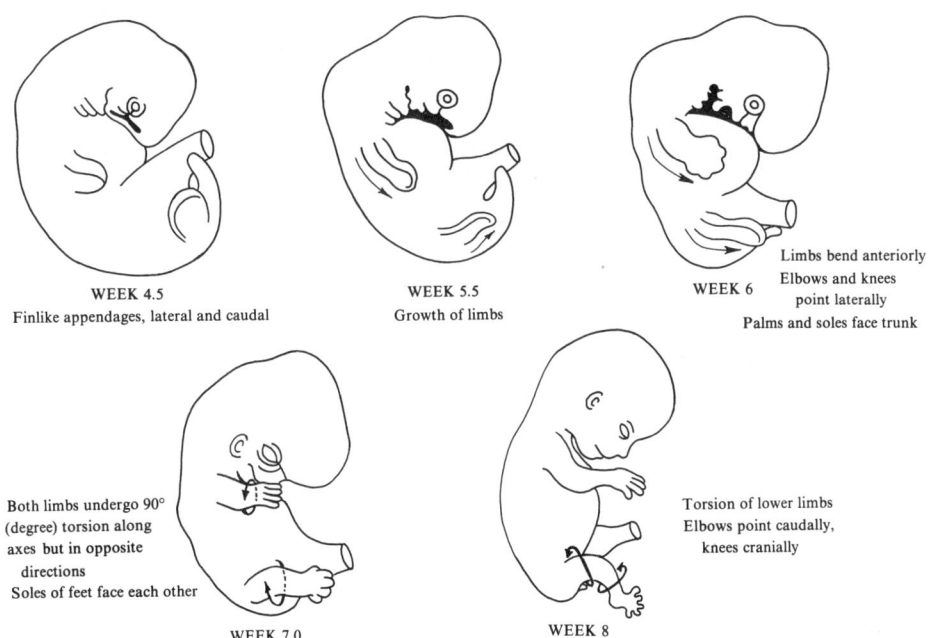

DEVELOPMENT OF LIMB DERMATOMAL PATTERNS

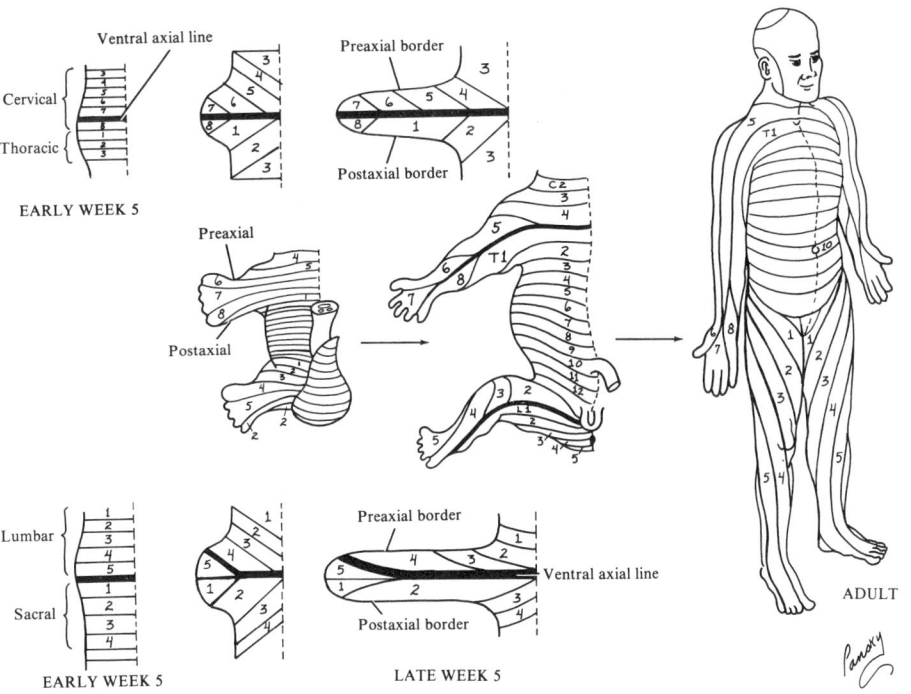

71. MALFORMATIONS OF THE APPENDICULAR SKELETON (THE LIMBS)

I. **Introduction:** abnormalities of the extremities vary greatly, and minor defects in development are relatively common. Major limb malformations are rare. Most are of a hereditary nature, such as chromosomal abnormalities. Environmental factors, such as drugs, have often caused an "epidemic" of limb deformities, e.g., thalidomide, which had been used as a sedative and antinauseant. Often, combinations of both environmental and hereditary factors have been involved. In addition, mechanical factors such as reduced amniotic fluid have been associated with limb deformities. In the extreme form, one or both extremities are absent (amelia) or are represented only by hands and feet which are attached to the trunk by small irregular bones (meromelia). Sometimes all segments of the extremities are present but are very short (micromelia)

II. **Types of deficiencies**
A. CLEFT HAND OR FOOT (lobster-claw deformities) are rare; 1 or more central digital rays or digits are absent, resulting in a hand or foot divided into 2 parts and opposing each other like a lobster claw. The remaining digits are usually fused (syndactyly)
B. CLUBHAND OR CONGENITAL ABSENCE OF RADIUS: the radius is partly or totally absent. The hand deviates to the radial side and the ulna is bowed
C. CLUBFOOT OR TALIPES EQUINOVARUS: is common; seen in males more frequently
  1. The sole of the foot is turned inwards, and the foot is adducted and plantar flexed at the midtarsal joint
D. BRACHYDACTYLY: not common; there is an abnormal shortness of the fingers or toes due to size reduction of the phalanges. This is an inherited dominant trait often associated with short stature
E. POLYDACTYLY OR SUPERNUMERARY DIGITS: extra fingers or toes are seen; is a common occurrence and due to an inherited dominant trait
  1. The extra digit may be incompletely formed, lack muscle fixation, and be useless
  2. In the hand, it is usually found on the radial or ulnar side (not central); in the foot, it is usually on the fibular side
F. SYNDACTYLY (FUSED OR WEBBED DIGITS): one of the most common malformations. It is the result of failure of differentiation between 2 or more digits. It is inherited as a simple dominant or recessive trait
  1. The webbing between the toes or fingers fails to break down during development
  2. In some cases, there is also fusion of bones
  3. It is most frequently seen between fingers 3 and 4 and toes 2 and 3
G. CONGENITAL HIP DISLOCATION: the capsule of the hip joint is abnormally relaxed at birth with underdevelopment of the acetabulum of the hip and head of the femur
  1. Dislocation usually takes place after birth
  2. It is a relatively common phenomenon and is more frequently seen in females
  3. It is associated with abnormal acetabulum development and joint laxity
H. SYMPODIA OR SIRENOMELIA: in this condition, the lower limbs are fused, a defect always associated with a disturbance in the formation of the pelvis
  1. The single lower extremity most commonly contains a single femur, 2 or 3 bones below the knee, and 5 or 6 digits attached to the foot
I. ARACHNODACTYLY (MARFAN'S SYNDROME): uncommon congenital and usually hereditary condition seen in tall slender individuals
  1. Characterized by thin fingers and toes with elongated phalanges, metacarpals, and metatarsals ("spider" fingers and toes)
  2. Asymmetry of skull seen in 80% of cases
  3. Marked atrophy and weakness of all muscles and laxity of ligaments
  4. 50% have eye deformities
  5. May see barrel-shaped thorax with kyphosis and marked pronation of feet

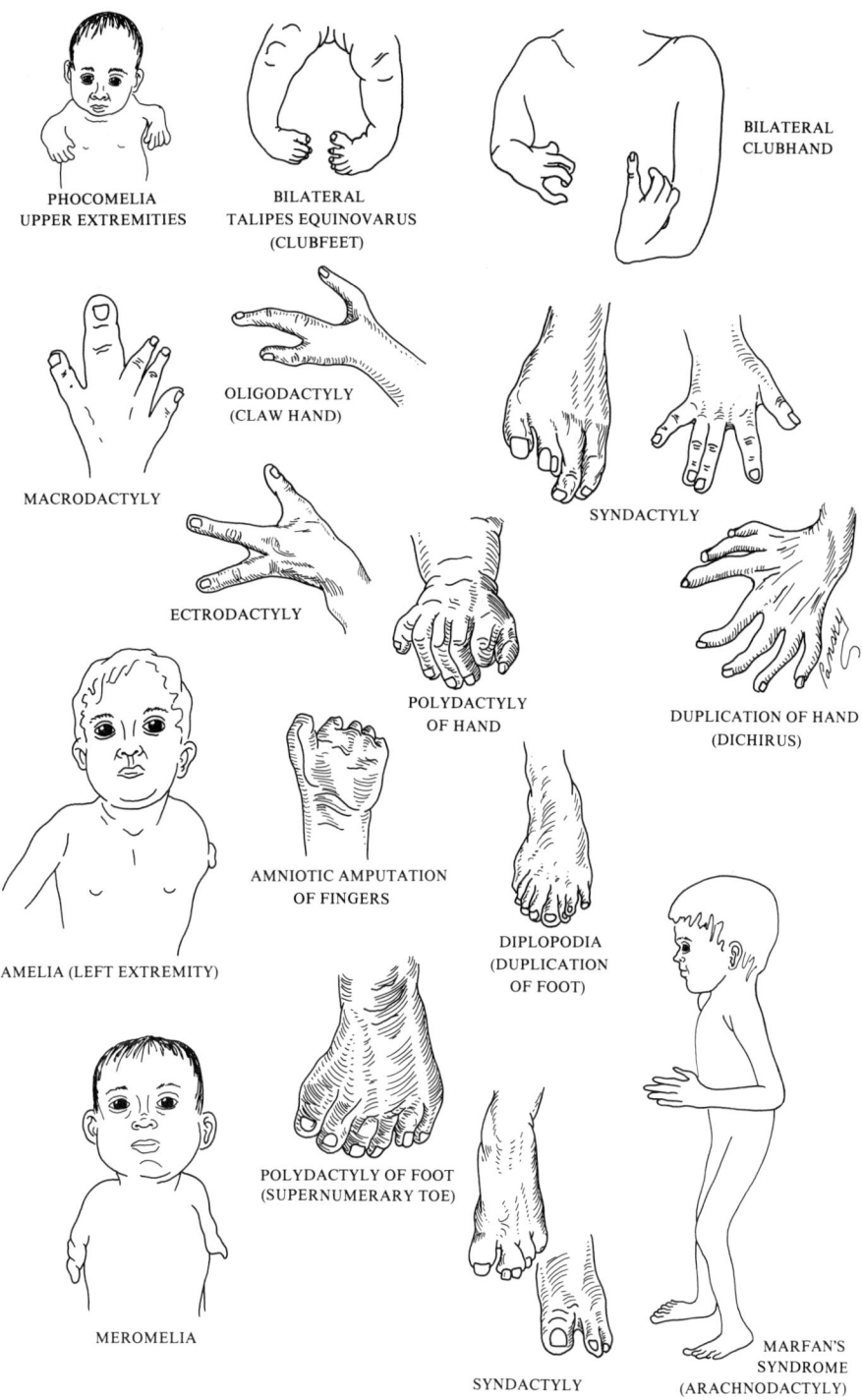

UNIT SIX

THE INTEGUMENTARY SYSTEM:

The Skin, Cutaneous Appendages, and Teeth

72. DEVELOPMENT OF THE INTEGUMENTARY SYSTEM: ECTODERMAL DERIVATIVES

I. **Integumentary system** develops from surface ectoderm and the underlying mesenchyme
A. THE SKIN has a twofold origin: a superficial layer, the *epidermis,* derived from surface ectoderm, and a deep, thick layer, the *dermis,* derived from mesenchyme
  1. Epidermis: initially the embryo's surface is covered by a single layer of ectodermal cells which, in month 2, divides to form a superficial protective layer of simple, flattened squamous epithelial cells, the *periderm or epitrichium*
     a. The cells of the periderm layer continually undergo keratinization and desquamation to be replaced by cells arising from the *basal layer*
        i. The basal layer of epidermis later becomes the *stratum germinativum* which produces new cells that are displaced into layers above
     b. The exfoliated cells form part of the *vernix caseosa,* a white, cheesy, protective substance that covers the fetal skin
        i. The vernix caseosa also includes sebaceous gland sebum, fetal hair, and desquamated amniotic cells
     c. By week 11, the basal layer (stratum germinativum) forms an intermediate skin layer, and by the end of month 4, all the epithelial layers of the adult epidermis of skin have acquired their definitive arrangement. Four successive layers are seen (bottom to top)
        i. Basal (stratum germinativum) layer: responsible for continuous development of new cells. It later forms genetically determined ridges and hollows which are filled by the underlying mesoderm. The patterns so formed are reflected on the surface of the skin (palms, fingers, and soles, including toes) in the form of fingerprints (dermatoglyphics)
        ii. Thick spinous (stratum spinosum) layer: large polyhedral cells, on top of the basal layer, connected by fine tonofibrils
        iii. Granular (stratum granulosum) layer: cells contain small keratohyaline granules, the first signs of keratinization
        iv. Horny (stratum corneum) layer: outermost layer which forms the scalelike hard surface of the epidermis and is loaded with keratin
     d. Replacement of the peridermal cells continues until about week 21 (the cells are lost into the amniotic fluid), thereafter the periderm normally disappears
     e. During the first 3 months of development, neural crest migrate and invade the epidermis, to form *melanoblasts* and then *melanocytes,* which synthesize *melanin pigment.* After birth, these cells cause skin pigmentation and are found in the epidermal-dermal junction
        i. In dark-skinned races, melanin granules are produced by fetal melanocytes; in white-skinned races, the fetal melanocytes contain very little to no melanin pigment
  2. The dermis is derived from mesenchyme of the somatic lateral mesodermal layer which underlies the surface ectoderm
     a. During months 3 and 4, the dermis forms many collagenous and elastic fibers; simultaneously, the superficial dermal layer or *corium* forms irregular papillary structures, the *dermal papillae,* which project into the epidermis
        i. Some papillae contain small capillary loops, and others have sensory nerve endings
     b. The deep dermal layer or *subcorium* is characterized by fatty tissue
  3. At birth, the skin is covered by the vernix caseosa, a whitish paste formed by sebaceous gland secretion, degenerated epidermal cells, and hairs. It protects the skin against the maceration action of the amniotic fluid
  4. If the superficial layers of the skin show excessive cornification, the skin has a scaly appearance, a condition spoken of as *icthyosis*

STAGES IN THE HISTOGENESIS OF SKIN

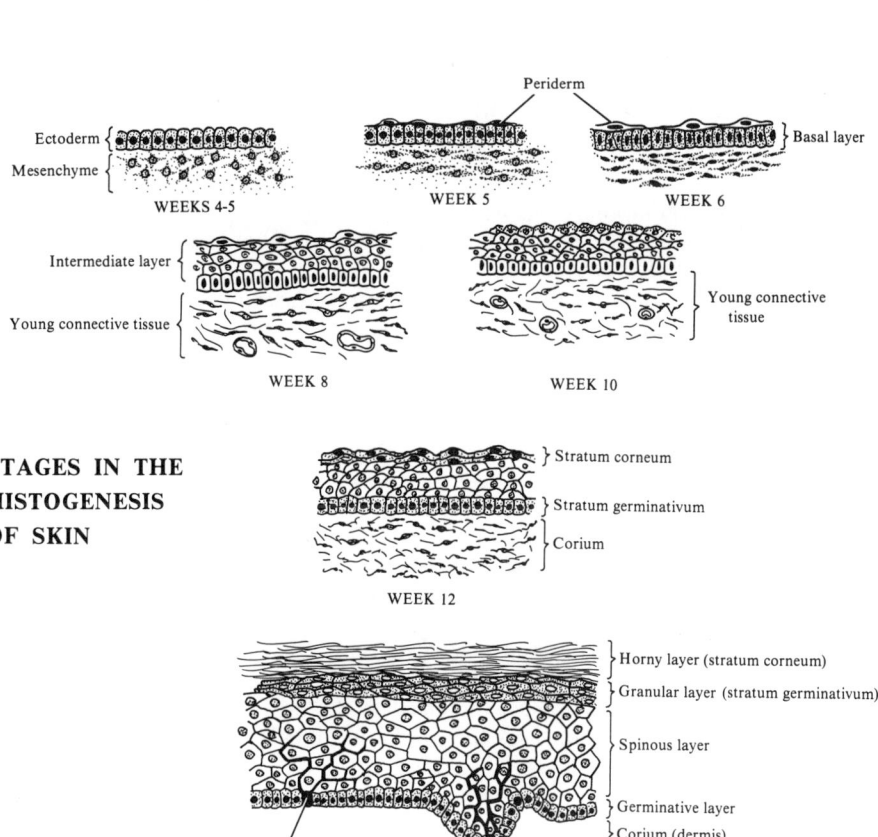

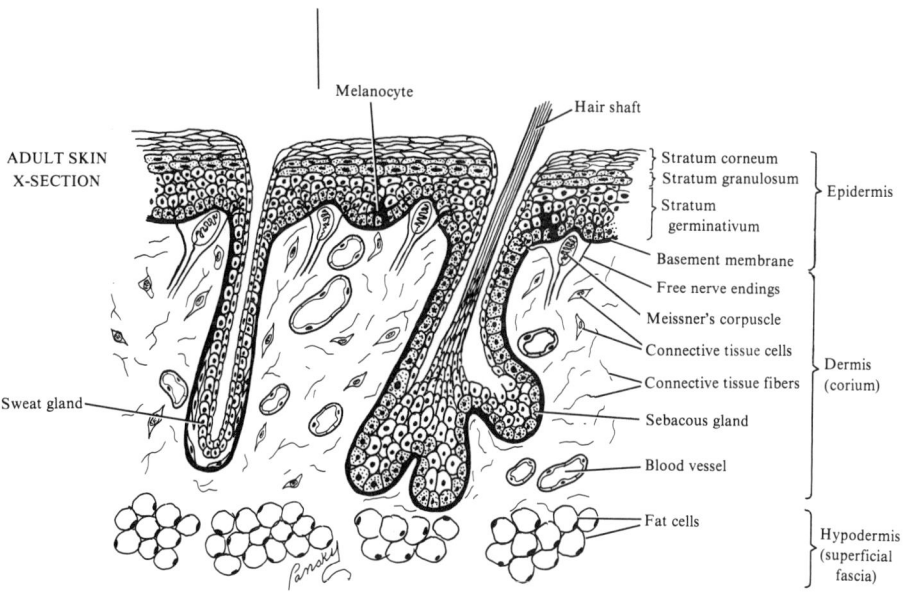

73. CONGENITAL MALFORMATIONS OF THE INTEGUMENTARY SYSTEM

I. Disorders of keratinization
A. ICTHYOSIS SIMPLEX: dry scaly skin with a decrease in sebaceous and sweat gland secretion. Develops during early infancy (not seen at birth) and is characterized by hyperkeratosis and thickening of the stratum corneum. Is transmitted by a single autosomal gene
B. ICTHYOSIS CONGENITA: the fetus is born with a smooth jellylike envelope over its skin. It is due to failure of the periderm to disappear during late fetal life. This layer is shed during early postnatal life, and the epidermis underneath is normal
C. CONGENITAL ECTODERMAL DYSPLASIA is a rare hereditary disorder displaying partial failure of the epidermis and appendages to develop

II. Neuroectodermal abnormalities
A. PIGMENTED NEVI: pigmented nevi or moles are common abnormal skin growths. Some originate from melanoblasts derived from neural crest cells, while others come from the surface ectoderm
  1. Some are seen in large vascular areas and are due to hypertrophy of blood and lymph vessels—capillary hemangiomata or port-wine stains
B. ALBINISM is an autosomal recessive trait in which the skin, hair, and retina lack pigment due to failure of the melanocytes to produce melanin as a result of the lack of the enzyme tyrosinase.

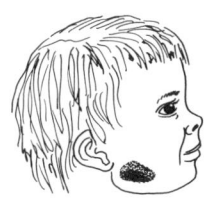

LOCALIZED PIGMENTED MACULAR LESION

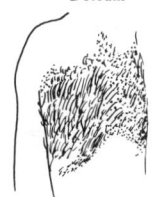

Dorsum

HAIRY, PIGMENTED NEVUS

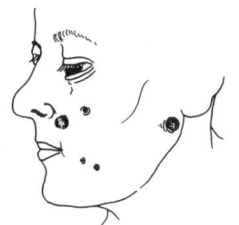

ELEVATED PIGMENTED MOLES

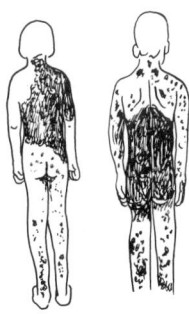

EXTENSIVE ("BATHING TRUNK") NEVUS

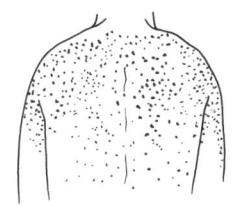

KERATOSIS FOLLICULARIS
Grouped papular lesions over trunk

VERRUCOUS NEVUS, EXTENSIVE

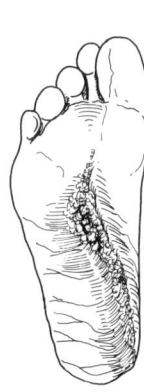

VERRUCOUS LINEAR NEVUS

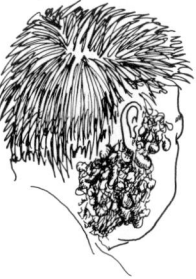

MIXED VASCULAR NEVUS
Macular and nodular

NEVUS FLAMMEUS ("PORT WINE" STAIN)
Vascular nevus

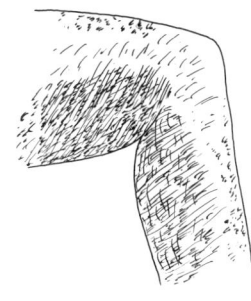

ICHTHYOSIS OF LEGS
Excessively dry skin with formation of scales

- 195 -

74. DEVELOPMENT OF THE HAIR AND ASSOCIATED STRUCTURES

I. **The hair** appears as a solid epidermal downgrowth of the stratum germinativum into the underlying dermis and is called the *hair bud*
  A. THE DEEPEST PORTION OF THE HAIR BUD becomes club-shaped and forms the *hair bulb*
    1. The epithelial cells of the hair bulb constitute the *germinal matrix,* which later will give rise to the hair
    2. The hair bulb is then invaginated by a small mesenchymal *hair papillae*
    3. As the cells of the germinal matrix, in the center of the hair follicles, proliferate, they are pushed upward and become keratinized forming the *hair shaft.* The peripheral cells of the developing hair follicle form the *epithelial root (hair) sheath*
      a. The surrounding mesenchymal cells differentiate into the *dermal (connective tissue) root sheath*
    4. The hair grows, penetrates the epidermis, and appears above the skin surface
    5. Melanoblasts invade the hair bulb and form melanocytes. They produce melanin which is transferred to the hair-forming cells in the germinal matrix before birth
  B. HAIRS BEGIN TO DEVELOP during early fetal life, but become visible at about week 20 on the eyebrows, upper chin, and lips and are called the *lanugo hairs*
    1. The lanugo are shed at about the time of birth and are later replaced by coarser hairs called *vellus hairs* which arise from new hair follicles
      a. The vellus persists over most of the body except in the axillary and pubic regions where, at puberty, they are replaced by coarse *terminal hairs* (seen also on the chest and face in males)
  C. THE ARRECTOR PILI MUSCLES are smooth muscle fibers which form from the surrounding mesenchyme and become attached to the connective tissue sheath of the hair follicle and dermal papillary layer

II. **The sebaceous glands** develop as buds from the side of the developing epithelial root sheath of the hair follicle
  A. THE BUDS GROW into the surrounding connective tissue and branch to form the primordia of the glandular alveoli and ducts
    1. The central cells of the alveoli break down and form an oily secretion, the *sebum,* which is extruded into the hair follicle and onto the skin surface to mix with the desquamated peridermal cells to help form the vernix caseosa
  B. INDEPENDENT GLANDS (not with hair follicles) also develop from the epidermis in the areas of the glans penis and the labia minora

III. **Sweat (eccrine or merocrine) glands** develop as a solid epidermal bud which grows down into the underlying dermis
  A. AS THE BUD ELONGATES, its end coils to form the primordium of the glands secretory portion, while the epithelial attachment to the epidermis forms the duct primordium
    1. The central cells of the primordia degenerate to form a lumen
    2. The peripheral cells of the secretory portion of the gland differentiate into *secretory and myoepithelial cells,* the latter being specialized ectodermal smooth muscle cells which aid in expelling the glandular secretion

IV. **Sweat (apocrine) glands** in humans, are confined to the axilla, pubic areas, and areola of the mammary glands
  A. THEY DEVELOP as downgrowths of the stratum germinativum of the epidermis
    1. Their ducts open into the hair follicles and not on the skin surface. They open just above the sebaceous glands
    2. Their chief function seems to be the production of small amounts of secretions which, on the surface, give rise to distinctive odors that enable animals to recognize each other. Human apocrine sweat glands have no odor in their secretion but contain substances readily degraded by bacteria into odiferous breakdown products

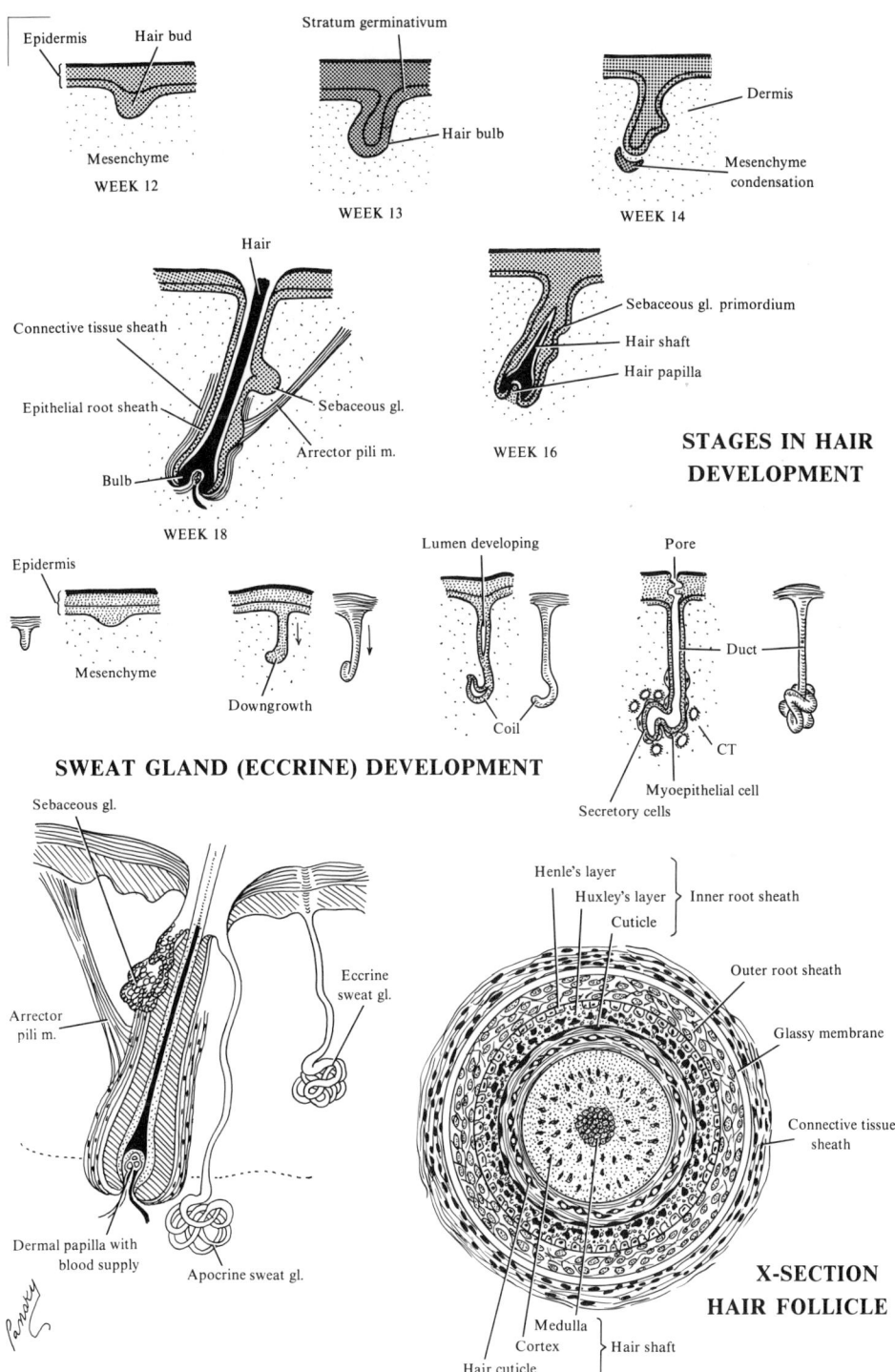

75. DEVELOPMENT OF THE NAILS

I. **Nails** are modifications of the epidermis and correspond to the claws and hoofs of lower animals
   A. THE FIRST INDICATION of a nail is foreshadowed at week 10 by a thickened area of epidermis, the *nail field,* seen on the dorsum of each digit
      1. The adjoining area, on each side and at the base of the field, tends to overgrow the field, giving rise to shallow *lateral nailfolds* which continue into a much deeper *proximal nailfold* that extends nearly to the proximal end of the terminal phalanx
      2. Development at the tips of the fingers precedes the development of the toenails
   B. THE MATERIAL of the true nail is developed within the underlayer of the proximal nailfold (although the primitive nail field undergoes some local cornification and forms a so-called false nail). This layer is named the *matrix*
      1. During month 5, specialized keratin fibrils differentiate in the matrix layer, without having passed through a keratohyalin or eleidin stage (ordinary method of cornification)
      2. The keratinized cells flatten and consolidate into the compact tissue of which the *nail plate* is composed
      3. Thus, the nail substance differentiates in the proximal nailfold as far distad as the outer edge of the *lunula* (the whitish crescent at the base of the exposed nail)
      4. Beyond the lunula, the nail plate merely shifts progressively over the *nail bed* and reaches the tip of the finger about 1 month before birth
      5. The dermis, beneath the nail, is thrown into parallel longitudinal folds to produce the characteristic ridges and grooves
   C. THE STRATUM CORNEUM AND PERIDERM of the epidermis, for a time, cover completely the free nail and are jointly referred to as the *eponychium*
      1. This layer, in late fetuses, is lost except for horny portions that continue to adhere to the nail plate along the curved rim of the nailfold (the *cuticle*)
   D. UNDERNEATH THE FREE END of the nail, the epidermal cells also accumulate to form a piled-up epidermal mass, the *hyponychium,* or substance beneath the nail

II. **Nail anatomy**
   A. THE HORNY ZONE of the nail is composed of hard keratin and has a distal, exposed part or *body,* and a proximal, hidden portion, the *root*
      1. The root is covered by a prolongation of the stratum corneum of the skin which is composed of soft keratin and is called the eponychium
      2. The lunula or "half-moon" lies distal to the eponychium and is a part of the horny zone which is opaque to the underlying capillaries
      3. The horny zone of the nail is attached to the underlying nail bed
      4. The matrix, or proximal part of the nail bed, produces hard keratin
      5. The fingernails reach the fingertips by week 32, and the toenails reach the toe tops by week 36
      6. On the average, after birth, the nail grows about 0.5 mm a week. They grow faster in the summer than in the winter; growth is also age dependent

III. **Malformations of the hair and nails**
   A. CONGENITAL ALOPECIA (atrichia congenita): fetal loss or absence of hair. may occur by itself or with other skin derivative abnormalities
   B. HYPERTRICHOSIS: excessive hairiness due to the development of supernumerary follicles or persistence of fetal hair that normally disappears
   C. ANONYCHIA: partial or complete absence of the nails due to a failure of the matrix to form or give origin to the nails
   D. MISSHAPEN NAILS are a common occurrence

DEVELOPMENT OF FINGERNAIL
Sections of fingertip

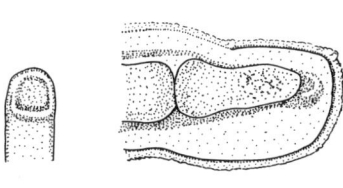

WEEK 10

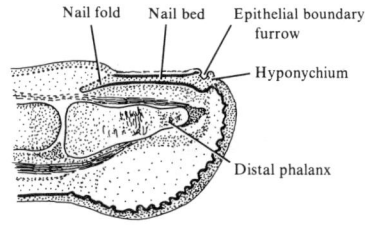

WEEK 11

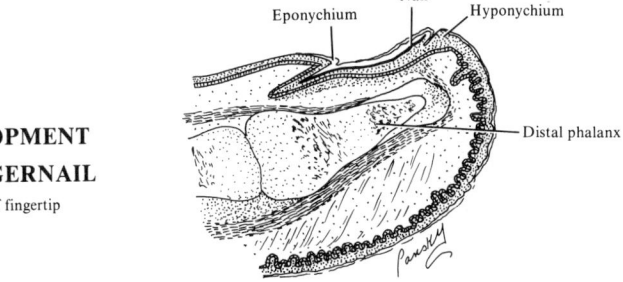

WEEK 14

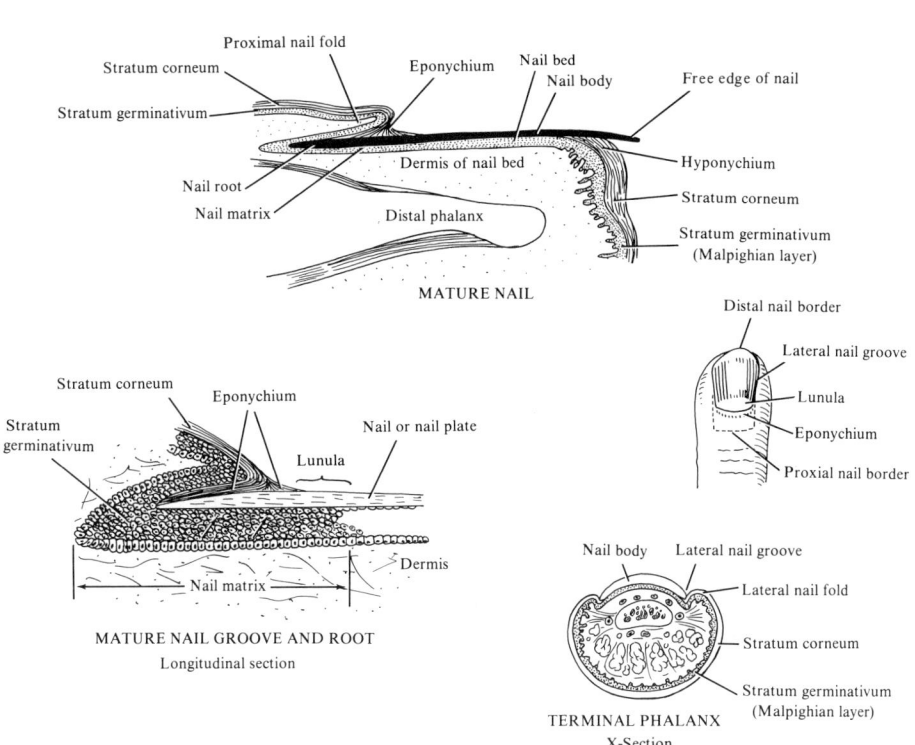

76. DEVELOPMENT OF THE MAMMARY GLANDS

I. **The mammary glands (breasts)** are derived from 2 thickened strips of epidermal ectoderm, the *primitive mammary ridges or milk lines,* which appear during week 6. The ridges extend from the axillae to the inguinal regions, but rapidly regress except in the thorax
   A. THE MAMMARY BUDS that persist in the thoracic region penetrate the underlying mesenchyme and give rise to several secondary buds which develop into *lactiferous ducts* and their branches. These are canalized by the end of the prenatal life
      1. The fibrous connective tissue and fat of the mammary gland develop from the surrounding mesenchyme. The lactiferous ducts form the small ducts and alveoli
      2. Only the main ducts are found at birth, and the gland remains undeveloped until puberty
   B. DURING THE LATE FETAL PERIOD, the epidermis, where the gland originated, becomes depressed to form a shallow *mammary pit* (epithelial pit) on which the ducts open
      1. The lactiferous ducts at first open onto this epithelial pit which is formed by the original mammary line
   C. THE NIPPLE itself forms during the perinatal period due to proliferation of the mesenchyme under the areola (circular area of skin around the nipple) in the area of the mammary pit. The nipple is often depressed and poorly formed during infancy
   D. THE MAMMARY GLANDS of both newborn males and females are often enlarged and may secrete "witches' milk" or *colostrum,* as a result of maternal hormones passing into the fetal circulation by way of the placenta
   E. AT PUBERTY, the female mammary glands enlarge rapidly as a result of the development of fat and connective tissue. The duct system also grows, stimulated by the estrogen and progesterone of the ovary
      1. The glandular tissue remains completely undeveloped until pregnancy when the intralobular ducts rapidly develop, form buds, and become alveoli
      2. The male glands undergo little postnatal development

II. **Malformations of the mammary gland**
   A. ABSENCE OF THE GLAND (AMASTIA) AND/OR NIPPLE (ATHELIA) is rare; may occur bilaterally or unilaterally, and is due to failure of development or complete disappearance of the mammary ridge(s). Also can be due to failure of the mammary bud to form.
   B. SUPERNUMERARY BREASTS (POLYMASTIA) AND NIPPLES (POLYTHELIA) are seen in about 1% of the female population and are usually inherited. They generally are found below the normal breast, but less commonly are seen in the axilla or abdominal area, developing along the mammary ridges
      1. Polythelia is uncommon, also may be seen in males
      2. In most cases, a single extra nipple or breast is seen, but in 30% of cases, 2 extra nipples or breasts are found
      3. Accessory breasts may have normal tissue and even function during lactation
   C. INVERTED NIPPLES: the nipple fails to develop normally and evert after birth. Probably due to a failure of the underlying mesenchyme to proliferate and push the nipple out
      1. Also may be caused by retraction of the nipple as a result of the presence of a fast-growing tumor in the gland

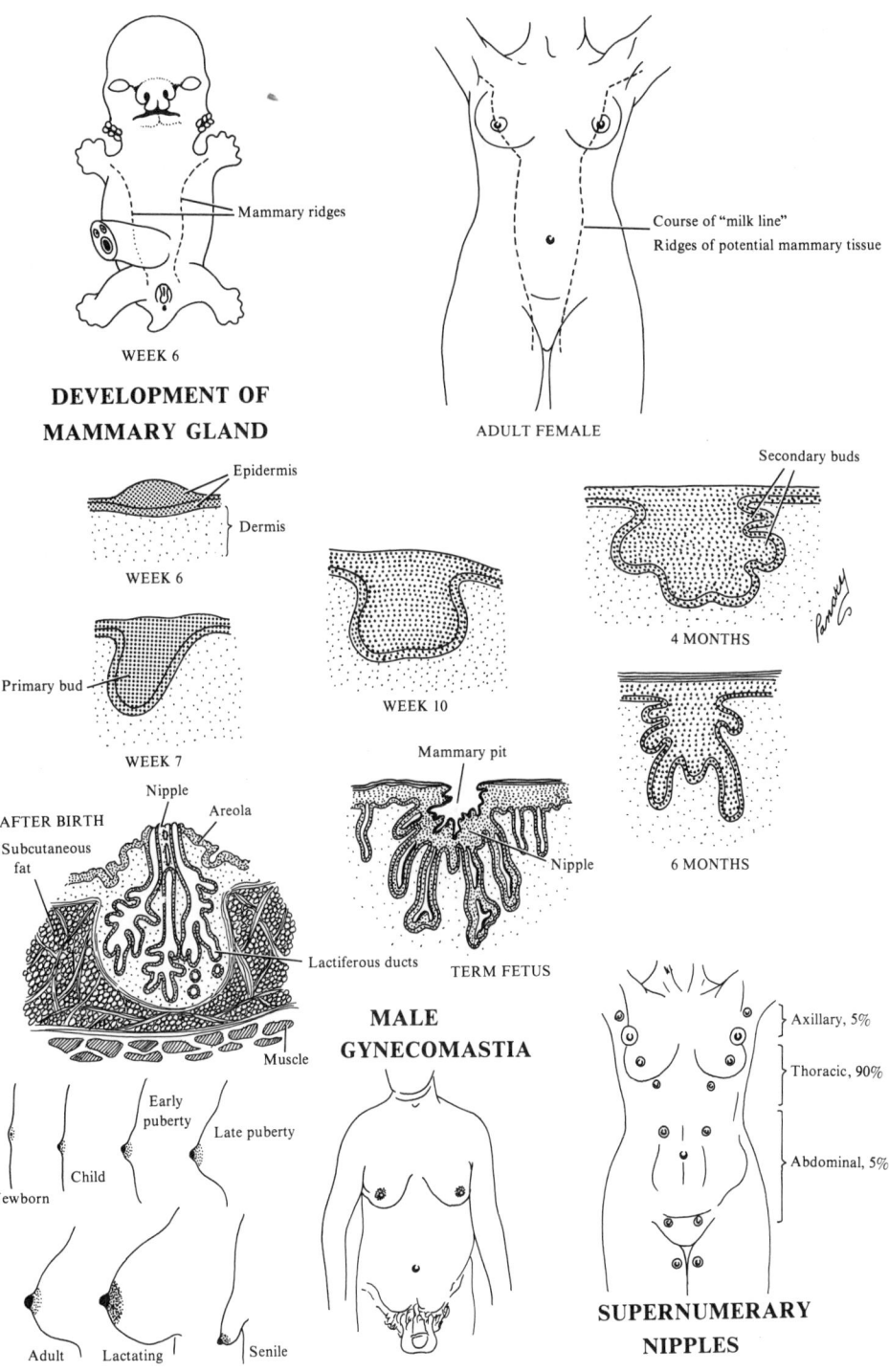

77. DEVELOPMENT OF THE TEETH

I. **Introduction:** the teeth develop from ectoderm and mesoderm: the enamel develops from ectoderm of the oral cavity, and all other tissues come from the associated mesenchyme. Not all teeth develop at the same time. The first tooth buds are seen in the anterior mandibular region, later in the anterior maxillary region, then posteriorly in both jaws. Development is in continuous stages

A. THE DENTAL LAMINA AND BUD STAGE: the dental laminae are seen early in week 6 as U-shaped thickenings or buds of the oral epithelium (surface ectoderm)
  1. Localized proliferation of cells in the dental laminae forms round or oval swellings, the *tooth buds,* which grow into the mesenchyme
  2. The tooth buds develop into the *deciduous or milk teeth* (shed during childhood). There are 10 tooth buds in each jaw, 1 for each tooth
  3. The tooth buds for the permanent teeth, with deciduous predecessors, are seen in the 10-week fetus, developing from deeper continuations of the dental lamina. They lie on the tongue or lingual side of the deciduous buds
  4. Tooth buds for the permanent teeth appear at different ages during the fetal period except for the second and third permanent molars, which appear after birth, at about 4 months and 5 years, respectively
  5. The permanent molars with no deciduous predecessors develop as buds from backward extensions of the dental laminae

B. CAP STAGE OF DEVELOPMENT: the deep surface of each ectodermal tooth bud becomes invaginated by mesenchyme called the *dental papilla,* which gives rise to the *dentin* and *dental pulp.* The ectodermal, cap-shaped covering over the papilla is called an *enamel organ* since it will produce the future enamel of the tooth
  1. The outer cellular layer of the ectodermal enamel organ is called the *outer enamel epithelium;* the inner layer lining the "cap" is the *inner enamel epithelium*
    a. The cell region between the above layers forms the core or bulk of the cap and is called the *stellate* or *enamel reticulum*
  2. As the enamel organ and dental papilla form, the surrounding mesenchyme condenses as the *dental sac,* which later forms the *cementum* and *periodontal ligament*

C. THE BELL STAGE: with invagination of the enamel organ, the tooth assumes a bell shape
  1. The mesenchymal cells in the dental papilla, adjacent to the inner enamel epithelium, differentiate into *odontoblasts,* which produce *predentin,* and deposit it adjacent to the inner enamel epithelium. The predentin later calcifies to form *dentin*
  2. As the dentin thickens, the odontoblasts regress toward the center of the dental papilla but odontoblastic processes remain embedded in the dentin and are called *Tomes' dentinal fibers* or *processes*
  3. Cells of the inner enamel epithelium near the dentin form *ameloblasts,* which produce *enamel* in the form of prisms or rods over the dentin layer, thus help form the outer layer of the tooth or the *crown.* As enamel increases, the ameloblasts regress
    a. Thus, both enamel and dentin help create the crown, which begins formation at the cusp or tip of the tooth and progresses, in development, to the future root
  4. The root begins after the enamel and dentin are well along in development
    a. The inner and outer enamel epithelia come together in the neck region and form an epithelial fold, the *epithelial root sheath,* which grows into the mesenchyme and begins the formation of the root
    b. The odontoblasts near the sheath form the dentin (continuous with that of the crown). As the dentin increases, the pulp cavity gets smaller and becomes a narrow canal for the vessels and nerves to enter the root
  5. The inner cells of the dental sac form *cementoblasts* which produce *cementum,* which is deposited over the root dentin and meets the enamel at the neck of the tooth
  6. As the teeth develop, the jaws ossify and the outer cells of the dental sac also become active in bone formation. Each tooth is soon surrounded by bone, except at its crown, and is held in its *bony socket* or *alveolus* by the peridontal ligament

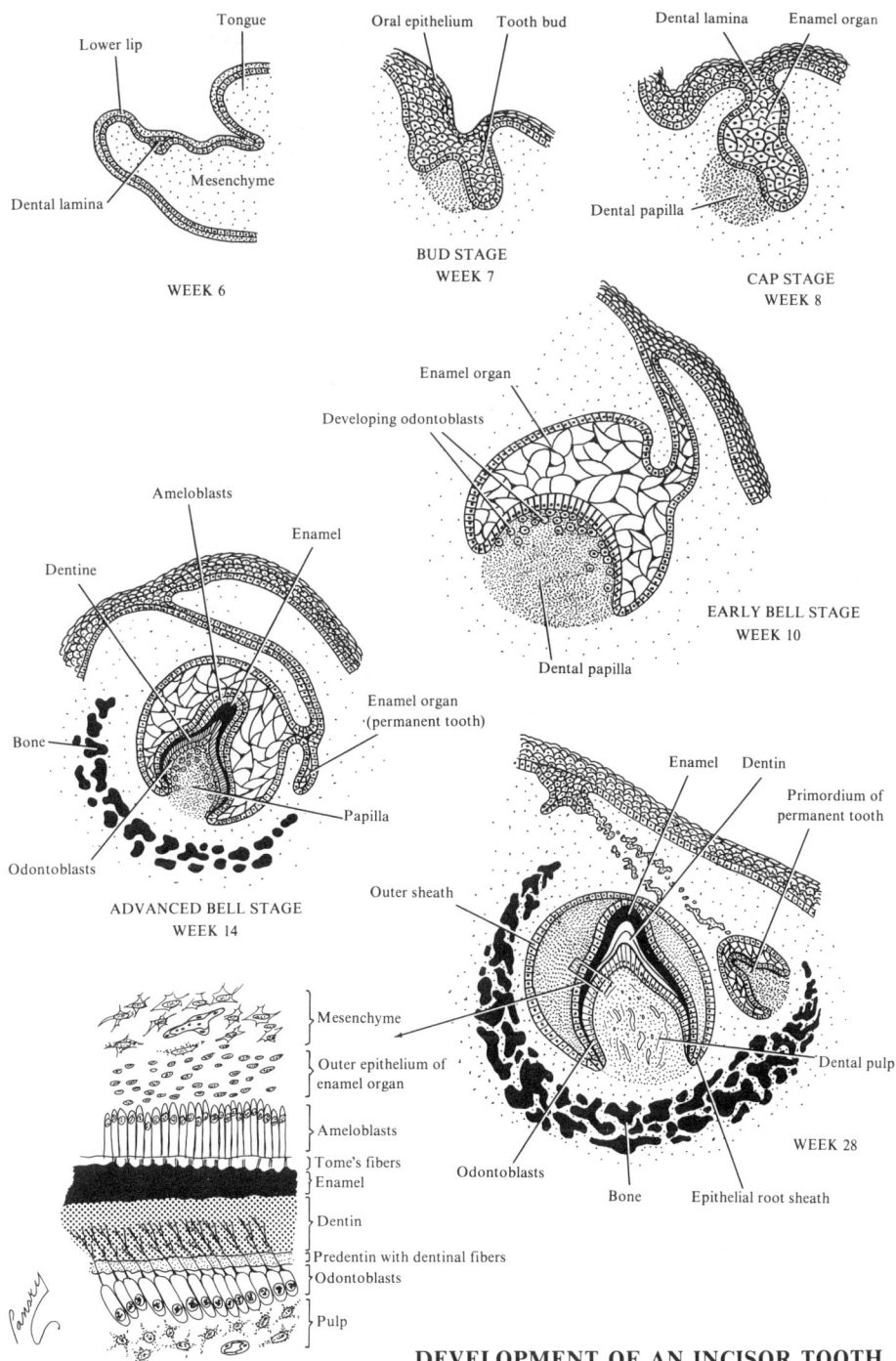

DEVELOPMENT OF AN INCISOR TOOTH

78. TOOTH ERUPTION AND MALFORMATIONS OF THE TEETH

I. **Tooth eruption:** as the tooth grows, the crown gradually erupts through the oral mucosa, and the mucosa around the crown becomes its *gum* or *gingiva*. The crown itself, as it emerges, consists of a nucleus of dentine covered by a layer of enamel
A. THE DECIDUOUS TEETH usually erupt between the 6th and 24th months after birth
B. THE PERMANENT TEETH develop later, but in a similar manner to the deciduous teeth, and as they grow, the root of the corresponding deciduous tooth is resorbed by osteoclasts. Thus, when the deciduous teeth are shed, the portion shed consists of only the crown and upper portion of the original root
   1. The permanent teeth erupt usually during one's sixth year and continue to appear until early adulthood

II. **Malformations of the teeth** generally are not visible at birth because the teeth do not erupt until after birth (usually)
A. ENAMEL HYPOPLASIA: defective enamel formation resulting in grooves, pits, and fissures on the enamel surface due to a disturbance in enamel formation. One of its most common causes is rickets, due to vitamin D deficiency
B. ABNORMALITIES IN SHAPE: quite common; due to aberrant groups of ameloblasts
C. NUMERICAL ABNORMALITIES: 1 or more extra teeth may develop or the teeth may not form at all
   1. Partial anodontia: 1 or more teeth are absent
   2. Total anodontia: no teeth develop. Is a very rare condition
D. NATAL TEETH AND CAPS: 1 or 2 mandibular incisors are found at birth
   1. Premature erupted teeth may be only small, loose enamel caps over a thin dentin sheet
E. FUSED TEETH: a tooth bud may divide or 2 buds may partly fuse to form a fused or joined tooth
F. AMELOGENESIS IMPERFECTA: the enamel of the tooth is soft and friable due to hypocalcification. In addition, the teeth are usually yellow to brown
   1. Probably due to an autosomal trait
G. DENTINOGENESIS IMPERFECTA: the teeth are brown to gray-blue and have an opalescent shine. The enamel wears down easily, and the dentin is exposed
   1. Due to an inherited autosomal dominant trait

TOOTH ERUPTION

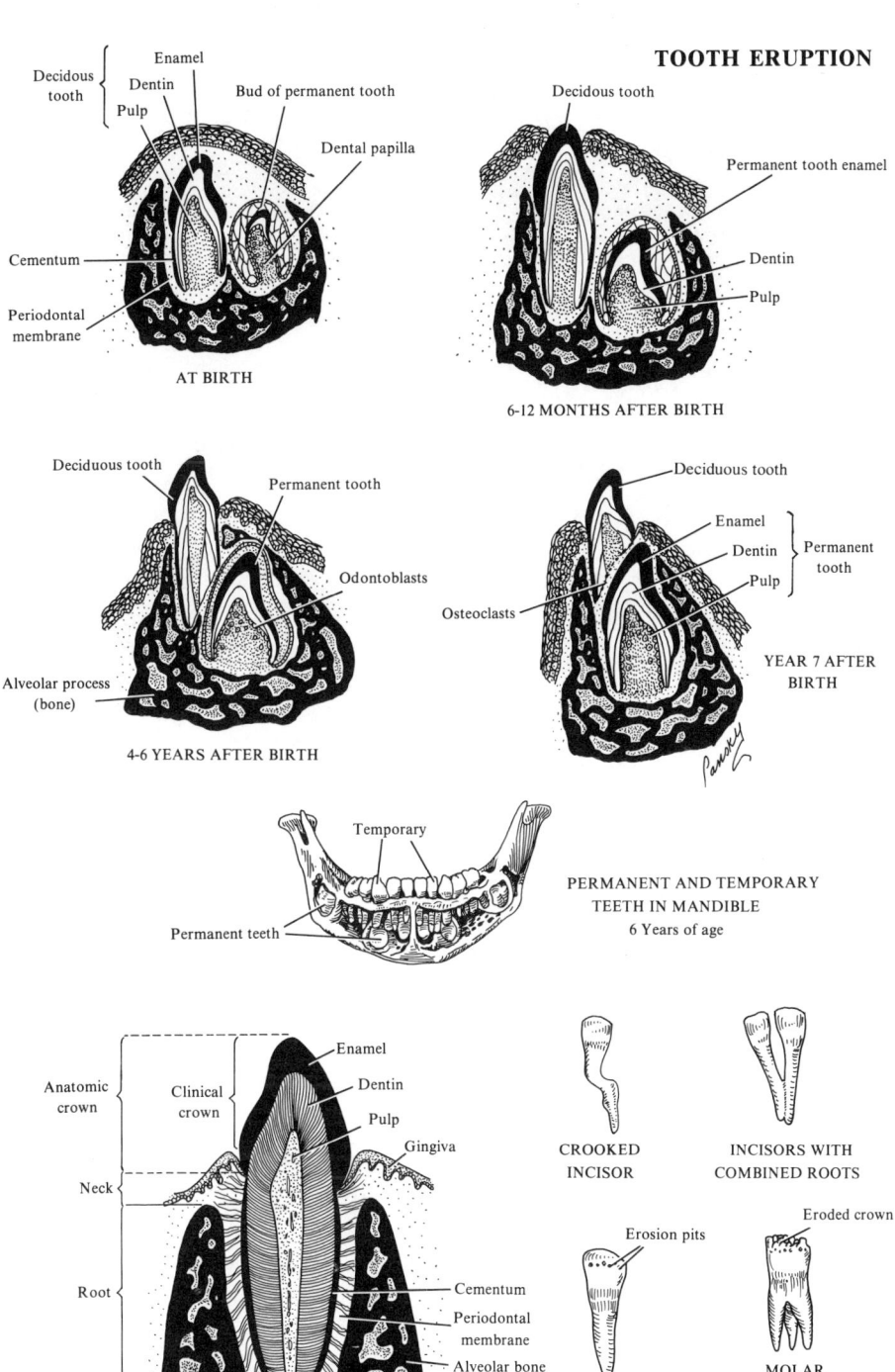

UNIT SEVEN
THE DIGESTIVE SYSTEM

79. THE DIGESTIVE SYSTEM: GENERAL INTRODUCTION*

I. Early development review
A. INTRODUCTION: the entoderm appears about day 8 and rapidly forms the *yolk sac* and the epithelial lining of the digestive tract and its associated glands. The process of body cylinder formation divides the yolk sac into 2 parts toward the end of month 1
  1. Part 1 is *extraembryonic* and is the yolk sac itself, which regresses early and disappears at about 3 months. The head, tail, and lateral folds incorporate the dorsal part of the yolk sac and the allantois into the embryo
  2. Part 2 is *intraembryonic* and is the *primitive gut* which is the early origin of the epithelium of the digestive tube and its accessory glands, the liver, the pancreas, and the biliary apparatus
      a. The epithelium at the cranial and caudal ends of the tract is derived from ectoderm of the stomodeum (primitive mouth) and the proctodeum (anal pit), respectively
B. THE SPLANCHNIC MESODERM, which is formed about day 15 during gastrulation, surrounds the entoderm and provides the digestive tract with its connective tissue, muscle, and serous (peritoneal) coverings

II. Major or principal stages of development
A. GENERAL INFORMATION
  1. The digestive tube is initially closed cephalically by the *buccopharyngeal* or *oropharyngeal membrane* (a bilaminar membrane: ectoderm externally and entoderm internally) and caudally by the *cloacal membrane*
      a. The buccopharyngeal membrane is resorbed at the beginning of week 4, connecting the amniotic cavity with the digestive tube
      b. The derivatives of the cloacal membrane open at the end of week 9
  2. The development of the digestive system consists of
      a. A very complex anterior pharyngeal portion, the *foregut*
          i. The foregut extends from the buccopharyngeal membrane to the duodenum where the liver bud arises (the anterior intestinal portal)
      b. A very extensive growth in length of its middle or abdominal portion, the *midgut*
          i. The middle portion of the midgut remains connected to the yolk sac via the vitelline or omphalomesenteric duct
          ii. The midgut extends from just caudal to the liver bud (the anterior intestinal portal) to a point where, in the adult, the right two-thirds and left one-third of the transverse colon are found (the posterior intestinal portal in the embryo)
      c. An intermingling and very close association with the urogenital system is found in relation to its terminal portion, the *hindgut*
          i. The hindgut extends from the posterior intestinal portal to the cloacal membrane
      d. The foregut, midgut, and hindgut are supplied by the celiac artery, the superior mesenteric artery, and the inferior mesenteric artery, respectively

---

*Development of the oral cavity, the tongue, the salivary glands, and the teeth is described in other units of the text.

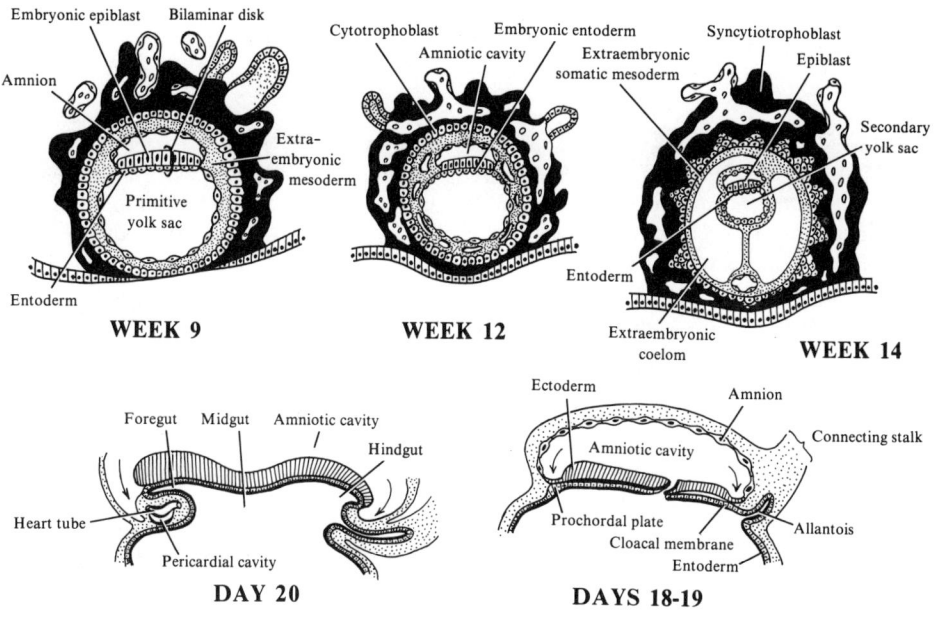
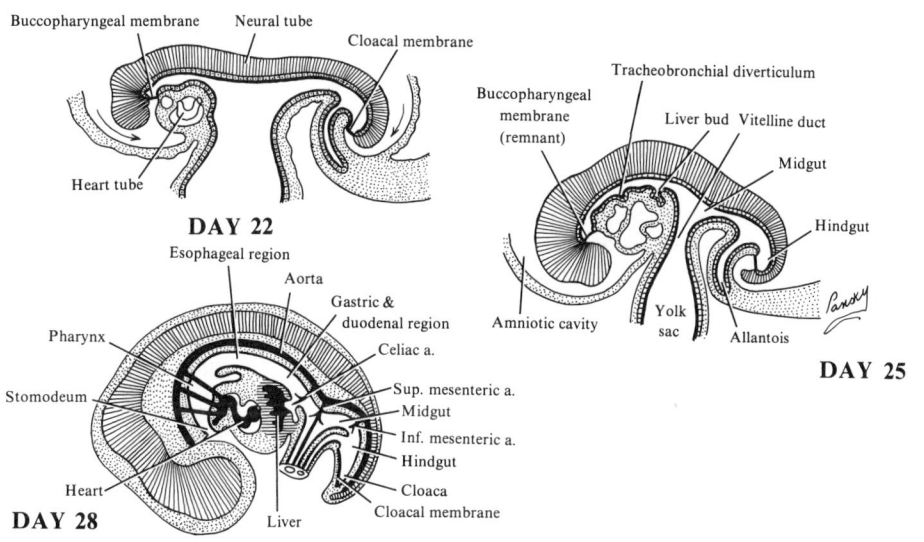
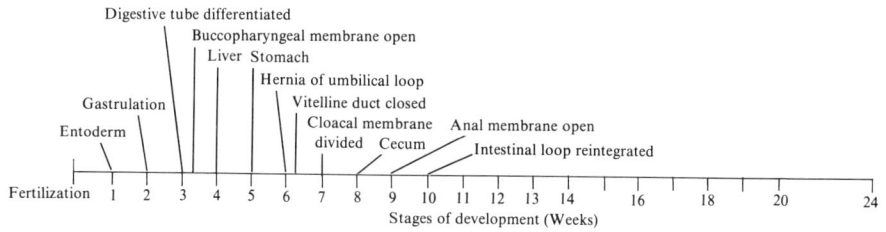

80. THE FOREGUT: ESOPHAGUS AND STOMACH

I. **Introduction:** the foregut derivatives are the pharynx and its derivatives (see Unit Two), the lower respiratory tract (see Unit Three), the esophagus, the stomach, the duodenum as far as the entrance of the common bile duct, the liver, the pancreas, and the biliary apparatus. All except for the pharynx, respiratory tract, and upper esophagus are supplied by the *celiac artery*

II. **The foregut** extends from the buccopharyngeal membrane to the duodenum, is initially located in the median sagittal plane, and is attached by *mesentery* to the anterior and posterior abdominal walls. It consists of a *cranial segment,* the pharyngeal gut or *pharynx,** which extends from the buccopharyngeal membrane to the tracheobronchial diverticulum; and a *caudal segment,* extending from the diverticulum as far caudally as the liver bud outgrowth from the duodenum
  A. FOREGUT (CAUDAL SEGMENT)
    1. The esophagus is partitioned from the trachea by the *tracheoesophageal septum*
       a. The esophagus is initially very short, but elongates rapidly, reaching its final relative length by about week 7
          i. Elongation is a result of cranial body growth (ascent of the pharynx), development of the heart, and retroflexion of the head
       b. The entoderm of the esophagus initially proliferates and almost obliterates the lumen, but recanalizes near the end of the embryonic period
       c. The striated muscle in the upper two-thirds of the esophagus is derived from the mesenchyme of the caudal branchial arches (innervated by cranial nerve X); the smooth muscle of the lower third of the esophagus develops from the surrounding splanchnic mesenchyme (innervated by the visceral nerve splanchnic plexus derived from neural crest cells)
    2. The stomach first appears as a fusiform dilatation of the caudal portion of the foregut in week 5. The primordium soon enlarges and broadens ventrodorsally. Its position and appearance change as a result of the different rates of growth in various regions of its walls, as well as changes in position of the surrounding organs
       a. The positional changes are explained most easily by assuming that the stomach rotates around a *longitudinal* and an *anteroposterior axis*
          i. Around the longitudinal axis, the stomach carries out a 90° clockwise rotation, causing its left side to face anteriorly and its right side posteriorly. (This explains why, in the adult, the left vagus nerve supplies its anterior or ventral wall and the right vagus nerve its posterior or dorsal wall)
          ii. Anteroposterior axis rotation displaces the pyloric part of the stomach to the right and upward and the cephalic or cardiac portion to the left and downward slightly, resulting in the future duodenum coming to be retroperitoneal
       b. The dorsal border of the stomach grows faster than the ventral one and produces the *greater and lesser curvatures of the stomach*
       c. Since at this stage of development, the stomach is attached to the posterior body wall by the dorsal mesogastrium, longitudinal rotation pulls the dorsal mesogastrium to the left and helps form the *omental bursa or lesser sac* (a peritoneal pouch found behind the stomach)
          i. As the embryo lengthens, the caudal part of the septum transversum thins and becomes the ventral mesentery or mesogastrium. It attaches the stomach and duodenum to the ventral wall of the abdominal cavity
       d. The stomach thus assumes its final position, and its long axis now runs from above left to below right. The greater curvature faces downward, and the lesser curvature faces upward and to the right

---

*The cranial segment or pharynx is discussed under the development of the branchial arches, clefts, and pouches, and lower respiratory system

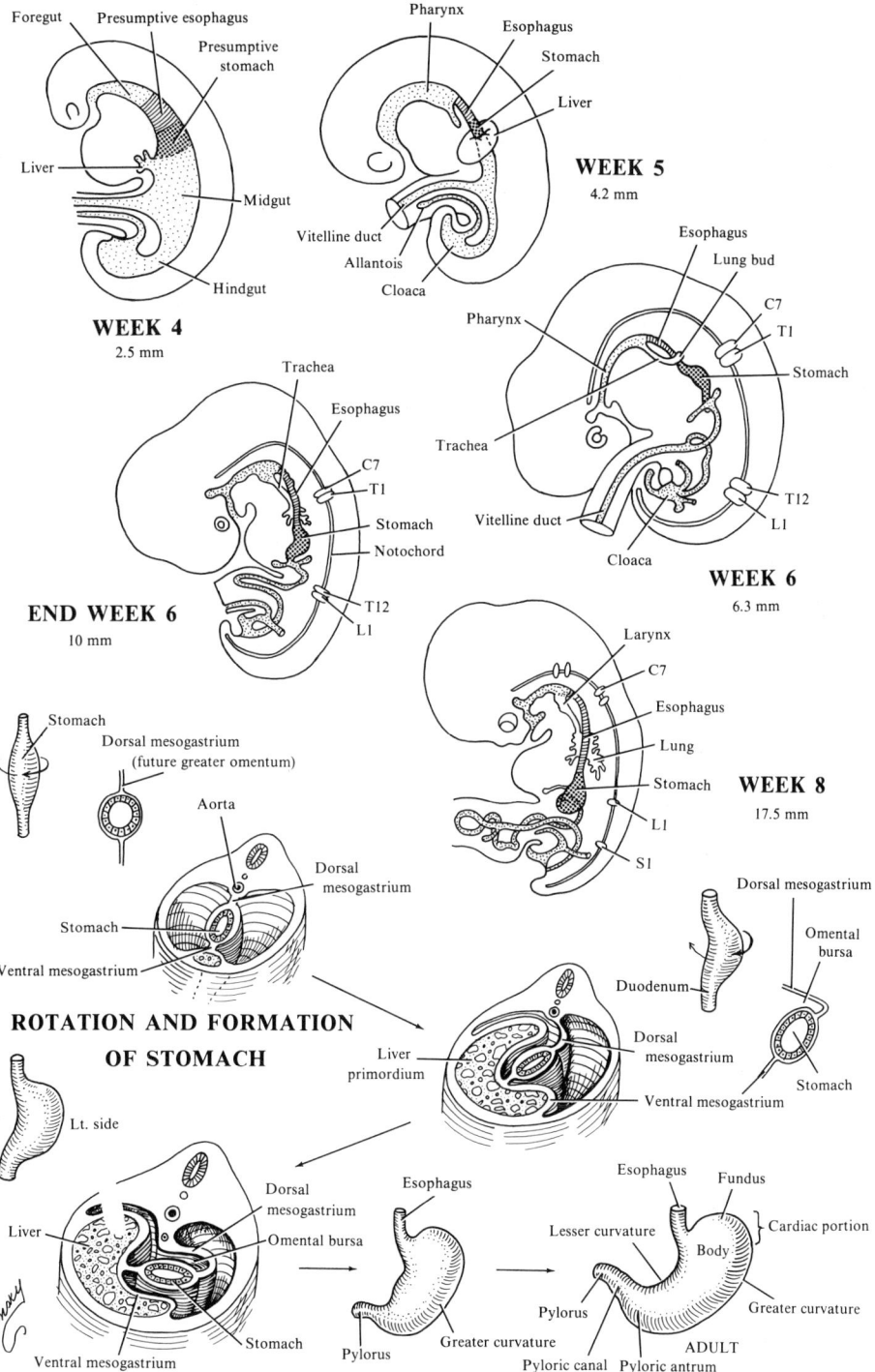

81. THE FOREGUT: THE OMENTAL BURSA AND DUODENUM

I. The omental bursa (lesser peritoneal sac)
A. CLEFTS develop between the cells of the dorsal mesogastrium which coalesce and eventually form a single cavity, the omental bursa
 1. The cavity expands in all directions and comes to lie behind the stomach and to the right of the esophagus
 2. The upper portion of the cranial extension of the sac is limited by the developing diaphragm, to form a closed space or sac called the *infracardiac bursa*
 3. The lower portion of the cranial extension of the sac persists as the *superior recess of the lesser sac*
 4. As the stomach enlarges, the lesser sac expands into an *inferior recess* which forms between the layers of the elongating dorsal mesogastrium (greater omentum)
    a. The 4-layer greater omentum overhangs the developing small intestines
    b. Most of the inferior recess of the lesser sac disappears as the layers of the greater omentum fuse
 5. The omental bursa communicates with the main peritoneal cavity or greater peritoneal sac by way of the *epiploic foramen* or *foramen of Winslow*

II. The duodenum forms from the terminal portion of the foregut and the cephalic or cranial portion of the midgut
A. THE 2 PARTS grow rapidly and form a C-shaped loop that projects ventrally
 1. The junction of the foregut and midgut is at the apex of this embryonic loop, just distal to the origin of the liver bud
B. AS A RESULT OF ITS DUAL ORIGIN, the duodenum is supplied by branches of both the celiac and superior mesenteric arteries
C. AS THE STOMACH ROTATES, the duodenum rotates to the right and comes to lie retroperitoneally
D. DURING WEEKS 5 AND 6, the duodenal lumen is reduced and temporarily may even be obliterated by epithelial cells. However, under normal conditions, the lumen recanalizes by the end of the embryonic period
E. A GREAT PORTION OF THE VENTRAL MESENTERY of the duodenum disappears. The free border of this mesentery which does remain lies between the duodenum and the liver and forms the ventral border of the epiploic foramen and the *duodenohepatic ligament*. This mesentery is also a portion of the so-called lesser omentum

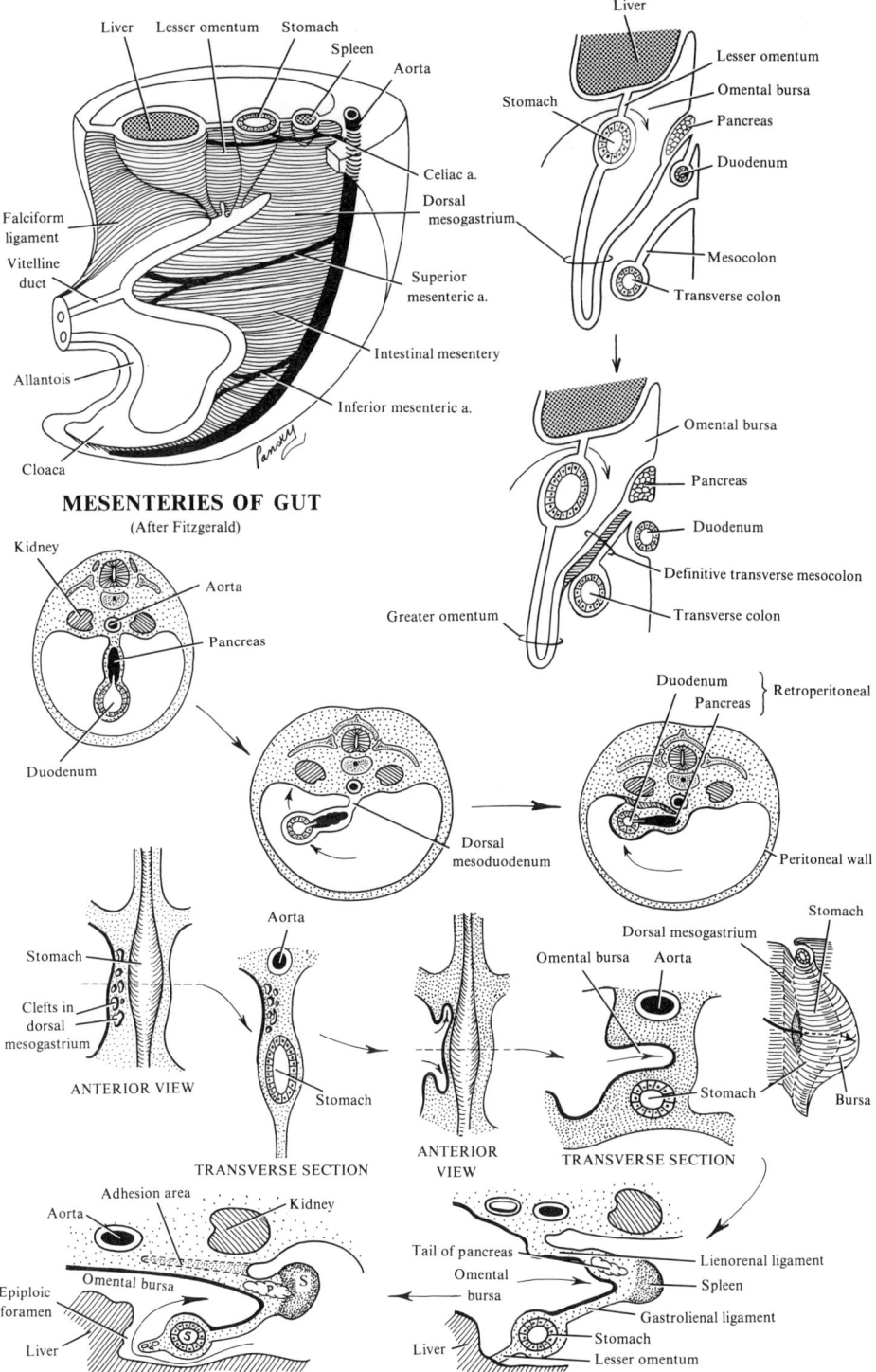

MESENTERIES OF GUT
(After Fitzgerald)

82. THE FOREGUT: THE LIVER AND BILIARY APPARATUS (GALLBLADDER AND DUCTS)

I. **The liver and biliary apparatus:** their primordium appears at about week 3 as an anterior thickening and outgrowth or bud from the entodermal epithelium of the most distal end of the foregut (future duodenum). This *hepatic diverticulum or bud,* consisting of rapidly proliferating cell strands, penetrates the septum transversum (the mesodermal plate between the pericardial cavity and the stalk of the yolk sac) where it enlarges and divides into 2 parts

  A. THE LARGER CRANIAL PART is the liver primordium
    1. The proliferating entodermal cells form interlacing cords of liver cells or *hepatic parenchyma* and the epithelial lining of the *intrahepatic portion of the biliary apparatus*
    2. As the cords penetrate the septum transversum, they cause fragmentation of the umbilical and vitelline veins supplying the liver, resulting in the formation of the *hepatic sinusoid network*
    3. The fibrous and hematopoietic tissue and Kupffer cells of the liver are derived from the splanchnic mesenchyme of the septum transversum
    4. The liver grows rapidly and soon fills most of the abdominal cavity. At first, the *right* and *left lobes* of the liver are equal in size, but the right lobe becomes larger, and the *caudate* and *quadrate lobes* develop as subdivisions of the left lobe
  B. THE SMALLER CAUDAL PORTION of the liver diverticulum expands to form the *gallbladder* and its stalk, the *cystic duct*
    1. Bile pigment begins to form during weeks 13 to 16 and enters the duodenum, making its contents (meconium) a greenish color
    2. At first, the extrahepatic biliary apparatus is occluded with entodermal cells, but it later recanalizes
    3. The stalk connecting the hepatic and cystic ducts to the duodenum becomes the *common bile duct*
      a. The common bile duct is initially attached to the ventral side of the duodenal loop, but with duodenal growth and rotation, the duct finally enters the duodenum on its dorsal side

II. **Hematopoiesis** begins during week 6 and the liver becomes a bright red. This function is responsible for the relatively large size of the liver during month 2. The liver weighs about 10% of the total fetal body weight by week 9

  A. LIVER HEMATOPOIESIS subsides gradually during the last 2 months of intrauterine life, and only small hematopoietic islands remain at birth, at which time, the liver is only about 5% of the total body weight

III. **The ventral mesentery** is a thin, double-layer mesodermal membrane which gives rise to 3 structures

  A. THE LESSER OMENTUM: lying between the liver and the ventral borders of the stomach (gastrohepatic ligament), and the liver and duodenum (duodenohepatic ligament)
    1. In the free margin of this omentum are the bile duct, the portal vein, and the hepatic artery
  B. THE FALCIFORM LIGAMENT: between the liver and the ventral abdominal wall
    1. Contains the umbilical vein (from umbilical cord to liver) in its inferior border
  C. THE VISCERAL PERITONEUM OF THE LIVER covers the liver except for the area in contact with the diaphragm, the *bare area of the liver*

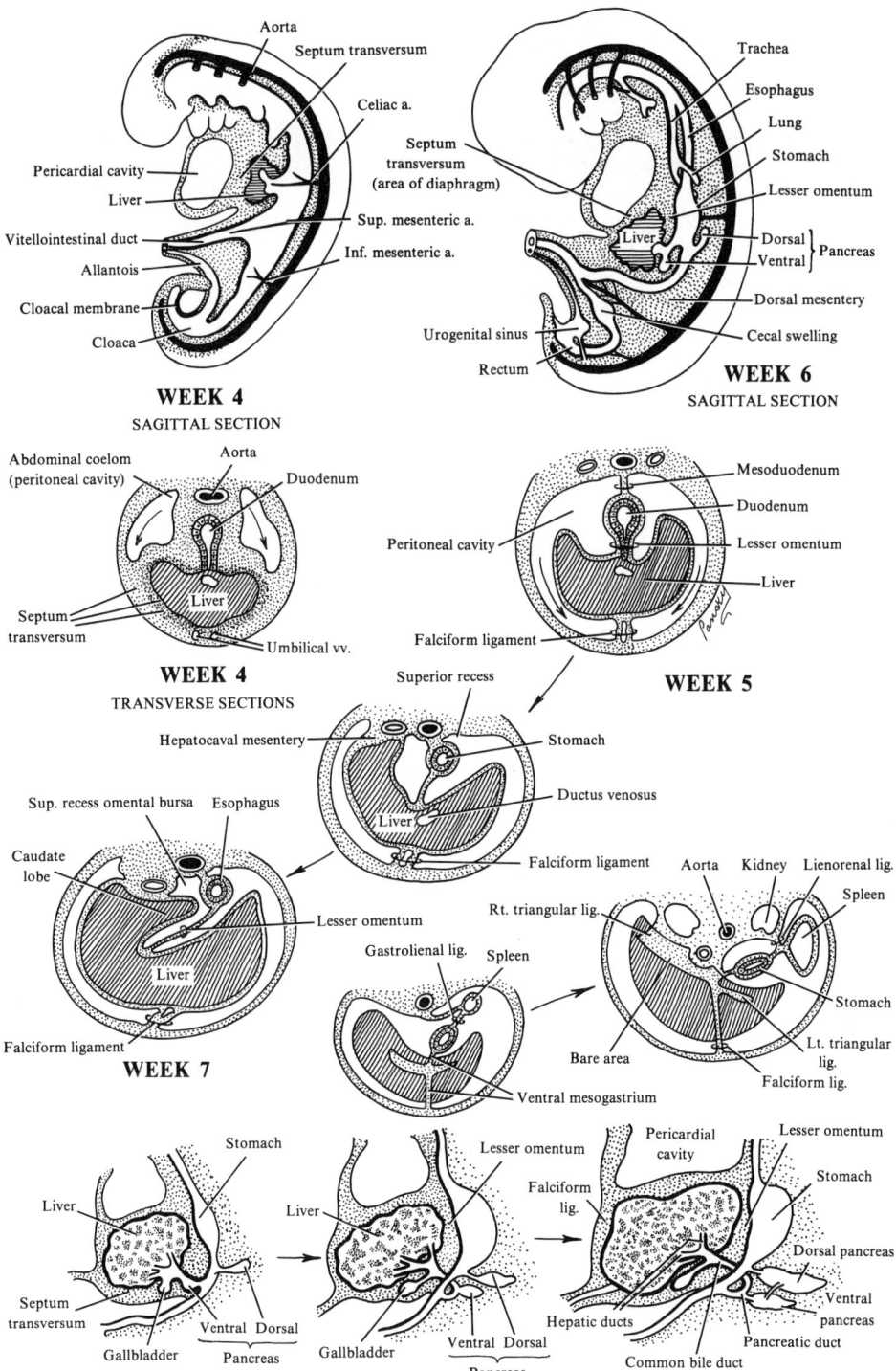

83. THE FOREGUT: THE PANCREAS AND SPLEEN

I. The pancreas makes its appearance in week 5 as 2 buds originating from the entodermal epithelium of the duodenum, namely, a *dorsal bud* (opposite and slightly above the hepatic diverticulum) and a *ventral bud* (in the angle below the hepatic rudiment)

A. THE LARGER DORSAL BUD appears first and rapidly grows into the dorsal mesentery
   1. The dorsal bud forms the major portion of the pancreas, namely, the *upper half of the head,* the *isthmus,* the *body,* and the *tail*

B. THE VENTRAL BUD develops near the entry of the common bile duct into the duodenum
   1. When the duodenum grows and rotates to the right (clockwise) and becomes C-shaped, the ventral pancreatic bud is carried dorsally along with the common bile duct and comes to lie in the mesoduodenum immediately below and behind the dorsal bud
      a. The parenchyma and the duct systems of both buds then fuse
   2. The ventral bud forms the *uncinate process* and the *inferior part of the head of the pancreas*

C. AS THE PANCREATIC BUDS FUSE, their ducts anastomose
   1. The main pancreatic duct (of Wirsung) forms from the duct of the ventral bud and the distal part of the duct of the dorsal bud
   2. The proximal portion of the duct of the dorsal part often persists as the accessory pancreatic duct (of Santorini), that opens just above the main duct
   3. The common bile duct and the duct of Wirsung open into the *ampulla of Vater* in the second part of the duodenum, either together or separately, with the bile duct above the pancreatic duct

D. PANCREATIC PARENCHYMA is of entodermal origin and forms a tubular primitive duct network
   1. Acini, early in the fetal period, develop from cell clusters around the ends of the tubules
   2. The *islets of Langerhans* develop in month 3 from the parenchymatous pancreatic tissue that separates from the tubules and lies between the acini
   3. Insulin secretion begins at about month 5
      a. Since fetal insulin levels are independent from maternal insulin levels, it is unlikely that insulin crosses the placenta
   4. The connective tissue covering and the pancreatic septa form from the surrounding splanchnic mesenchyme

II. The spleen is derived from mesenchymal cells found between the layers of the dorsal mesogastrium

A. THE SPLEEN obtains its characteristic shape early in the fetal period

B. AS A RESULT OF ROTATION OF THE STOMACH, the left surface of the mesogastrium fuses with the peritoneum over the left kidney, accounting for the dorsal attachment of the *lienorenal ligament* (connecting the spleen to the left kidney) and the passage of the splenic artery behind the peritoneum on its way to the spleen

C. MESENCHYMAL CELLS DIFFERENTIATE to form the splenic capsule, the connective tissue framework of the spleen, and its parenchyma

D. THE SPLEEN serves as a hematopoietic organ until late in fetal life; however, lymphocyte and monocyte production continues through life

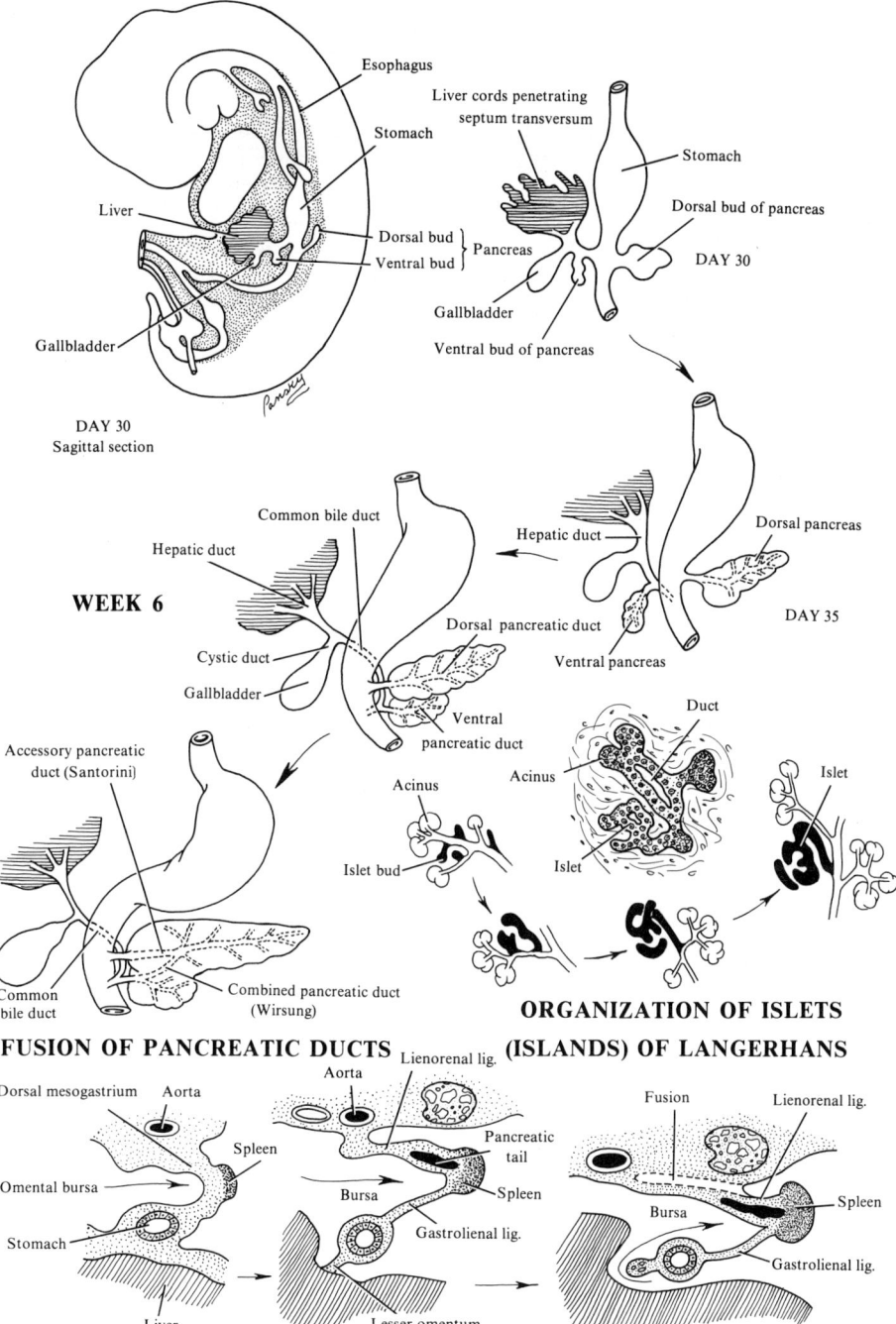

84. DEVELOPMENT OF THE MIDGUT: GENERAL INTRODUCTION

I. **The midgut:** in the 5-mm embryo, it is seen beginning just caudal to the entrance of the bile duct into the duodenum and terminating at the beginning of the last third of the transverse colon (from anterior to posterior intestinal portals). It is suspended from the dorsal abdominal wall by a short mesentery and communicates with the yolk sac via the vitelline duct. Its derivatives consist of the *small intestines* (except the first part of the duodenum to the common bile duct entrance); *the cecum, the appendix, the ascending colon;* and the right one-half to two-thirds or *proximal part of the transverse colon.* It is supplied by the superior mesenteric artery and vagus nerve

II. **Midgut development** is characterized by a rapid elongation of the gut and its mesentery
  A. ROTATION AND FIXATION initially, the midgut communicates with the yolk sac, but this connection narrows to the yolk stalk or vitelline duct. Elongation of the gut occurs faster than elongation of the embryo's body, thus, a series of intestinal changes takes place, usually in 3 stages
  1. Physiologic herniation of the midgut
     a. As it elongates, the midgut forms a ventral U-shaped umbilical loop of gut, the *primary intestinal loop*, which projects into the umbilical cord
     b. This "herniation" takes place at weeks 6 to 10 and is a normal migration of the midgut into the extraembryonic coelom. It occurs because there is not enough room in the abdomen to accommodate the fast-growing midgut due to the space occupied by the massive liver and the kidneys
        i. Thus at this stage, the intraembryonic and extraembryonic coeloms communicate at the umbilicus, and the midgut develops entirely outside the abdominal cavity
     c. The midgut has 2 limbs: a proximal (cranial) and a distal (caudal) limb. The yolk stalk is attached to the apex of the loop at their junction. If the duct persists, it is called *Meckel's diverticulum*
        i. The proximal limb grows rapidly to form the small intestinal coils (distal duodenum, jejunum, and ileum)
        ii. The caudal limb changes little except for developing the lower ileum, the cecal diverticulum, the appendix, the ascending colon, and proximal two-thirds of the transverse colon
     d. In the umbilical cord, the midgut loop rotates 90° counterclockwise (seen from in front) around the axis of the superior mesenteric artery, bringing the proximal limb of the loop to the right and the distal limb to the left
        i. From the artery arise the colic branches for the caudal limb and jejunoileal branches for the proximal limb
  2. Return of the midgut (reduction of the midgut hernia)
     a. During week 10, the intestines return to the abdomen
     b. The proximal limb (jejunum of the small intestines) returns first and passes behind the superior mesenteric artery to the left side. The later returning loops settle more to the right
        i. The cecal swelling is seen in the 12-mm embryo as a small conical swelling on the caudal limb of the primitive intestinal loop and is the last part to reenter the cavity
     c. As they return, the gut undergoes another 180° counterclockwise rotation, placing the cecum and appendix near the right lobe of the liver, from where they descend into the right iliac fossa at a later date
        i. Total rotation is thus: $90° + 180° = 270°$
     d. Return to the abdomen is related to a decrease in relative size of the liver and mesonephric kidneys, as well as abdominal enlargement and expansion
  3. Fixation of the midgut: see next page

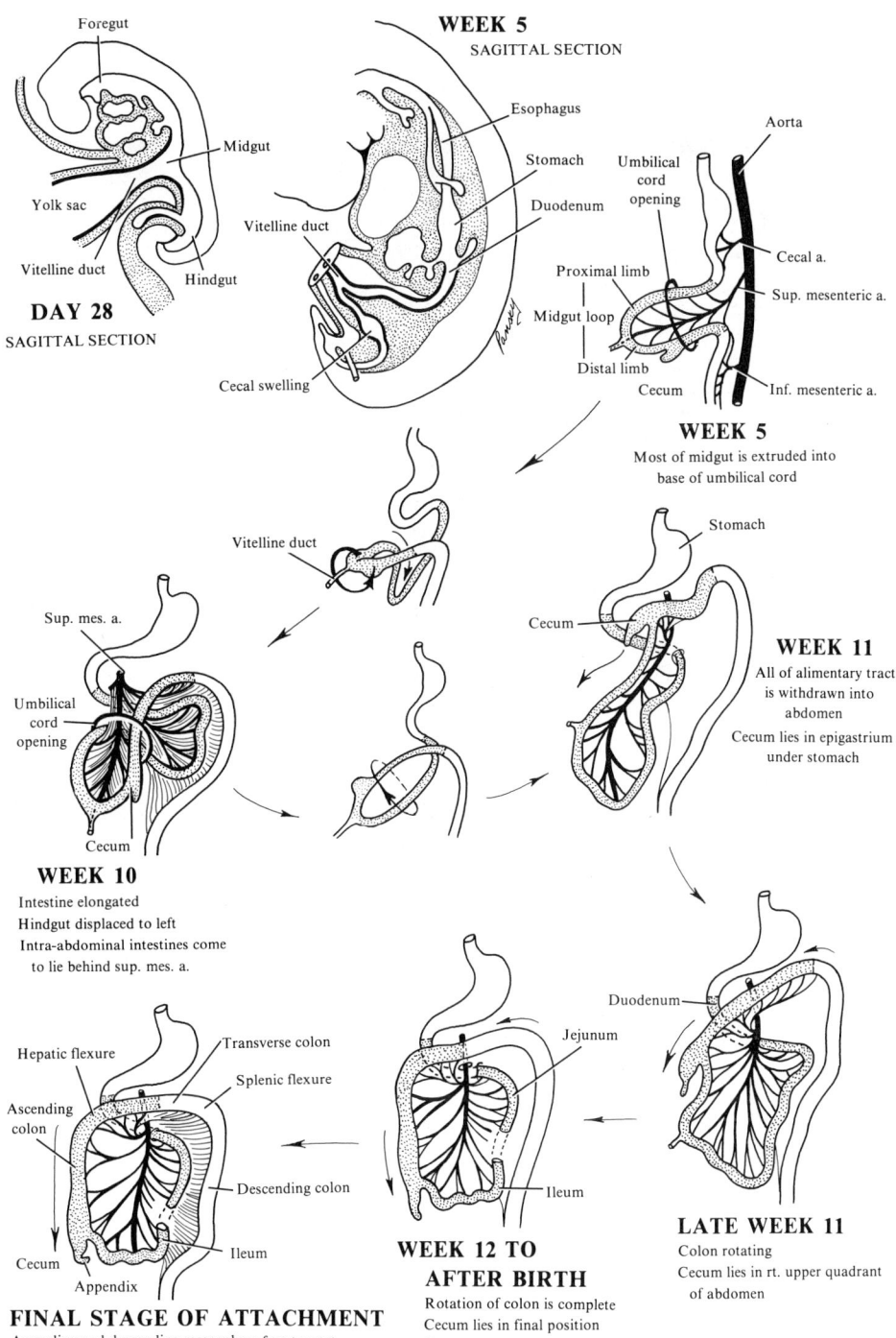

85. THE MIDGUT: FIXATION, THE CECUM AND APPENDIX

II. Midgut development (cont.)
A. ROTATION AND FIXATION (cont.)
  3. Fixation of the midgut
     a. The proximal portion of the colon lengthens, giving rise to the hepatic flexure and ascending colon as the cecum descends from the upper to the lower right side of the abdomen into the right iliac fossa
     b. As the intestines settle into their final positions, their mesenteries press against the back wall of the abdominal cavity
        i. The mesentery of the ascending colon fuses with parietal peritoneum and disappears; thus, the ascending colon becomes retroperitoneal
        ii. The duodenum (except for its foregut portion) also becomes retroperitoneal
        iii. The rest of the midgut loop derivatives (jejunum and ileum) keep their mesenteries, which at first are attached to the midline of the posterior abdominal wall, but during midgut rotation they twist around the origin of the superior mesenteric artery. When the mesentery of the ascending colon disappears, the intestinal mesentery gets a new line of attachment from the duodenojejunal junction down to the ileocecal junction
B. DUODENAL FIXATION: with rotation of the duodenum and stomach, the duodenum and pancreas fall to the right and meet the dorsal abdominal wall; the adjacent layers of peritoneum fuse and disappear; and most of the duodenum and head of the pancreas become retroperitoneal
C. THE CECUM AND APPENDIX
  1. The cecal diverticulum appears in week 6 and is the primordium of the cecum and the vermiform appendix
     a. The diverticulum is seen as a conical pouch on the antimesenteric border of the caudal limb of the midgut loop just beyond the apex of the loop
     b. The distal end of the blind sac does not grow as fast, thus the appendix, which is a vestige of the incomplete development of the cecum, develops
     c. As the proximal portion of the colon elongates, the cecum and the appendix descend on the right side of the abdomen. The position of the appendix can be variable
        i. Retrocecal appendix: behind the cecum
        ii. Retrocolic appendix: behind the ascending colon
        iii. Pelvic appendix: appendix descends into the pelvis
     d. The appendix grows in length, so that at birth, it is long and worm-shaped, or vermiform
     e. After birth, the cecal wall grows unequally and the appendix comes to lie on its medial side

WEEK 8

3 MONTHS

5 MONTHS

WEEKS 7-8

FULLY DEVELOPED (ADULT)

ADULT

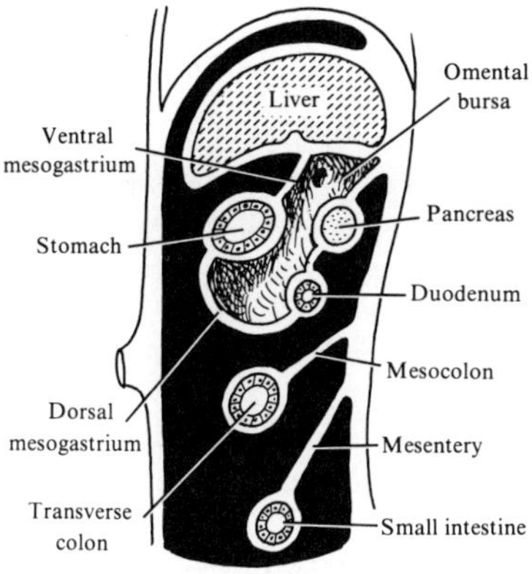

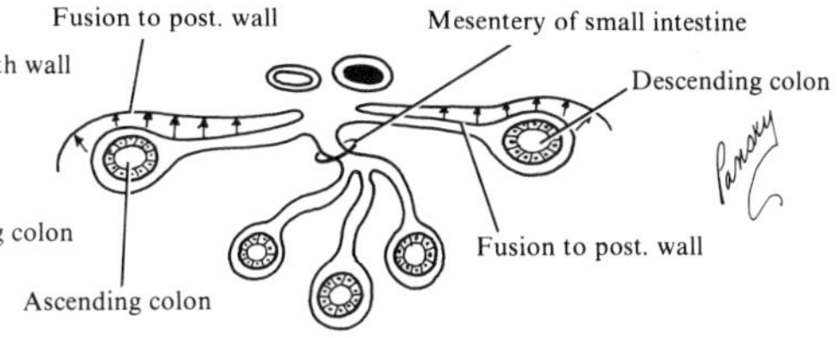

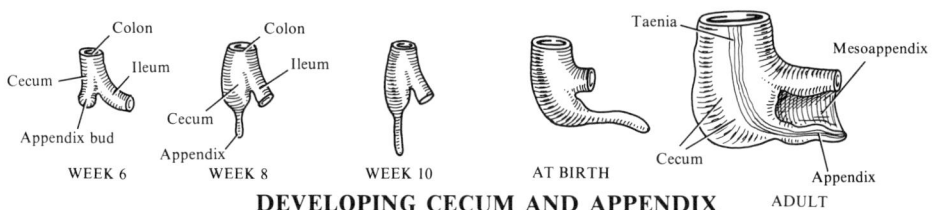

ARRANGEMENT OF MESENTERIES

WEEK 6 WEEK 8 WEEK 10 AT BIRTH ADULT

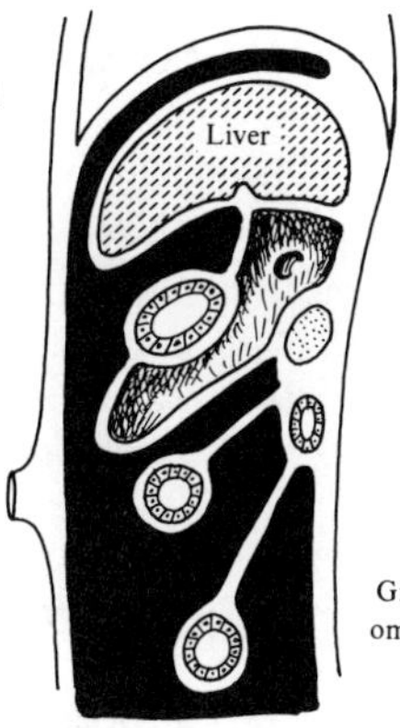

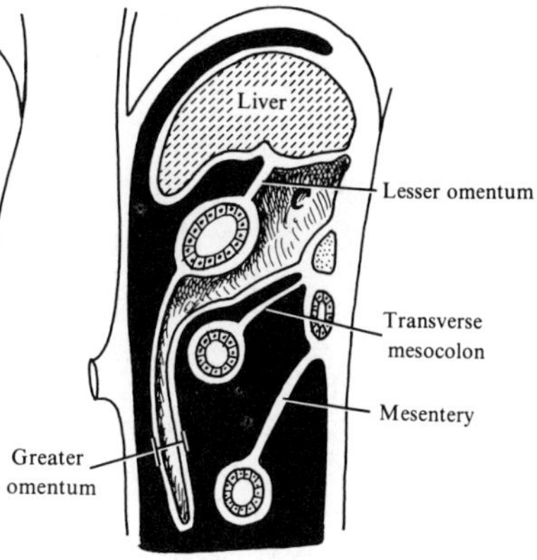

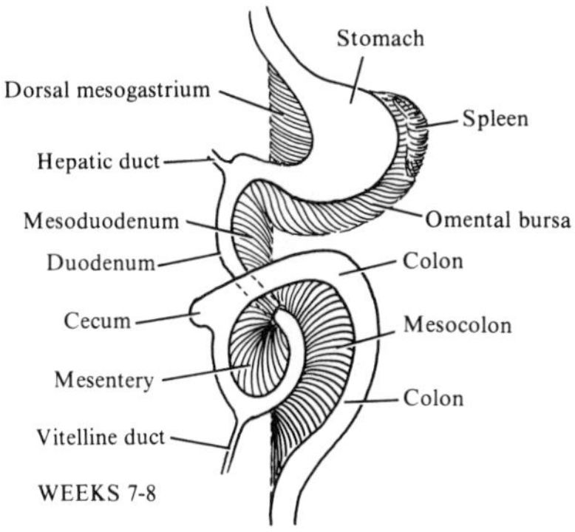

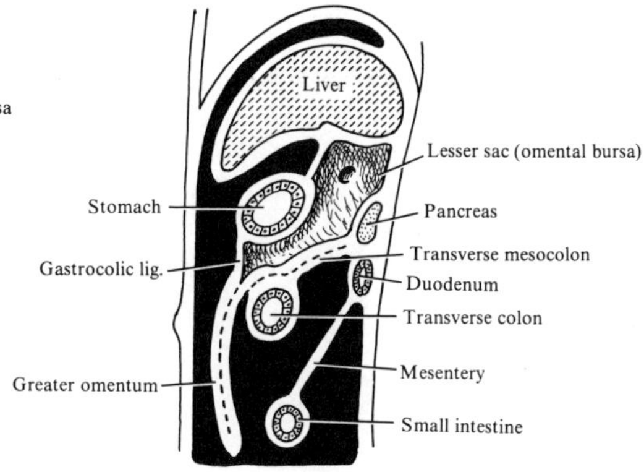

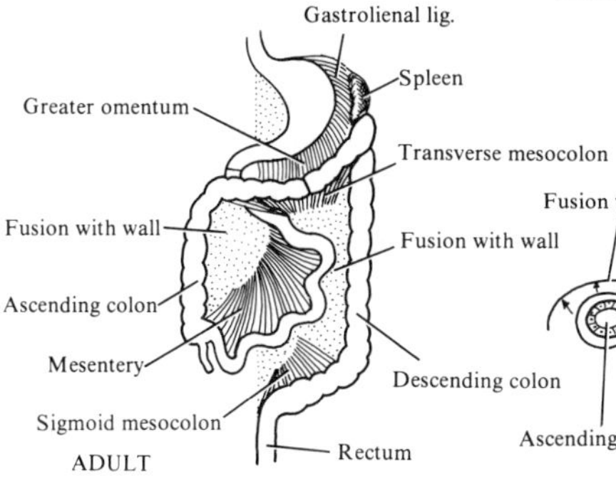

DEVELOPING CECUM AND APPENDIX

86. DEVELOPMENT OF THE HINDGUT

I. **The hindgut** follows the midgut, in the embryo, and extends from the posterior intestinal portal to the cloacal membrane. It gives rise to the left one-third to one-half or distal portion of the transverse colon, the descending colon, the sigmoid or pelvic colon, the rectum, the upper portion of the anal canal, and part of the urogenital system (e.g., the bladder and urethra). It is supplied by the inferior mesenteric artery. The terminal part of the hindgut enters into the cloaca, which is an entoderm-lined cavity that is in direct contact with the surface ectoderm
   A. THE CLOACAL MEMBRANE is composed of entoderm of the cloaca and ectoderm of the proctodeum or anal pit
   B. THE TERMINAL PART OF THE HINDGUT, the cloaca, receives the allantois ventrally and the mesonephric ducts laterally

II. **Fixation of the hindgut:** when the mesentary fuses with the peritoneum of the left dorsal abdominal wall and then disappears posteriorly, the descending colon becomes retroperitoneal
   A. THE MESENTERY OF THE SIGMOID COLON, however, persists, although diminished

III. **Partitioning of the cloaca**
   A. DURING DEVELOPMENT, a coronal wedge or ridge of mesenchyme, the *urorectal septum*, forms in the angle between the allantois and the hindgut
      1. As the septum grows caudad toward the cloacal membrane, it divides the cloaca into an anterior portion, the *primitive urogenital sinus,* and a posterior part, the *anorectal canal*
      2. By 7 weeks of age, the urorectal septum reaches the cloacal membrane and fuses with it. Thus, the membrane is divided into a posterior *anal membrane* and a larger anterior *urogenital membrane.* The area of fusion of the urorectal septum and the cloacal membrane becomes the *primitive perineum or the perineal body*
      3. In week 9, proliferation of mesenchyme around the anal membrane raises the surrounding ectoderm to form a shallow pit, the *anal pit* or *proctodeum.* The surrounding swellings are called the *anal folds*
         a. Soon after, the anal membrane, at the bottom of the anal pit, ruptures to establish the *anal canal,* an open pathway from the rectum to the outside (actually, to the amniotic cavity, at this time of development)

IV. **The anal canal:** its upper two-thirds is derived from hindgut and its lower third from the anal pit
   A. THE JUNCTION of the anal pit ectoderm and the hindgut entoderm is indicated by the *anatomic anorectal* or *pectinate (dentate) line* which is at the level of the anal or semilunar valves. This is the former site of the anal membrane and where the epithelium changes from columnar to stratified squamous
   B. AT THE ANUS, the epithelium (anoderm) is keratinized and is continuous with the surface skin of the perineum. The surrounding tissue is derived from splanchnic mesenchyme
   C. THE HINDGUT PORTION of the anal canal is supplied by the inferior mesenteric artery; whereas, the power portion (anal pit) is supplied by the internal pudendal branch of the internal iliac artery
   D. AS A RESULT OF THE DIFFERENT EMBRYOLOGIC ORIGINS of the upper and lower parts of the anal canal, venous (external and internal hemorrhoidal or rectal plexi) and lymphatic drainage (via the iliac and inguinal nodes) and nerve supply (autonomic and peripheral) differ in the various parts of the canal

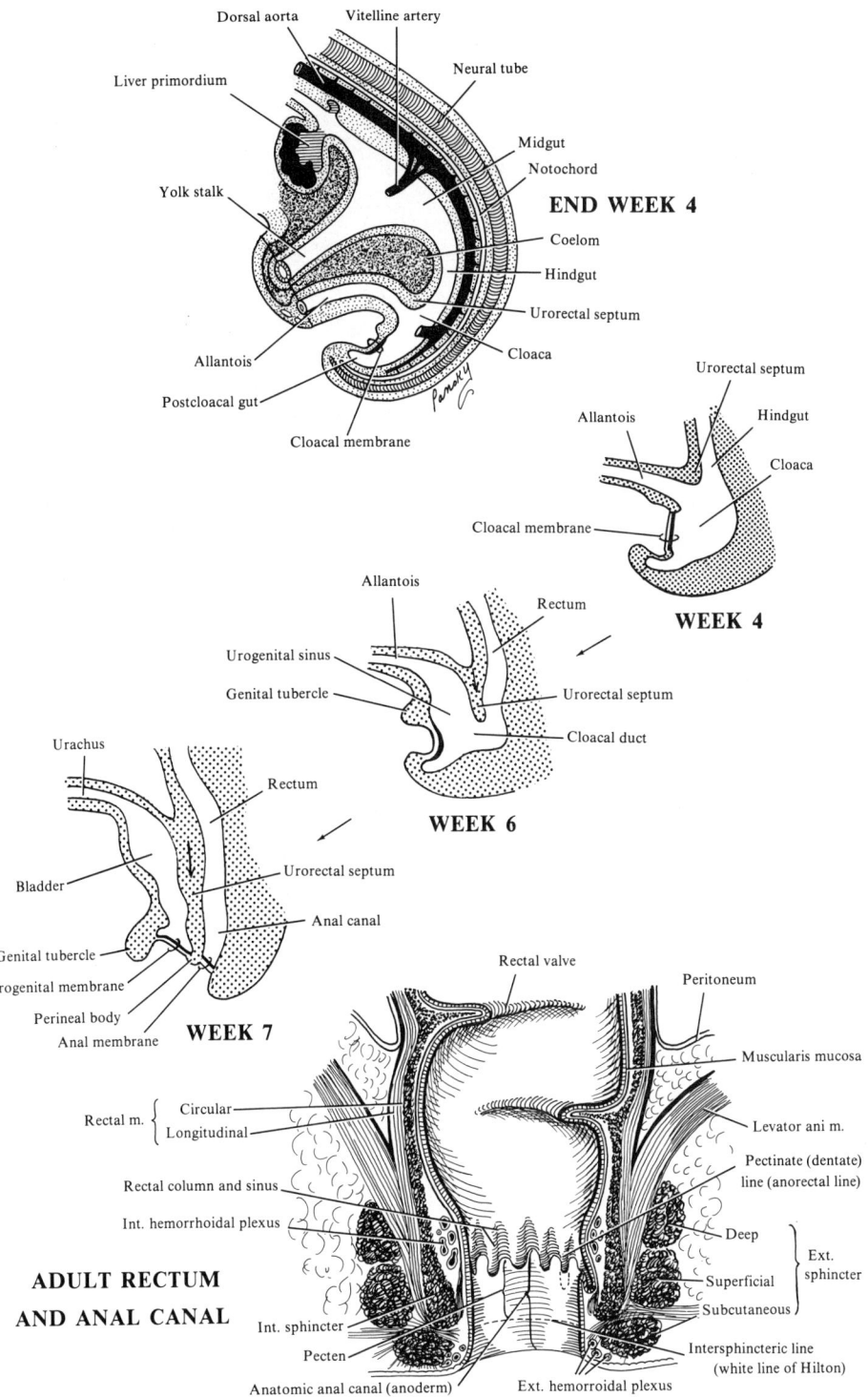

87. CONGENITAL MALFORMATIONS OF THE DIGESTIVE SYSTEM: FOREGUT MALFORMATIONS

I. Foregut malformations

A. ESOPHAGEAL ATRESIA AND STENOSIS: esophageal atresia often occurs with tracheoesophageal fistula but may occur independently (rare)
  1. Atresia or stenosis is due to unequal division of the foregut into respiratory and digestive parts or failure of esophageal recanalization
  2. In esophageal atresia, amniotic fluid cannot enter the intestine for absorption and transfer to the placenta for removal, resulting in *polyhydramnios* (excess amniotic fluid)

B. PYLORIC STENOSIS: reduced lumen of the pyloris due to hypertrophy of the sphincter muscle layers. Its cause is unknown but may be genetic
  1. It is the most common abnormality of the stomach in infants, develops in fetal life, and occurs in about 1/200 males and 1/1000 females
  2. Extreme narrowing of the pyloric lumen obstructs the passage of food, resulting in severe progressive vomiting

C. ATRESIA OF THE GALLBLADDER AND BILE DUCTS is the most serious malformation of extrahepatic biliary system and is seen in 1/20,000 births
  1. Results from persistence of the solid stage of the duct and gallbladder development. The bladder remains atretic, and the ducts appear as narrow, fibrous cords
  2. Atresia may be limited to only a small part of the common bile duct leading to distention of the bladder and the hepatic duct, resulting in a severe, steadily increasing jaundice after birth
  3. Duplication, partial subdivision, and diverticula of the gallbladder are also seen

D. LIVER MALFORMATIONS: variations in lobulation are common, but gross malformations are rare. Variations of the hepatic ducts, common bile ducts, and cystic ducts are common

E. PANCREATIC MALFORMATIONS
  1. Accessory pancreatic tissue (heterotopic pancreatic tissue) may be found anywhere from the distal esophagus to the tip of the primary intestinal loop. It is seen most frequently in the wall of the stomach or duodenum or in a Meckel's diverticulum
  2. Pancreatic bladder: a part of the ventral pancreatic bud grows out with the liver bud and forms a pancreatic nodule
  3. Annular pancreas is a rare malformation consisting of a thin flat band of pancreatic tissue surrounding the second portion of the duodenum
     a. It may be symptomless but may also constrict the duodenum and result in an obstruction
     b. Probably caused by the growth of a bifid ventral pancreatic bud around the duodenum which fuses with the dorsal bud to form a ring

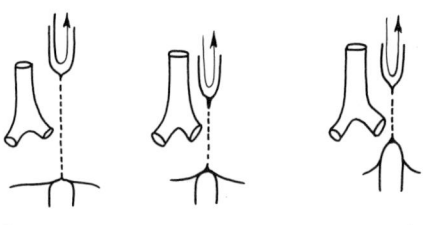
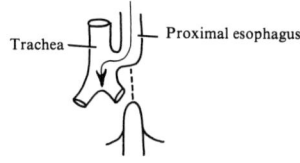

ESOPHAGAL ATRESIA ONLY
Excessive salivation

ATRESIA WITH PROXIMAL FISTULA
Coughing, choking, cyanosis

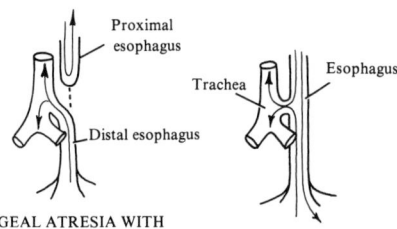
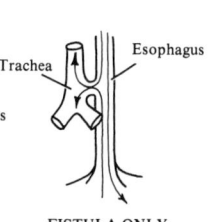
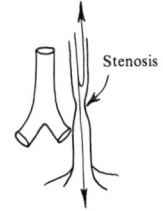

ESOPHAGEAL ATRESIA WITH DISTAL FISTULA
Excessive salivation, coughing, choking, and cyanosis

FISTULA ONLY
Episodic coughing, choking, and cyanosis

ESOPHAGEAL STENOSIS
Partial regurgitation and dysphagia

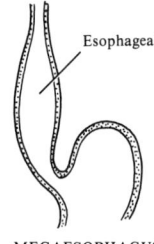
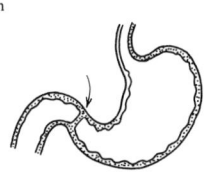
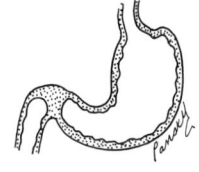

MEGAESOPHAGUS

GASTRIC ATRESIA
(MEMBRANOUS)

COMPLETE, SOLID GASTRIC ATRESIA

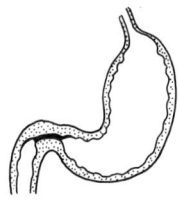
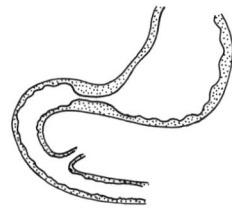
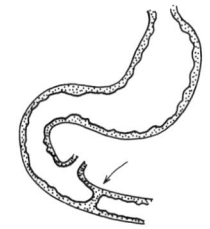

LUMINAL ATRESIA

INFANTILE HYPERTROPHIC PYLORIC STENOSIS

DUODENAL ATRESIA

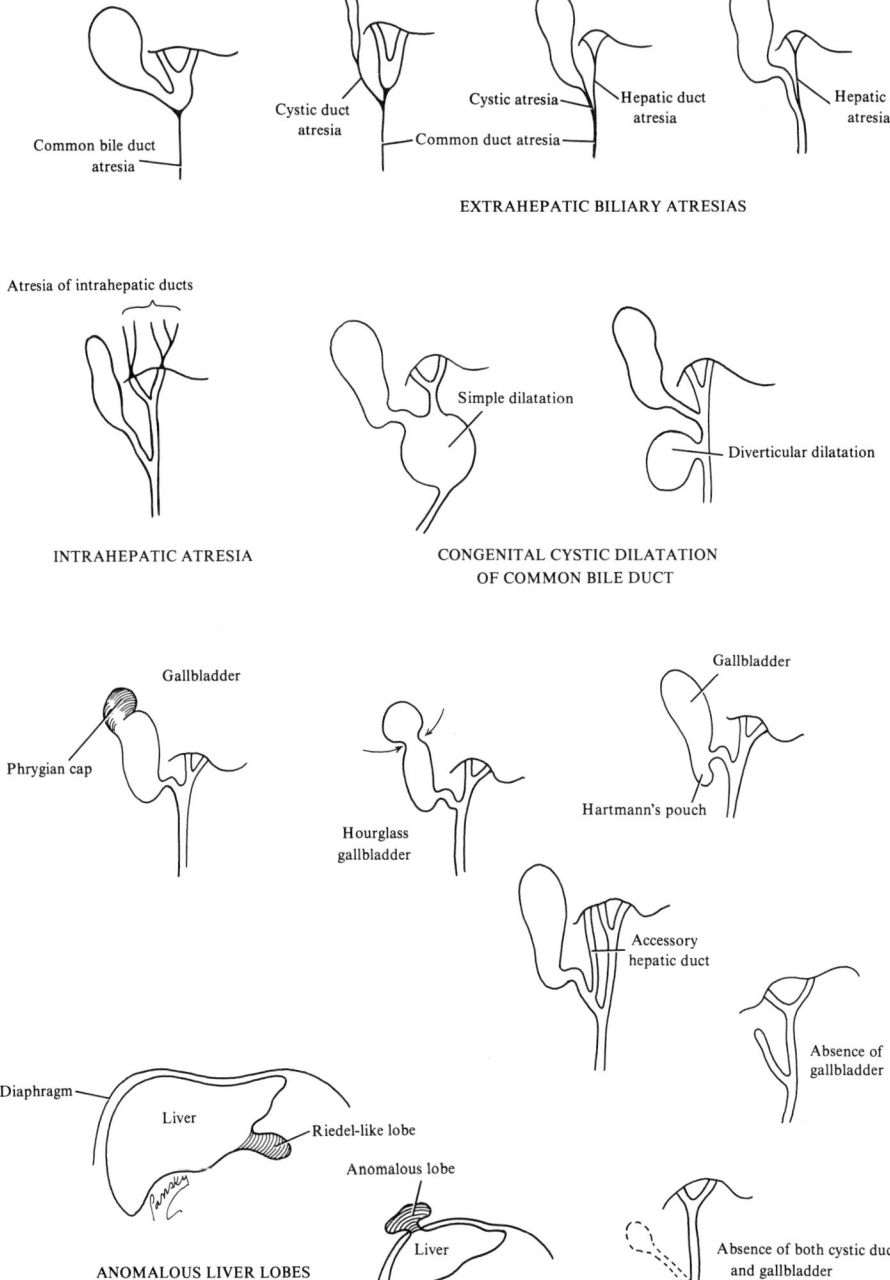

FIGURE 35. **Congenital anomalies and malformations of the biliary apparatus.**

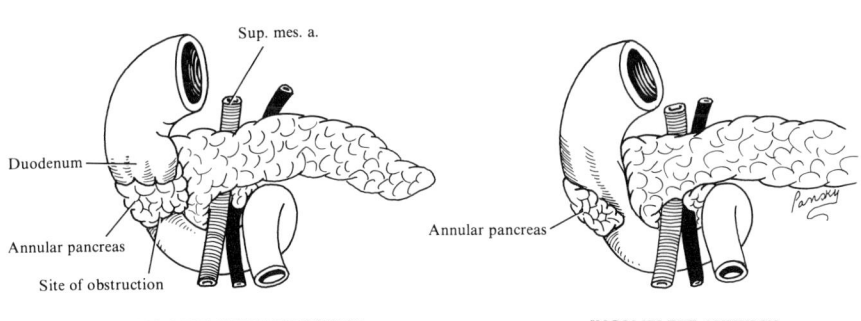

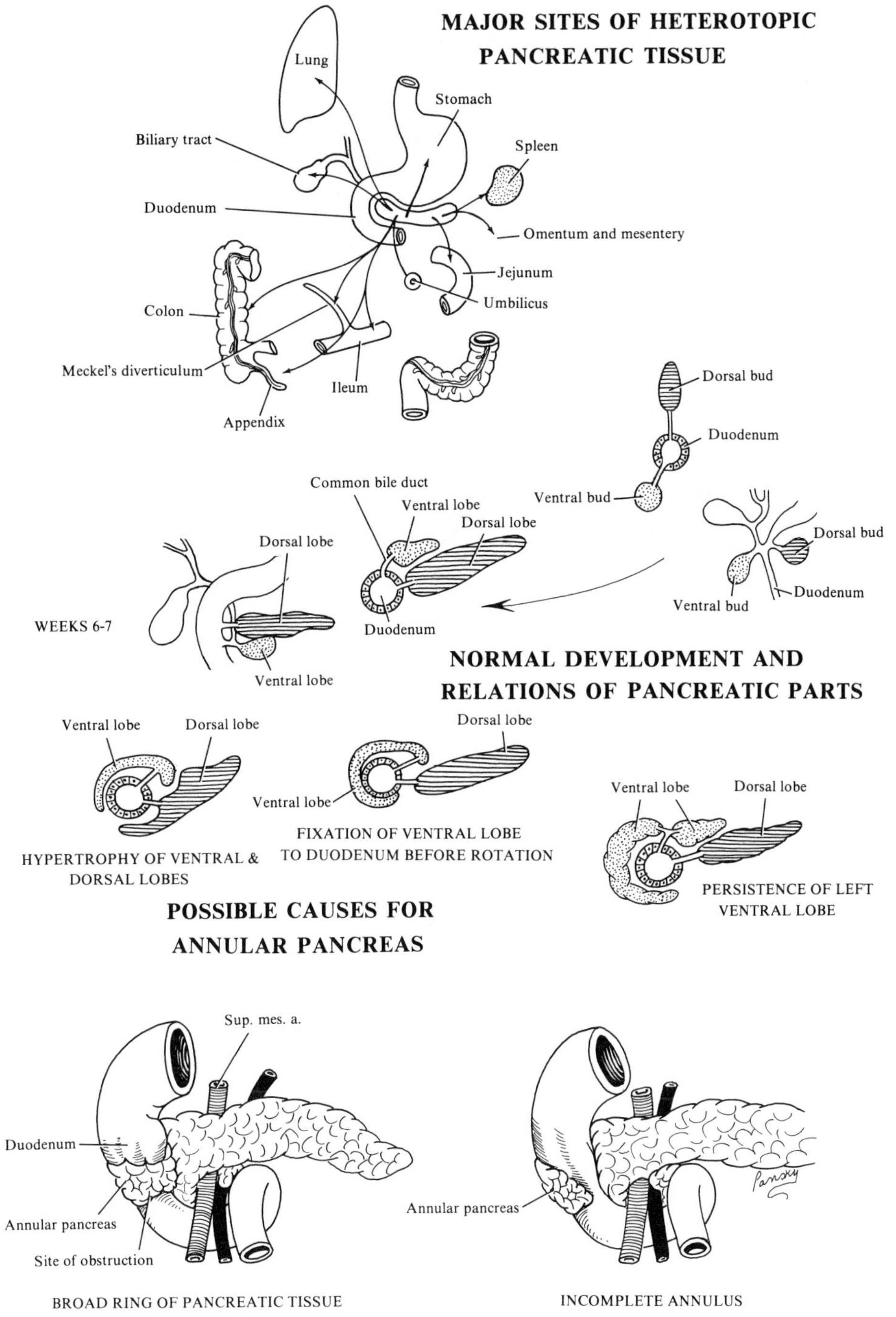

FIGURE 36. **Annular pancreas and heterotopic pancreatic tissue.**

88. CONGENITAL MALFORMATIONS OF THE DIGESTIVE SYSTEM: MIDGUT MALFORMATIONS

I. **Midgut malformations** are common and may result from abnormal development of the digestive tube, incomplete rotation, and/or failure of fixation, from abnormalities in its location and arrangement, or from defective development of neighboring organs. More than one cause may be interconnected. The major clinical manifestation of malformation is a syndrome of neonatal intestinal obstruction
   A. UNDERDEVELOPMENT
      1. Agenesis: complete absence of an intestinal segment (incompatible with survival if it is very extensive)
      2. Atresia and stenosis: an intestinal segment is, and remains, narrow and constricted (obstructs the passage of food)
         a. Seen most often in the ileum
         b. Duodenal atresia: distal to the duodenal papilla; vomitus always has bile
         c. Polyhydramnios may occur with duodenal atresia
         d. Cause often failure of recanalization or interruption of the blood supply
      3. Aplasia: where the contracted segment does have a mucosa and lumen
      4. Mucosal narrowing: often associated with other anomalies
   B. OVERDEVELOPMENT
      1. Duplications: range from simple *diverticulae* to almost complete doubling of the digestive tube. Also may include many varieties of *cystic malformations*
         a. Commonly found on the dorsal (mesenteric) border of the intestine
         b. All duplications are caused by failure of normal recanalization and formation of two lumina
   C. OMPHALOCELE OR EXOMPHALOS: seen in 1/6500 births and results from failure of the intestines to return to the abdomen during stage 2 of midgut loop rotation. The loop remains in the extraembryonic coelom of the umbilical cord
      1. Hernia can be a single loop of bowel or may contain most of the intestine as well as the spleen, liver, and pancreas
         a. The hernial sac is covered by the amnion of the umbilical cord
      2. Eventration of the abdominal viscera or congenital umbilical hernia (type of omphalocele) is due to faulty closure of the lateral body folds during week 4 of embryonic life. The abdominal viscera develop outside the embryo in a sac of amnion.
         a. Is often associated with exstrophy of the urinary bladder
      3. Gastroschisis is uncommon; due to a defect of the anterior abdominal wall (not stomach) with extrusion of abdominal contents without involving umbilical cord
         a. The viscera protrude into the amniotic sac and float in fluid
         b. Usually seen on the right side and is more common in males
   D. ANOMALIES OF POSITION
      1. Nonrotation: quite common; called "left-sided colon" and generally is asymptomatic, but twisting or volvulus can occur
         a. Midgut does not rotate after it enters the abdomen. Thus, the caudal limb enters before the cranial limb
         b. Small intestines lie on the right side and the entire large intestines on left. May cause obstruction of vessels and gut if kinking or twisting occurs
      2. Volvulus and mixed rotation: cecum lies below the pyloris and is fixed to the posterior abdominal wall by peritoneal bands that cross over the duodenum
         a. Usually causes duodenal obstruction
         b. Due to a failure of the midgut loop to complete final 90° of rotation, thus, terminal ileum enters the abdominal cavity first
      3. Reversed rotation: rare, clockwise rotation (not counterclockwise)
         a. Duodenum lies in front of the superior mesenteric artery and transverse colon behind it, which may obstruct the latter due to pressure from the artery
         b. Small intestines lie on the left; large intestines lie on the right, and cecum is found in the center

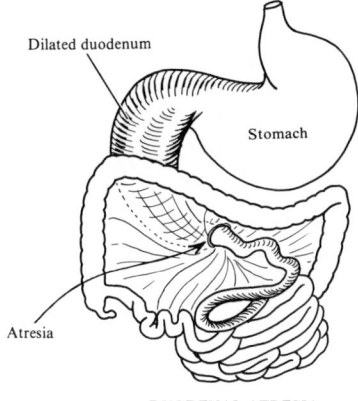

DUODENAL ATRESIA

DUODENAL STENOSIS

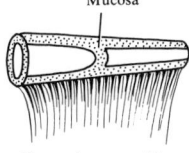

Closure by mucosal layer

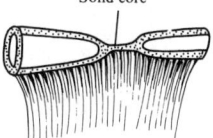

Closure by solid core

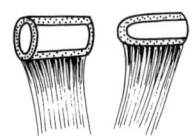

Complete segmental absence

INTESTINAL ATRESIA

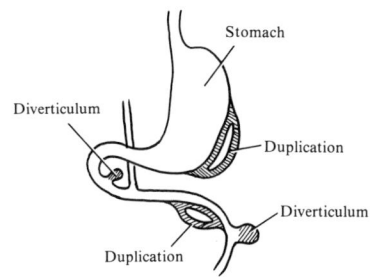

VOLVULUS AROUND FIBROUS CORD ATTACHED TO ANT. ABDOMINAL WALL

TRUE ANTIMESENTERIC DIVERTICULUM

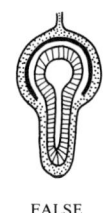

FALSE DIVERTICULUM
Missing some intestinal coats

INTESTINAL DUPLICATION

INTESTINAL DUPLICATION

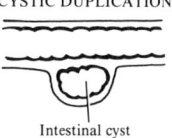

CYSTIC DUPLICATION

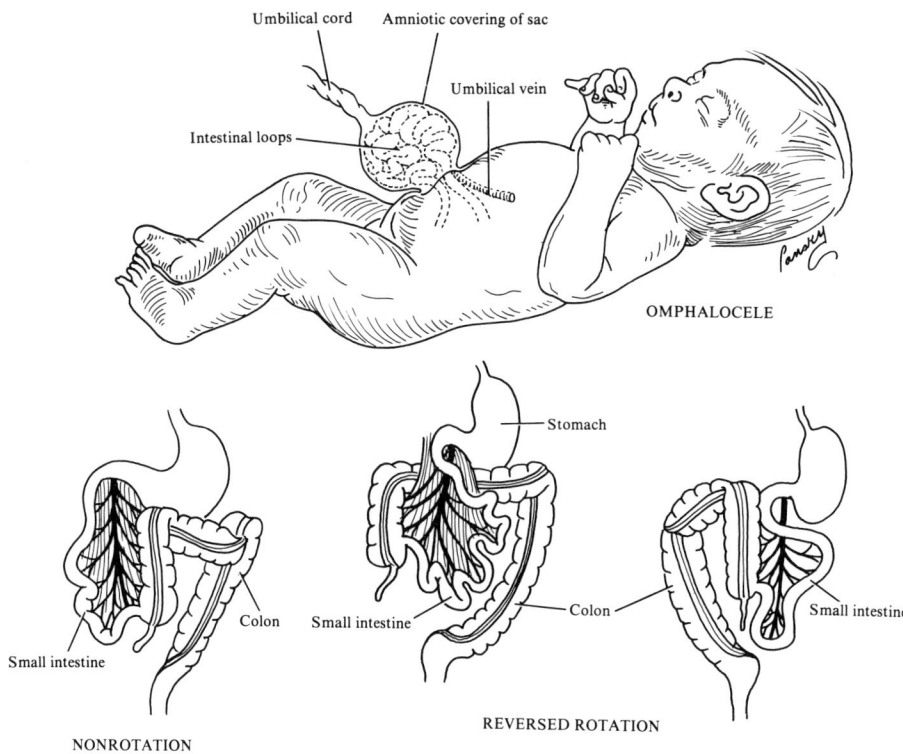
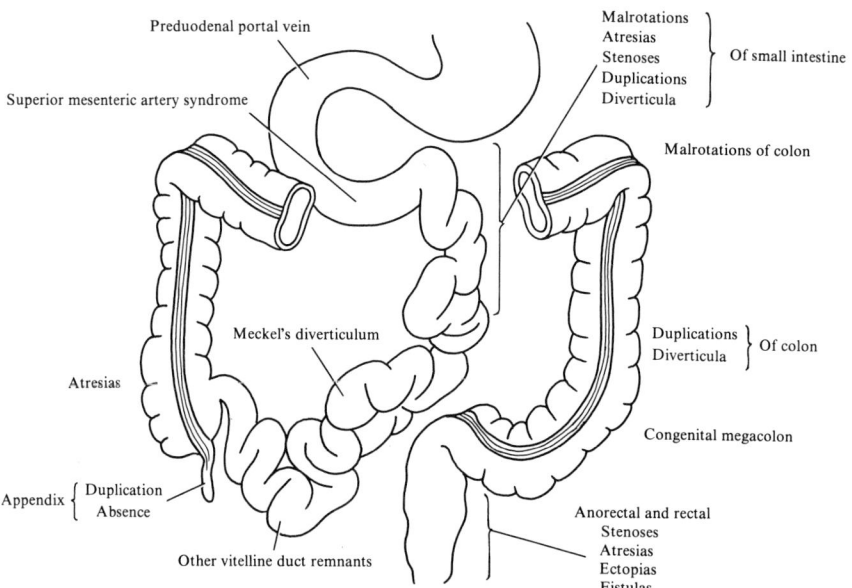

FIGURE 37. **Anomalies of the small intestine and colon.**

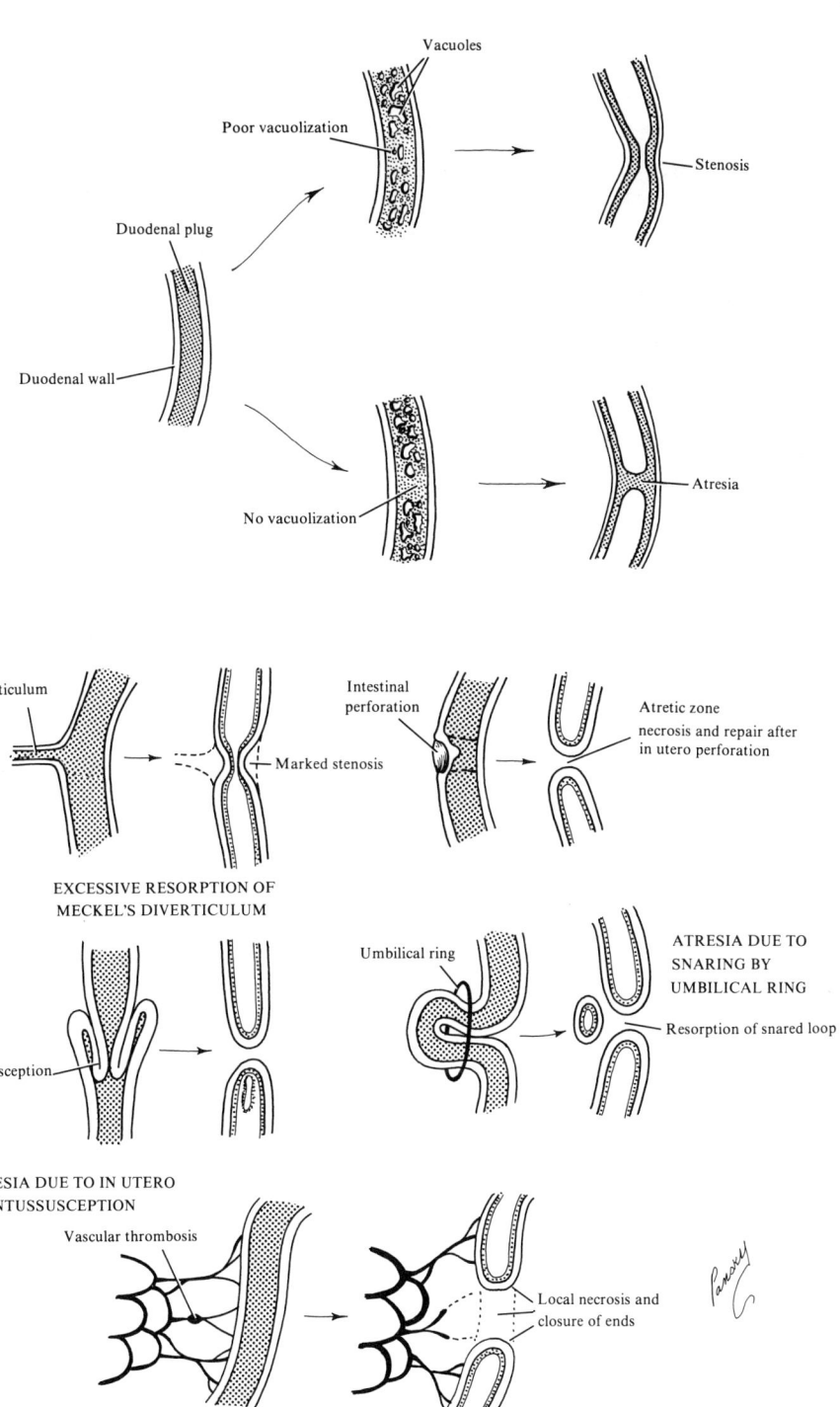

FIGURE 38. **Atresia, stenosis, and Meckel's diverticulum.**

89. MIDGUT MALFORMATIONS

I. Midgut malformations (cont.)
  D. ANOMALIES OF POSITION (cont.)
   4. Subhepatic cecum: failure of the proximal colon to elongate during stage 3 of rotation, thus, cecum ends up near the liver
   5. Mobile cecum: due to incomplete fixation of the ascending colon. Results also in a mobile and variable appendix and even volvulus of the cecum
   6. Midgut volvulus: mesenteries fail to undergo normal fixation, and the intestines twist with incomplete rotation of the midgut loop
       a. The small intestines hang by a narrow stalk of the superior mesenteric vessels and twist around it, thus obstructing at or near the duodenojejunal junction
  E. REMNANTS OF THE VITELLINE DUCT: the duct usually disappears at 6 weeks
   1. Meckel's diverticulum is an *ileal diverticulum* and the most common malformation of the digestive tract (2–4% of people). It is clinically significant because it can become inflamed and cause symptoms mimicking appendicitis
       a. It is located about 2 to 3 ft (0.6–1.0 m) from the ileocecal valve and is a finger-like pouch about 3 to 6 cm long arising from the antimesenteric border of the ileum
       b. Its walls contain all the layers of the ileum, but may also contain gastric and pancreatic tissue. The gastric mucosa may secrete acid and produce ulceration, bleeding, and perforation
   2. Umbilical or vitelline fistula: vitelline duct remains patent over its entire length, thus, connects the umbilicus and intestinal tract. It may lead to fecal discharge at the umbilicus or ileal prolapse through the fistula
       a. One may find the duct closed at both ends and the formation of a *vitelline cyst* or *enterocystoma* in its midportion
  F. HISTOLOGIC ANOMALIES
   1. Congenital aganglionic dystony or megacolon (Hirschsprung's disease) is rare; may affect the colon, small intestine, or duodenum, but especially the rectum and its internal and external sphincters
       a. Causes a portion of the colon to dilate due to the absence of ganglion cells of the myenteric plexus distal to the dilated segment, as a result of failure of migration of the neural crest cells. The dilated portion itself has a normal population of ganglion cells
       b. Dilatation is caused by failure of the distal segment to move the intestinal contents onward
       c. The severity of the condition is directly proportional to the length of the gut segment involved
   2. Mucoviscidosis: adherence of meconium to the intestinal wall secondary to a deficiency of trypsin secretion by the pancreas as a result of the latter being invaded by interstitial fibrosis of unknown origin

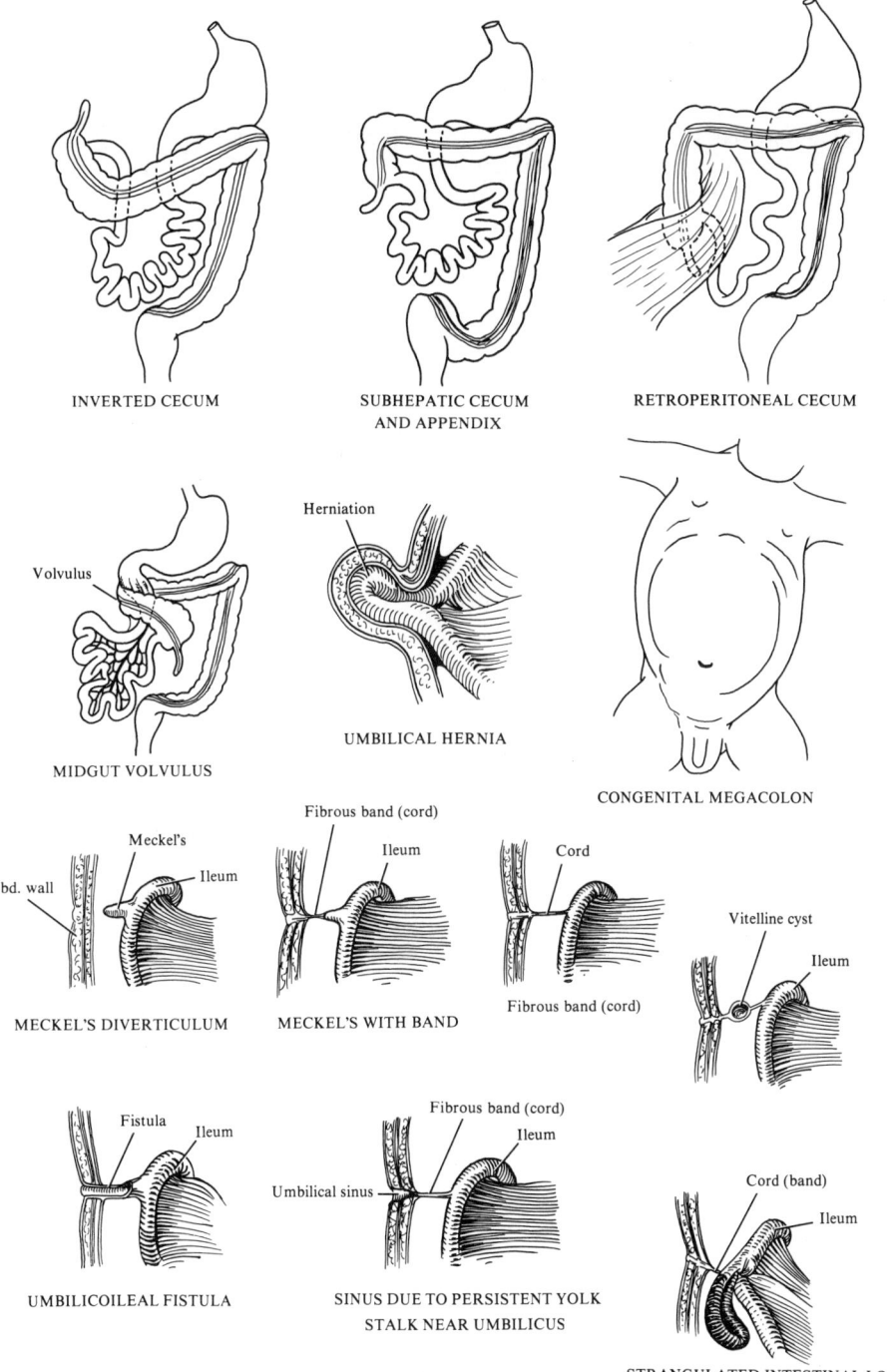

90. CONGENITAL MALFORMATIONS OF THE DIGESTIVE SYSTEM: HINDGUT MALFORMATIONS

I. **Malformations of the hindgut:** some form of imperforate anus is usually seen, occurs in 1/5000 births, and is more common in males. Most anorectal malformations result from an abnormal development of the urorectal septum, resulting in an incomplete separation of the cloaca into the urogenital and anorectal parts

A. ANORECTAL MALFORMATIONS
   1. Clinically 3 major types are seen
      a. Imperforate anus or absent anus: classical anal imperfections include many varieties, and all require surgical intervention
      b. Insufficient anus leads to problems of meconium evacuation and should be treated without delay
      c. Ectopic sinus: less serious than a or b since it does allow some intestinal transport, but it, too, is usually functionally insufficient
   2. Developmental problems
      a. Superficial malformations: due to anomalies in formation and fixation at superficial perineal levels. Here we see
         i. Anal agenesis or insufficient anus (with or without fistula): the canal may end blindly, and there may be an ectopic opening (ectopic anus) or fistula opening in the perineum or vulva or male urethra. Due to incomplete separation of the cloaca by the urorectal septum
         ii. Membranous atresia or covered anus (with or without fistula): very rare; anus is in normal place, but a thin layer of tissue separates the anal canal from the exterior. Due to a failure of the anal membrane to perforate at the end of week 8
         iii. Anorectal agenesis with or without fistula: rectum ends blindly above the anal canal, but there is usually a fistula to the urethra in the male or vagina in the female. The defect is similar to i
      b. Deep malformations: where the anomaly affects septation of the cloaca by aberrant migration of the urorectal septum
         i. Pure rectal atresia: complete failure of the formation of the inferior part of the rectum and anal canal
         ii. Rectal atresia with fistula: always insufficient. The length and degree of anastomosis differentiate the various types of anomaly
      c. Mixed malformations include all forms of ectopic anus
         i. May involve abnormal anastomoses of the anus to the perineum and reflect both perineal and cloacal abnormalities

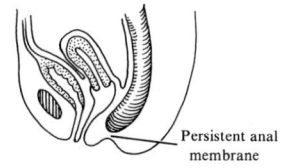
IMPERFORATE ANUS
Membranous atresia

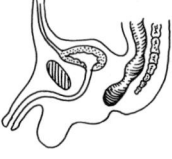
ANAL AGENESIS
WITHOUT FISTULA

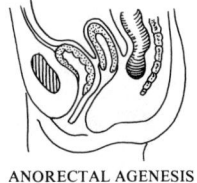
ANORECTAL AGENESIS
WITHOUT FISTULA

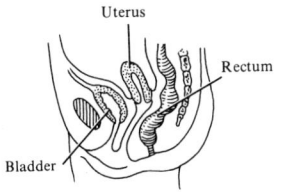
ANAL STENOSIS

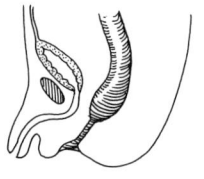
RECTAL STENOSIS

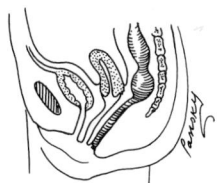
ANORECTAL AGENESIS
WITH FISTULA
Female

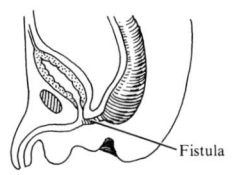
URORECTAL FISTULA
WITH ANAL AGENESIS

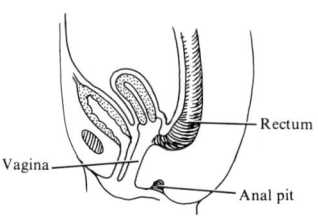
RECTOVAGINAL FISTULA

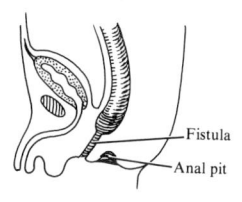
ANOPERINEAL FISTULA

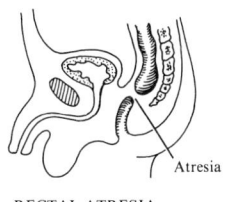
RECTAL ATRESIA
Male

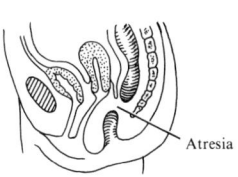
RECTAL ATRESIA
Female

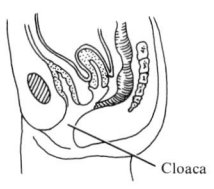
PERSISTENT CLOACA

UNIT EIGHT
THE URINARY SYSTEM

91. THE URINARY OR EXCRETORY SYSTEM: INTERMEDIATE PLATE, NEPHROGENIC CORD, AND PRONEPHROS

I. **Introduction:** embryologically and anatomically, the *urinary system* (excretes waste products and excess water via an intricate tubular system in the kidneys) and the *genital system* (assures continuity of the race by production of germ cells) are closely associated, especially in early stages of development. Both develop from a common mesodermal ridge along the posterior abdominal wall, and the excretory ducts of both systems initially enter a common cavity, the cloaca. In the male, the urethra conveys both urine and semen; although separate in the female, the urethra and vagina both open into a common vestibule. The genital system is discussed elsewhere

II. **The intermediate plate:** the urinary system is derived from the *intermediate mesoderm* or *plate* [lying between the paraxial (somite-forming) mesoderm and the lateral plate] and the *cloaca*
   A. THE URINARY SYSTEM develops in a craniocaudal direction in successive chronologic steps
      1. The *definitive kidney* or *metanephros* is preceded by 2 transitory structures
         a. The *pronephros* may be thought of as a "rough draft" and is rapidly replaced
         b. The *mesonephros* reaches complete development but later predominantly regresses. Remnants of this system are incorporated into the urogenital system

III. **The nephrogenic cord:** the intermediate mesoderm migrates ventrally and loses its connections with the somites. This longitudinal mass of nephrogenic mesoderm on each side of the body becomes the nephrogenic cord
   A. THE NEPHROGENIC CORD is at first continuous with the paraxial mesoderm (internally) and the lateral plate (externally), but it later separates yet remains close to the intraembryonic coelom
   B. LIKE SOMITE MESODERM, the nephrogenic cord undergoes *metameric segmentation* into *nephrotomes*
      1. Metamerization is clear at the cranial end of the embryo, rudimentary in its middle portion, and almost nonexistent at its caudal end, where the nephrogenic mesoderm remains undivided
      2. The nephrogenic cords give rise to the renal tubules of the kidney
   C. THE NEPHROGENIC CORDS produce bilateral longitudinal bulges, the *urogenital ridges*, on the dorsal wall of the coelomic cavity. The ridges give rise to both nephric and genital structures

IV. **Development of the nephrogenic cords:** arising from the long ribbons of nephrogenic cords, each transitory kidney develops from 3 primordia which succeed each other, not only in time but in space, and differentiate progressively from the cervical to the caudal region of the embryo. The primordia are referred to as the pronephros, mesonephros, and metanephros, from cervical to caudal region, respectively
   A. THE PRONEPHROS (forekidney) differentiates at the end of week 3 in the cervical region and is nonfunctional. It disappears at the end of week 4. Cranialcaudal development is especially true of this transitory kidney
      1. First, the nephrogenic cord cleaves into *nephrotomes*
      2. Second, each nephrotome hollows out into a *nephrotomal vesicle,* which becomes oval in shape
      3. Third, the union of vesicles forms the beginning of the *pronephric duct,* which progresses toward the cloaca
      4. Fourth, while above is taking place, the pronephros degenerates and disappears
      5. Currently, it is thought that the pronephros disappears completely, leaving no vestiges, thus, it is the mesonephros that forms its own collecting duct
         a. The classical conception was that the pronephric duct persisted and was used by the mesonephros

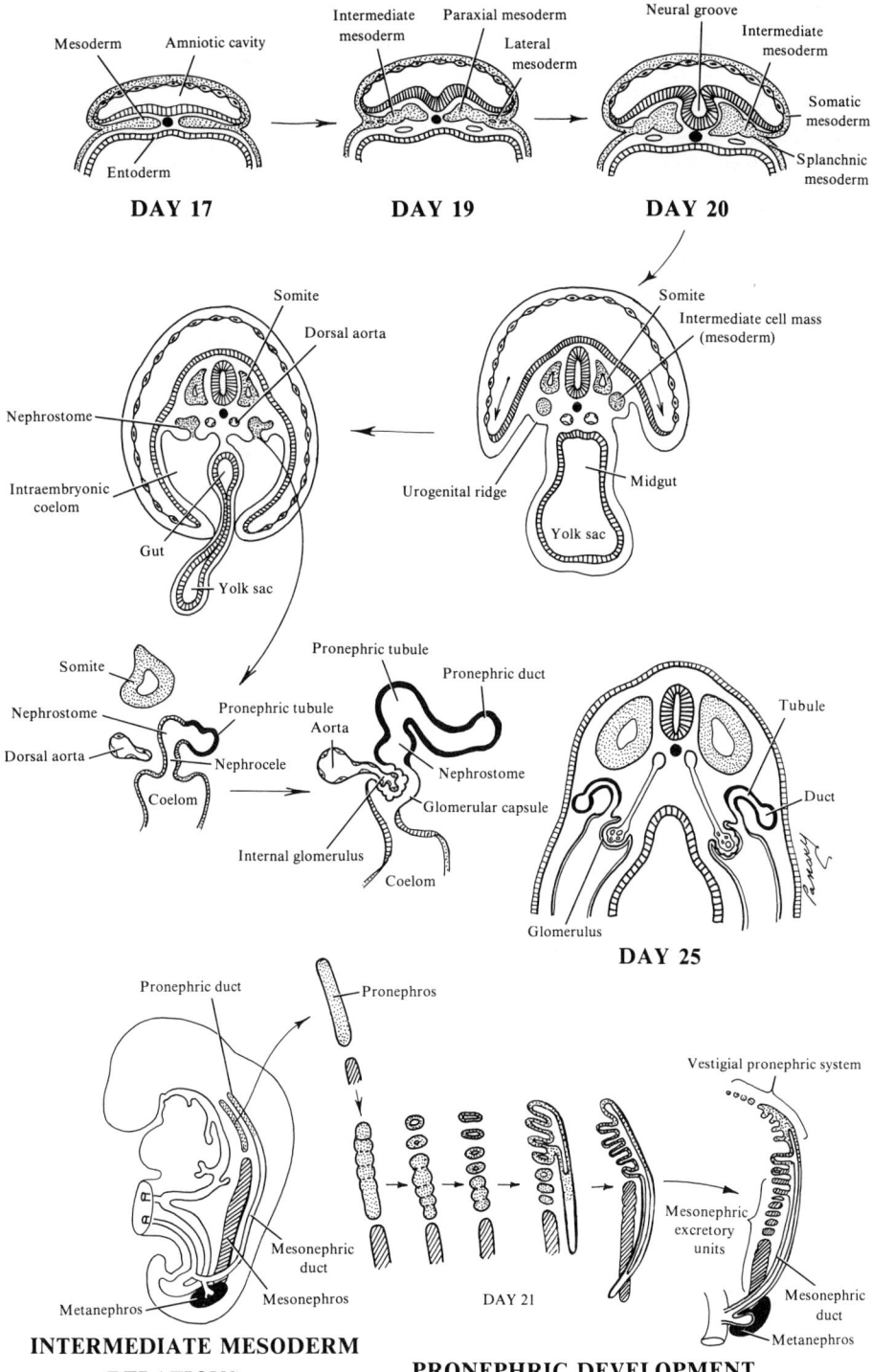

92. THE URINARY OR EXCRETORY SYSTEM: THE MESONEPHROS

I. **The mesonephros (midkidney) or wolffian body*** differentiates in week 4 caudal to the rudimentary pronephros and may function as an interim kidney until the permanent kidney is established. The mesonephros regresses in week 8
   A. As in the development of the pronephros, the mesonephric nephrotomes pass through the following stages, beginning in week 4
      1. Nephrotomal (mesonephric) vesicle formation
      2. The vesicle grows into an S-shaped mesonephric tubule
      3. The tubules lengthen rapidly
      4. There is an enlargement of the vesicle's internal or medial end to form a glomerular chamber opposite an arterial loop from the aorta
      5. Arrangement of the glomerular chamber around the arterial capillary cluster (glomerulus)
         a. The chamber forms a double-layer cup, the *glomerular* or *Bowman's capsule*
         b. The capsule plus the glomerulus forms the *mesonephric* or *renal corpuscle*
      6. The external end of the tubule opens into the mesonephric or wolffian duct (which may or may not be preexistent as the pronephric duct)
   B. The characteristic segmentation of the mesonephric nephrotomes is evident at the cranial portion of the mesonephric mass. The caudal portion regresses before its differentiation is complete.

II. **The anatomic relations of the mesonephros (wolffian body)**
   A. The mesonephros appears, in cross-section, to be a mass projecting into the peritoneal cavity
      1. It is attached by the *urogenital mesentery* and on its anteroexternal edge one sees the *urogenital cord*, containing the *mesonephric (wolffian)* and *müllerian ducts*
      2. On its anterointernal side, one sees the *gonadal primordium* which is attached by the *gonadal mesentery*
   B. Posteriorly, the mesonephros (wolffian body) is attached to the posterior (dorsal) body wall by the *mesonephric (wolffian body) mesentery* which is wide and close to the root of the mesentery near the aorta
      1. The coelomic epithelium on the inside surface of this mesentery gives rise to the adrenal cortex

III. **Regression of the mesonephros (wolffian body)**
   A. The mesonephros begins to regress toward the end of month 2
      1. No more than 40 mesonephric tubules are present in each kidney at any one time
      2. By the beginning of the fetal period, most of the mesonephros has degenerated and disappeared except for its duct and a few tubules near the gonads, which persist and are annexed by them as genital ducts in the male or form vestigial remnants in the female

---

*The gonadal primordium develops at the internal, anterior side of the mesonephros (wolffian body).

STAGES IN DEVELOPMENT OF THE MESONEPHRIC TUBULE

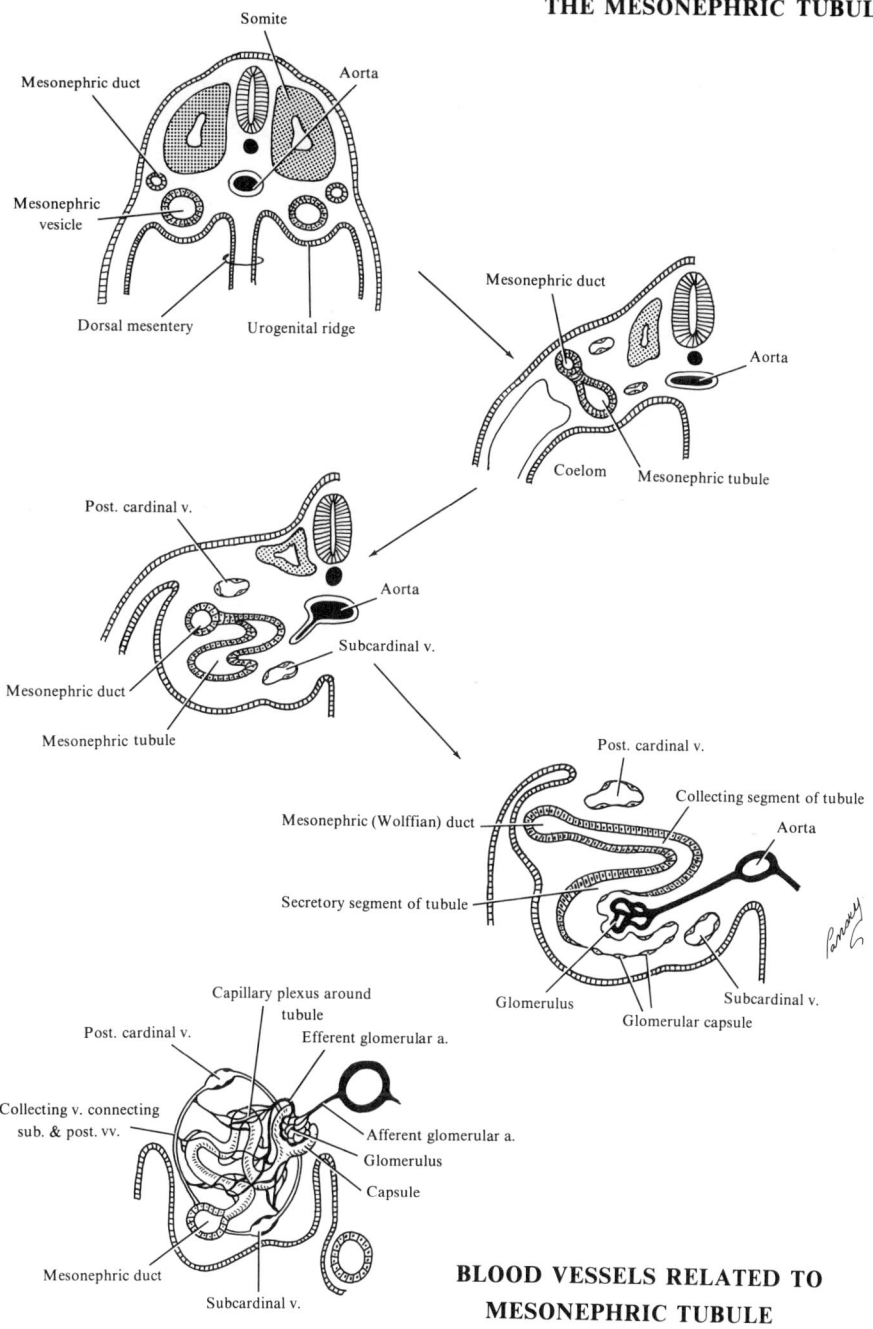

BLOOD VESSELS RELATED TO MESONEPHRIC TUBULE

DEVELOPING MESONEPHROS

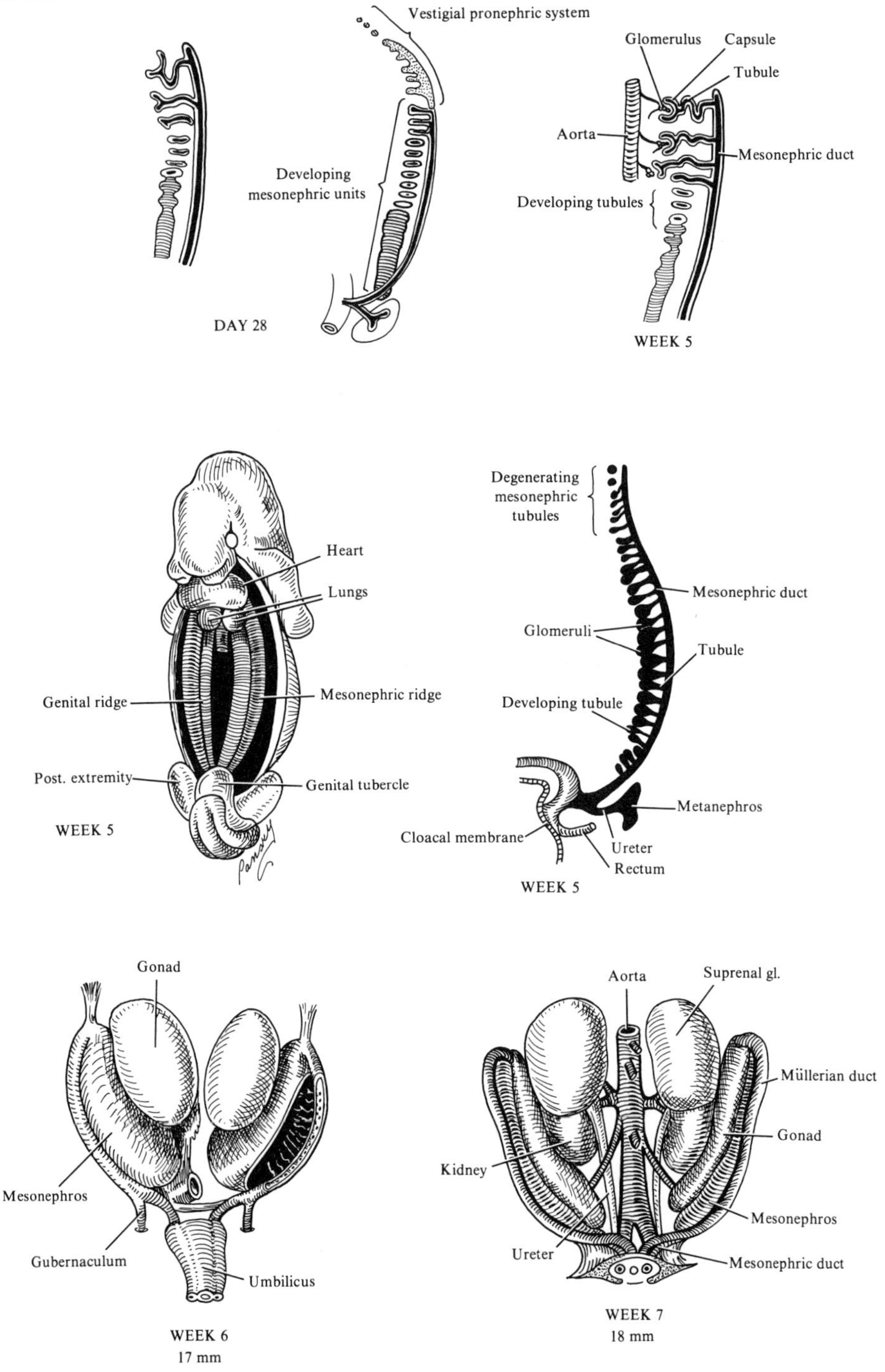

FIGURE 39. **Developing mesonephros.**

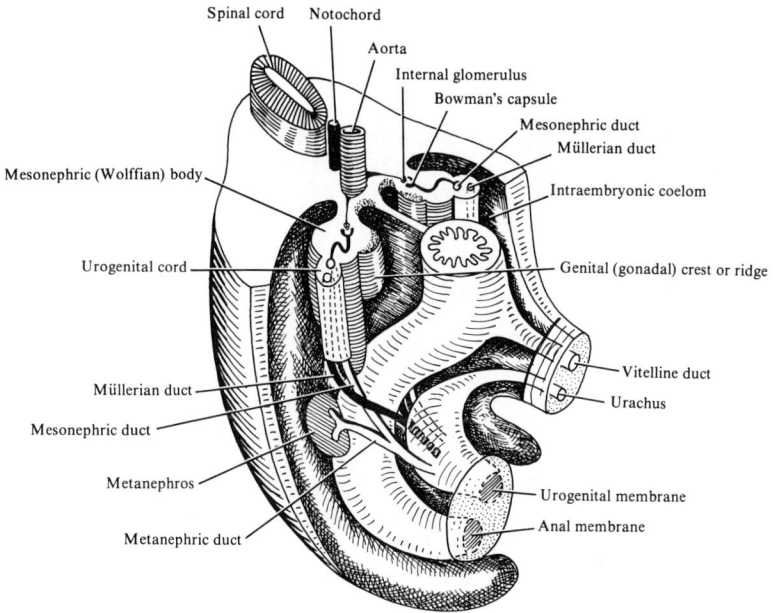

RELATIONSHIP OF MESONEPHROS TO SURROUNDING DEVELOPING STRUCTURES
(After Kollmann)

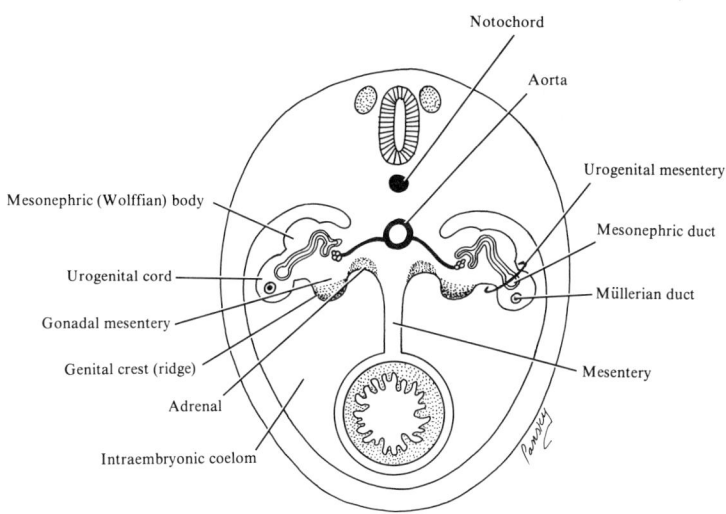

FIGURE 40. Relationship of mesonephros to surrounding developing structures.

93. THE URINARY OR EXCRETORY SYSTEM: THE METANEPHROS

I. **The metanephros (hind kidney) or permanent kidney:** the metanephric blastema, a nonsegmented homogenous mesodermal mass, enters its active phase of differentiation at the beginning of month 2 (week 5) and begins to function about 3 weeks later
   A. URINE FORMATION continues actively throughout the fetal life
      1. Urine combines with amniotic fluid, which the fetus drinks, and is absorbed by the intestine. The fetal kidneys regulate amniotic volume, and if the kidneys are absent, amniotic fluid volume is abnormally small (oligohydramnios)
   B. THERE IS NO VITAL NEED FOR PRENATAL FUNCTION of the kidneys because metabolic wastes are transferred via the placenta to the mother, but the fetus would die perinatally without the kidneys

II. **Metanephric differentiation** begins when the metanephric blastema (mass of mesoderm) is penetrated by the ureteric bud (metanephric diverticulum) which arises from the lower portion of the mesonephric (wolffian) duct
   A. THE URETERIC BUD arises as a dorsal bud, at the beginning of month 2, from the lower end of the mesonephric duct, and rapidly reaches and penetrates the metanephric blastema which, in turn, forms a "cap" over the bud
      1. The bud then enlarges and forms the beginning of the *renal pelvis* in week 5. It then develops two diverticula: a cranial and caudal portion, the future *major calyces*. The stalk itself becomes the *ureter*
      2. The renal pelvis divides into the major and then minor calyces, the latter giving rise to the collecting ducts or tubules, which penetrate farther into the metanephric mass
         a. The collecting tubules, in turn, further subdivide and, in the process, compress the blastema eccentrically. They subdivide until about 13 or more generations of tubules (ducts) are formed. Differentiation of the collecting tubules depends on an induction stimulus from the ureteric bud and its derivatives
         b. A complete system of branching tubules (ducts), forming a *renal lobe,* results from each tubule of the first generation
      3. While more and more new tubules arise on the periphery of the blastema, the major calyces absorb the ducts of the third and fourth generations, which are transformed into the *minor calyces* of the renal pelvis. The tubules of the fifth and successive generations form the definitive collecting tubules of the adult kidney
         a. The minor calyces receive the *papillary ducts* of the tubular system, the terminal end of the collecting system opening at the apex of the pyramid
      4. The ureteric bud gives rise to the entire collecting system of the kidney, namely, the *ureter,* the *renal pelvis,* the *calyces,* the *papillary ducts,* and the *collecting ducts*
   B. THE METANEPHRIC BLASTEMA develops from the intermediate mesoderm, which forms a solid mass of tissue. The development of its collecting ducts differs from that of the pronephros and mesonephros systems in that it is formed by the developing metanephric (ureteric) bud. The tubular branching from the bud compresses the blastema eccentrically and fragments it into tissue caps or clusters of mesenchymal cells
      1. At each end of the cap, a metanephric spheroid arises which undergoes the same transformations described for the nephrotomes of the pronephros and mesonephros; namely, there is a spheroid stage, vesicle stage, stage where the vesicle elongates in an S-shape, and finally the vesicle opens into the collecting tubule. Here differentiation is elaborate and leads to the formation of *functional units* or *nephrons* (the glomerulus plus the tubular system). The latter include Bowman's capsule, the proximal and distal convoluted tubules, and the nephronic (Henle's) loop. The glomerulus and Bowman's capsule make up the *renal corpuscle*
      2. Distal convoluted tubules join arching collecting tubules and become confluent
      3. The metanephric blastema thus gives rise to the *entire excretory system*
      4. Each uriniferous tubule has 2 parts: a nephron from the metanephric mass of mesoderm and a collecting tubular system from the ureteric bud

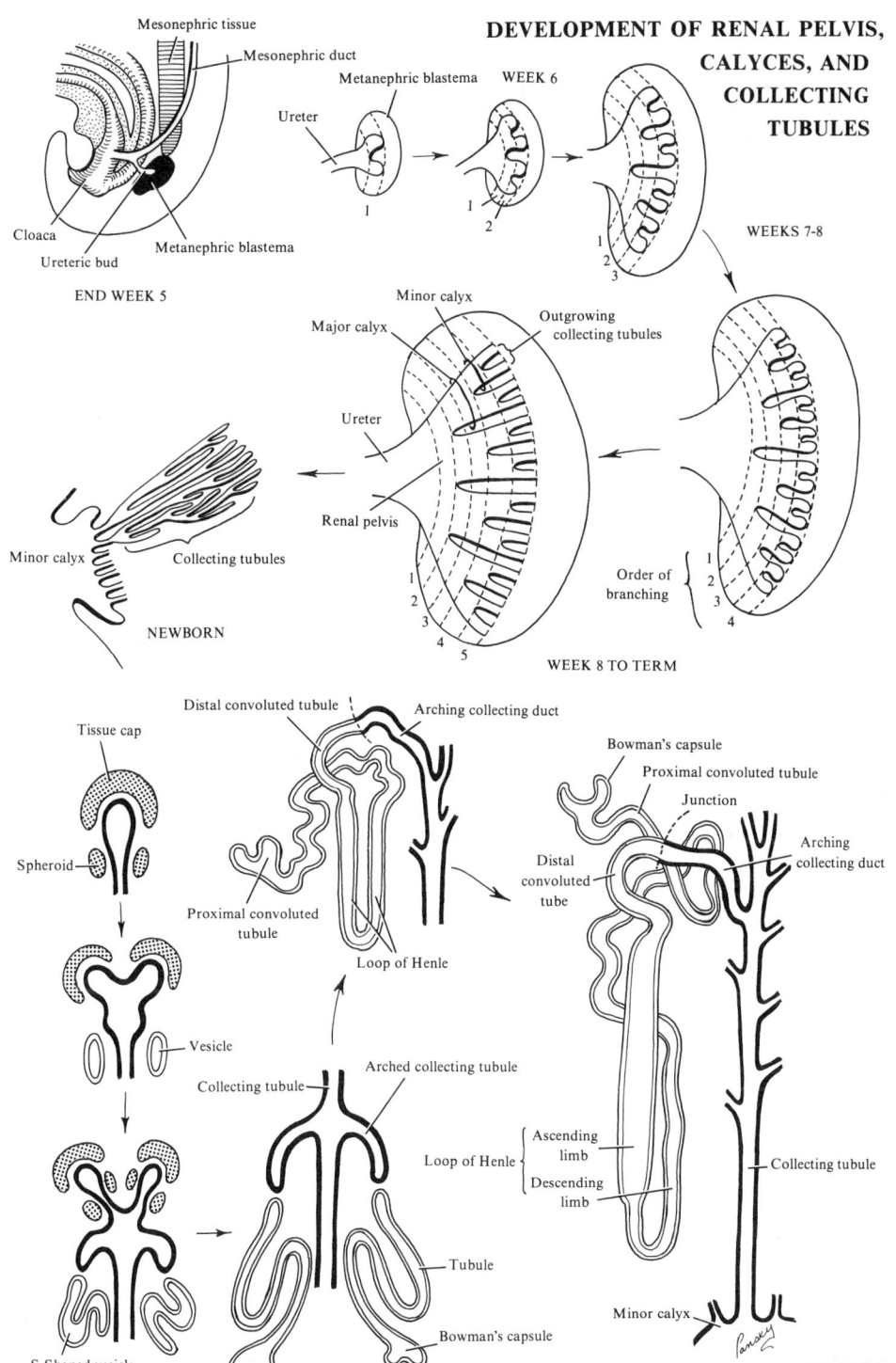

94. THE URINARY OR EXCRETORY SYSTEM: THE DEFINITIVE KIDNEY

I. The definitive kidney
A. BY MONTH 3, THE NEPHRONS already formed are anatomically and histologically identical to those of the adult kidney, and at this stage of development, the kidney begins to make urine, even though it has not acquired as yet its total number of approximately one million functional units
   1. The functional units grow, in number, by concentric layers throughout the prenatal life and are laid down until about term
   2. No nephrons usually are developed after birth, except in premature infants, but existing nephrons complete their differentiation during infancy and increase in size until adulthood
   3. Thus, the increase in kidney size is due to hypertrophy and not to an increase in the number of nephrons after birth
B. THE FETAL KIDNEY normally has a polylobar appearance due to the manner of development of the ureteric bud in the metanephric blastema
   1. The lobular form diminishes at birth by means of progressive filling in of the interlobular grooves
   2. Although the adult kidney is smooth and regular, the prenatal appearance sometimes persists, and we refer to a *polylobed kidney or fetal-like kidney*
C. CHANGES IN KIDNEY POSITION: the metanephros initially is located in the pelvic region but shifts later to a more cranial position in the abdomen
   1. This so-called "ascent" of the kidney is probably due to a diminution of the body curvature as well as growth of the body in the lumbar and sacral regions
   2. The kidney hilum initially faces ventrally, but ascent and rotation of 90° turn the hilum so that it faces medially
D. VASCULAR SUPPLY
   1. In the pelvis, the metanephros receives its arterial supply from the pelvic branches of the aorta
   2. During ascent to the abdomen, the kidney is vascularized by arteries that originate from the aorta at continuously higher levels
   3. The lower vessels usually degenerate, except for vascular variations and anomalies

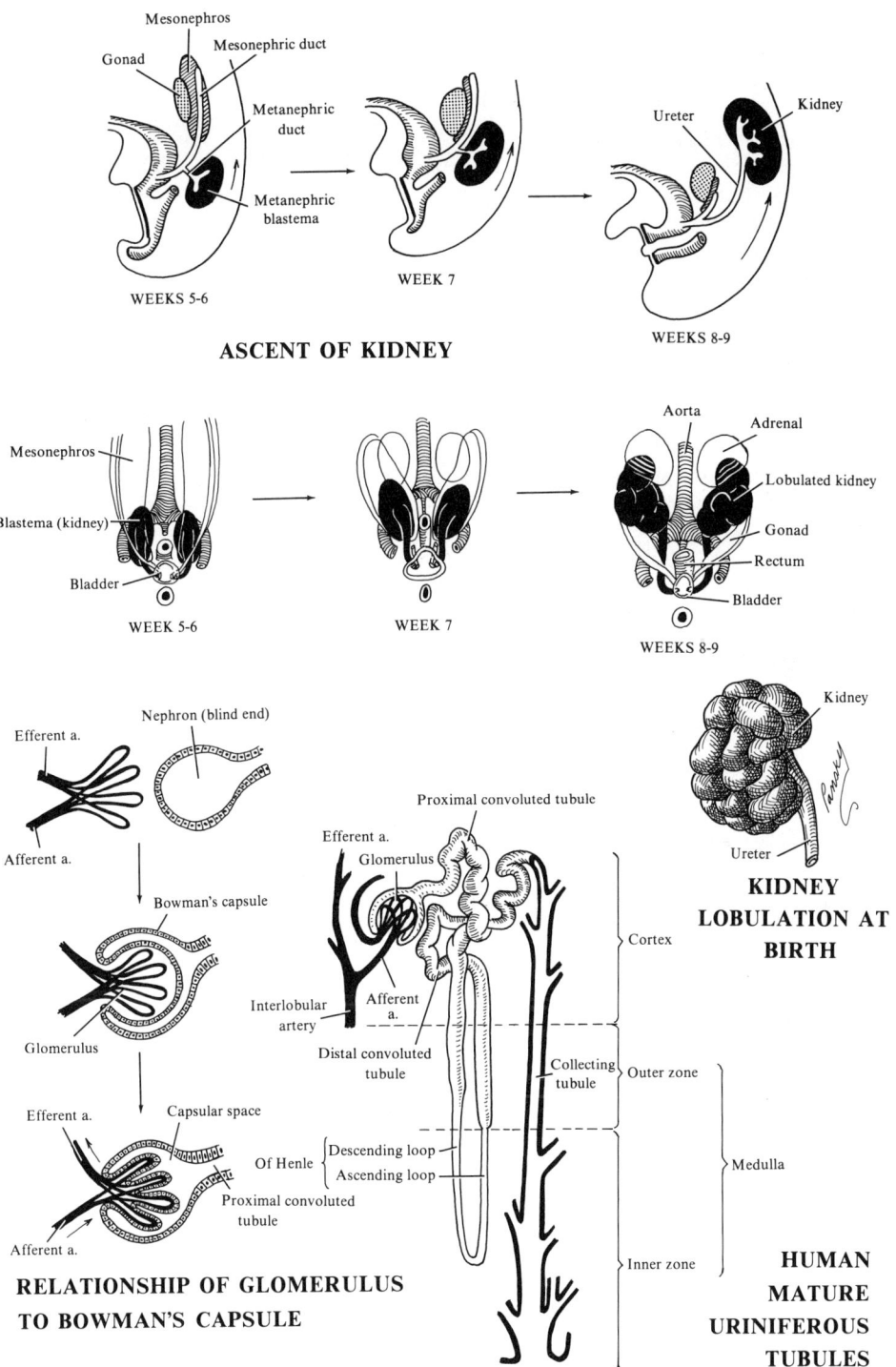

95. THE URINARY OR EXCRETORY SYSTEM: THE URINARY BLADDER AND URETHRA

I. **The urogenital sinus and anorectal canal** are formed during septation of the entodermal cloaca by the urorectal septum, during weeks 4 to 7. Simultaneously, the distal portions of the 2 mesonephric (wolffian) ducts, undergo complex development
   A. THE UROGENITAL SINUS, according to the position of the opening of the mesonephric ducts, can be divided into 3 distinct zones
      1. The vesicourethral canal or urinary zone: the upper region, above the entrance of the mesonephric ducts, is continuous with the allantois
      2. The middle or pelvic zone
      3. The caudal or genital (phallic) zone: the lower region, below the entrance of the mesonephric ducts, closed by the urogenital membrane
   B. EACH MESONEPHRIC DUCT forms a diverticulum, the *ureteric bud,* near the end of week 5
      1. The part of the mesonephric ducts between the ureteric buds and the posterior wall of the urogenital sinus enlarges into the *ampullae* or *horns of the urogenital sinus.* The mesonephric ducts and the ureters open side by side into these ampullae
      2. Selective development of the posterior wall of the urogenital sinus absorbs the 2 ampullae at about week 7, resulting in the ureters opening separately and directly into the urogenital sinus, just outside the mesonephric ducts
   C. THE POSTERIOR WALL OF THE UROGENITAL SINUS continues to develop, and by week 8 the orifices of the ureters have moved farther cranially and laterally, while those of the mesonephric ducts have remained relatively fixed
      1. Further remodeling of the urogenital sinus causes the ureters to open into the urinary bladder and the caudal end of the mesonephric ducts (future *ejaculatory ducts*) to open into the urethra (in the male) beneath the bladder
      2. The caudal ends of the ducts, in the female, subsequently degenerate
      3. The portion of the posterior urogenital sinus wall between the openings of the ureters and the mesonephric ducts appears triangular in shape and becomes the future *trigone of the bladder* (trigonal vesicle of mesonephric origin)
         a. The mesodermal epithelium of the trigone, derived from the mesonephric ducts, is soon replaced by the entodermal epithelium of the urogenital sinus

II. **The urinary bladder:** as the bladder forms, the allantois is progressively obliterated and forms the *urachus,* a thick tube. It becomes the median umbilical ligament after birth

III. **In the male:** the vesicourethral canal of the urogenital sinus gives rise to the bladder (vesical) and the upper part of the prostatic urethra (proximal to the ejaculatory ducts). All are derived from entoderm. The lamina propria, smooth muscle, and serosa (adventitia) develop from adjacent splanchnic mesenchyme
   A. MUCOSA of the cranial prostatic urethra, originally mesoderm is replaced by entoderm
   B. THE LOWER PORTION of the prostatic urethra and the membranous urethra are derived from entoderm of the pelvic portion of the urogenital sinus
   C. THE EPITHELIUM of the penile urethra, except for its glandular part, comes from cells of the phallic or genital portion of the urogenital sinus
      1. The epithelium of the glandular part of the penile urethra (terminal or navicular fossa) develops by canalization of an ectodermal cord of cells that extends into the glans from its tip. The connective tissue and smooth muscle develop from adjacent splanchnic mesenchyme
   D. AT THE END OF 3 MONTHS, the epithelium of the cranial part of the urethra proliferates and forms outbuddings that penetrate the surrounding mesenchyme. In the male, these buds form the *prostate gland;* in the female the *urethral and paraurethral glands*

IV. **In the female:** the entodermal vesicourethral canal of the urogenital sinus gives rise to the bladder (vesical) and the entire urethra. The connective tissue and smooth muscle come from the adjacent splanchnic mesenchyme

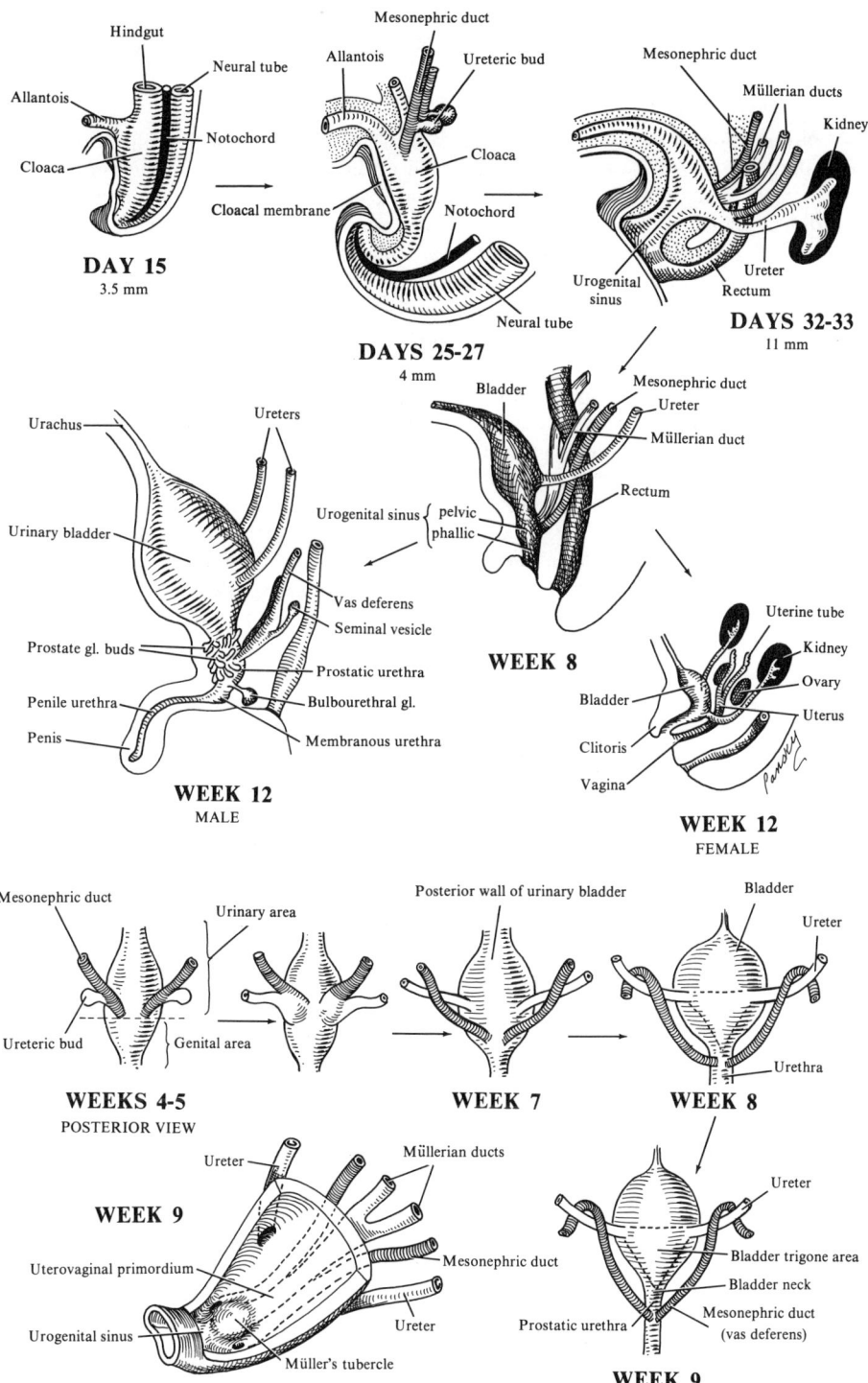

96. MALFORMATIONS OF THE URINARY SYSTEM

I. **Introduction:** many anomalies rise as a result of disturbances of renal development. Abnormalities of the kidneys and ureters are seen in 3–4% of the population. Morphologically, anomalies involving the excretory part, the collecting part, or both, can be distinguished. Modifications of shape, position, and blood supply are also seen

  A. HYPOPLASIA: underdevelopment of tissue or an organ usually caused by a decrease in cell numbers or atrophy due to a destruction of some elements

  B. APLASIA: defective development or congenital absence of the organ

  C. AGENESIS: unilateral absence of a kidney is relatively common; bilateral agenesis is rare and incompatible with postnatal life
   1. The result of failure of the ureteric bud to develop, failure of the bud to grow into the metanephric mass of mesoderm and induce nephron formation, or failure of the mesonephric ducts to descend to the area of the ureteric bud
   2. In the female, the müllerian ducts are also affected and may result in an absence of the uterus and part of the vagina

  D. HORSESHOE KIDNEY: seen in 1/600 births; U-shaped kidney that lies at the level of the lower lumbar vertebrae. Its ascent is prevented by the inferior mesenteric artery
   1. During migration from the sacral region, the 2 metanephric blastemas can come in contact with each other, most often at the lower poles and fuse on the median line, forming the classic form of the horseshoe kidney
   2. The ureters, which pass in front of the zone of kidney fusion, are usually normal
   3. Often no symptoms exist, but it may be associated with abnormalities of the renal pelvis and kidney and favor obstruction and infection

  E. POLYCYSTIC KIDNEYS: common, death occurs at or shortly after birth if bilateral, with severe renal insufficiency
   1. This form of kidney is caused by defective junction between the metanephric origin of the kidney and the derivatives of the ureteric bud. As a result of the absence of a connection between the collecting and excretory parts of the tubules, urine cannot be evacuated, pressure increases in the glomerulotubular system, and distention develops. Cystic degeneration follows, and the kidney rapidly loses its ability to function
     a. Other causes may be cysts from remnants of the first rudimentary nephrons, which normally degenerate, or abnormal development of the collecting tubules

  F. RENAL DUPLICATIONS are rather frequent, but may remain completely latent without any effect on urinary function. Different types, as a result of precocious division of the ureteric bud into 2 branches, result in the formation of
   1. Complete double ureters (unilateral or bilateral)
   2. Partial double ureters
   3. Double kidney: where the division also involves the metanephric blastema

  G. SUPERNUMERARY KIDNEYS: rare; the result of complete division of the ureteric bud; often seen with a bifid ureter or separate ureters
   1. When 1 of the 2 ureters remains very short and the second kidney remains low, we refer to the abnormally placed kidney as supernumerary rather than a double kidney

  H. ECTOPIC URETERAL ORIFICES: the ureter opens anywhere, except into the trigone of the urinary bladder, and the complaint is usually incontinence
   1. In the male, it usually opens into the prostatic urethra, prostatic utricle, or seminal gland
   2. In the female, it usually opens into the urethra, the vagina, or vestibule
   3. Caused by ureteric bud developing more cranially than usual from the mesonephric duct

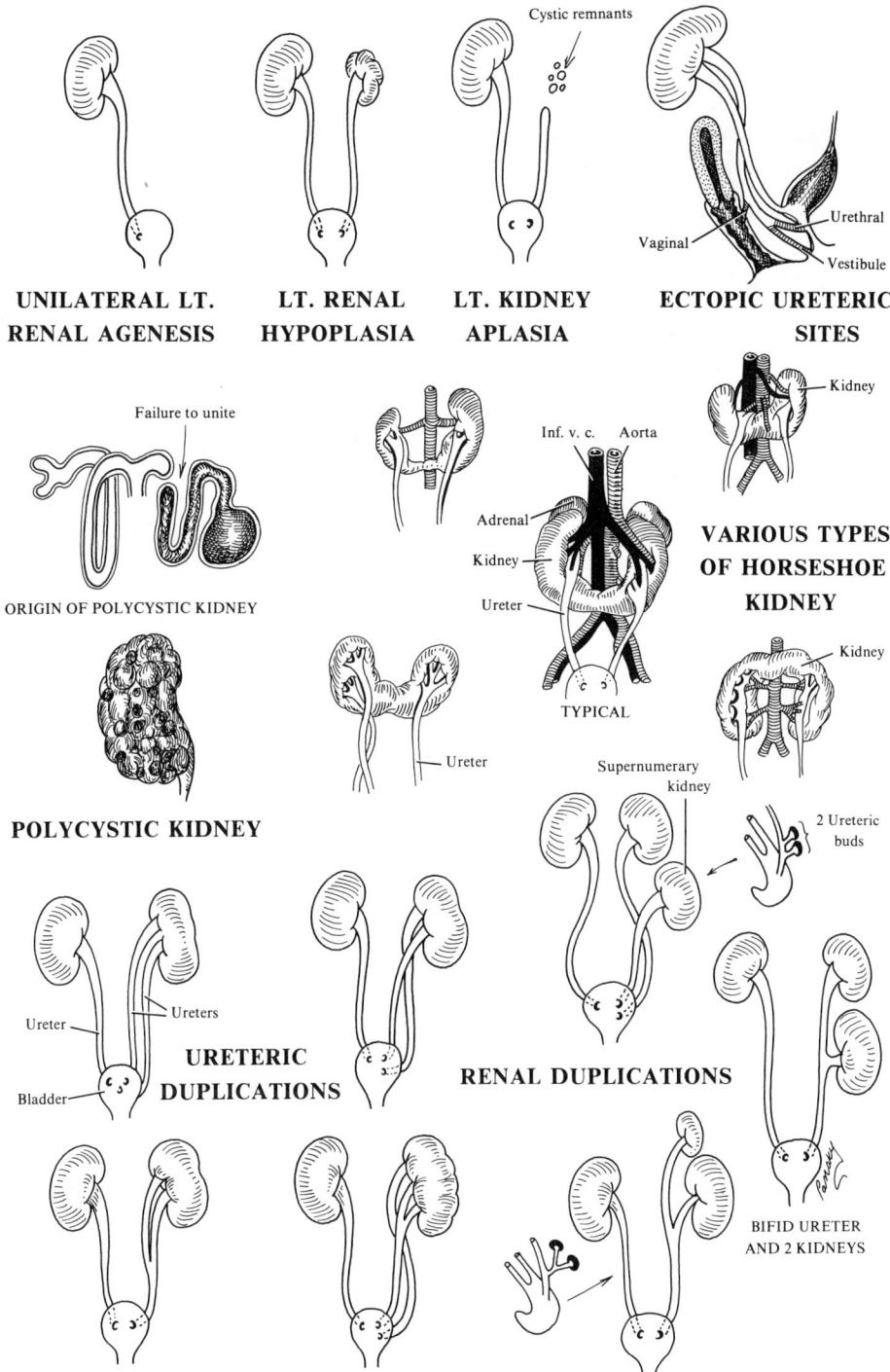

97. MALFORMATIONS OF THE URINARY SYSTEM

I. Introduction (cont.)

I. MULTIPLE RENAL VESSELS are common; about 3% of population show variations in the number of renal arteries and their position with respect to the renal veins
   1. Supernumerary arteries (2 or 3) are twice as common as extra veins and usually arise at the level of the kidneys
   2. They are caused by the persistence of embryonic vessels, which normally disappear

J. ABNORMAL ROTATION OF KIDNEYS: uncommon; kidney hilum does not face medially
   1. Can be caused by no rotation (hilum faces ventrally), overrotation (hilum faces dorsally), or malrotation (hilum faces laterally)
   2. Often associated with ectopic kidneys

K. SIMPLE RENAL ECTOPIA: one or both kidneys are in an abnormal position. One is usually lower than the other and malrotated. Majority are seen in pelvis or lower abdomen
   1. The pelvic kidney is due to a failure of the kidney to ascend. It may fuse to form a round mass or *discoid* or *pancake kidney*
   2. These kidneys receive their blood supply from nearby vessels

L. CROSSED RENAL ECTOPIA: in its ascent, the kidney crosses to the opposite side and may fuse with the other kidney, resulting in one large organ. Note: one ureter descends on the right side, the other on the left (differs from duplication of kidney)

M. URACHAL MALFORMATIONS: the median umbilical ligament is the adult derivative of the allantois and urachus. The latter lies between the umbilical arteries and connects the umbilicus and bladder in the fetus. The lumen in the lower urachus persists in 50% of people and is continuous with the bladder cavity. Malformations include
   1. Lower urachus may dilate: forms a *urachal sinus* or *diverticulum* opening into bladder.
   2. Dilatation of the upper urachus may form a *urachal sinus* that opens at umbilicus
   3. Urachal fistula: entire urachus remains patent
   4. Urachal cysts: remnants of the allantois may persist

N. EXSTROPHY OF BLADDER OR ECTOPIA VESICAE: uncommon; chiefly seen in males; seen in about 1/50,000 births as exposure and protrusion of the posterior bladder wall
   1. Associated with epispadias and wide separation of pubic bones, division of penis and clitoris, and wide separation of the labial and scrotal halves
   2. Due to an incomplete midline closure of the lower abdominal wall and anterior wall of urinary bladder as a result of failure of mesenchymal cells to migrate between the surface ectoderm and the urogenital sinus in week 4
      a. No muscle in anterior abdominal wall over bladder, and the epidermis and anterior bladder wall rupture to expose the cavity of the bladder

O. EXSTROPHY OF THE CLOACA: rare; due to entire infraumbilical body wall rupture and failure of the cloaca to divide into a urogenital sinus and rectum, exposing the posterior wall of the cloaca
   1. In addition, all the viscera (including liver) may be outside the body cavity (eventration of the abdominal viscera)

P. RECTOURINARY FISTULAS: connecting rectum and lower urinary tract (rectovesical and rectourethral fistulas, etc.) or vagina. Due to an abnormal division of the cloaca into the rectum and urogenital sinus

Q. CONGENITAL HYDRONEPHROSES involve dilatation of the renal pelvis, the calyces, and papillary ducts, resulting in a thinning of the cortex. Anomaly is attributed to a high ureteral obstruction due to a problem of urine elimination. May be result of
   1. Fusion of the pyeloureteral junction (normal in embryo, but should not persist)
   2. Ureteral compression by an aberrant vessel or an abnormal position of ureter
   3. Mucosal narrowing
   4. An anomaly of development of the ureteric bud, resulting in a high insertion of the ureter on the pelvis
   5. In an abnormal persistence of the ureteral membrane, *low ureteral obstruction* develops, the ureter is dilated, and a *ureterohydronephrosis* is created

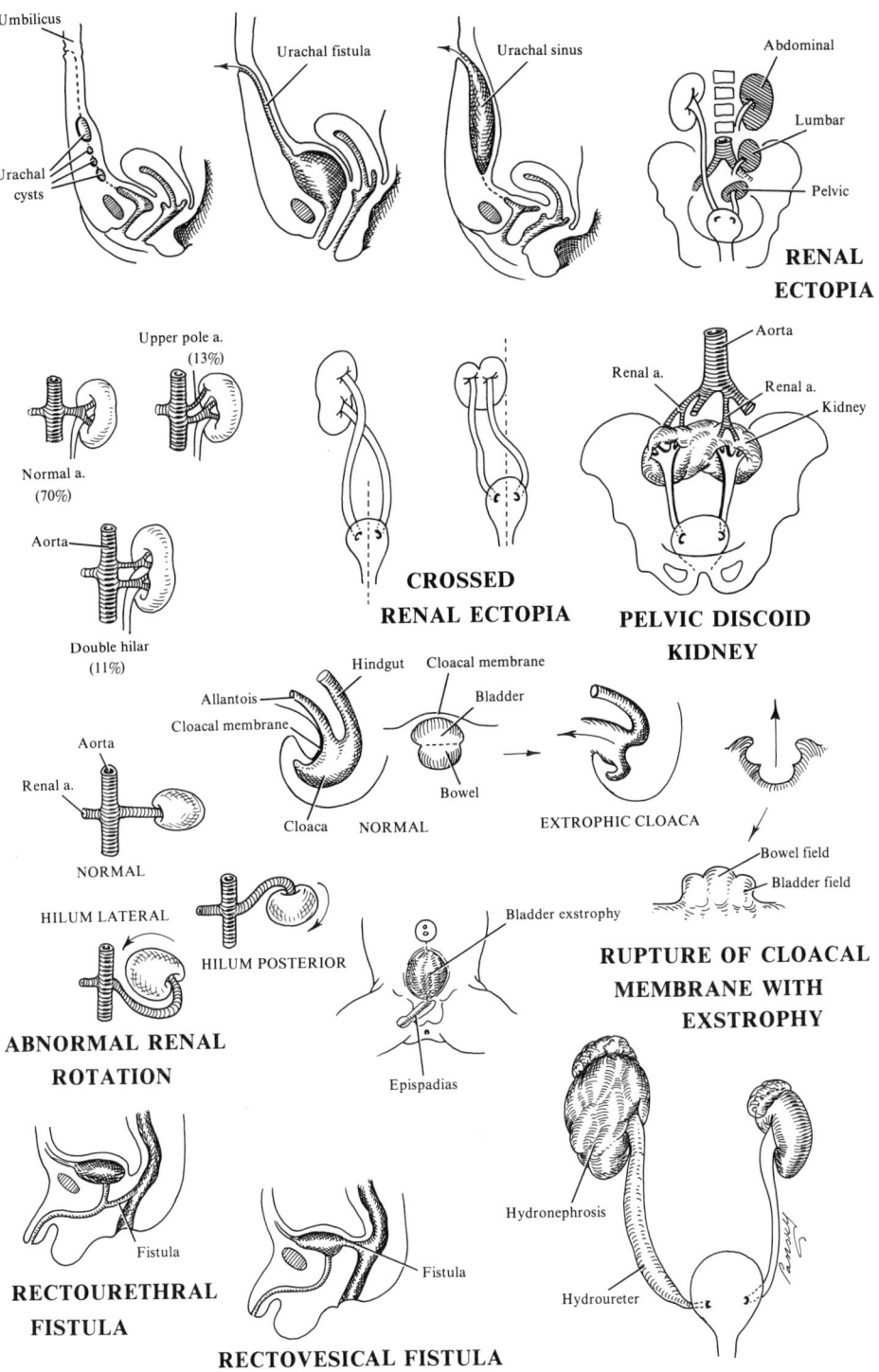

UNIT NINE
THE GENITAL OR REPRODUCTIVE SYSTEM

98. THE GENITAL OR REPRODUCTIVE SYSTEM: THE PRIMITIVE GENITAL SYSTEM

I. **Introduction:** the genetic sex of an embryo is determined at fertilization by the sperm that fertilizes the oocyte, but the gonads do not acquire male or female morphologic characteristics until week 7 of development. The early genital system is similar in both sexes, and in the beginning all human embryos are potentially bisexual. The period of early genital development is called the *indifferent* or *primitive stage* of the reproductive organs

II. **Primitive genital system**
  A. PRIMORDIAL GERM CELLS: the genital glands or gonads, the testes and ovaries, are formed from 2 types of cells, the *reproductive germinal cells* or *primordial germ cells* and the *nutrient supporting cells* (follicular cells of the ovary; the Sertoli cells of the testis)
   1. The primordial germ cells are large, spherical primitive sex cells of about 25 to 30 μm, with a granular cytoplasm, rich in lipids, and containing a large attraction sphere or *idiozome* consisting of 2 centrioles surrounded by Golgi apparatus
      a. The human primordial germ cells are discernible at about day 21 of embryonic life and are seen among the entodermal cells in the wall of the yolk sac near the origin of the allantois. Thus, they are at first at some distance from their eventual definitive location in the genital or gonadal ridge
  B. GONADAL PRIMORDIUM (indifferent gonad)
   1. The primordial germ cells migrate, by ameboid movement, along the dorsal mesentery of the hindgut during week 5 and reach the lumbar region of the developing embryo, the future gonadal ridge
      a. The coelomic epithelium which lines the anterior internal side of the mesonephric (wolffian) body, thickens to form the *genital* or *gonadal ridge* and provides the nutrient supporting cells of the gonad.
      b. If the cells fail to reach the ridges, the gonads do not develop and gonadal dysgenesis occurs, a well-known syndrome in the female
   2. In week 6, the primordial germ cells invade the genital ridges and are incorporated into the primary sex cords, which proliferate and grow from the coelomic epithelium into the underlying mesenchyme to form the *primary sex cords*
      a. The gonad is called "indifferent" or undifferentiated at this stage because it has the same morphologic appearance in both male and female
      b. The indifferent gonad now consists of an *outer cortex* and an *inner medulla*
         i. In embryos with an XX sex chromosome complex, the cortex forms an ovary, and the medulla regresses; in one with an XY chromosome complex, the medulla differentiates into a testis, and the cortex regresses
      c. The sex cords eventually become the seminiferous tubules in the male and the medullary cords in the female
   3. The primary sex cords continue to proliferate actively, anastomose deep in the mesenchyme, and produce a complex network called the *rete,* which is seen as a bulge under the coelomic epithelium on the anterointernal side of the mesonephric (wolffian) body.
  C. UROGENITAL CONNECTIONS: the rete anastomoses with the adjacent part of the mesonephric proximal convoluted tubules establishing the initial or first urogenital connections
   1. Toward the end of month 2, the mesonephric (wolffian) body begins to regress, the glomeruli disappear, and only the mesonephric tubules remain linked with the genital gland

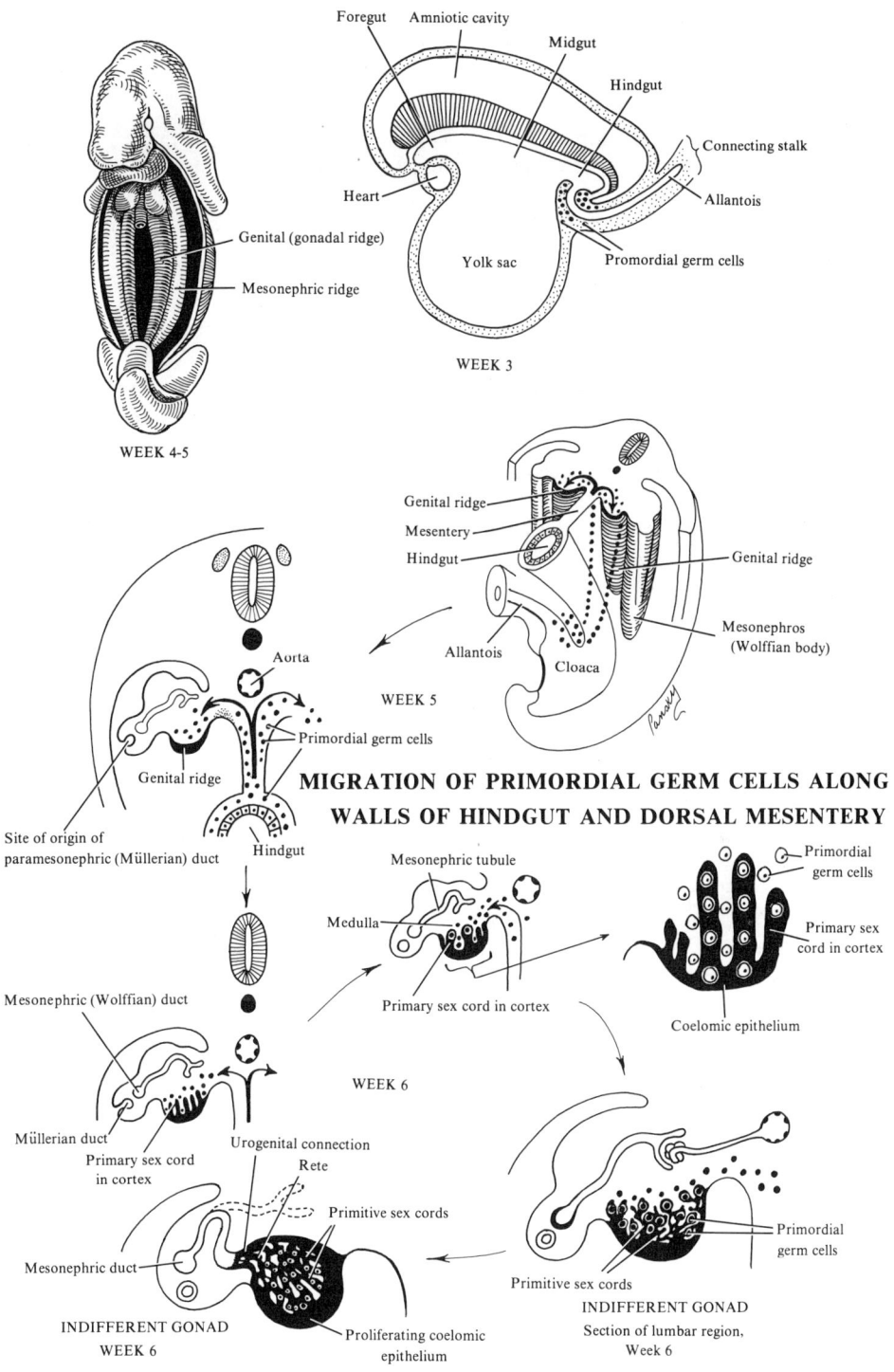

MIGRATION OF PRIMORDIAL GERM CELLS ALONG WALLS OF HINDGUT AND DORSAL MESENTERY

99. DEVELOPMENT OF THE TESTIS

I. Testicular differentiation. The male gonad develops into the testis toward week 7 of development. Until then, it is undifferentiated. The differentiation is determined by the XY genetic constitution and proceeds as follows
  A. THE PRIMARY SEX CORDS, proliferating from the coelomic epithelium, condense and extend into the medulla of the gonad.
    1. In the medulla, the cords branch, their deep ends anastomose, and they form the *rete testis*
    2. The prominent sex cords become the *seminiferous* or *testicular cords* which soon lose their connections with the germinal epithelium because of the development of a thick fibrous capsule, the *tunica albuginea*
      a. The tunica albuginea is a layer of connective tissue which is interposed early, between the coelomic epithelium (parietal peritoneum) and the rest of the gland. It produces partitions which compartmentalize the gland, closing off the seminiferous ducts, about day 50, into testes cords
      b. Development of the tunica albuginea is a characteristic and diagnostic feature of testicular development
    3. The seminiferous or testicular cords develop into the *seminiferous tubules,* whose deep portions narrow to form the *tubuli recti,* which converge on the rete testis
    4. The seminiferous tubules become separated by mesenchyme which gives rise to the *interstitial cells of Leydig*
      a. It is here that the androgenic hormones are secreted which help in the differentiation of the genital tract and the external genital organs
      b. The interstitial cells of Leydig reach their maximum development between $3\frac{1}{2}$ and 4 months
    5. The walls of the seminiferous tubules, as a result of their cellular duality of origin, are composed of 2 types of cells: *supporting* or *sustentacular cells of Sertoli,* derived from the germinal epithelium and the *spermatogonia,* derived from the primordial germ cells (unlimited in number, in contrast to the oogonia)
      a. The cells of Sertoli make up most of the seminiferous epithelium in the fetal testis
    6. Gradually, the enlarging testis separates from the regressing mesonephros and is suspended by its own mesentery, the *mesorchium*
  B. IN LATER DEVELOPMENT, THE GERMINAL EPITHELIUM flattens to form the mesothelium on the surface of the testis and the rete testis becomes continuous with the 15 to 20 adjacent persistent mesonephric tubules
    1. The persistent mesonephric tubules, after regression of the mesonephric (wolffian) body, participate in the formation of the excretory tracts of the testis, forming the *vasa efferentia or efferent ductules*
    2. The efferent ductules open into the adjacent mesonephric duct which becomes the *ductus epididymidis (epididymis)*
    3. Thus, the vasa efferentia and the epididymis are of mesonephric origin

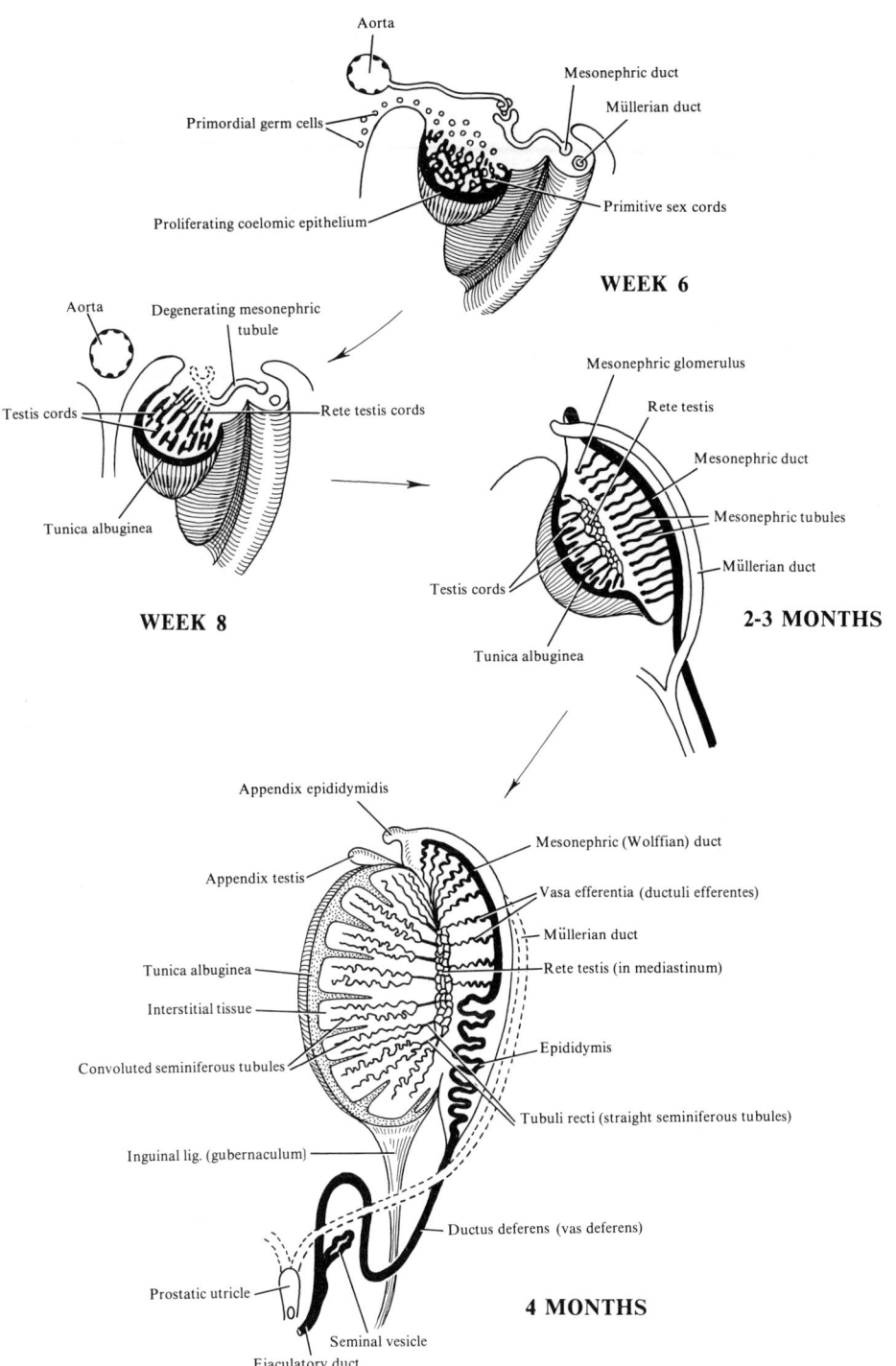

100. THE GENITAL OR REPRODUCTIVE SYSTEM: PRIMITIVE GENITAL TRACTS AND SEX DETERMINATION

I. The early or primitive genital tracts
A. THE GENITAL TRACTS have the same appearance in both male and female embryos until week 7 of development, consisting of the 2 *paramesonephric* or *müllerian ducts* and the 2 *mesonephric* or *wolffian ducts*
   1. The mesonephric ducts drain the mesonephric kidneys, but persist as the male genital ducts when the mesonephric system undergoes degeneration
   2. The paramesonephric ducts develop bilaterally from invaginations of the coelomic epithelium, on the lateral aspects of the mesonephroi
      a. In the 10-mm embryo, the paramesonephric (müllerian) ducts induce an invagination of the coelomic epithelium opposite the cranial end of each mesonephric duct which creates an epithelial bud that penetrates the mesenchyme and progresses caudally along the mesonephric duct
      b. The bud hollows out at the same time it grows and thus becomes an open paramesonephric duct in the coelomic or peritoneal cavity
   3. The paramesonephric ducts cross in front of the mesonephric (wolffian) ducts at the lower pole of the mesonephric body and then run alongside it
      a. The terminal parts of the paramesonephric ducts fuse to form a small, single median duct (uterovaginal primordium or canal) which ends blindly at the posterior surface of the urogenital sinus
      b. The blind ending projects into the dorsal wall of the sinus to create an elevation, the *sinus* or *müllerian tubercle* which is located between the openings of the mesonephric ducts into the urogenital sinus
   4. The mesonephric and paramesonephric ducts are located in the *urogenital cord*, which is attached to the anterior external edge of the mesonephric (wolffian) body by the *urogenital mesentery*. The latter attaches the urogenital cord to the body wall, below the mesonephric body. Furthermore, the 2 urogenital mesenteries (right and left) join below the median line
   5. The mesonephric body is attached to the abdominal wall by the mesonephric mesentery throughout its entire length. Above the body, the urogenital mesentery and mesonephric mesentery extend upward and form the *diaphragmatic ligament*
      a. The lower pole of the mesonephric body is attached at the inquinal region by the *inguinal ligament*

II. Sex determination: before week 7 of embryonic life, the gonads of both sexes are identical in appearance (undifferentiated). Genetic sex is determined by fertilization. Gonadal sex is determined by the sex chromosome complex: XX in the female embryo; XY in the male embryo
A. THE Y CHROMOSOME has a testis-determining effect on the medulla of the undifferentiated gonad
   1. Primary sex cords differentiate into seminiferous tubules in the male
B. WITHOUT THE Y CHROMOSOME, an ovary forms
C. THE TYPE OF SEX CHROMOSOME COMPLEX AT FERTILIZATION determines gonad type, and the latter determines the type of sexual differentiation seen in the genital ducts and external genitalia
D. WITH ABNORMAL SEX CHROMOSOME COMPLEXES
   1. The number of X chromosomes is unimportant in sex determination
      a. If a Y is present, the embryo becomes a male
      b. If there is no Y, a female develops
   2. Loss of a sex chromosome (XO females) causes ovarian dysgenesis
   3. Loss of an X chromosome does not interfere with migration of the primordial germ cells to the gonadal ridges because the germ cells are seen in the XO females
   4. Two X chromosomes, however, are needed for complete ovarian development

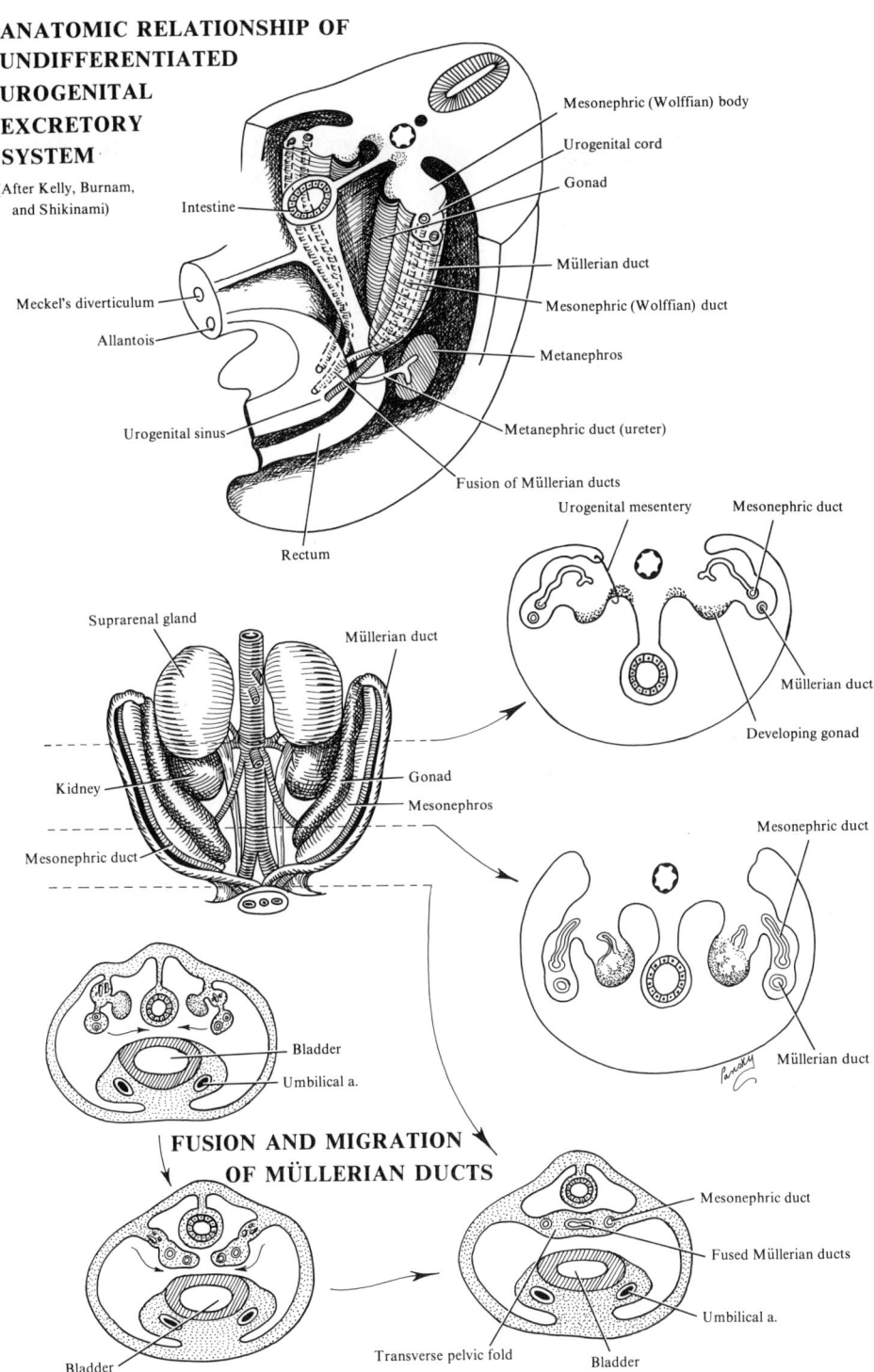

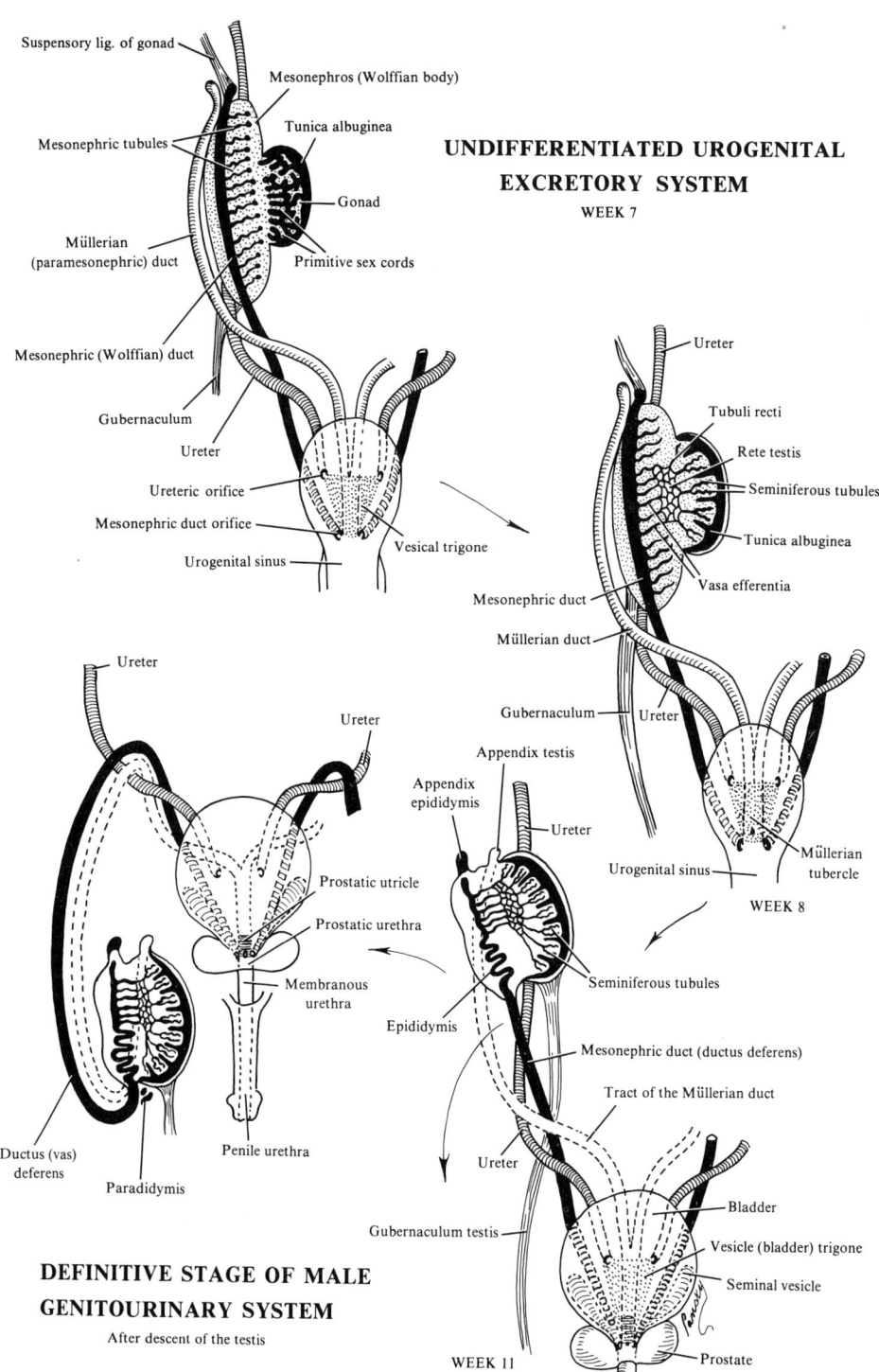

FIGURE 41. **Stages in development of the definitive male genitourinary system.**

- 262 -

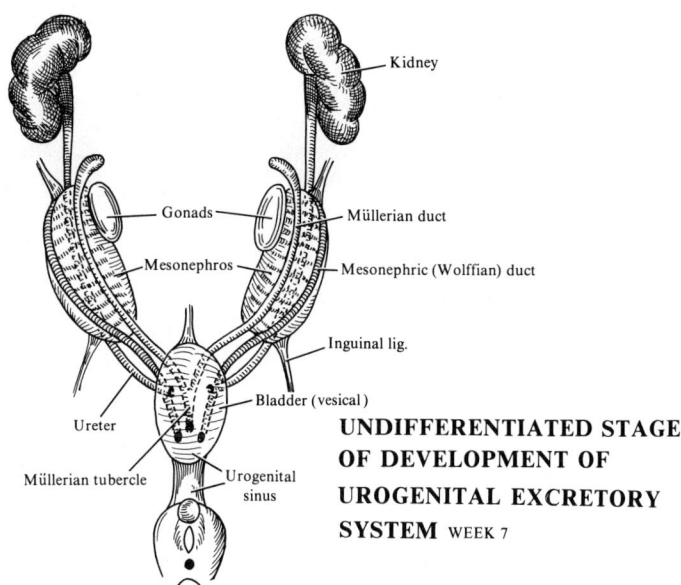

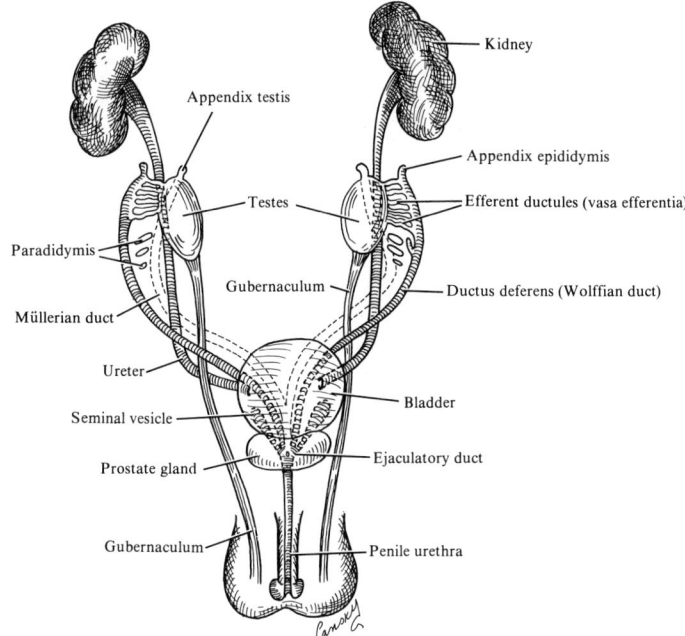

FIGURE 42. Differentiation and development of the male excretory system.

101. DIFFERENTIATION OF THE MALE GENITAL TRACTS AND AUXILIARY GLANDS

I. **Introduction:** development of the male urogenital system is caused by action of androgenic fetal hormones produced by the fetal testes. They take effect at the end of month 2 as seen by regression of the paramesonephric (müllerian) structures and differentiation of the mesonephric (wolffian) structures

II. **Development of upper portion of male genital tract:** after regression of the mesonephric (wolffian) body, the inguinal ligament inserts at the inferior pole of the testis (above) and in the inguinal region (below), forming the so-called *gubernaculum testis*. The diaphragmatic ligament disappears
   A. BOTH PARAMESONEPHRIC DUCTS regress completely, in the male, by week 11
      1. The *appendix testis* persists at the superior pole of each testis, and the *utriculus prostaticus* (*prostate utricle*) persists from the early joining of the paramesonephric stays at the back of the urogenital sinus between the mesonephric ducts
   B. THE 2 MESONEPHRIC (WOLFFIAN) DUCTS persist and develop as follows
      1. Their most cranial portion forms the *appendix of the epididymis*
      2. The segment of the mesonephric duct opposite the testis forms the *epididymis* or *ductus epididymidis*, with its cranial end connected to the vasa efferentia
         a. Below the cranial end, the epididymis is highly convoluted and descends along the testis, still receiving mesonephric tubules (ducts of Haller) which do not connect to the rete testis. Below, isolated mesonephric tubules form the *paradidymis*
      3. Below the testis, the mesonephric ducts acquire a thick investment of smooth muscle and form the *ductus deferens*
         a. As each ductus deferens joins the posterior side of the urogenital sinus, 2 lateral outgrowths form the *seminal gland* or *vesicle*
      4. The ductus deferens finally becomes the *ejaculatory duct*, that duct portion between the seminal vesicle duct and the urethra
      5. The ejaculatory ducts open on the posterior wall of the urogenital sinus on an elevation, called the *verumontanum* or *seminal colliculus*
         a. The seminal colliculus, a small elevation in the prostatic urethra, is the remnant of the müllerian tubercle and is homologous to the hymen
         b. The prostatic utricle, seen between the ejaculatory duct openings, opens on the colliculus in the prostatic urethra and is probably homologous to the vagina

III. **Development of the lower portion of the male genital tract:** the urogenital sinus caudal to the union of the mesonephric ducts consists of a vertical *pelvic segment* and a horizontal *phallic segment* in the genital tubercle. The latter, after resorption of the urogenital membrane, opens to the exterior. The 2 segments undergo reorganization, from month 3, related to secretion of androgenic hormones
   A. THE PELVIC SEGMENT: epithelial (entodermal) buds begin to detach themselves from the posterior aspect of the urogenital sinus, on both sides of the verumontanum, during month 3. They penetrate the adjacent mesenchyme and form the glandular epithelium of the *prostate gland*, which usually is well differentiated at 4 months
      1. The prostate eventually encloses the ejaculatory ducts and the prostatic utricle and completely surrounds the prostatic urethral area of the urogenital sinus
         a. The prostatic urethra thus is composed of a cephalic half, between the bladder and verumontanum, which belongs to the urinary region of the urogenital sinus; and a caudal half, between the verumontanum and the cranial part of the pelvic segment of the urogenital sinus
      2. The remaining pelvic segment forms the *membranous urethra*. The latter is continuous with the phallic segment of the urethra (described under the external organs)
         a. The pea-sized bulbourethral (Cowper's) glands develop from paired entodermal outgrowths from the membranous urethra. Adjacent mesenchyme contributes both smooth muscle fibers and stroma

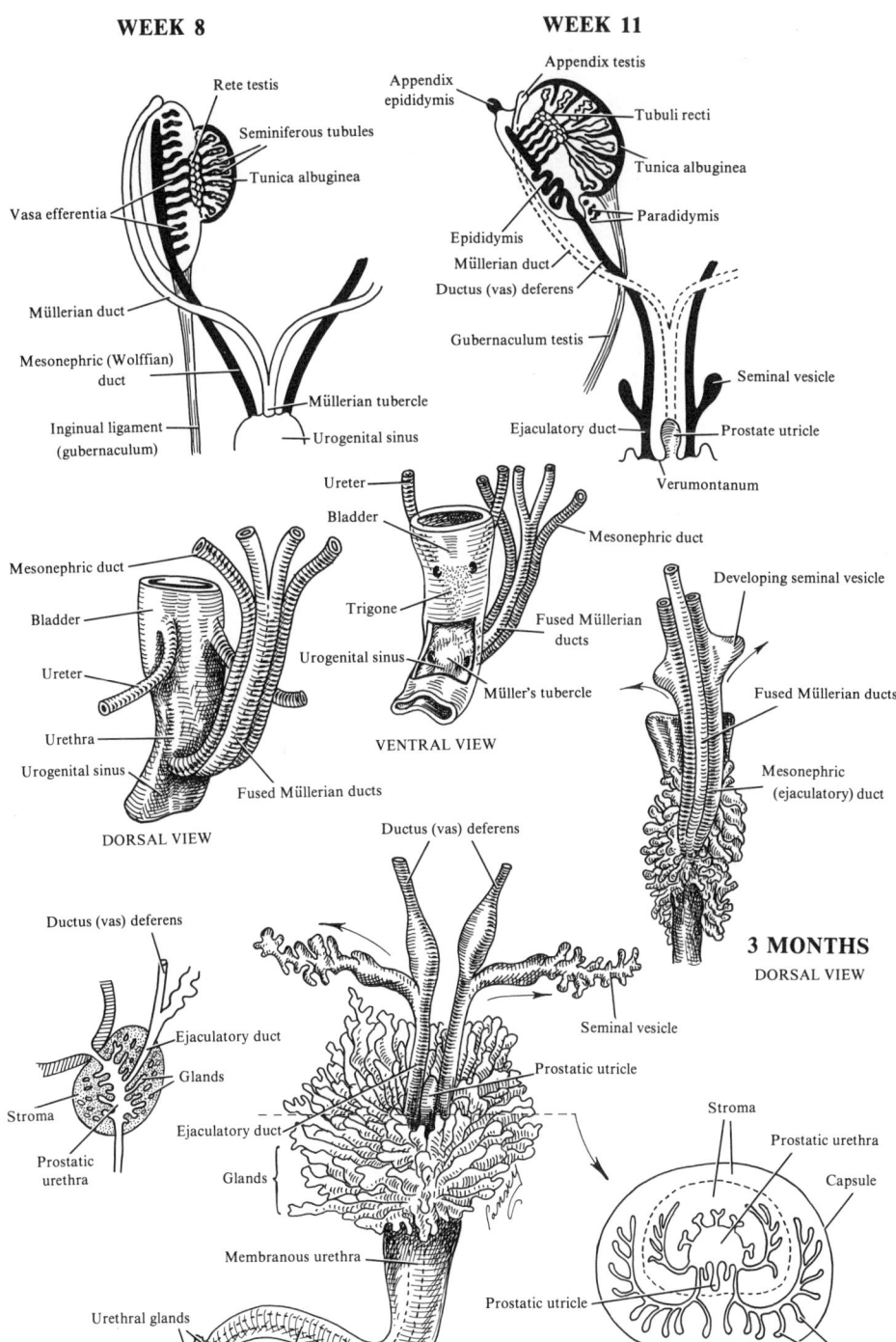

102. DEVELOPMENT OF THE MALE EXTERNAL GENITAL ORGANS

I. **The indifferent stage:** stage that is indistinguishable as either male or female
A. THE CLOACAL MEMBRANE, in week 3, is gradually surrounded by mesenchyme from the primitive streak. This mesenchyme forms a pair of elevations, the *cloacal folds*, which fuse with each other in front of the cloacal membrane to form the *cloacal eminence*
   1. The cloacal membrane, in week 6, is subdivided by the urorectal septum into the *urogenital* and *anal membranes;* the cloacal swellings also are split into the *genital* or *urethral folds*. The membranes rupture a week later to form the urogenital and anal openings, respectively
   2. The cloacal eminence elongates, by week 4, to form the *genital tubercle*
B. ANOTHER PAIR OF ELEVATIONS, THE LABIOSCROTAL OR GENITAL SWELLINGS, are seen on each side of the genital folds
   1. In the male, the swellings differentiate into the *scrotal swellings;* in the female, they become the *labia majora*
C. THE MALE AND FEMALE EXTERNAL GENITALIA are similar until week 9, even though distinguishing external sexual characteristics are seen during the early fetal period. The final form is usually not seen until week 12

II. **External male genital organs:** development and differentiation are evident beginning at month 3 and are related to the action of androgens produced by the testes
A. THE GENITAL TUBERCLE, initially seen in week 4, elongates by week 7 to form a *phallus,* which in turn will form the future penis. The phallus is as large in the female as in the male at this stage
   1. The phallus carries with it (pulls forward), the *genital* or *urethral (urogenital) folds* surrounding the phallic segment of the urogenital sinus. The folds elongate on the ventral side of the penis to form the lateral walls of the *urethral (urogenital) groove*
      a. The urethral groove is lined by an extension of the entoderm from the phallic portion of the urogenital sinus and is continuous with the urogenital opening
      b. At the base of the groove, the entoderm thickens into a *urethral plate*
   2. The posterior portion of the genital swellings thickens to form the *scrotal swellings*
B. THE GENITAL FOLDS, which circumscribe the median urethral (urogenital) groove, fuse along the ventral (under) side of the penis, from behind forward, at about 3 months, changing the groove into a duct, the definitive *penile urethra*
   1. The penile urethra ends blindly just before the end of the penis and is surrounded by a mass of erectile tissue of mesenchymal origin, the *corpus cavernosum urethrae* or *spongiosum*. This erectile tissue forms the end of the penis, the *glans penis*
      a. The external urethral opening moves progressively toward the glans
   2. The paired *corpora cavernosa penis,* also of mesenchymal origin, complete the erectile tissue in the shaft of the penis
C. THE LABIOSCROTAL OR GENITAL SWELLINGS grow toward each other and fuse in the midline to form the *scrotum*
   1. Both the scrotum and penis bear the signs of their early formation through closure of the urogenital groove as is evidenced by the median raphé
D. DURING MONTH 4, THE EPITHELIUM at the end of the penis forms 2 invaginations
   1. At the tip of the glans, an ectodermal ingrowth forms a cellular cord, the *glandular epithelial plate*. Splitting of this plate forms a groove, the *glandular urethra,* on the ventral part of the glans that is continuous with the urethral groove in the body of the penis. Closure of the groove in the glans moves the urethral opening to the tip of the glans and joins the 2 urethral parts
   2. The second invagination is circular and is called the *preputial epithelial plate*. Cleavage of this plate before birth separates the glans penis from the *prepuce* or *foreskin*. The latter is a fold of skin at the tip of the penis which, during week 12, grows over the glans and surrounds it by week 14. It is fused to the glans and not retractable at birth, but breakdown of the fused surfaces normally occurs during infancy

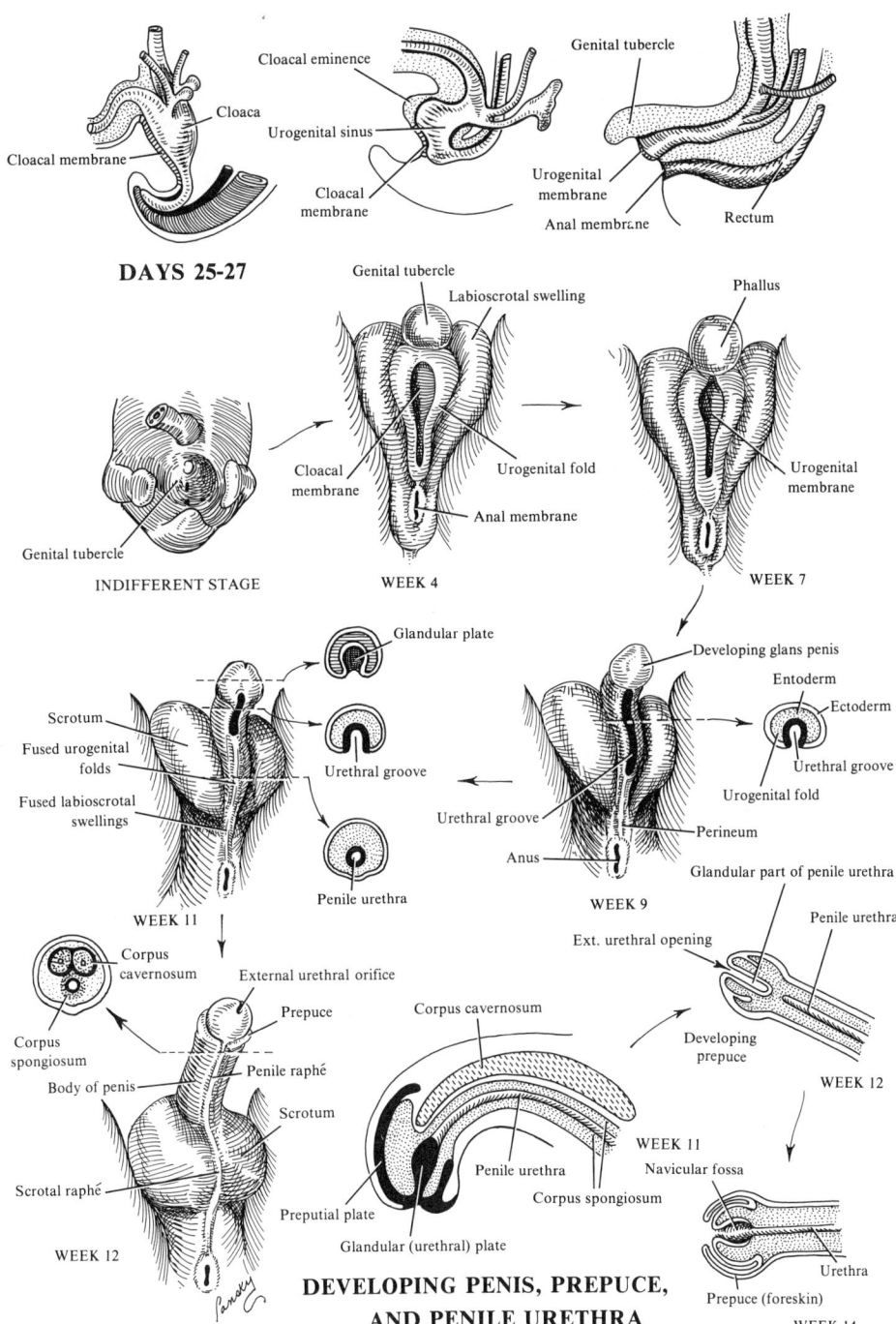

DEVELOPING PENIS, PREPUCE, AND PENILE URETHRA

103. INGUINAL CANAL DEVELOPMENT AND TESTICULAR MIGRATION

I. **Inguinal canal development:** the canals form the pathways for the testes to descend through the abdominal wall into the scrotum. They develop in the female embryo as well as in the male, even though the ovaries (except in rare cases) do not enter the canals
   A. AS THE MESONEPHROS degenerates, the gubernaculum (a ligament) descends on each side from the lower poles of the gonads, passes obliquely through developing abdominal wall, and attaches to the labioscrotal swellings (future scrotum or labia majora)
   B. THE PROCESSUS VAGINALIS (peritoneal sac) develops later, on each side, ventral to the gubernaculum, and herniates through the lower abdominal wall along the pouch formed by the gubernaculum
      1. Each processus carries extensions of layers of the abdominal wall before it, and together they form the walls of the inguinal canal. In the male, they also form the coverings for the testes and the spermatic cord
      2. The opening produced in the transversalis fascia by the processus is the *deep inguinal ring* and that in the external oblique aponeurosis becomes the *external* or *superficial inguinal ring*. Between the rings is the *inguinal canal*.

II. **Testicular migration:** the testes migrate from their early developing lumbar position, in the dorsal abdominal wall, to the deep inguinal rings above the scrotum between month 3 and term. Migration takes place as the pelvis enlarges and the embryonic trunk enlarges and, because the gubernaculum does not grow as fast as the body wall, the testes descend. Their descent through the inguinal canals into the scrotum is probably due to hormones (androgens and gonadotropins)
   A. THE VAGINAL PROCESS, a bilateral prolongation of the coelomic cavity peritoneum, is protruded into the scrotum
      1. The vaginal process at first is mostly open, but progressively narrows, and finally its proximal portion is entirely obliterated. The remainder of the process, in the scrotum, then exists as a double serous envelope around the testes, called the *tunica vaginalis*
      2. The vaginal process (processus vaginalis) is parallel with the inferior ligament of the testis, which becomes the gubernaculum testis. The migration of the testes follows the line of the gubernaculum, although the role of the gubernaculum is uncertain
      3. Descent through the inguinal canal begins during week 28 and takes 2 to 3 days. The testes move beneath the peritoneum (retroperitoneal) and behind the processus vaginalis
         a. The testes reach the orifice of the inguinal canal around month 6, cross the canal during month 7, and arrive at their definitive intrascrotal position near the end of month 8
         b. After the testes pass into the scrotum, the inguinal canal contracts around the spermatic cord. In full-term newborn boys, over 97% have bilateral descended testes. Some, however, may descend during the first 3 months after birth
   B. ANOMALIES OF TESTICULAR MIGRATION are many and are not necessarily related to a hormonal deficit since they are often unilateral. They range from *simple ectopy*, where the testis may be inguinoscrotal or inguinal, to *cryptorchism*, where the testis is pelvic, iliac, or even lumbar
      1. The gland also may be in an aberrant location such as crural or perineal
      2. Cryptorchism usually results in an alteration in spermatogenesis or, if bilateral, even sterility
   C. ANOMALIES OF VAGINAL PROCESS CLOSURE are also frequent. They may be (not always) associated with problems of testicular migration
      1. Cysts of the spermatic cord are signs of incomplete closure
      2. A complete failure of closure may result in *congenital oblique external hernia* or the so-called *communicating hydrocele*

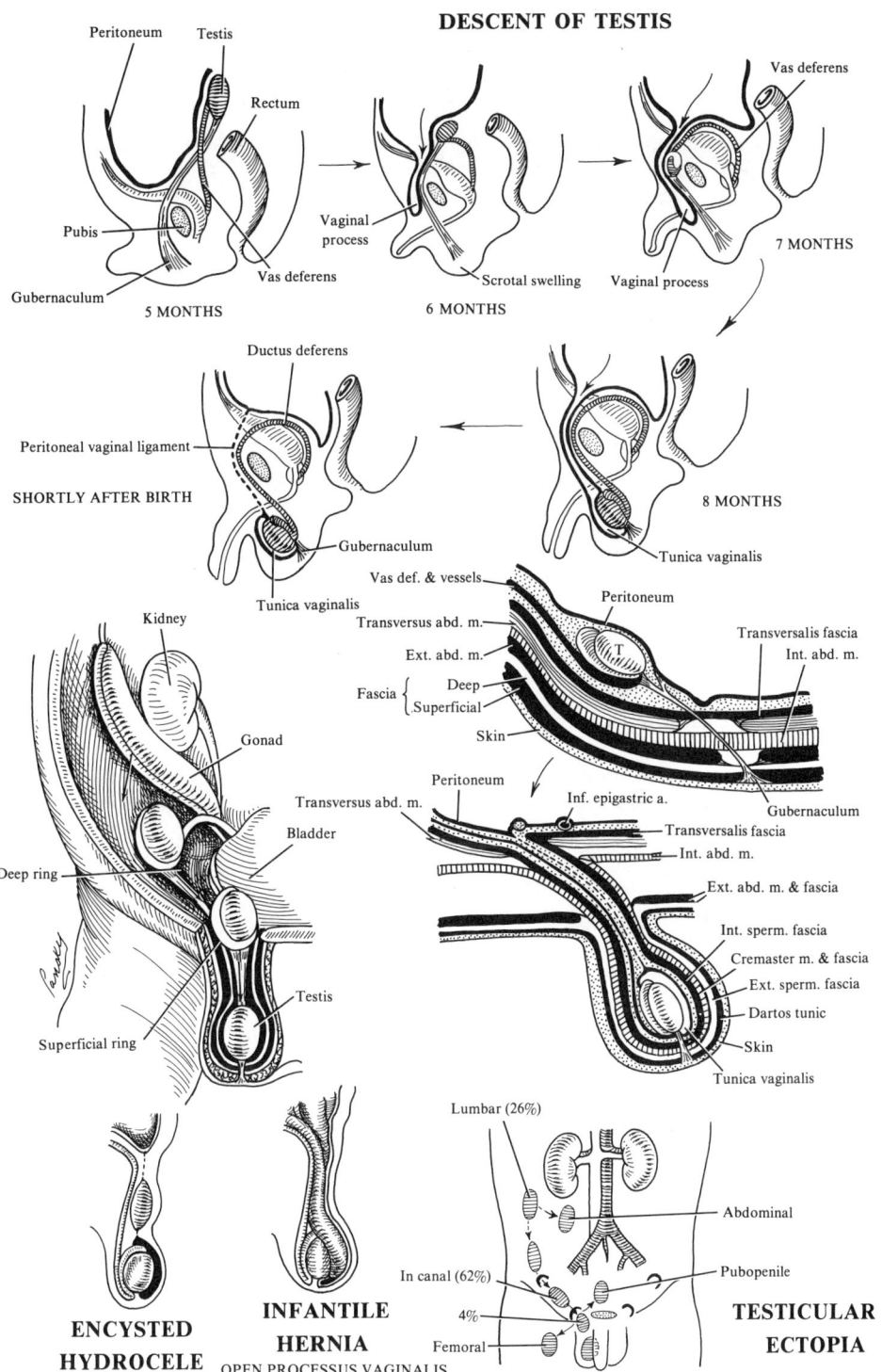

104. DEVELOPMENT OF THE FEMALE GENITAL SYSTEM: OVARIAN DIFFERENTIATION

I. **Ovarian differentiation:** see page 256 for description of indifferent gonads
A. IN EMBRYOS THAT LACK Y-CHROMOSOMES, development occurs slowly in the areas of the gonadal ridges, and the female gonad, which is initially undifferentiated, develops into an identifiable ovary at about week 10 of embryonic life, when its characteristic cortex becomes evident. The testes, on the other hand, are recognizable by week 7. The development and differentiation of the ovary are determined by the genetic constitution of its XX sex chromosomes and are characterized by the following
   1. The primary sex cords are not prominent in the embryonic female gonads, but extend into the medulla to form a rudimentary *rete ovarii*. Both the rete ovarii and the primary sex cords normally degenerate and disappear
      a. The first proliferation of sex cords of the indifferent period are reoriented toward the center of the gland and form the *medullary cords*. The latter, as well as the rete ovarii and its mesonephric connections, regress to eventually constitute Rosenmüller's body or the so-called epoöphoron
   2. During fetal life, there are proliferation of the coelomic germinal epithelium into the underlying mesenchyme and the appearance of a second outgrowth of cords, the *cortical* or *secondary sex cords* (Pflüger's tubes). These eventually occupy the cortex of the gland
   3. As the cortical cords increase in size, primordial germ cells are incorporated into them
   4. At about week 16 (early month 5), the cortical cords, which were originally trabecular, begin to break up into isolated cell clusters called *primordial follicles*. These follicles consist of large cells with clear cytoplasm, the *oogonia*, which are derived from primordial germ cells that migrated to this location from the yolk sac wall, and *supporting nutrient cells* that form a single layer of flattened follicular cells derived from the secondary cortical or sex cords
      a. The primordial follicles are distributed in a connective tissue stroma and demonstrate the cellular duality of the undifferentiated gonadal primordium
         i. Their number is limited, and about 300,000 to 2,000,000 are present at birth
         ii. They contain a specific reproductive cell, the oocyte, at the primary stage, with 46 chromosomes
         iii. Of the initial stock of primordial follicles, approximately 300 develop between puberty and menopause to produce fertilizable ova
   5. Active mitosis of oogonia occurs during fetal life, producing the thousands of primitive germ cells. No oogonia form postnatally in full-term humans. Although many oogonia degenerate before birth, those that do remain enlarge to become primary oocytes.
   6. When the primary oocyte is surrounded by 1 or 2 layers of cuboidal or low columnar follicular cells, it is called a *primary follicle*. Most follicles remain quiescent until puberty
   7. The mesenchyme around the follicles becomes the *ovarian stroma*
   8. After birth, the germinal epithelium flattens out to a single cuboidal layer of cells that is continuous with the mesothelium of the peritoneum at the ovarian hilum. The germinal epithelium is separated from the follicles in the cortex by a thin fibrous capsule, the *tunica albuginea*
   9. As the ovary separates from the regressing mesonephros, it is suspended by its own mesentery, the *mesovarium*

THE DEVELOPING OVARY

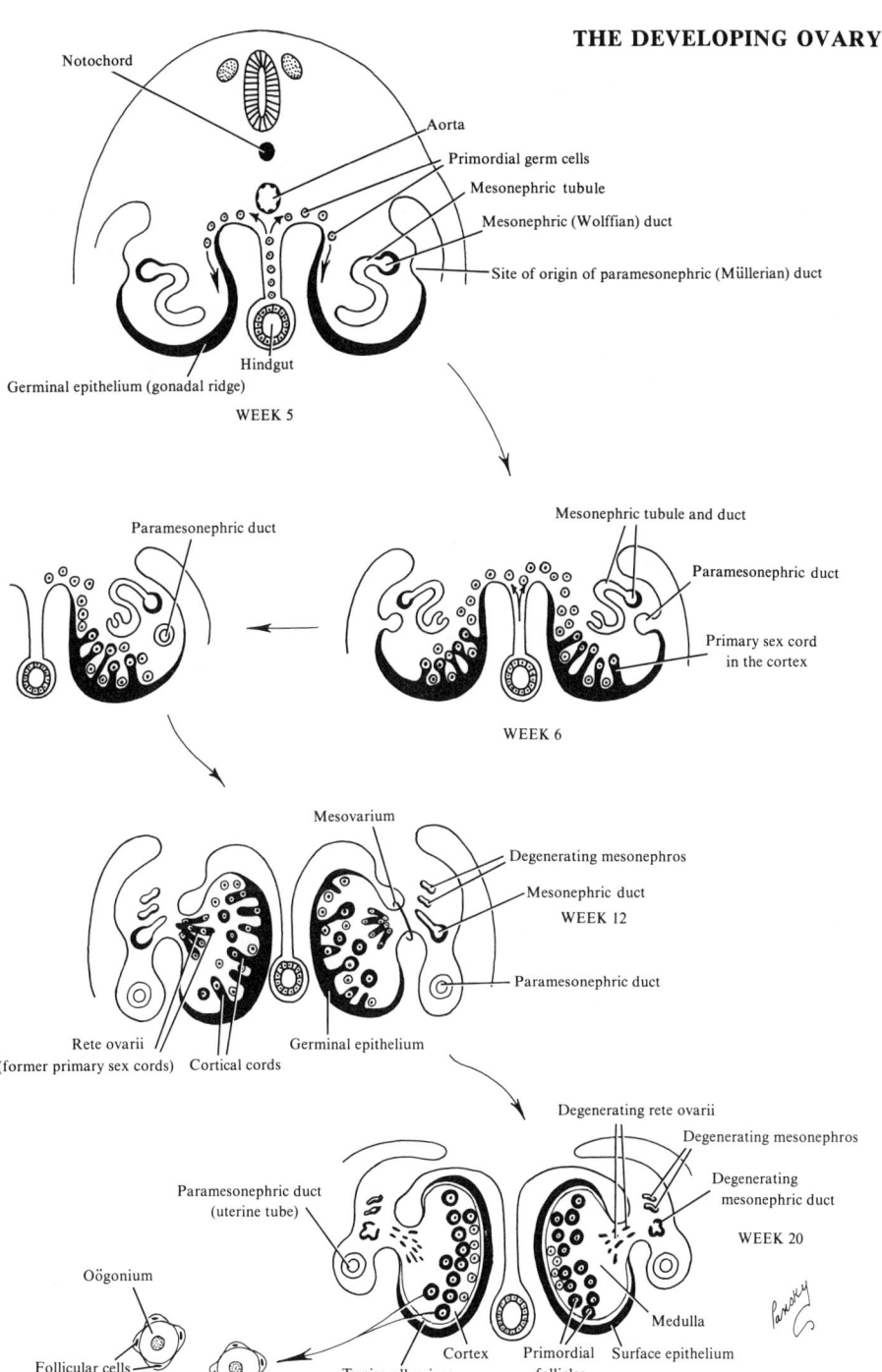

105. DIFFERENTIATION OF THE FEMALE GENITAL TRACTS: UTERUS, VAGINA, AUXILIARY GLANDS, MESENTERIES

I. **Female genital tracts:** see page 256 for indifferent stages of development
   A. IN EMBRYOS WITH OVARIES the mesonephric ducts regress, and the paramesonephric (müllerian) ducts develop to form the female genital tracts. The cranial, unfused parts of the paramesonephric ducts form the uterine tubes and the caudal fused parts form the single median uterovaginal primordium (about week 8) which gives rise to the epithelium and glands of the uterus. The endometrial stroma and myometrium form from adjacent mesenchyme
      1. The fusion of the caudal parts of the paramesonephric ducts commences caudally and progresses up to the future uterine tubes
         a. The median septum of their union disappears by the end of month 3
      2. The two mesonephric (wolffian) ducts regress, but persist as vestigial structures, *Gärtner's ducts*
   B. FUSION OF THE PARAMESONEPHRIC DUCTS brings together 2 peritoneal mesenchymal folds, which form the right and left *broad ligaments* (attached to the abdominal wall on each side and are a continuation of the urogenital mesentery); and 2 peritoneal compartments of pelvic cavity, the *uterorectal pouch* (*of Douglas*) and *uterovesical pouch*
      1. Between the layers of the peritoneal broad ligaments, on each side of the uterus, the mesenchyme proliferates and differentiates into loose connective tissue and smooth muscle to form the *parametrium*
      2. The inguinal ligaments caudal to the uterine tubes become the *round ligaments of the uterus*
   C. FORMATION OF THE VAGINA
      1. The terminal end of the primitive uterovaginal canal touches the posterior wall of the urogenital sinus and forms *Müller's tubercle* or *sinus*. The mesonephric ducts enter the urogenital sinus on each side of this tubercle
      2. The posterior wall of the urogenital sinus thickens opposite the tubercle and with it forms the *vaginal epithelial plate*
      3. From the plate, 2 solid evaginations, the *sinovaginal bulbs,* grow and encircle the caudal end of the uterovaginal primordium (uterine canal)
      4. Canalization of the vaginal plate (begins at week 11) proceeds from caudal to cranial end, producing the lumen of the vagina, with the peripheral cells remaining as the vaginal epithelium. Canalization is completed by month 5
      5. The sinovaginal bulbs surround the uterine cervix to form the *fornices of the vagina*
      6. Thus, the vagina is of entodermal origin derived from the wall of the urogenital sinus. The paramesonephric ducts form the body and cervix of the uterus
         a. The vaginal lumen remains separated from the urogenital sinus until late fetal life by a membrane, the *hymen*. The latter usually ruptures during perinatal life

II. **Auxiliary genital glands**
   A. BUDS FROM THE URETHRA grow into the mesenchyme and form the *urethral glands* and the *paraurethral glands* (*of Skene*). Both correspond to the male prostate gland
   B. OUTGROWTHS FROM THE UROGENITAL SINUS form the *greater vestibular glands* (*Bartholin*) and are homologous to the male bulbourethral (Cowper's) glands

III. **Ovarian migration**
   A. THE UTERINE (FALLOPIAN TUBES, which were the cranial parts of the paramesonephric (müllerian) ducts, are initially vertical. During development of the uterus, they move toward the interior of the abdominal cavity and become horizontal
   B. THE OVARY is temporarily found cranial to the uterine tube on the posterior side of the regressing mesonephric body, but moves to a position posterior to the uterine tube
   C. THE ASSOCIATED MESENTERIES follow the positional changes. Their arrangement in the pelvis forms the broad ligament of the uterus with its 3 "flanges": the *round ligament of the uterus;* the *mesovarium,* posteriorly; and the *mesosalpinx,* cranially

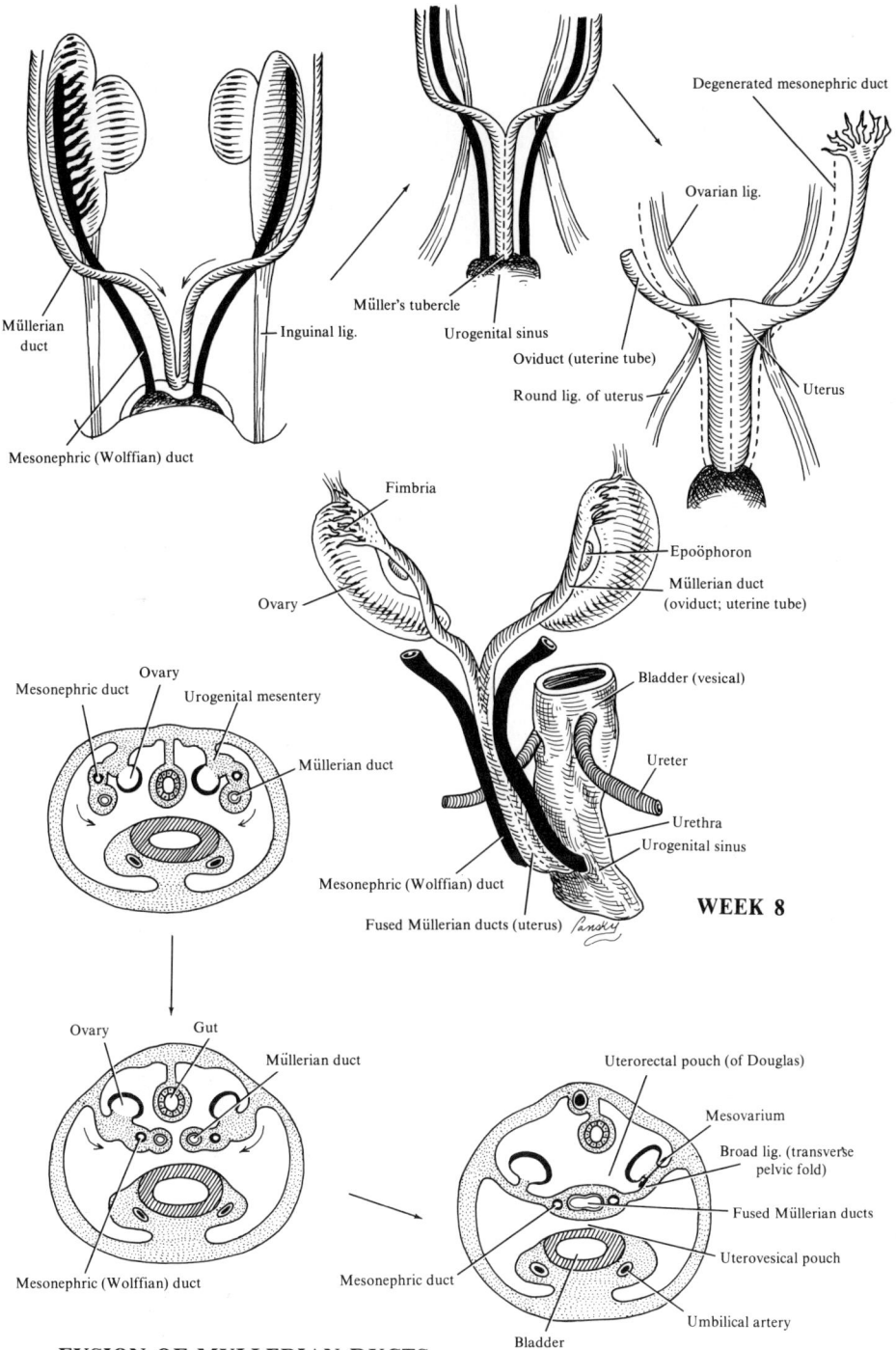

FUSION OF MULLERIAN DUCTS

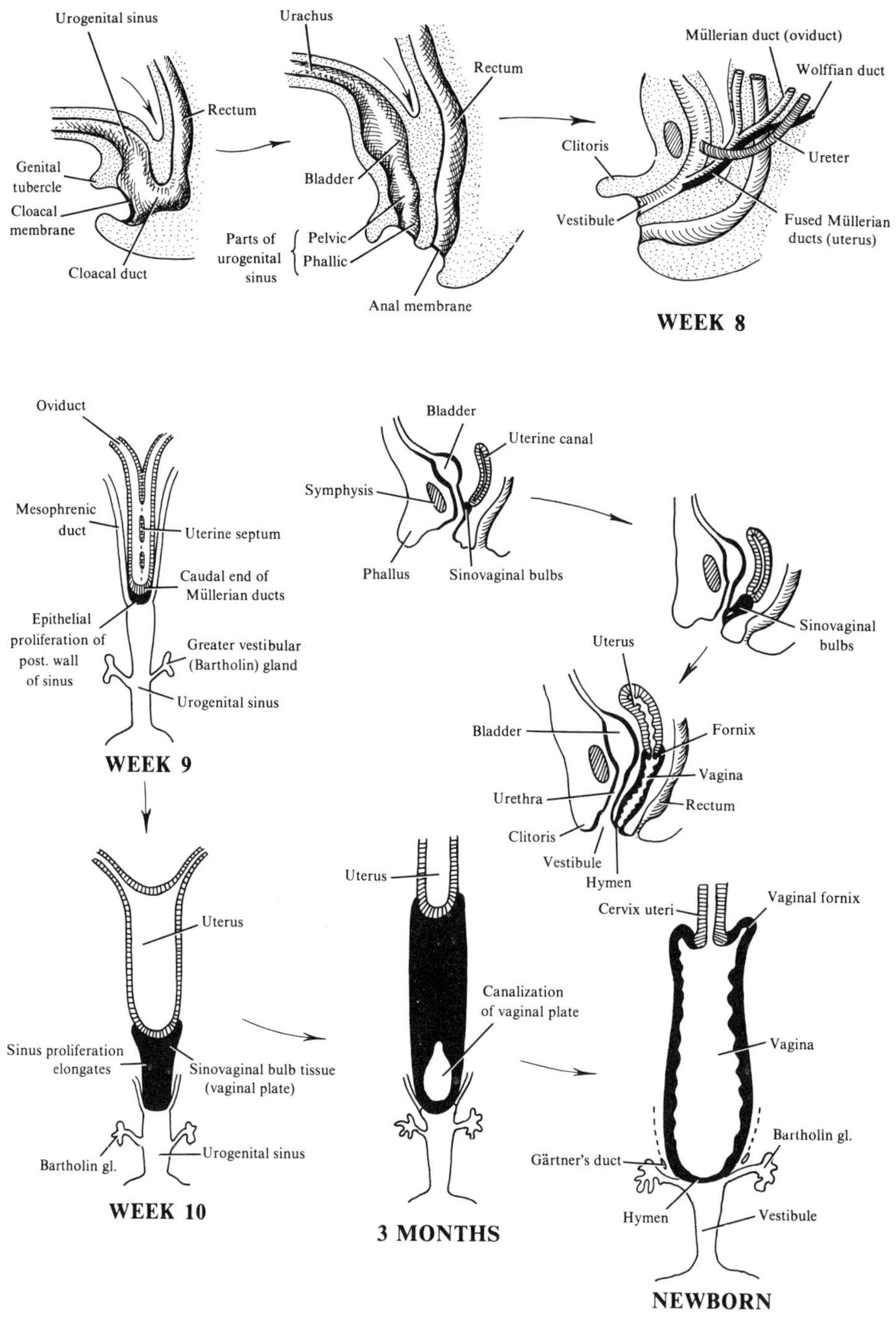

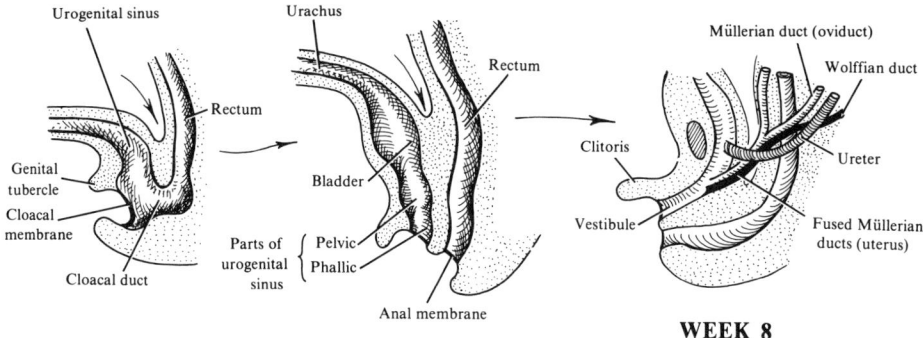

FIGURE 43. **Development of uterus and vagina.**

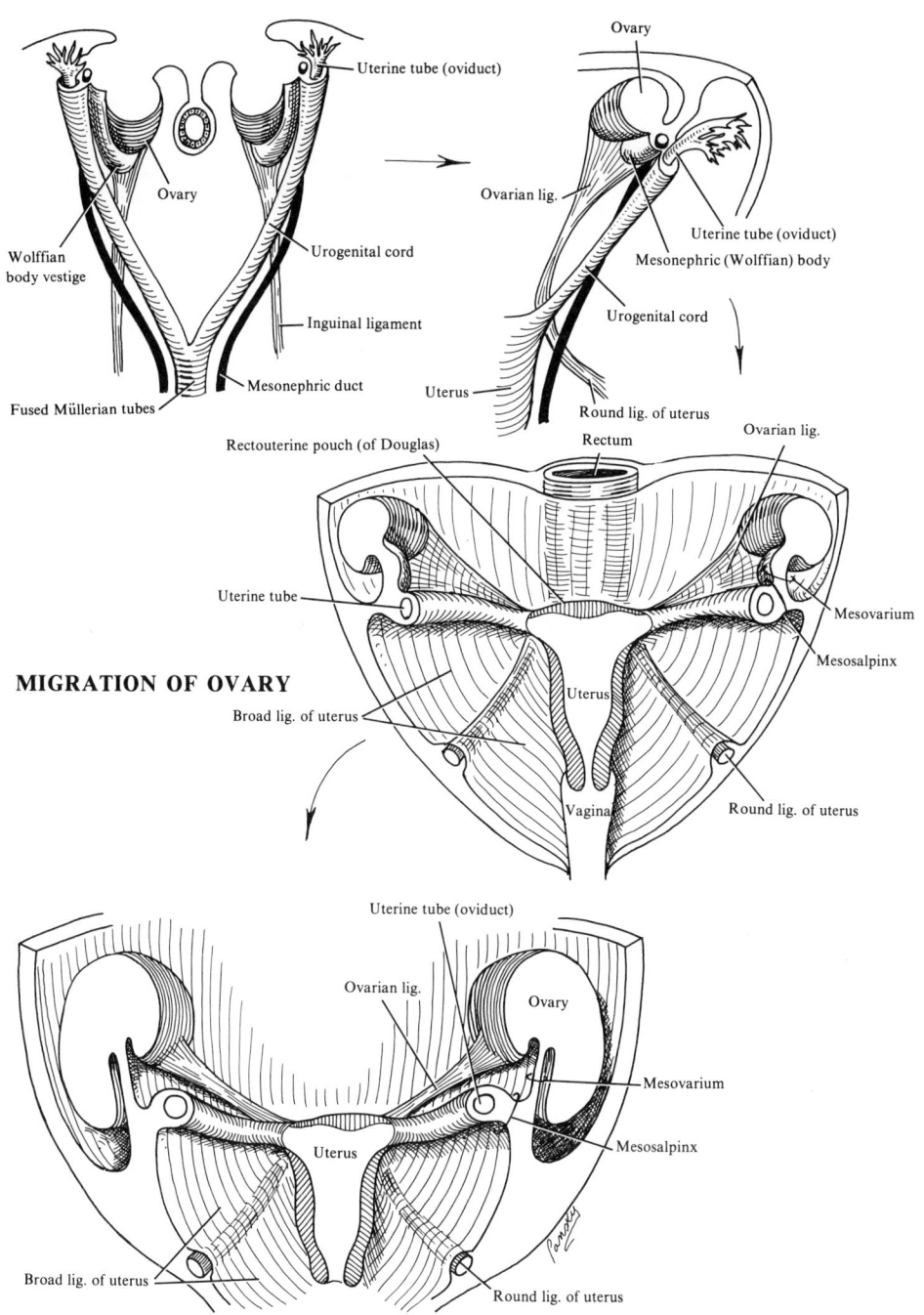

FIGURE 44. **Migration of ovaries and ligaments of the uterus, ovaries, and uterine tubes.**

106. DEVELOPMENT OF THE UPPER AND LOWER PORTIONS OF THE FEMALE GENITAL TRACT

I. The upper portion of the genital tract
A. AFTER FORMATION OF THE UTEROVAGINAL CANAL, the segment of each paramesonephric (müllerian) duct, which is cranial to the junction of the inguinal ligament and the urogenital cord, becomes the *uterine (fallopian) tubes*
   1. The cranial opening of the paramesonephric duct, which is open to the peritoneal cavity, becomes the *fimbriae* of the uterine tube
   2. The uterine tube eventually moves caudally and becomes horizontal
B. THE MESONEPHRIC (WOLFFIAN) DUCTS REGRESS almost completely, leaving only a short segment connected with the mesonephric tubes adjacent to the ovary which persists as the *epoöphoron or Rosenmüller's body*
   1. Other mesonephric tubules caudal to the ovary form the *paroophoron*
C. AFTER THE MESONEPHRIC DUCT REGRESSES, the diaphragmatic ligament is attached directly to the ovary at its cephalic pole and becomes the *suspensory ligament of the ovary*
   1. The inguinal ligament, which is also attached to the ovary but at its caudal pole, becomes the *proper ligament of the ovary*
D. THE MESONEPHRIC BODY flattens and atrophies beginning in month 3, and only a vestige persists between the urogenital and the mesonephric mesenteries. These 3 structures together form the *mesosalpinx,* which connects the urogenital cord and the uterine tube to the abdominal wall
   1. The *mesovarium* is thus attached definitively to the inside of the mesosalpinx and is also continuous caudally with the *mesometrium,* forming with it, the *broad ligament*

II. The lower portion of the genital tract
A. TWO REGIONS CAN BE DISTINGUISHED on either side of the openings of the mesonephric and paramesonephric ducts, in the urogenital sinus
   1. The urinary or cranial region (described with the urinary system)
   2. The genital or caudal region, which in turn, has 2 major parts
      a. The interior or deep vertical portion is the *pelvic portion of the urogenital sinus*
         i. In the female fetus, its cranial end contains the urethral orifice ventrally, and the termination of the vaginal epithelial plate dorsally
         ii. The pelvic portion of the urogenital sinus becomes less deep, bringing to the surface the urethral orifice (or urinary meatus) and the hymen
         iii. The pelvic portion is finally incorporated with the phallic portion, to form with it the *vestibule,* which is surrounded by the external genital organs
      b. The superficial or horizontal portion is the *phallic portion of the urogenital sinus*
         i. It is bordered by the genital tubercle, cranially; and the urogenital membrane, caudally
         ii. After resorption of the membrane in week 9, the phallic portion of the urogenital sinus becomes open to the exterior
   3. Differentiation of the genital portion of the urogenital sinus in the female follows that of the urogenital excretory system and begins in month 3, paralleling the formation of the vagina and the external genital organs

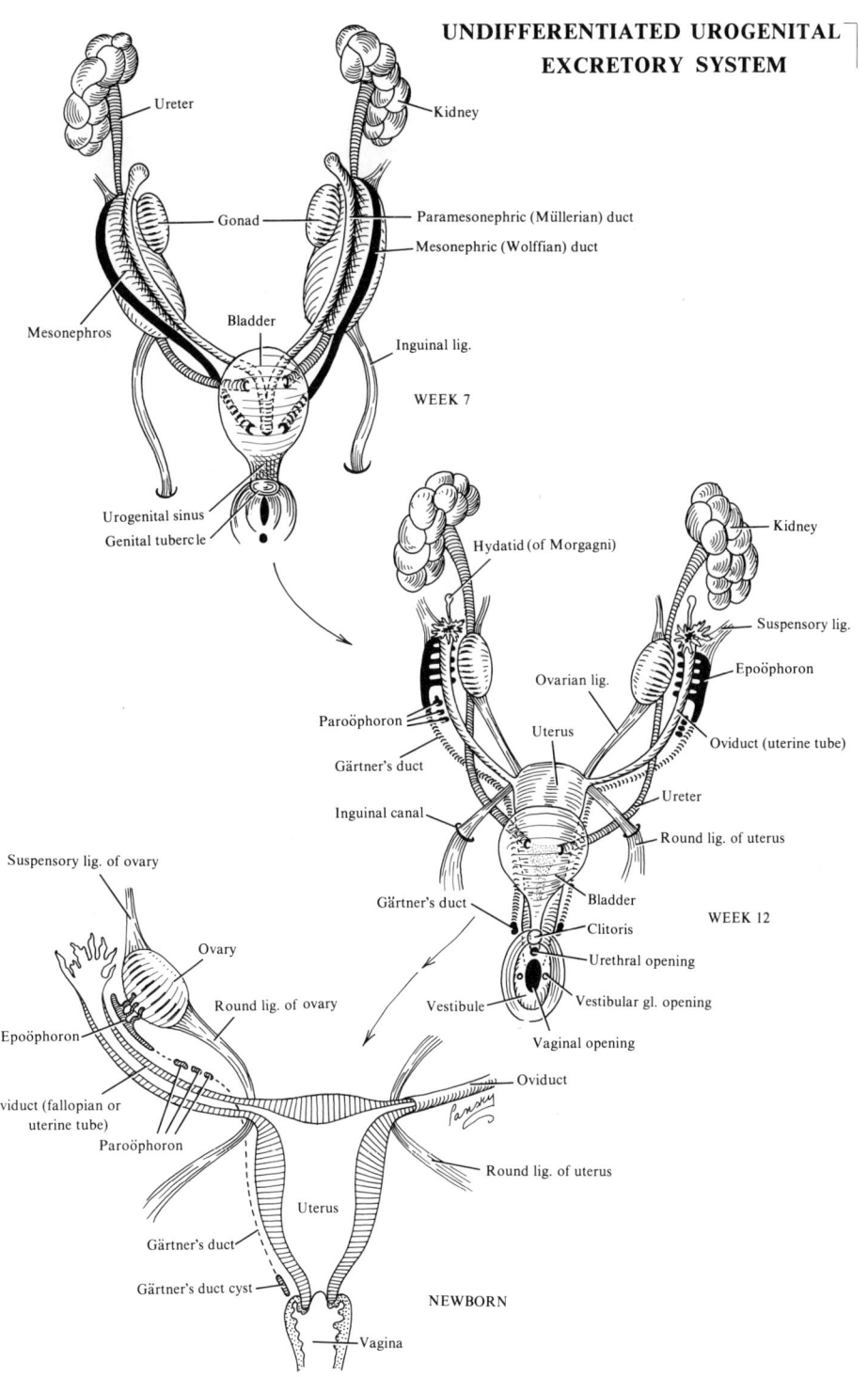

107. DEVELOPMENT OF THE FEMALE EXTERNAL GENITAL ORGANS

I. **The indifferent stage:** see page 256 for description

II. **Definitive development:** in the absence of androgens, feminization of the indifferent external genitalia takes place
   A. THE CLOACAL MEMBRANE, by week 3, is very extensive, and its anterior end is level with the base of the umbilical cord
      1. It is at this stage that the cloacal membrane is bordered laterally by 2 mesenchymal projections, covered by ectoderm, which are the paired primordia of the genital tubercle
      2. At week 4, the anterior end of the cloacal membrane retracts from the base of the umbilical cord, permitting formation of the anterior body wall below (caudal to) the umbilicus, and the paired primordia of the genital tubercle come together in the midline to form the *genital tubercle*
      3. The cloacal fold, which surrounds the cloacal membrane, further prolongs the genital tubercle, and about this time new swellings appear, the *genital swellings,* on either side of the cloacal folds and surround the genital tubercle and the cloacal folds
      4. At week 7, the cloacal membrane is divided into the *urogenital membrane* (anteriorly) and the *anal membrane* (posteriorly), separated by the *perineal body* (perineum)
         a. The cloacal fold is divided into the *genital fold,* which surrounds the urogenital membrane, and the *anal fold,* which circumscribes the ectodermal depression of the proctodeum covered by the anal membrane
   B. THE UROGENITAL MEMBRANE, in week 9, disappears, thus, the phallic segment of the urogenital sinus becomes open to the exterior
   C. DIFFERENTIATION OF THE EXTERNAL GENITALIA takes place during month 3 and closely follows the pattern of the primitive structures
      1. The genital tubercle elongates only slightly and forms the *clitoris,* where erectile tissues develop
      2. The urogenital sinus remains open with the *urethra* opening anteriorly and the *vagina* posteriorly, within the interior of the vestibule portion of the sinus
      3. The vestibule is bordered laterally by the genital folds, which become the *labia minora*
      4. The genital (labioscrotal) swellings largely remain unfused and form the *labia majora*
         a. They do fuse posteriorly to form the *posterior labial commissure* and anteriorly to form the elevation known as the *mons pubis*

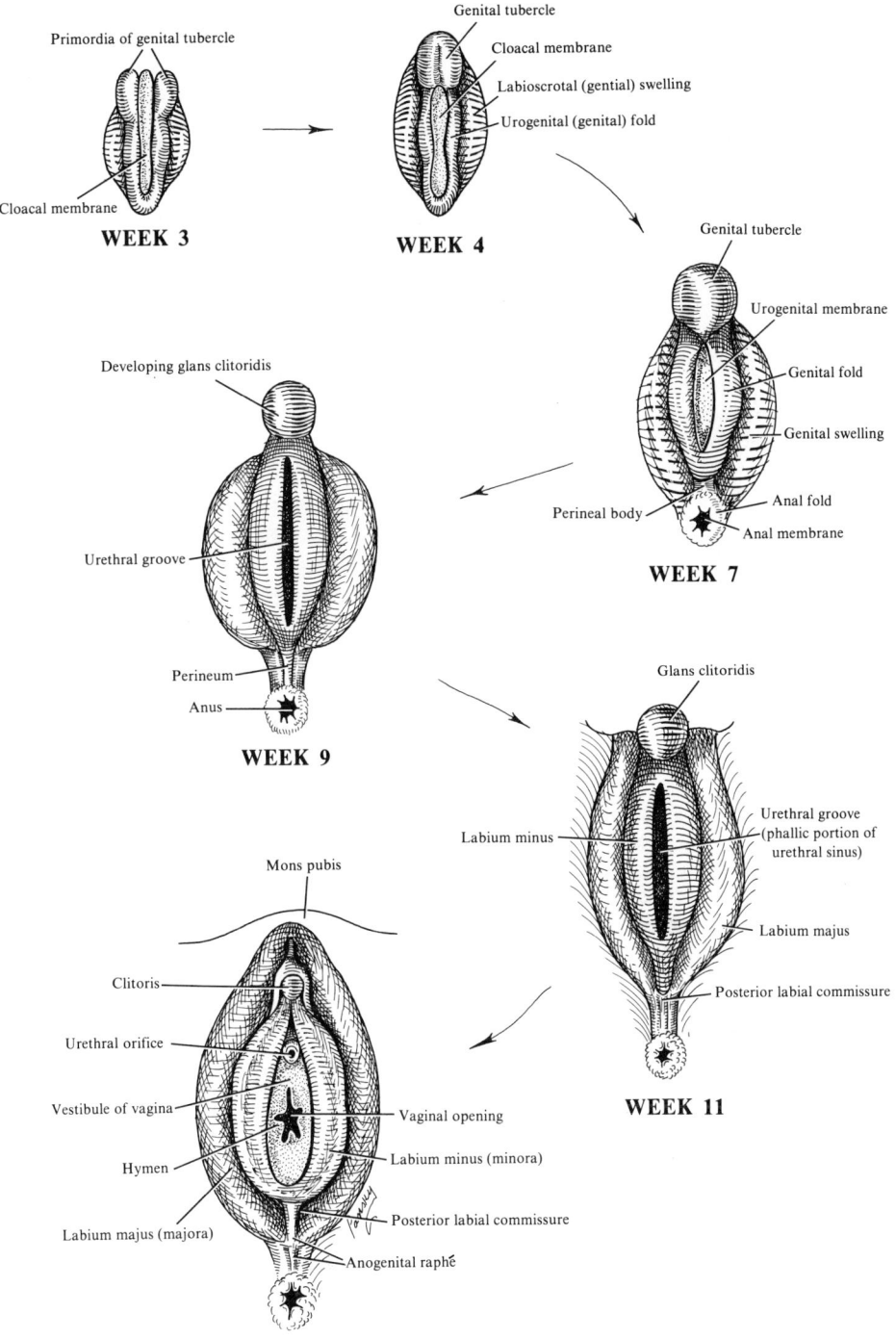

108. SEXUAL ANOMALIES OF GENETIC AND HORMONAL ORIGIN

I. **Sexual anomalies of genetic origin:** sex chromosome alterations may be transmitted by either parent (gonadal dysgeneis) or may occur in the embryo itself, even though the egg initially had a normal chromosome count (true hermaphroditism)
   A. GONDAL DYSGENESIS is due to a failure of disjunction of the sex chromosomes, during gametogenesis, in one of the parents. It may result in
      1. Testicular dysgenesis of Klinefelter, with male morphology and karyotype XXY
      2. Ovarian dysgenesis of Turner, with female morphology and a karyotype of only 1 sex chromosome which is an X and is thus called XO
      3. Mixed gonadal dysgenesis: rare; chromatin-negative nuclei; a testis on one side and an undifferentiated gonad on the other
         a. Internal genitalia are female, but male derivatives of the mesonephric ducts are present
         b. External genitalia range from normal female to intermediate to normal male, but at puberty, neither breast development nor menstruation occurs. One sees varying degrees of virilization
      4. Testicular feminization: rare; appears as a normal female despite the presence of testes and XY sex chromosomes
         a. There is normal breast development at puberty, but no pubic hairs and the vagina ends blindly. Other internal genitalia are also absent or rudimentary
         b. The testes are intra-abdominal or inguinal, but may be in the labia majora
         c. This is an extreme form of male pseudohermaphroditism, but is not intersex because female external genitalia are normal. The testes develop and secrete androgens, but masculinization of the genitals fails because the indifferent external genitalia are insensitive to the androgens
   B. TRUE HERMAPHRODITISM: rare; chromatin-positive nuclei; is secondary to failure of disjunction of sex chromosomes during the first cleavage mitosis of the egg and results in sex mosaics such as XY/XX or XY/XO
      1. The relative quantitative importance of the male component (Y) helps explain the ultimate variations in androgenic effects, including varying degrees of the differentiation of the external genitalia and some subtle variations in somatic morphology

II. **Sexual anomalies of hormonal origin** are seen in people with normal genetic makeup. The phenomenon of primary and secondary sexual characteristics leading to pseudohermaphroditism is related to the somatic effects of abnormal androgen secretion
   A. MALE PSEUDOHERMAPHRODITISM: a result of insufficient androgen secretion of an otherwise normal testis with chromatin-negative nuclei and a male 46,XY karyotype
      1. A slight deficiency affects only the last stages of differentiation of the external genitalia, resulting in a small penis, hypospadias, and a valviform appearance of the scrotum. The general morphology is masculine
      2. A severe deficiency shows persistence of the paramesonephic system, and thus, a vagina and uterus coexist with 2 normal vas deferens ducts. The testes are ectopic, and the external genitalia, as well as general morphology, are female type
   B. FEMALE PSEUDOHERMAPHRODITISM is the result of an *abnormal virilization* of a female fetus that has normal ovaries and a 46,XX karyotype
      1. Virilization may be of endogenous origin, or due to excessive androgen secretion by the fetal adrenal, or of exogenous origin, related to administration of synthetic progesterones or anabolic hormonal medications containing androgens
      2. The paramesonephic system develops normally, but the androgens cause persistence of the mesonephric system and differentiation of the external genitalia toward the male type, resulting in a peniform clitoris and a tendency to closure of the urogenital sinus. Coalescence of the labia majora is completely latent and becomes known by chance through complications of pregnancy or by hysterography
      3. Only severe atresias are found early; they may create difficult treatment problems

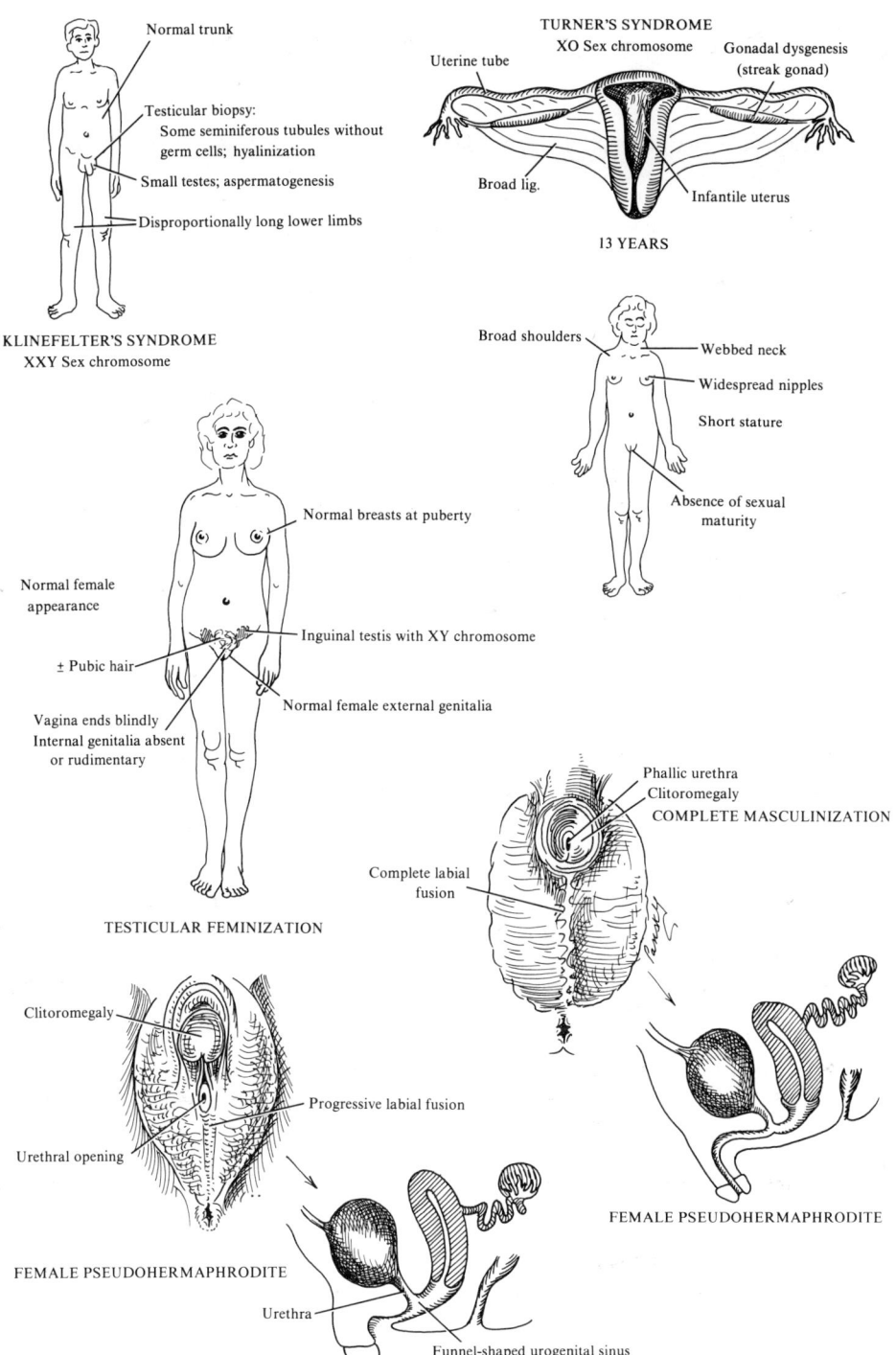

109. GENITAL MALFORMATIONS IN THE MALE

I. **Epispadias:** seen in 1/30,000 infants where the urethra opens on the dorsal surface of the penis. It often is associated with exstrophy of the bladder. It may involve the glans or the entire penis. It also is seen in the female but is rare and appears as a fissure of the upper urethral wall that opens on the dorsum of the clitoris

II. **Hypospadias** seen in 1/300 males where the external urethral opening is on the ventral (under) surface of the penis rather than at the tip of the glans
   A. THE PENIS is often underdeveloped, curves downward (chordee), and we speak of 4 types of hypospadias: *glandular, penile, penoscrotal,* and *perineal*
   B. IT IS THE RESULT of an inadequate production of androgens by the fetal testes, resulting in failure of fusion of the urogenital folds and incomplete formation of the urethra

III. **Cryptorchidism, or undescended testes:** seen in about 30% of premature infants and in about 3% of full-term infants. Normal descent into scrotum is complete by first year
   A. IF BOTH TESTES STAY IN or just outside the abdominal cavity, they fail to mature histologically, and sterility is almost certain
   B. THE TESTES MAY BE FOUND in the abdominal cavity or anywhere along the path of testicular descent, but they usually lie in the inguinal canal. May be unilateral or bilateral
   C. CAUSE OF UNDESCENT is unknown, but is probably a failure of normal androgen production

IV. **Ectopic testis:** rare; in its descent, it deviates from the normal pathway and lodges in abnormal locations, such as the external aponeurosis of the external oblique muscle (interstitial), the thigh (femoral triangle area), dorsal to the penis, or on the opposite side (crossed ectopia)

V. **Hydrocele:** abdominal end of the processus vaginalis remains open but is too small for a hernia. Peritoneal fluid passes into processus, and we get a hydrocele of the testis and spermatic cord
   A. IF THE MIDDLE OF THE CANAL OF THE PROCESSUS VAGINALIS stays open, fluid may accumulate and form a *hydrocele of the spermatic cord*

VI. **Congenital inguinal hernia:** a persistent processus vaginalis. Loops of intestine may herniate into scrotum or labium majora. More common in males than in females and is associated with cryptorchidism

VII. **Agenesis, or absence of penis:** failure of the genital tubercle to develop. The urethra opens into the perineum near the anus. Testes and scrotum are normal

VIII. **Bifid penis and double penis:** failure of fusion of the 2 parts of the genital tubercle results in a bifid condition; formation of 2 tubercles creates a double penis

IX. **Micropenis:** penis is very small and is hidden by the suprapubic fat. Probably due to a functional hormonal deficiency of the fetal testes. Is commonly associated with hypopituitarism

X. **Retroscrotal penis, or transposition of penis and scrotum:** the penis is found behind the scrotum. Due to a failure of the labioscrotal folds to shift caudally as they fuse to form the scrotum. They may even develop in front of the genital tubercle and then fuse

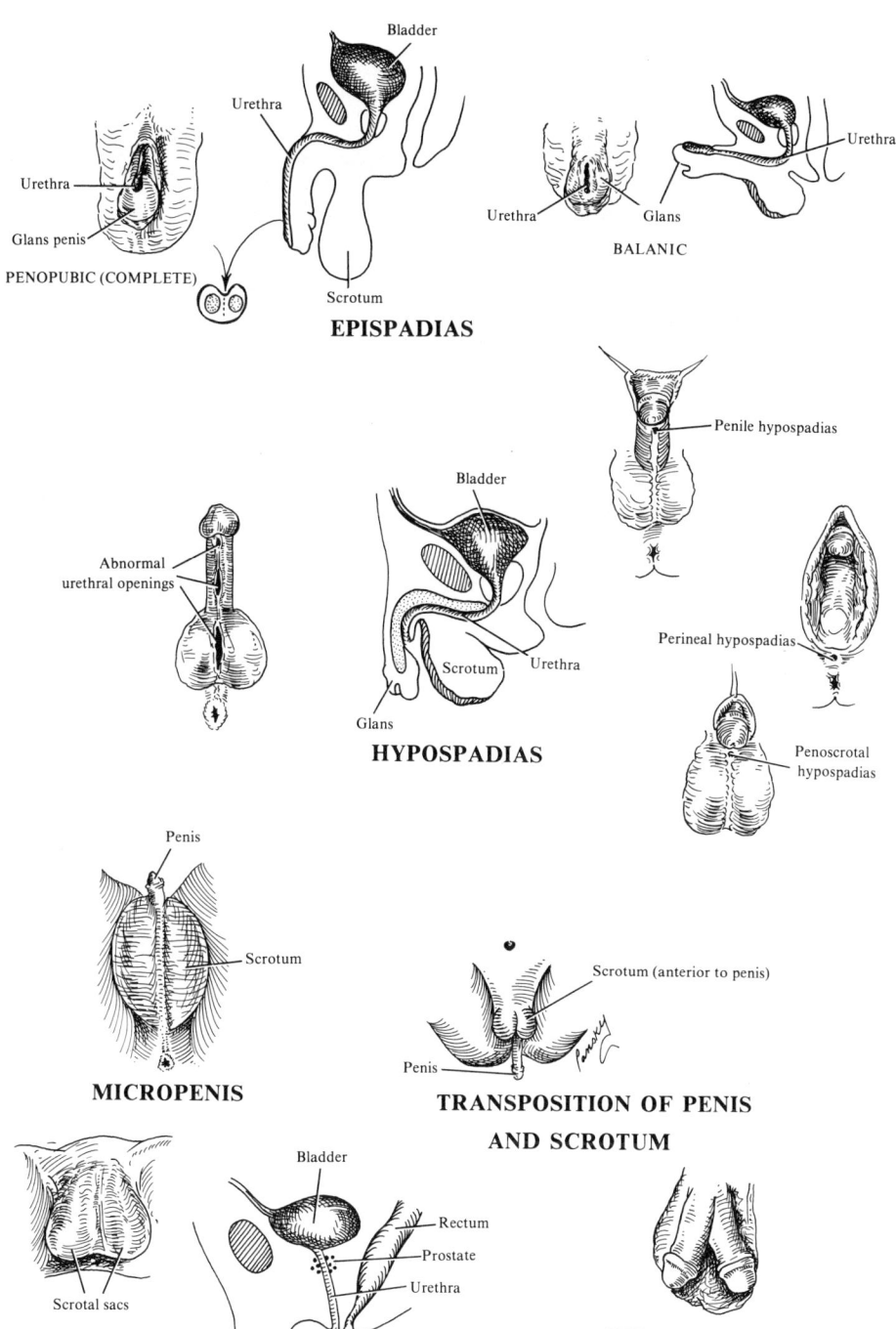

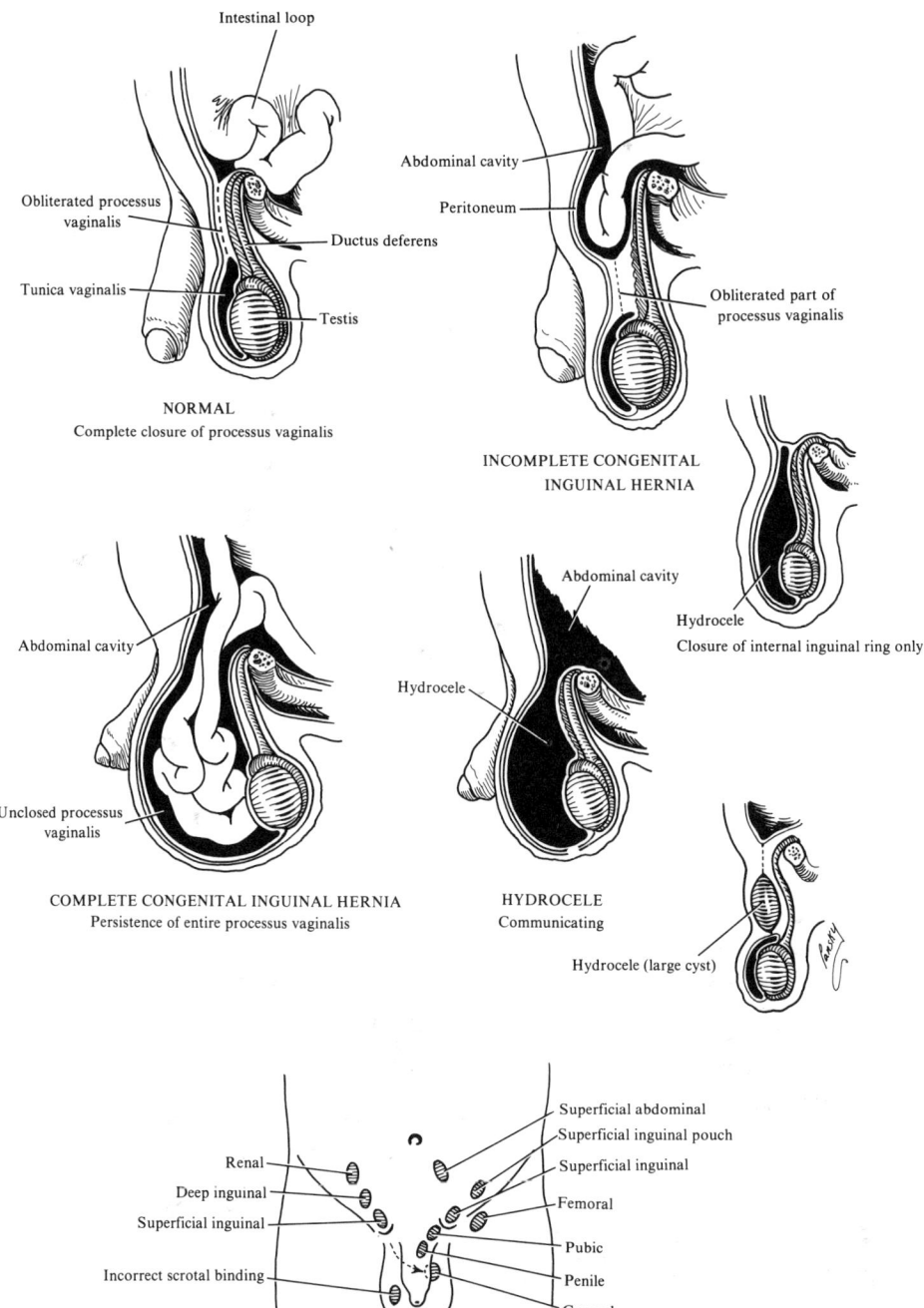

FIGURE 45. **Hydrocele, inguinal hernia, and cryptorchid testes.**

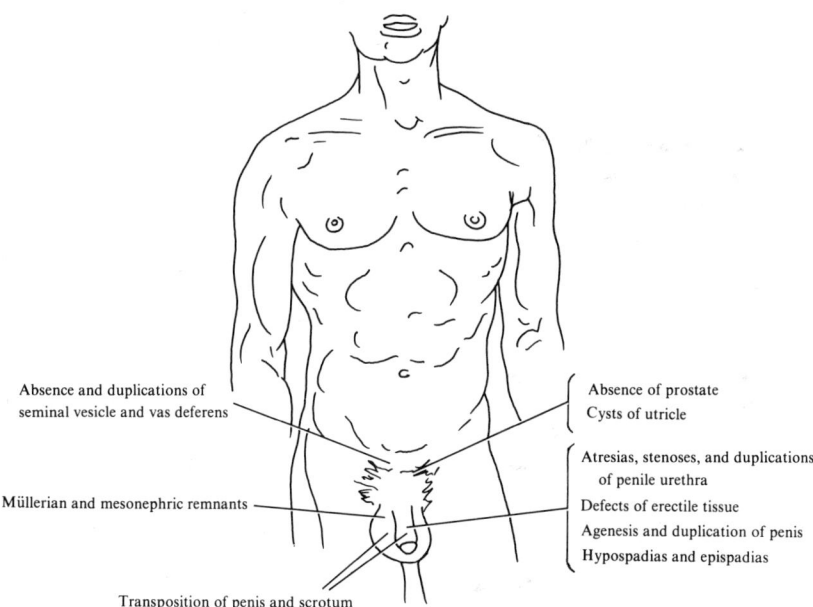

ANOMALIES OF MALE REPRODUCTIVE TRACT

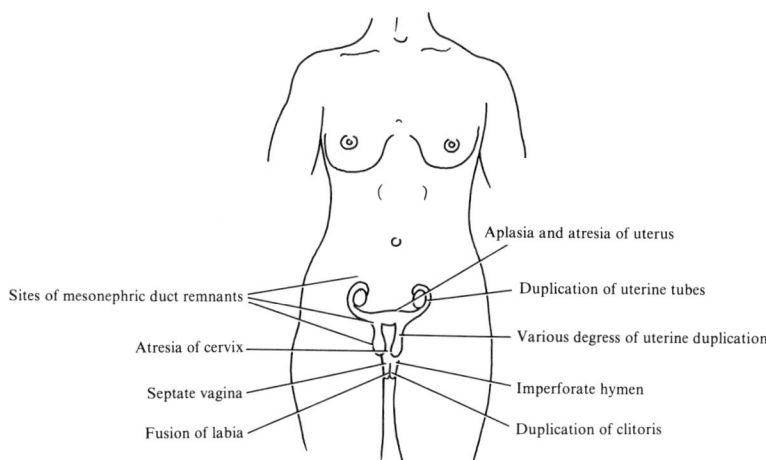

ANOMALIES OF FEMALE REPRODUCTIVE TRACT

FIGURE 46. Anomalies of the male and female reproductive tracts.

110. UTEROVAGINAL MALFORMATIONS OF THE FEMALE

I. Uterovaginal malformations
  A. AS A RESULT OF PARTIAL OR TOTAL FAILURE OF FUSION of the terminal portion of the paramesonephric (müllerian) ducts, we may see
   1. Uterus didelphis: double uterus with a double cervix and double vagina
   2. Uterus bicornis (bicornuate or bifid): uterus is more or less divided into 2 lateral horns.
      a. The cervix may be single: uterus bicornis unicollis
      b. The cervix may be double: uterus bicornis bicollis
   3. Uterus biforis: cervix is divided into 2 by a septum
   4. Uterus cordiformis: an incomplete uterus bicornis with a wedge-shaped depression at the fundus (uterus becomes heart-shaped)
   5. Uterus parvicollis: abnormal disproportionately small cervix
   6. Absence of uterus
  B. BY PARTIAL OR TOTAL ATRESIA of the terminal portion of 1 or both paramesonephric (müllerian) ducts, we may see
   1. Unilateral atresia with uterus bicornis unicollis or rudimentary horn
      a. May also see a uterus unicornis: one-horned
   2. Partial bilateral atresia with atresia of the cervix
   3. Partial bilateral atresia with atresia of the vagina
   4. Absence of the vagina: seen in 1/4000 females
  C. BY FAILURE OF RESORPTION OF THE UTEROVAGINAL SEPTUM after fusion of the paramesonephric (müllerian) ducts, we may see
   1. Complete bilocular uterus: a uterus having 2 cavities (also called U. duplex)
   2. Bilocular unicervical uterus
   3. Bilocular bicervical uterus

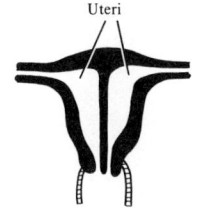
COMPLETE BILOCULAR UTERUS

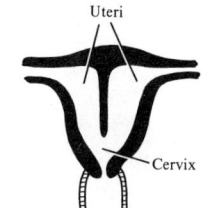
BILOCULAR UNICERVICAL UTERUS

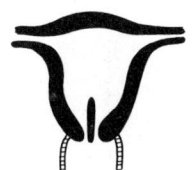
BILOCULAR BICERVICAL UTERUS

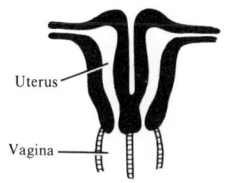
DOUBLE UTERUS (DIDELPHYS) AND DOUBLE VAGINA

BICERVICAL UTERUS BICORNIS

UNICERVICAL UTERUS BICORNIS

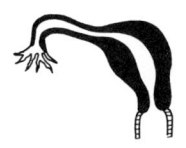
UNICORNUATE UTERUS

UNILATERAL ATRESIA

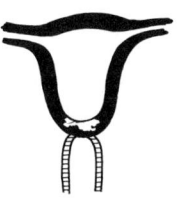
CERVICAL ATRESIA

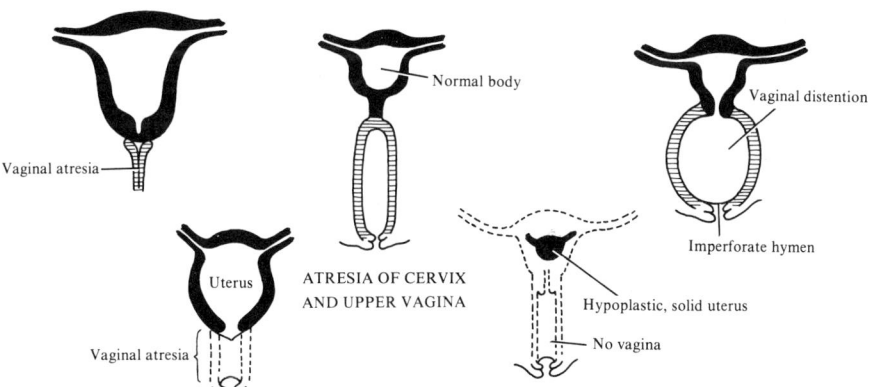

111. ADULT DERIVATIVES OF EMBRYONIC UROGENITAL STRUCTURES

Embryonic Structure	Male	Female
Indifferent gonad	Testis	Ovary
Cortex	Seminiferous tubules	Ovarian follicles
Medulla	Rete testis	Rete ovarii
Gubernaculum	Gubernaculum testis	Ovarian ligament and round ligament of uterus
Mesonephric tubules	Ductuli efferentes	Epoöphoron
	Paradidymis	Paroöphoron
Mesonephric duct (wolffian duct)	Appendix of epididymis	Appendix vesiculosa
	Ductus epididym	Duct of epoöphoron
	Ductus deferens	Duct of Gärtner
	Ureter	Ureter
	Ureteric pelvis	Ureteric pelvis
	Calyces	Calyces
	Collecting tubules	Collecting tubules
	Ejaculatory duct and seminal vesicle	
Paramesonephric duct (müllerian duct)	Appendix of testis	Hydatid (of Morgagni)
		Uterine tube
		Uterus
Urogenital sinus	Urinary bladder	Urinary bladder
	Urethra (except glandular portion)	Urethra (all)
	Prostatic utricle	Vagina
	Prostate gland	Urethral and paraurethral glands
	Bulbourethral glands	Greater vestibular glands
Sinus or müllerian tubercle	Seminal colliculus	Hymen
Phallus	Penis	Clitoris
	Glans penis	Glans clitoridis
	Corpora cavernosa penis	Corpora cavernosa clitoridis
	Corpus spongiosum	Bulb of vestibule
Urogenital folds	Ventral (under) aspect of penis	Labia minora
Labioscrotal swellings	Scrotum	Labia majora

DIFFERENTIAL DIAGNOSIS OF PATIENTS WITH AMBIGUOUS EXTERNAL GENITALIA*

Sex chromatin pattern
- Chromatin positive
  - 17-ketosteroid output normal
    - No history of maternal hormone ingested
    - Nonandrogenic female pseudohermaphrodite
    - True hermaphrodite
    - Chromatin-positive male pseudohermaphrodite
    - Exploratory laporotomy, gonadal biopsies, and chromosome studies needed
    - History of maternal hormone ingestion
    - Female pseudohermaphrodite due to maternal androgens or progestins
    - Exploratory laporotomy not needed
  - 17-ketosteroid output elevated
    - Congenital virilizing adrenal hyperplasia
- Chromatin negative
  - Male pseudohermaphrodite
  - True hermaphrodite
  - Mixed gonadal dysgenesis
  - Gonadal dysgenesis with phallic enlargement
  - Gonadal dysgenesis with male pseudohermaphroditism
  - Exploratory laporotomy, gonadal biopsies, and chromosome studies usually needed

*After Federman (1967), Hamerton (1971), Gray and Skanakakis (1972), Schlegel and Gardner (1975), Page et al. (1976), and Villee (1975).

UNIT TEN

THE CIRCULATORY SYSTEM:

Cardiovascular and Lymphatic Systems

112. HEMATOPOIESIS AND GENERAL DEVELOPMENT OF THE CIRCULATORY SYSTEM

I. Hematopoiesis

A. THE YOLK SAC CAPILLARIES form a network which is drained by the vitelline veins. The latter flow directly into the venous sinus of the heart, forming first an anastomotic network around the duodenum, and then crossing the septum transversum (described later)

B. THE HEPATIC PRIMORDIUM, of entodermal origin, begins to invade the septum transversum in the middle of week 3 at about the same time that vascular primordia appear

C. THE HEPATIC EPITHELIAL CORDS proliferate and surround the vitelline veins and fragment them into many *sinusoidal capillaries*. Extension of this proliferation to the entire septum transversum carries the same process to the umbilical veins

D. THE HEPATIC CELLS or hepatocytes are next arranged into cords which surround the sinusoidal capillaries

E. HEPATIC HEMATOPOIESIS begins during month 2 and attains its maximum in month 3. It then decreases and ceases about month 7 when it is assumed by the bone marrow, which has already become functional beginning with month 4
   1. The megaloblasts (primitive nucleated red cells) are replaced by *erythroblasts* (*normoblasts*) and finally by *mature erythrocytes*, without nuclei

F. PATHOLOGY: in fetal-maternal blood incompatibility, where there is immunization of the mother against the red blood cells of her fetus, the maternal antibodies destroy the fetal red cells. The fetus reacts against the anemia by an intense erythropoiesis.
   1. The liver retains its hematopoietic function beyond the usual month 7, and the red cells appear in the circulation even before they are completely mature
   2. Thus, the presence of erythroblasts in the blood of the newborn is one characteristic of this form of *hemolytic anemia*

II. General development of the circulatory system

A. INTRODUCTION: the many important rearrangements which the circulatory system undergoes during its development and the complex modifications of the general vascular plan are usually related to function. Thus, the organs of nutrition and excretion which are necessary for survival have a great deal of metabolic activity and show priority in development. The transition from fetal life to that after birth are not only marked by physiologic changes but also show marked circulatory modifications

B. THE ESSENTIAL STAGES OF CIRCULATORY DEVELOPMENT
   1. The vitelline stage: where the embryo lives on its own small reserves, from week 3 to the beginning of month 2
      a. The vitelline circulation of the yolk sac is predominant, and the primitive intraembryonic circulation and allantoic circulation are first forming
   2. The placental stage: an intermediary organ, the placenta, develops between the fetus and the mother
      a. The vitelline circulation disappears at the end of month 2, but from its only vestige arise the superior mesenteric vessels
      b. The allantoic circulation becomes placental and is predominant after day 30
      c. The umbilical vessels accomplish the change of circulation and become responsible for oxygenation, nutrition, and filtration
      d. In addition, the above intraembryonic circulation development is marked by special enlargement of the liver, the brain, and the mesonephros
   3. The neonatal stage: the organism assumes its postbirth or postnatal autonomous existence. Placental circulation is interrupted. The placental circulation now is taken over by specialized organs
      a. The lungs begin to function for oxygenation
      b. The metanephros, or adult kidney, begins to function at about month 3 of fetal life and then takes over all renal function and filtration
   4. The mesenteric network, which drains the digestive tract, takes over nutritional responsibilities

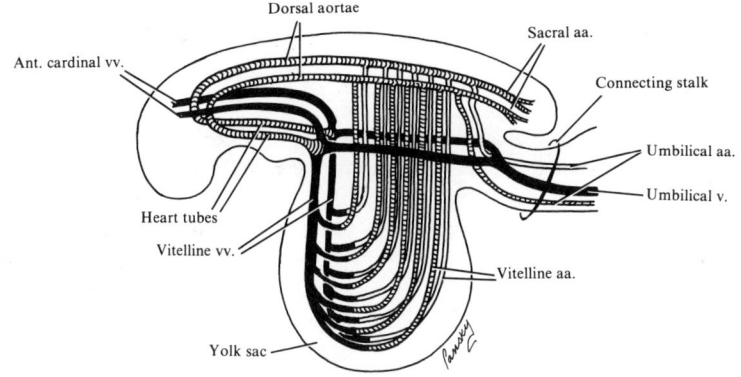
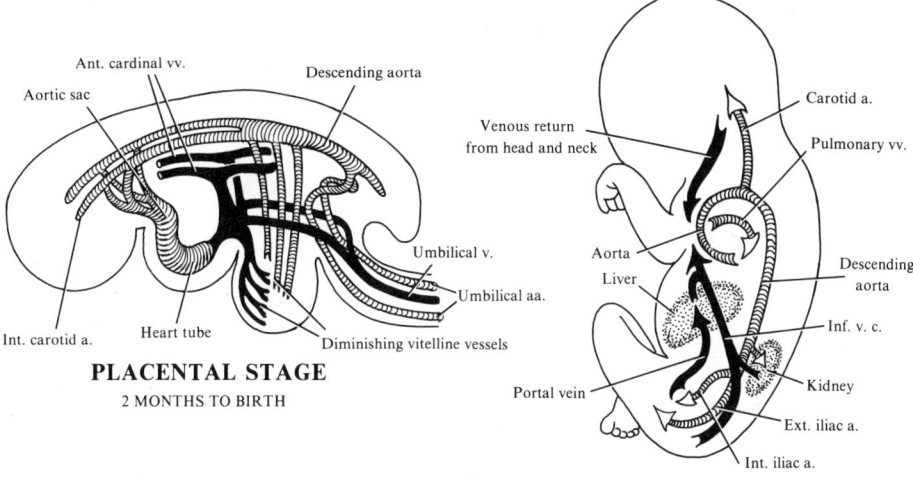
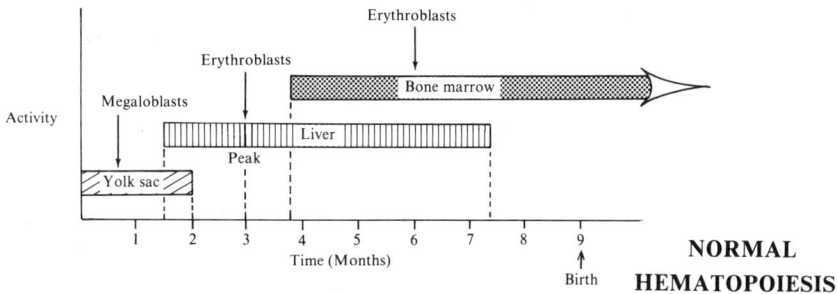

113. CARDIOVASCULAR CIRCULATORY AND LYMPHATIC SYSTEMS: EARLY DEVELOPMENT

I. **Introduction:** the blood and cardiovascular system are derived from mesoderm. They both develop at the same time, beginning about the middle of week 3, when the embryo can no longer satisfy its nutritional needs by diffusion alone. Their development is correlated with the absence of a significant amount of nutritive yolk in the oocyte and yolk sac. The first blood and vascular elements appear near the exterior of the embryo in the mesenchyme lining the yolk sac. This extraembryonic network rapidly blends with the intraembryonic circulation which appears a little later, beginning week 4. As the yolk sac regresses, near the end of month 2, so do the blood-forming islands, and the hematopoietic function is then assumed by the liver

II. **The primitive cardiovascular system and blood islands**
  A. EXTRAEMBRYONIC BLOOD VESSELS: angiogenesis begins in the extraembryonic mesoderm as clusters of mesenchymal cells which differentiate in the chorion, the connecting stalk, and the yolk sac wall, toward days 13 to 15
   1. These angiogenetic clusters, made up of mesenchyme angioblasts, give rise to blood- and vascular-forming structures called the *blood islands of Wolff and Pander* (about 2 days later) when a lumen is formed in the clusters by the appearance and confluence of intercellular clefts
   2. Cells on the periphery of an island flatten and form *endothelial cells* which outline the vessels. Mesenchymal cells surrounding the primitive endothelial vessels differentiate into the muscular and connective tissue elements of the vessels
   3. The central cells of the islands become free and give rise to the blood cells. These early "parent" cells are called *hemocytoblasts* and represent the origin of 3 lines of blood cells. However, at this stage, they give rise essentially to *nucleated red cells called megaloblasts*
      a. Blood formation does not begin in the embryo until week 5. It is seen first in the various mesenchymal areas, namely, in the liver and later in the spleen, the bone marrow, and the lymph nodes
   4. The blood islands approach each other by sprouting endothelial cells, fuse, and form a plexiform network. The latter is transformed into small blood vessels under the influence of hemodynamic factors
      a. With continuous budding, the extraembryonic vessels gradually penetrate the embryo proper
  B. INTRAEMBRYONIC BLOOD VESSELS develop independently from angiogenetic cell clusters that appear in the splanchnic mesoderm layer of the late preomite embryo
   1. The clusters, at first, are seen on the lateral sides of the embryo, but soon spread rapidly in a cephalic direction
      a. In time, they acquire a lumen, unite, and form a plexus of small blood vessels which gradually becomes horseshoe-shaped
      b. In addition, other clusters of angiogenetic cells appear bilaterally, parallel and close to the midline of the embryonic shield
         i. These also acquire a lumen and form a pair of longitudinal vessels, the *dorsal aortae*. These, at a later date, connect up with the horseshoe-shaped plexus (the latter becomes the heart tube)
   2. The intraembryonic coelomic cavity found over the central portion of the horseshoe-shaped plexus later develops into the pericardial cavity. Thus, the pericardial cavity lies anterior to the prochordal plate

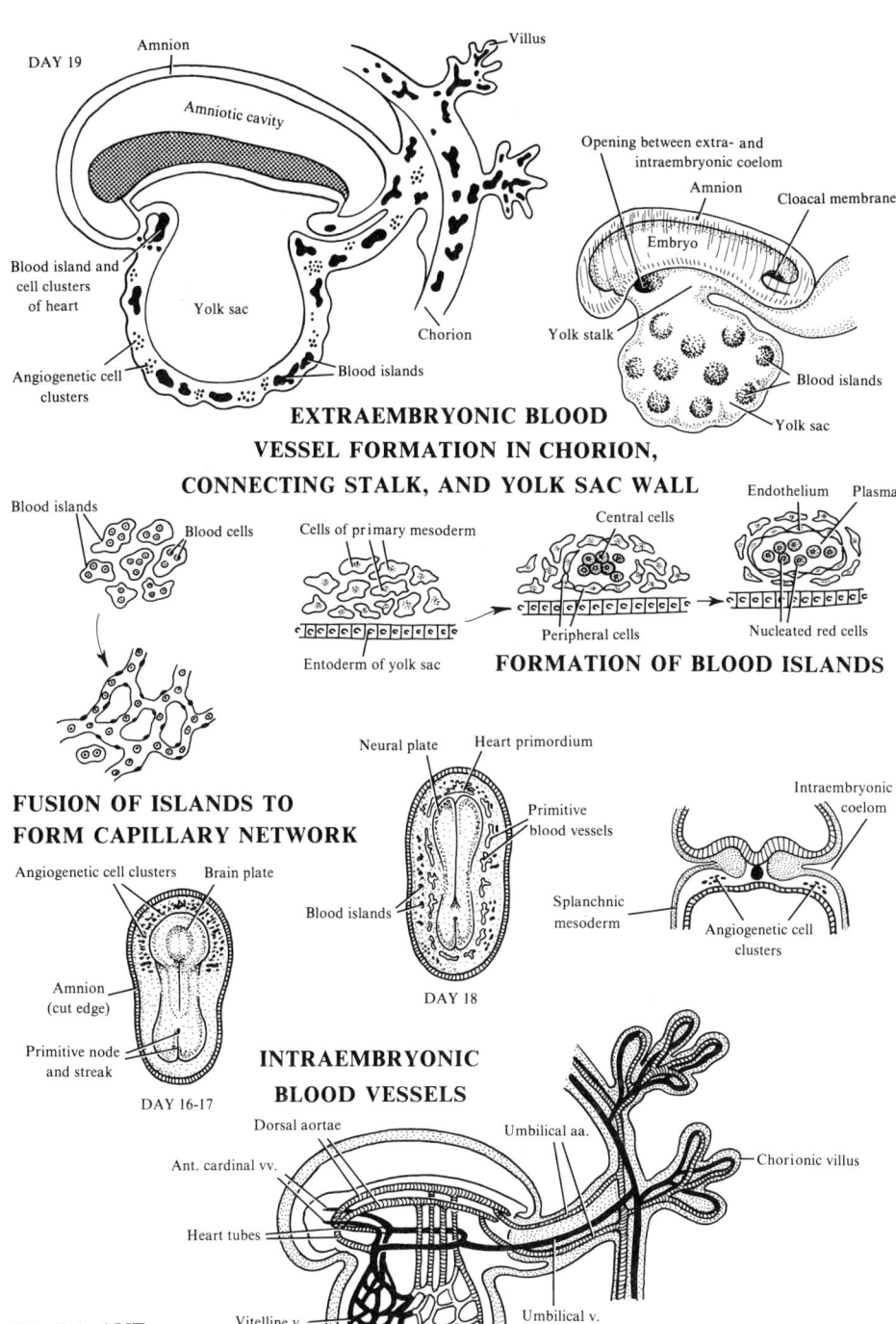

114. DEVELOPMENT OF THE HEART: CARDIAC TUBE DEVELOPMENT

I. **Development of the heart** is first indicated at day 18 or 19, in the cardiogenic area

A. CARDIAC TUBE DEVELOPMENT
1. During the stage of gastrulation, angiogenetic cell clusters from the splanchnic mesoderm layer of the late presomite embryo migrate from the primitive streak to the area of the oropharyngeal membrane. They unite with comparable migrating mesoderm from the opposite side to form a horseshoe-shaped plexus, the *cardiac primordium* or *cardiogenic cords*
    a. Initially, the central portion of the horseshoe-shaped plexus is found anterior to the prochordal plate and neural plate. With closure of the neural plate and formation of the brain vesicles, the CNS grows so rapidly in a cephalic direction that it extends over the central cardiogenic area and the future pericardial cavity
    b. During this growth, the expanding brain pulls the prochordal plate (future buccopharyngeal membrane) and the central part of the cardiogenic plate forward, rotating the plate and pericardial part of the intraembryonic coelomic cavity so that the central portions of the cardiogenic plate and pericardial cavity, initially rostral to the buccopharyngeal plate, are now ventral and caudal to it
2. Cleavage of the lateral plate by the coelom reaches this region, resulting in the differentiation of the splanchnopleure and somatopleure. These form the walls of the future pericardial cavity
3. Islands appear in the splanchnic mesoderm after day 20, and then by confluence, the cords become 2 thin-walled endothelial tubes which are called the *endocardial heart tubes*
4. As the embryo undergoes cephalocaudal flexion, the endocardial tubes approach each other in the midline. Closure of the foregut moves both tubes to a ventral position
5. The 2 endocardial tubes come together about day 22 and fuse, beginning at the cephalic end of the original horseshoe-shaped structure and extending in a caudal direction. Thus, a single endocardial tube is formed
    a. The developing primitive heart tube, located in the splanchnic mesoderm of the pericardial cavity, bulges gradually more and more into the pericardial cavity and continues until the heart tube, with its investing layer, lies completely in the cavity
6. The fusion of the 2 tubes is followed by disappearance of the *ventral mesocardium* and a temporary attachment to the dorsal side of the pericardial cavity by a fold in mesodermal tissue, the *dorsal mesocardium*
    a. As the heart tubes fuse, the mesenchyme around them thickens to form the *myoepicardial mantle*, which at first is separated from the endothelial wall of the tube by the *cardiac jelly* (gelatinous connective tissue substance). The jelly later is invaded by mesenchymal cells
    b. The inner endocardial tube will become the internal endothelial lining of the heart, the *endocardium;* the myoepicardial mantle gives rise to the *myocardium* (heart muscle) and *epicardium* or *visceral pericardium*
    c. The embryo is now about 23 days old, has 7 somites, and is 2.2 mm long
7. The time between the first appearance of the intraembryonic vessels and the heart tube formation is about 3 days. The resulting single median endocardial tube begins to beat about day 22
8. True embryonic circulation is established between days 27 and 29

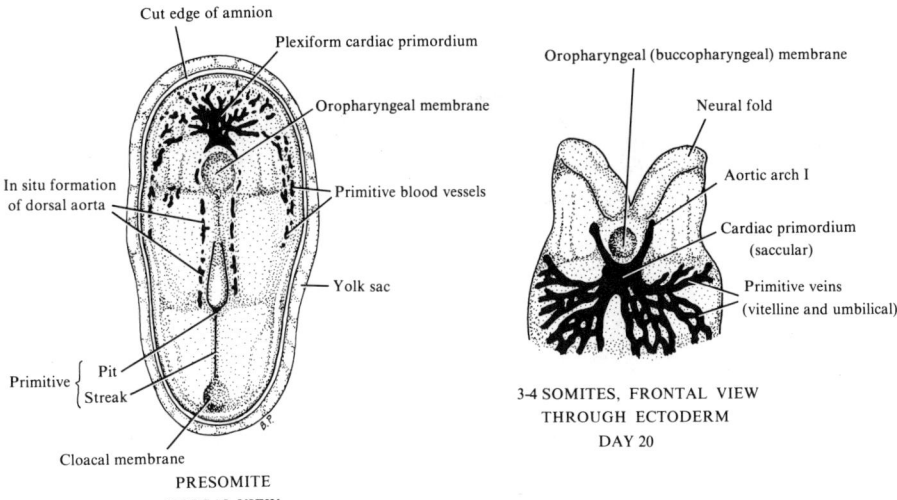
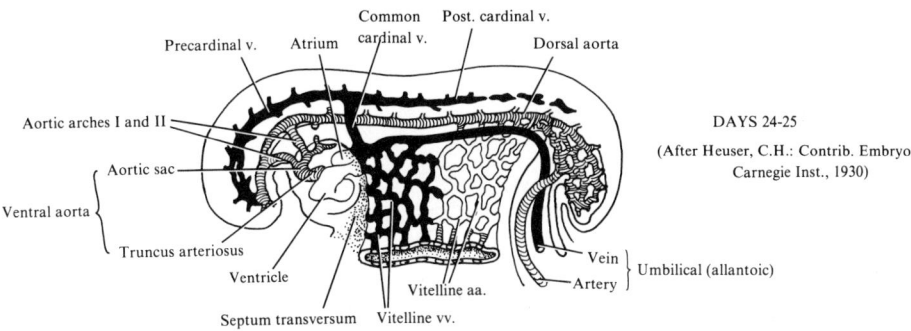
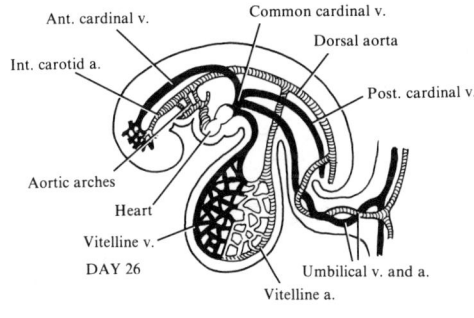

PRIMITIVE VASCULAR SYSTEM

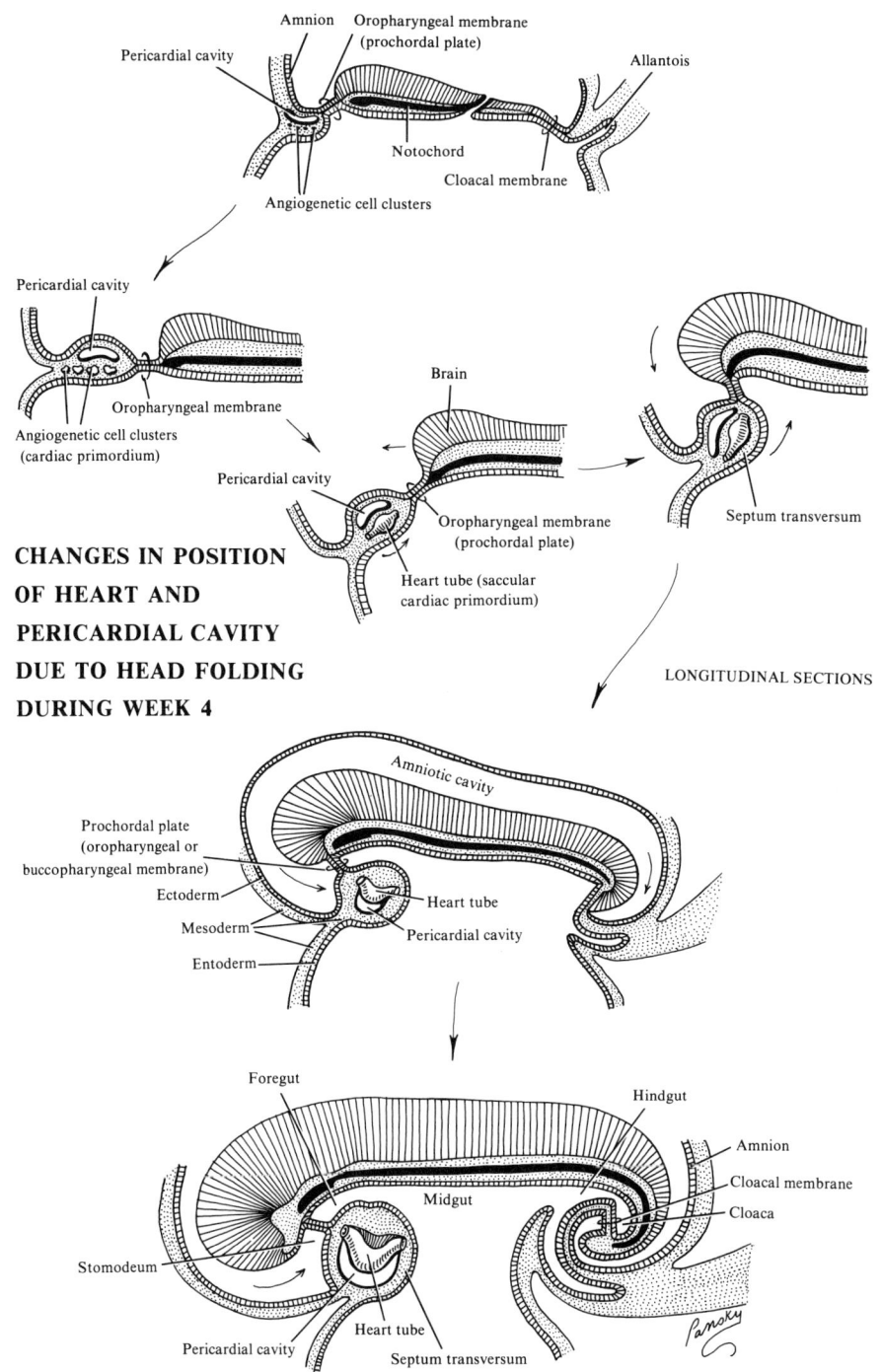

FIGURE 47. **Longitudinal sections: changes in position of the heart and pericardial cavity due to head folding.**

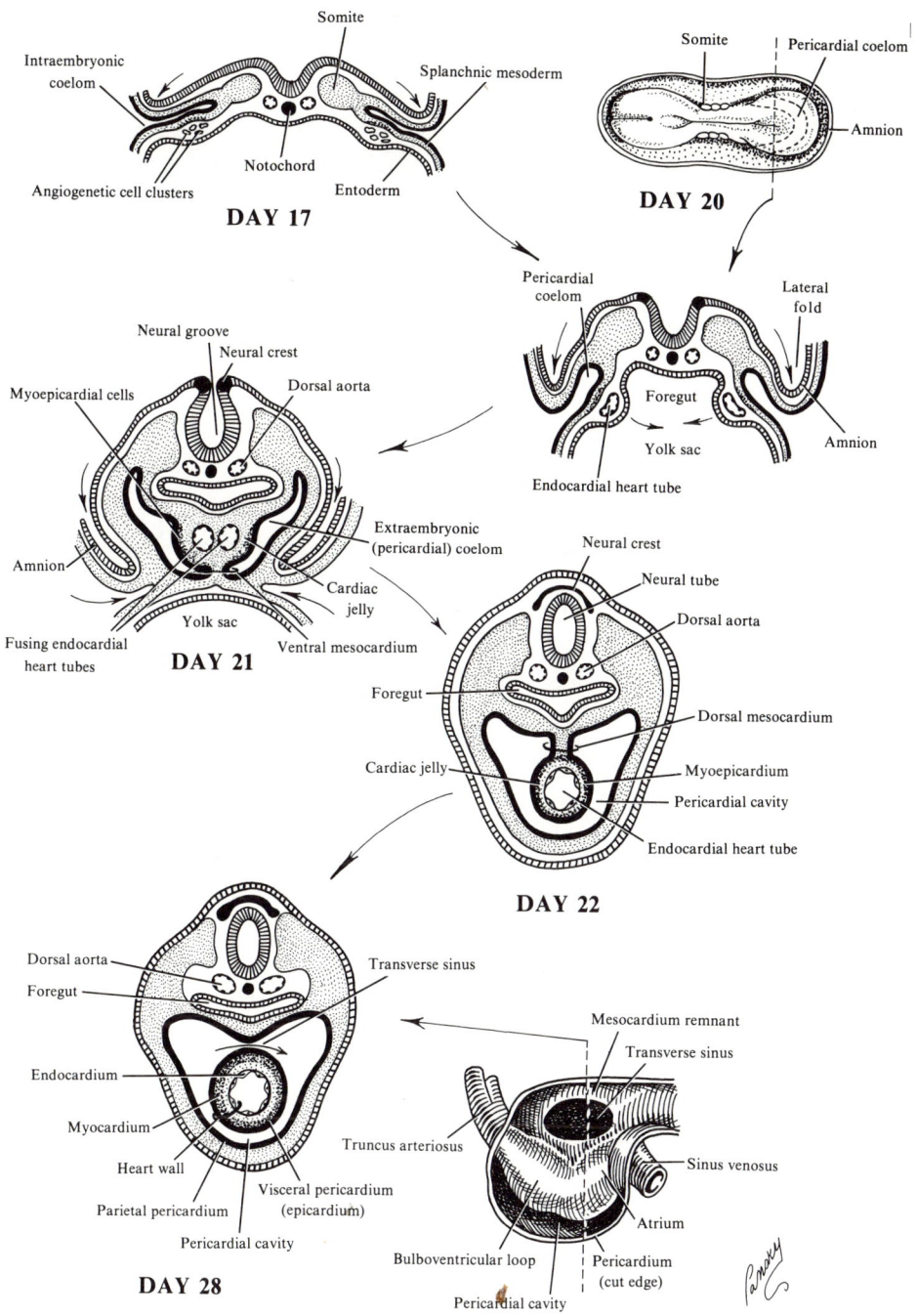

FIGURE 48. Coronal sections: changes in position of the heart and pericardial cavity due to head folding.

115. DEVELOPMENT OF THE HEART: FORMATION OF THE HEART LOOP

I. **Formation of the heart loop:** since the cardiac tube grows more rapidly than the pericardial cavity, it must undergo a series of complex foldings to be accommodated
  A. THE SINGLE TUBULAR HEART elongates and develops dilatations and constrictions
   1. The intrapericardial part consists of the future *bulboventricular portion* (bulbus cordis plus ventricle), whereas the *atrial part* and *sinus venosus* are still paired and lie outside the pericardium in mesenchyme of the septum transversum
     a. The bulbus cordis, ventricle, and atrium appear first (day 22±1). The truncus arteriosus and sinus venosus are seen a day later (day 23±1). The truncus is continuous caudally with the bulbus cordis and cranially with the aortic sac and aortic arches
     b. The sinus venosus (a large venous sinus) receives the umbilical, common cardinal, and vitelline veins from the primitive placenta, body of the embryo proper, and the yolk sac, respectively
   2. During development, the bulboventricular part grows faster than the pericardial cavity, and because its 2 ends are fixed to the surrounding tissue outside the pericardial cavity (arterial end to the branchial arches, venous end to the septum transversum), elongation cannot take place in a longitudinal direction
  B. THE CEPHALIC END OF THE LOOP bends ventral and caudal and slightly to the right. As a result, the bulboventricular sulcus (between bulbus cordis and ventricle) becomes visible on the outside. Internally, it remains narrow as the *primary interventricular foramen,* with a fold being formed, the *bulboventricular fold*
   1. Because the bulbus cordis and ventricles grow faster than the other regions, the heart tube bends on itself, forming a U-shaped bulboventricular loop
   2. As the heart bends, the atrium and sinus venosus come to lie dorsal to the bulbus cordis, truncus arteriosus, and ventricle. In addition, the sinus venosus at this stage also develops lateral expansions, the *right and left sinus horns*
  C. AS A SECONDARY SEQUENCE OF BENDING AND TORSION, the atrioventricular junction comes to lie on the left side of the pericardial cavity, while the right side is occupied by the greatly elongated bulbus cordis
   1. The cardiac loop thus consists of a cephalic or ascending limb (the bulbus) and a descending limb formed by the embryonic ventricle
   2. The bulbus cordis is narrow except for its proximal one-third which will form the *trabeculated part of the right ventricle.* Its midportion, the *conus cordis,* forms the outflow tracts of both ventricles. Its distal part, the *truncus arteriosus,* forms the roots and proximal parts of the aorta and pulmonary artery
  D. WHILE THE CARDIAC LOOP IS FORMING, changes occur throughout the length of the tube
   1. The atrial part, a paired structure outside the pericardial cavity, forms a common atrium by fusion of the right and left sides. During fusion, it is incorporated into the pericardial cavity and moves in a dorsocranial direction. The atrioventricular junction assumes a cranial position and forms the *atrioventricular canal,* connecting the left side of the common atrium with the embryonic ventricle
  E. WHEN LOOP FORMATION ENDS, the smooth-walled heart tube forms primitive trabeculae in 2 distinct areas: just proximal and distal to the primary interventricular foramen. The atrial part and other parts of the bulbus remain temporarily smooth-walled
  F. THE PRIMITIVE VENTRICLE, now trabeculated, is called the *primitive left ventricle* since it will form the major part of that definitive structure. The trabeculated proximal third of the bulbus cordis is called the *primitive right ventricle*
  G. THE CONUS-TRUNCUS PART of the heart tube, initially on the right side of the pericardial cavity, shifts to a medial position as a result of the formation of 2 transverse dilatations of the atrium, which bulge on each side of the bulbus cordis
   1. As a result, the truncus is found in a depression between the right and left atria, and the conus takes an oblique position between the roof of the primitive left ventricle and the anteromedial wall of the atrium

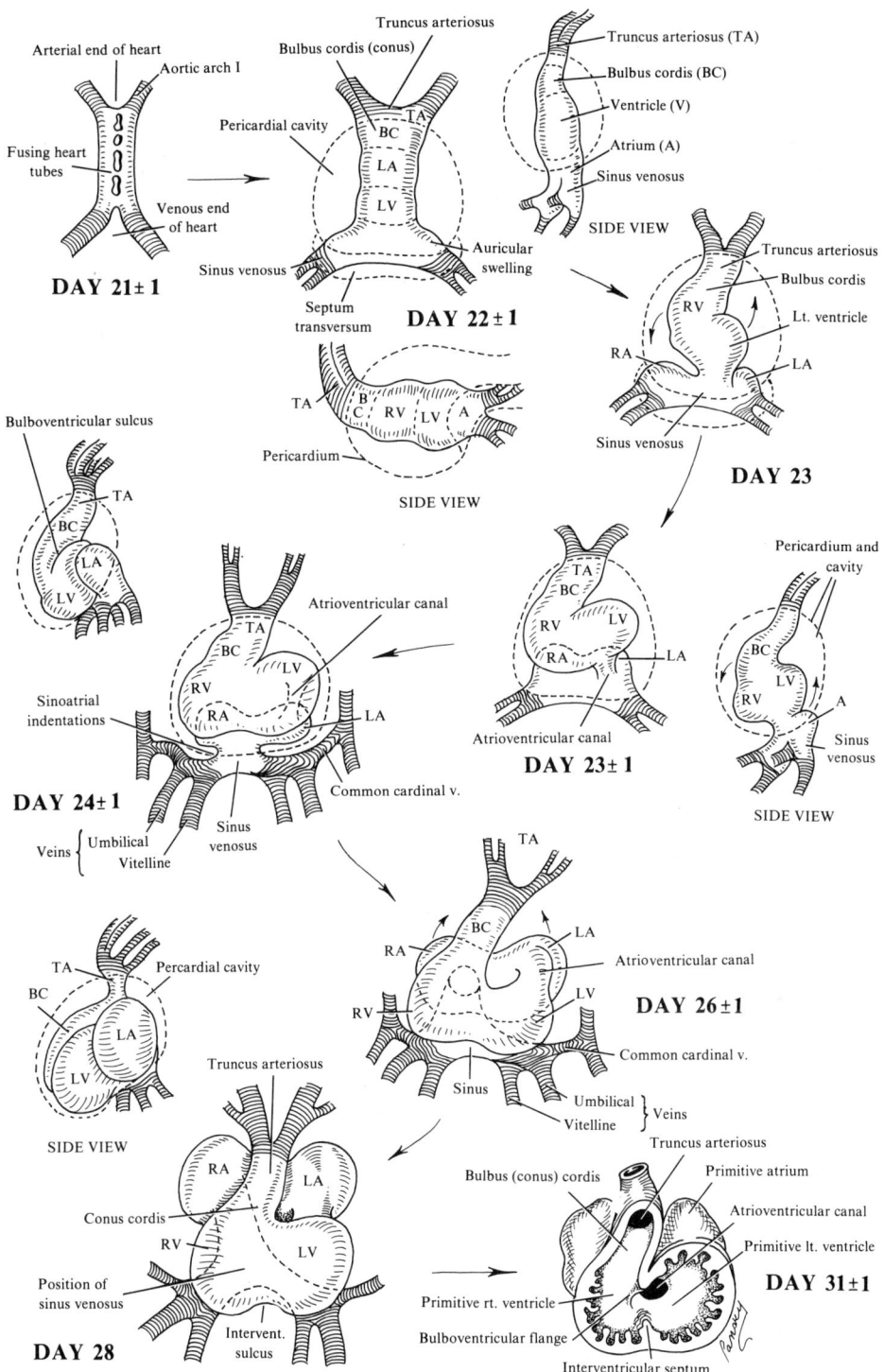

- 301 -

116. PERICARDIAL CAVITY DEVELOPMENT AND PRIMITIVE HEART CIRCULATION

I. Pericardial cavity development

A. THE FIRST SIGN OF HEART FORMATION is found at the end of week 3. The first heartforming cells appear as irregular clusters and cords in the cephalic part of the embryo between the entoderm of the yolk sac and the splanchnic mesoderm. These cell clusters form solid strands across the midline in front of the neural plate and extend down on each side of the embryo by the time the first somites appear

1. As the head end of the embryo grows forward and folds off from the yolk sac, the 2 solid strands approach each other ventrally and also acquire a lumen lined by endothelial cells. Thus, the 2 endocardial tubes are formed
2. The lumen of each of the 2 tubes gradually extends cranially into the midline cell strands and finally the 2 meet
3. With further lateral folding of the embryo, the fusion of the 2 endocardial tubes then progresses from the cephalic point in a caudal direction, thus forming a single endocardial tube

B. SIMULTANEOUSLY, WITH LATERAL FOLDING and the medial migration and fusion of the tubes, the intracoelomic cavities, right and left, also approach each other in the midline. Initially, at the 4-somite state (about day 21), the primitive heart tubes are connected to the anterior and posterior walls, between the right and left coelomic cavities, by the *dorsal* and *ventral mesocardium*

1. Whereas the ventral part disappears immediately after its early formation, the dorsal mesocardium persists a little longer
2. As the heart tube elongates, bends, and loops, it slowly sinks into the dorsal wall of the pericardial cavity, which is formed from a fusion of the right and left intraembryonic coelomic cavities
3. Eventually, beginning at the cranial end, the dorsal mesocardium also breaks down and has entirely disappeared at the 16-somite stage; and the heart tube is then freely suspended in the pericardial cavity and is attached to the surrounding tissues only at its cephalic and caudal ends. The newly formed passage, dorsal to the primitive heart tube, is the future *transverse sinus of the pericardial cavity*

C. IN WEEK 5, THE INTRAEMBRYONIC COELOM consists of a thoracic and abdominal component, connected by a canal found on each side of the foregut. In the adult, the intraembryonic coelom is divided into 3 well-defined compartments: the pericardial cavity with the heart, the pleural cavities with the lungs, and the peritoneal cavity with the viscera below the diaphragm. The *diaphragm* forms the septum between the thorax and abdomen; the *pleuropericardial membrane* forms between the pericardial and pleural cavities (see discussion on coelomic cavities and mesenteries)

II. Primitive heart circulation

A. CONTRACTIONS OF THE HEART begin by day 22 and are of myogenic origin. The muscles of the atrium and ventricle are continuous, and contraction, in peristaltic waves, begins in the sinus venosus

1. The circulation through the embryo and heart, at first, is an ebb-and-flow type, but by the end of week 4, coordination of heart contractions creates a unidirectional flow

B. BLOOD RETURNS to the sinus venosus from
1. The embryo proper via the common cardinal veins
2. The developing placenta via the umbilical veins
3. The yolk sac via the vitelline veins

C. THE SINOATRIAL VALVES control the blood flow from the sinus venosus into the atrium, and the blood then passes through the atrioventricular canal into the ventricle. When the latter contracts, blood is pumped through the bulbus cordis and truncus arteriosus into the aortic sac and aortic arches of the branchial arches. Blood then passes to the dorsal aortae for distribution to the embryo, yolk sac, and the placenta

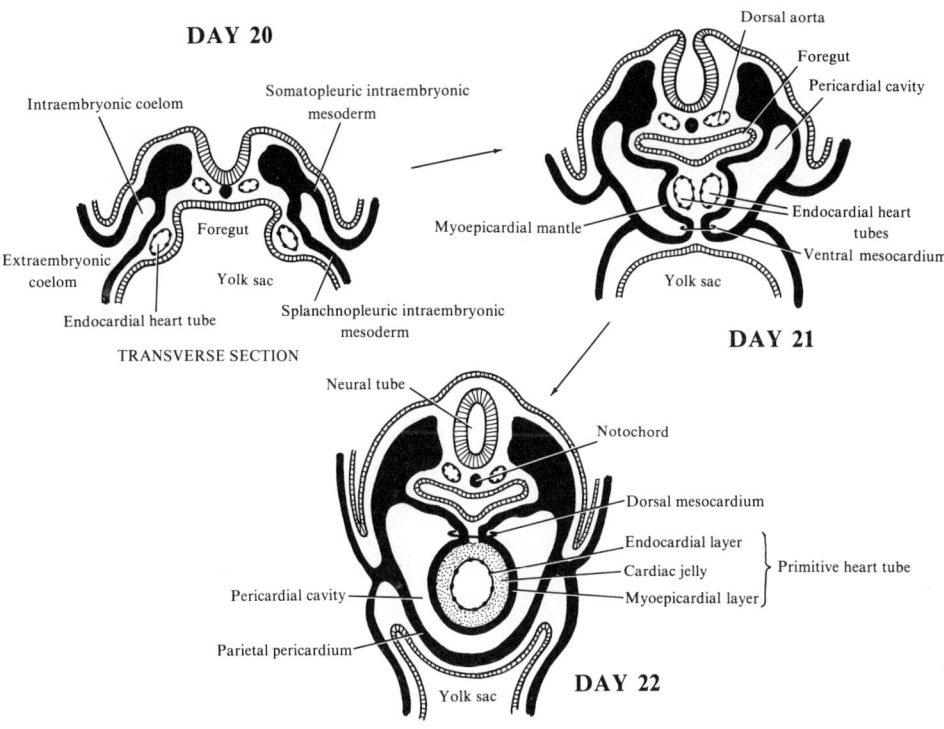
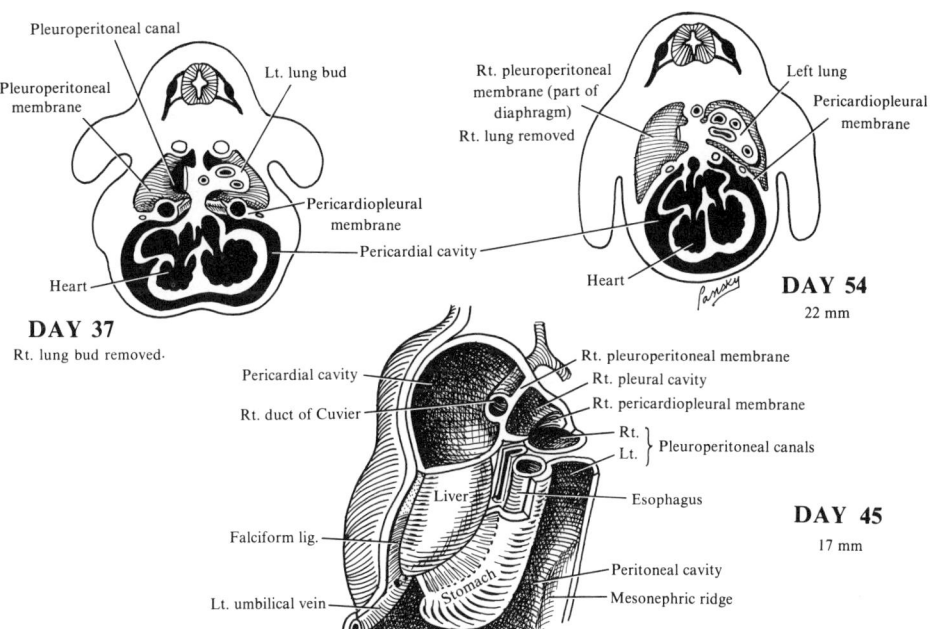

PERICARDIAL, PLEURAL, PERITONEAL CAVITIES

117. ATRIOVENTRICULAR AND INTERATRIAL SEPTATION AND DEVELOPMENT

I. **Introduction:** the cardiac tube is folded in the pericardial cavity by day 28 and consists of
   A. THE SINUS VENOSUS, into which enter the vitelline veins, the umbilical veins, and the common cardinal veins
   B. THE ATRIAL REGION, which communicates with the ventricle via the atrioventricular canal
   C. THE VENTRICULAR REGION
   D. THE BULBUS CORDIS, which is a prolongation of the ventricle and is continuous with the truncus arteriosus and gives rise to the aortic roots
   E. PARTITIONING of the atrioventricular canal, the atrium, and the ventricle begins about the middle of week 4 and is essentially complete by the end of week 5 (days 27 to 37) when the embryo grows in length from 5 mm to about 16 or 17 mm. Although described separately, the processes take place concurrently

II. **Atrioventricular canal septation:** by day 28, the atrial region forms a large cavity dorsal (behind) to the ventricular region and becomes divided into the right and left atria. Thus, at this time, the ventricle is bounded ventrally by the bulbus cordis and the atria dorsally. The fold between the ventricle and the bulbus rapidly disappears
   A. THE SEPARATION BETWEEN ATRIA AND VENTRICLE increases, resulting in a shrinkage of the atrioventricular canal. On the ventral and dorsal walls of the canal, thickenings of subendocardial tissue now appear, the 2 *endocardial cushions*, move toward each other, and finally fuse (between days 35 and 40) to form the *primitive interventricular septum*
      1. By day 40, the atrioventricular canal is divided into *right* and *left atrioventricular canals*. The mesenchyme around each canal proliferates and forms the atrioventricular valves (mitral valve at left and tricuspid valve at right)

III. **Primitive interatrial septation** begins during week 5. At this stage, one sees a single atrium, common cardinal, vitelline, and umbilical veins
   A. A THIN, SICKLE-SHAPED MEMBRANE, the *septum primum*, appears on the posterosuperior wall of the primitive atrial chamber and grows toward the endocardial cushions. A large, temporary opening exists between the lower free edge of the septum primum and the endocardial cushions called the *foramen primum,* which rapidly gets smaller
   B. BEFORE CLOSURE OF THE FORAMEN PRIMUM, small openings or perforations appear in the upper central part of the septum primum, which merge to form another opening, the *foramen secundum.* At the same time, the free edge of the septum primum fuses with the left side of the fused endocardial cushions to obliterate the foramen primum
      1. Thus, when the foramen (ostium) primum is closed, the foramen (ostium) secundum remains patent and affords free access between the 2 atria
   C. A NEW CRESCENTERIC MEMBRANE appears to the right of the "delicate" septum primum, on the antero-superior wall of the atrium, near the end of week 5. It converges toward the endocardial cushions as the *septum secundum*
      1. The septum secundum enlarges, covers the foramen secundum in the septum primum, but remains as an incomplete partition which results in an oval-shaped passageway, the *foramen ovale,* in the interatrial septum directly in the path of the blood coming from the inferior vena cava
      2. The upper portion of the septum primum, which is attached to the roof of the left atrium, gradually disappears, but the rest of the septum becomes the *valve of the foramen ovale*
   D. COMPLETE FUSION OF THE SEPTUM PRIMUM to the septum secundum forms the definitive interatrial septum obliterating the foramen ovale
      1. Traces of the former passage are often seen
      2. A depression, the *fossa ovalis,* is evident on the right side of the interatrial septum
      3. The crest of the septum secundum becomes the *limbus of the fossa ovalis*

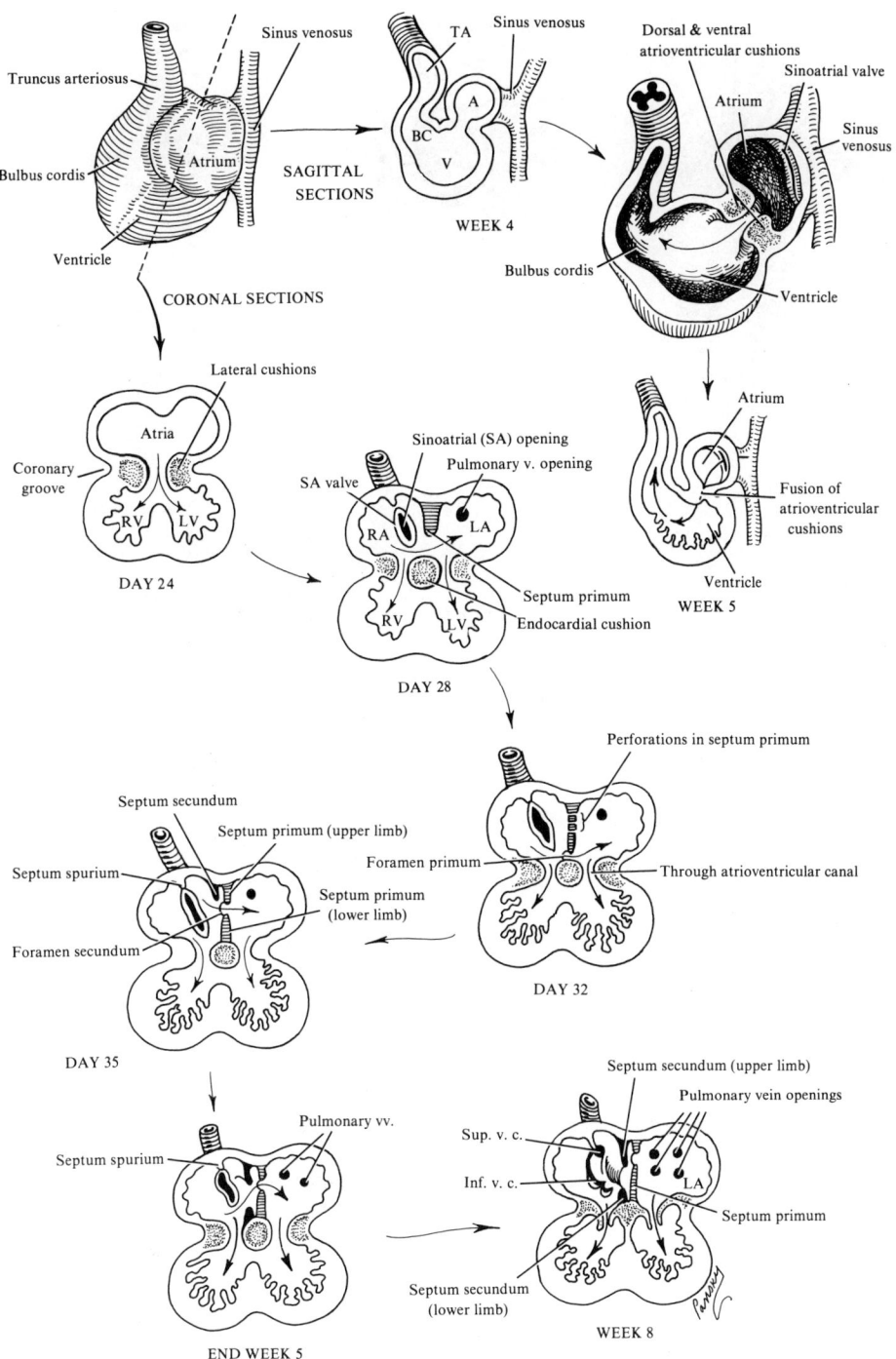

ATRIOVENTRICULAR SEPTATION

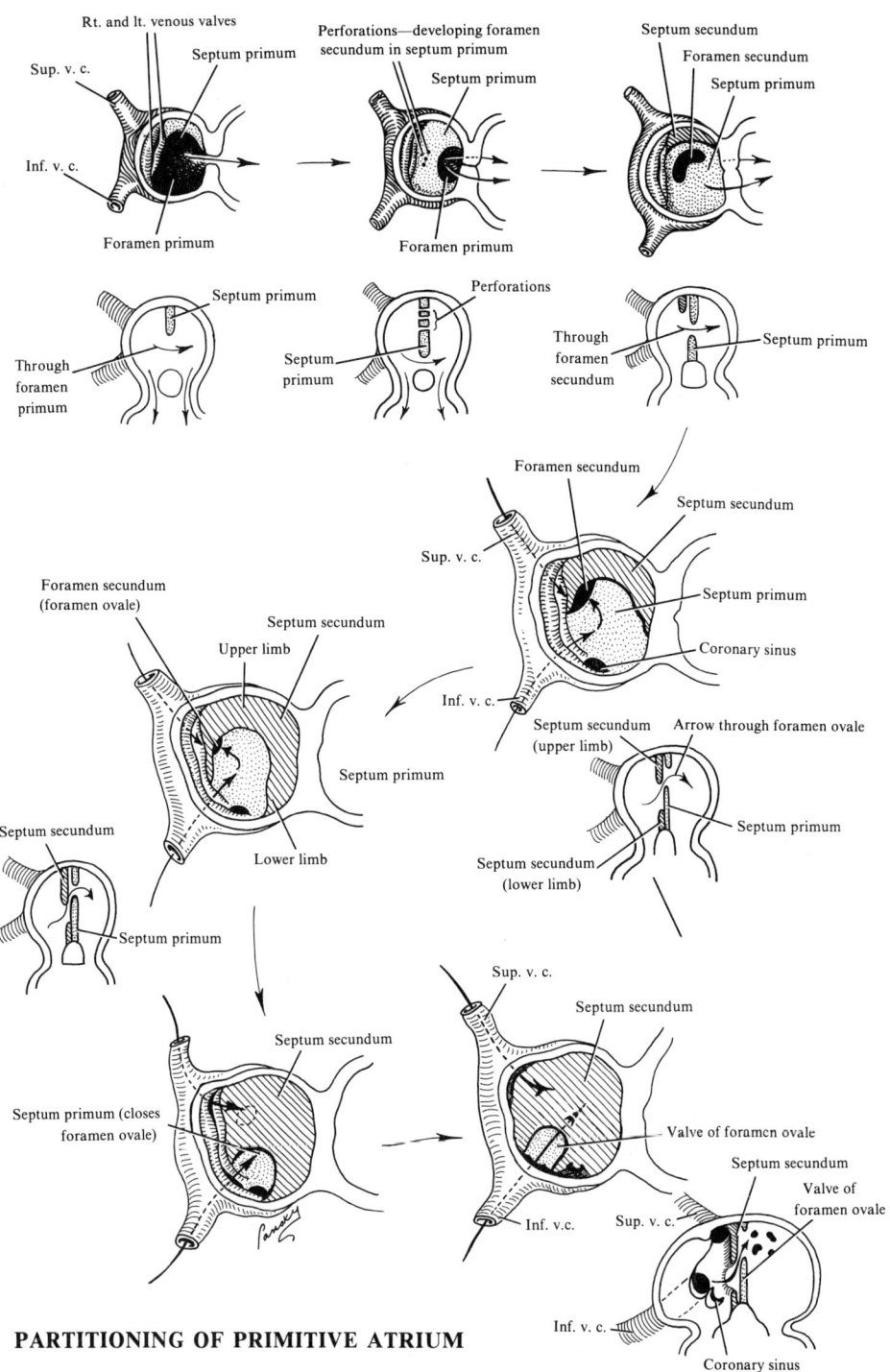

FIGURE 49. **Partitioning of the primitive atrium.**

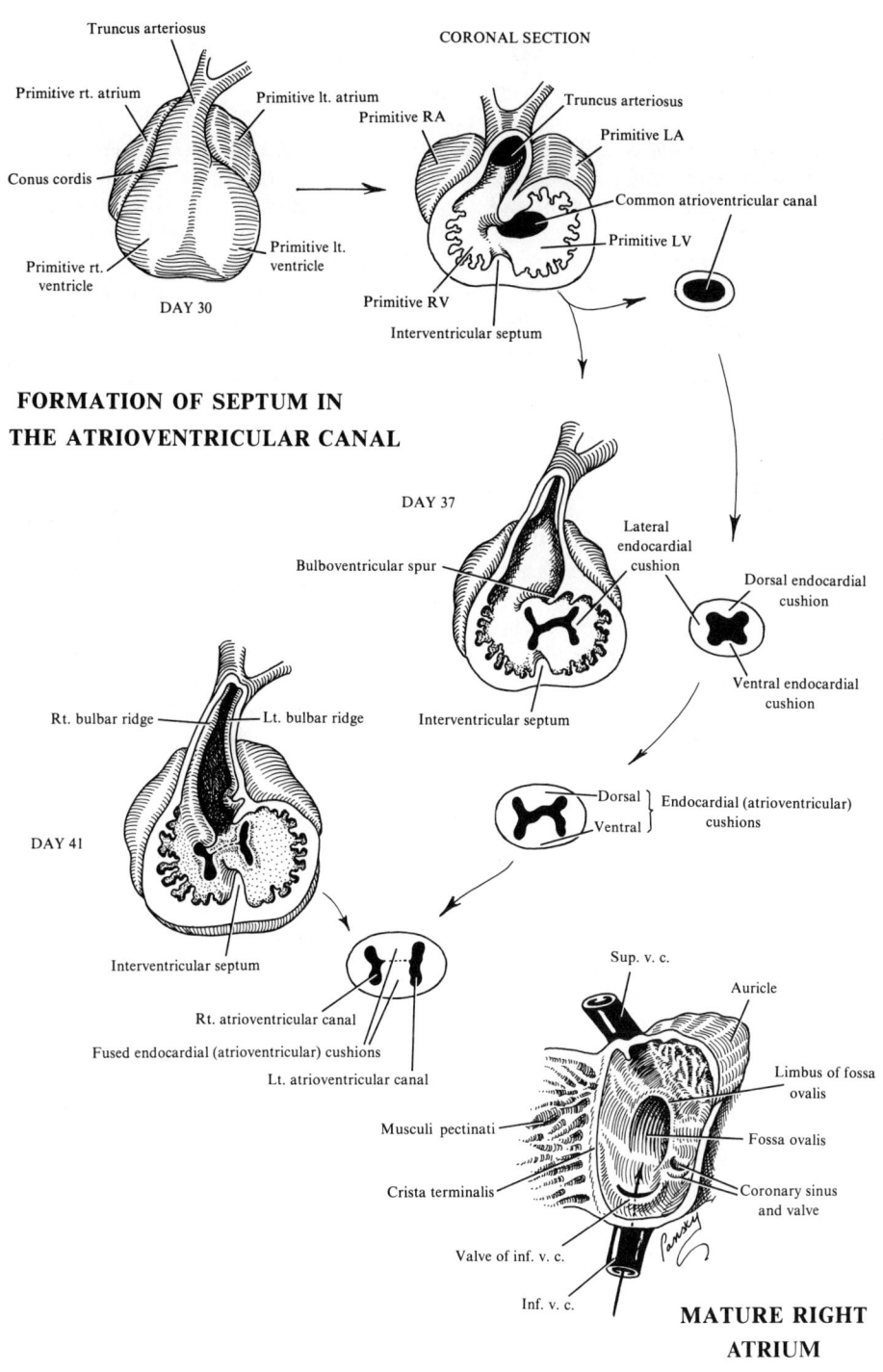

FIGURE 50. **Formation of the septum in the atrioventricular canal.**

118. DEVELOPMENT OF THE SINUS VENOSUS AND ASSOCIATED VEINS

I. **The sinus venosus** contributes to the definitive form of the atrium. It maintains its paired condition longer than than any other part of the heart tube, and in the 4-mm embryo consists of a *small transverse part,* opening into the center of the primitive atrium, and the *right* and *left sinus horns.* Each horn receives blood from 3 major veins: the vitelline (omphalomesenteric), umbilical, and common cardinal veins

   A. THE COMMUNICATION BETWEEN THE SINUS VENOSUS AND ATRIUM, which at first is wide, becomes narrow and shifts to the right as a result of development of a deep fold, the *sinoatrial fold,* which separates the left part of the sinus from the left side of the atrium. In addition, the right horn enlarges due to 2 left-to-right shunts of blood, and by the end of week 4, the right horn is much larger than the left. Thus, the sinoatrial opening shifts to the right and opens into the future right atrium

   1. The first left-to-right shunt of blood results from transformation of the vitelline and umbilical veins

      a. The vitelline veins enter the embryo with the yolk stalk, pass through the septum transversum, and enter the sinus venosus. Between the yolk sac and the septum, the paired vitelline veins are connected via anastomoses. In the septum, the veins are broken up into sinusoids by the developing cords of liver cells, which later become the hepatic sinusoids

      b. The *terminal part of the inferior vena cava* forms from the right vitelline vein between the liver and right horn of the sinus venosus. The *hepatic veins* form from the remains of the right vitelline vein in the area of the developing liver. The *portal vein* forms from the anastomotic network formed around the duodenum by the vitelline veins

      c. The umbilical veins

         i. The right umbilical vein and the part of the left between the liver and the sinus venosus degenerate

         ii. The persistent part of the left umbilical vein carries all the blood from the placenta to the fetus

         iii. The *ductus venosus* forms in the liver and connects the left umbilical vein with the inferior vena cava. The ductus serves as a bypass through the liver, enabling blood to bypass the liver and go from the placenta to the heart

   2. The second left-to-right shunt of blood occurs when the anterior cardinal veins become connected by an oblique anastomosis which shunts blood from the left to right anterior cardinal vein. The anastomosis becomes the *left brachiocephalic vein*

      a. The right anterior cardinal and right common cardinal veins become the *superior vena cava*

      b. The right posterior cardinal vein forms the *root of the azygos vein*

      c. The posterior cardinal vein contributes to the formation of the *left superior intercostal vein*

      d. The left anterior cardinal vein vanishes

      e. The left common cardinal vein is greatly reduced at week 10 to form the *oblique vein of the left atrium*

   B. DUE TO THE SHUNTS AND OBLITERATION of the left umbilical vein at the 5-mm stage and the left vitelline vein at the 7-mm stage, the left horn of the sinus decreases greatly in size and loses its importance. The right horn enlarges to receive all the blood from the head and neck via the superior vena cava and from the lower body region and the placenta via the inferior vena cava. When the left common cardinal vein is finally obliterated at week 10, the distal part of the left sinus horn remains as the *oblique vein of Marshall,* while the proximal part of the horn and transverse part of the sinus become the *coronary sinus*

   1. Due to obliteration of the veins on the left side, the right sinus horn and veins greatly enlarge. Subsequently, the right horn, the only connection between the original sinus venosus and atrium, is slowly incorporated into the right atrium

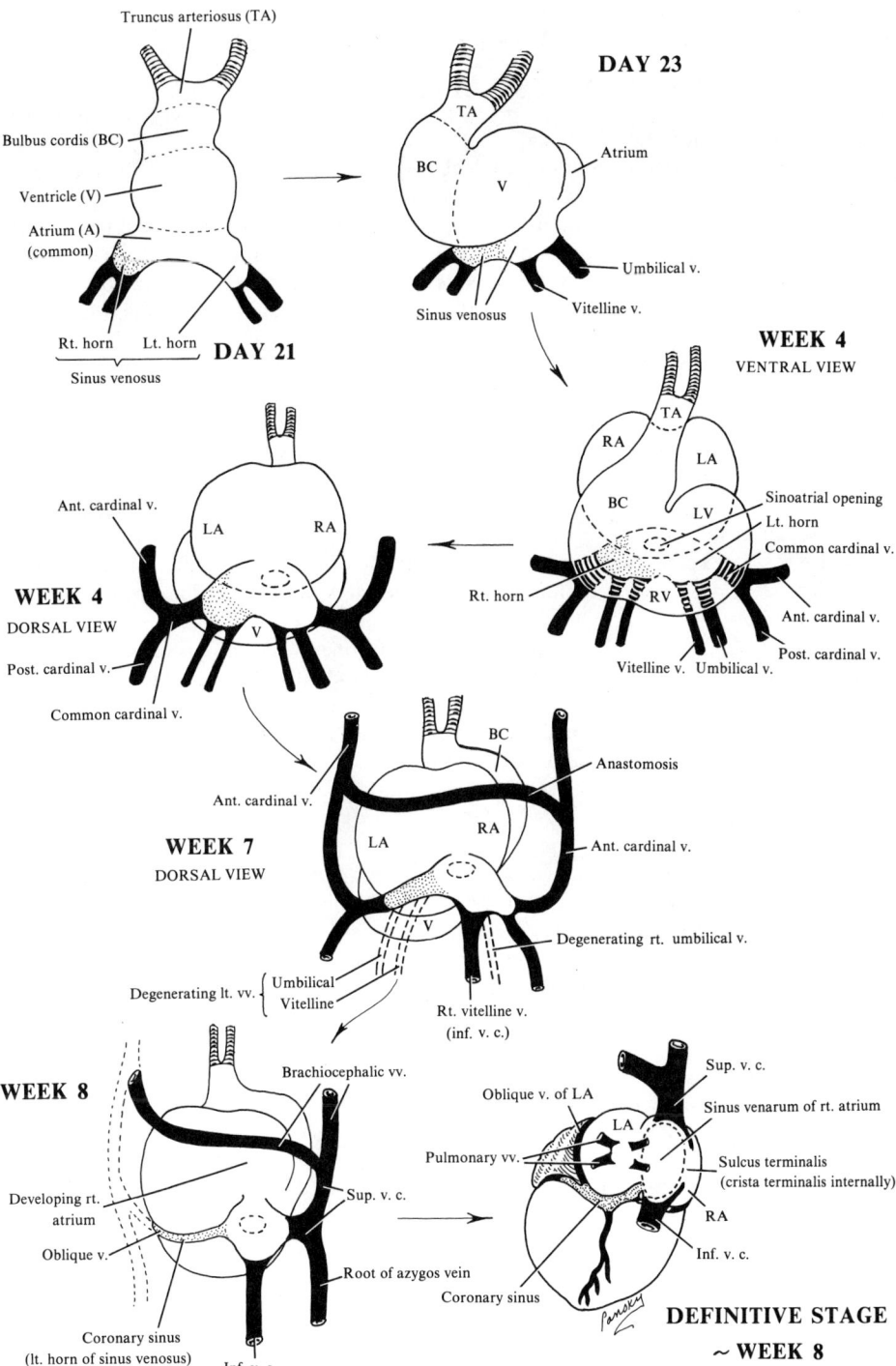

119. THE RIGHT AND LEFT ATRIAL WALLS AND THE VENOUS VALVES

I. The right atrial walls
  A. THE SINUS VENARUM (smooth part of the wall of the right atrium into which the great veins open) is derived from the sinus venosus
   1. The rest of the atrium and its muscular extension, the *auricle,* have a rough trabeculated surface and are derived from the primitive right atrium
  B. THE SINUS VENARUM AND THE PRIMITIVE ATRIUM are demarcated internally by a vertical ridge, the *crista terminalis,* and externally, by an inconspicuous groove, the *sulcus terminalis*
   1. Thus, the crista represents the cranial part of the right sinoatrial valve
     a. The lower portion of the right sinoatrial valve forms the valves of the inferior vena cava and coronary sinus
   2. The left sinoatrial valve fuses with the septum secundum and is incorporated into the interatrial septum

II. The left atrial walls
  A. MOST OF THE LEFT ATRIUM is smooth and is derived from the primitive pulmonary vein, which develops as an evagination from the dorsal wall of the atrium in the sinoatrial region
   1. Initially, the single common pulmonary vein opens into the primitive left atrium, but as the latter expands, parts of the vein are gradually absorbed into the wall of the left atrium
   2. Progressively, the proximal parts of the branches of the pulmonary vein are also absorbed, thus, the 4 pulmonary veins all open independently into the left atrium
   3. Only the left auricle (derived from the primitive atrium) has a rough, trabeculated appearance

III. The venous valves
  A. THE ENTRANCE OF THE SINOATRIAL OPENING is flanked on each side by a valvular fold, the *right* and *left venous valves*
   1. On the right, this fold is formed by a sinoatrial fold
   2. On the left, there is a smaller fold, called the left venous valve
  B. DORSOCRANIALLY, THE VALVES FUSE to form a ridge called the *septum spurium*
  C. INITIALLY, THE VALVES ARE LARGE, but when the right horn is incorporated into the atrial wall, the left sinus venosus valve and septum spurium fuse with the developing atrial septum
   1. The superior portion of the right venous valve disappears completely, while its inferior part fuses with the septum that develops between the orifice of the right vitelline vein (inferior vena cava) and the orifice of the coronary sinus. The remainder of the valve is divided into 2 parts: valve of the inferior vena cava and valve of the coronary sinus

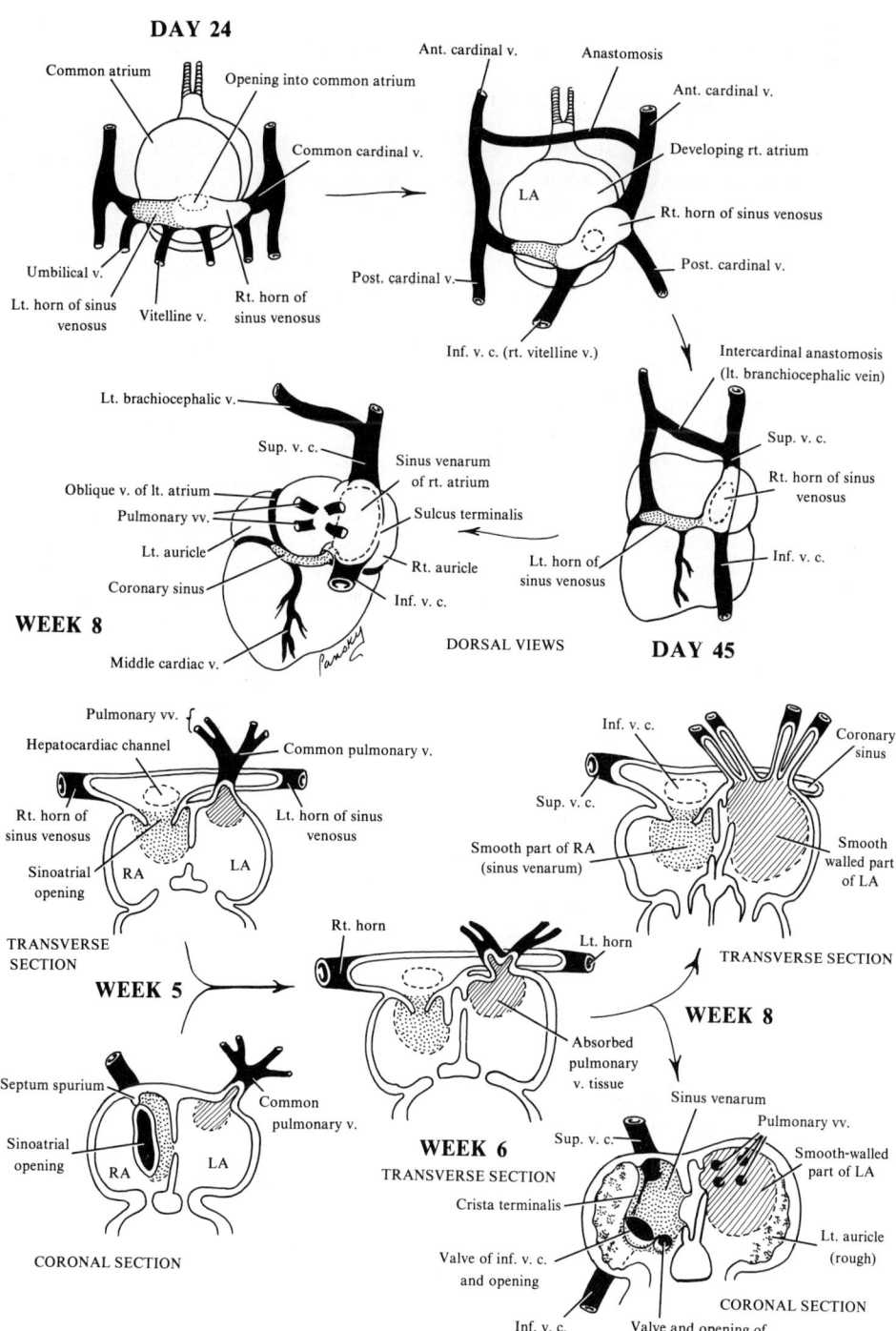

120. SEPTATION OF VENTRICLES, TRUNCUS ARTERIOSUS, AND CONUS CORDIS

I. Interventricular septation

A. A MUSCULAR CREST (ridge or fold) appears on the anterior ventricular wall near its floor and almost in a median plane near the apex, at the same time that the interatrial septum is forming, at about week 5. This is the *interventricular septum primordium*
   1. The interventricular septum is incomplete and has an upper free concave edge
   2. Most of its increase in length, at first, is the result of dilatation of the ventricles on each side of the septum, which produces an *external interventricular groove*
   3. With active growth, the septum *forms the muscular portion* of the interventricular septum
   4. A crescenteric *interventricular foramen* is seen between the free edge of the septum and the fused endocardial cushions, allowing for communication between the right and left ventricles, until about the end of week 7

B. SEPTUM FORMATION in the truncus arteriosus and bulbus cordis
   1. During week 5, aortic arch VI appears and contributes to the formation of the pulmonary arteries. Just cephalic to arch VI, the subendocardial tissue in the bulbus cordis thickens into 2 opposing ridges called the *truncoconal* or *bulbar ridges*. Semilunar ridges also form in the truncus arteriosus and are continuous with those in the bulbus cordis
   2. The bulbar ridges soon fuse with those of the truncus arteriosus
      a. The fusion takes a spiral orientation, possibly due to the streaming of blood from the ventricles, and forms the *aorticopulmonary septum,* which definitively separates the aorta and the pulmonary artery
   3. As a result of the spiral orientation of the septum, the aorta and pulmonary artery twist around each other
      a. Blood from the aorta now passes into the third and fourth parts of the aortic arches
      b. Blood from the pulmonary trunk flows into the sixth pair of aortic arches
   4. The bulbus cordis is gradually incorporated into the walls of the ventricles
      a. In the adult right ventricle, the bulbus cordis is seen as the conus arteriosus or *infundibulum*; and in the left ventricle, it is seen as the *aortic vestibule*

C. CLOSURE OF THE INTERVENTRICULAR FORAMEN and completion of the interventricular septum
   1. Ventricular septation is completed by closure of the interventricular communication (foramen) around the end of week 7, as the bulbar ridges fuse. Closure from fusion of subendocardial tissue is from 3 sources
      a. A proliferation of the right bulbar ridge near the tricuspid orifice
      b. A proliferation of the left bulbar ridge near the mitral orifice
      c. A proliferation of the posterior (atrioventricular) endocardial cushion
   2. Fusion of the two ridges and the endocardial cushion outgrowth forms the *membranous portion of the interventricular septum* and is completed toward the end of month 2.
      a. This tissue fuses with the aorticopulmonary septum and muscular portion of the interventricular septum, thus, the pulmonary trunk communicates with the right ventricle and the aorta with the left ventricle

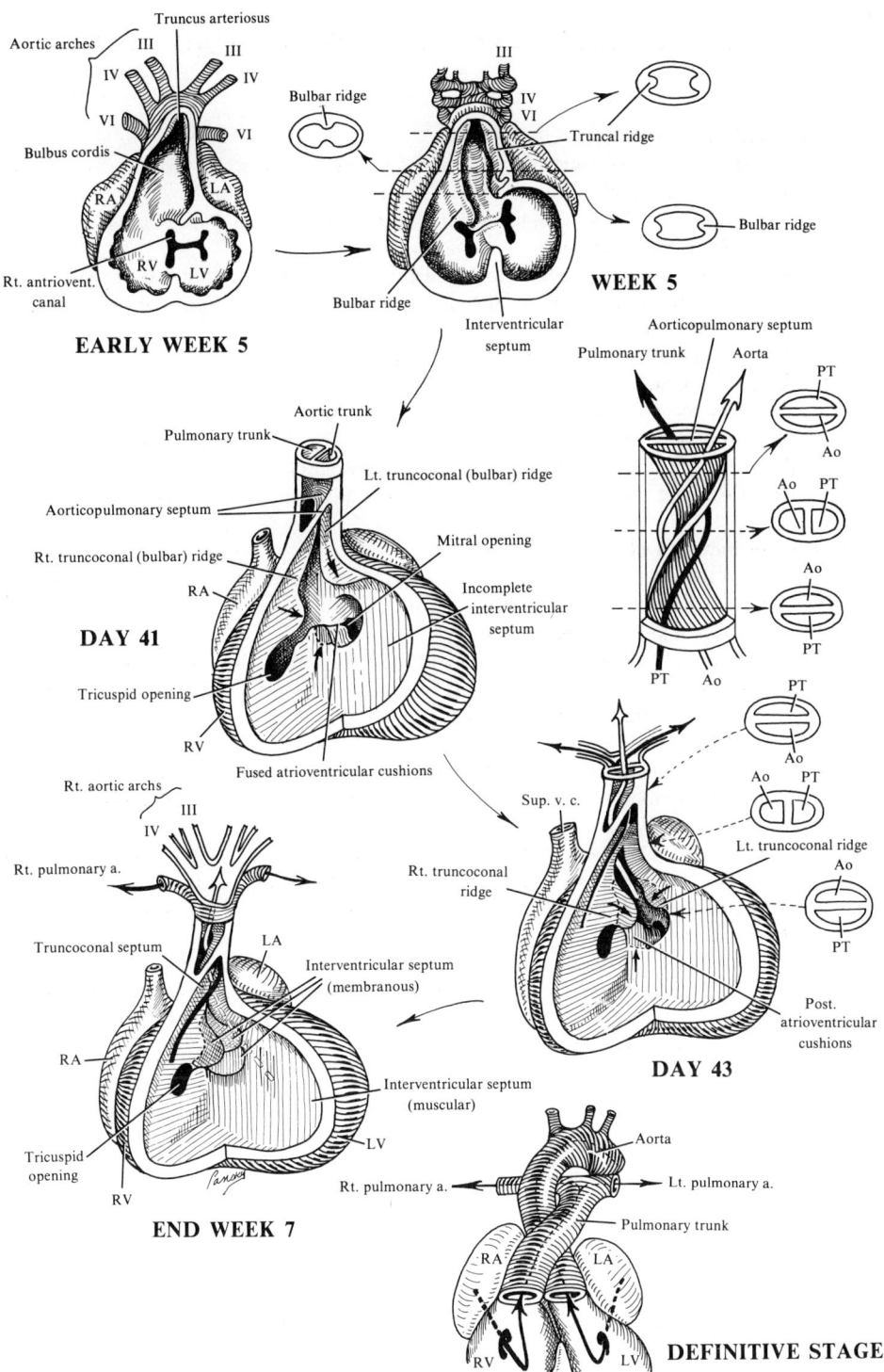

121. THE CARDIAC VALVES AND CONDUCTING SYSTEM

I. The cardiac valves
A. THE SEMILUNAR VALVES form from 3 valve swellings or ridges of subendocardial connective tissue at the openings of the aorta and pulmonary trunk
  1. The swellings hollow out and reshape themselves to form 3 thin cusps, which are covered by endocardium
B. THE ATRIOVENTRICULAR VALVES [the tricuspid (rt. AV valve) and mitral (lt. AV valve)] develop simultaneously from localized proliferation of subendocardial tissue around the atrioventricular canals
  1. They are hollowed out on their ventricular sides

II. The ventricular walls
A. CAVITATION OF THE VENTRICULAR WALLS forms a spongelike mass of muscle bundles
  1. Some muscle remains as the *trabeculae carneae* of the ventricular walls
  2. Other muscles form the *papillary muscles* and modify to form the *chordae tendineae* which connect the ventricular walls with the atrioventricular valves

III. The conducting system
A. THE MUSCULAR LAYERS (myocardium) of the atrium and ventricle are initially continuous, and the primitive atrium serves as a pacemaker for the primitive heart until the sinus venosus takes over that function
  1. The sinoatrial (SA) node is originally in the right wall of the sinus venosus, but as it is incorporated with the sinus venosus into the wall of the right atrium, it comes to lie near where the superior vena cava enters the right atrium
  2. After the sinus venosus is incorporated into the right atrium, cells from its left wall, which are found in the base of the interatrial septum just anterior to the opening of the coronary sinus, and cells along the atrioventricular canal make up the *atrioventricular (AV) node* and the *bundle of His*
  3. As the 4 heart chambers form, a band of connective tissue grows in from the epicardium and separates the atrial muscle from the ventricular muscle to form the *cardiac skeleton*. Thus, the conducting system remains as the only pathway from the atria to the ventricles
  4. The sinoatrial (SA) node, the atrioventricular (AV) node, and the atrioventricular bundle of His are soon supplied by external nerves (the parasympathetic vagus and sympathetic fibers). Histologic differentiation of these specialized tissues is continued up to, and after, birth

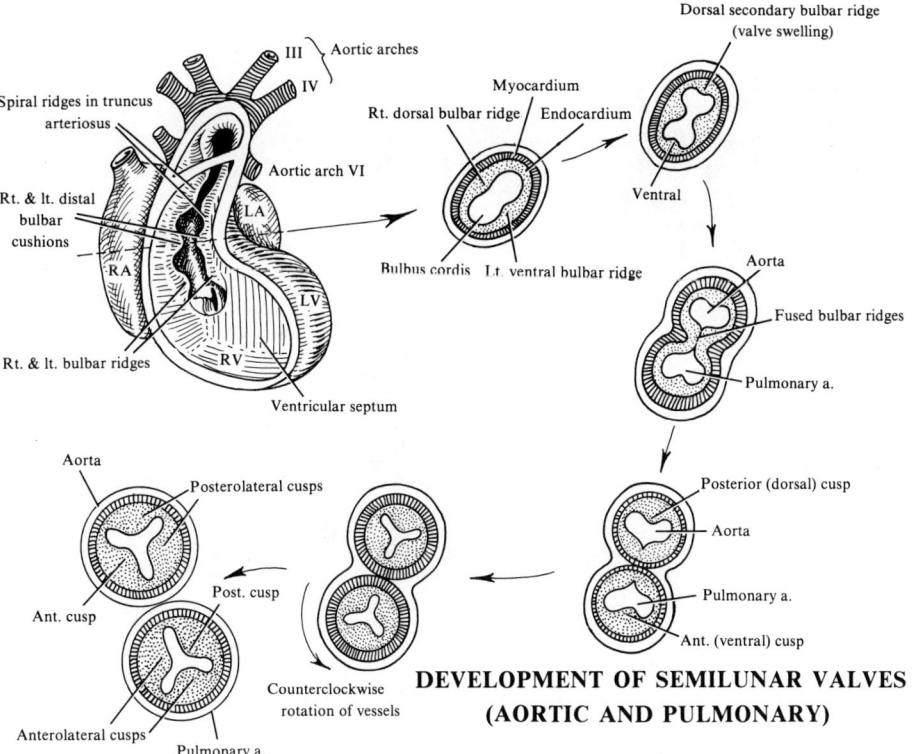

DEVELOPMENT OF SEMILUNAR VALVES
(AORTIC AND PULMONARY)

LONGITUDINAL SECTIONS THROUGH SEMILUNAR VALVES

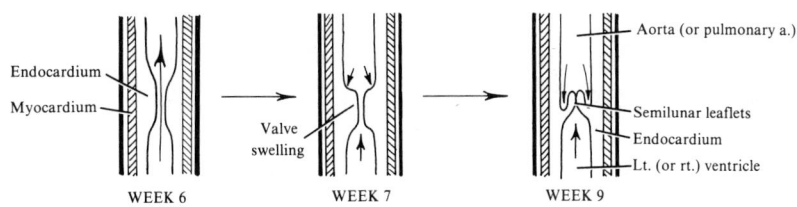

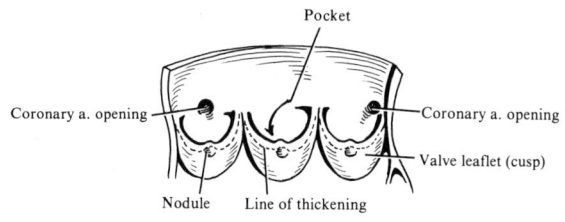

THREE LEAFLETS OF AORTIC VALVE
VALVE CUT OPEN AND SPREAD FLAT

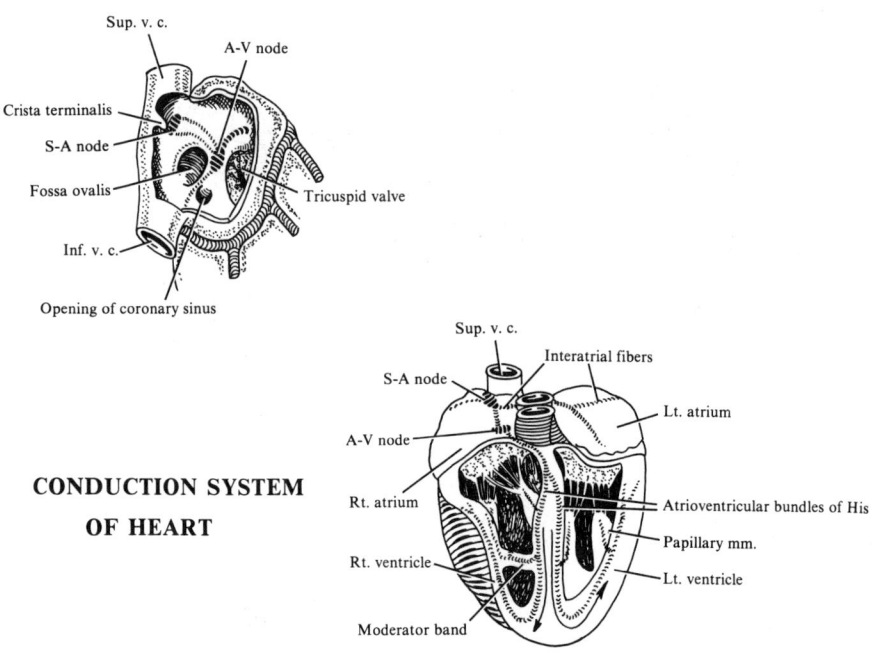
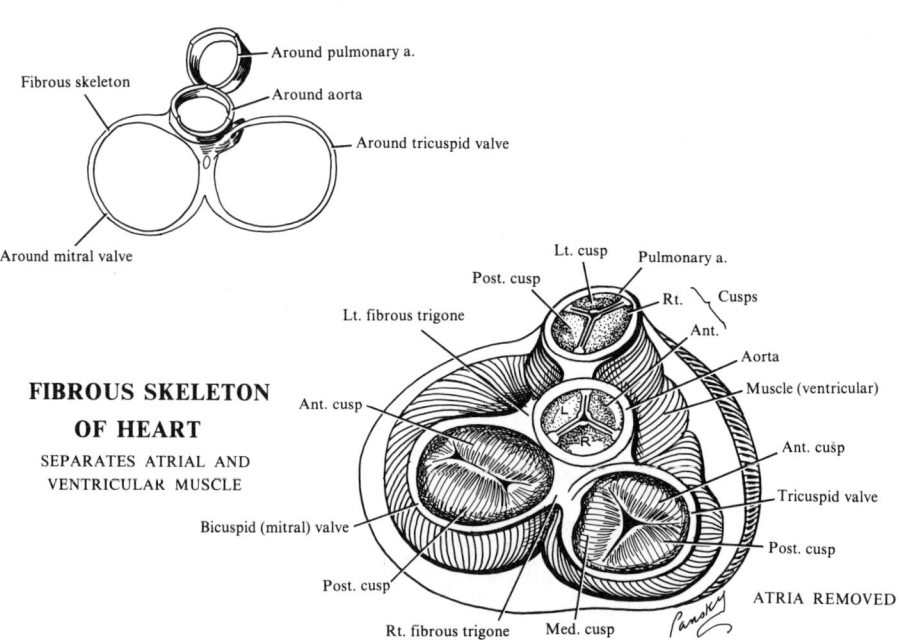

FIGURE 51. **Conduction system and fibrous skeleton of the heart.**

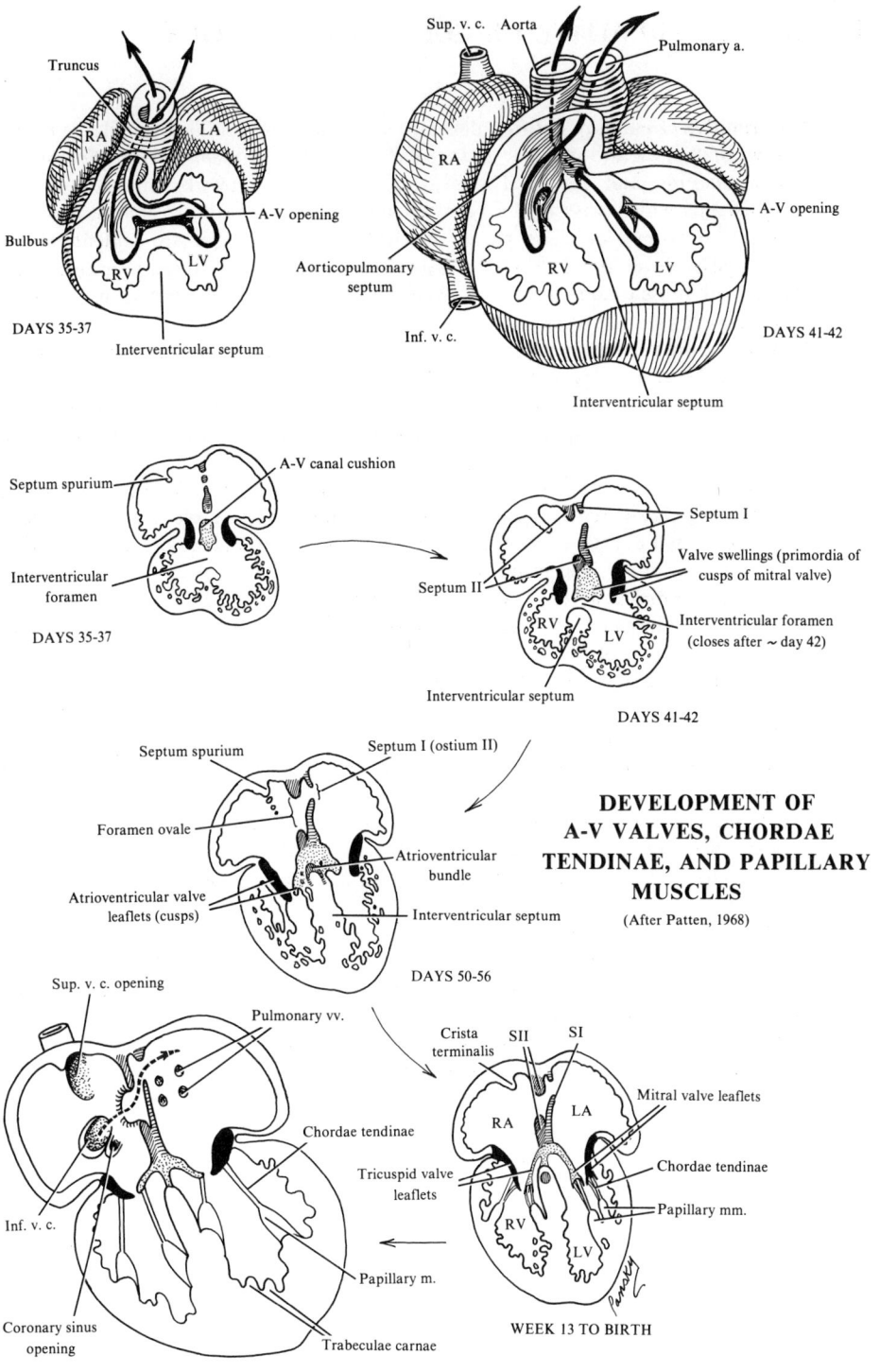

FIGURE 52. **Development of the atrioventricular valves, chordae tendineae, and papillary muscles.**

122. THE PRIMITIVE CIRCULATORY NETWORK

I. **The primitive circulatory network** develops as a result of the almost simultaneous formation of the heart and the 3 vascular networks: *intraembryonic, placental* or *umbilicoallantoic,* and *vitelline.* These vascular embryonic networks form from vascular blood islands which arise in the mesenchyme. The islands hollow out and combine (coalesce) to form a capillary plexus in which certain branches are evident from an early stage of development. *Only these persist.* At a later stage, these primitive vessels derive their muscle and connective tissue layers from the neighboring mesenchyme
   A. INTRAEMBRYONIC VASCULAR NETWORK
      1. Arteries: the *ventral arteries,* the *first aortic arches,* and the *dorsal aortas* are continuous. In the anterior region of the embryo, in each branchial arch, 5 pairs of aortic arches are formed successively and join the ventral to the dorsal aortas. The anterior arches eventually disappear, whereas the posterior ones are those that develop further
         a. The dorsal aortas extend from the cranial to the caudal region of the embryo and develop *paired segmental arteries* to the somites. The paired segmental arteries consist of a dorsal series, which vascularizes the neural tube, and a ventral series, which surrounds the primitive gut
         b. Some ventral arteries make a junction with the extraembryonic network
            i. The omphalomesenteric arteries continue as the vitelline arteries in the vascular system of the yolk sac
            ii. The allantoic or umbilical arteries feed the placental network
      2. Veins: the *paired anterior* and *posterior cardinal veins* develop in the same way as do the aortas, but at a slightly later time. In the heart, they join to form the *common cardinal veins,* which open into the sinus venosus, close to the vitelline and umbilical veins
   B. THE PLACENTAL OR UMBILICOALLANTOIC VASCULAR NETWORK, which develops in the mesenchyme around the allantois, forms 4 large vessels
      1. The 2 allantoic or umbilical arteries are really 2 of the posterior segmental arteries of the aorta
      2. The 2 umbilical veins run into the sinus venosus. The 2 umbilical veins form a single trunk at the level of the umbilical cord
   C. THE VITELLINE VASCULAR NETWORK develops on the yolk sac surface, especially on its caudal half and forms 4 large vessels
      1. The 2 vitelline arteries, whose proximal parts are the omphalomesenteric arteries
      2. The 2 vitelline veins, which run ventrally in the embryo into the sinus venosus

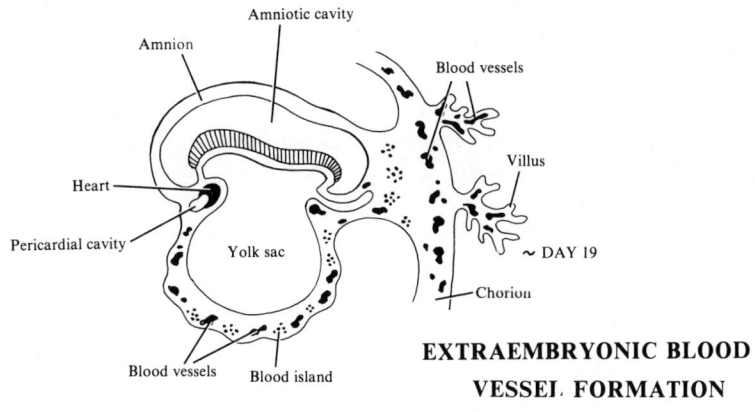

EXTRAEMBRYONIC BLOOD VESSEL FORMATION

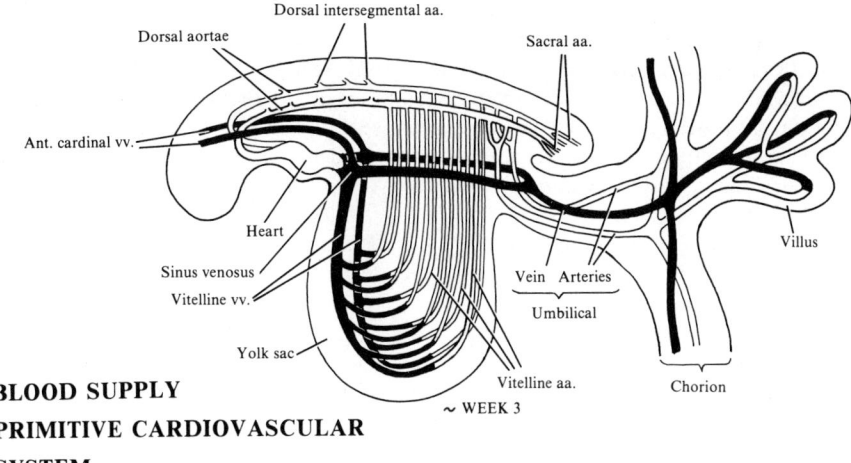

BLOOD SUPPLY PRIMITIVE CARDIOVASCULAR SYSTEM

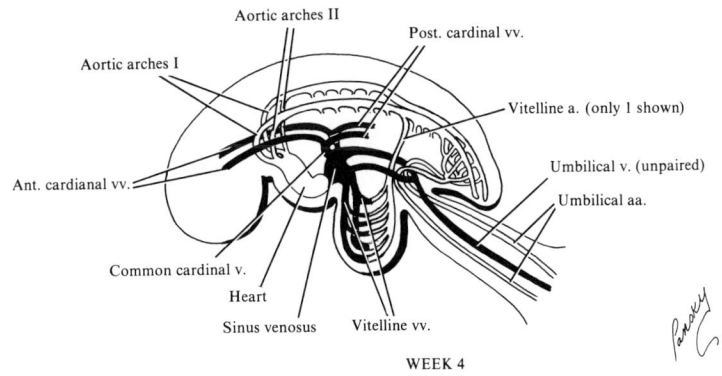

123. DEVELOPMENT OF THE ARTERIAL SYSTEM

I. **The arterial system:** the primitive arterial system undergoes numerous modifications in its development, which are minimal in the embryonic posterior region, basic in its middle region where fusion of the 2 aortas forms a single aorta, and quite complex in the anterior region where the aortic arches are formed
  A. POSTERIOR REGION OF THE EMBRYO
    1. The paired umbilical arteries arise from the dorsal aorta at the posterior end of the ventral segmental arteries
      a. The umbilical arteries remain paired during the growth of the embryo, although their point of origin moves slightly in a caudal direction, and they do give off a small external branch to the lower limbs
      b. When placental circulation is interrupted at birth, the umbilical arteries become fibrosed over most of their length. Their proximal portion gives rise to the primitive *iliac, hypogastric,* and *superior vesical arteries.* The branch to the lower limb becomes the *external iliac artery*
  B. MIDDLE REGION OF THE EMBRYO
    1. The 2 dorsal aortas approach each other and fuse at about week 4. They fuse initially in the middle portion, then cranially up to the eighth segmental artery and caudally to the posterior end. By the middle of week 5, the single dorsal aorta is formed
    2. The paired ventral segmental arteries approach each other in the midline in the mesentery, and some fuse into median vessels creating 3 visceral arterial systems
      a. The celiac trunk
      b. The superior mesenteric artery, which is derived from a special segmental artery, the omphalomensenteric artery, after regression of its vitelline portion (the left one disappears entirely)
      c. The inferior mesenteric artery
    3. Temporary longitudinal anastomoses result in the caudal movement of the origins of the above arteries, and they reach their final levels at about the end of month 2
    4. The dorsal segmental arteries remain unpaired (in contrast to the ventral). Initially, they only supply the neural tube. Their somatic branches, however, grow considerably and finally predominate. Some of the dorsal segmental arteries persist to form the *intercostal arteries*
    5. The lateral segmental arteries, which exist in two symmetric series, provide vascularization to the mesonephros and the gonads
  C. ANTERIOR REGION OF THE EMBRYO
    1. Six pairs of aortic arches are theoretically formed; however, the fifth pair is essentially only a temporary doubling of the fourth pair
    2. The aortic arches form successively, thus are never all present at the same time
      a. The aortic arches arise from the dilated region of the truncus arteriosus, known as the *aortic sac,* and terminate in the dorsal aorta of the corresponding side
      b. During weeks 6 to 8, the primitive aortic arch pattern is transformed into the basic adult arterial arrangement

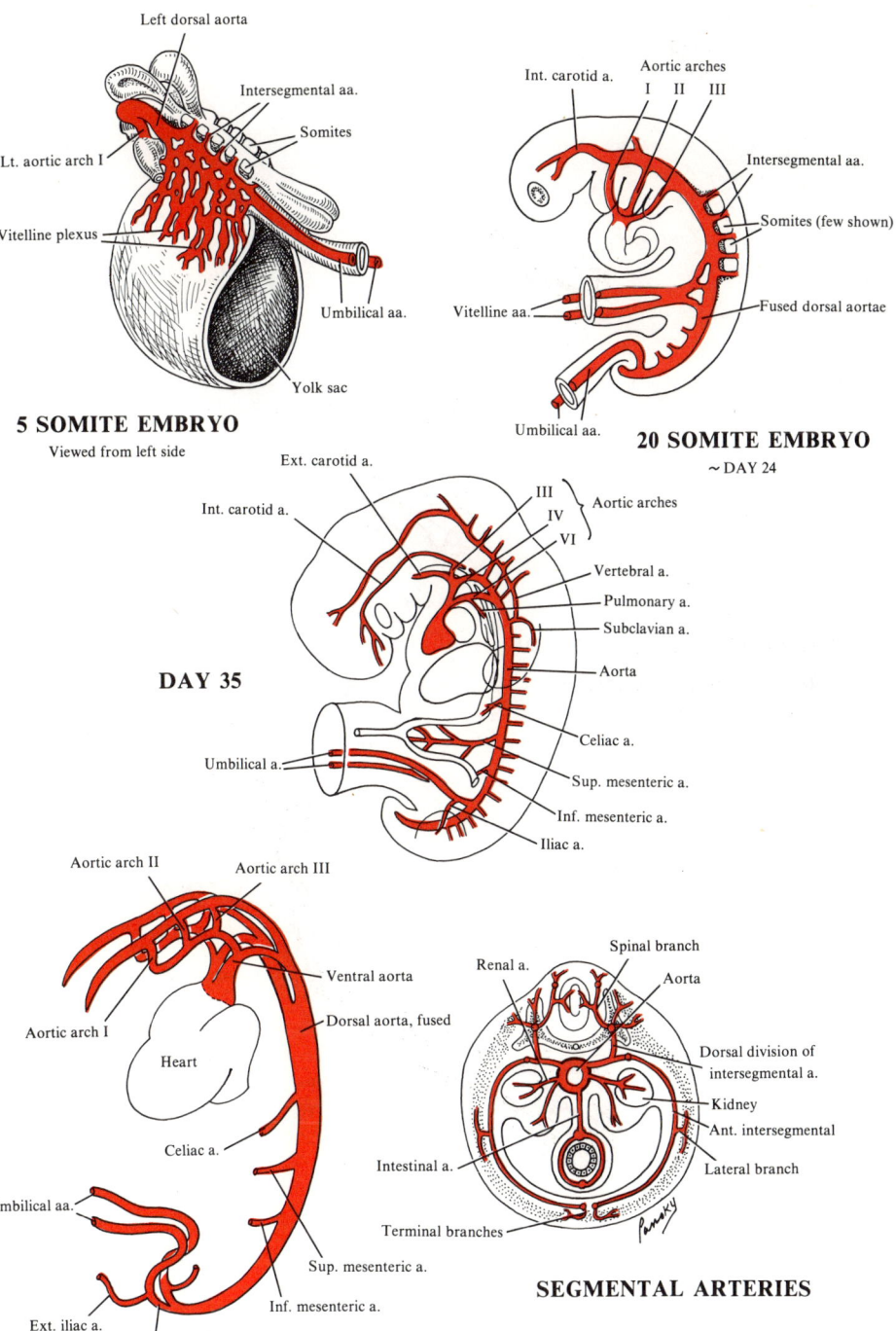

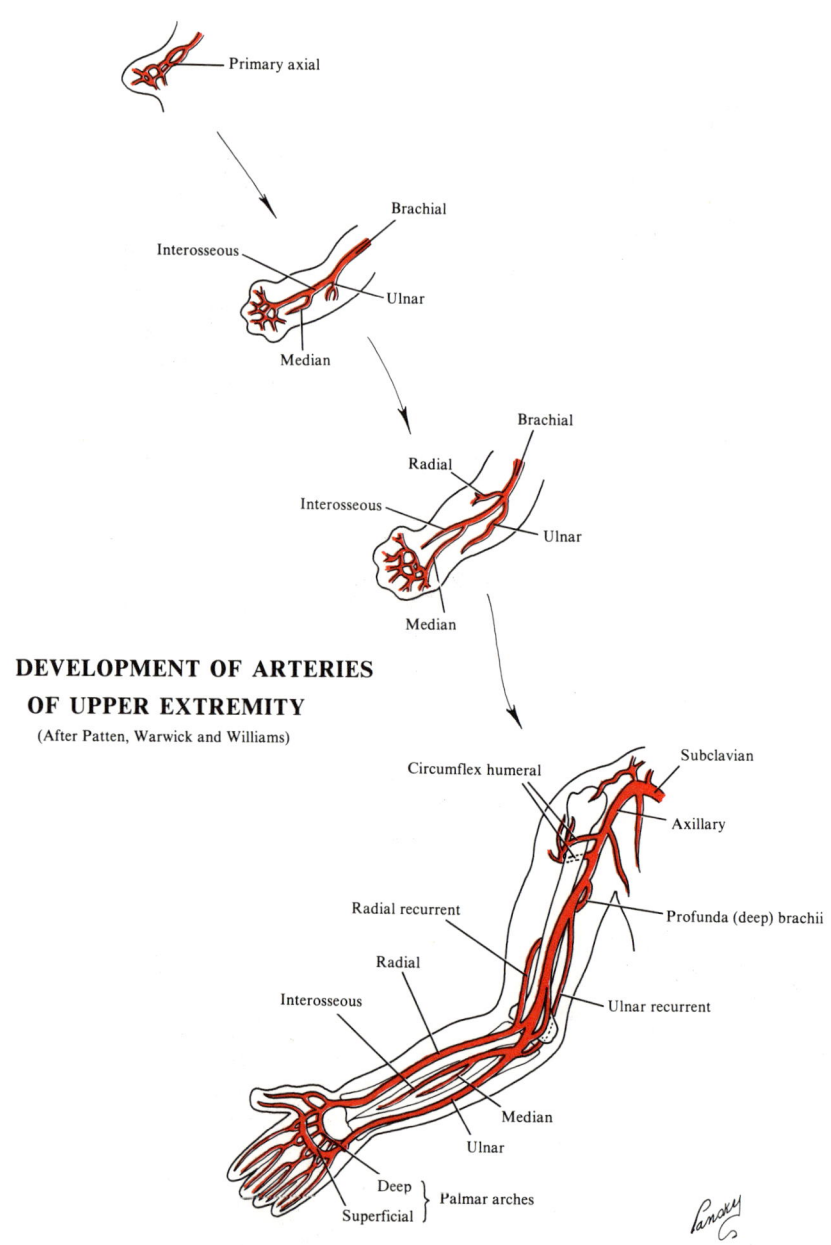

FIGURE 53. Development of the arteries of the upper extremity.

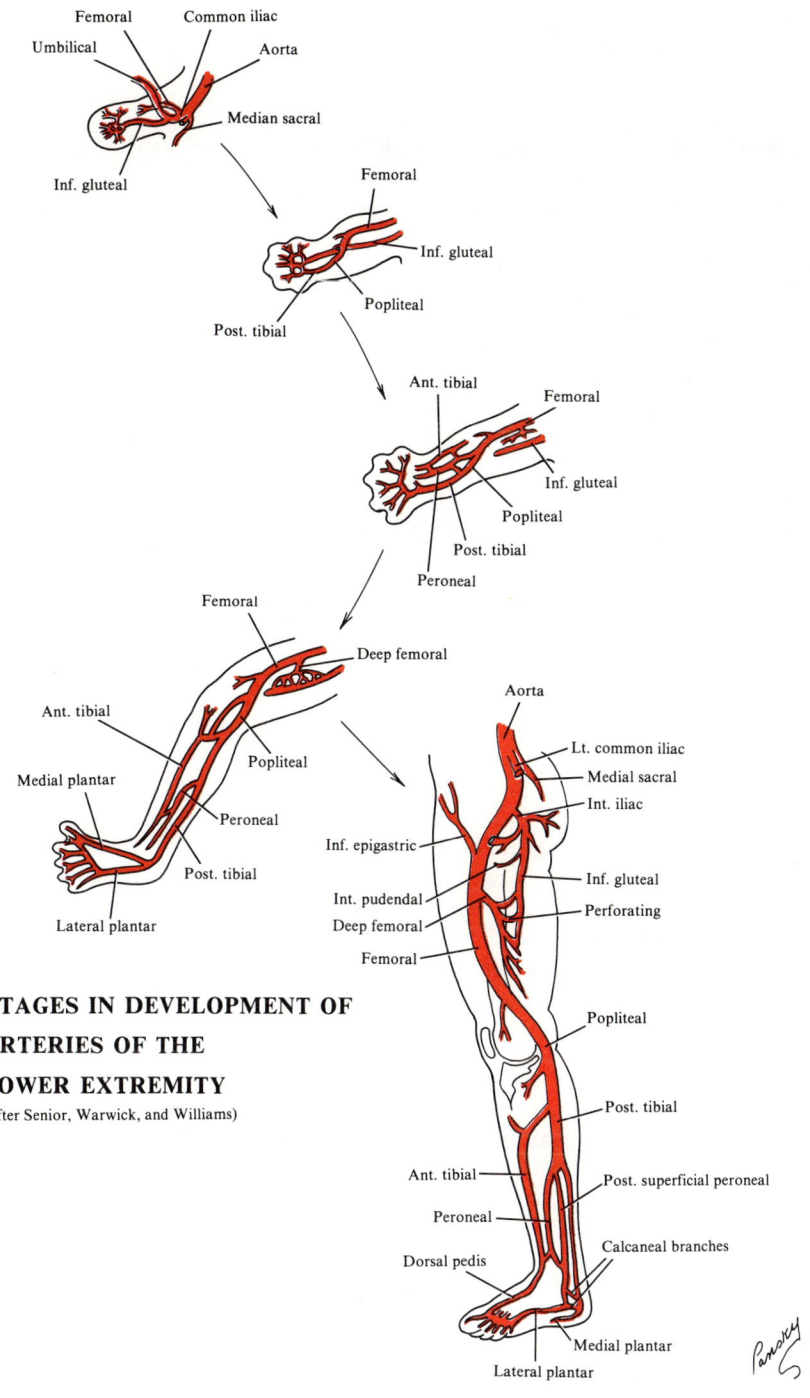

FIGURE 54. **Development of the arteries of the lower extremity.**

124. THE AORTIC ARCHES

I. **The first pair of aortic arches** is formed by the curving of the ventral aorta into the primitive dorsal aorta. This arch is hidden in the mandibular arch and participates in formation of the *maxillary artery,* and contribute to the *external carotid artery*

II. **The second pair of aortic arches** make their appearance in the middle of week 4. They cross the second branchial arches and give rise to the *stapedial* and *hyoid arteries.* (It should be noted that arches I and II regress rapidly and are not seen after day 31)

III. **The third pair of aortic arches** make their appearance at the end of week 4. They give rise to the *common carotids* and *proximal portions of the internal carotid arteries.* The latter are the short cephalic prolongations of the primitive dorsal aortas and are associated with development and supply of the brain
   A. THE INTERNAL CAROTID ARTERIES are secondarily attached to the cranial portions of the dorsal aortas, which form the remainder of the carotid artery
   B. THE ORIGIN OF THE EXTERNAL CAROTID ARTERIES is controversial, but in later stages of development, they are found to sprout from aortic arch III. (Arch I, however, has been implicated in its developmental contribution)

IV. **The fourth pair of aortic arches** make their appearance shortly after the third arches, at the end of week 4. Their development is different for the right and left sides
   A. ON THE RIGHT SIDE arch IV forms the proximal portion of the *right subclavian artery* and is continuous with the seventh segmental artery
      1. The caudal portion of the right primitive dorsal aorta disappears
      2. The distal portion of the subclavian artery forms from the right dorsal aorta and the right seventh intersegmental artery
   B. ON THE LEFT SIDE arch IV persists as the *arch of the aorta,* which grows significantly and is continuous with the primitive left dorsal aorta.
      1. The *left subclavian artery* (or seventh segmental) arises directly from the aorta
   C. THE SHORT PORTION of the right primitive ventral aorta, which persists between arches IV and VI, forms the *brachiocephalic arterial trunk* and the *first portion of the aortic arch*

V. **The fifth pair of aortic arches:** in 50% of embryos, these arches are rudimentary vessels that degenerate with no derivatives. In fact, they may never even develop

VI. **The sixth pair of aortic arches** make their appearance in the middle of week 5 and give rise to the *right* and *left pulmonary arteries.* After pulmonary vascularization is established, the communication with the corresponding primitive dorsal aorta regresses
   A. REGRESSION is total and complete on the right side. The proximal portion of the right arch forms the proximal part of right pulmonary artery; its distal portion degenerates
   B. THE PROXIMAL PORTION OF THE LEFT ARCH persists as the proximal part of the left pulmonary artery
      1. The distal portion of the left arch, in which communication persists with the dorsal aorta until birth, forms the *ductus arteriosus* and diverts blood from the pulmonary artery to the aorta. Closure of the ductus arteriosus takes place in the neonatal period, and the functional duct becomes the anatomic *ligamentum arteriosum*
   C. THE DISTAL PORTIONS OF THE PULMONARY ARTERIES are derived from buds of the sixth aortic arches that grow into the developing lungs. After partitioning of the truncus arteriosus, the pulmonary arteries arise from the pulmonary trunk

VII. **Summary of the aortic arches:** arch I regresses; arch II regresses; arch III forms the carotid system; arch IV forms the aortic arch (on the left) and the subclavian (on the right); arch V disappears; and arch VI forms the pulmonary arteries and the ductus arteriosus (on the left)

DEVELOPING ARTERIES IN HEAD REGION OF EARLY EMBRYO

BRANCHIAL (PHARYNGEAL) ARCH ARTERIES

ADULT FORM

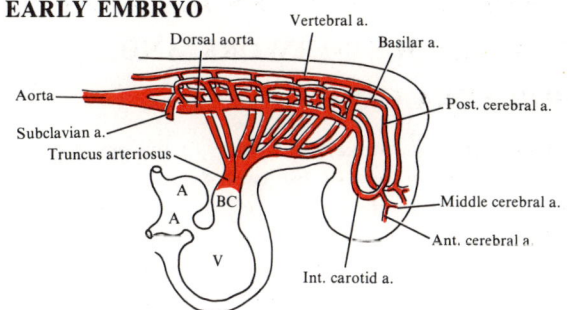

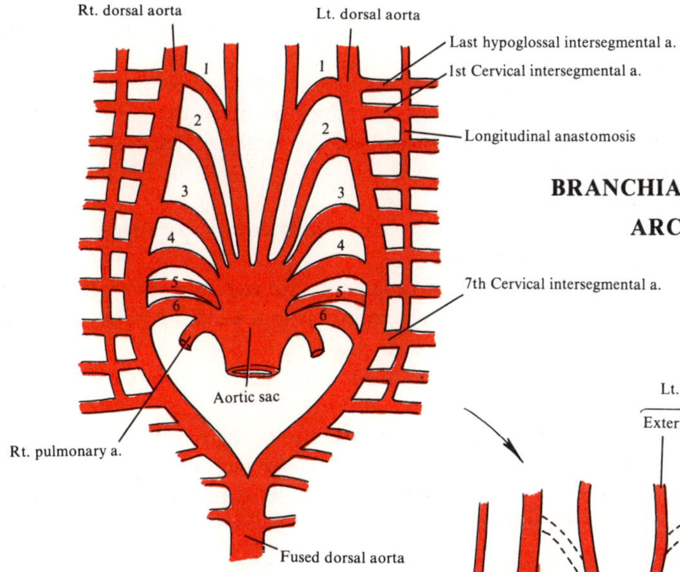

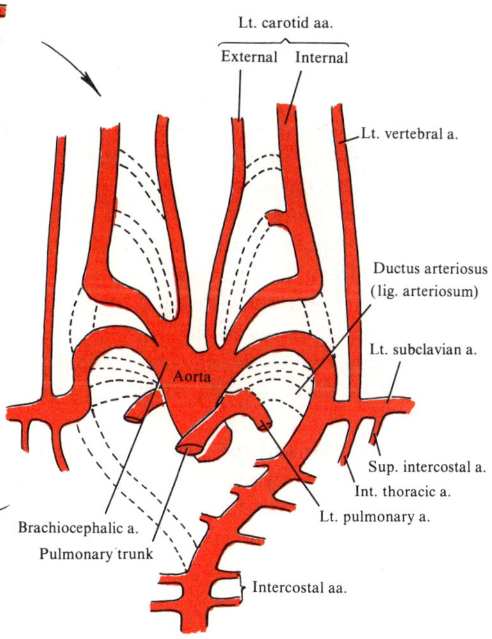

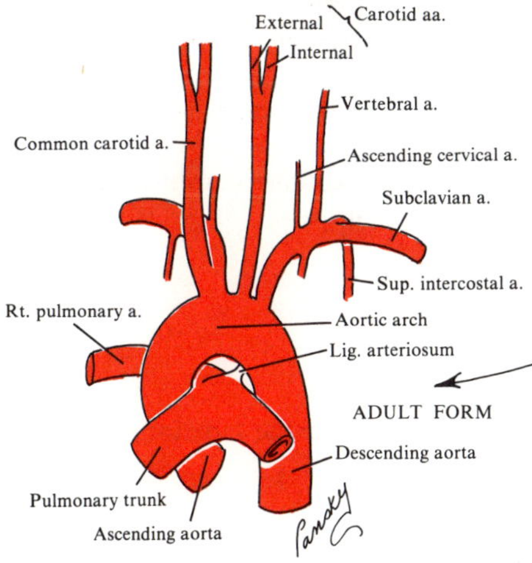

125. DEVELOPMENT OF THE VENOUS SYSTEM: PRIMITIVE VENOUS NETWORK AND SUPERIOR VENA CAVA

I. **The primitive venous network:** the venous system of the embryo consists of a *dorsal systemic* and *double nutritional network* by week 4
   A. THE DORSAL SYSTEMIC NETWORK drains the body of the embryo proper and carries all the *intraembryonic blood*. It is formed by the *anterior* and *posterior cardinal veins* which reach the sinus venosus through the *common cardinal veins*
   B. THE DOUBLE NUTRITIONAL NETWORK carries the extraembryonic blood by way of
      1. The vitelline or omphalomesenteric system carrying blood from the yolk sac toward the heart
      2. The umbilicoallantoic system carrying blood (oxygenated) from the placenta to the embryo
   C. BOTH NETWORKS are initially paired and symmetric, but by a series of cross or transverse anastomoses, they are converted into single major trunks in the right half of the embryo. The left-sided vessels diminish in size and are largely obliterated

II. **Development of the superior vena cava**
   A. THE SUPERIOR VENA CAVA has a relatively simple development compared to the inferior vena cava, but is formed somewhat later
      1. In week 8, a large anastomosis (derived from the thymic and thyroid veins) channels the blood from the left anterior (superior) cardinal vein toward the right. This gives rise to the future *left brachiocephalic venous trunk*
         a. Above the anastomosis, the anterior cardinal veins become the *internal jugular veins*
      2. The anterior veins of the mandibular region give rise to the *external jugular veins*
      3. The venous plexuses of the upper limb fuse to form the *subclavian vein*. The latter originally opens into the posterior cardinal vein, but as the heart shifts somewhat caudally in its development, the subclavian vein finally shifts to open into the anterior cardinal vein
      4. The left anterior cardinal vein, below the anastomosis, loses its connection with the left common cardinal vein. The part that persists is a short segment which forms the *left superior intercostal vein*
      5. The left common cardinal vein persists as a very short segment which forms the *coronary sinus venosus*
      6. The *superior vena cava* itself is finally formed by the right common cardinal vein and the proximal portion of the right anterior cardinal vein

III. **Malformations of the superior vena cava**
   A. MALFORMATIONS are rare. Some of those that are seen are
      1. Left superior vena cava
      2. Double superior vena cava
      3. Abnormal pulmonary venous return which drains into either the superior vena cava or the right atrium

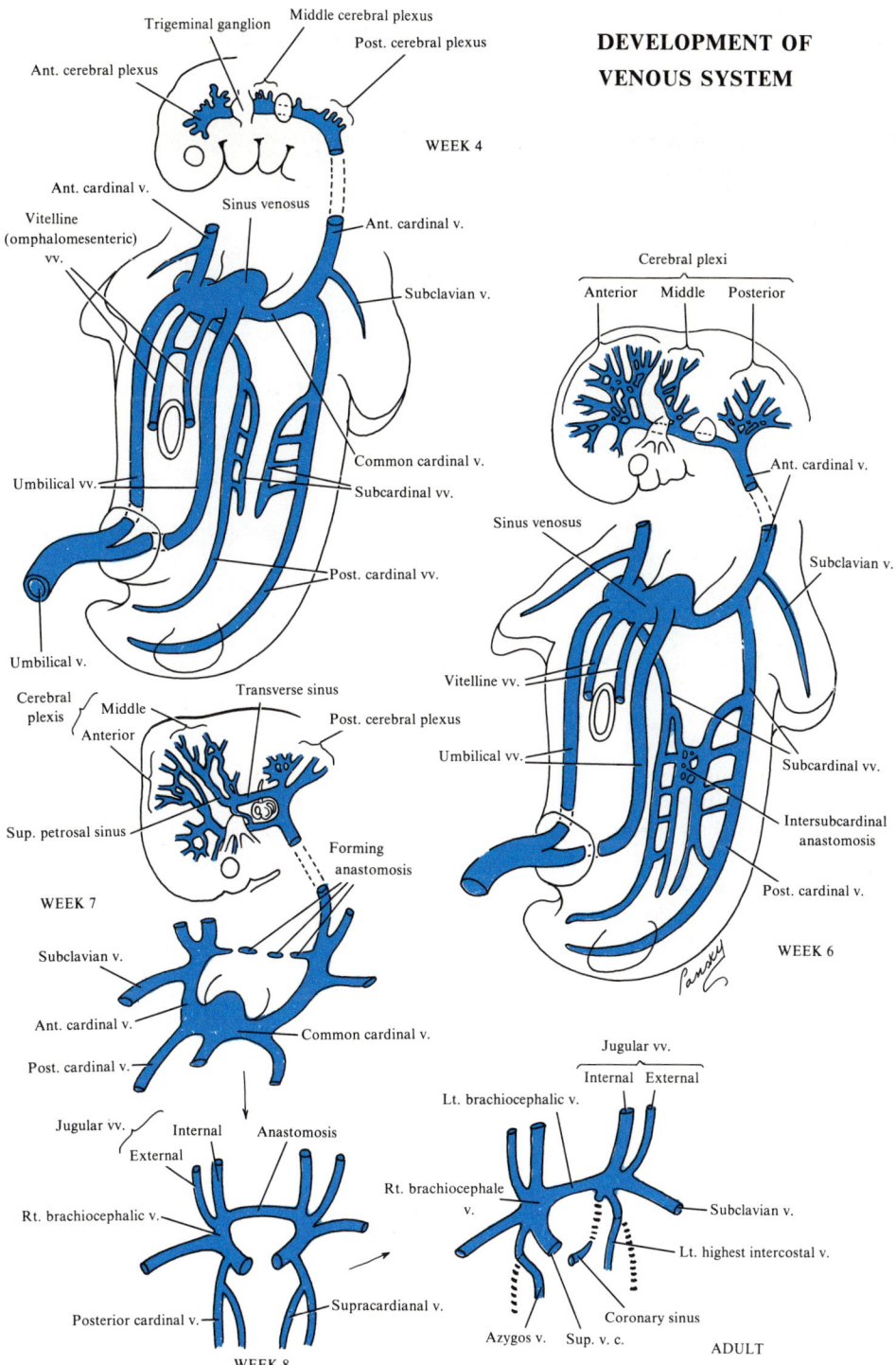

126. DEVELOPMENT OF THE VENOUS SYSTEM: THE INFERIOR VENA CAVA

I. **The inferior vena cava:** a series of successive venous networks take part in the formation of the inferior vena cava. Each predominates temporarily, then regresses, and remains only partly in the final definitive system
  A. THE MESONEPHROS grows considerably and becomes very highly vascularized during week 4. Although it is drained initially only by the posterior cardinal veins, a new system takes over after week 4, the *subcardinal network,* which is formed by the *internal veins of the wolffian body*
    1. The internal veins of the wolffian body are widely anastomosed with the initial posterior cardinal network and with each other, forming the *median subcardinal network,* which soon predominates. It takes over the posterior cardinal system, which disappears in the middle region of the embryo
    2. The *subcardinal sinus* persists as the *left renal vein*
    3. The anterior segment of the left subcardinal vein disappears, but its posterior segment forms the *left gonadal vein*
    4. The right subcardinal vein forms the *right gonadal vein* and the pararenal portion of the definitive inferior vena cava
       a. Cranially, it continues with the mesenteric segment and the hepatic segment derived from the hepatic vein (proximal right vitelline) and hepatic sinusoids
  B. DURING WEEKS 6 AND 7, a supplementary dorsal network develops, called the *supracardinal system* which runs parallel to the paravertebral sympathetic chain and opens into the proximal segment of the posterior cardinal veins. Anastomoses are formed
    1. Between the two supracardinal veins
    2. Between the supracardinals and the subcardinals, on the right side
    3. Between the extremities of the posterior cardinal veins
    4. The left supracardinal vein becomes the *hemiazygos vein* and is drained toward the right by the transverse anastomosis which forms an *interazygos communication*
    5. The right supracardinal vein becomes the *azygos vein,* which opens into the right anterior cardinal vein
       a. From below, the azygos vein drains the 2 iliac veins, thus becomes the prerenal portion of the definitive inferior vena cava
  C. IN SUMMARY: the definitive inferior vena cava is composed of (from caudal to cranial)
    1. The posterior intercardinal anastomosis
    2. The caudal portion of the right supracardinal vein
    3. The right anastomosis between the supracardinal and the subcardinal veins
    4. A segment of the right subcardinal vein
    5. The anastomosis between the right subcardinal and right vitelline veins
    6. The terminal portion of the right vitelline vein

II. **Malformations of the inferior vena cava**
  A. MALFORMATIONS of the inferior vena cava may be due to the complexity of formation of the system, yet, even in cases of severe malformation, one of the constituent networks invariably substitutes some form of venous flow
    1. Agenesis (absence) of the inferior vena cava is the most conspicuous. The right subcardinal vein has failed to make its connection with the liver and shunts blood directly into the right supracardinal vein. Thus, blood from the caudal part of the body reaches the heart via the azygos and superior vena cava. The hepatic vein enters the right atrium at the site of the inferior vena cava
    2. An abnormality of position of the vein may affect the adjacent organs, such as the ureter, compressing it and causing a hydronephrosis
    3. Double inferior vena cava at the lumbar region: the left sacrocardinal vein has failed to lose its connection with the left subcardinal, and the left common iliac vein may or may not be present. The left gonadal vein is usually present and in a normal condition

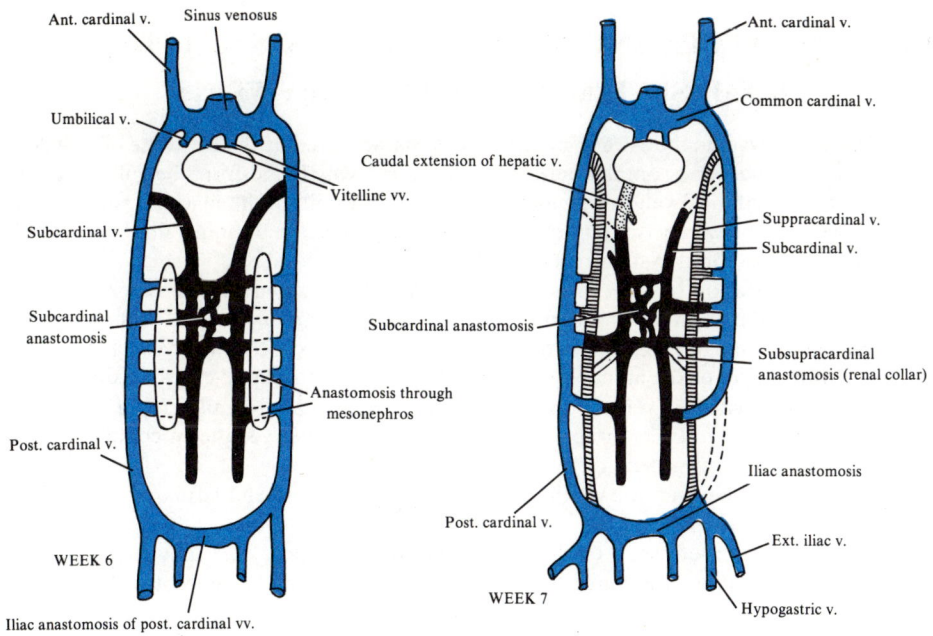

DEVELOPMENT OF INFERIOR VENA CAVA
(After McClure, Butler, and Arey)

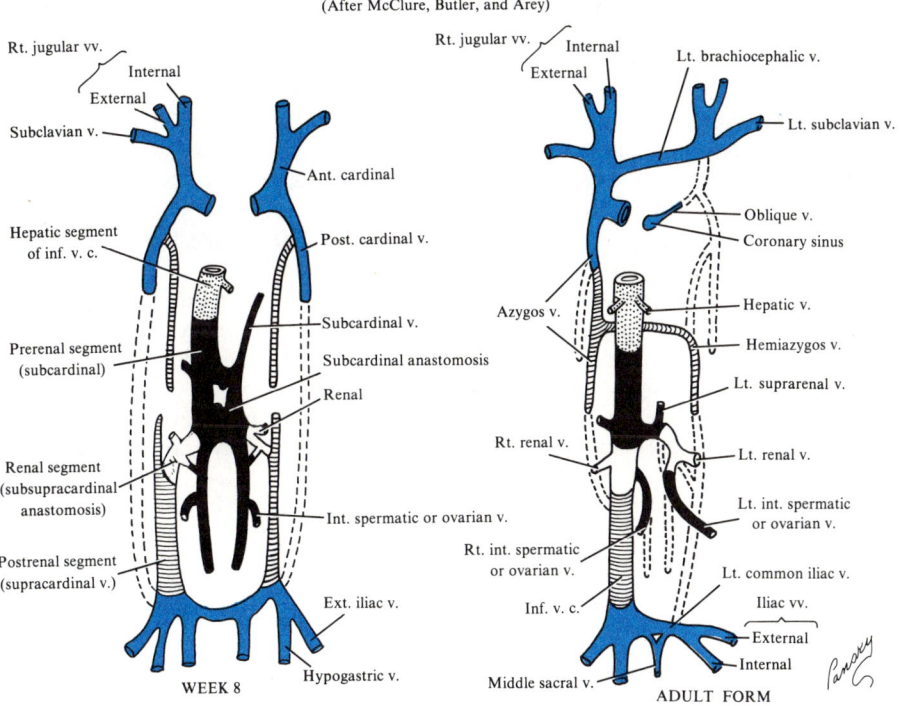

127. DEVELOPMENT OF THE VENOUS SYSTEM: THE PORTAL SYSTEM AND PULMONARY VEINS

I. **The portal system:** the 3-week-old embryo is characterized by the existence of 3 pairs of veins: the vitelline or omphalomesenteric veins carrying blood from the yolk sac to the heart, the umbilical veins originating in the chorionic villi of the placenta and carrying oxygenated blood to the embryo, and the cardinal veins draining the body of the embryo. The portal system relates to the first 2

A. THE VITELLINE OR OMPHALOMESENTERIC VEINS enter the body of the embryo via the yolk sac stalk, form an anastomotic network around the duodenum of the digestive tract, and then enter the septum transversum, which they cross on their way to the heart
1. Proliferation of the entodermal liver cords, which form the liver primordium, fragments the vitelline veins to form a vascular labyrinth, the so-called *hepatic sinusoids*
   a. After the yolk sac disappears, the vitelline veins regress almost completely and persist only in their mesenteric branches
2. Cranial to the liver, the vitelline veins open into the right and left horns of the sinus venosus
   a. When the left horn of the sinus venosus disappears, the right vitelline trunk receives the anastomosis of the inferior vena cava and becomes its terminal portion or suprahepatic portion of the inferior vena cava
3. Caudal to the liver, the vitelline vein anastomotic network around the duodenum fuses to form a single trunk, the *portal vein,* partly by obliteration and partly by growth of different portions
   a. The *superior mesenteric vein* which drains the primitive intestinal loop is considered to be the successor of the right vitelline vein
   b. The distal portion of the left vitelline vein disappears completely

B. THE UMBILICOALLANTOIC (UMBILICAL) VEINS enter the embryo by way of the connecting stalk and then course through the mesoderm of the septum transversum toward the heart. They lie more lateral than the vitelline veins but are also fragmented by the development of the liver and are connected to the hepatic sinusoids
1. The proximal portions of both umbilical veins as well as the remainder of the right umbilical vein ultimately disappear, so that only the left umbilical vein continues to drain the blood coming from the placenta to the liver
2. As a result of the marked increase of the placental circulation during further development, a short circuit (direct communication) is temporarily established between the left umbilical vein and the inferior vena cava, namely, the *ductus venosus,* which bypasses the sinusoidal plexus of the liver
   a. After birth, the left umbilical vein and the ductus venosus are obliterated and form the *ligamentum teres hepatis* and the *ligamentum venosum,* respectively

II. **The pulmonary veins:** at about the 4-mm stage, the common pulmonary vein is visible as an evagination of the dorsal wall of the atrium

A. THE BUD subsequently grows out into the dorsal mesocardium in the direction of the primitive foregut, which gives rise to the lung buds
1. As the atrial cavity continues to develop, the stem of the pulmonary vein is progressively incorporated into the left atrial wall
2. The incorporation of the pulmonary veins continues until 2 right and 2 left branches of the pulmonary stem enter the atrial cavity

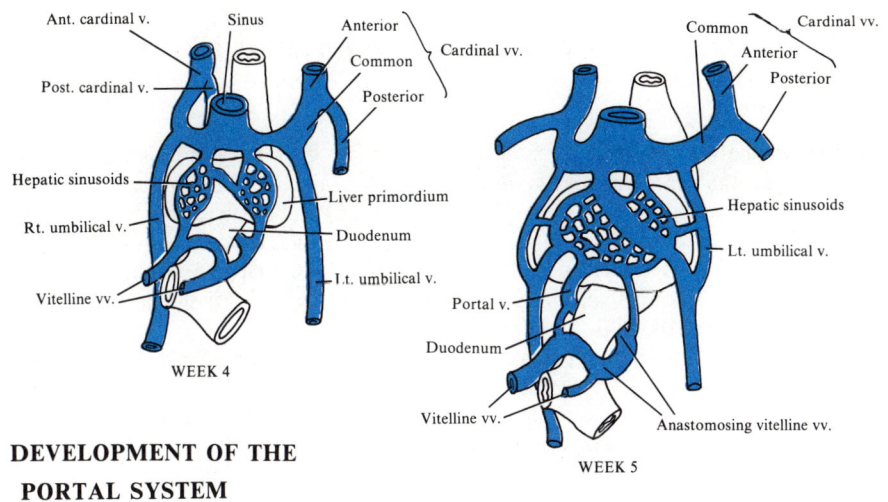

DEVELOPMENT OF THE PORTAL SYSTEM

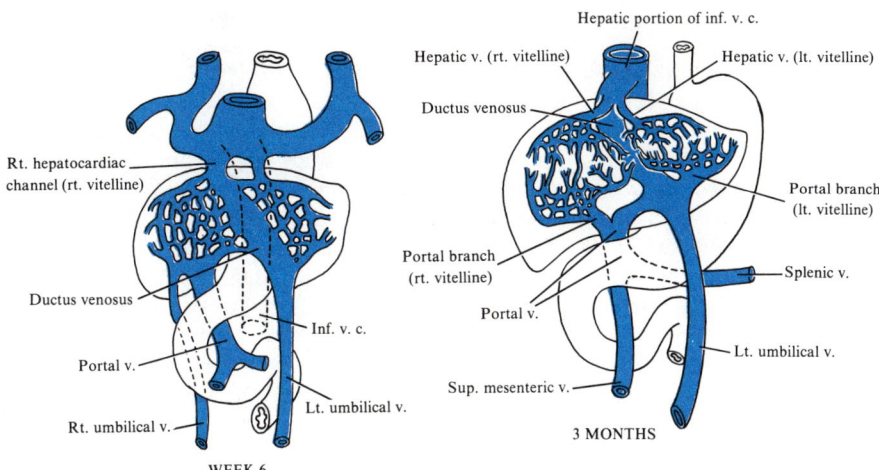

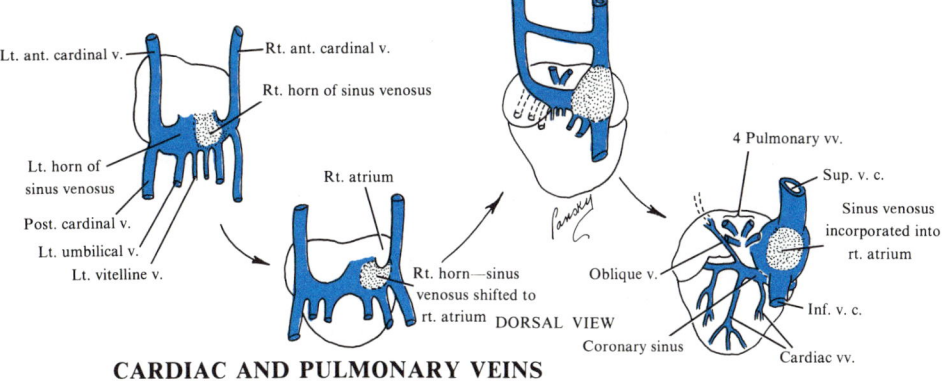

CARDIAC AND PULMONARY VEINS

128. DEVELOPMENT OF THE LYMPHATIC SYSTEM

I. **The lymphatic system** begins to develop at the end of week 5, approximately 2 weeks later than the cardiovascular system. One view states that the lymphatics develop as diverticulae of the endothelium of veins; whereas another states that like other blood vessels they develop from clefts in the mesenchyme that connect with the venous system secondarily. Thus, the cells lining the mesenchymal clefts assume an endothelial shape, and subsequent sprouting of these cells causes the clefts to fuse and form the lymphatic channels
  - A. IN WEEKS 6–9, LOCAL DILATATIONS of the lymphatic channels form 6 primary lymph sacs
    1. Two jugular lymph sacs near the junction of the subclavian veins with the anterior cardinals (future internal jugular vein)
    2. Two iliac lymph sacs near the junction of the iliac veins with the posterior cardinal veins
    3. One retroperitoneal lymph sac in the root of the mesentery on the posterior abdominal wall
    4. One so-called *cisterna chyli* dorsal to the retroperitoneal lymph sac, at the level of the adrenal glands
  - B. LYMPH VESSELS GROW OUT from the lymph sacs, along the major veins, to the head, neck, and arms from the jugular sacs; to the lower trunk and legs from the iliac sacs; and to the gut from the retroperitoneal and cisternal sacs
    1. The cisterna chyli is connected to the jugular lymph sacs by 2 large channels, the *right* and *left thoracic ducts*. An anastomosis forms between the 2 ducts, thus, the definitive thoracic duct is formed by the caudal portion of the right thoracic duct, the anastomosis, and the cranial portion of the left thoracic duct
    2. The right lymphatic duct is derived from the cranial part of the right thoracic duct
    3. Both the right and left thoracic ducts join the venous system at the angle of the subclavian and internal jugular veins at the base of the neck

II. **Lymph node development**
  - A. EXCEPT FOR THE UPPER PORTION OF THE CISTERNA CHYLI, which persists, the lymph sacs are transformed into groups of lymph nodes during early fetal life, at about month 3.
    1. Surrounding mesenchymal cells invade each sac and break it up into lymphatic channels or *sinuses*. The mesenchymal cells give rise to the lymph node capsule and the connective tissue framework of the node
    2. The lymphocytes seen in the node before birth come from the thymus gland
    3. The lymph nodule and germinal centers of lymphocyte production do not appear in the nodes until just before or after birth
    4. Lymph nodes also develop along the course of other lymph vessels

III. **Other lymphatic tissues**
  - A. THE SPLEEN develops from an aggregation of mesenchymal cells in the dorsal mesentery of the stomach
  - B. THE PALATINE TONSILS form from the second pair of pharyngeal pouches
  - C. THE TUBAL (PHARYNGOTYMPANIC) TONSILS develop from aggregations of lymph nodules around the openings of the auditory tubes
  - D. THE PHARYNGEAL TONSILS (adenoids) develop from an aggregation of lymph nodules in the nasopharyngeal wall
  - E. THE LINGUAL TONSILS develop from aggregations of lymph nodules in the root of the tongue
  - F. LYMPH NODULES also are seen in the mucosa of the digestive tract and respiratory tract

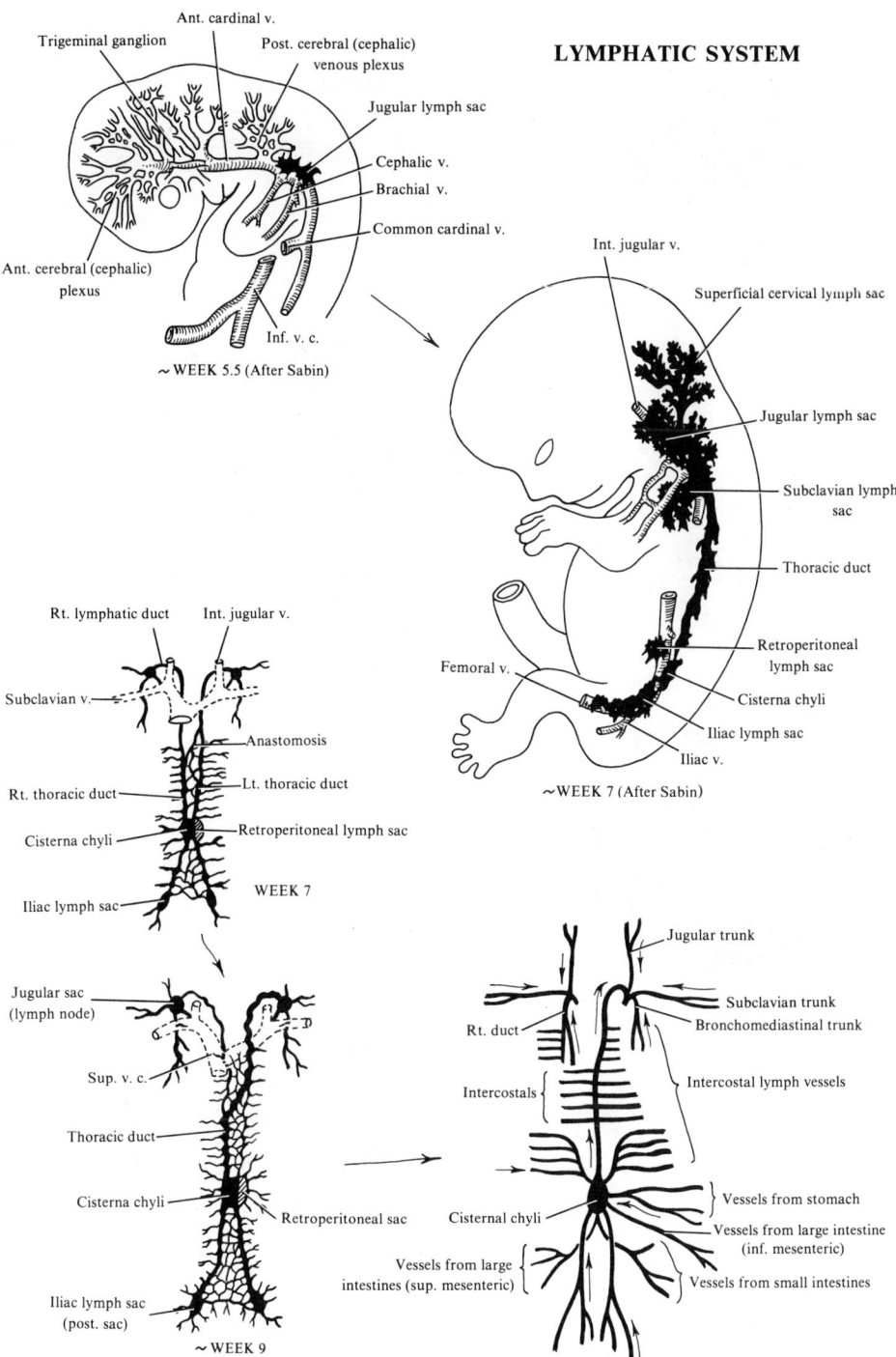

LYMPHATIC SYSTEM

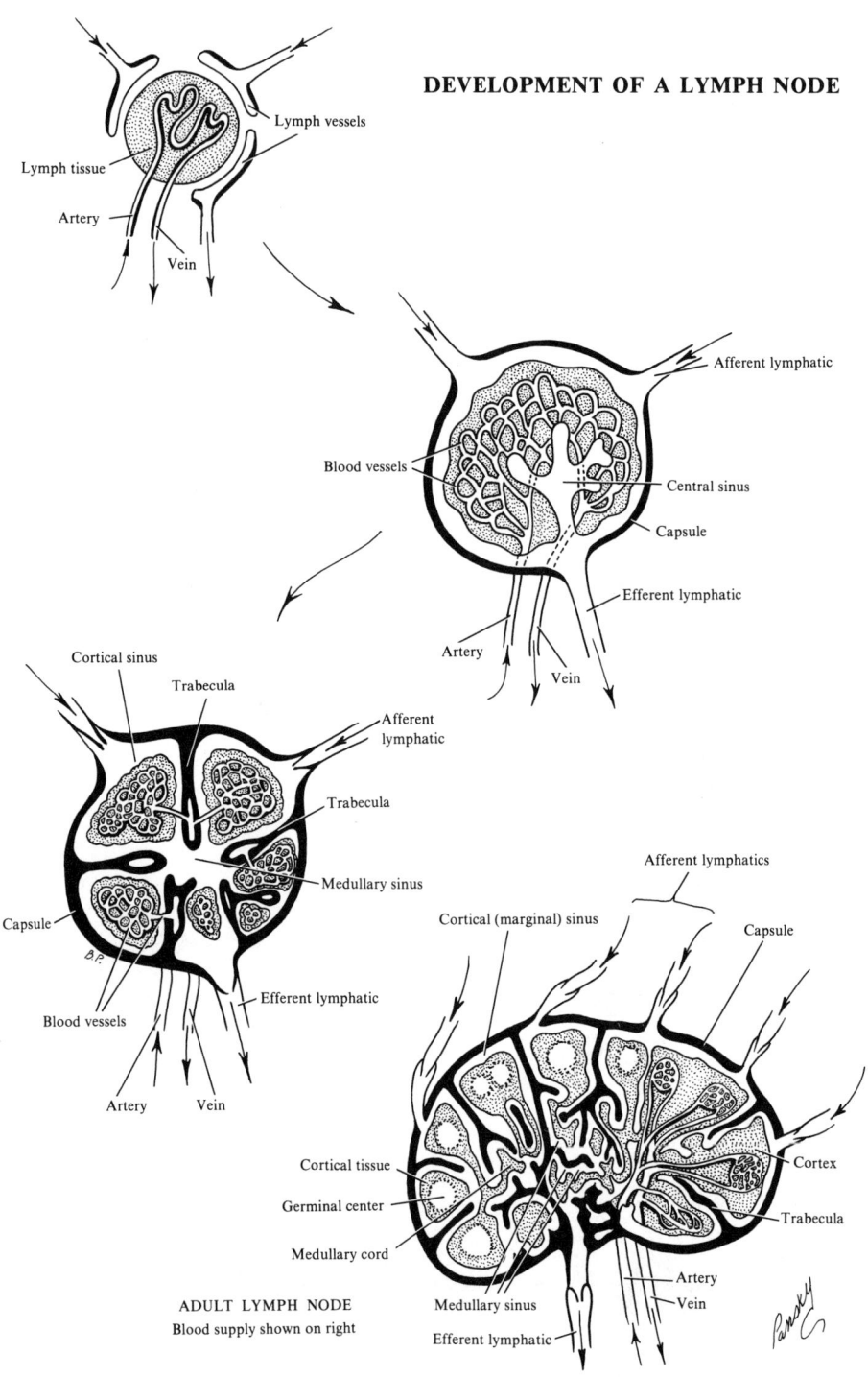

FIGURE 55. **Development of a lymph node.**

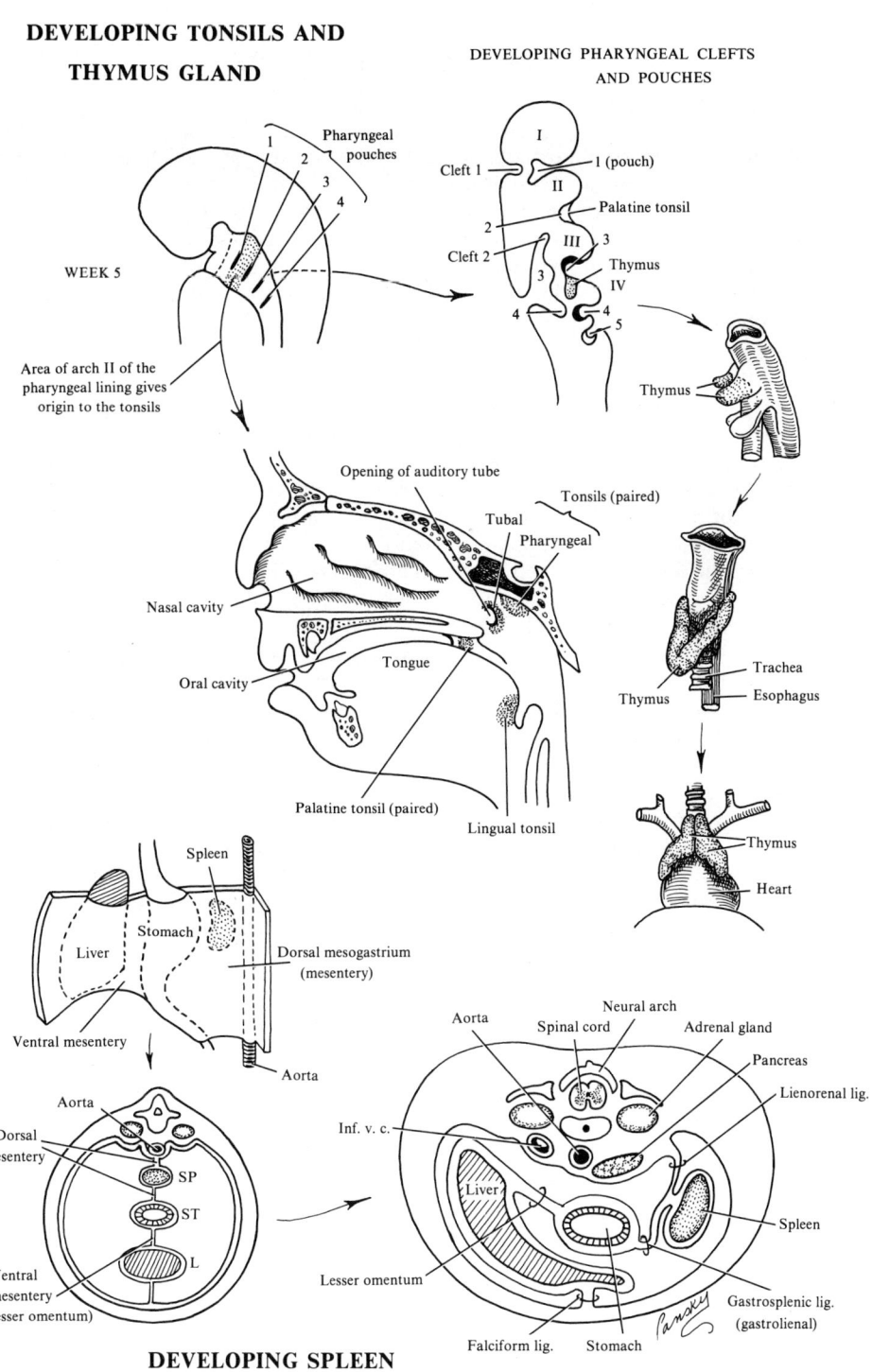

FIGURE 56. **Development of the tonsils, the thymus, and the spleen.**

129. THE CIRCULATORY SYSTEM BEFORE AND AFTER BIRTH

I. **The fetal blood** reaches the placenta via the umbilical arteries, which are branches of the caudal aortic system. Well-oxygenated blood from the placenta is returned by the umbilical veins, about one-half passing through the liver sinusoids, and the rest bypassing the liver via the ductus venosus into the inferior vena cava system. The blood flow is regulated by a muscular sphincter in the ductus venosus, close to the umbilical veins. The blood then passes to the right atrium. Since the inferior vena cava also receives deoxygenated blood from the lower portion of the body, the blood entering the right atrium is not as well oxygenated as that in the umbilical veins

II. **The minor circulation (circulation through the lungs):** although existing anatomically, the minor circulation is almost completely short-circuited by 2 major mechanisms
   A. THE BLOOD FROM THE INFERIOR VENA CAVA is directed largely by the lower border of the septum secundum, through the *foramen ovale,* which directs blood from the right to the left side of the heart. Here it mixes with small amounts of deoxygenated blood returning from the lungs via the pulmonary veins
      1. From the left atrium, blood goes to the left ventricle and out of the heart via the aorta to vessels of the head and neck, upper limbs, and the rest of the body. The former receive richer, well-oxygenated blood
      2. Some oxygenated blood from the inferior vena cava stays in the right atrium, mixes with deoxygenated blood from the superior vena cava and coronary sinus, and passes to the right ventricle. Blood leaves via the pulmonary trunk and passes through the *ductus arteriosus,* which diverts it from the pulmonary to the aortic system
         a. Little goes to the lungs ($\pm 10\%$) due to pulmonary vascular resistance
         b. The mixed blood in the descending aorta (58% saturated with oxygen) passes to the umbilical arteries and is returned to the placenta for reoxygenation
   B. THE FETUS cannot use its pulmonary system since it lives in a liquid environment. The lungs, nevertheless, have been ready and prepared to fulfill their role from month 6 of pregnancy. It becomes effectively functional only when the fetus is born

III. **Circulation at birth:** placental circulation is interrupted. The abrupt drop in intrathoracic pressure brought about by the first respiration (aeration of the lungs) helps contribute to the initial pulmonary circulation
   A. BLOOD PRESSURE decreases in the pulmonary artery even though its flow is increased since it supplies a capillary network considerably enlarged by expansion of the pulmonary parenchyma
      1. As a result, blood flow decreases (even reverses momentarily) in the ductus arteriosus; its walls contract, and in a few days, it closes off completely
         a. Closure appears to be mediated by *bradykinin,* released from the lungs during their initial inflation. Its action depends on the high oxygen content of the aortic blood resulting from ventilation of the lungs at birth
      2. Similarly, influx of pulmonary blood into the left atrium causes the septum primum to be pressed against the septum secundum, and the foramen ovale is closed
      3. The umbilical arteries also constrict at birth and prevent blood loss
   B. THE CIRCULATORY SYSTEM at this time resembles its adult form, with separation of the minor and major circulations. Nevertheless, occlusion of the 2 circuits is still, for some time, only physiologic (functional)
   C. THE CIRCULATORY SYSTEM becomes fully anatomic after several weeks due to the proliferation of endothelial and fibrous tissue
      1. Fibrous degeneration of the ductus arteriosus forms the ligamentum arteriosum, between the left pulmonary artery and the concave inferior surface of the aortic arch
      2. Complete fusion of the septum primum to the septum secundum forms a definitive interatrial septum. Traces of the former passage, the foramen ovale, are seen as a depression, the fossa ovalis, in the right interatrial wall

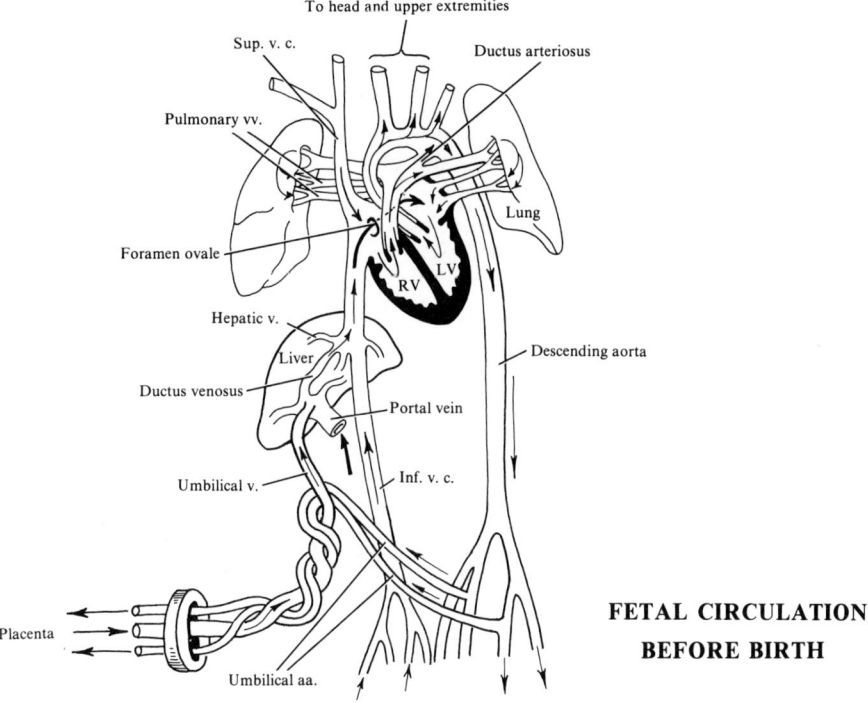

FETAL CIRCULATION BEFORE BIRTH

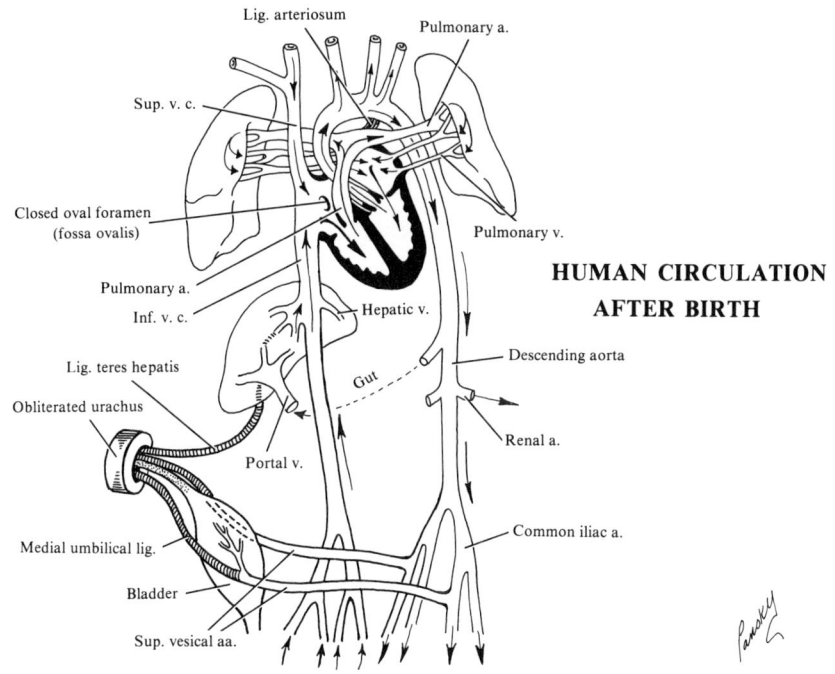

HUMAN CIRCULATION AFTER BIRTH

130. ADULT DERIVATIVES OF FETAL STRUCTURES

I. **The umbilical vein** forms the ligamentum arteriosum and passes from the umbilicus to the porta hepatis where it attaches to the left branch of the portal vein

II. **The ductus venosus** becomes the ligamentum venosusm which passes through the liver from the left branch of the portal vein to the inferior vena cava

III. **The umbilical arteries (intra-abdominal portion)** form the median umbilical ligaments; the proximal portions persist as the superior vesicle arteries

IV. **The foramen ovale** becomes the fossa ovalis with adhesion of the septum primum to the left margin of the septum secundum. The lower edge of the latter forms the limbus of the fossa ovalis (anulus ovalis)

V. **The ductus arteriosus** becomes the ligamentum arteriosum passing from the left pulmonary artery to the aortic arch. This closes anatomically by the end of the third postnatal month

MAJOR STAGES IN CARDIAC DEVELOPMENT

Weeks	3				4						5							6							7	
Days	20	21	22	23	24	25	26	27	28	29	30	31	32	33	34	35	36	37	38	39	40	41	42	43--------49
Somites	1	4	7	10	13	15	20		25		28														
Length (mm)	1.5				2.0		3.0		4.0				5.0		6.5		8.0					13.0			20.0

Cardiogenic plate
Endocardial tubes
Fusion of endocardial tubes
Single median tube
— Cardiogenic loop
Single atrium
First contractions (ineffective)
Bilobed atrium
Beginning of circulation
Septum primum (1°)
A-V orifices (3-chamber heart)
Septium secundum (2°)
Complete inferior septum
Septation of bulbus and ventricle
Divided truncus arteriosus
4-Chambered heart
Absorption of pulmonary vv.

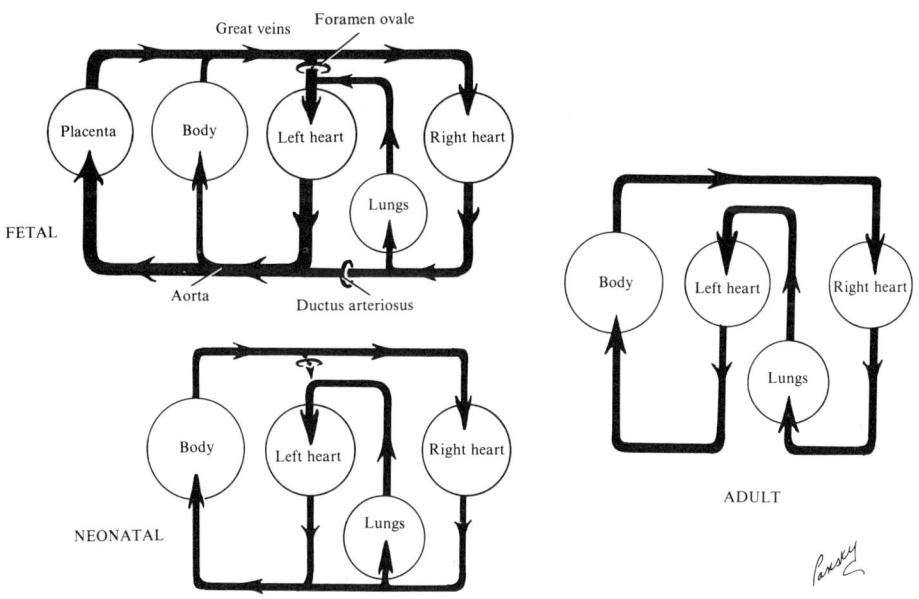

FETAL

NEONATAL

ADULT

-339-

131. MALFORMATIONS OF THE CARDIOVASCULAR SYSTEM

I. Aortic arch malformations
A. PATENT DUCTUS ARTERIOSUS is 2 to 3 times more common in the female than in the male. Is due to a failure of the distal part of the left sixth aortic arch to involute after birth and become ligamentous
B. COARCTATION OF THE AORTA: common; a narrowing of the aorta either below or above the ductus arteriosus. May occur directly opposite the ductus, but usually is classified into 2 types
   1. Preductal type: less common; constriction is above the level of the ductus. The ductus usually stays open, and a communication exists between the descending aorta and the pulmonary artery
   2. Postductal type: common; constriction is below the level of the ductus. The latter is usually closed and forms a ligament. Condition permits collateral circulation in the fetal period and helps circulation in lower parts of the body
C. DOUBLE AORTIC ARCH: rare; characterized by a *vascular ring* that compresses the trachea and esophagus. Due to failure of involution of distal part of the right dorsal aorta. Both a right and left aortic arch arise from the ascending aorta
D. RIGHT AORTIC ARCH: the entire right dorsal aorta persists, and the distal segment of the left dorsal aorta involutes. Two types are described
   1. Right aortic arch without a retroesophageal component: the ductus (ligamentous) arteriosus passes from the right pulmonary artery to the right aortic arch. No vascular ring. It is usually asymptomatic, but other cardiac defects are seen
   2. Right aortic arch with retroesophageal component: the right aortic arch is behind the esophagus. The normal left ductus arteriosus (ligamentum arteriosum) attaches to the descending aorta and forms a vascular ring that may constrict the trachea and esophagus
E. ABNORMAL ORIGIN OF THE RIGHT SUBCLAVIAN ARTERY (retroesophageal subclavian): the right subclavian artery arises from the ascending aorta and passes behind the trachea and esophagus to the right arm
   1. Is the result of the right fourth aortic arch and right dorsal aorta involuting cranial to the seventh intersegmental artery. The right subclavian forms from the right seventh intersegemental artery and distal part of the right dorsal aorta

II. Venous anomalies
A. DOUBLE SUPERIOR VENA CAVA (persistent left superior vena cava): due to an absence of or inadequate anastomosis between the anterior cardinal veins, resulting in the left anterior cardinal vein persisting as part of the left superior vena cava
   1. The abnormal left superior vena cava may open into the coronary sinus or, less commonly, into the left atrium
B. ANOMALOUS PULMONARY VENOUS CONNECTIONS (drainage or return)
   1. Total: none of the pulmonary veins connects with the left atrium, but they connect to the right atrium or to 1 of the systemic veins or both
   2. Partial: 1 or more (not all) of the pulmonary veins have anomalous connections. Others enter the left atrium of the heart normally
C. DOUBLE INFERIOR VENA CAVA AT THE LUMBAR REGION: the left sacrocardinal vein has failed to lose its connection with the left subcardinal, and the left common iliac vein may or may not be present. The left gonadal vein is normal
D. ABSENCE OF THE INFERIOR VENA CAVA: the right subcardinal vein fails to make connection with the liver and shunts its blood to the right supracardinal vein. Blood from the caudal part of the body reaches the heart via the azygos and superior vena cava, and the hepatic vein enters the right atrium at the site of the inferior vena cava
E. LEFT SUPERIOR VENA CAVA: persistence of left anterior cardinal vein and obliteration of common cardinal and proximal part of anterior cardinal vein on the right. Blood from the right is channeled to the left by the brachiocephalic vein

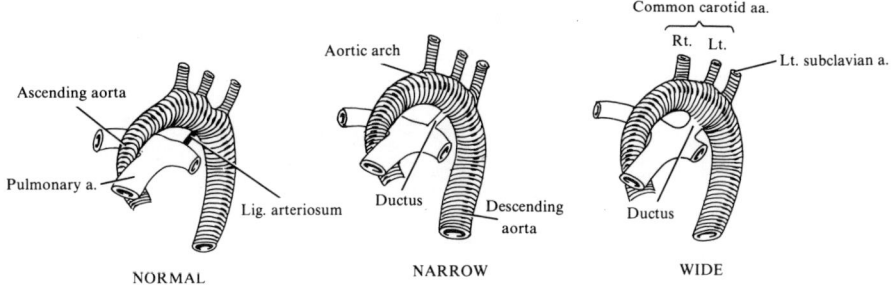

PATENT DUCTUS ARTERIOSUS

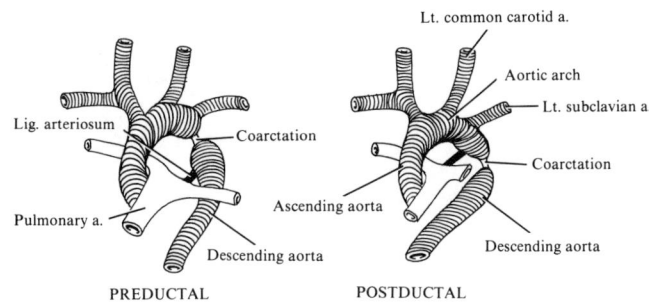

AORTIC COARCTATION

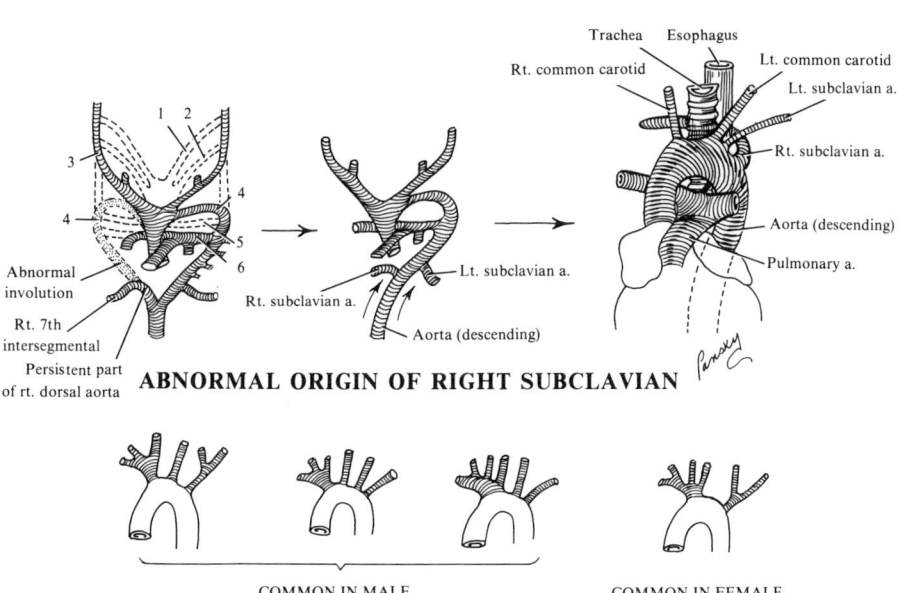

ABNORMAL ORIGIN OF RIGHT SUBCLAVIAN

VARIATIONS IN BRANCHES OF AORTIC ARCH (After Adachi, McDonald, and Anson)

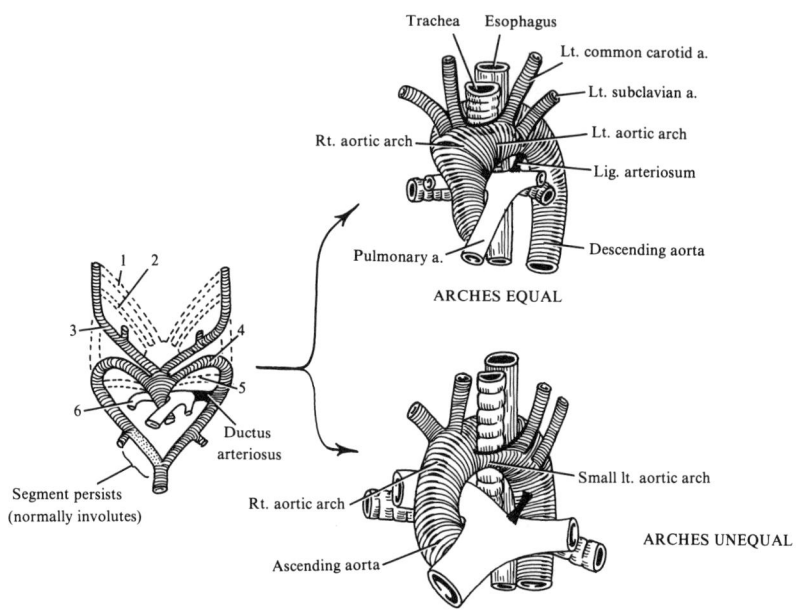

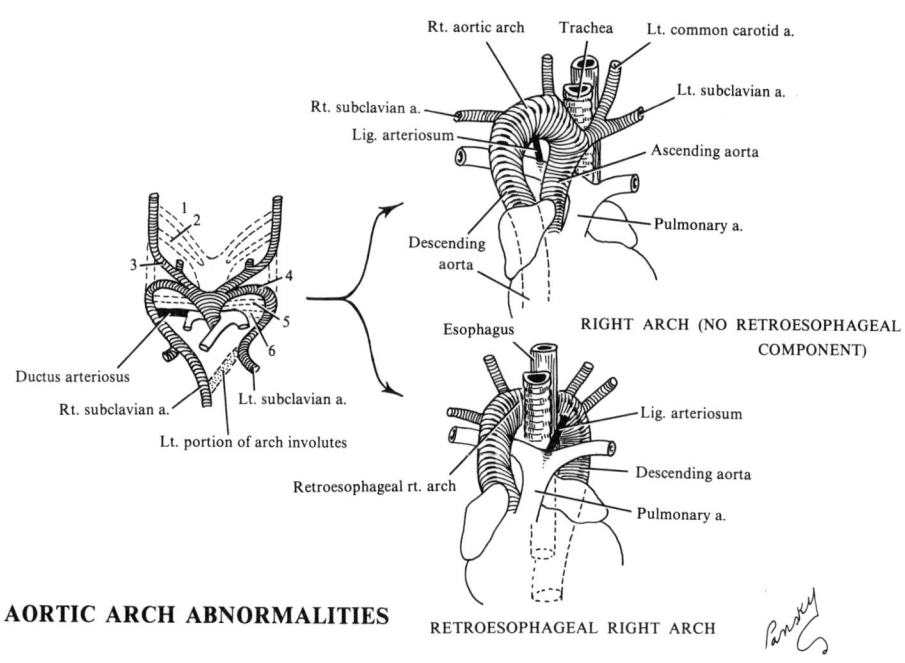

FIGURE 57. **Aortic arch abnormalities.**

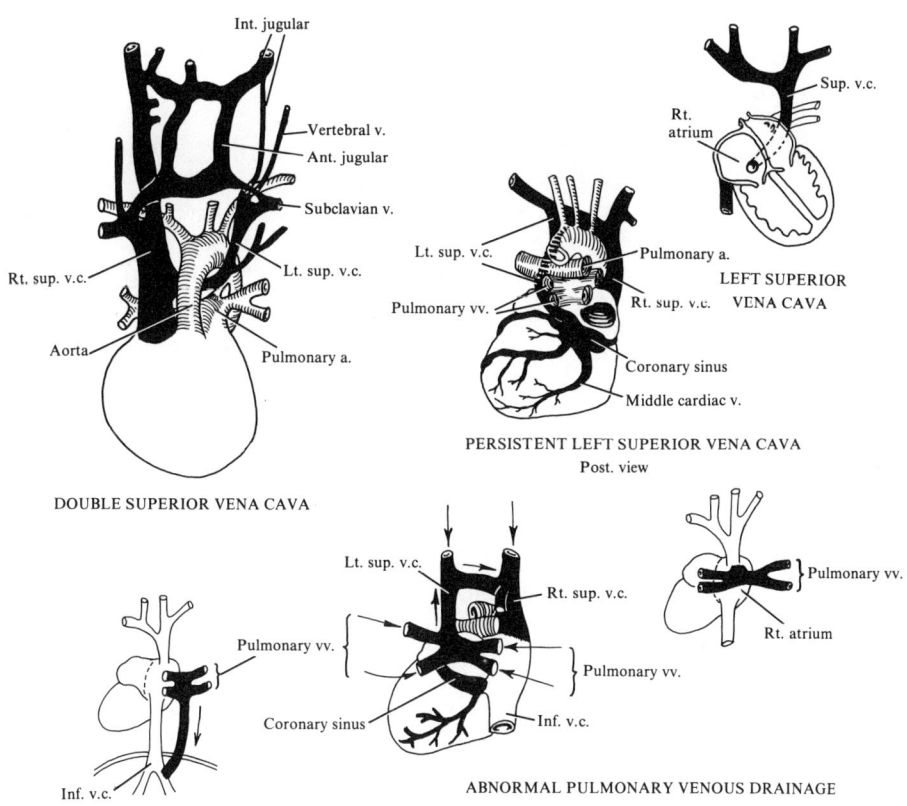
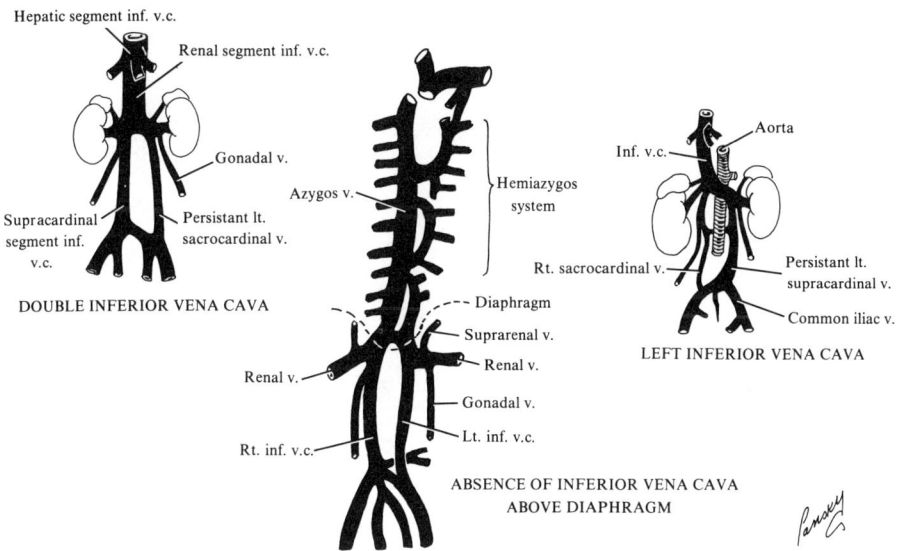

FIGURE 58. **Abnormal development of the superior and inferior venae cavae and pulmonary venous drainage.**

132. MALFORMATIONS OF THE HEART AND GREAT VESSELS

I. **Malformations of the heart and great vessels** are common because the development of the structures is very complex. The incidence is about 0.7% of live births and 2.7% of stillbirths. Only the major conditions are described and outlined here
   A. DEXTROCARDIA (displacement of the heart to the right): if the heart tube bends to the left instead of the right, there is transposition, in which the heart and its vessels are reversed left to right, as in a mirror image. It is uncommon, yet is the most frequent of positional abnormalities
      1. Isolated dextrocardia: abnormal heart positioning without other visceral displacement; however, one may see other heart defects
      2. Dextrocardia with situs inversus includes transposition of the viscera, such as the liver being on the left side and the heart on the right. Few other heart defects accompany
   B. ECTOPIA CORDIS: very rare; heart lies partly or completely outside the body
      1. Extrathoracic ectopia cordis: most common; heart protrudes through a sternal defect caused by failure of the lateral folds to fuse in the thorax during week 4 of development. Other congenital defects also are seen. Survival rate is low
      2. Incomplete ectopia cordis: the heart lies within the body but sternal and mediastinal defects cause abnormal positioning of the heart in the chest
   C. ATRIAL SEPTAL DEFECTS are one of the most common of congenital defects. The clinically significant atrial defects are discussed here
      1. Patent foramen ovale (secundum type): most common congenital heart defect; includes both the septum primum and septum secundum defects. Usually results from an abnormal resorption of the septum primum (foramen ovale valve) during formation of the septum secundum
         a. If resorption occurs in an abnormal location, the septum primum is fenestrated
         b. If excessive resorption of the septum primum occurs, there is a short septum which fails to cover the foramen ovale
         c. If a large foramen ovale results from a defective development of the septum secundum, the normal septum primum will not close the foramen ovale at birth
         d. Large atrial septal defects may result from excessive resorption of the septum primum and a large foramen ovale
         e. A patent foramen ovale without atrial septal defects may be due to abnormal intra-atrial pressure after birth
      2. Endocardial cushion defect with primary type of atrial septal defect: relatively common. The septum primum fails to fuse with the endocardial cushions, leaving a *patent foramen primum*
         a. In the complete type: seen in 20% of people with Down's syndrome, but otherwise is uncommon; fusion of the cushions does not occur, resulting in a large hole in the center of the heart (atrioventricularis communis or persistent atrioventricular canal). This condition is usually accompanied by a patent foramen primum and a ventricular septal defect
      3. Sinus venosus type of atrial septal defect: very rare; seen high in the interatrial septum, peripheral to the fossa ovalis. It is due to incomplete absorption of the sinus venosus into the right atrium and failure or abnormal development of the septum secundum
         a. Associated with partial anomalous pulmonary connections
      4. Common or single atrium: very rare; no interatrial septum. Due to failure of the septum primum and septum secundum to develop

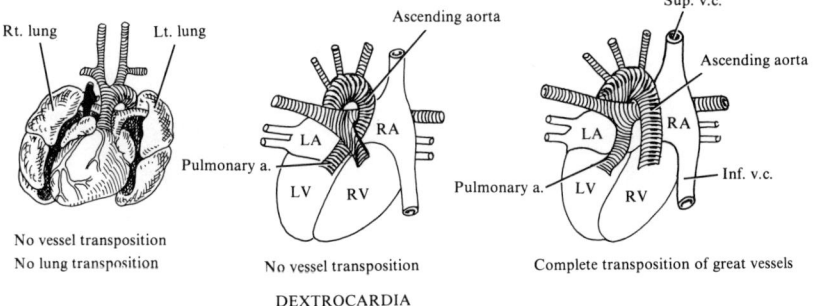

DEXTROCARDIA

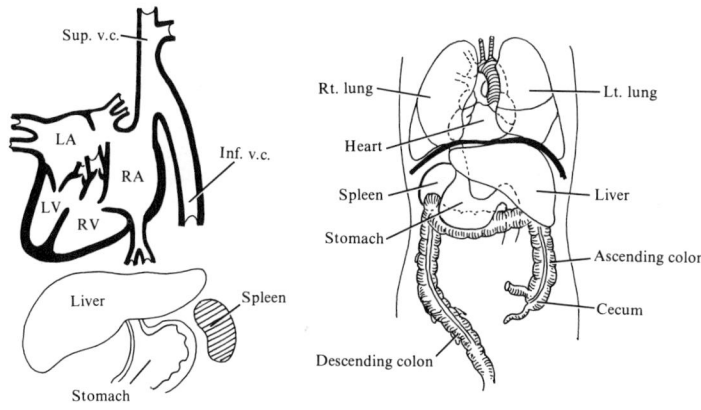

DEXTROVERSION OF SITUS SOLITUS OF HEART WITHOUT VISCERAL INVERSION

SITUS INVERSUS VISCERUM TOTALIS

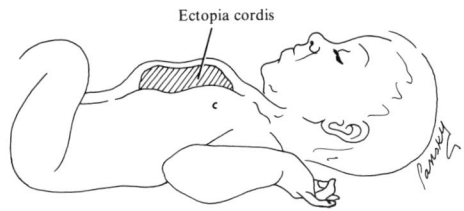

THORACIC ECTOPIA WITH ABSENCE OF STERNUM

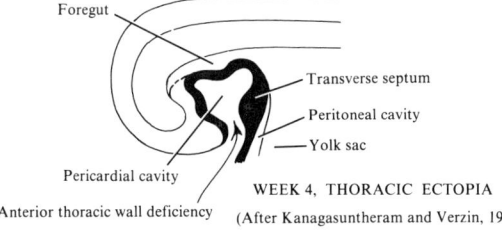

WEEK 4, THORACIC ECTOPIA
(After Kanagasuntheram and Verzin, 1962)

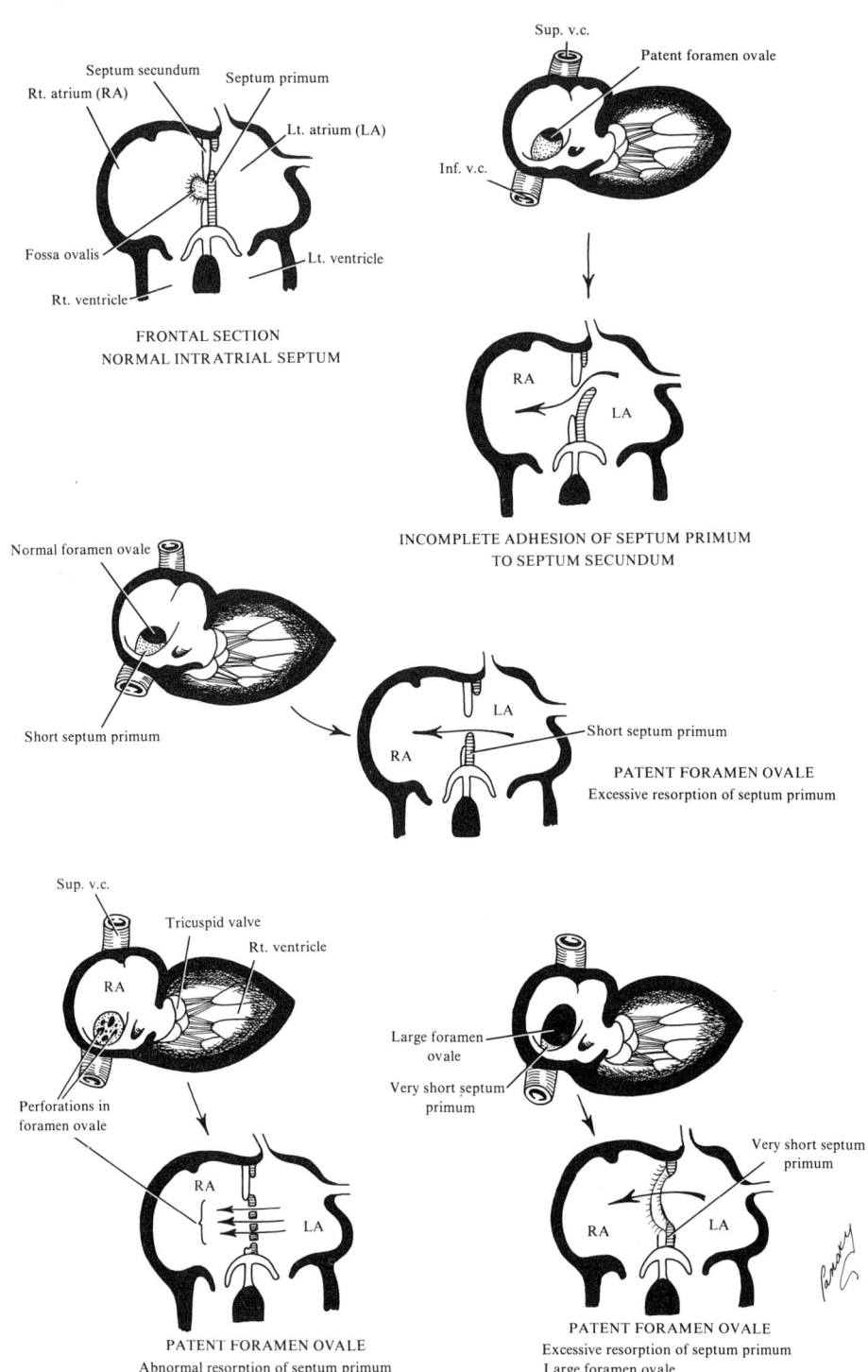

FIGURE 59. **Interatrial septal abnormalities.**

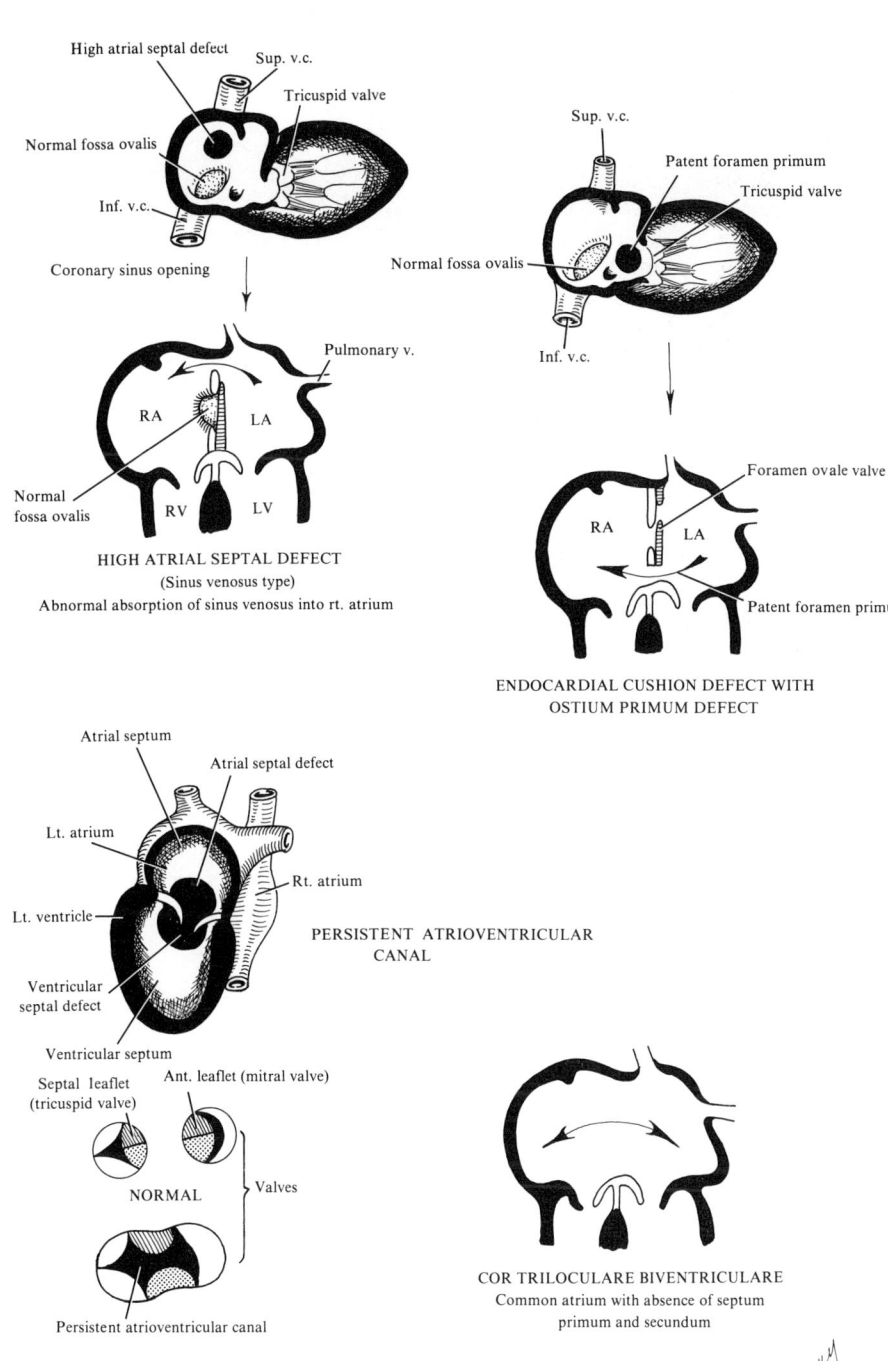

FIGURE 60. **Interatrial septal abnormalities.**

133. MALFORMATIONS OF THE HEART AND GREAT VESSELS

I. Malformations of the heart and vessels (cont.)

D. ANOMALIES OF THE INTERVENTRICULAR SEPTUM: common; isolated ventricular defects are detected in about 10 to 12/10,000 infants between birth and the age of 5 years
  1. High membranous type: most common. Incomplete closure of the interventricular foramen and failure of the membranous septum to develop due to failure of extensions of subendocardial tissue to grow from the right side of the fused endocardial cushions and fuse with the aorticopulmonary septum and muscular part of the interventricular septum
  2. Low, muscular type: less common; seen anywhere in the muscle portion of the septum (single or multiple). Probably a result of excessive resorption of myocardial tissue during the formation of the muscular part of the septum
  3. Single ventricle (absence of interventricular septum): very rare; results in a 3-chamber heart (cor triloculare biatriatum)

E. ANOMALIES OF SEPTATION OF THE BULBUS CORDIS
  1. Persistent truncus arteriosus: failure of the aorticopulmonary septum to develop and divide the truncus into an aorta and pulmonary trunk. Also may see defective fusion of the bulbar ridges. Most commonly see 1 vessel giving rise to an aorta pulmonary trunk, but other variations are possible
  2. Aorticopulmonary septal defect (or fistula): very rare; consists of a round or oval opening between the aorta and pulmonary arterial trunk near the aortic valve. Due to localized defect in the aorticopulmonary septum
  3. Unequal division of the truncus arteriosus: unequal partitioning of the truncus above the valves, resulting in 1 great artery being large and the other small or stenotic. The aorticopulmonary septum may not be aligned with the interventricular septum, and one sees ventricular septal defects. The larger of the 2 vessels usually overrides the septal defect

F. COMPLETE TRANSPOSITION OF THE GREAT VESSELS
  1. Typically, the aorta lies anterior to the pulmonary trunk and arises from the right ventricle, and the pulmonary trunk arises from the left ventricle
  2. For survival, there is an associated septal defect or patent ductus arteriosus to allow for interchange between pulmonary and systemic blood
  3. Its possible cause: during partitioning of the bulbus cordis and truncus arteriosus, the aorticopulmonary septum fails to follow a spiral course
     a. The straight septum results from failure of the conus arteriosus or pulmonary infundibulum to develop during incorporation of the bulbus into the ventricles

G. PULMONARY VALVE STENOSIS: the pulmonary valve cusps are fused together to form a narrow central opening
  1. Infundibular pulmonary stenosis: one sees an underdeveloped infundibulum in the right ventricle
  2. The degree of right ventricular hypertrophy depends on the degree of obstruction of the blood flow through the pulmonary trunk

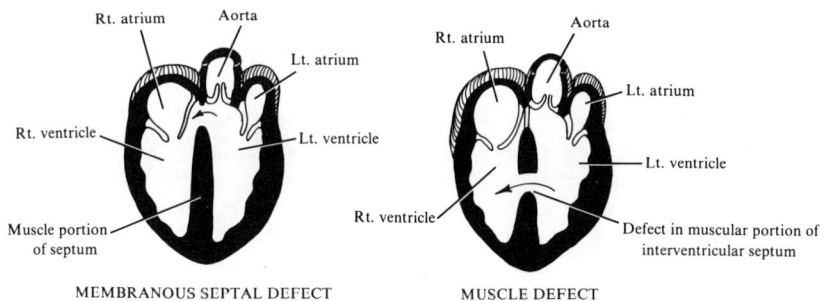
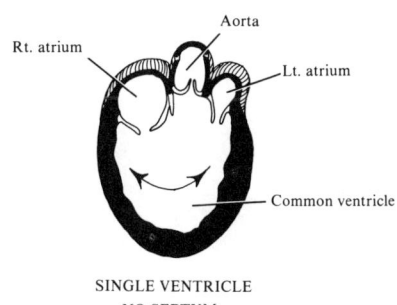
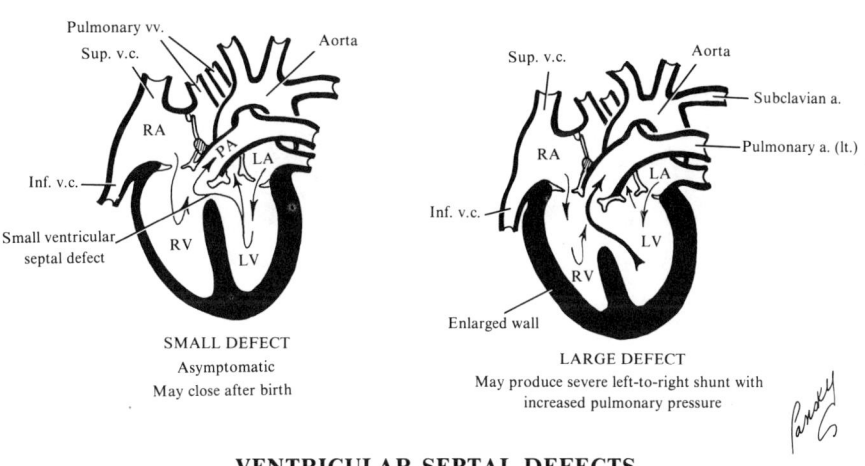

VENTRICULAR SEPTAL DEFECTS
(After Edwards et al., 1957 and 1965)

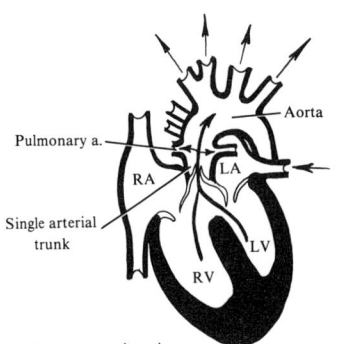

Pulmonary arteries arise
from lateral walls of undivided
arterial trunk

PERSISTANT TRUNCUS ARTERIOSUS

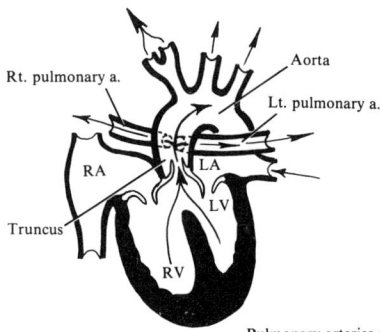

Pulmonary arteries arise close together
from posterior wall of truncus

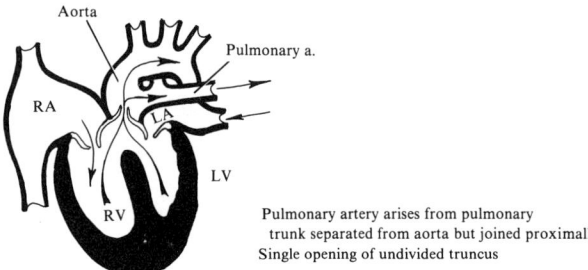

Pulmonary artery arises from pulmonary
trunk separated from aorta but joined proximally
Single opening of undivided truncus

INTERVENTICULAR SEPTAL DEFECT

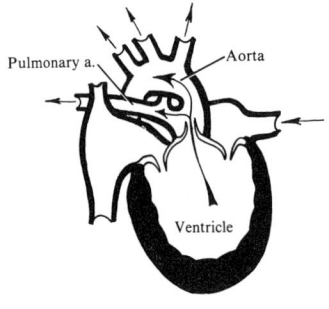

SINGLE VENTRICLE AND RIGHT AORTIC ARCH
Pulmonary artery arises from pulmonary trunk separated
from aorta but joined proximally
Single opening of undivided truncus

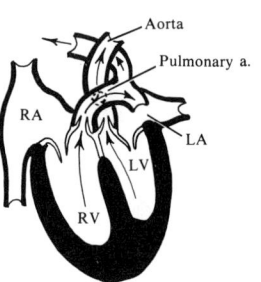

AORTICUPULMONARY SEPTAL DEFECT
Partial persistent truncus arteriosus

FAILURE OF SEPARATION OF AORTIC AND PULMONARY TRUNKS

FIGURE 61. **Failure of separation of aortic and pulmonary trunks.**

TRANSPOSITION OF GREAT VESSELS

(After Kanjuh and Edwards, 1964; Edwards *et. al.* 1965)

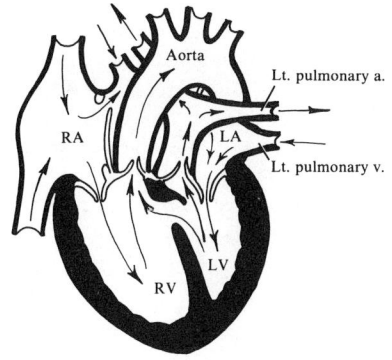

VENTRICULAR SEPTAL DEFECT AND
SUBPULMONARY STENOSIS WITH PATENT
FORAMEN OVALE

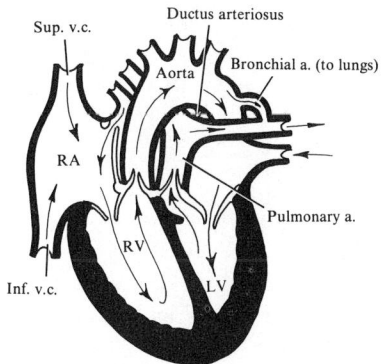

INTACT VENTRICULAR SEPTUM AND CLOSED
DUCTUS ARTERIOSUS (Interatrial septal defect)

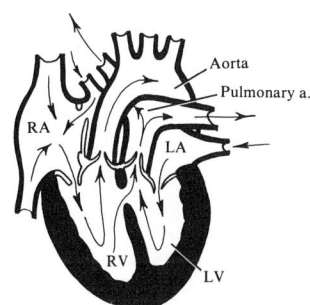

VENTRICULAR AND ATRIAL SEPTAL DEFECTS

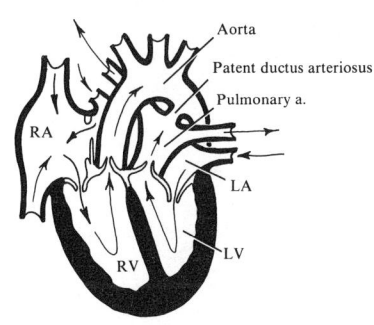

INTACT INTERVENTRICULAR SEPTUM
ATRIAL SEPTAL DEFECT
PATENT DUCTUS ARTERIOSUS

PULMONARY STENOSIS

With intact ventricular septum
and patent foramen ovale
(After Edwards, 1964)

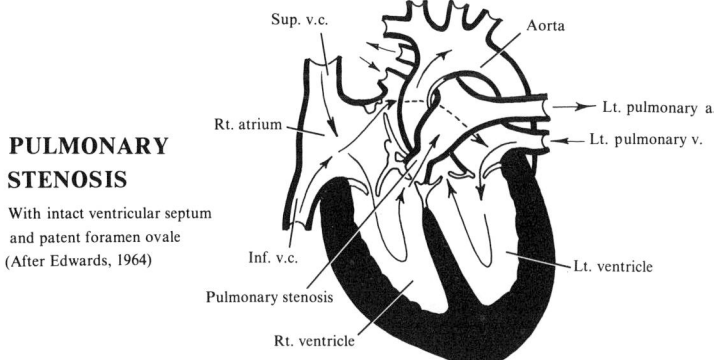

FIGURE 62. **Transposition of the great vessels and pulmonary stenosis.**

134. COMPLEX CARDIAC MALFORMATIONS

I. **Complex cardiac malformations** involve more than one anatomic defect. Surgical treatment does not always correct the anatomic lesion, and the methods available may only be palliative with the hope of improving oxygenation
  A. TETRALOGY OF FALLOT is a classic and common group of cardiac defects thought to result from underdevelopment of the pulmonary infundibulum. The basic defect in this cardiac malformation is an unequal division of the conus arteriosus due to an anterior displacement of the aorticopulmonary septum. This results in
    1. Pulmonary stenosis or narrowing of the region of the right ventricular outflow (infundibular stenosis)
    2. Overriding aorta: the aorta arises directly above the septal defect from both ventricular cavities
    3. Hypertrophy of the right ventricle: result of high pressure on the right side
    4. Ventricular septal defect
    5. This abnormality is regarded as the most important type of malformation causing cyanosis, but compatible with life, and as a result of the tetralogy there is *severe dyspnea* (primarily) and often *moderate cyanosis* (secondarily). Children have a tendency to assume a crouching position in an effort to get better oxygenation
  B. PULMONARY ATRESIA: division of the truncus arteriosus is unequal and the pulmonary trunk has no lumen. Thus, there is no orifice at the level of the pulmonary valve. There is often an associated ventricular septal defect
  C. AORTIC STENOSIS AND ATRESIA
    1. Aortic valvular stenosis: edges of the valve are usually fused to form a dome with a narrow opening (sometimes pinpoint). Often, only 2 valves fuse and we have an abnormal bicuspid valve. The size of the aorta itself is usually normal
      a. Subaortic stenosis: a ring of fibrous tissue is found circling the outflow tract of the left ventricle just below the aortic valves
    2. Aortic valvular atresia: fusion of the semilunar aortic valves is complete, resulting in the absence of an aortic orifice. This results in
      a. Intense cyanosis
      b. Atrophy of the left ventricle
      c. Hypertrophy of the right ventricle
      d. Perfusion of the coronary arteries by the aorta
      e. Survival is possible only if it is associated with an interatrial connection, an interventricular connection, or a patent ductus arteriosus
  D. TRICUSPID ATRESIA (absence of the right atrioventricular orifice) results in
    1. Intense cyanosis
    2. Atrophy of the right ventricle
    3. Hypertrophy of the left ventricle
    4. Survival is possible only if associated with interatrial communication, interventricular communication, or a patent ductus arteriosus

II. **Abnormalities of the lymphatic system:** congenital malformations are rare
  A. THERE MAY BE A DIFFUSE SWELLING of a portion of the body or an extremity due to dilatation of the primitive lymphatic channels
    1. More rarely, a diffuse cystic dilatation of channels over the entire body
  B. CYSTIC LYMPHANGIOMA OR HYGROMA: swellings in the lower third of the neck
    1. Large single or multilocular fluid-filled cavities due to a failure of the lymphatic channels to communicate, or a pinching off of the jugular lymph sac

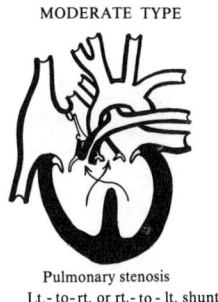
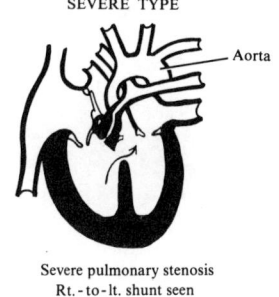

TETRALOGY OF FALLOT

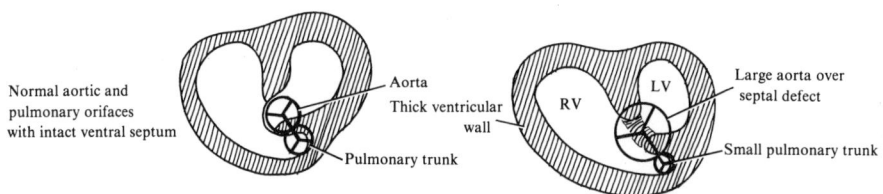

RELATIONSHIP OF AORTA AND PULMONARY ORIFACES TO VENTRICULAR SEPTAL DEFECT

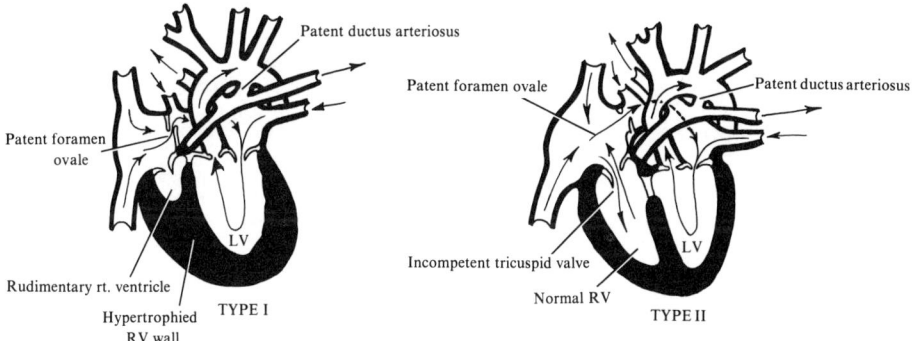

PULMONARY ATRESIA
(After Edwards, 1960)

VALVULAR ATRESIA AND STENOSIS

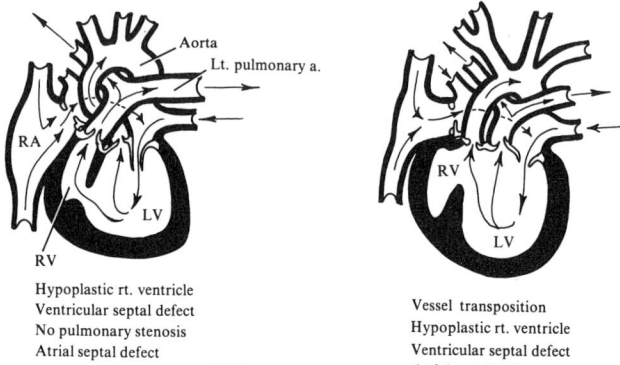

TRICUSPID ATRESIA
Hypoplastic rt. ventricle
Ventricular septal defect
No pulmonary stenosis
Atrial septal defect

Vessel transposition
Hypoplastic rt. ventricle
Ventricular septal defect
Atrial septal defect

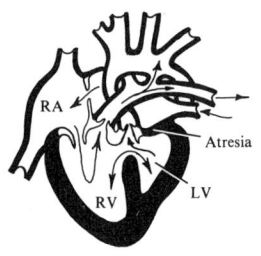

MITRAL ATRESIA
Aortic stenosis
Hypoplastic lt. ventricle
Ventricular septal defect
Atrial septal defect

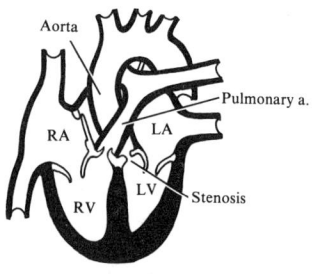

AORTIC VALVULAR STENOSIS

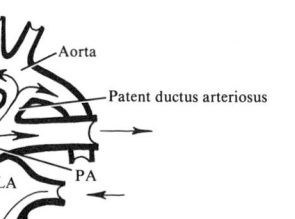

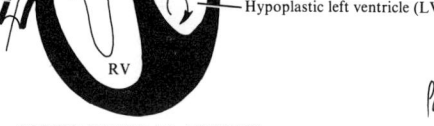

AORTIC VALVULAR STENOSIS

FIGURE 63. **Valvular atresia and stenosis.**

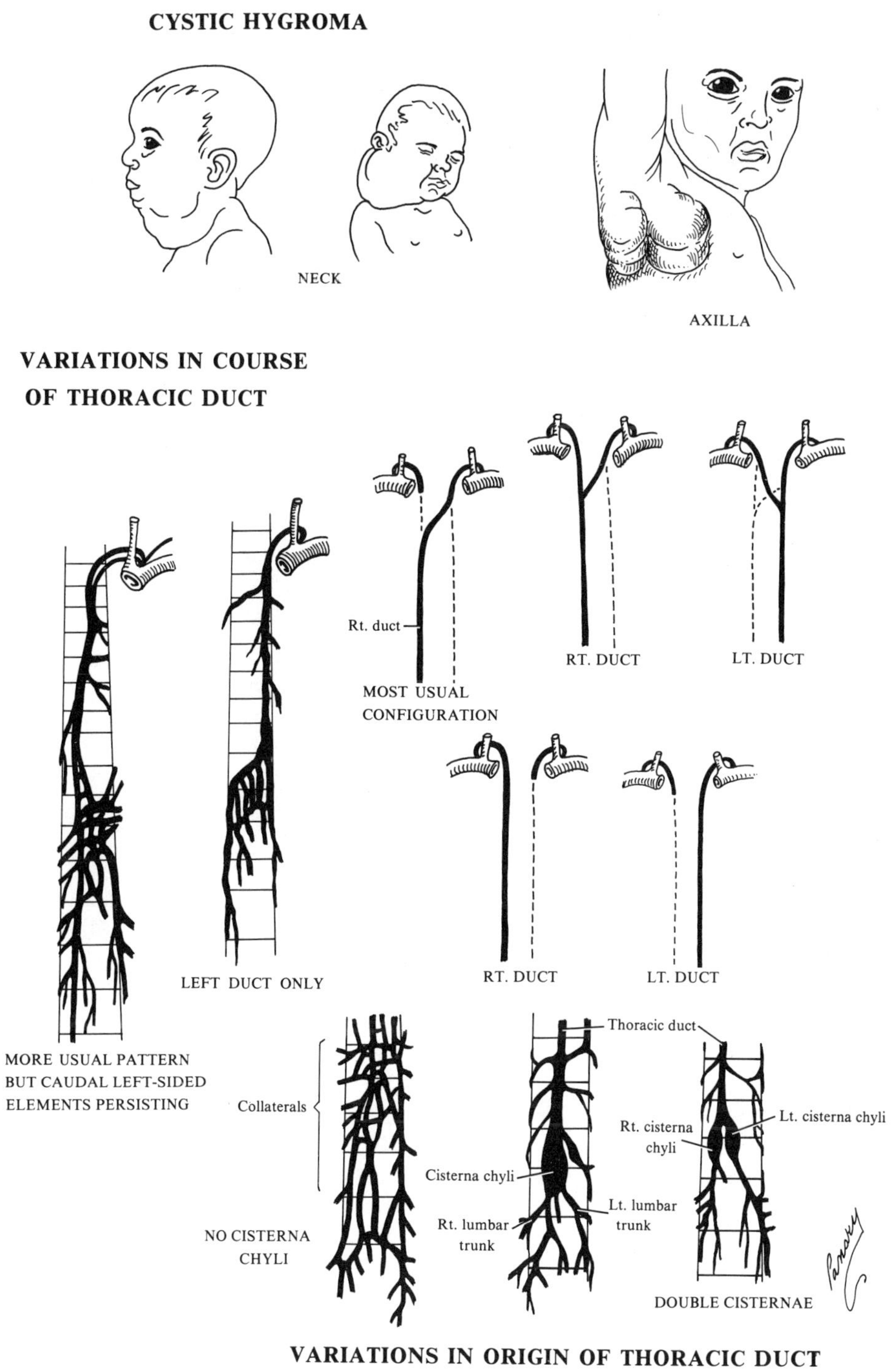

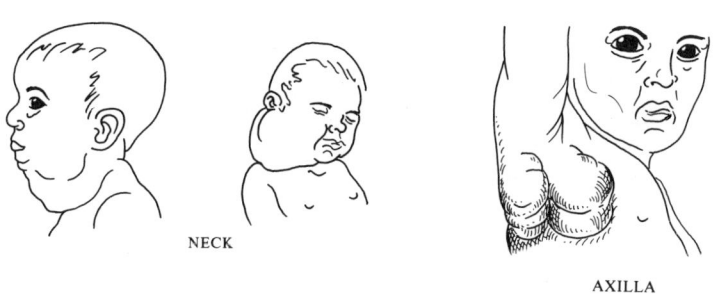

FIGURE 64. **Abnormalities in the development of the lymphatic system.**

UNIT ELEVEN

THE NERVOUS SYSTEM

135. EARLY DEVELOPMENT OF THE NERVOUS SYSTEM

I. **General considerations**
A. THE NERVOUS SYSTEM, AS A WHOLE, including the spinal cord, brain, and peripheral nerves, is derived from *ectoderm*
   1. The primordial structure, which gives rise to the nervous system or *neural ectoderm,* appears very early in development, around day 17. It develops from the ectoderm in the dorsomedian region of the embryo, above the mesoderm, and cephalic to Hensen's node
      a. The mesoderm, formed during gastrulation, induces formation of the neural ectoderm from the overlying ectoderm
   2. During development, the neural ectoderm and the remaining ectoderm separate. The latter forms the surface ectoderm, which gives origin to the epidermis and certain sense organs

II. **Neurulation** is the transformation of the ectoderm overlying the notochord into a *neural tube,* which is flanked by 2 longitudinal formations, the *neural crests*
A. INDUCTION OF THE NERVOUS SYSTEM
   1. The various phases of neurulation are induced by the notochord and the parachordal mesoderm (inductors) from the overlying competent ectoderm. Induction is also essential in the subsequent development of the nervous system
   2. The activity of the inductors is not uniform, and there is a *cephalocaudal gradient*
      a. Thus, the trunk areas of the notochord and parachordal mesoderm induce the formation of the spinal cord, while their extremities are inducing the middle and posterior parts of the brain (the mesencephalon and rhombencephalon)
      b. The prosencephalon, on the other hand, is induced by the *prochordal plate,* a small circular area of columnar entodermal cells found in front of the notochord
         i. The prochordal plate is an entodermal structure firmly attached to the overlying ectoderm, forming the oropharyngeal or buccopharyngeal membrane
         ii. The prochordal plate indicates the site of the mouth and serves as an organizer of the head region, giving rise to mesenchyme in the head region and to the entodermal layer of the oropharyngeal membrane
   3. During its development, the neural tube, in turn, induces the formation of the posterior arch of the vertebrae and of the cranial vault
      a. It also plays an important role in the development of the face, the eye, and the nose
   4. Since either induction or competence can be defective at each stage of the developmental process, many abnormalities can occur as a result of such a complex mechanism of development

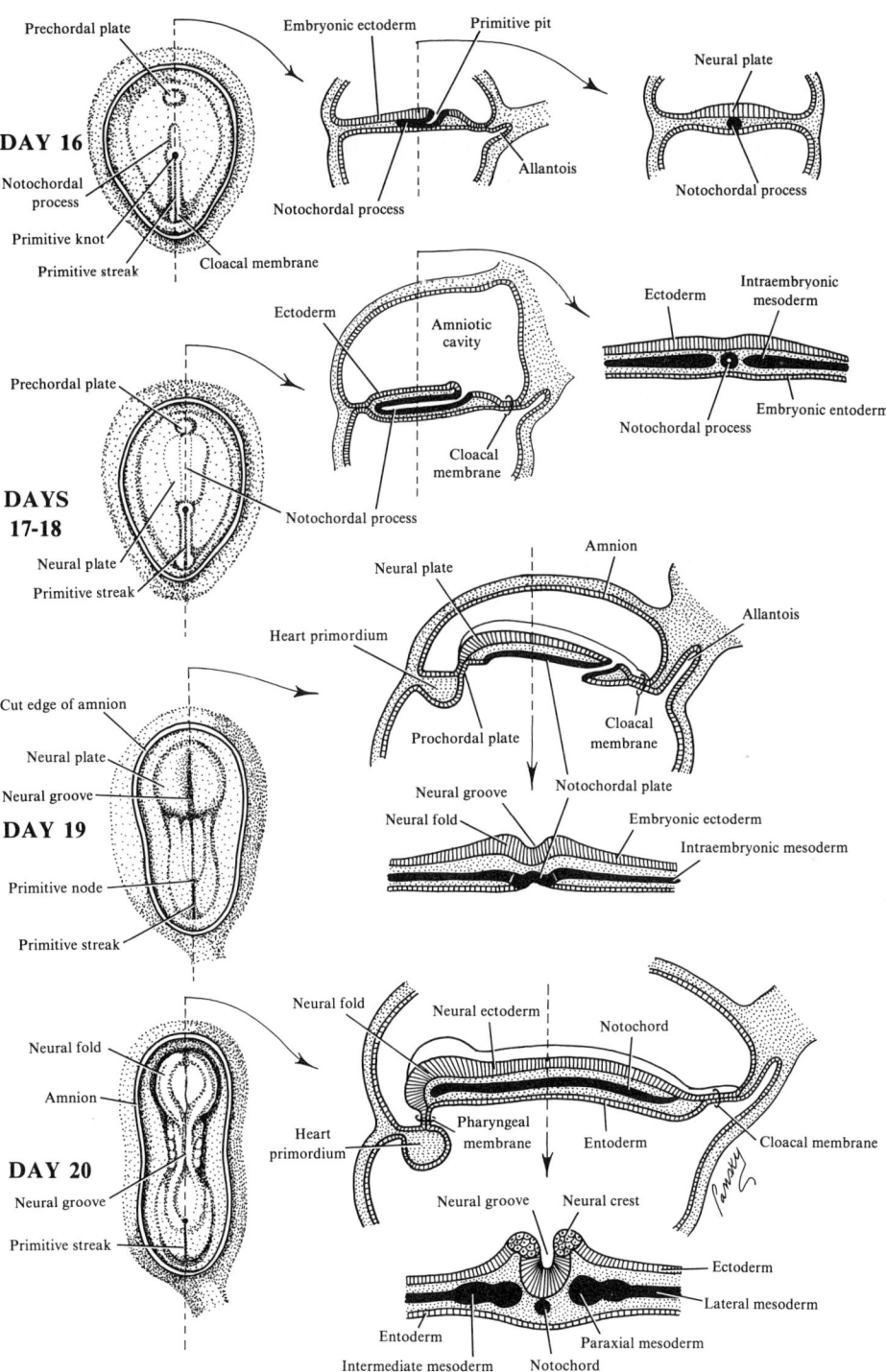

136. EARLY NERVOUS SYSTEM DEVELOPMENT: THE NEURAL TUBE AND NEURAL CREST

I. **Formation of the neural tube** consists of 3 successive stages of development
A. THE NEURAL PLATE results from thickening and differentiation of ectoderm overlying the notochord. The ectoderm of the plate is called neuroectoderm and will give rise to the central nervous system, consisting of the brain and spinal cord
   1. The plate is wider at its cephalic end than at its caudal end and is formed, in humans, before the appearance of the first somite or about day 18 of development
   2. The ectoderm remains relatively thin at the periphery
   3. The plate is first seen cranial to the primitive knot and dorsal to the notochordal process. As the notochordal process elongates, the plate broadens and eventually extends cranially to the oropharyngeal membrane
   4. The notochordal process and paraxial mesoderm act as primary inductors for the differentiation of the embryonic ectoderm into the neural plate
B. THE NEURAL GROOVE AND NEURAL FOLDS: the lateral edges of the neural plate elevate to form the neural folds. The neural groove develops as a result of invagination of the neural plate along its central axis, between the neural folds (about day 18)
C. The neural tube: the neural folds move together, fuse by the end of week 3, and convert the plate into a tube (also called the neurula stage)
   1. The ectoderm thus is separated from the nervous tissue to form the surface ectoderm. Between the 2, mesenchymal cells gradually infiltrate

II. **The 3 stages (plate, groove and folds, and tube)** all coexist at the same time in different regions of the embryo
A. CLOSURE OF THE NEURAL GROOVE, at about day 21, in humans, takes place near the fourth somite (future cervical region), which is in the middle portion of the embryo, and progresses toward both its cephalic and caudal ends

III. **Two orifices at the ends of the neural tube** persist for a relatively short period of time and are called the *anterior* and *posterior neuropores*. They temporarily form open connections between the neural tube lumen and the amniotic cavity
A. THE ANTERIOR NEUROPORE indicates the area of the lamina terminalis and closes about day 25 or the 18-to-20-somite stage
B. THE POSTERIOR NEUROPORE closes at about day 28 or the 25-somite stage

IV. **The neural crests:** when the neural folds fuse and, before the neural groove closes completely, some neuroectodermal cells detach themselves from the region where the neural tube borders on the ectoderm (lateral edges of the neural plate). These groups of cells are not incorporated into the neural tube, but form a neural crest over the tube
A. THE CRESTS are distinct from both the neural tube as well as the overlying ectoderm after the neural groove is closed. They form longitudinal tracts which extend from the middle brain regions of the neural tube to its caudal end
B. THE NEURAL CRESTS soon fragment into segments that give rise to the *primordia of the ganglia*. The segmentation corresponds to that of the somites, with each ganglionic primordium corresponding to a muscle primordium. The ganglia later give rise to general sensory innervation, and the corresponding level of the spinal cord gives motor innervation to the muscles
   1. Derivatives of the neural crests
      a. Ectomesenchyme is the result of mesenchymal differentiation of neuroblastic cells and is therefore of ectodermal origin. It gives rise to the dermis of the head, the meninges, the skeleton, and perhaps the musculature of the branchial arches
      b. Cells of the spinal ganglia, ganglia of the sympathetic and parasympathetic systems, and adrenal medulla
      c. Schwann cells that form the myelin sheaths of peripheral nerves
      d. Pigment cells, particularly of the skin

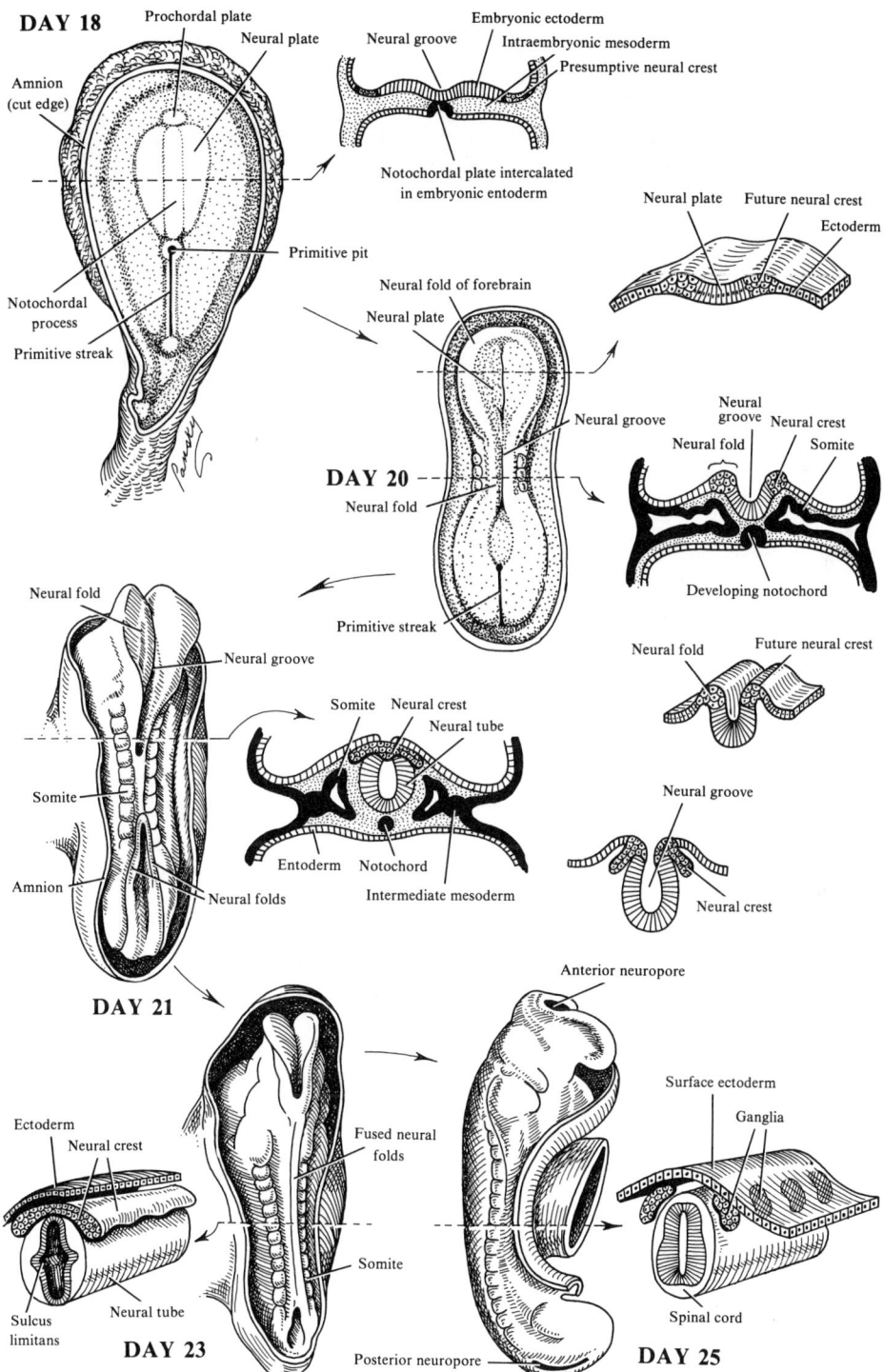

137. GENERAL DEVELOPMENT OF THE CENTRAL NERVOUS SYSTEM

I. **The somites:** alongside the developing neural tube are strips of mesoderm which begin to show segmentation into somites near the end of week 3 of development
   A. THE FIRST SOMITE appears behind the cephalic tip of the notochord
   B. SUCCESSIVE SOMITES appear in a craniocaudal sequence, and approximately 41 to 44 pairs of somites are present by day 31 of human development
   C. MOST OF THE AXIAL SKELETON AND MUSCULATURE develop from the somites (see the units on skeletal and muscular development)

II. **Divisions of the neural tube:** the cephalic (head) end of the neural tube is larger than its caudal (tail) end, and even before the neural tube is closed cephalically, the cephalic end shows 3 distinct dilatations, the *primary brain vesicles*
   A. THE PROSENCEPHALON OR FOREBRAIN is the most anterior vesicle
   B. THE MESENCEPHALON OR MIDBRAIN is the central vesicle
   C. THE RHOMBENCEPHALON OR HINDBRAIN is the most posterior vesicle

III. **Unequal growth rates and cell migration** result in flexures, constrictions, thickenings, invaginations, and evaginations
   A. NEURULATION contributes to the cephalocaudal flexion of the embryo. The extensive proliferation of the nervous tissue causes curvature of the embryo on its long axis. Progressive dorsal flexion results in raising and isolating the embryo from its membranes. With the appearance of the vesicles (about a 5-mm embryo stage), the neural tube bends ventrally to form 2 flexures
      1. A *cervical flexure:* at the junction of the spinal cord and hindbrain
      2. A *cephalic flexure:* in the midbrain
      3. Later, between the above 2 major flexures, unequal growth in the hindbrain produces the *pontine flexure,* in the opposite direction

IV. **Five components in the developing brain at week 5**
   A. THE PROSENCEPHALON OR FOREBRAIN now consists of 2 parts
      1. An anterior *telencephalon* or *endbrain* which will give rise to 2 anterolateral expansions called the *primitive cerebral hemispheres*
      2. An *intermediate brain* or *diencephalon,* characterized by the outgrowth of the optic vesicles
         a. Dorsal and ventral evaginations from the diencephalon become the primordia of the pineal gland and the posterior hypophysis, respectively
   B. THE MESENCEPHALON OR MIDBRAIN undergoes little change by this age
   C. THE RHOMBENCEPHALON consists of 2 parts
      1. The anterior *metencephalon,* which later forms the pons and the cerebellum
      2. The posterior *myelencephalon* which later forms the medulla oblongata
      3. The boundary between the metencephalon and myelencephalon is marked by the third or pontine flexure, compensatory to the cervical and cephalic flexures

V. **The lumen of the spinal cord, *the central canal*** is continuous with the brain vesicles, permitting cerebrospinal fluid to circulate freely between the brain and cord
   A. THE CAVITY OF THE RHOMBENCEPHALON is the fourth ventricle
   B. THE CAVITY OF THE DIENCEPHALON is the third ventricle
   C. THE CAVITIES OF THE CEREBRAL HEMISPHERES are the lateral ventricles
   D. THE LUMEN BETWEEN THE THIRD AND FOURTH VENTRICLES is the narrow aqueduct of Sylvius
   E. THE LATERAL VENTRICLES communicate with the third ventricle via the foramina of Monro
   F. THE FOURTH VENTRICLE opens into the subarachnoid space via the foramina of Luschka (2) and Magendie (1)

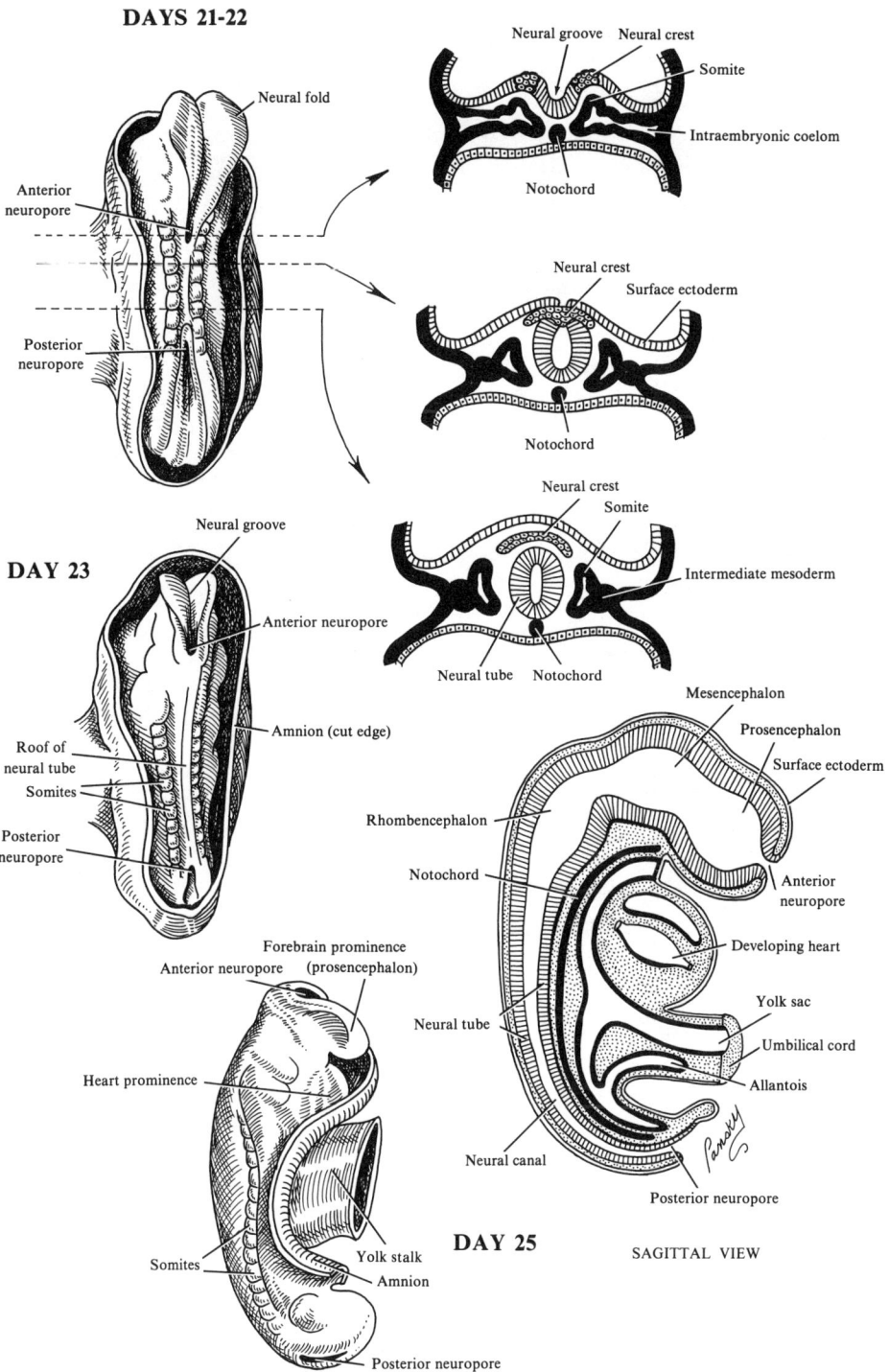

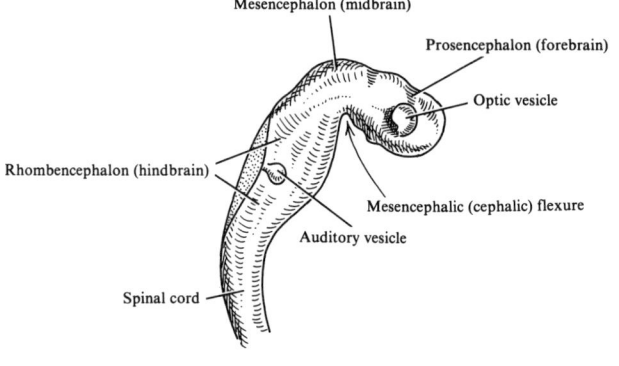

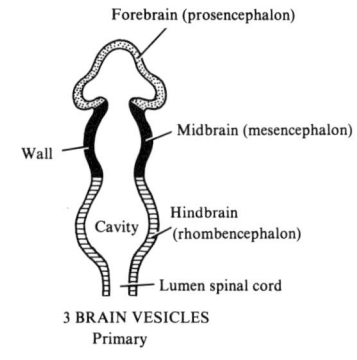

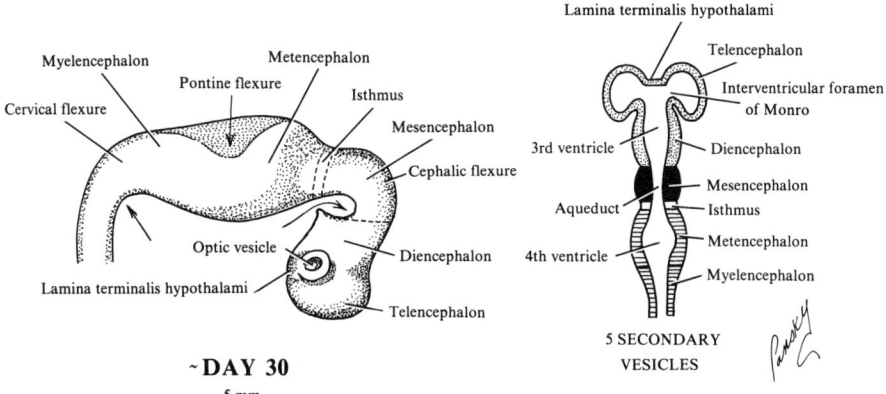

FIGURE 65. **Primary brain vesicles and neural tube flexures: days 23–30.**

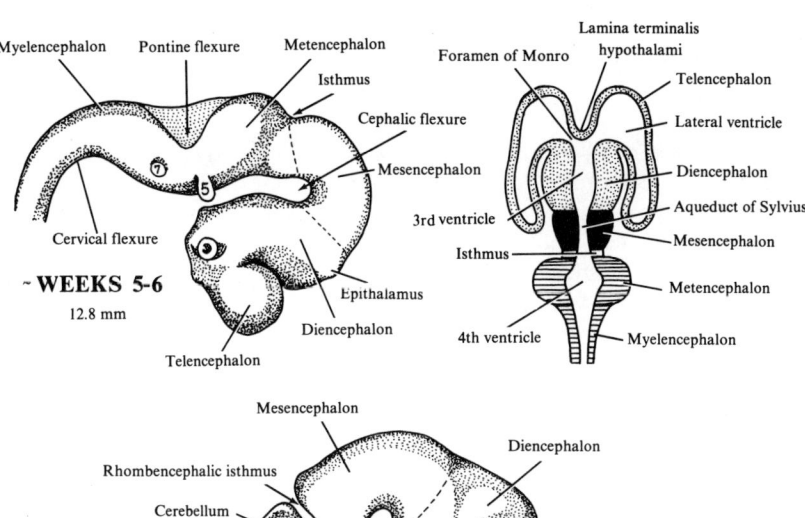
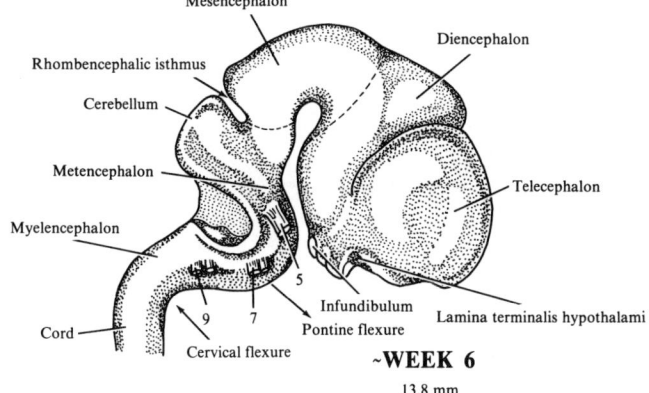
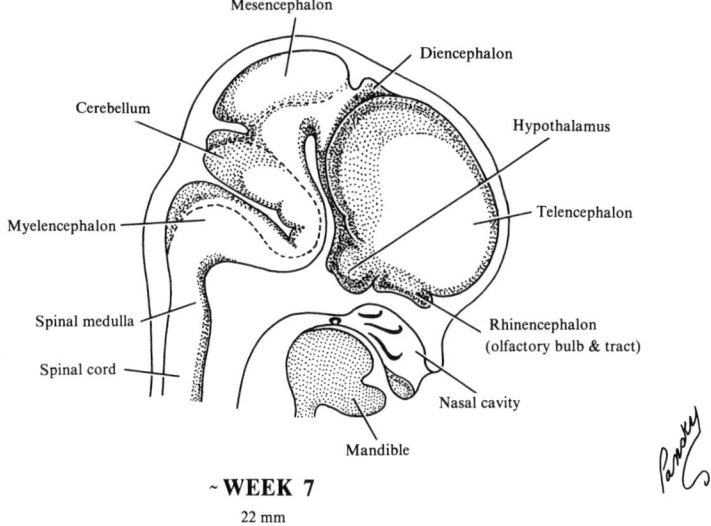

FIGURE 66. **Major components of the developing brain: weeks 5-7.**

138. PHYLOGENESIS OF THE NERVOUS SYSTEM

I. **Introduction:** vertebrates are distinguished from other groups of animals by certain characteristics, principally related to their spinal cords and nervous systems, namely, the presence of a *dorsal notochord* surrounded by mesodermal tissue, which gives rise to a vertebral column in adults of most species, and a *tubular nervous system* lying dorsal to the cord, concentrated in the midline, and very well developed at its cephalic end (cephalization). The extent and pathways of nervous system development differs between invertebrates and vertebrates, even though, in all animals that have a nervous system, it is derived predominantly from ectoderm

II. **Comparative embryology and anatomy** have demonstrated various phases of the nervous system evolution. Specific aspects of evolutionary differences are
   A. PROGRESSIVE CONCENTRATION of the nervous system and beginning of cephalization
   B. THE APPEARANCE AND CHANGES IN SEGMENTATION seen in lower and superior annelids
   C. THE APPEARANCE OF A NOTOCHORD AND NERVOUS SYSTEM in vertebrates

III. **Invertebrate nervous system development:** before the evolutionary appearance of the spinal cord, neural organization apparently depends on *sensory surface ectoderm*
   A. IN COELENTERATES (sea anemones, hydra, jellyfish) the cells are diffuse and the nervous system consists of a diffuse neural network
   B. IN PRIMITIVE WORMS (Convoluta, Platyhelminthes, or flatworms) the sensory cells are concentrated in certain regions and the nervous system has a parallel development. It approaches a central nervous system organization, but it is ventral and without a cavity
      1. At its anterior end, sensory concentration is clear, and there is the beginning of cephalization with cerebral ganglia
      2. Nervous concentration is greatest anterior, and strict parallelism is seen between the anterior concentration of sensory and nervous cells
   C. IN LOWER AND LARVAL FORMS OF ANNELIDS (Serula, Dinophilus) concentration has increased. Segmentation (metamerization) appears. The central nervous system is ventral (bilateral ventral cords), has no cavity and localization of sensory cells seems to be responsible for the segmental concentration which forms the metameric ganglia
   D. THE SUPERIOR AND ADULT ANNELIDS (nereis, Lumbricus): the central nervous system is still ventral (single cord and ganglia) without a cavity, and there is transverse concentration. The cerebral ganglia are specialized for vision and olfaction
   E. THE HEMICHORDATES (enteropneusts): a dorsal tubular and a solid ventral nervous system. There is no cord, but there is a cephalodorsal diverticulum of the intestine corresponding to a dorsal nervous tube. The diverticulum is incompletely separated from the overlying ectoderm from which it is derived
   F. THE PROCHORDATES (Chephalochordata, *Amphioxus*): a dorsal cord and a dorsal tubular central nervous system entirely developed from dorsal ectoderm. A ventral system is no longer seen. The anterior vesicle represents a true rudimentary brain

IV. **Vertebrate nervous system (Chordata):** mesoderm determines the development of the nervous system and sense organs. The appearance of mesoderm groups together the inductive capabilities and marks an important stage in the genesis of development
   A. THE NERVOUS SYSTEM is relatively independent of the sensory organization. However, the eye and general sensory receptors, are directly derived from the nervous system
      1. During evolution, relationships between the nervous system and the sense organs are modified. In invertebrates, the nervous system appears to be only an accessory structure to the sensory system. In vertebrates, the importance of the nervous system becomes so great that everything else is organized for it and around it
   B. IN VERTEBRATES SUCH AS FISH, the 5 neural vesicles presage the cephalization of the higher vertebrates. During the slow progression of evolution, the role of the prochordal plate becomes essential since, in humans, it induces brain formation

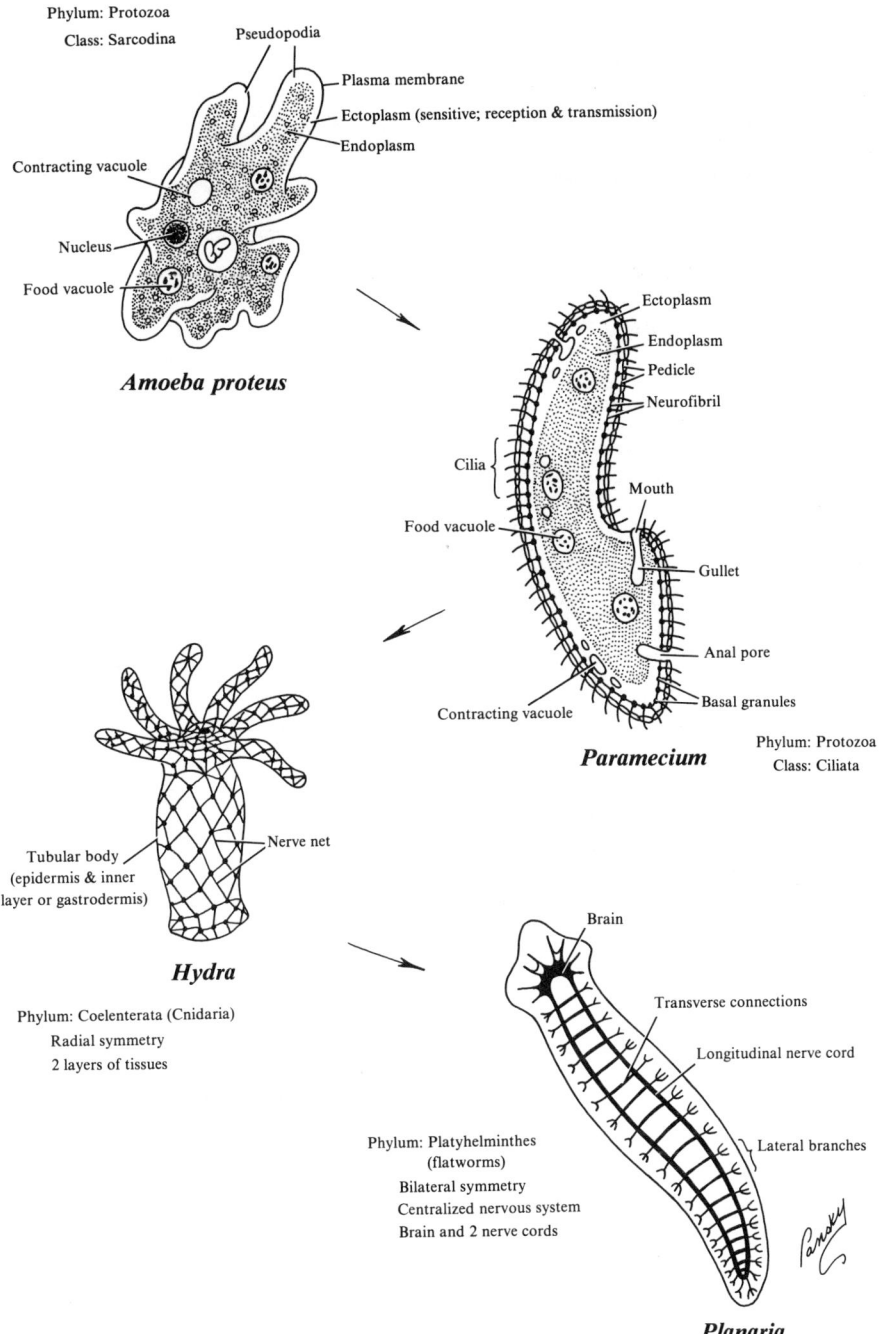

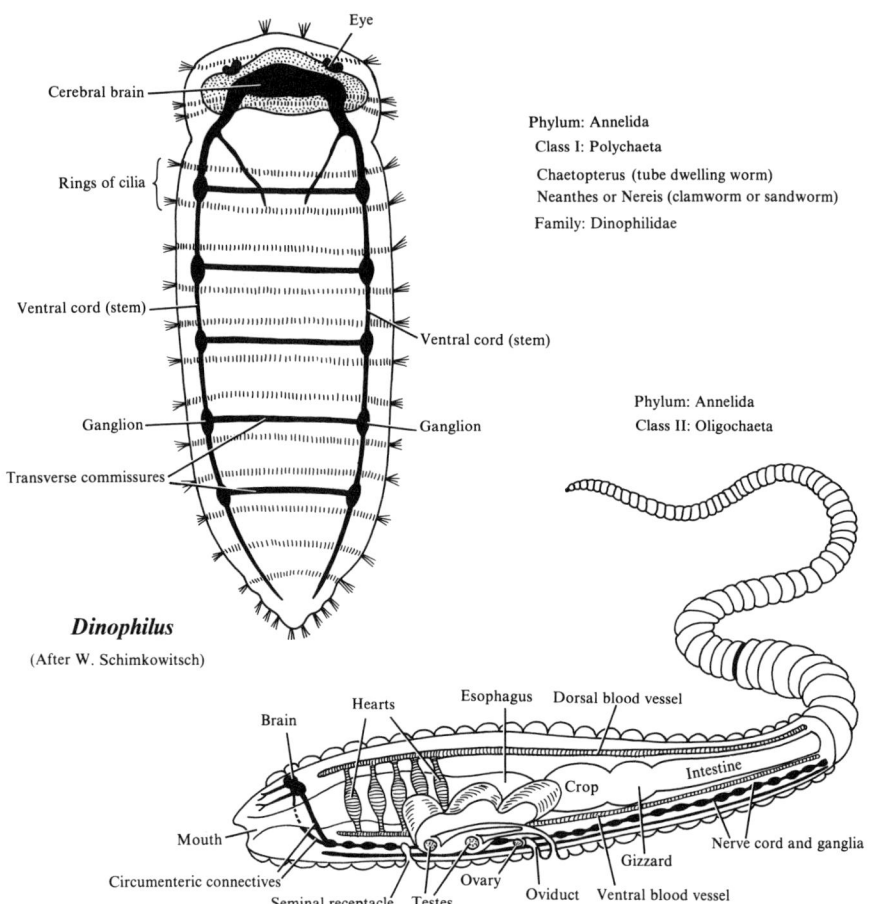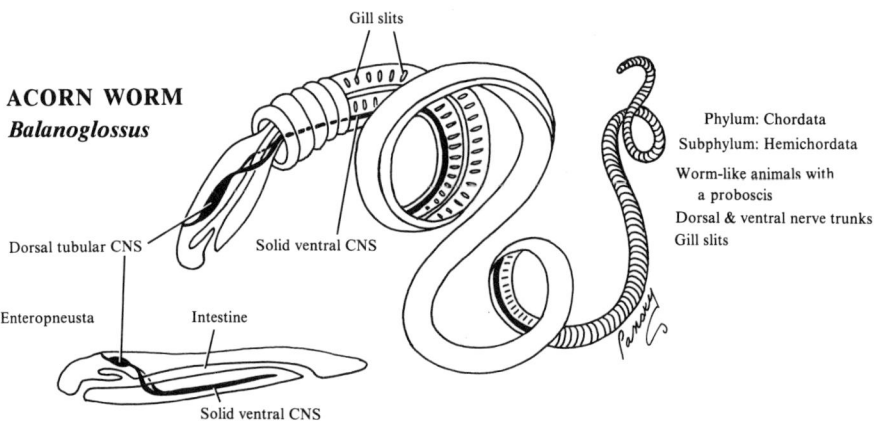

FIGURE 67. **Annelid and worm nervous systems (phyla Annelida and Chordata).**

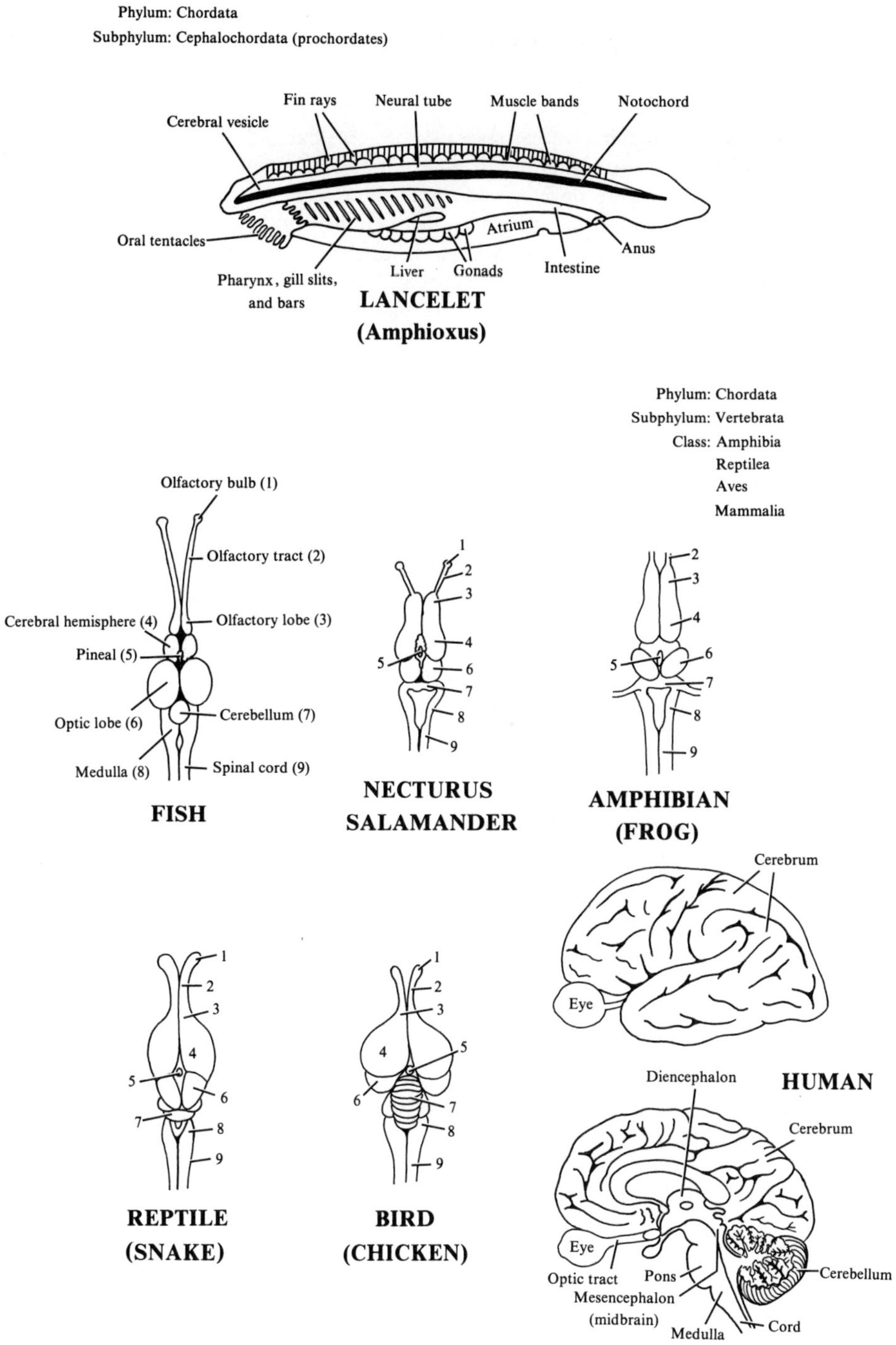

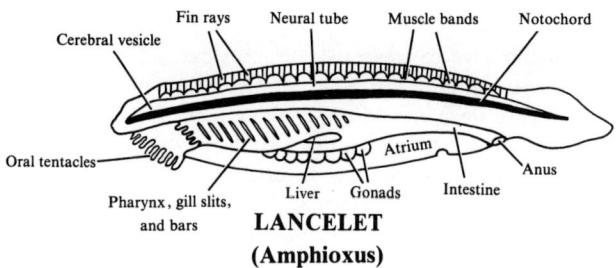

FIGURE 68. **Comparative anatomy of the chordate brain.**

139. METAMERIC ORGANIZATION OF THE NERVOUS SYSTEM

I. **Introduction:** each metamere* carries a pair of nerve ganglia, a cutaneous component, a group of muscles, and a number of other mesodermal derivatives, such as visceral, vascular, etc. Thus, there develops an anatomic and physiologic unit which is more advanced in organization than the diffuse structures of primitive invertebrates
   A. AS WE ASCEND THE PHYLOGENETIC SCALE of invertebrates, toward the insects, segmental independence is sacrificed for the welfare of the organism as a whole, and neighboring ganglionic pairs tend to fuse, and corresponding metameres interweave and coordinate their functions

II. **In vertebrates and humans** the neural tube and neural crest systems replace what was essentially the ganglionic system of the invertebrates. Metamerization, however, is still obvious, especially in the trunk region. A metamere consists of dermatomes, ganglia, neuromere, and myotomes
   A. METAMERIZATION manifests itself by the aggregation of paraxial mesoderm into somites
      1. The first somites make their appearance in the posterior part of the cephalic region, and segmentation, from that area, progresses toward the caudal end
   B. THE NEURAL CRESTS, in a parallel manner, break up and produce the ganglionic primordia which correspond, one to one, to the lateral somites
   C. THE CUTANEOUS AREAS opposite the somites contribute the *dermatomes*
   D. THE SPINAL CORD itself does not divide; however, the cord can be considered to be virtually segmented since a medullary level or neuromere corresponds functionally to each ganglion and somite level

III. **Medullary metamerization**
   A. EACH NEUROMERE receives sensory fibers from a specific cutaneous area, the *dermatome,* and sends out motor fibers to a specific group of muscles, the *myotome.* Afferent sensory fibers also pass to the viscera, and the response from the latter is made via the motor pathways of the autonomic nervous system
   B. THE VARIOUS ELEMENTS OF THE METAMERE and its organization are embryologically interdependent

IV. **Variations in segmentation:** in humans, and in mammals segmentation persists in the sensory pathways, but is less well defined in the motor pathways
   A. SENSORY PATHWAYS: the *dermatome* is a cutaneous zone which corresponds precisely to the sensory root that it innervates. There are a number of clinical applications related to this relationship
      1. In herpes zoster, the ganglion level affected can be determined from the pattern of the cutaneous lesion
      2. In cases of medullary compression by a tumor, the level of compression can be determined by the loss of sensitivity in all the subjacent dermatomes.
   B. MOTOR PATHWAYS: segmentation persists and is clearly seen in the intercostal spaces. It is much less clear in the limbs and the girdles since the primordia of these structures are not located opposite a single metamere, but opposite several. Thus, the same muscle group may be innervated by motor nerves coming from different levels
      1. Functionally, segmental medullary activities, such as osteotendinous or micturition reflexes, are progressively controlled by the higher centers as one ascends the evolutionary scale of the vertebrates. Man can regulate many of these activities at will
   C. INVOLUNTARY PATHWAYS
      1. Segmentation is diminished in the involuntary pathways
      2. Organs are innervated by nerves from various levels, as in most of the somatic muscle groups

*A metamere is one of a series of homologous segments in the body. Metameric organization is seen phylogenetically for the first time in annelids.

7 SOMITES

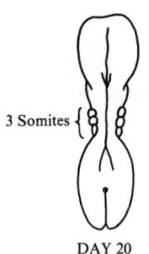

DAY 20

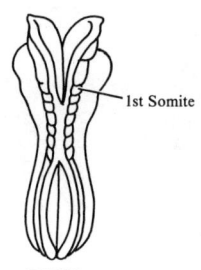

DAY 22

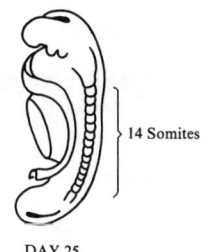

DAY 25

25 SOMITES

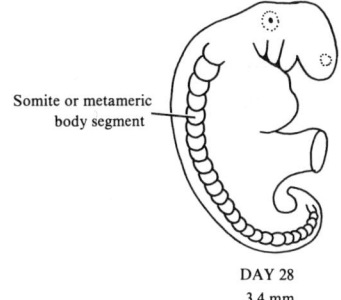

DAY 28
3.4 mm

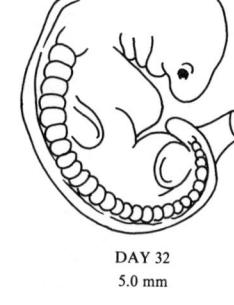

DAY 32
5.0 mm

30 + SOMITES

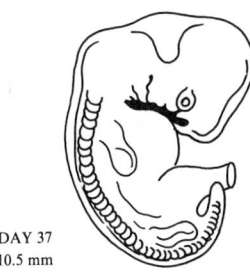

DAY 37
10.5 mm

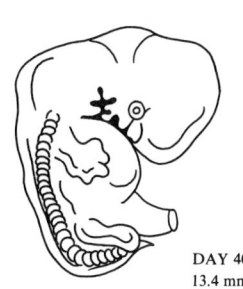

DAY 40
13.4 mm

DAY 46
17.0 mm

NO SOMITES

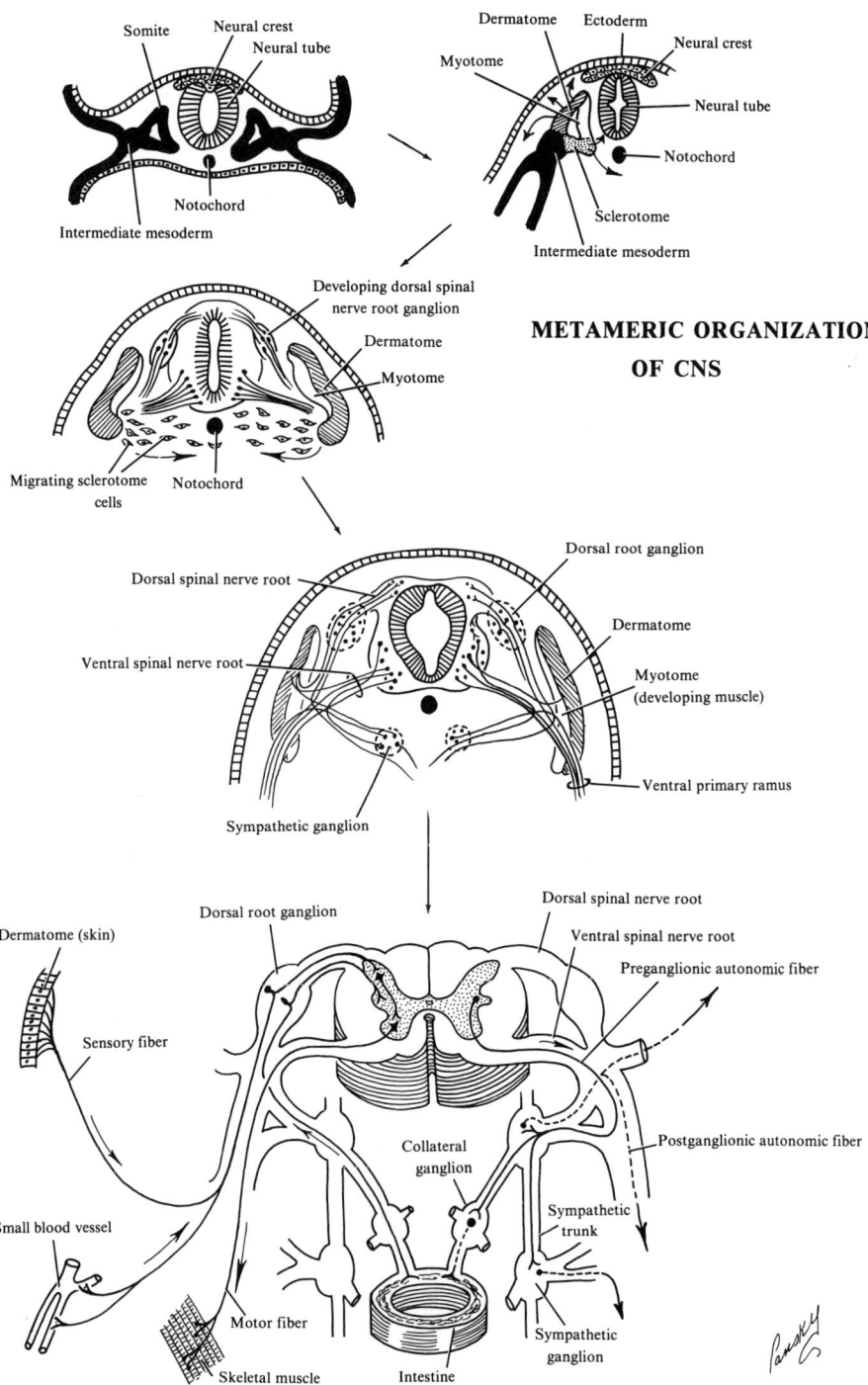

FIGURE 69. **Metameric organization of the central nervous system.**

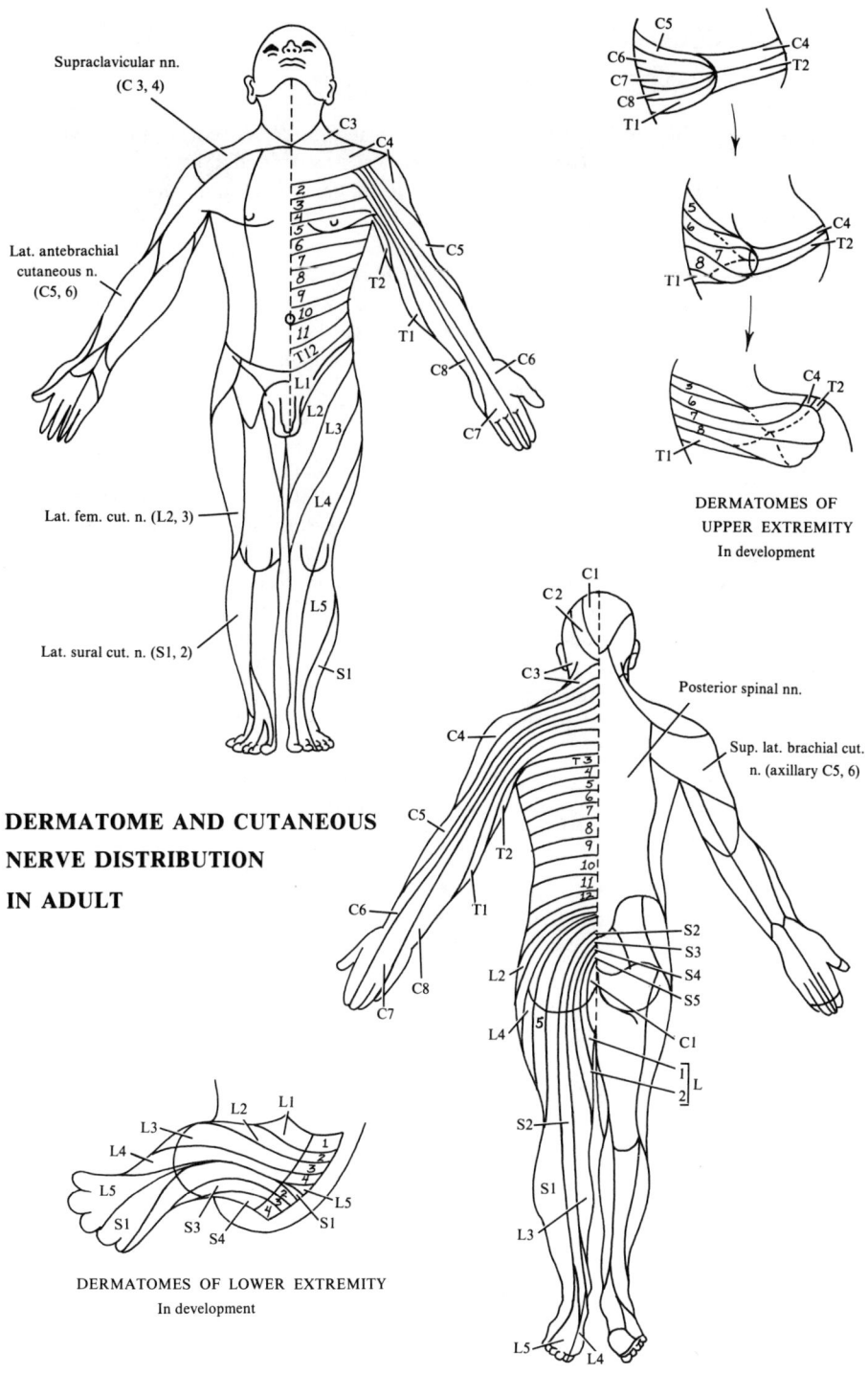

FIGURE 70. **Dermatomes and cutaneous nerve distribution in the adult.**

140. GENERAL CONSIDERATIONS RELATED TO THE ANATOMY OF THE SPINAL CORD

I. **Introduction:** the spinal cord occupies the dorsal and median portions of the embryo and is directly caudal to (behind) the medulla. It is surrounded by membranes called meninges and is lodged in the bony vertebral canal. Its axial cavity is called the spinal or central canal

II. **Functional synthesis:** the spinal cord of vertebrates is formed by the combination of neuromeres connected to each other, on the one hand, and to the brain on the other. Many interneuronic associations give the spinal cord its unity and coordination. These associations and the great influence of the brain in mammals, particularly in humans, greatly reduce the autonomy of each metamere

III. **General characteristics of the spinal cord**
   A. THE SIZE of the spinal cord varies with the age of the fetus
   B. THE RELATIVE VOLUME of the cord, which at first is large, diminishes progressively in relation to the total volume of the central nervous system, especially in relation to the total body size
   C. THE NEURAL TUBE AND VERTEBRAL CANAL initially develop in an almost parallel way. The spinal cord occupies the entire length of the canal, and the spinal nerves emerge between the vertebral bodies through the intervertebral foramina
   D. GROWTH OF THE NEURAL TUBE slows down considerably by month 4 of fetal life, but that of the vertebral canal continues. As a result of this fact
      1. The spinal cord no longer occupies the entire length of the vertebral canal
      2. The roots of the lumbar and sacral nerves, which originally were horizontal, are pulled down by the vertebral canal. They become long and vertical and form the *cauda equina* (horse's tail) below the spinal cord, which ends at about the second lumbar vertebra in the adult

IV. **Composition of the spinal cord:** the cord consists of
   A. A SERIES OF NERVE CENTERS, the gray matter, formed by
      1. Neural cells derived from neuroblasts
         a. Somatic and visceral motor neurons, emission centers derived from the basal plates (anterior and lateral horns)
         b. Somatic and visceral sensory neurons, reception centers derived from the alar plates (posterior and lateral horns)
         c. Metameric interneurons
         d. Cell bodies of the intermetameric neurons, derived from the alar plates and particularly numerous in the posterior horns
      2. The neuroglial cells
      3. Fine nerve fibers, slightly myelinated or unmyelinated, coming from the cells listed above. All these elements develop in the mantle layer
   B. A GROUPING OF PATHWAYS OF TRANSIT or of association, the white matter (marginal layer), is formed of myelinated fibers and consists of
      1. Motor fibers from the brain
      2. Sensory fibers from the periphery, passing through the spinal ganglia, making connections in the posterior horn and ascending toward the brain
      3. Axons of intersegmental association neurons
      4. The bundles of fibers in the marginal layer are compressed against each other. The white matter thus is formed from the pasage of myelinated fibers in the marginal layer

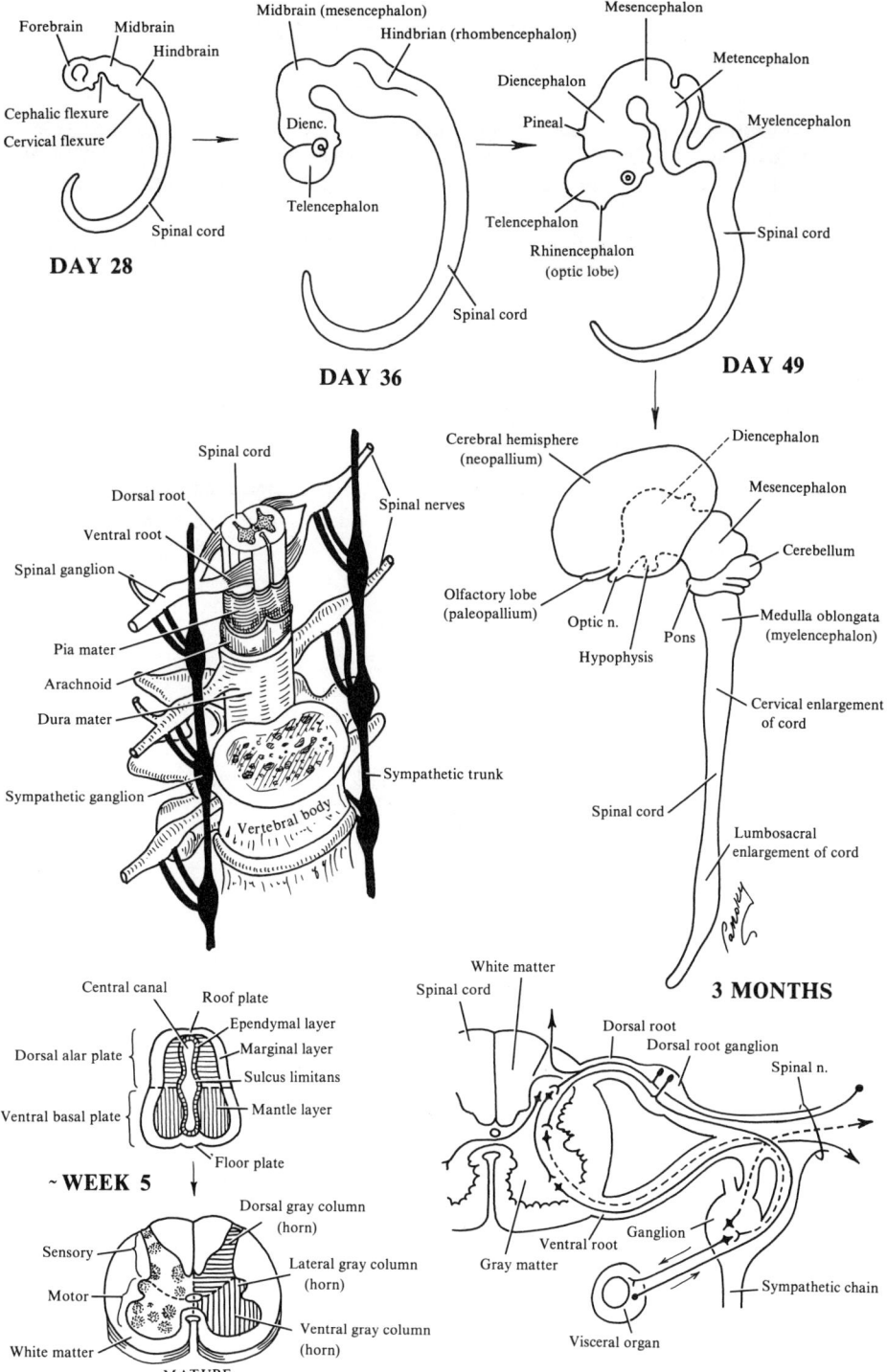

141. THE SPINAL CORD: NORMAL DEVELOPMENT

I. Neuroepithelial, mantle, and marginal layers
A. THE WALL OF A RECENTLY CLOSED NEURAL TUBE consists of only 1 cell type, the *neuroepithelial cells,* which extend over the entire thickness of the wall and form a thick pseudostratified epithelium
   1. The cells are connected to each other by terminal bars at the lumen
   2. During interphase, when DNA synthesis takes place, the cells are wedge-shaped, with the broader portion containing the nucleus in the outer zone of the wall and a slender cytoplasmic portion extending toward the lumen
   3. Just after DNA synthesis, the nucleus begins to move toward the lumen, while the cell contracts toward the terminal bars
   4. During metaphase, the cells are round and in broad contact with the lumen, squeezing the thin cytoplasmic processes of neighboring nondividing cells
B. DURING THE NEURAL GROOVE STAGE and just after tube closure, the neuroepithelial cells divide rapidly, resulting in the production of more cells, and we now refer to the thickened epithelium in the recently closed neural tube as the *neuroepithelial layer* or *neuroepithelium*
C. ONCE THE TUBE IS CLOSED, the neuroepithelial cells give rise to another cell type characterized by a round nucleus with pale cytoplasm and a dark-staining nucleus, the *primitive nerve cells* or *neuroblasts*
   1. The neuroblasts form a zone that surrounds the neuroepithelial layer, called the *mantle layer* (the future gray matter of the spinal cord)
   2. The outermost layer of the cord contains the nerve fibers emerging from the neuroblasts in the mantle layer and is called the *marginal layer*
      a. As a result of myelination of the nerve fibers, the marginal layer takes on a "white" appearance and is called the *white matter of the cord*

II. Basal, alar, roof and floor plates
A. WITH THE CONTINUAL ADDITION OF NEUROBLASTS to the mantle layer, each side of the neural tube shows a ventral and dorsal thickening
   1. The ventral thickenings, the *basal* or *motor plates,* contain the anterior motor horn cells and form the motor areas of the spinal cord
   2. The dorsal thickenings, the *alar* or *sensory plates,* form the sensory areas of the cord
   3. A longitudinal groove, the *sulcus limitans,* is found bilaterally on the inner surface of the tube. It marks the boundary between the anterior motor and posterior sensory areas of the cord and ends in the region of the mamillary recess in the ventral portion of the diencephalon
      a. One should not expect, therefore, any truly motor nerves emerging from the brain rostral to the mamillary recess
      b. Since all neural tissue rostral to this point is an extension of the alar plate, its function is regarded as sensory or associational
B. THE THIN DORSAL PORTION AND VENTRAL MIDLINE PARTS of the tube are the *roof* and *floor plates,* respectively. They contain no neuroblasts and serve primarily as pathways for nerve fibers crossing from one side of the cord to the other
C. THE BASAL PLATES bulge ventrally on each side of the midline as a result of the continuous enlargement of the neuroblasts, creating a deep longitudinal groove called the *ventral fissure,* which later will contain the anterior spinal artery
D. THE ALAR PLATES expand predominantly in a medial direction, compressing the dorsal portion of the lumen of the neural tube
   1. The *posterior median septum* is formed where the 2 alar plates fuse in the midline
E. ACCUMULATION OF NEURONS between alar and basal plates causes the formation of the *intermediate horn,* which contains motor neurons of the autonomic nervous system
F. THE SPINAL CORD acquires its definitive form: motor horns anteriorly, sensory horns posteriorly, intermediate horns laterally, and a small lumen, the *central canal*

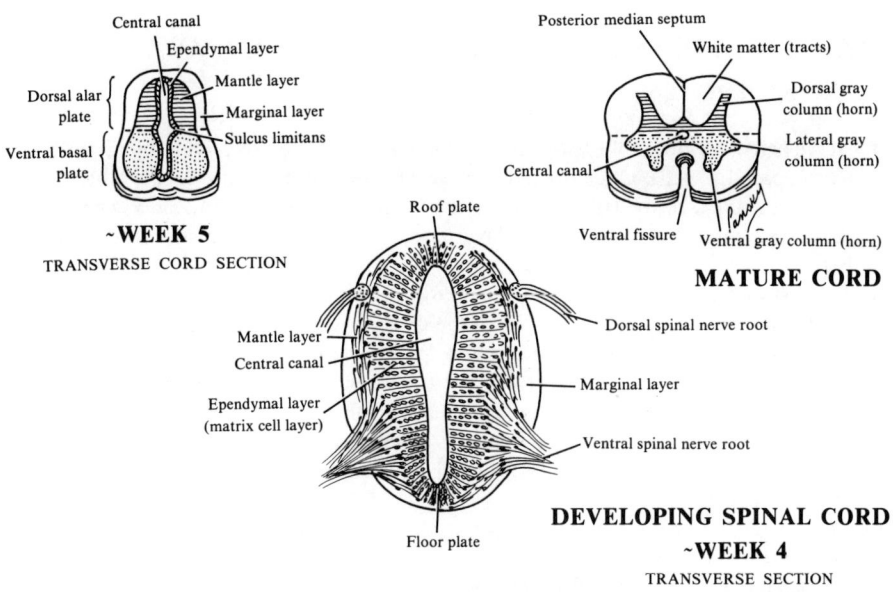

142. THE SPINAL CORD: DIFFERENTIATION OF NERVE AND GLIAL CELLS

I. **Nerve cells:** the neuroblasts or primitive nerve cells arise exclusively by division of the neuroepithelial cells. Once the neuroblasts are formed, they lose their ability to divide
   A. THE NEUROBLASTS OF THE ANTERIOR HORN are formed first, and only when most of these have migrated to the mantle layer does formation of nerve cells for the alar plate begin
   B. THE NEUROBLASTS INITIALLY have a central process extending to the lumen, the transient dendrite, but it disappears when the cells migrate to the mantle zone. The neuroblasts are temporarily round and are called *apolar neuroblasts*
   C. THERE IS FURTHER DIFFERENTIATION and 2 new cytoplasmic processes appear on opposite sides of the cell body. Thus, the *bipolar neuroblasts* are formed
   D. THE PROCESS AT ONE END ELONGATES rapidly to form the *primitive axon,* whereas that at the other end develops a number of cytoplasmic arborizations which are known as the *primitive dendrites*
   E. THE CELL IS NOW A MULTIPOLAR NEUROBLAST and with further development becomes the *adult nerve cell* or *neuron*
   F. THE AXONS OF THE NEURONS in the posterior sensory horn behave differently than those in the anterior horn
      1. Those in the posterior horn penetrate the marginal layer of the cord and then ascend or descend to a higher or lower level (association neurons)
         a. The dendrites of the posterior horn grow peripherally to form the sensory part of the mixed nerve. Their axons grow toward the posterior aspect of the tube to form the *sensory roots.* Their cell bodies lie in the dorsal root ganglia
      2. Those in the anterior horn break through the marginal zone and are seen on the ventral aspect of the cord where they form the *anterior motor root* of the spinal nerve (they conduct motor impulses from the cord to the muscles)

II. **Glia cells:** the majority of primitive supporting cells are called *glioblasts* or *spongioblasts* and are formed by the neuroepithelial cells after the production of neuroblasts has ceased
   A. THE GLIOBLASTS migrate from the neuroepithelial layer to the mantle layer (some even to the marginal layer) where they differentiate into the *fibrillar* and *protoplasmic astrocytes*
   B. ANOTHER TYPE OF SUPPORTING CELL, possibly of glioblast origin, is the *oligodendroglia* cell
      1. This cell is mainly found in the marginal layer and forms the myelin sheaths around the ascending and descending axons in that layer
      2. They may be derived from mesenchyme cells which have penetrated into the central nervous system, thus, their origin from neuroepithelial cells is in doubt
   C. IN THE SECOND HALF OF DEVELOPMENT, a third type of supporting cell, the *microglia cell,* appears in the central nervous system. It is believed that its origin is from mesoderm which surrounds the neural tube
      1. They are macrophagic cells of the histiocyte system, with a neural localization, brought into the central nervous system by blood vessels. There is also some evidence, however, that the microglia could be ectomesenchymal.

III. **Fibers crossing over from both sides of the spinal cord** form the so-called *commissures,* which connect the right and left portions of the gray matter

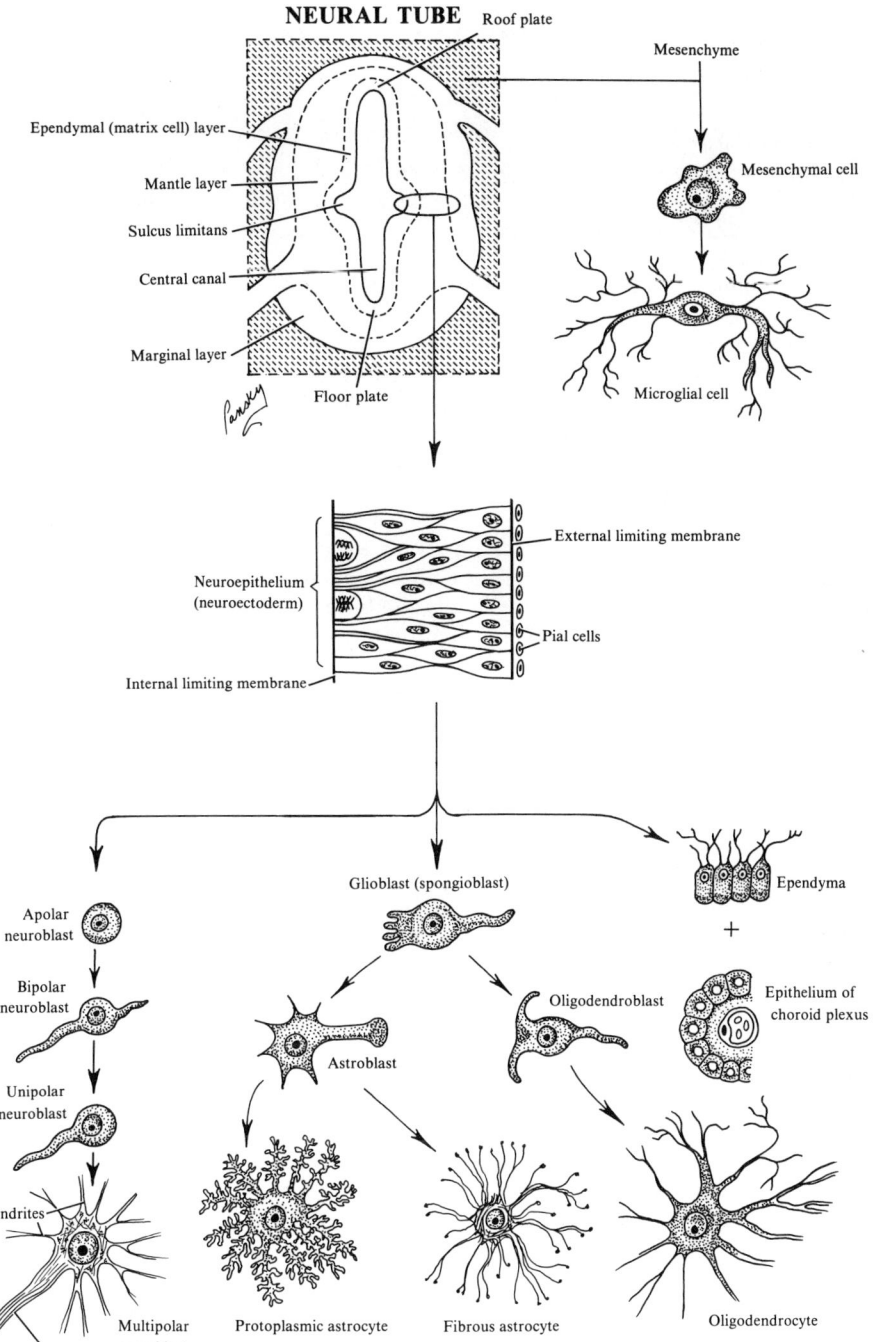

143. THE SPINAL CORD: NEURAL CREST CELLS AND MYELINATION

I. The neural crest cells
A. DURING INVAGINATION OF THE NEURAL PLATE, a distinct group of cells appears along each edge of the neural groove, are ectodermal in origin, and are called the neural crest
  1. The cells temporarily form an intermediate zone between the tube and surface ectoderm
  2. The zone extends from the mesencephalon to the level of the caudal somite, and in time divides into 2 parts, each of which migrates to the dorsolateral aspect of the neural tube
     a. Here the cells of the neural crest form a series of bipolar cell clusters that give rise to the sensory or dorsal root ganglia of the spinal and cranial nerves (V, VII, IX, and X), and multipolar cells of the future sympathetic ganglion cells
B. DURING DEVELOPMENT, the neuroblasts of the sensory ganglia form 2 processes
  1. One grows centrally and penetrates the dorsal portion of the neural tube
     a. In the spinal cord, they either end in the dorsal horn or ascend through the marginal layer to one of the higher brain centers
     b. Collectively, these processes are called the dorsal sensory root of the spinal nerve
  2. The other process grows peripherally and forms fibers of the ventral motor root and participates in the formation of the trunk of the spinal nerve
     a. These processes eventually terminate in the sensory receptor organs
C. IN ADDITION TO FORMING THE SENSORY GANGLIA, the cells of the neural crest differentiate into sympathetic neuroblasts, Schwann cells, pigment cells, odontoblasts, meninges, and cartilage cells of the branchial arches
  1. Removal of neural crest cells of the trigeminal region results in facial abnormalities, including clefts of the primary palate

II. Myelination follows histogenesis and tends to occur later and persist longer as the systems are phylogenetically more recent
A. MYELINATION OF THE PERIPHERAL NERVE is brought about by the *neurilemma cells* or *cells of Schwann,* beginning in month 4 of fetal life*
  1. These cells originate from the neural crest, migrate peripherally, and wrap around the axons to form the neurilemma sheath
  2. Axons, varying in number from 1 to 20, can be enwrapped by one neurilemma cell
B. AT MONTH 4 OF FETAL LIFE, the nerve fibers gradually obtain a whitish appearance as a result of myelin deposition between the axon and the neurilemma
  1. This substance is formed by repeated coiling of the membrane around the axon
C. BOTH THE NEURILEMMA AND THE MYELIN SHEATH of the peripheral nerve fibers are formed by the cells of Schwann
D. THE MYELIN SHEATH OF NERVE FIBERS within the spinal cord is of different origin, being formed by the oligodendroglia cells
E. ALTHOUGH MYELINATION OF NERVE FIBERS IN THE CORD generally begins at about month 4 of fetal life, some motor fibers that descend from higher brain centers to the cord do not become myelinated until the first year of postnatal life
  1. The tracts in the nervous system apparently become myelinated at about the time they begin to function
     a. It appears in the vestibulospinal tract in month 6, and in the rubrospinal tract in month 7
  2. Some motor fibers coming from the upper cerebral centers, the pyramidal tract, for example, are myelinated during the 2 years after birth
  3. The slowness of this process and the correlation between myelination and the final development of functional ability partly explain the long duration of psychomotor development in the child

---

*Myelination in the interior of the CNS is furnished by the neuroglial cells (oligodendrites) derived from the neural tube.

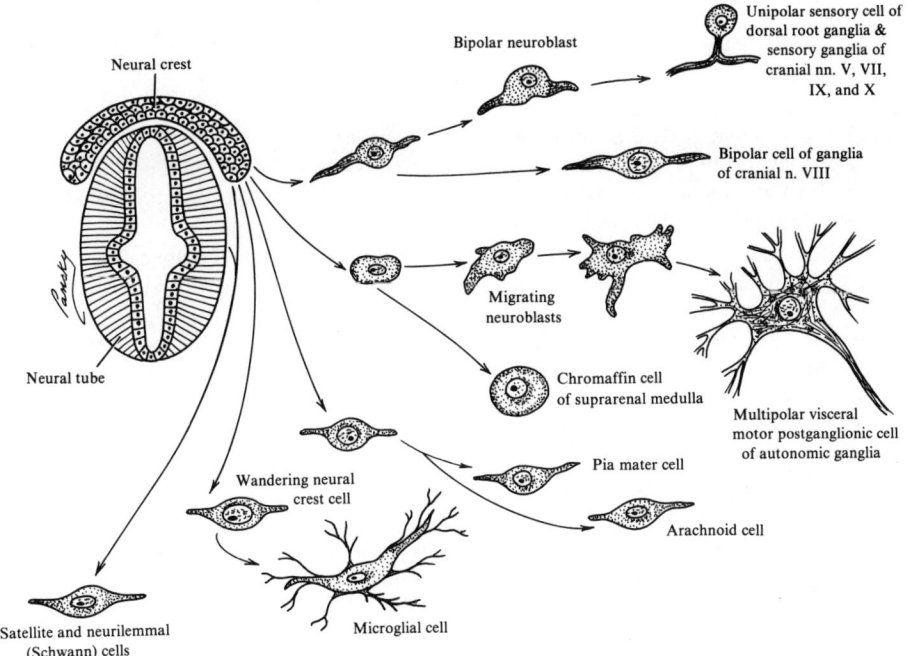
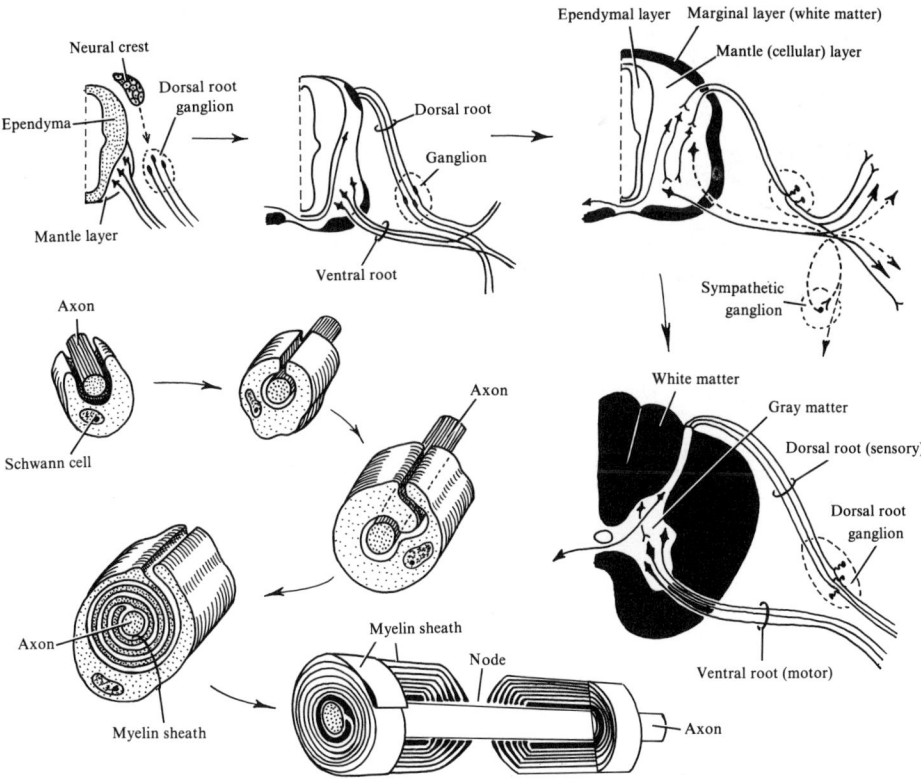

144. SPINAL CORD LENGTH AND SPINAL MENINGES

I. **Changes in spinal cord position during development**
   A. WHEN THE CROWN-RUMP LENGTH of the embryo is about 30 mm (in month 3 of development), the spinal cord extends the entire length of the embryo, and the spinal nerves pass through the intervertebral foramina at their level of origin
   B. WITH INCREASING AGE, the vertebral column and dura mater lengthen more rapidly than the neural tube, and the terminal end of the spinal cord gradually shifts to a higher level
   C. AT BIRTH, THE END OF THE CORD is located at the level of the third lumbar vertebra (L3)
      1. Due to disproportionate growth, the spinal nerves run obliquely from their segment of cord origin to the corresponding level of the vertebral column
      2. The dura remains attached to the bony column at the coccygeal level
   D. IN THE ADULT, THE SPINAL CORD terminates at the level of lumber 2 (L2)
      1. Below this point, the CNS is represented by only the *filum terminale internum* marking the tract of spinal cord regression
      2. Nerve fibers (roots of lumbar and sacral nerves) below the terminal end of the cord are collectively called the *cauda equina (horse's tail)*
   E. WHEN CEREBROSPINAL FLUID is taken in a lumbar puncture, the needle is inserted at the lower lumbar level to avoid injuring the lower end of the spinal cord
      1. At this level, there is no spinal cord, only meninges, cerebrospinal fluid, and the mobile nerves of the cauda equina, through which the needle will slip
   F. THE SPINAL CORD WIDENS in 2 regions to form the *cervical* and *lumbar enlargements* which are located at the level of origin of nerves that give rise to the *brachial plexus* for the upper extremity and the *lumbosacral plexus* for the lower extremity

II. **Spinal meninges**
   A. THE PRIMITIVE MENINX is a covering or membrane of condensed mesenchyme that surrounds the neural tube
      1. The inner cell layer of the meninx is probably derived from neural crest and remains there to form the pia-arachnoid
         a. The pia mater and the arachnoid membrane together make up the so-called leptomeninges
      2. The outer cell layer of the meninx thickens to form the dura mater
   B. FLUID SPACES IN THE LEPTOMENINGES form and then coalesce to create the subarachnoid space which, in the adult, is traversed by fine, delicate strands of connective tissue (the arachnoid trabeculae). Thus, the trabeculae pass between the pia mater and the arachnoid membrane

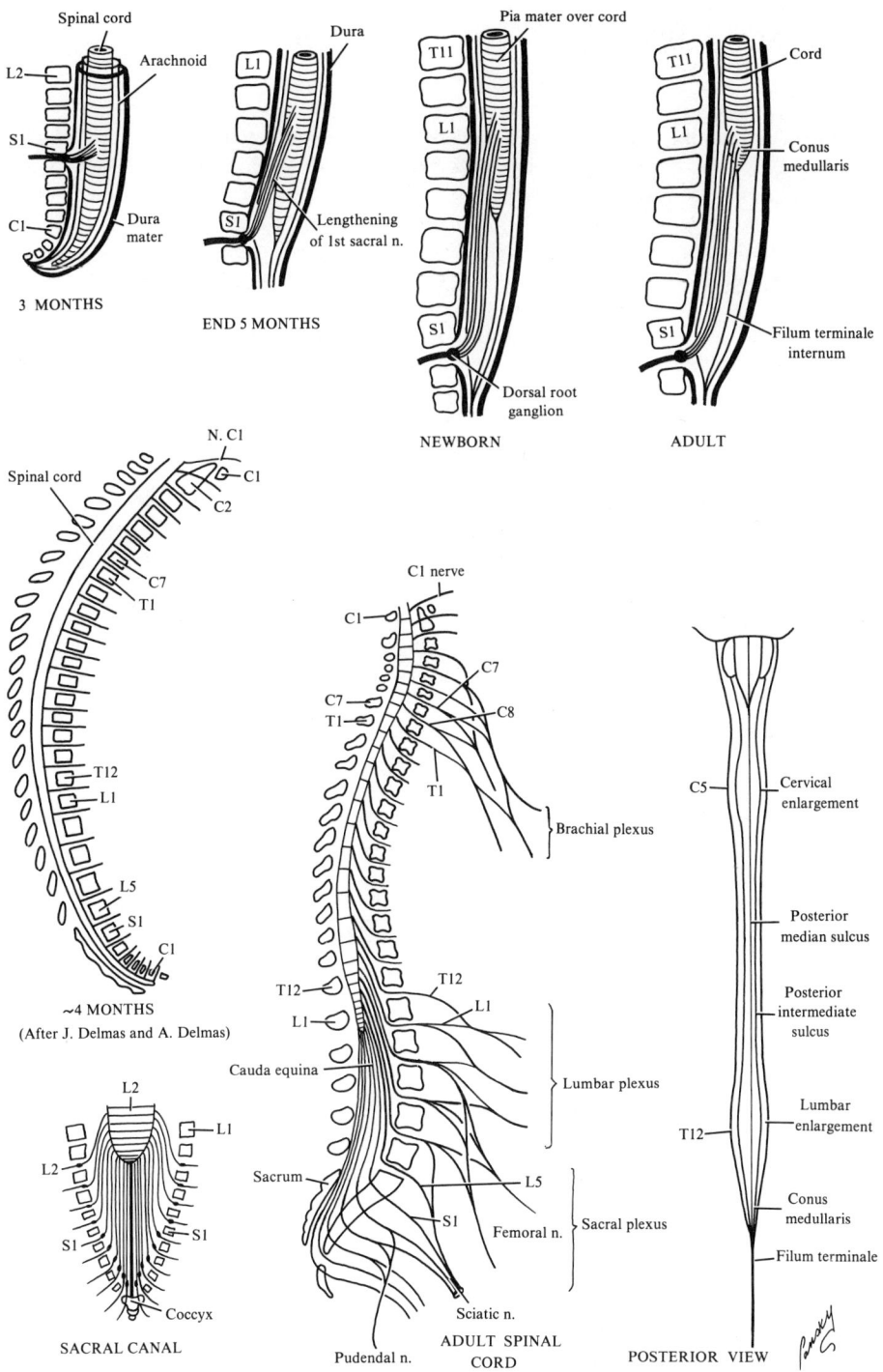

145. MALFORMATIONS OF THE SPINAL CORD

I. **Introduction:** the central nervous system forms a closed tubular structure which is detached from the overlying ectoderm by the end of week 4. Malformations of the spinal cord are usually the result of problems related to the closure of the neural groove, the causes of which are still obscure. It may be the result of faulty induction by the underlying notochord, or action of environmental teratogenic factors acting on the neuroepithelial cells. Malformations involve defects in early embryogenesis since neural groove closure, in humans, usually takes place between days 21-28 of gestation
   A. FORMATION OF THE NEURAL TUBE also affects formation of the posterior arch of the vertebra, thus, anomalies of the tube may produce anomalies of the spine as well
      1. The most classic form of these malformations is *spina bifida*. The term "spina bifida" results from the fact that the posterior arch of one or more vertebrae is not formed, thus each vertebral body ends with 2 bony spines, rather than an arch, which frame the spinal cord
   B. THE LUMBOSACRAL REGION is the most common site of spina bifida. When the spinal cord is open from top to the bottom, the condition is called *rachischisis,* which may be complete or partial. Failure of closure in the cephalic region is called *anencephaly*

II. **Major types of malformations**
   A. THE SPINAL CORD IS OPEN. The nervous tissue is in direct continuity with skin, thus producing a condition similar to the general structure of the neural plate or neural groove stages of neurulation
   B. SPINA BIFIDA refers to a wide range of defects, but is usually localized in the cord region
      1. In its most simple form, it is seen as a failure of the dorsal portions of the vertebrae to fuse with one another (bifid spine)
         a. Usually localized in the lumbosacral region, covered by skin and not noticeable on the surface except for the presence of a small tuft of hair over the affected area. It is usually discovered by radiography
         b. It is the most frequent type of spina bifida, affecting about 15 to 20% of the population and is referred to as *spina bifida occulta*
         c. The cord (closed spinal cord) and nerves are normal, and there are usually no neurologic symptoms
      2. If more than 1 or 2 vertebrae are involved in the defect, the meninges of the cord bulge through the opening, and a sac covered with skin is seen on the surface. This is called *meningocele* and surgical correction is feasible
         a. If the sac is so large that it contains not only the meninges but also the spinal cord and its nerves, we refer to this abnormality as *meningomyelocele*
            i. Usually covered by a very thin, easily torn membrane
            ii. Neurologic symptoms usually are present in this condition
      3. If there is a failure of the neural groove to close, the nervous tissue is widely exposed to the surface and we have a condition called a *myelocele* or *rachischisis*
         a. Occasionally, the neural tissue shows much overgrowth, but the excess tissue invariably becomes necrotic before or after birth
      4. *Myelomeningoceles* generally are associated with a caudal displacement of the medulla oblongata and a part of the cerebellum into the spinal canal
         a. The myelomeningocele is frequently combined with hydrocephaly and we have what is called *Arnold-Chiari malformation*
   C. DERMAL SINUS: closure of the levels above the cord is almost complete. Only a narrow opening connecting the meninges and the surface ectoderm opens to the outside. The latter may close secondarily. Lined by epidermis and skin appendages

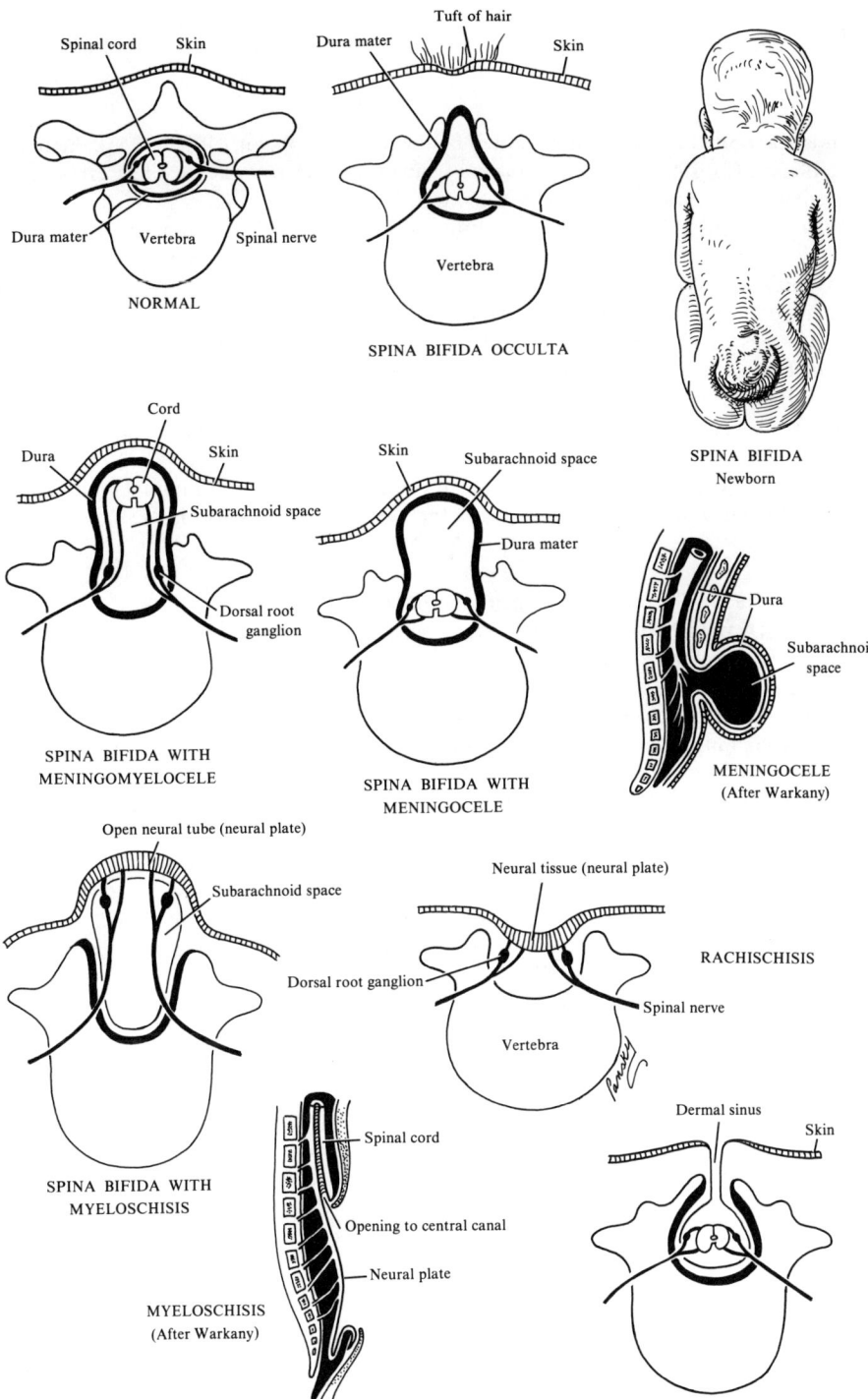

146. INTRODUCTION TO BRAINSTEM DEVELOPMENT

I. **Introduction:** the brainstem is formed by the myelencephalon (fifth vesicle), the metencephalon (fourth vesicle), and the mesencephalon (third vesicle). The cerebellum is derived from and is attached to the brainstem
  A. IN THE BRAINSTEM, THE GRAY AND WHITE MATTER are not in the same relationships to each other as they are in the spinal cord, nor does the gray matter have the same potential
  B. CEPHALIZATION, through which these suprasegmental structures are found, begins in the brainstem. Vesicle development reveals how these changes occur
  C. THERE IS ALWAYS A ROOF PLATE derived from the dorsal part of the neural tube and a floor plate from its ventral part. The plates and their lateral boundaries delineate the cavities of the neural tube, the future ventricles, and the interventricular communications
  D. THE MYELENCEPHALIC CAVITY, the fourth ventricle, remains in communication with the spinal canal, at its caudal end, and with the mesencephalic cavity via the metencephalic cavity

II. **The derivatives of the third, fourth, and fifth vesicles** consist of
  A. GRAY MATTER, derived from the alar and basal plates, includes,
    1. Segmental nuclei: similar to the medullary centers and consist of nuclei of cranial nerves as well as autonomic nuclei
    2. Suprasegmental structures: indicating cephalization. These structures are relay or association centers and head up the spinal cord. Even though they are integrated into a motor system (the extrapyramidal tract), they are *derived in whole or in part from the alar plates* which, even in the spinal cord, give rise to the synaptic relay and association centers
      a. Development of the suprasegmental structures is characterized by cellular migrations which are more extensive than those in the spinal cord. These migrations increase with the complexity of the organs (i.e., the cerebellum)
        i. Some of these structures, in each vesicle, are rather discrete, such as the olivary nuclei, the nuclei of Goll and Burdach, the nuclei of the pons and cerebellum, the red nuclei, and substantia nigra, and the nuclei of the corpora quadrigemina (or colliculi)
        ii. Others are diffuse, such as the reticular formation, which occupies the ventral portion of the 3 vesicles. The reticular formation is formed from many small nuclei and is derived from either the alar or basal plates, or both, in variable proportions. Its mesencephalic portion activates the extrapyramidal tracts. Its pontine-medullary portion activates the ascending sensory tracts and inhibits the pyramidal motor tracts
  B. WHITE MATTER is formed from myelinated tracts which thicken the marginal layer ventrally or just pass through the gray nuclei
    1. Segmentary association tracts make the brainstem a functionally homogeneous whole and connect it to its subjacent and suprajacent structures
    2. Cerebromedullary or medullocerebral pathways use the brainstem only as a crossover (e.g., the pyramidal tracts, the spinothalamic pain and heat pathways) or make connections at the level of its suprasegmentary centers (e.g., the extrapyramidal pathways, the spinobulbothalamic pathways of proprioception)
    3. Brainstem pathways, e.g., the geniculate tract, connecting the cerebral cortex to the nuclei of the cranial nerves
    4. Spinocerebellar pathways, e.g., Flechsig's tract of deep unconscious sensory perception

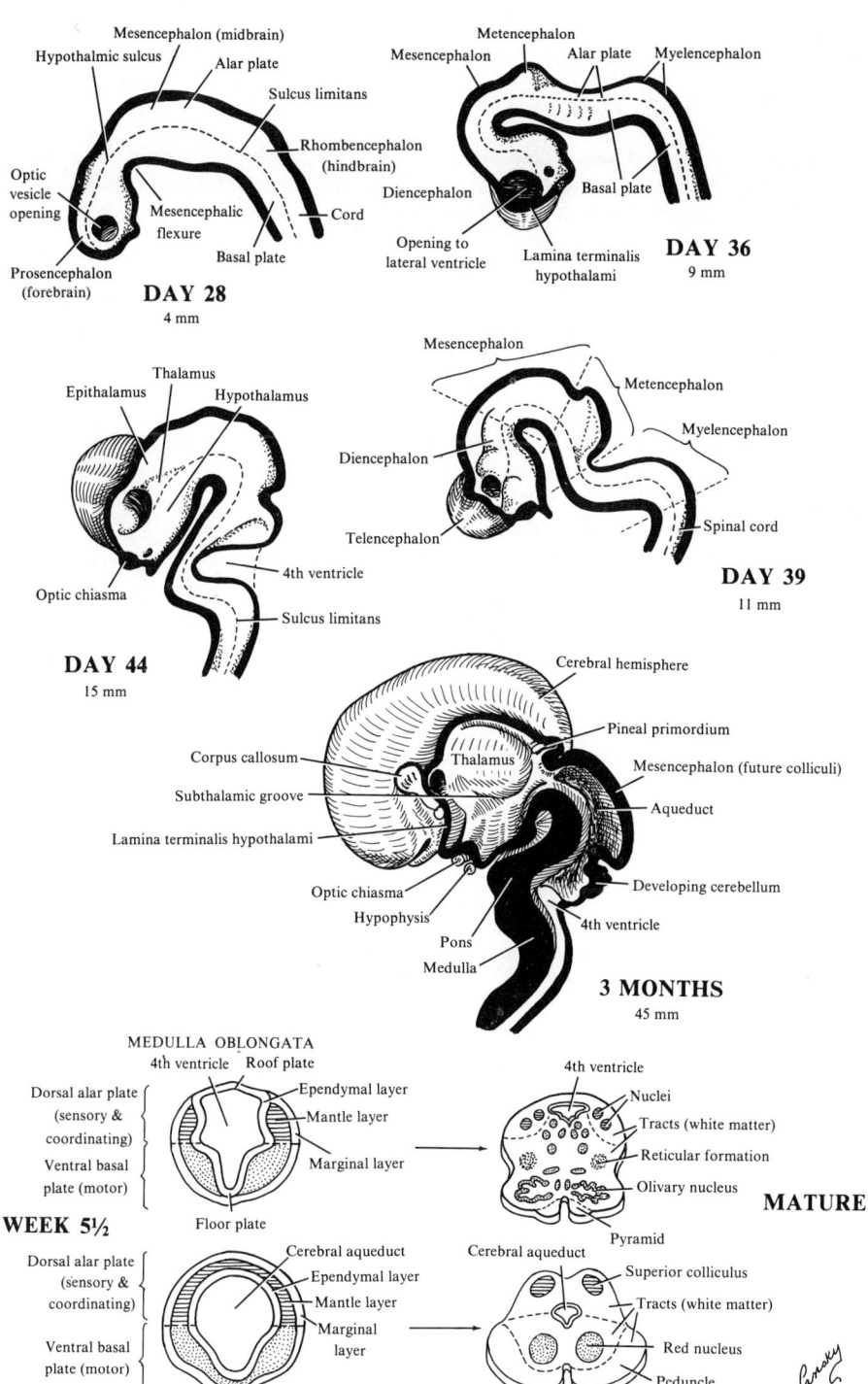

147. THE BRAINSTEM: MYELENCEPHALON (FIFTH VESICLE)— BASAL MOTOR PLATE

I. **Introduction:** the external shape of the cephalic portion of the neural tube changes markedly with the appearance of brain vesicles and the development of cervical and cephalic flexures. Despite changes, some morphologic characteristics seen in the spinal cord are recognizable in most of the brain vesicles
  A. DISTINCT BASAL AND ALAR PLATES, representing the motor and sensory areas, respectively, are seen on each side of the midline in most of the brain vesicles
  B. THE SULCUS LIMITANS, which formed the boundary line between the alar and basal plates in the cord, is present in the rhombencephalon and mesencephalon, where it also forms the divider between sensory and motor areas

II. **The myelencephalon** is the most caudal brain compartment and extends from the first spinal nerve to the pontine flexure and gives rise to the medulla oblongata
  A. THE MEDULLA differs from the cord in that its lateral walls rotate around an imaginary long axis in the floor plate, like opening a textbook, and as a result, the roof plate stretches and becomes a single layer of cells covering an enlarged central cavity called the fourth ventricle
    1. The lateral wall structure is similar to that of the spinal cord, and one sees alar and basal plates separated by the sulcus limitans
  B. THE BASAL MOTOR PLATE of the myelencephalon
    1. Like the spinal cord, the basal plate contains the motor nuclei, but these are divided into 3 groups
      a. A medial somatic efferent group forms the cephalic continuation of the anterior horn cells containing the motor neurons which innervate striated muscle (derived from myotomes in the cephalic region)
        i. Since the somatic efferent group continues rostrally into the mesencephalon (through the metencephalon), it is referred to as the somatic efferent motor column
        ii. It is represented in the myelencephalon by the neurons of the hypoglossal (XII) cranial nerve which supplies the 4 occipital myotomes for the tongue musculature
        iii. In the metencephalon and mesencephalon, the column is represented by the neurons of the abducens (VI), trochlear (IV), and oculomotor (III) cranial nerves, supplying eye muscles thought to be derived from preoptic myotomes
        iv. The neurons of nerves III, IV, VI, and XII are all located near the midline
      b. An intermediate special visceral efferent group extends into the metencephalon and forms the special visceral efferent motor column which contains motor neurons supplying striated muscles derived from the mesenchyme of the pharyngeal or branchial arches
        i. In the myelencephalon, the column is represented by neurons of the accessory (XI), vagus (X), and glossopharyngeal (IX) cranial nerves
        ii. In the adult, motor neurons of the above nerves are formed by the nucleus ambiguus and the bulbar portion of the accessory nerve
      c. A lateral general visceral efferent group contains the neurons whose axons grow out as preganglionic fibers to synapse in the parasympathetic ganglia supplying involuntary muscles of the heart, respiratory tract, and intestinal tract as well as innervating the salivary glands
        i. In the myelencephalon, this group is represented by the dorsal nucleus of the vagus (X) and the inferior salivatory nucleus which, by way of the cranial nerve IX (glossopharyngeal), innervates the parotid gland

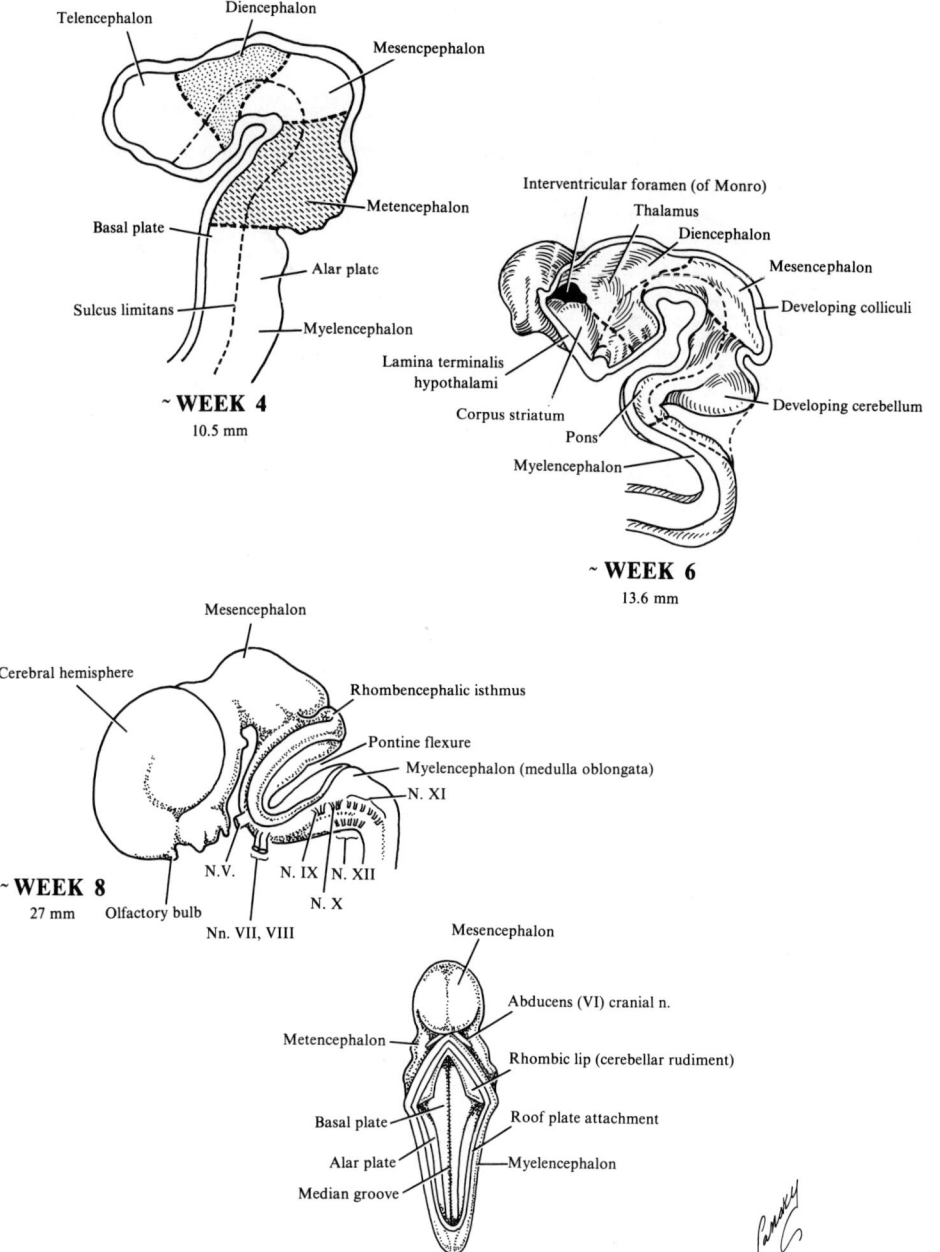

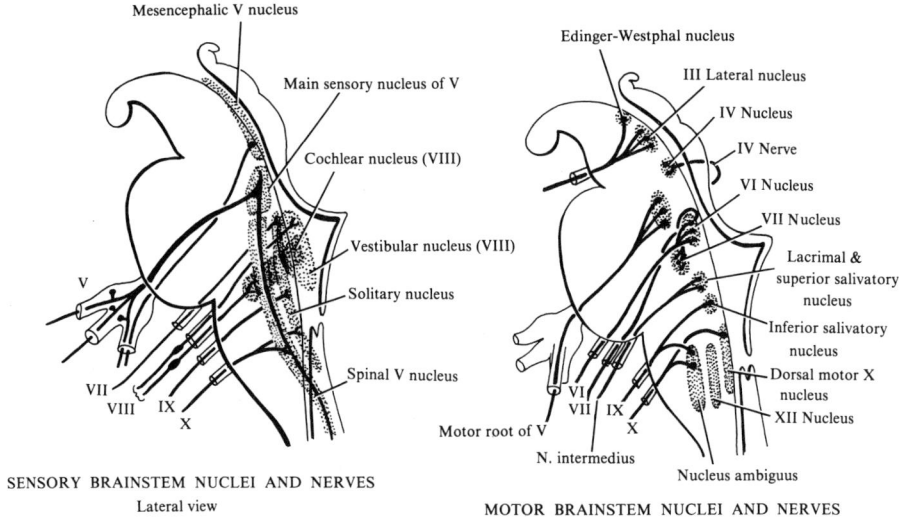
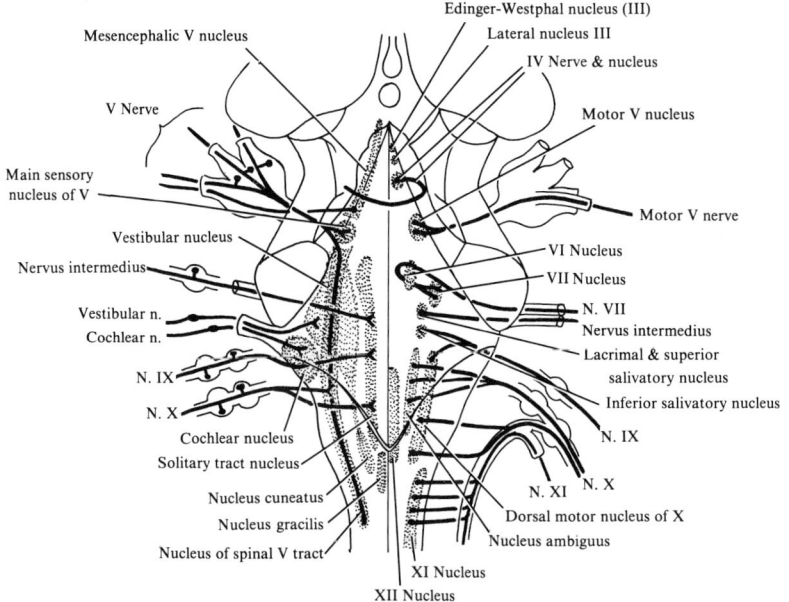

FIGURE 71. **Brainstem nuclei and nerves (dorsal view).**

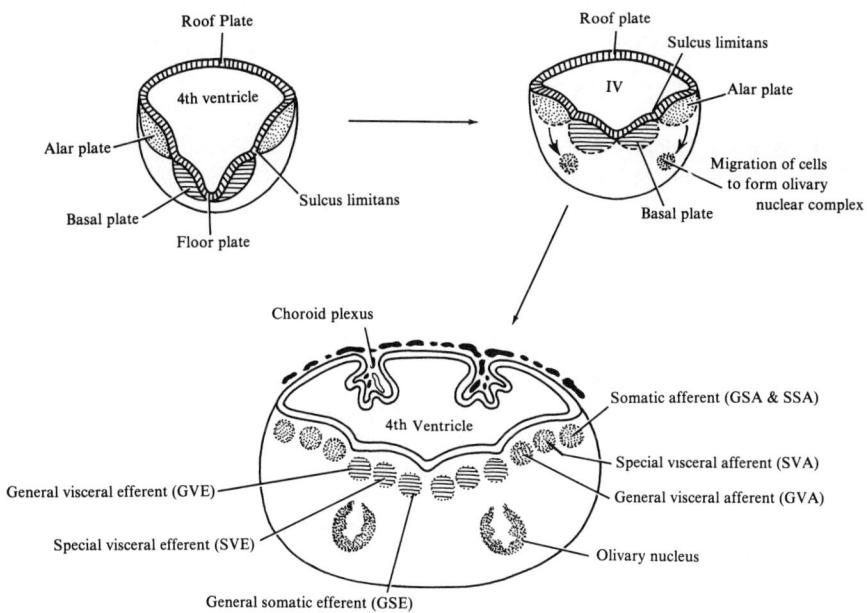

POSITION AND DIFFERENTIATION OF BASAL AND ALAR PLATES OF MYELENCEPHALON AT DIFFERENT STAGES OF DEVELOPMENT

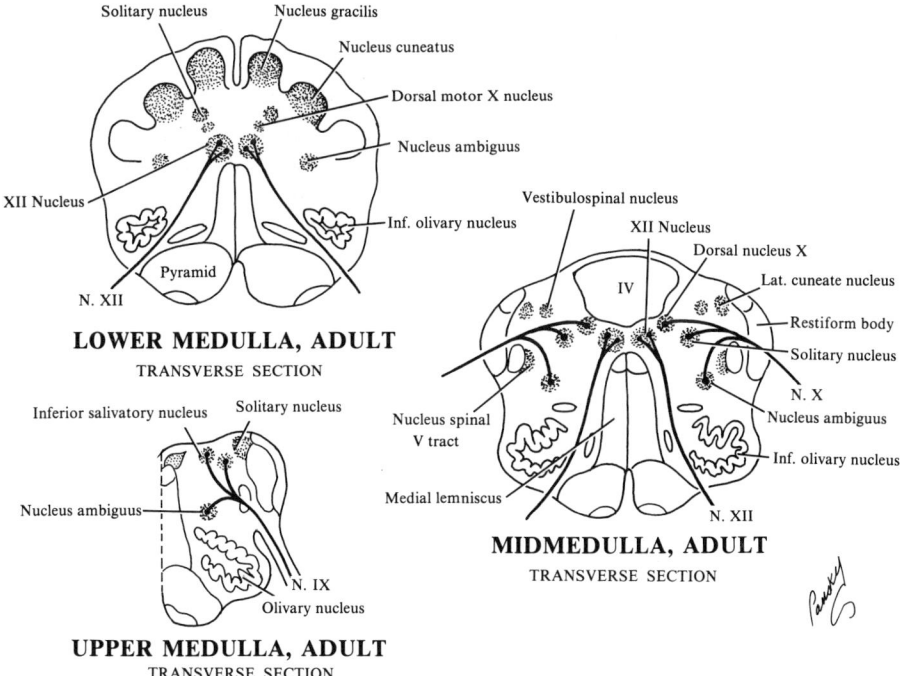

FIGURE 72. Basal and alar plate development of myelencephalon and transverse sections of medulla at various levels.

148. THE BRAINSTEM: MYELENCEPHALON (FIFTH VESICLE)—ALAR SENSORY AND ROOF PLATES

I. The myelencephalon

A. THE ALAR SENSORY PLATES contain the sensory relay nuclei which, like the basal plate, are divided into 3 groups
   1. The most lateral is the somatic afferent group: receives impulses from the ear and surface of the head via the staticoacoustic (VIII) and bulbospinal part of the trigeminal (V) nerves
   2. The intermediate is the special visceral afferent group: receives impulses from the taste buds of the tongue and from the palate, oropharynx, and epiglottis
      a. These neurons later form the nucleus of the solitary tract
   3. The medial is the general visceral afferent group: represented by the dorsal sensory nucleus of the vagus (X) nerve with its neurons receiving interoceptive information from the heart and gastrointestinal tract
   4. In addition to the sensory relay nuclei, other cells of the alar plate migrate downward to be ventrolateral to the basal plate and form a part of the olivary nuclear complex

B. THE ROOF PLATE, CHOROID PLEXUS, AND THE FORAMINA OF LUSCHKA AND MAGENDIE
   1. The roof plate of the myelencephalon consists of a single layer of ependymal cells which is later covered by vascular mesenchyme, the pia mater. Together they make up the *tela choroidea*
      a. As a result of active proliferation of vascular mesenchyme, the tela choroidea forms a series of saclike invaginations that project into the underlying ventricular cavity in the region of the pontine flexure, forming the *choroid plexus*. The latter are tuftlike invaginations that produce the cerebrospinal fluid of the central nervous system
   2. At about month 4, areas of the roof plate of the rhombencephalon thin out, bulge outward, and finally disappear. The apertures formed are the 2 lateral foramina of Luschka and a median foramen of Magendie which allow the cerebrospinal fluid to move freely between the ventricles and the surrounding subarachnoid space

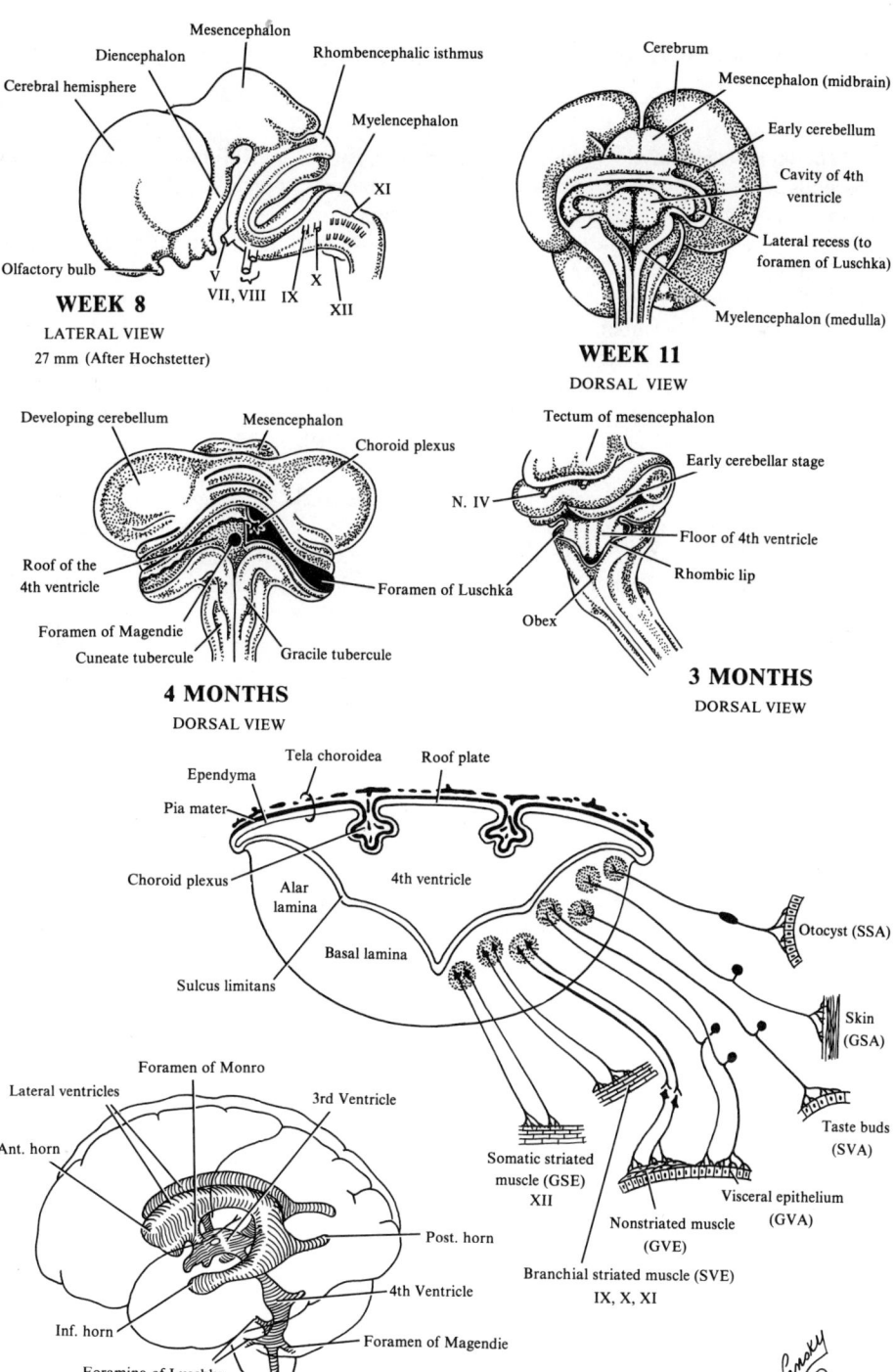

149. THE BRAINSTEM: METENCEPHALON (FOURTH VESICLE)

I. **The metencephalon** develops from the anterior part of the rhombencephalon and extends from the pontine flexure to the rhombencephalic isthmus. It differs from the myelencephalon in forming 2 specialized components
   A. THE DORSAL PORTION OR ROOF PLATE forms the cerebellum, which functions as a coordination center for posture and movement
   B. THE VENTRAL PORTION OR FLOOR PLATE, which becomes the pons, functions as the pathway for nerve fibers between the cord and the cerebral and cerebellar cortices

II. **The basal or floor plate and pons:** the major morphologic features of the metencephalon do not change even though the lateral walls reapproach each other, and the basal motor and sensory alar plates are seen easily. Each basal plate contains 3 groups of motor neurons
   A. THE MEDIAL SOMATIC EFFERENT GROUP gives rise to the abducens (VI) cranial nerve
   B. THE SPECIAL VISCERAL EFFERENT GROUP contains the nuclei of the trigeminal (V) and facial (VII) cranial nerves which innervate the muscles of branchial arches I and II.
      1. The nuclei of nerves V and VII pass across the level of both the metencephalon and myelencephalon
   C. THE GENERAL VISCERAL EFFERENT GROUP contains the superior salivatory nucleus, axons of which grow out into the facial nerve to supply the submandibular and sublingual glands as well as the nasal and lacrimal glands
   D. THE MARGINAL LAYER OF THE BASAL PLATES expands and thickens to serve as a bridge for the nerve fibers connecting the cerebral cortex and cerebellar cortex with the spinal cord and is known as the *pons*
      1. In addition to nerve fibers, the pons contains pontine nuclei which originate in the sensory alar plates of the metencephalon and myelencephalon
         a. These nuclei serve as relay stations in the extrapyramidal pathways that connect the cortex of the telecephalon with the cerebellum
      2. The axons of the pontine nuclei grow toward the cerebellum and form the *middle cerebellar peduncles*

III. **The alar plate and rhombic lip:** development is very complicated
   A. THE VENTROMEDIAL PORTION of the alar plate contains 3 groups of sensory nuclei
      1. The lateral somatic afferent group contains neurons of the pontine portion of the trigeminal (V) nerve and a small part of the vestibulocochlear complex (VIII)
      2. The special visceral afferent group is represented by the cranial portion of the nucleus of the solitary tract
      3. The general visceral afferent group is represented by the most cranial part of the dorsal sensory nucleus of the vagus (X) nerve
   B. THE DORSOLATERAL PARTS of the alar plates bend medially and form the *rhombic lips* which project partly into the lumen of the fourth ventricle and partly above the attachment of the roof plate (extraventricular portion) to give rise to the *cerebellum* (see next unit for development of the cerebellum)

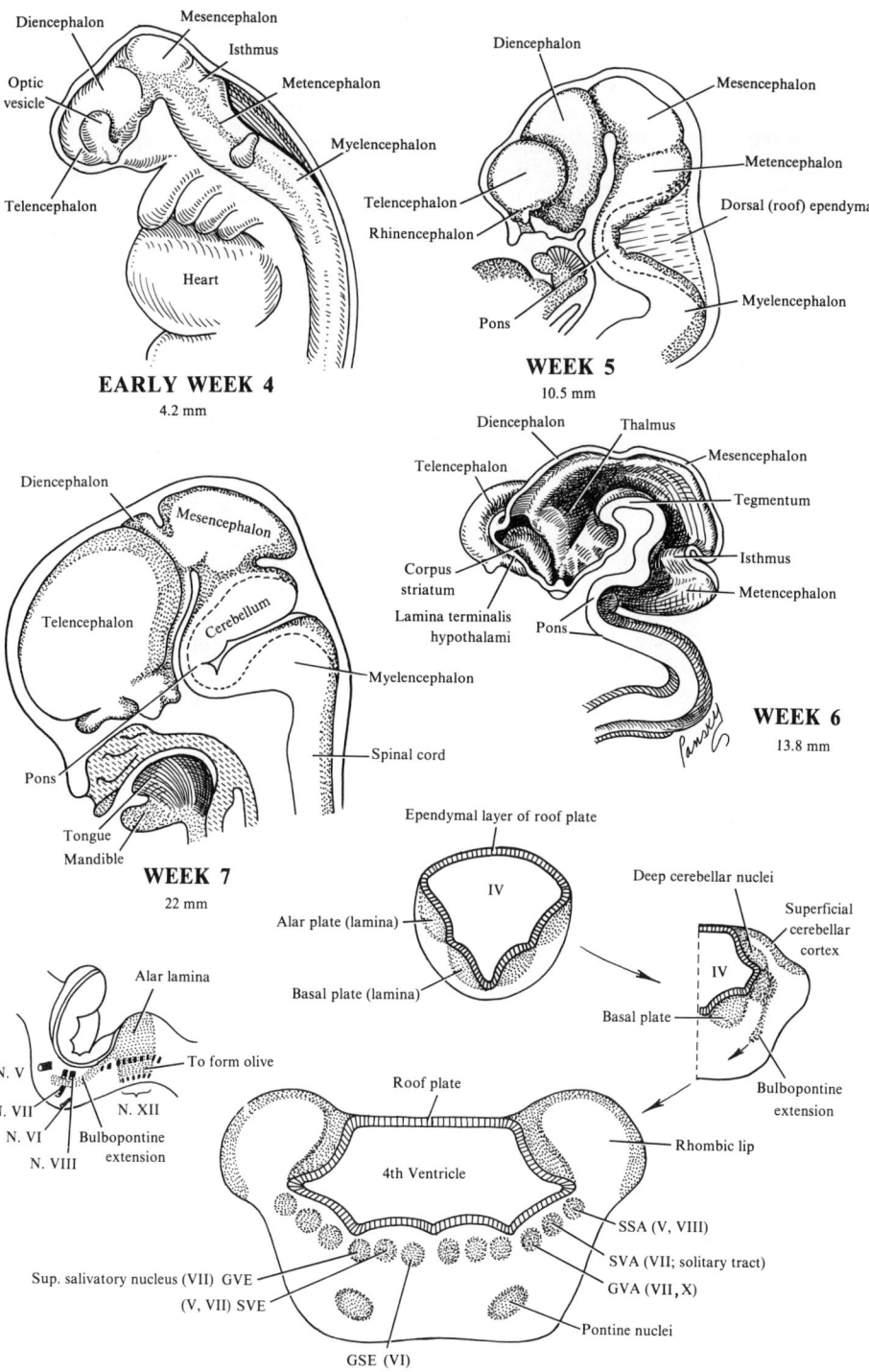

150. THE BRAINSTEM: METENCEPHALON (FOURTH VESICLE)—THE CEREBELLUM

I. **The roof plate** is derived from the dorsal part of the alar plate and thickens to form the cerebellum, which is developed increasingly as one ascends the phylogenetic scale of vertebrates. It begins development between days 40 and 45
   A. THE RHOMBIC LIPS, in the caudal part of the metencephalon, are widely separated, but just below the mesencephalon, they approach each other in the midline to form a transverse thickening which extends to the thin roof plate. With further deepening of the pontine flexure, the rhombic lips become compressed in a cephalocaudal direction to form the *cerebellar plate*
      1. In the 12-week embryo, the plate shows a small midline portion, the *vermis,* and 2 bulging lateral masses, the *lateral lobes*
      2. A transverse fissure soon separates the nodule (arising from the vermis) and the flocculus (derived from the lateral lobes) to form the *flocculonodular lobe,* which is the most primitive part of the cerebellum, maintains connections with the vestibular system, and is concerned with subconsciously controlled equilibrium

II. **Other transverse fissures:** many appear with development, giving the cerebellum its characteristic adult appearance
   A. THE CEREBELLUM is divided into 2 fundamental parts: the *flocculonodular lobe* and the *corpus cerebelli,* separated by the *posterolateral fissure.* The corpus cerebelli is further subdivided into an *anterior* and *middle lobe* by the *primary fissure*
      1. The anterior lobe consists of the lingula, the central lobule, the culmen, the alae of the central lobules, and the quadrangular lobules
      2. The middle lobe consists of the declive, folium, tuber, pyramid, uvula, lobus simplex, biventral lobule, semilunar lobules, and the tonsils
   B. THE ARCHICEREBELLUM consists of the flocculonodular lobe (vestibular) and the lingula (has spinocerebellar and vestibular connections)
   C. THE PALEOCEREBELLUM consists of the anterior lobe (minus the lingula) with the pyramid and uvula of the middle lobe. Phylogenetically, it appears after the archicerebellum and is of spinocerebellar function (for sensory limb information)
   D. THE NEOCEREBELLUM consists of the middle lobe (minus the pyramid and uvula) and is mostly corticopontocerebellar in function (of selective limb control). It is well developed in mammals, especially in humans, and parallels cerebral neocortical development. The neocerebellum matures slowly after birth, whereas the paleocerebellum is completed before term

III. **Histogenesis of the cerebellum**
   A. THE CEREBELLAR PLATE initially consists of a neuroepithelial, a mantle, and a marginal layer. With development, cells of the neuroepithelium migrate through the mantle layer to the marginal layer to form the *external granular or superficial cortical layer.* Simultaneously, some neuroepithelial cells, in a second wave, migrate into the marginal layer to form the *Purkinje cells*
   B. DURING MONTH 6, THE EXTERNAL GRANULAR LAYER releases various cell types, which migrate inward toward the differentiating Purkinje cells and form the granule cells, the basket cells, the stellate cells, and the Golgi cells. The layer eventually becomes the *definitive cerebellar cortex,* with all these cells. Production and migration of the cells continue postnatally for about $1\frac{1}{2}$ years
   C. THE DEEP PARAVENTRICULAR CEREBELLAR NUCLEI, i.e., the dentate, are formed by nonmigrating neuroblasts in the mantle layer and reach their definitive position long before birth. Their axons constitute the major part of the superior cerebellar peduncles containing the cerebellovestibular and cerebellorubrothalamic tracts
   D. THE GREATER PART OF THE ORIGINAL ROOF PLATE of the fourth ventricle forms the pia mater on the cerebellar surface. However, the part in front of and behind the cerebellum specializes to form the *anterior* and *posterior medullary velum,* respectively

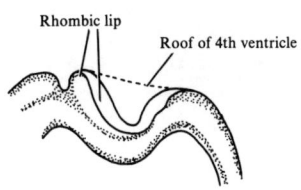
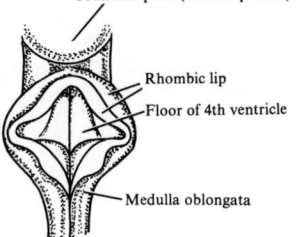

WEEK 5

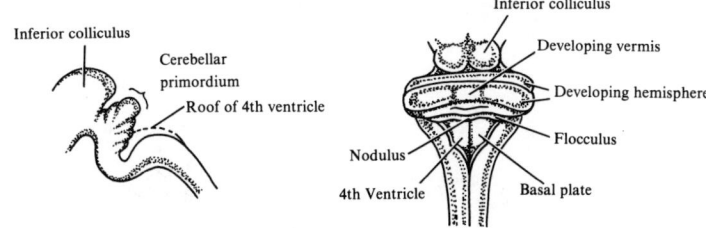

~ 3 MONTHS

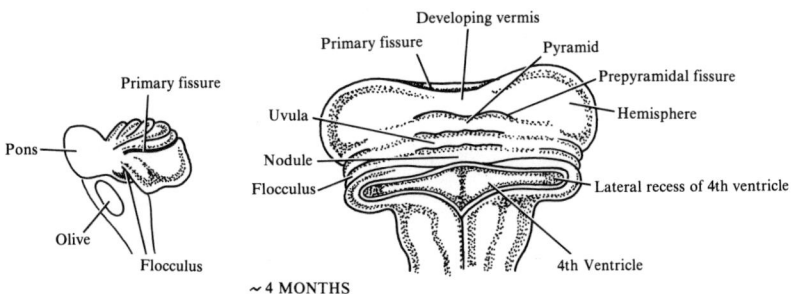

~ 4 MONTHS

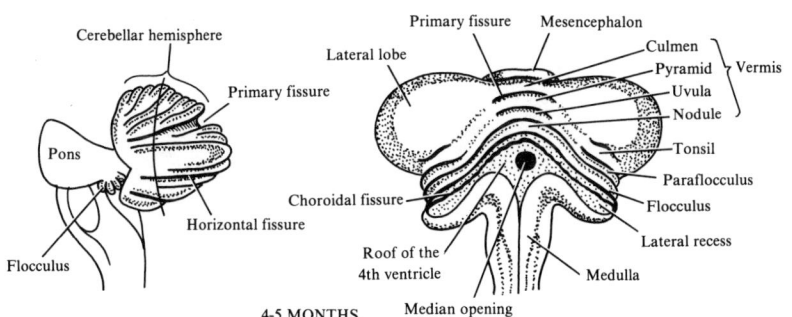

4-5 MONTHS

DEVELOPING CEREBELLUM

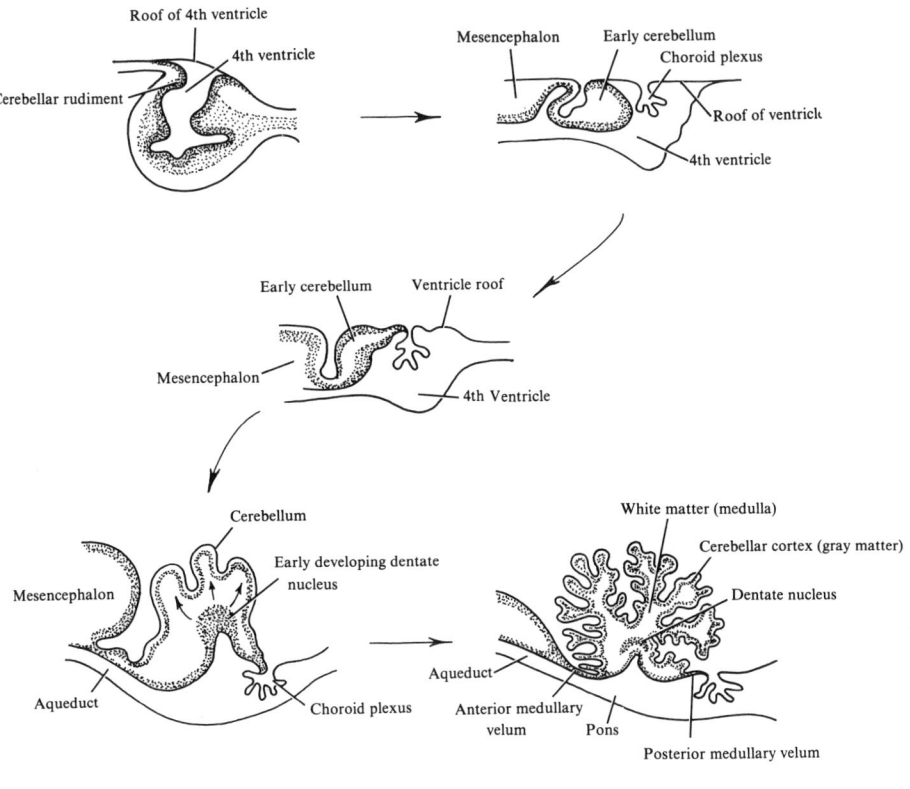
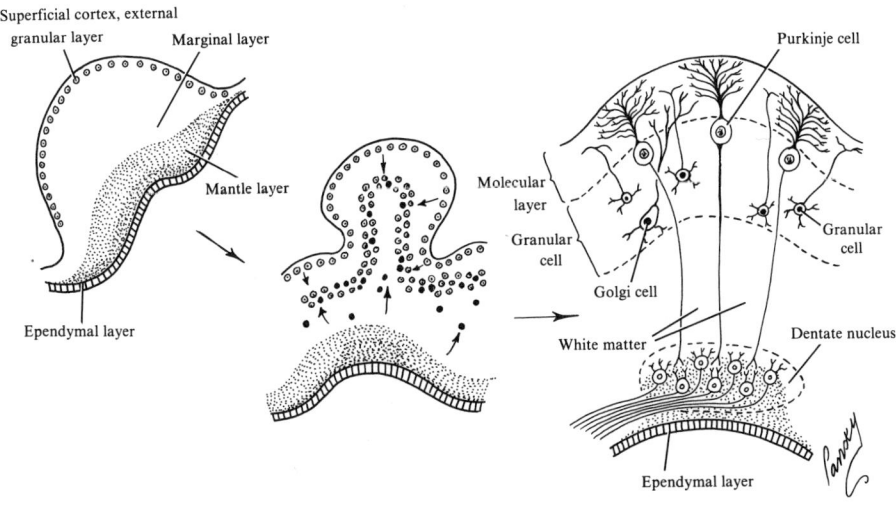

FIGURE 73. **Cerebellar development and histogenesis.**

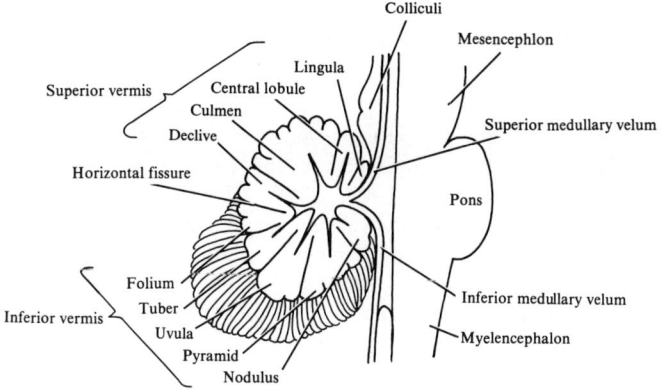

MIDSAGITTAL SECTION THROUGH ADULT CEREBELLUM (VERMIS)

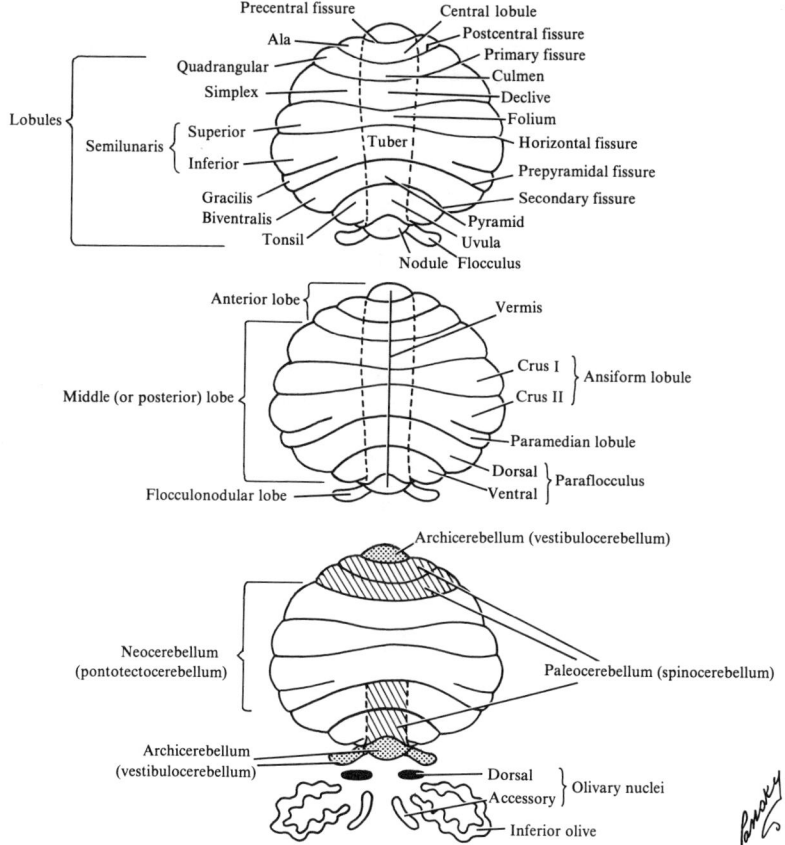

ANATOMIC TERMINOLOGY OF CEREBELLAR COMPONENTS

FIGURE 74. Adult cerebellum—components and terminology of description.

151. THE BRAINSTEM: MESENCEPHALON (THIRD VESICLE)

I. **The mesencephalon** is morphologically the most primitive of the brain vesicles. Its basal and alar plates, separated by the sulcus limitans (as seen in transverse section), are easily identified
   A. THE BASAL OR FLOOR PLATE AND CRUS CEREBRI (BASIS PEDUNCULI)
      1. Each basal plate contains 2 groups of motor nuclei
         a. A medial somatic efferent group is represented by the oculomotor (III) and trochlear (IV) cranial nerves, which innervate the preoptic (eye) muscles
         b. A small general visceral efferent group is represented by the Edinger-Westphal nucleus, also associated with the oculomotor (III) nerve, and innervates the sphincter pupillary muscle
      2. The marginal layer of each basal plate enlarges and forms the basis pedunculi (crus cerebri or cerebral peduncles)
         a. The crura serve as pathways for the nerve fibers descending from the cerebral cortex to the lower centers in the pons and spinal cord
         b. In the adult, the crura contain the corticospinal, the corticorubral, and the corticopontine tracts
   B. THE ALAR OR ROOF PLATE AND THE COLLICULI
      1. The alar plates initially appear as 2 longitudinal elevations separated by a shallow midline depression
         a. With development, a transverse groove divides each longitudinal elevation into an *anterior* (*superior*) and a *posterior* (*inferior*) *colliculus*
            i. The nuclei of the posterior colliculus serve as synaptic relay stations for the auditory reflexes
            ii. The nuclei of the anterior colliculus serve as correlation and reflex centers for the visual impulses
            iii. The colliculi are formed by waves of neuroblasts produced by the neuroepithelial cells that migrate into the overlying marginal zone and become arranged in stratified layers
            iv. In higher vertebrates that have a visual and auditory neocortex, the colliculi tend to regress somewhat and become only oculo- and auditory motor reflex centers, independent of conscious perception. In lower vertebrates, the colliculi are true sensory "brains"
      2. Some feel that the cells of the alar plate also give rise to the tegmental nuclei, the nucleus ruber (red nucleus), and the substantia nigra. Others, however, feel they develop *in situ* from the basal plates
         a. These nuclei are considered to be suprasegmental structures and are part of the extrapyramidal motor pathways. The red nuclei have a dual structure consisting of large phylogenetic ancient cells and many new small cells, also suggesting a dual origin
   C. THE CAVITY OF THE MESENCEPHALON grows smaller, as a result of the growth of its walls, and eventually forms the so-called *aqueduct of Sylvius* which connects the third and fourth ventricles
   D. THE CEREBRAL PEDUNCLES become more prominent with development as more descending groups of fibers pass through the developing midbrain on their way to the brainstem and spinal cord

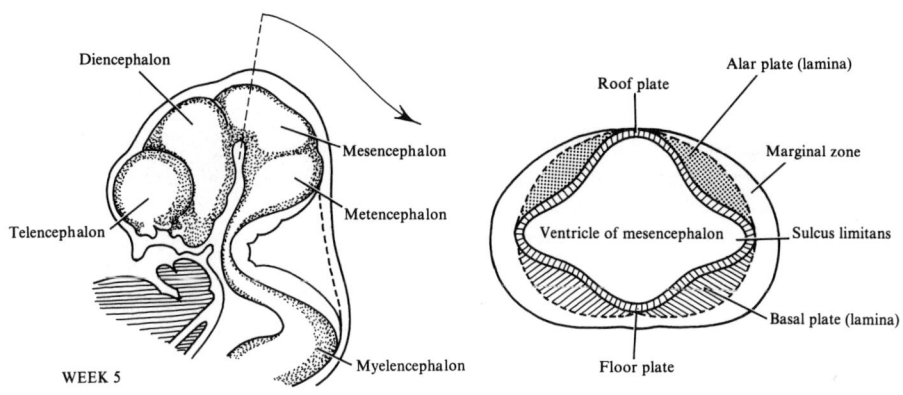
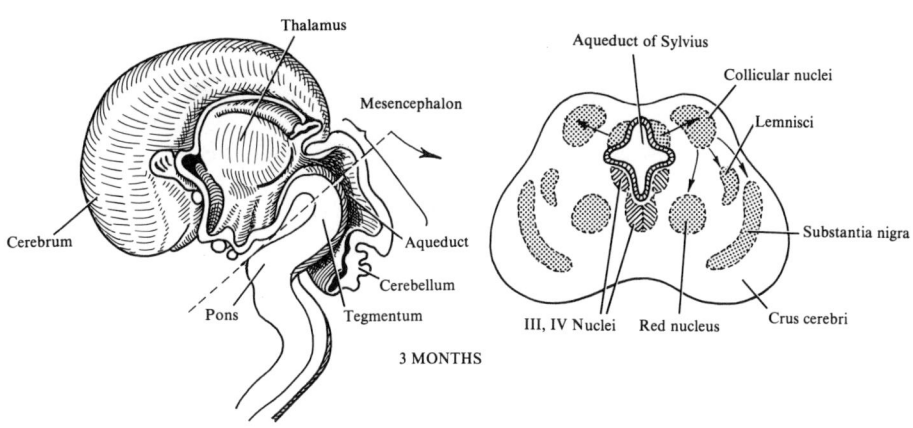
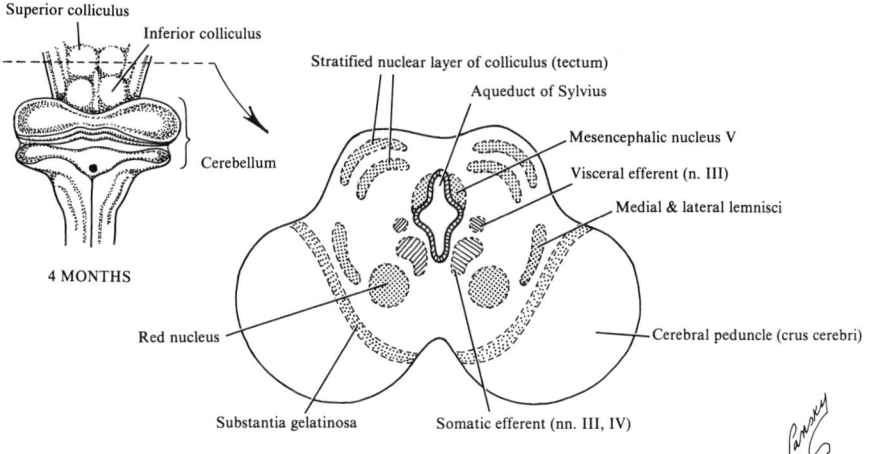

DEVELOPMENT OF MESENCEPHALON

152. THE PERIPHERAL NERVOUS SYSTEM AND CRANIAL NERVES

I. **Introduction:** the peripheral nervous system consists of the cranial, spinal, and visceral nerves as well as the cranial, spinal, and autonomic ganglia. The peripheral nervous system develops from a variety of sources
   A. ALL SENSORY CELLS, both somatic and visceral, of the peripheral nervous system are derived from *neural crest cells,* and the cell bodies of these cells are found outside the central nervous system
      1. All the peripheral sensory cells are at first bipolar (except for the cells in the spiral ganglion of the cochlear and in the vestibular ganglion of cranial nerve VIII). However, the 2 processes soon unite to form a single process and a unipolar type neuron
         a. The process has peripheral and central branches or processes
         b. The peripheral process terminates in an afferent ending; the central process enters the spinal cord or brain
         c. The sensory cells of the ganglion of the vestibulocochlear (VIII) cranial nerve remain bipolar
   B. THE CELL BODY OF EACH AFFERENT NEURON is invested by a capsule of *satellite cells,* also derived from the neural crest. The capsule is continuous with the neurilemmal sheath of Schwann cells (of neural crest origin) that surrounds the axon of the afferent neuron
      1. A layer of connective tissue, continuous with the endoneurial sheath of the nerve fiber and derived from mesenchyme, surrounds the satellite cells

II. **The cranial nerves**
   A. THE NEURAL CREST CELLS in the region of the brain migrate to form sensory ganglia only in relation to the trigeminal (V), the facial (VII), the glossopharyngeal (IX), the vagus (X), and the accessory (XI) cranial nerves. These are the branchial mixed nerves
   B. THE PURELY SENSORY NERVES are the olfactory (I), the optic (II), and the vestibulocochlear (VIII) cranial nerves
      1. The neural crest cells in the brain region form the sensory ganglia for the vestibulocochlear (VIII) nerve
      2. The olfactory and optic nerves, however, are not typical sensory nerves since the olfactory bulbs and the eyes are outgrowths from the brain. Thus, these nerves are really tracts of the brain
   C. THE REMAINING CRANIAL NERVES, the oculomotor (III), the trochlear (IV), the abducens (VI), the accessory (XI), and the hypoglossal (XII) are motor nerves
      1. They all, however, contain sensory fibers of proprioception, except for cranial nerve XI. The cell bodies of these proprioceptive fibers do not form ganglia but remain scattered among the axons of the motor neurons as they leave the brainstem
   D. THE CRANIAL ROOT OF THE ACCESSORY (XI) NERVE is actually an isolated root of the vagus (X) cranial nerve that becomes enclosed with the fibers of the so-called spinal root of the accessory nerve. After traveling with the spinal root for a short time, the cranial root enters the foramen magnum and joins the vagus nerve and is enclosed with it
      1. The spinal portion of the accessory (XI) nerve is actually a cervical spinal motor nerve arising from the lateral surface of the upper 5 or 6 cervical segments of the spinal cord. It innervates the sternocleidomastoid and trapezius muscles.

III. **Cells of the neural crest** also differentiate into multipolar neurons of the autonomic ganglia: sympathetic ganglia, collateral or prevertebral ganglia (cardiac, celiac, mesenteric, etc.), and parasympathetic or terminal ganglia (Meissner's or submucosal). They also give rise to melanoblasts and cells of the adrenal medulla
   A. CELLS OF THE PARAGANGLIA are called chromaffin cells (also of neural crest origin)

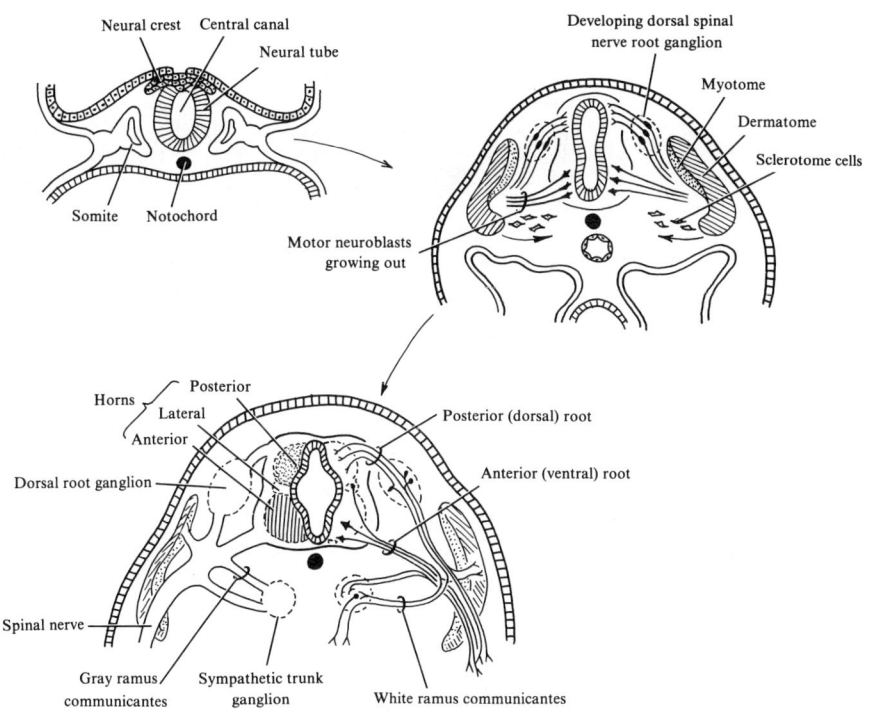
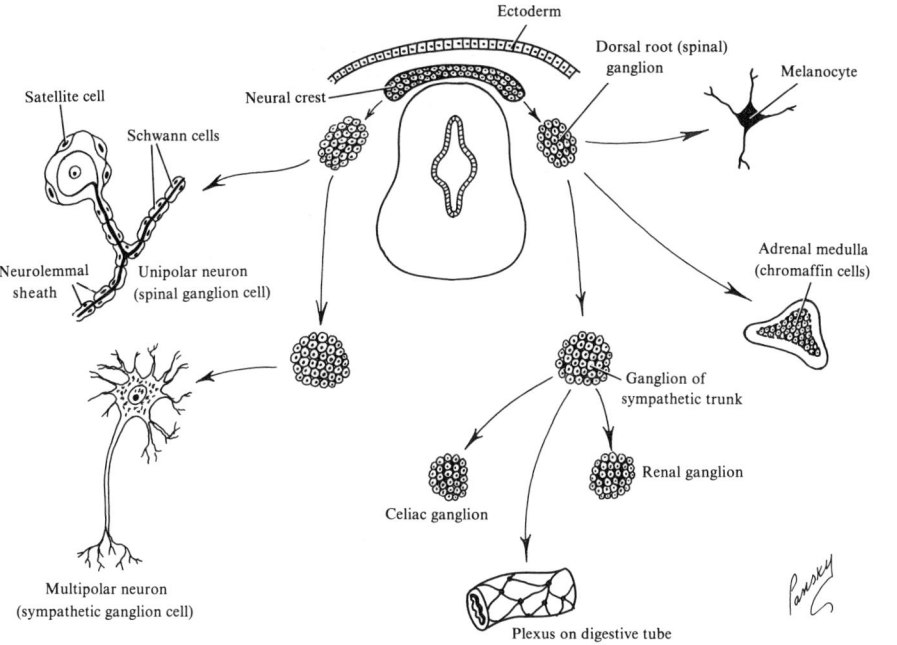

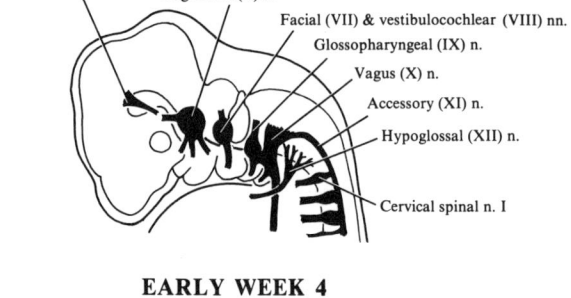
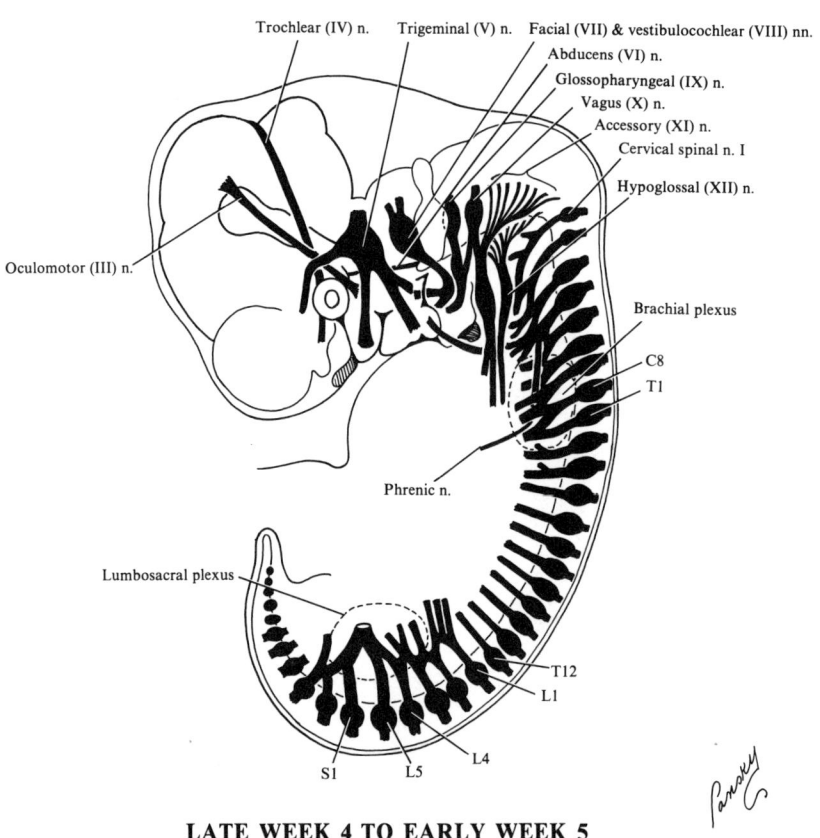

FIGURE 75. **Cranial nerves (4- to 5-week embryo).**

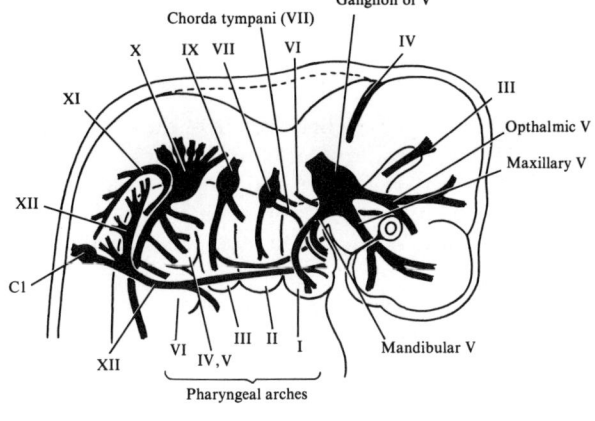
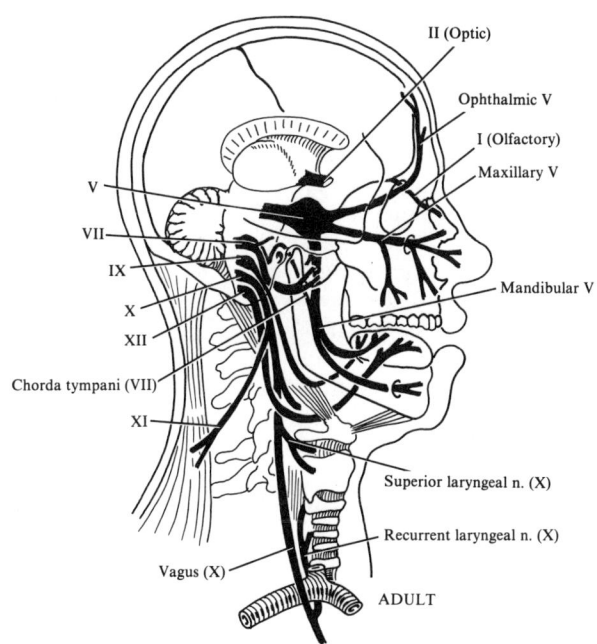

DISTRIBUTION OF ADULT CRANIAL NERVES
(After Corning and Kollman)

FIGURE 76. **Cranial nerves (fetal and adult).**

153. THE DIENCEPHALON (SECOND VESICLE)

I. **The diencephalon** develops from the median portion of the prosencephalon and consists of a roof plate, 2 alar plates, and the third ventricle. There is a question of whether it has a floor or basal plate. It is bounded posteriorly by a plane passing behind the pineal gland and mamillary bodies and anteriorly by a plane passing just rostral to the optic chiasma and encircling the foramen of Monro. The lamina terminalis is considered to be a part of the telecephalon
   A. THE ROOF PLATE consists of a single layer of ependymal cells covered by vascular mesenchyme (meninges). The 2 layers later combine to form the *choroid plexus* of the third ventricle, which closes it from above
      1. The most caudal part of the roof plate does not take part in formation of the choroid plexus but develops into the *pineal body* or *epiphysis*. The latter initially appears as an epithelial thickening in the midline, begins to evaginate by week 7, and eventually forms a solid organ. Its structure and function are both neural and glandular
      2. Occasionally, the roof plate forms another evagination near the foramina of Monro, called the *paraphysis,* which sometimes persists into postnatal life and may give rise to small cysts. It is usually seen in lower vertebrates
      3. The roof plate is also thought to give rise to the *epithalamus,* a group of nuclei located on each side of the midline close to the pineal gland. It may, however, arise from the alar plates
         a. The epithalamic region is originally large, but it regresses to a small area where the *habenular nuclei* are seen. The latter form a link in the olfactory conduction path and are connected to each other, across the midline, by a group of nerve fibers collectively called the *habenular commissure*
         b. Just posterior to the pineal body, a small commissure, the *posterior commissure,* connects the paramedian epithalamic nuclei
   B. THE ALAR PLATES form both the lateral walls and the floor of the diencephalon
      1. A distinct longitudinal groove, the *hypothalamic sulcus,* divides the alar plate into a dorsal and ventral region, the *thalamus* and *hypothalamus,* respectively
         a. This sulcus is of a different nature than the sulcus limitans since it does not form a dividing line between motor and sensory areas
      2. The thalamus is important in evolution. It begins as a simple relay station in the opticomesencephalic pathways, but gradually becomes a polysensorial connection, interposed between the sensory receptors and the cerebral cortex. Its major role is in humans
         a. After proliferation, the thalami gradually bulge into the diencephalic lumen, and the 2 may fuse in the midline to form the *massa intermedia* or *interthalamic connexus*
         b. The thalamic nuclear areas eventually form 2 distinct nuclear groups
            i. A dorsal nuclear group: important for the reception and transmission of visual and auditory impulses
            ii. A ventral nuclear group: a passage and relay station for higher centers
      3. The hypothalamus is the coordinating and effector receptor center of autonomic function in all vertebrates. It differentiates into a number of nuclear groups that serve as regulation centers of visceral functions, i.e., sleep, digestion, body temperature, emotional behavior, etc.
         a. One of the groups, the *mamillary bodies,* forms rounded elevations on the ventral surface of the hypothalamus on either side of the midline and connects the hypothalamus to higher centers of the rhinencephalon and lower brainstem centers
         b. The lateral wall of the diencephalon also provides the *pallidum* which is the only striated intercerebral structure having a diencephalic origin
      4. The floor of the diencephalon gives rise to the neural primordia of the eyes and between the 2, the funnel-shaped *infundibulum* and future *neural lobe of the hypophysis*

WEEK 4
4 mm (After Hochstetter)

WEEK 6
(9 mm)

WEEK 11
65 mm

WEEK 8
25 mm

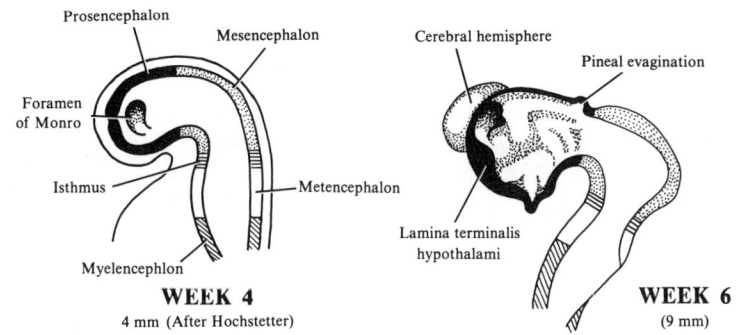

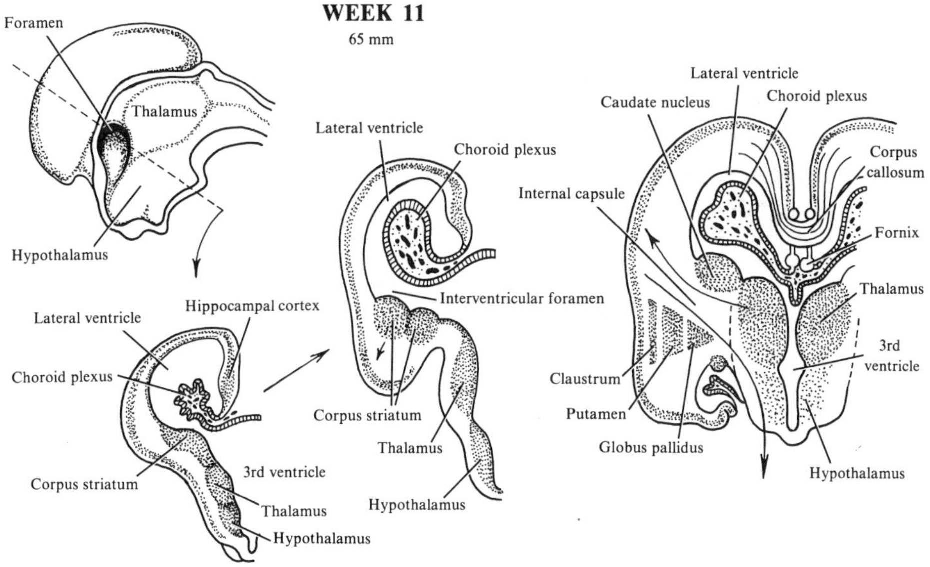

SECTIONS AT LEVEL OF INTERVENTRICULAR FORAMEN
Early to late stages

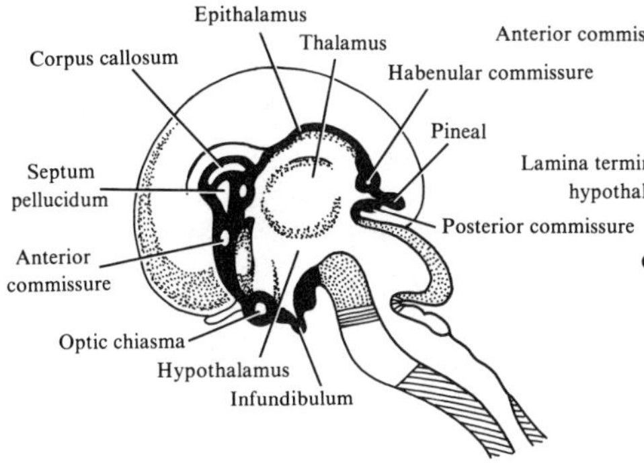

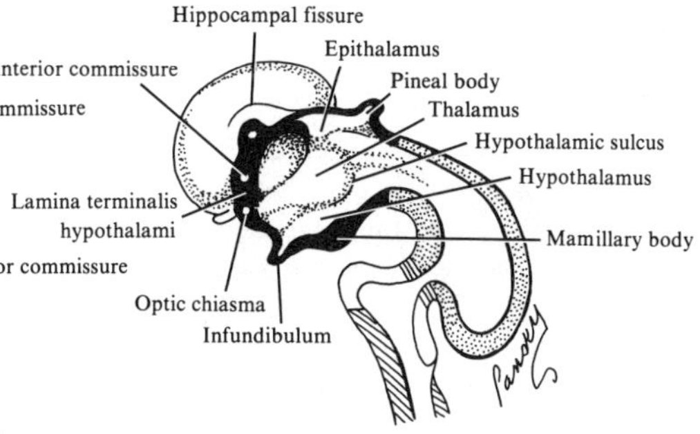

154. THE TELENCEPHALON (FIRST VESICLE): PHYLOGENESIS

I. **Development of the telencephalon** becomes more and more extensive as one goes up the evolutionary scale of vertebrates
  A. IN CYCLOSTOMES AND FISH the telencephalic vesicle remains simple, and its cavity consists of a *single ventricle*. Only the olfactory bulbs are differentiated
  B. IN AMPHIBIANS the telencephalon gives rise to 2 evaginations or hemispheres, both of which enclose a lateral ventricle
  C. UP TO THE REPTILES development is slightly modified, and one sees 5 vesicles
  D. IN MAMMALS the hemispheres grow extensively, both laterally and caudally, and gradually engulf the diencephalon and surround the dorsal structures of the mesencephalon and part of the metencephalon
    1. Centers which were once dominant lose their autonomy and come under control of the telencephalon
    2. Special areas related to the hemispheres are differentiated, such as the neocerebellum, neorubrum, etc.
    3. With growth of the telencephalon, the olfactory bulbs slowly regress and essentially become appendages (in contrast to development in lower vertebrates)
    4. The hemispheres extend progressively backward in the human fetus and increase in volume and in cortical surface, related to the association areas (±70% in man; ±20% of cortical surface in rabbits)
  E. THERE IS AN INCREASED POTENTIAL FOR ASSOCIATION CENTERS, corresponding to the extension of the alar plates
    1. In cyclostomes the pallium* is seen as a thin cellular layer without special structure or importance. The cells are paraventricular. The telencephalon is essentially olfactory and regulates rudimentary behavior
    2. In fish the pallium is thicker, but the cells are still paraventricular and lack a clearcut organization. The telencephalon is not exclusively olfactory since it receives nonolfactory afferent fibers via the striatum and the septum
    3. In amphibians the cells of the pallium migrate toward the surface and differentiate into 2 cell types: small granular receptor cells and large effector cells. Together they form the *archipallium*
       a. A poorly differentiated paleopallial layer is seen on the outer surface of the ventricle which connects the olfactory afferent fibers and the archipallium
    4. In reptiles the 2 cellular types, described above, are stratified and form the *archeocortex*. The paleopallial zones migrate toward the surface and show a stratification to form the *paleocortex*, rich in receptor and association cells
       a. The rhinencephalon now consists of archeocortex, paleocortex, striatum, and septum and becomes the center of behavior, since it connects many afferent as well as olfactory and visual areas
    5. In mammals the major development is a 6-layer neocortex (from neopallium), accompanied by extensive cellular migration to establish an outer gray matter and an inner white matter, reversing the primitive medullary structure. With evolution, the neocortex is extended and compresses the archeocortex on the inside and the paleocortex below
    6. In primates (man) the neocortex is so extensive that it invades most of the former cortices. The human rhinencephalon is composed of neocortical zones integrated into a new system which is no longer the "principal" brain but is superceded by the *neoencephalon,* which consists of neocortex plus the neostriated structures. Here all sensory afferent fibers converge. Olfactory functions of the rhinencephalon regress
       a. The neoencephalon receives much information, and as a result of its being formed predominantly of association areas, its potential for integration of afferent impulses is immense, leading to thought processes and complex associations

*Pallium [L. cloak]. Mantle; brain mantle; the cerebral cortex with the subjacent white substance.

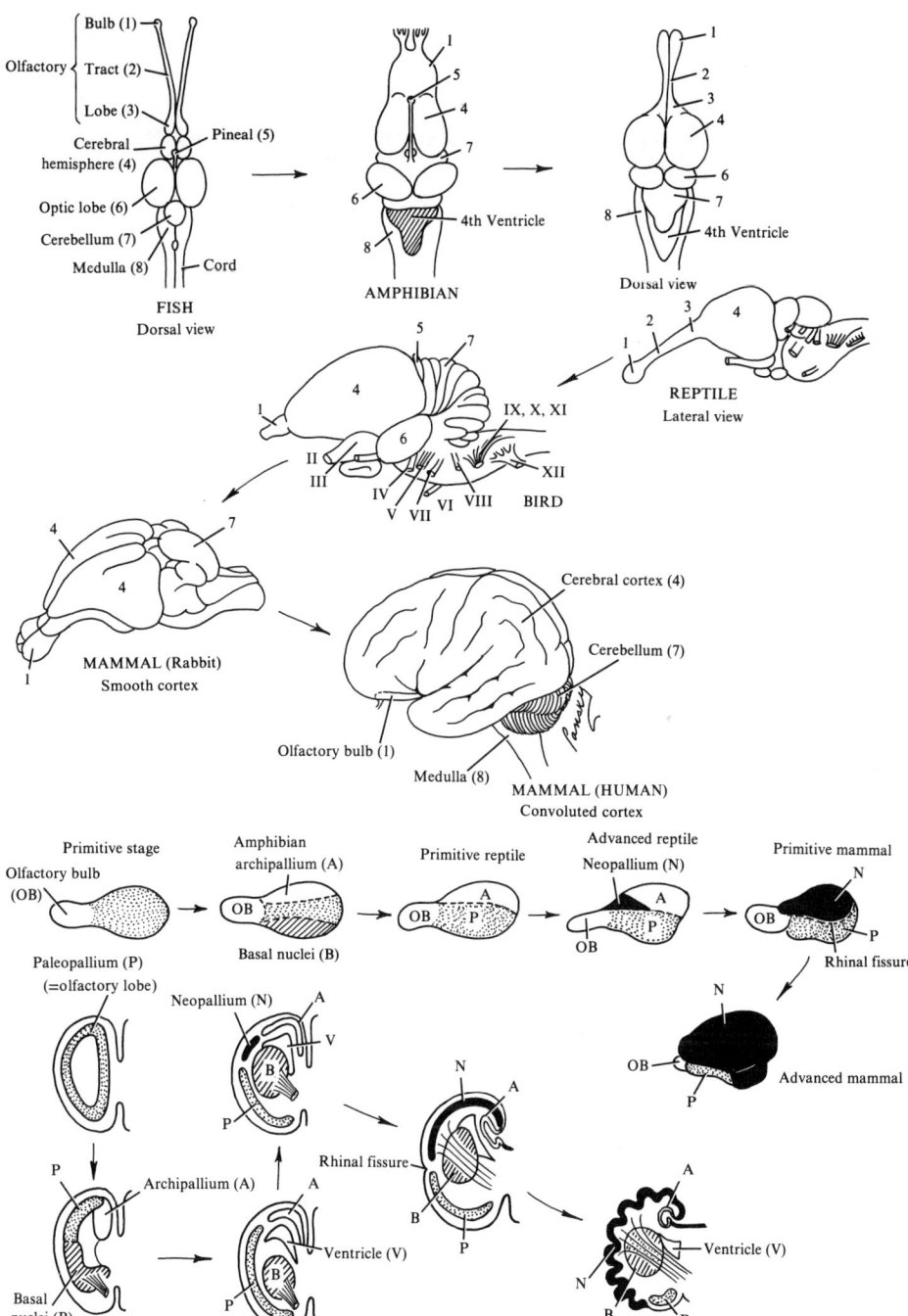

DEVELOPMENT OF CORPUS STRIATUM (BASAL NUCLEI) AND CORTEX

155. THE BRAIN: THE TELENCEPHALON (FIRST VESICLE)

I. **Introduction:** the telencephalon is the most rostral part of the brain vesicle and consists of 2 lateral outpocketings, the *cerebral hemispheres,* and a median part, the *lamina terminalis.* The hemispheres grow simultaneously in lateral, longitudinal, and parietal directions. The growth is dominated by the great development of the neocortex, which, in humans, occupies about 90% of the cerebral surface. The *pallium* or vault of each hemisphere forms the *cortex.* The floor gives rise to the *striated bodies,* and between the cortex and the bodies, the *lateral ventricles* are formed

II. **Development during the first 2 months of embryonic life**
   A. 30- TO 32-DAY EMBRYO: each hemisphere contains an evagination consisting of a vault (pallium or future cortex) and a lateroventral region (floor or future striatum)
      1. The 2 lateral cavities of the telencephalon communicate with the cavity of the diencephalon via the *2 foramina of Monro*
   B. 45- TO 50-DAY EMBRYO: the vault and floor begin to differentiate. The pallium thickens slightly, and one can recognize the archeo-, paleo-, and neopallial regions which will give rise to the corresponding areas of the cortex
      1. The floor becomes greatly thickened as a result of marked activity of its germinating zone and gives rise to the primordia of the *striated nuclei,* the *lateral striated body* on the outside, and the *median striated body on the inside*
      2. The *interhemispheric or choroid fissure* is formed. The junction between the 2 vaults is very thin, and along the fissure the *choroid plexuses* invaginate
      3. The ventricular cavities narrow and differentiate as a result of parietal growth. One sees the lateral ventricles, the foramina of Monro, and the third ventricle
   C. 50- DAY TO 2-MONTH EMBRYO: the neopallium extends and engulfs the paleopallium on the ventral side and the archeopallium on the dorsal side so that the telencephalon progressively surrounds the diencephalon
      1. In a parallel manner, the various cells of the striatum develop and contribute to the thickening of the area of junction of the telencephalon and diencephalon. This permits the 2 structures to become continuous, and a wide area is formed that helps join the hemispheres to the rest of the neural axis
         a. Efferent and afferent fibers to and from the cortex arrange themselves in this junctional zone to form the *internal capsule*

III. **Development after the second month**
   A. PALLIAL DEVELOPMENT: the neopallium grows and compresses the archeopallium on the inside and the palleopallium below. The striated bodies become paramedian. The pallial zones give rise to distinct cellular layers which are the primordia of the cortex
      1. The parietal growth produces some volume reduction of the ventricular cavities
      2. The mass of white matter between the cortex, the ventricles, and the central gray nuclei also increases, and myelinated fibers come to and leave the cortex
   B. FLOOR DEVELOPMENT is characterized by the appearance of the striated nuclei. The latter are traversed by fibers of the internal and external capsules
      1. The lateral nucleus striatum gives rise to the *neostriatum* which forms the *caudate nucleus* and *putamen.* The *claustrum* also may be derived from it
      2. The median nucleus striatum gives rise to the *paleostriatum* or *globus pallidus.* This nucleus is responsible for most of the strioencephalic connections. It fuses laterally with the putamen to form the *lenticular nucleus*
      3. The *amygdaloid nucleus* or *archeostriatum* differentiates from the most ventral region of the floor, below the lenticular nucleus
      4. The septal formations are seen in humans but have lost their great importance. They are derived from the median telencephalic parts that blend ventrally with the diencephalon. They form an important junction where the hippocampal, olfactory, and neocortical pathways connect. They send efferent fibers to the hypothalamus and the reflex centers of the brainstem

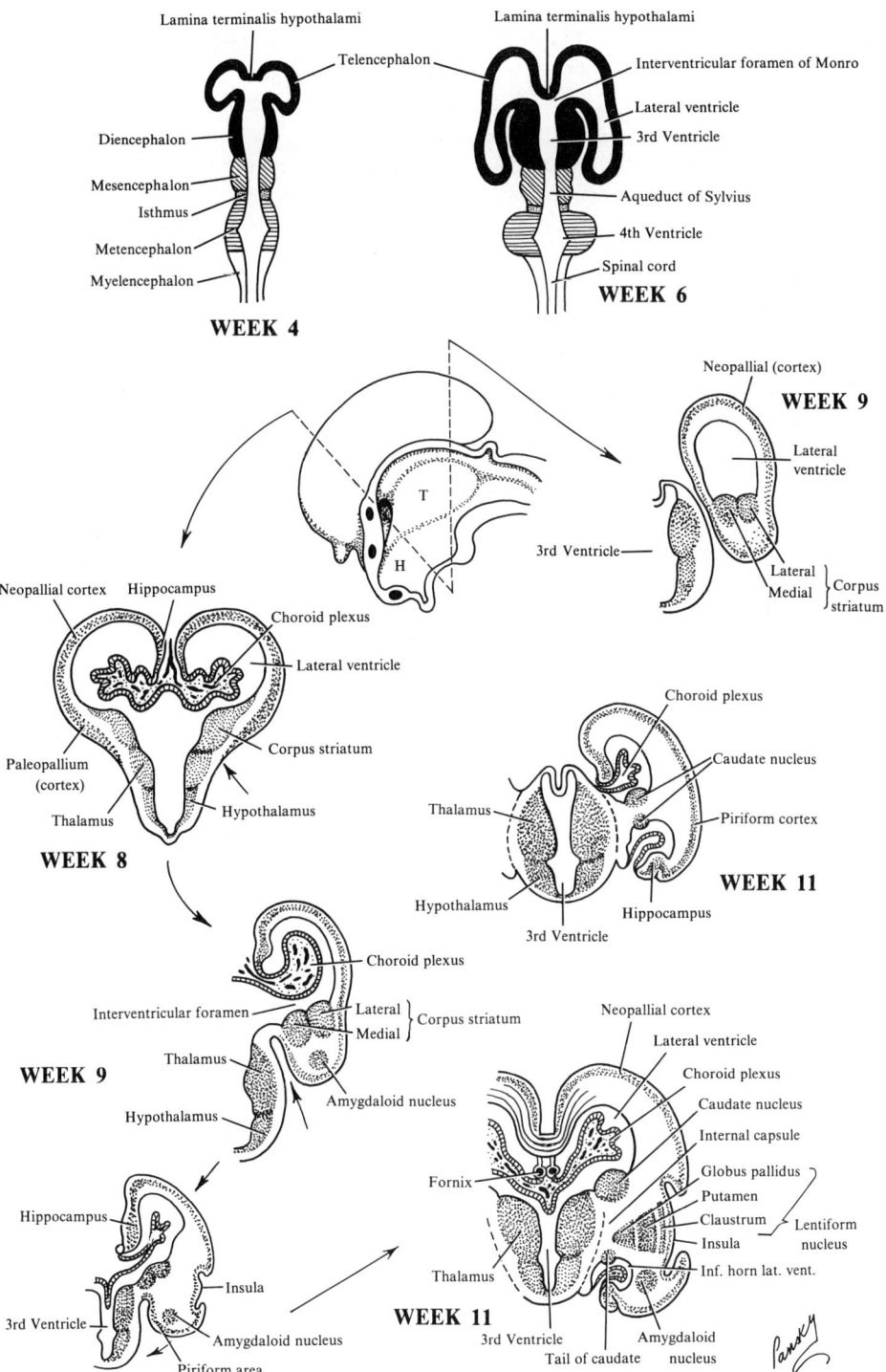

156. THE BRAIN: THE TELENCEPHALON (FIRST VESICLE)—LOBES AND PALLIAL DEVELOPMENT

I. Development of the lobes of the telencephalon

A. THE 2 PRIMITIVE HEMISPHERES GROW FORWARD, producing the *frontal lobes,* and backward where they cover the diencephalon and form the *parietal* and *occipital lobes.* The posterior pole then expands and grows in a forward direction to form the future *temporal pole*
   1. Thus, the vaults of the hemispheres take on a horseshoe shape" and delimit the *fossa of Sylvius* (sylvian fissure), by month 9
   2. In a parallel manner, the posterior areas continue to grow and gradually cover the mesencephalon and part of the metencephalon

B. RESULTS OF DEVELOPMENT OF THE PALLIUM
   1. The ventricular cavity sends forth a process into the temporal lobe to form the *inferior horn of the lateral ventricle* and another into the occipital lobe to form the *posterior horn of the lateral ventricle*
   2. In the paraventricular areas of the floor plate, the caudate nucleus, which protrudes into the cavity of the lateral ventricle, follows the turning movement of the posterior pole of the hemisphere and thus becomes horseshoe-shaped.
   3. The putamen rotates very little and is arched in an anteroposterior direction. It is surrounded progressively by the caudate nucleus
   4. The central globus pallidus does not move at all
   5. The surface of the pallium folds increasingly during its transverse and longitudinal development. Toward month 6, the *parieto-occipital fissure of Rolando* makes its appearance. From month 7 on, grooves form that separate various convolutions well established at birth

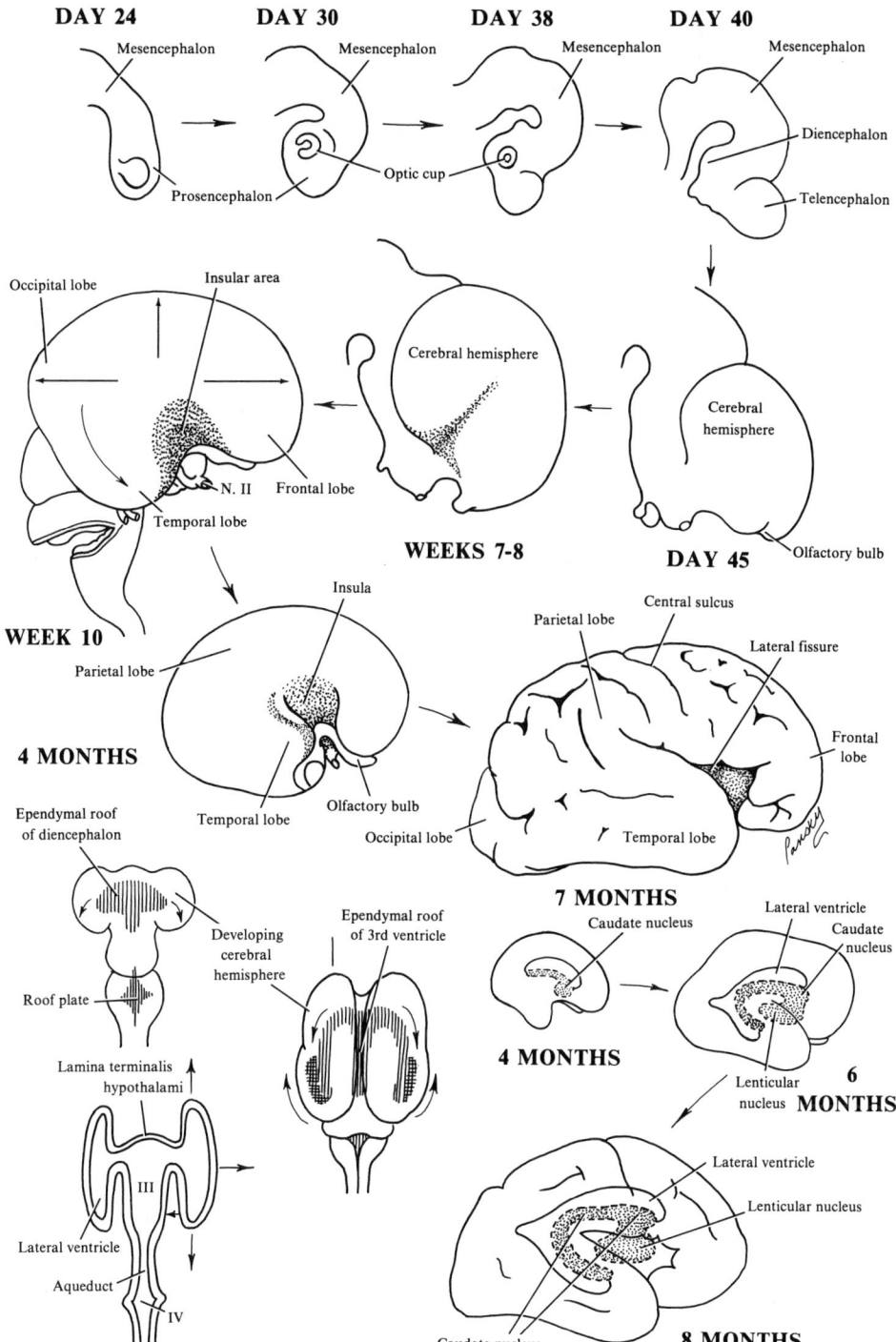

157. THE TELENCEPHALON (FIRST VESICLE): DEVELOPMENT OF THE RHINENCEPHALON

I. **The rhinencephalon** is derived from the archeo- and paleopallium and consists of a number of intracerebral structures, namely, the *limbic lobe* and *attached structures,* and the *olfactory bulbs*
   A. THE RHINENCEPHALON is the essentially olfactory brain of the lower vertebrates without a neocortex and regulates their behavior
      1. In mammals, growth of the neocortex brings about the relative regression of the rhinencephalon and, in humans, where regression is maximal, olfactory function is markedly reduced. Nevertheless, with its neocortical parts, the rhinencephalon still regulates a great deal of fundamental behavior
   B. DEVELOPMENT OF THE LIMBIC STRUCTURES
      1. The hemispheric vesicle has a simple form at about $2\frac{1}{2}$ months. The archipallium is located at the internal face of the hemispheres and the paleopallium at their ventral face, below and outside the striated bodies
      2. Between months 3 and 5, the topography of the cortex is greatly modified by both the longitudinal and transverse growth of the neocortex to form the *temporal lobes*
      3. As it grows, the neocortex invades most of the dorsal archeocortex to form the *convolution of the corpus callosum.* It also invades most of the paleocortex to form the *fifth temporal convolution.* Both convolutions merge posteriorly to form the *limbic lobe*
         a. The structure of the limbic lobe is intermediate between that of the neocortex and that of the corresponding cortices, except anteriorly where the limbic lobe is directly related to the afferent fibers of the olfactory bulbs. In these areas, it remains an archeocortical or paleocortical organization and give rise to the olfactory areas of the cortex
      4. The structures attached to the limbic lobe consist of
         a. Archeocortical elements which are generally spared by the neocortex and more or less are regressed in man
         b. The dorsal hippocampus (hippocampal cortex) consists of a thin, atrophied band of gray matter on the dorsal surface of the corpus callosum (the hippocampal rudiment)
         c. The ventral hippocampus is more developed and projects into the ventricular lumen of the inferior horn of the lateral ventricle and forms the intraventricular hippocampus and the dentate gyrus
         d. The intra- or interhemispheric fiber systems of association
            i. The hippocampus sends projection fibers via the fimbria, the fornix system, and its commissures to the septum pellucidum, the anterior nucleus of the thalamus, the premamillary region of the hypothalamus, the nuclei of the mamillary body, and to the hippocampus of the opposite side
      5. The olfactory areas of the cortex connect with lower structures (hypothalamus and brainstem) and with upper structures (neocortex of the limbic lobe and frontal cortex) by way of the amygdaloid nucleus, the nuclei of the septum, and the hippocampus
   C. DEVELOPMENT OF THE OLFACTORY BULBS is the result of a cortical formation which takes place under the inductive influence of nerve fibers from the olfactory epithelium of the nasal cavities
      1. At first, it is a hollow evagination, which gradually fills and elongates under the frontal lobe of the cerebrum. The enlarged end of the evagination forms the olfactory bulb
      2. The narrower stem by which the bulb is attached to the hemisphere forms the olfactory tract
      3. The olfactory bulb connects with the entorhinal paleocortex and the archeocortex of the subcallosal convolution

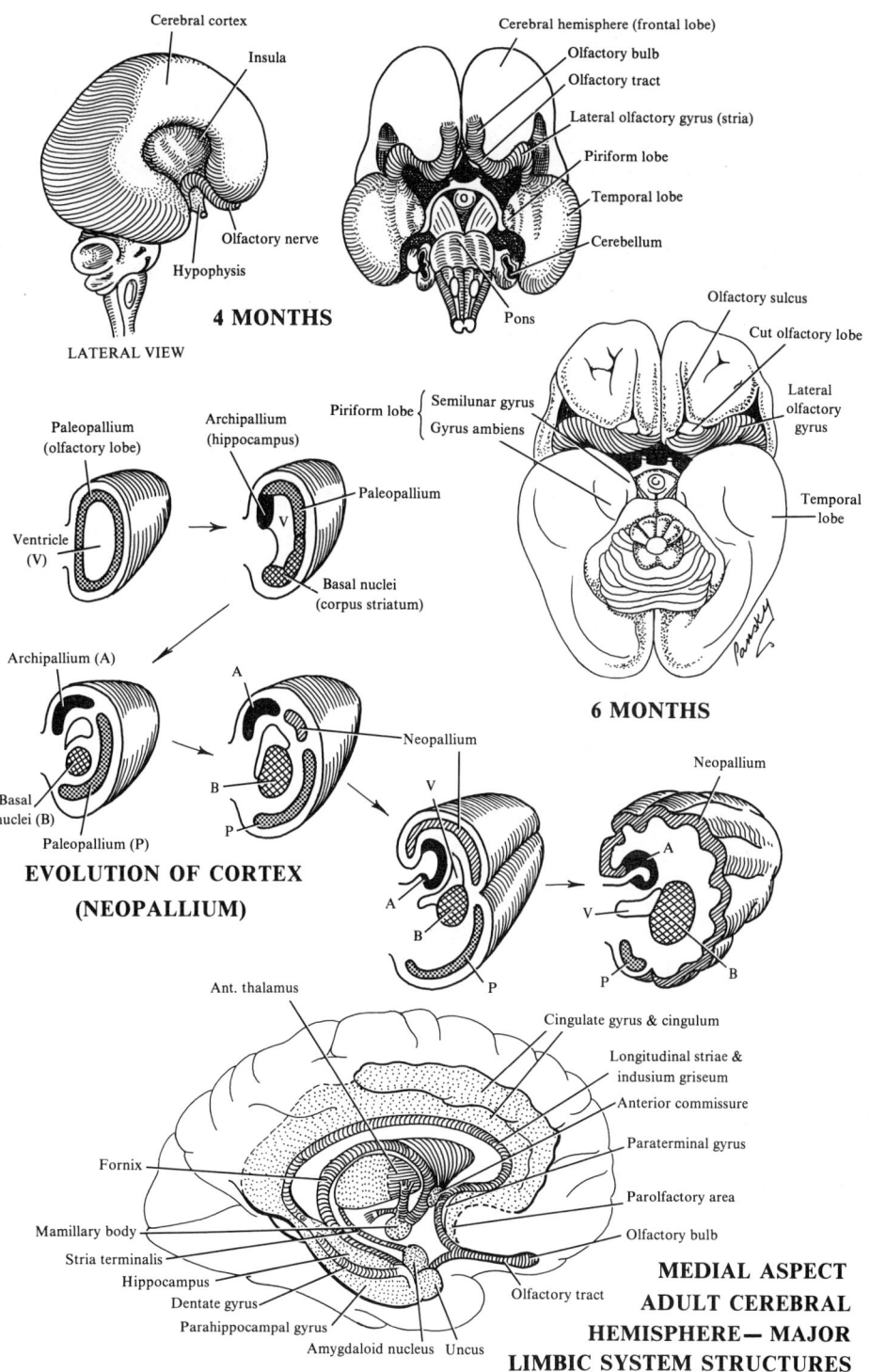

158. HISTOGENESIS OF THE CEREBRAL CORTEX

I. **By the end of month 1,** the telencephalic pallium (cortex), in the human fetus consists of a germinal layer and an ependymal layer, which together form the stratified cellular wall of the neural tube

II. **By the end of month 2,** medullary differentiation is well advanced, and the cells migrate toward the surface to form the mantle layer. Thus, the wall of the hemisphere has the general primitive structure of the neural axis with the germinal or mantle layer on the interior and the marginal or future molecular layer on its exterior

III. **During month 3,** the cortical layer is formed by cells from the mantle layer migrating toward the surface. This cortical layer is thin in the archi- and paleopallial regions and thick in the neopallial region. Various cortical regions differentiate from the cortical layers
   A. THE PRIMORDIAL OLFACTORY CORTEX begins first between 2 and 3 months, and we see the *hippocampus* (*archeocortex*) and *paleocortex*
      1. A 3-layer cortex forms in these areas
   B. DIFFERENTIATION OF THE NEOCORTEX extends from the beginning of month 3 to the end of month 6. It is characterized by extensive cellular migrations, resulting in the formation of *6 cellular layers*
      1. In the adult, the cerebral cortex forms a layer of gray matter approximately 3 to 5 mm thick
      2. The cellular migration and layering of cells are first seen at the level of the insula and parietal cortex. Thus, the somesthetic system, which ends in the ascending parietal convolutions of the brain, is functional in the fetus very early, before the special senses like sight and hearing
      3. Cellular migrations then appear at the level of the frontal and occipital cortices
      4. In month 6, the neurons form their processes and in month 7, the various types of cortical structures are established, e.g., motor, sensory, associative, and intermediary, according to the proportion of specialized cells they contain
   C. IN CONTRAST TO OTHER REGIONS of the nervous system, the germinal layer is active for only several months after birth

IV. **Cortical organization:** 5 or 6 major types have been described, as well as their distribution in various areas of the cerebral hemispheres
   A. THIS HAS PROVED VALID in functional terms even though variations do occur

V. **Cellular migrations** are increasingly evident as they relate to the phylogenetically more recent structures
   A. THEY ARE DISCRETE in the spinal cord, notable in the brainstem, important in the cerebellum and in the archeocortex, and at their maximum in the neocortex
      1. As a result of their degree of development in the brainstem, cerebellum and archeocortex, and neocortex, the gray matter in these areas is peripheral and the white matter central, which is just opposite to that in the spinal cord, where the white matter is peripheral and the gray matter is central
   B. THE MATURE INFANT is born with most of its cortical neurons, approximately about 9 to 14 billion, besides nerve fibers, neuroglia, and blood vessels
      1. Only the neuroglial cells continue to multiply actively and separate the neurons
      2. The connections of each neuron increase progressively and may reach a huge number, in the order of 10,000 synapses per cell
      3. The total surface of the adult cerebral cortex has been estimated at about 285,000 mm^2, with a volume of about 300 cm^3
      4. The cerebral cortex varies in thickness from 1.55 to 3.5 mm

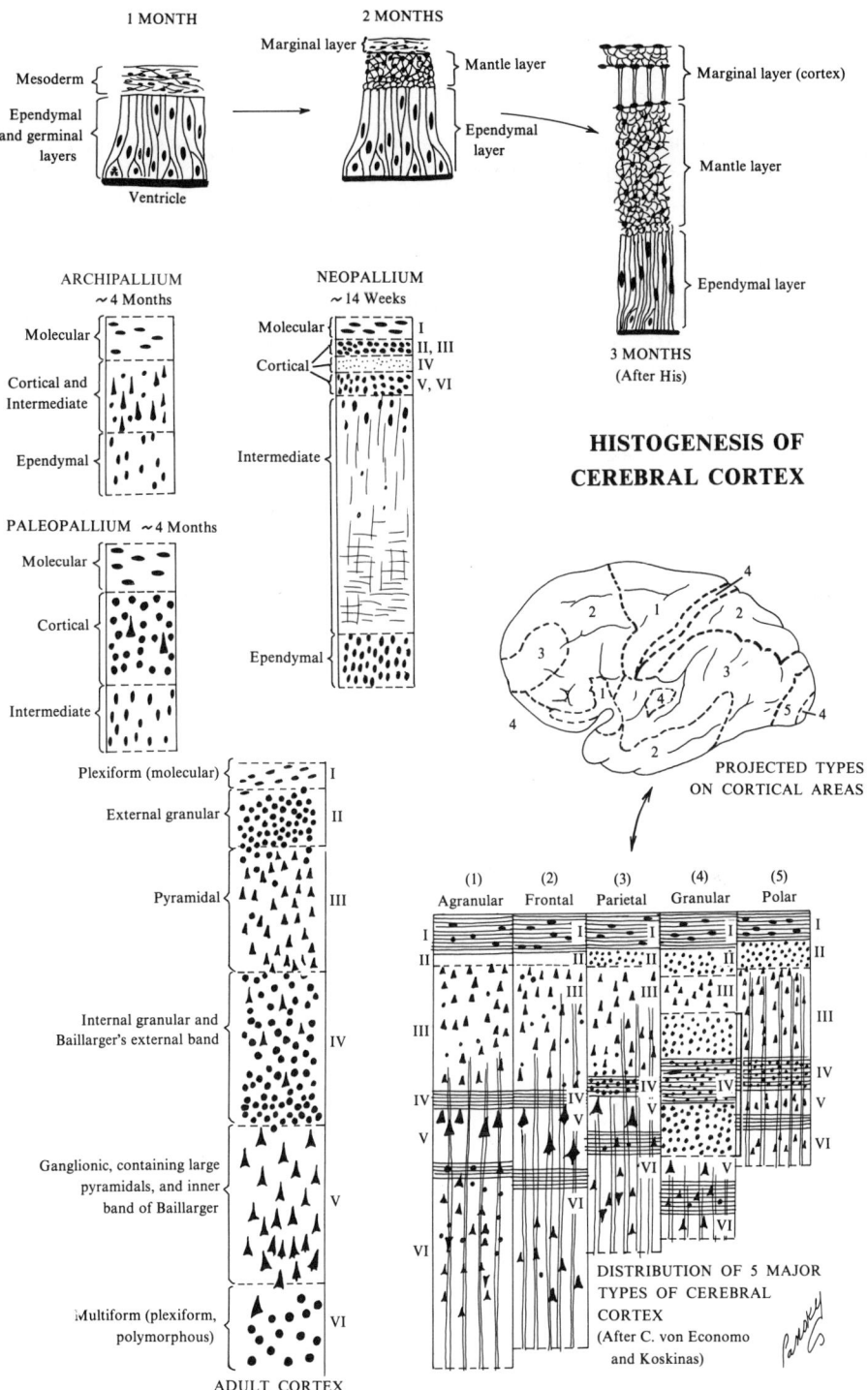

HISTOGENESIS OF CEREBRAL CORTEX

159. COMMISSURES OF THE TELENCEPHALON

I. **Introduction:** the commissures of the telencephalon are the fiber systems that cross the midline and connect homologous portions of the right and left halves of the cortical hemispheres. The most important make use of the *lamina terminalis* which is the zone of junction of the 2 telencephalic vesicles, corresponding to the closure of the anterior neuropore, and extends from the roof plate of the diencephalon to the optic chiasma
   A. THE RHINENCEPHALON COMMISSURES appear before those of the neocortex, probably as a result of progressive maturation of various cortices as well as phylogenetic order
   B. FUNCTIONAL UNITY OF THE 2 HEMISPHERES, particularly of the 2 halves of the symmetric central nervous system, is assured by the systems of association, thus necessitating commissures in the subjacent layers and even in the spinal cord

II. **The anterior commissure** is the first to appear and is seen at the end of month 2, connecting the 2 olfactory bulbs by its anterior processes and the 2 convolutions of the hippocampus (right and left temporal lobes) by its posterior processes. Its fibers pass in the inferior part of the lamina terminalis

III. **The hippocampal or fornix commissures** connect the two hippocampi and the 2 paleocortices in the area of the *trigonum.* It appears during month 3
   A. THE COMMISSURAL FIBERS arise in the hippocampus, follow the posterior pillars of the fornix, and converge on the lamina terminalis near the roof plate of the diencephalon and in the middle of the lamina terminalis. It tends to regress somewhat after the corpus callosum appears
      1. The above fibers then continue to form an arching system, just outside the choroid fissure as the anterior pillars, which connect to the mamillary bodies
      2. The anterior and posterior pillars of the same side join the hippocampus and the homolateral mamillary bodies. Thus, in the true sense, they are not commissures but rather a system of intrahemispheric association fibers

IV. **The corpus callosum** appears by week 10 and connects nonolfactory areas of the 2 hemispheres (neocortex). It is not fully formed until about month 6
   A. INITIALLY, THE CORPUS CALLOSUM forms a small bundle in the lamina terminalis just rostral to the hippocampal commissure, but with expansion of the neopallium, it expands anteriorly and posteriorly, arching over the thin roof of the diencephalon
      1. Growth of the corpus callosum in an anterior direction pulls the area of the lamina terminalis away from the fornix commissure, thus out from the lamina terminalis locally, forming the *septum pellucidum* (may contain a small cavity)
         a. It has been suggested that the septum represents the apposed walls of the 2 hemispheres anterior to the lamina terminalis

V. **Appearance of several other commissures in the lamina terminalis**
   A. TWO APPEAR JUST BELOW AND ROSTRAL to the stalk of the pineal gland: the *posterior commissure* and the *habenular commissure*
   B. THE THIRD, THE OPTIC CHIASMA, seen in the rostral wall of the diencephalon contains fibers from the medial halves of the retinae which cross on their way to the lateral geniculate bodies and the anterior colliculi

VI. **Pathology related to the commissures:** the most frequently affected development is that of the corpus callosum and may involve total or partial agenesis or hypogenesis, with an abnormally thin corpus. Possible causes for the derangement are
   A. A DEFICIENCY OF THE NEOCORTICAL REGIONS in producing fibers
   B. A DEFICIENCY OF THE PATHWAY, the lamina terminalis
   C. AGENESIS OF THE CORPUS CALLOSUM, seen in 2% of cases with cephalic pathology, usually results in debility and epilepsy, however, some types have no clinical symptoms because the callosal fibers reach the neocortex via the anterior commissure

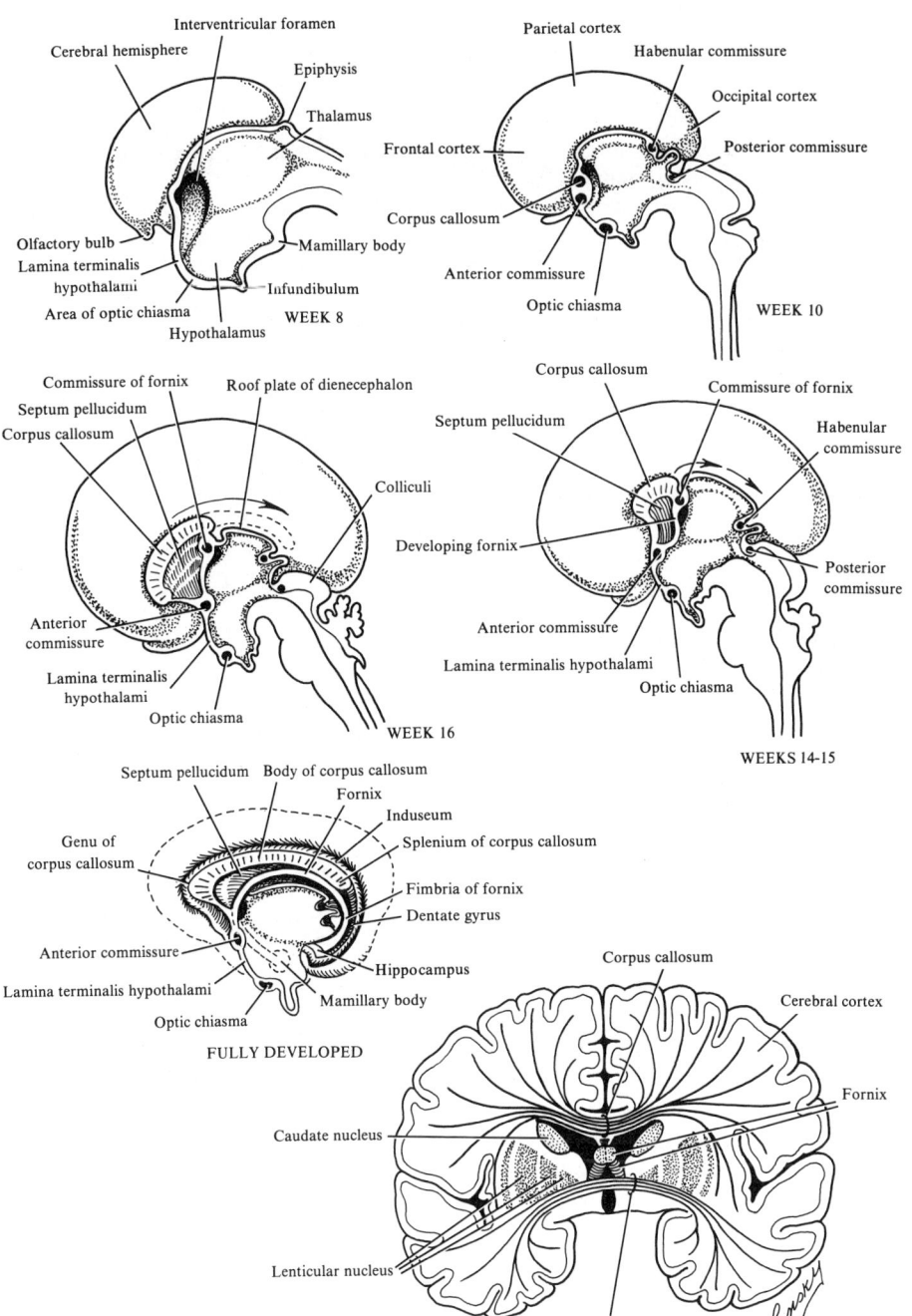

COMMISSURES OF TELENCEPHALON   TRANSVERSE SECTION

160. THE COVERINGS AND VASCULARIZATION OF THE BRAIN

I. **The meninges** are membranes interposed between the bones of the skull and spine and the central nervous system (CNS) which completely enclose the CNS. Normal development of the meninges depends on that of the CNS. The loose mesenchyme around the neural tube condenses to form a covering or membrane called the *primitive meninx*. The innermost cells of this layer may be of neural crest origin
   A. THE LEPTOMENINGES (soft meninges) are the *pia mater* and the *arachnoid*
      1. They are derived from ectomesenchyme from neural crest cells
      2. The pia mater (innermost layer) lies directly on the nervous tissue
      3. The arachnoid, just outside the pia mater, surrounds the blood vessels
      4. The arachnoid trabeculations between the pia mater and the arachnoid attest to the fact that the leptomeninges form from a single layer
   B. THE DURA MATER (hard meninges) is the outermost layer and is derived from ordinary mesenchyme and differentiates after the leptomeninges
   C. FLUID-FILLED SPACES in the leptomeninges coalesce to form the *subarachnoid space*

II. **Role of the meninges** is generally protection against mechanical shock
   A. VASCULAR ROLE: the pia mater and the arachnoid follow the vessels that penetrate the nervous tissue. They sheath the vessels at the level of the capillaries which are then in direct contact with the neuroglial cells. The combination of capillary wall and neuroglial cells make up the so-called *blood-brain barrier*
      1. The barrier is impermeable to some microbes and some compounds, like antibiotics, but is permeable to others, like sulfonamides
      2. The barrier is not present in the young embryo, but develops slowly
   B. CLEANING OR CARRYING OFF WASTES takes place in the subarachnoid space

III. **The choroid plexuses:** primordia are found in the regions where the ependymal wall is thin, such as the roof of the third and fourth ventricles and the internal part of the lateral ventricles. Here the leptomeninges push the wall into the ventricles
   A. THE PLEXUSES form a highly vascularized meningeal axis covered with thin cuboidal ependymal epithelium. The first choroid plexus develops in the fourth ventricle between days 48 and 50 in a 20-mm embryo
   B. ROLE OF THE PLEXUSES is the production of cerebrospinal fluid
      1. In the fetus, the cerebrospinal fluid may furnish proteins to the CNS
      2. The cerebrospinal fluid circulates toward the fourth ventricle and the spinal canal. It then passes through the foramina of Magendie and Luschka and is resorbed in the subarachnoid space. The main site of absorption of the fluid into the venous system is through the *arachnoid villi* (Pacchionian bodies) projecting into the dural sinuses. The villi are a thin cellular layer derived from the arachnoid epithelium and endothelium of the sinus

IV. **Vascularization of the brain**
   A. CEPHALIC CIRCULATION begins to be established very early, at about week 3, even before closure of the neural tube. It develops rapidly, along with the brain
      1. The prosencephalon is vascularized first, by the *internal carotid arteries* which come off the dorsal aortas. Vascularization of the rhombencephalon and mesencephalon takes place later and is derived from the *basilar artery,* which is formed from the confluence of *vertebral arteries*
      2. The major arteries join together to form the *circle of Willis,* which is definitively formed by about $7\frac{1}{2}$ to 8 weeks. It has its own regulatory system to protect it from wide fluctuations and provides the brain with about 15% of its total blood
         a. The circle of Willis is nourished by blood richer in oxygen than that of other parts of the body since it passes through the heart via the foramen ovale to the left side of the heart and then to the aorta and primitive carotids to the brain

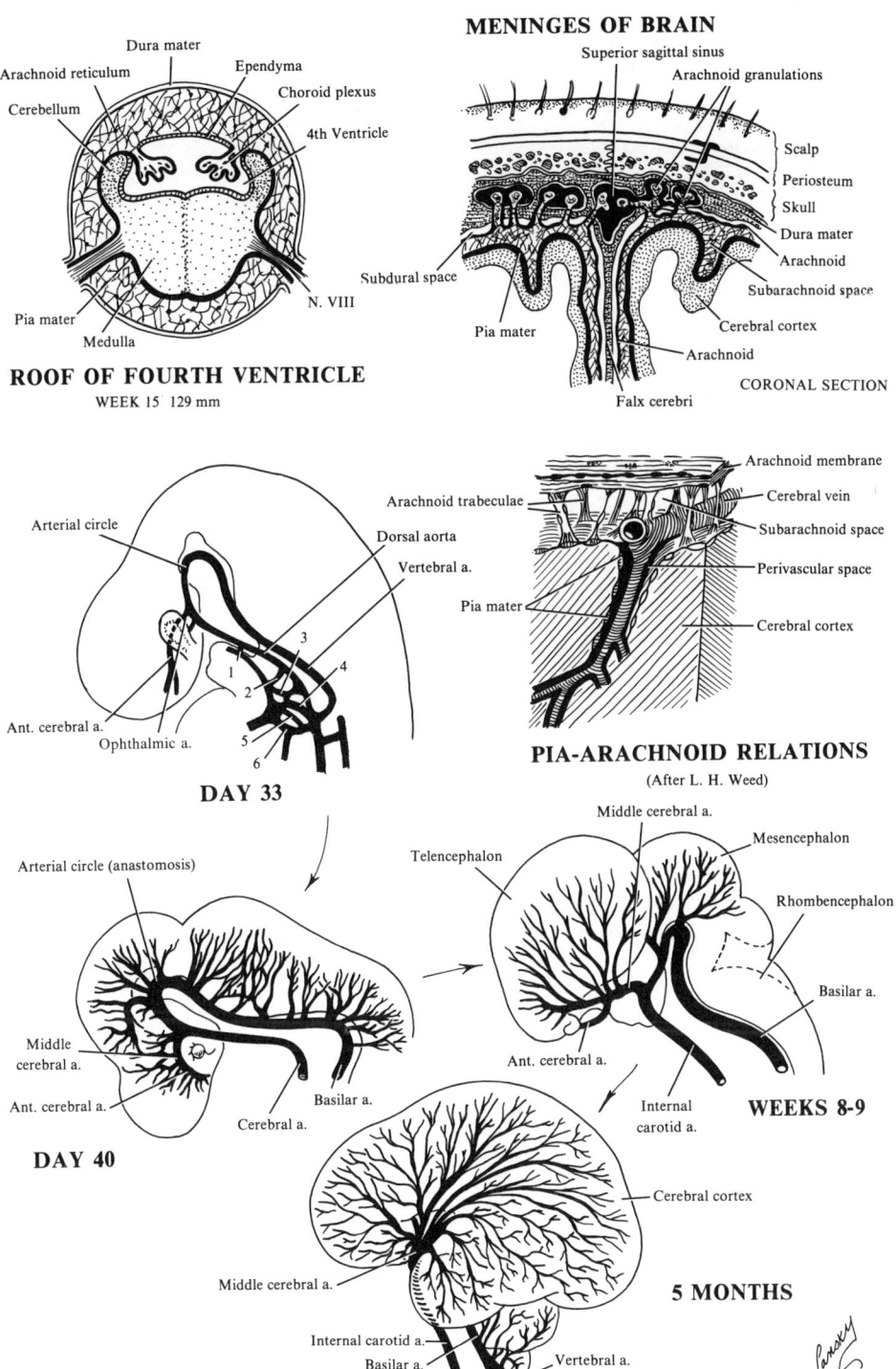

161. MALFORMATIONS OF THE BRAIN

I. **Introduction:** malformations of the brain occur in about 0.51% of live births and in about 3.0% of stillbirths. In females, malformations of the nervous system are more numerous than in other organ systems, whereas in males, the incidence of malformations of the nervous system fall somewhat between those involving the cardiac and digestive systems
  A. BRAIN MALFORMATIONS ARE THE RESULT OF
    1. Exogenous causes: nutritional, physical (especially x-rays), viral, parasitic, chemical, and medications
    2. Endogenous causes: hereditary (often poorly defined in some cases)
    3. Interaction of exogenous and endogenous causes: these have been particularly shown in experimental animals and are a result of differences in sensitivity to exogenous factors as a result of species and animal strain
  B. TERATOGENIC FACTORS can exert their influence for a very long time since brain development takes place over an extended period of time

II. **Malformations of squamous occipital bone:** the most frequently affected bone of the skull is the squamous occipital, which may be partly or totally missing and the resultant opening confluent with the foramen magnum
  A. IF THE SKULL OPENING IS SMALL, only meninges bulge through forming a *meningocele*
  B. IF THE DEFECT IS LARGE, part of the brain also bulges, forming a *meningoencephalocele*
  C. IF THE DEFECT IS VERY LARGE, part of the brain and even ventricle may penetrate through the opening into the meningeal sac to form a *meningohydroencephalocele*

III. **Problems of neural groove closure**
  A. ANENCEPHALY: common abnormality (1/1,000); 4 times more frequent in the female and 4 times more frequent in whites; results from a failure of the cephalic part of the neural groove to close in the brain. It is a disorder of early embryogenesis and is caused by defective induction of the prochordal plate or the parachordal mesoderm or poor receptivity of the competent neural plate
    1. The overall brain structure is disturbed, and at birth the brain is a mass of degenerated tissue exposed to the surface
    2. The nervous tissue, as in typical spina bifida, is not covered with bone or skin since the cranial vault differentiates only under the influence of a normal neural tube
    3. The skin is in continuity with the nervous tissue, as in the neural groove stage
    4. Normal histogenesis takes place until the beginning of neocortical differentiation, but several weeks before term, vascular problems lead to general necrosis so that only choroid plexus, some nerves, and degenerating nervous tissue persist
    5. Anencephaly results in neonatal death, can be recognized by x-rays since the skull vault is missing, and in the last 2 months of pregnancy is characterized by hydramnios since the fetus lacks the control mechanism for swallowing
    6. *Craniorachischisis* is a particularly extreme form of anencephaly. The neural groove remains open throughout its length
  B. ENCEPHALOCELE is the most serious type of defect involving closure of the encephalic neural groove. Part of the brain is herniated under the skin through a hole in the skull. It is most often untreatable
  C. MENINGOCELE is a hernia of the meninges, is compatable with life, and can be treated surgically. As a result of its origin, it is usually midline
    1. Localization of the groove defect enables the bone to form normally in its lateral parts. Similarly, the small extent of the opening allows the nervous tissue and the skin at the edges to join more or less at a late stage; however, the skin is always abnormal, fine parchmentlike, and hairless. In some cases, the abnormality of closure is evident
    2. Meningoceles have been produced experimentally by x-rays, anoxia, nutritional deficiencies, hypervitaminosis A, drugs, etc.

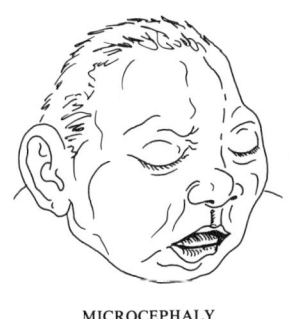
MICROCEPHALY

ANENCEPHALY

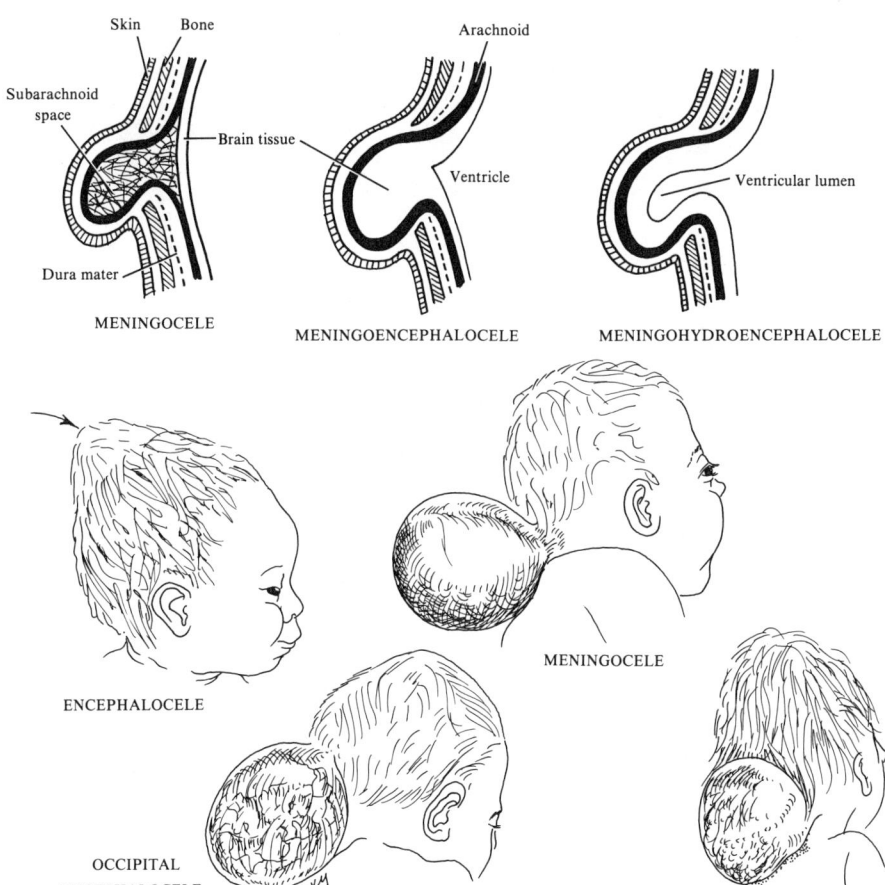

- 423 -

162. MALFORMATIONS OF THE BRAIN

I. **Ethmocephaly** is a rare malformation involving only prosencephalic derivatives
   A. THE TELENCEPHALON FAILS TO SUBDIVIDE into 2 hemispheres and has a single cavity which communicates with the third ventricle by a single opening. Since the neural groove is closed, the cranial vault and skin are normal. However, there is only 1 median eye (cyclopia), usually no hypophysis, and the nasal apparatus is more or less atrophied and often replaced by a trunk with a single median opening instead of 2 nasal openings
   B. THIS MALFORMATION is the result of a defective induction of the prochordal plate
      1. In humans, a chromosomal defect has been suggested as a possible cause
      2. It has been produced experimentally by excision of the prochordal plate, x-rays, and hypervitaminosis A
   C. ETHMOCEPHALY usually results in death; however, some children with minor types of malformations have been known to live for several years

II. **Arrhinencephaly** falls into the general category of ethmocephaly and is characterized by dysgenesis of the rhinencephalon and a general atrophy of the telencephalon
   A. MALFORMATIONS OF THE INTERMAXILLARY SEGMENT, whose development is related to that of the rhinencephalon, are also seen and result in
      1. Cleft plate
      2. Agenesis of the nasal septum with a median union of the nostrils and nasal fossa
      3. Hypotelorism
   B. HYPOGENESIS or absence of the rhinencephalon is due to a disorder of the olfactory placodes which fail to produce fibers or induce the formation of the olfactory bulbs
      1. Another cause is a genetic defect or defective induction of the prochordal plate

III. **Microcephaly:** one sees a small brain in a small cranium
   A. IT CAN BE THE RESULT of a genetic abnormality, pelvic x-rays during pregnancy, or toxoplasmosis, just to name a few
   B. DEVELOPMENTAL ARREST takes place at late stages of gestation and may involve problems of cellular multiplication or migration. Cellular densities may be abnormally low, or the cortical layers may be less numerous than normal
   C. THIS CONDITION usually is accompanied by mental deficiency and convulsions

IV. **Dysgenesis** involves agenesis or degeneration of certain cellular groups
   A. IN PORENCEPHALY (collective term for variety of cerebral defects involving cortical tissue) a lateral ventricle communicates with the subarachnoid spaces or cortical tissue
      1. These anomalies are probably the result of a defect of vascularization and often result in debility, convulsions, and cerebral motor disability
   B. AGENESIS OF THE CORPUS CALLOSUM
   C. HEMISPHERIC OR CEREBELLAR AGENESIS OR HYPOGENESIS is rarely seen and is probably also caused by early vascular problems

V. **Heterotopias**
   A. ROSETTES are small, accessory paraependymal cavities which may or may not be connected to the main cavity. They are surrounded by a germinal layer which can give rise to nervous cells
      1. These disorders vary according to localization and have been produced experimentally by x-rays or vitamin deficiencies
   B. CONGENITAL MYXEDEMA: thyroid deficiency can result in defective cellular migration, resulting in heterotopias that cause problems

VI. **Anomalies of the cortical surface** are always accompanied by debility or idiocy and are of unknown pathogenesis. They may include lissencephaly or absence of cortical convolutions and micro- or macrogyria (too small or too large convolutions)

MICROCEPHALY

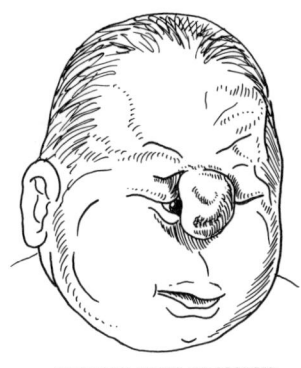

CYCLOPS WITH PROBISCIS
AND EXOMPHALOS

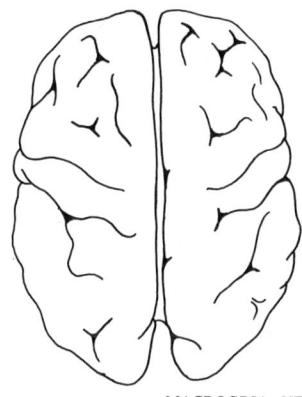

MACROGRIA, NEWBORN

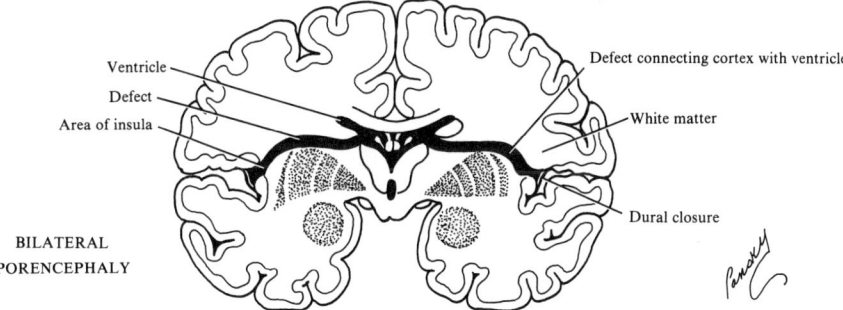

BILATERAL
PORENCEPHALY

163. MALFORMATIONS OF THE BRAIN: HYDROCEPHALUS

I. **Hydrocephalus** is a frequent anomaly, but is also compatible with life. It is characterized by an abnormal accumulation of cerebrospinal fluid (CSF) in the cerebral ventricles or in the subarachnoid spaces in the case of external hydrocephalus
   A. THE 3 MAJOR CAUSES of hydrocephalus are
      1. Excess production of cerebrospinal fluid (communicating hydrocephalus)
      2. Obstruction of circulation of cerebrospinal fluid (noncommunicating hydrocephalus)
      3. Defective resorption of cerebrospinal fluid (communicating hydrocephalus)
   B. HYDROCEPHALUS IN THE NEWBORN, in most cases, is thought to be the result of an obstruction in the aqueduct of Sylvius due to inflammation, a tumor, or an anomaly
   C. SOME FORMS OF HYDROCEPHALUS are the result of obliteration of the openings of the fourth ventricle (Dandy-Walker deformity)
      1. In some cases of spina bifida, the posterior part of the brain sinks into the occipital opening. With caudal elongation of the trunk, the spinal cord follows the movement and pulls the brain posteriorly. This results in the *Arnold-Chiari syndrome*
   D. HYDROCEPHALUS MAY BE DUE to a recessive, sex-linked genetic anomaly in some cases
      1. The hydrocephalus does not appear in the female who transmits it, but is seen in 50% of the males
      2. One usually sees an overproduction of cerebrospinal fluid
   E. MANIFESTATIONS OF HYDROCEPHALUS
      1. An increased size of the head
      2. An enlargement of the cranial sutures
      3. Progressive thinning of the cranial bones
      4. Lamination of the cerebral cortex in severe cases
      5. It is often accompanied by psychologic retardation, often severe enough to cause debility, convulsions, or cerebral motor disability
   F. HYDROCEPHALUS is often associated with spina bifida cystica, even though the hydrocephalus is not obvious at birth
      1. The condition of hydrocephalus after birth usually results from meningitis which causes a postinflammatory obstruction
   G. SURGICAL TREATMENT: arachnoid-ureterostomy or ventricle-cardiac, ventricle-pleura, or ventricle-vena cava shunt may stabilize the condition by removing the obstruction. This, however, has no effect on the nervous lesions that may have occurred before surgery. Prognosis is guarded

HYDROCEPHALUS

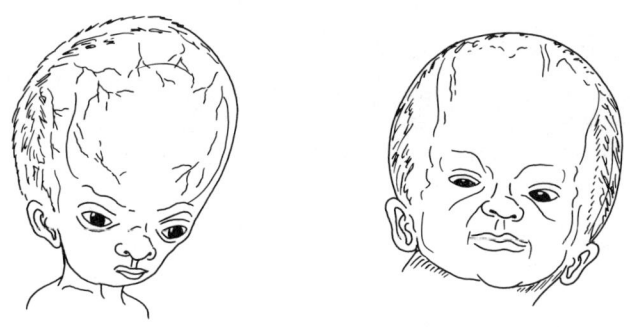

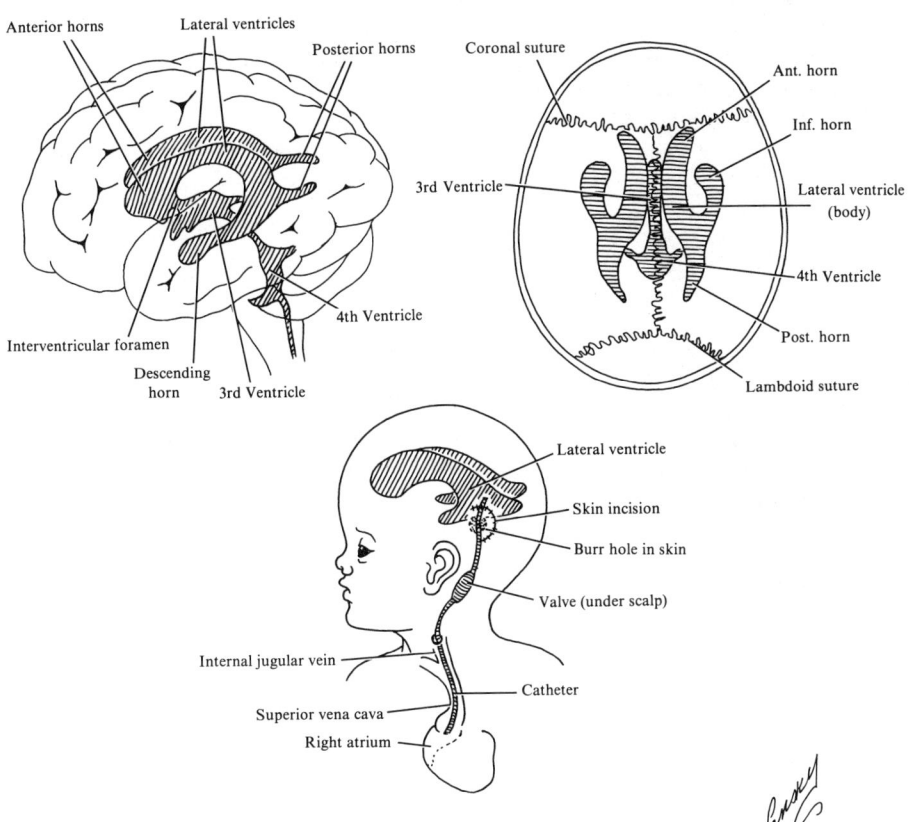

SURGICAL TREATMENT FOR HYDROCEPHALUS

164. THE AUTONOMIC NERVOUS SYSTEM: THE SYMPATHETIC SYSTEM

I. **Introduction:** the autonomic nervous system (ANS) involves those processes that are normally beyond voluntary control and, for the most part, beneath consciousness. In this way, it differs from the voluntary central nervous system. However, it is under the control of centers in the central nervous system and cannot function as an independent unit

A. THE AUTONOMIC NERVOUS SYSTEM is composed of 2 major portions which are anatomically and physiologically distinct: the *sympathetic* (thoracolumbar) and *parasympathetic* (craniosacral) *systems*. These systems are essentially motor systems since the sensory afferent nerves, with but a few exceptions, follow the ordinary sensory pathways. They are also essentially a 2-chain system of *pre- and postganglionic fibers*

II. **The sympathetic nervous system**

A. CELLS FROM THE NEURAL CREST and ventral portion of the neural tube of the thoracic region migrate on either side of the spinal cord, toward the region just behind the dorsal aorta, at about week 5 of development. These are to become the *sympathetic neuroblasts* or future sympathetic cells
  1. Some detach themselves from the tube and arrange themselves along the motor root
B. THE MIGRATING CELLS form *2 chains of sympathetic ganglia* on either side of the vertebral column
  1. The ganglia are segmental or metameric, but in contrast to the spinal ganglia, they are interconnected to each other by longitudinal nerve fibers or axons of some of the cells. The resulting interconnected ganglia form the *lateral vertebral sympathetic chains*
  2. From their thoracic portion, the neuroblasts migrate and extend the sympathetic system into both the cervical (neck) and lumbosacral region
    a. An upward extension into the neck forms the *superior, middle,* and *inferior cervical ganglia,* which exist to supply structures of the head and neck
C. SOME OF THE SYMPATHETIC NEUROBLASTS migrate even farther ventrally to form *preaortic ganglia* such as seen in the *solar (celiac)* and *mesenteric plexuses,* the *visceral* or *gastrointestinal ganglia of the myenteric plexus of Auerbach,* and in the *submucous plexus of Meissner*
  1. Still other sympathetic cells migrate to the heart and lungs where they give rise to the sympathetic organ plexuses
D. WHILE THE GANGLIA ARE FORMING, fibers coming from the visceral motor areas of the medulla and spinal cord make synapses with the sympathetic neuroblasts of 1 of the 3 ganglionic levels to form the *preganglionic fibers*
  1. The preganglionic fibers are myelinated, and their paths from the spinal nerve to the sympathetic ganglia are thus called the *white rami communicantes*
E. THE AXONS OF SYMPATHETIC NEUROBLASTS, found in the ganglia, constitute the *unmyelinated postganglionic fibers*
  1. These fibers leave the lateral chain ganglion system at 1 of its 3 levels to join the spinal nerves and are called the *gray rami communicantes*
  2. The postganglionic fibers innervate diffuse structures such as *smooth muscle, cardiac (heart) muscle* and *glands*
  3. The fibers innervating the eye, heart, and lungs, as well as the digestive system, originate in the 3 ganglion levels (cervical, thoracic, and preaortic, respectively)

FORMATION OF SYMPATHETIC GANGLIA

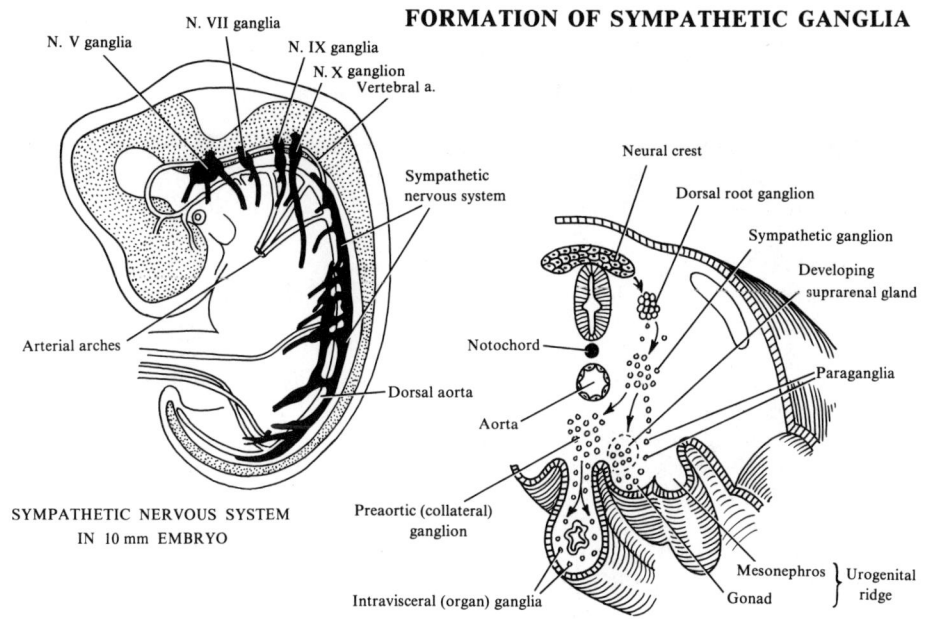

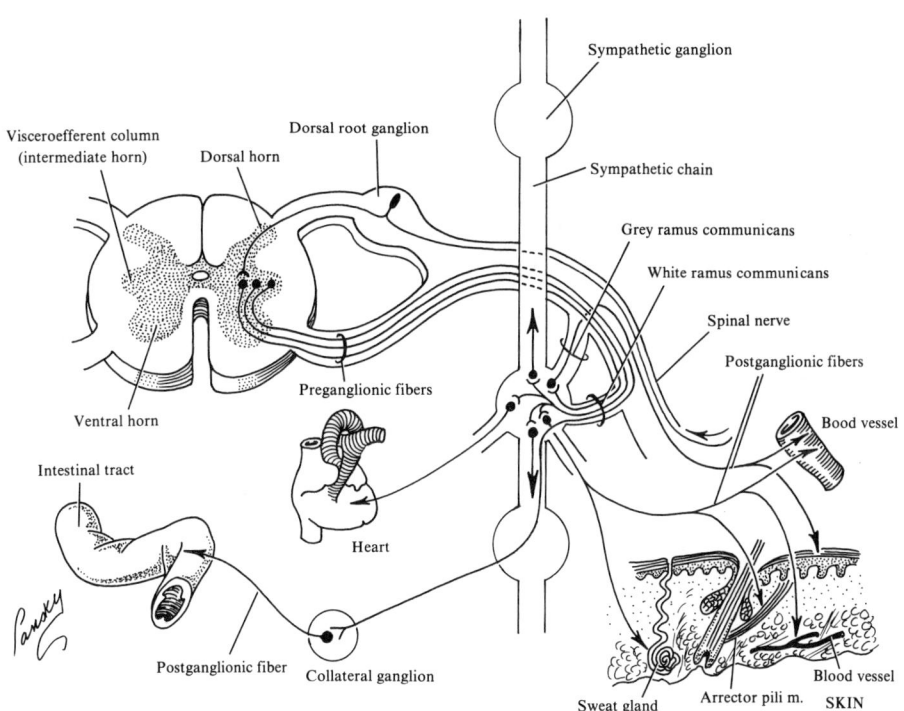

165. THE AUTONOMIC NERVOUS SYSTEM: THE PARASYMPATHETIC SYSTEM

I. **The parasympathetic system** is less extensive than the sympathetic system. Preganglionic fibers arise only in certain centers of the cerebral trunk and in the sacral portion of the spinal cord. Thus it is called the *craniosacral* portion of the autonomic nervous system

   A. THE PREGANGLIONIC FIBERS of the parasympathetic system follow the path of specific cranial nerves, namely, the oculomotor (III), the facial (VII), the glossopharyngeal (IX), and the vagus (X). In addition, they follow the sacral spinal nerves arising from segments S2, 3, and 4 of the cord

   B. LIKE THE SYMPATHETIC NEUROBLASTS, their ganglion cells come from the neural crest and neural tube, but only at the level of the preganglionic fibers. Their long migrating preganglionic fibers take them to the viscera; and the short postganglionic fibers pass to the branchial arches and the cardiac, pulmonary, and intestinal plexuses

   C. ALL THE PARASYMPATHETIC GANGLIA are preaortic or visceral and do not appear in the chain ganglia

   D. SOME SPECIALIZED PARASYMPATHETIC RECEPTORS
      1. The carotid and aortic bodies are mesenchymal chemoreceptors, innervated by the glossopharyngeal (IX) cranial nerve. The neurosensory cells which make up these structures are of parasympathetic origin, and they migrate along nerve IX from the neural crest or neural tube

II. **Physiologic significance of the autonomic nervous system**

   A. THE POSTGANGLIONIC NEURONS of the sympathetic nervous system are *adrenergic,* while those of the parasympathetic nervous system are predominately *cholinergic*
      1. Antagonism of the sympathetic and parasympathetic systems helps to maintain equilibrium of involuntary functions, although in some instances the 2 systems may work together
      2. The entire autonomic nervous system is under the control of the hypothalamus, which coordinates information relating to involuntary body functions

III. **Pathology**

   A. ABNORMAL ORGANOGENESIS of the autonomic nervous system is responsible for certain problems, especially of the digestive system, such as is seen in *Hirschsprung's disease or megacolon,* which results in congenital dilatation of the colon with anomalies of Meissner's and Auerbach's plexuses

CRANIOSACRAL PARASYMPATHETIC NERVES

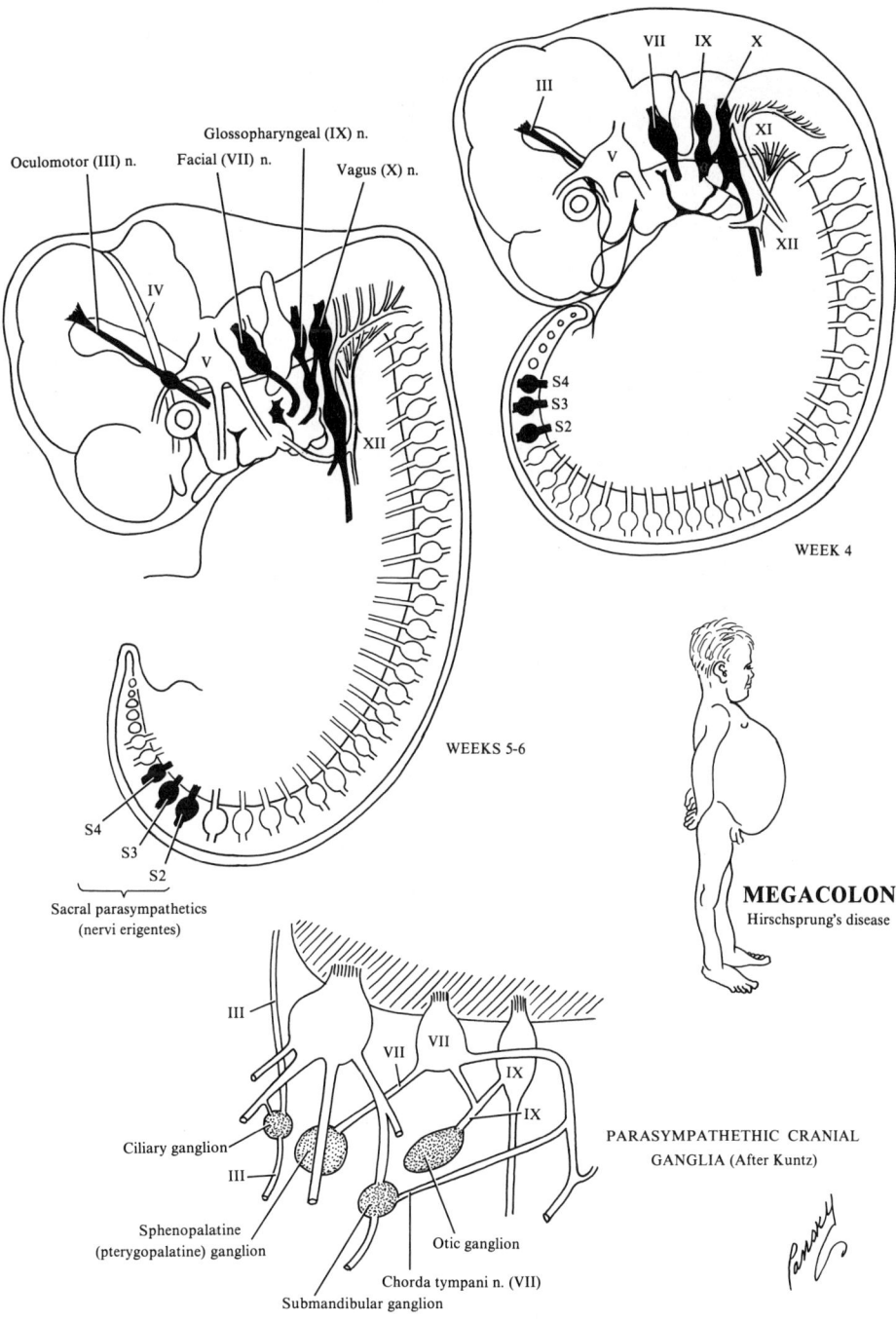

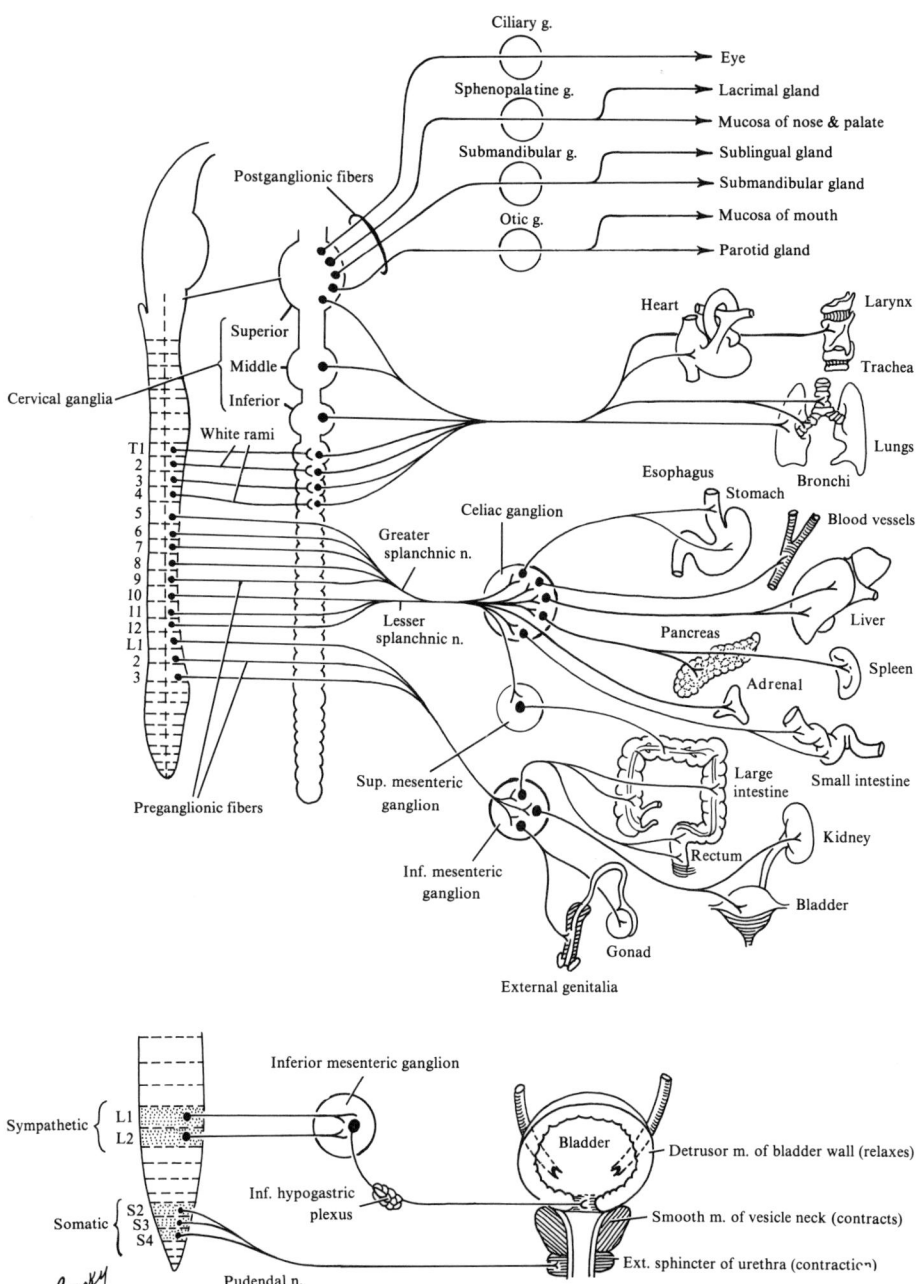

FIGURE 77. **The sympathetic nervous system and urinary retention mechanism.**

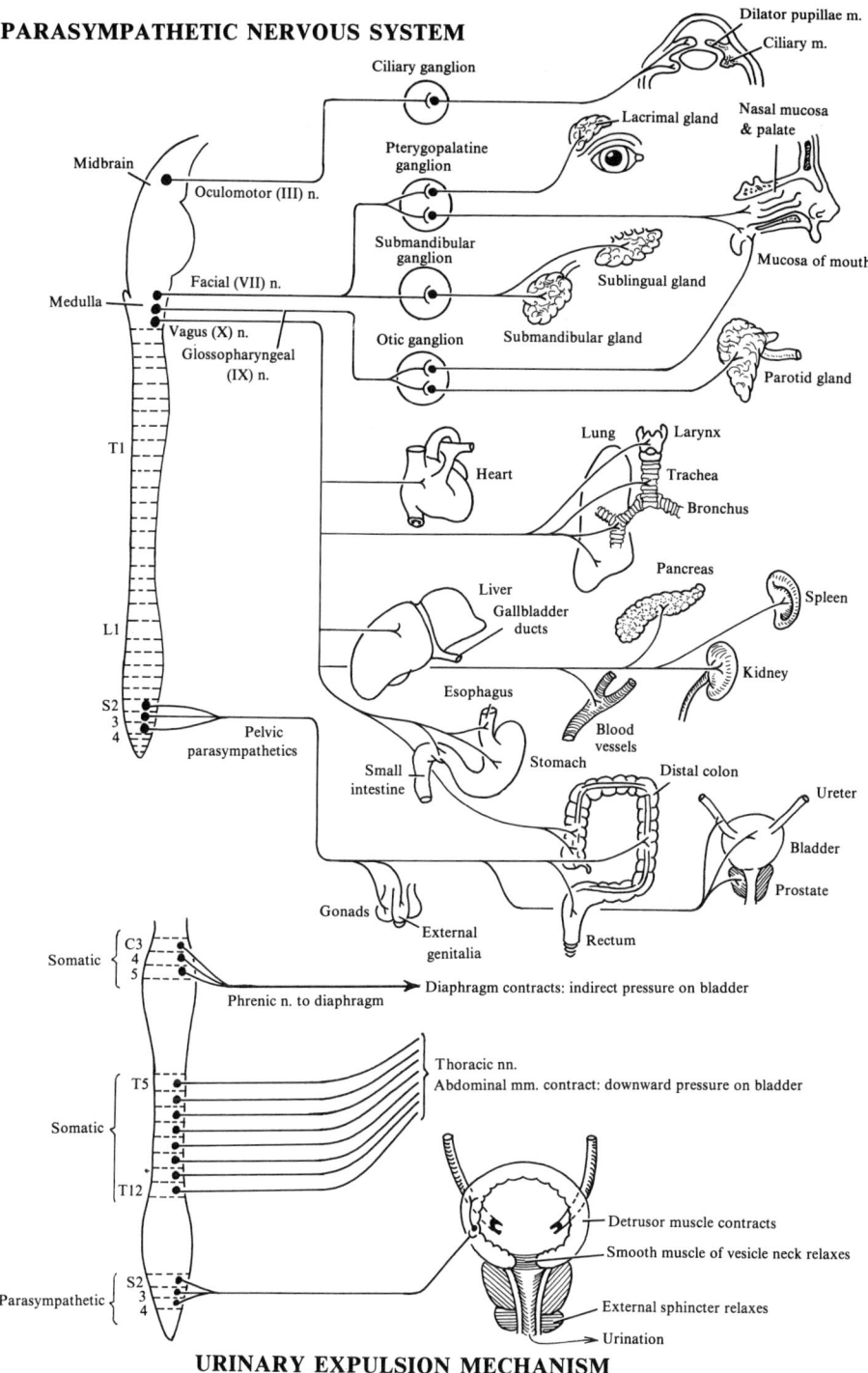

FIGURE 78. **The parasympathetic nervous system and urinary expulsion mechanism.**

166. THE OLFACTORY SYSTEM

I. **The primordia of the olfactory system** consist of 2 placodes on the right and left of the anteroinferior portion of the frontal prominence, situated above the stomodeum and below and lateral to the forebrain
   A. THE PRIMORDIA appear about day 30, after those of the optic and otic placodes. At this time, too, the neural tube is completely closed
   B. THE PLACODES are primarily induced by the adjacent mesoderm and secondarily by the ventral surface of the prosencephalon
   C. EACH PLACODE INVAGINATES, in the direction of the adjacent brain, to form the *olfactory pits*
      1. The stratified placodal base of the invagination forms the *olfactory epithelium*
      2. The lateral walls around the invaginating pits form the surface ectodermal covering of the nasal cavities
   D. THE PLACODAL CELLS of the olfactory epithelium differentiate into neurosensory cells within the thickness of the epithelium and eventually give origin to olfactory nerve fibers
      1. At about 1.5 months, the deep pole of the superficial cells gives rise to an axon that crosses the epithelium and the mesenchyme and contacts the olfactory area of the cerebral hemisphere (telencephalon)
      2. The arrival of these fibers at the telencephalon induces formation of the *olfactory bulb*
      3. The axons then connect with the specialized structures of the central nervous system corresponding to the olfactory system, namely, the bulbs
   E. NEAR THE END OF MONTH 3, the mesenchyme between the sensory epithelium and the bulb gives rise to a cartilaginous structure, the *lamina cribrosa of the ethmoid bone* which is eventually organized around the olfactory nerve networks and separates them into a number of bundles. The cartilage ossifies here to form the cribriform plate of the ethmoid through which the nerves pass to enter the olfactory bulbs
   F. THE OLFACTORY BULB elongates, and eventually the extension of the ventricular cavity into it becomes obliterated
      1. Cells in the bulb, around which the olfactory nerve fibers terminate and synapse, give origin to secondary olfactory fibers which grow centrally and form the *olfactory tract*
         a. The olfactory tract terminates in the region of the piriform cortex
   G. FIBERS OF THE OLFACTORY NERVES are entirely of placodal origin and their cell bodies remain in the olfactory epithelium
      1. In lower mammals, a special part of the olfactory nerve is distributed to Jacobson's organ as the vomeronasal nerve. The organ is found in the lower part of the nasal septum

II. **Malformations of the olfactory system**
   A. MALFORMATIONS are usually quite serious because they are always accompanied by anomalies of the central nervous system as well as the face. They are classified in the category of the more general malformations. Examples are
      1. Ethmocephalus: in these malformations, the nose is replaced by a proboscis with a single canal, as a result of convergence of both nasal primordia on the midline
         a. The sensory epithelium is much reduced or even totally absent
         b. The olfactory bulbs also may be absent
         c. In some cases, there may be 2 proboscises or even trunks
         d. There are also minor types of this malformation, in which the nose is more or less cylindric and the nasal fossae are totally closed
      2. Arrhinencephaly (see unit on Brain Malformations p. 424)
   B. IN HUMANS, many of the causes of these malformations are apparently genetic in origin

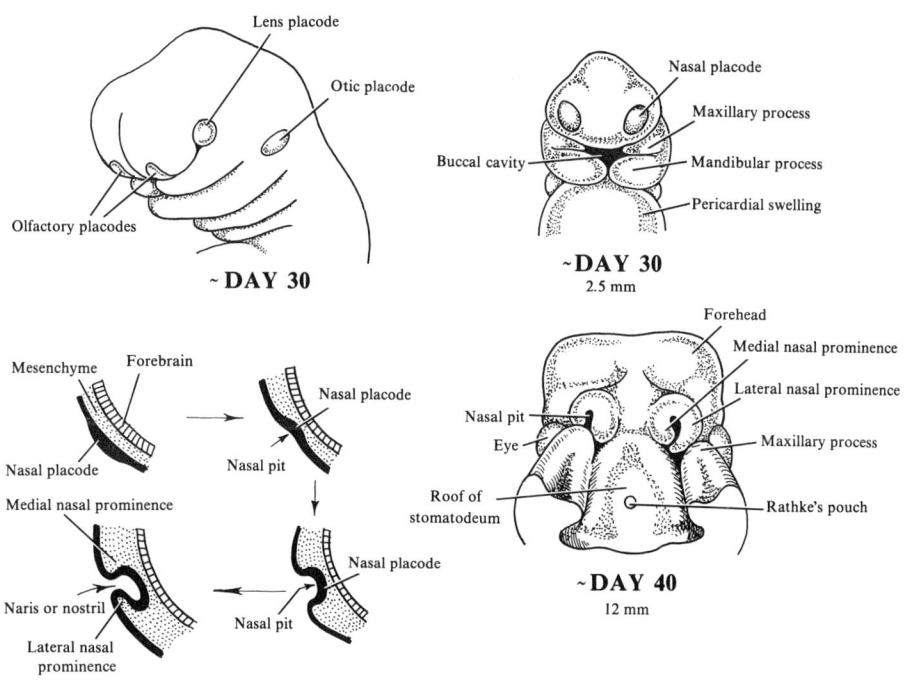

DEVELOPMENT OF NASAL SAC

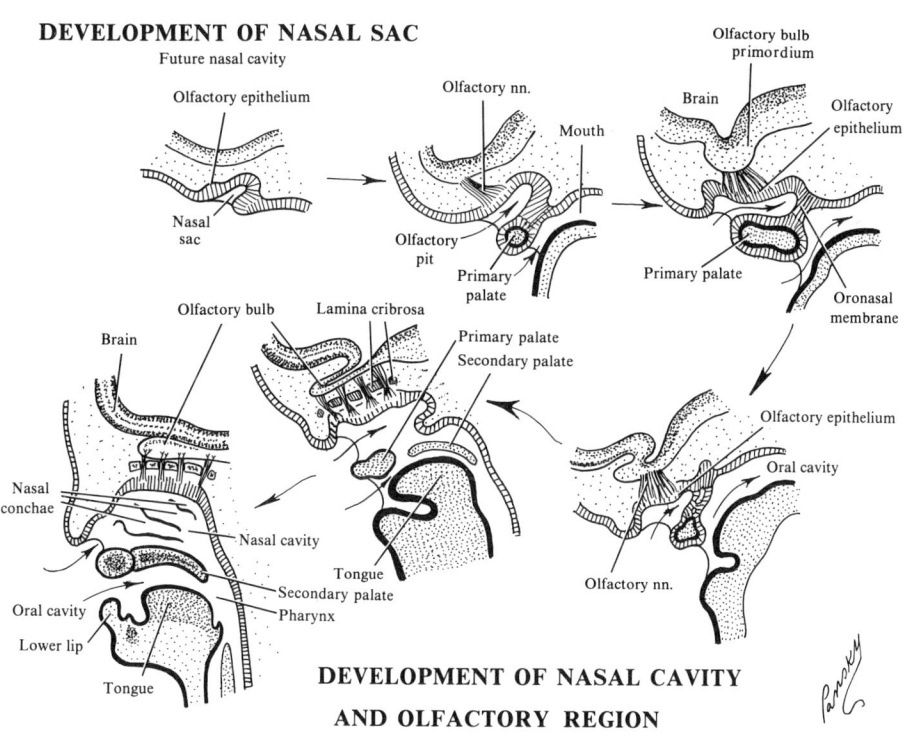

DEVELOPMENT OF NASAL CAVITY AND OLFACTORY REGION

167. THE EYE: OPTIC CUP AND LENS VESICLE, RETINA, IRIS, AND CILIARY BODY

I. The optic cup and lens vesicle

A. THE EARLIEST SIGN OF EYE DEVELOPMENT is seen in the 22-day embryo as a pair of shallow grooves on each side of the invaginated forebrain
 1. With closure of the neural tube, the grooves form outpocketings of the forebrain, the *optic vesicles*, which are in contact with surface ectoderm
 2. The vesicles cause chemical changes in the surface ectodermal cells needed for lens formation, and the optic vesicles invaginate and form the double-wall *optic cup*
    a. The inner and outer wall of the cup are initially separated by the *intraretinal space*, but with development, the lumen disappears and the walls oppose each other
 3. The invagination not only is centrally restricted in the cup, but also involves its ventral rim where the *choroid fissure* is formed, which extends along the undersurface of the *optic stalk* where it tapers off
    a. Fissure formation allows the *hyaloid artery* to reach the inner eye chamber
    b. During week 7, the lips of the choroid fissure fuse and the mouth of the optic cup becomes a rounded opening, the future *pupil*
 4. In the meantime, the cells of the surface ectoderm, initially in contact with the optic vesicle, elongate and form the *lens placode*
    a. The placode subsequently invaginates and develops into the *lens vesicle*
    b. During week 5, the lens vesicle loses surface contact and is then located in the mouth of the optic cup

II. Retina, iris, and ciliary body

A. DEVELOPMENT OF THE OUTER LAYER of the optic cup is characterized by the appearance of small pigment granules during week 5 to form the pigment layer of the retina
B. THE POSTERIOR FOUR-FIFTHS of the inner layer of the optic cup (*pars optica retinae*)
 1. This part thickens and undergoes a series of changes similar to those seen in the wall of the brain vesicle
 2. Bordering the intraretinal space, the ependymal layer differentiates into the light-receptive elements, the *rods and cones*
 3. Adjacent to the photoreceptive layer, the "mouth" layer, as in the brain, gives rise to neurons and supporting cells
    a. In the adult, the outer nuclear layer, the inner nuclear layer, and the ganglion cell layer are all distinguished
 4. On the surface of the mantle layer, the marginal zone contains the axons of the nerve cells of the deeper layers; and the nerve fibers in this marginal layer converge toward the optic stalk, which gradually forms the *optic nerve*
C. THE ANTERIOR ONE-FIFTH of the inner layer of the optic cup (*pars caeca retinae*)
 1. Changes very little and remains 1-cell layer thick but later divides into the *pars iridica retinae*, forming the inner layer of the iris, and the *pars ciliaris retinae*, which participates in the formation of the ciliary body
    a. The pars ciliaris retinae is recognized by its marked folding
       i. Externally, it is covered by a layer of mesenchyme which forms the *ciliary muscle*
       ii. Internally, it is connected to the lens by a network of elastic fibers which form the *suspensory ligament* or *zonula*
       iii. Contraction of the ciliary muscle changes the tension in the ligament and controls the curvature of the lens
D. IN HUMANS, the sphincter and dilator pupillae muscles develop from the underlying ectoderm of the optic cup in the area between the optic cup and surface epithelium
E. THE IRIS IN THE ADULT is formed by the pigment-containing internal and external layers of the optic cup and by a layer of vascularized connective tissue which also contains the pupillary muscles

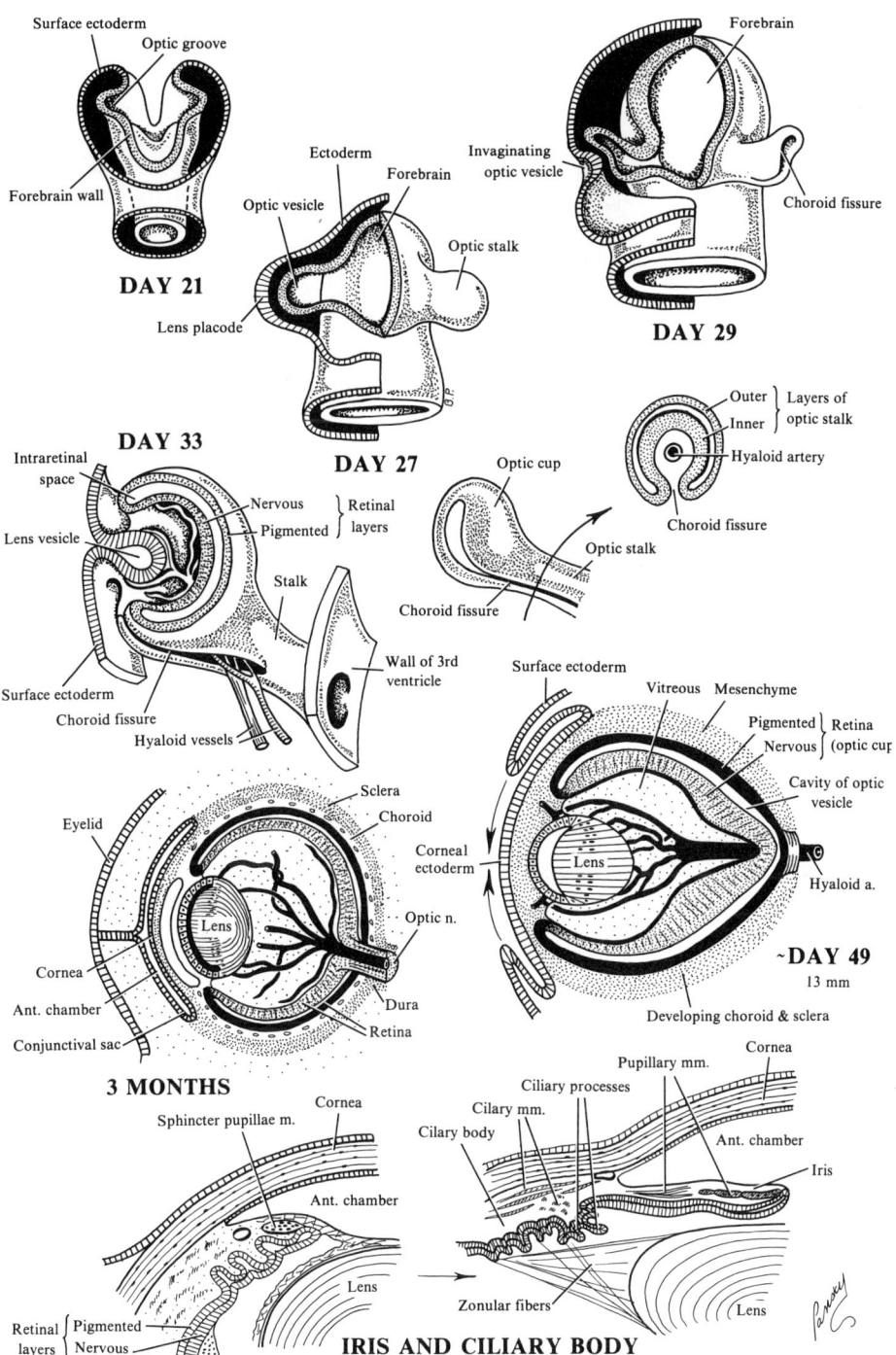

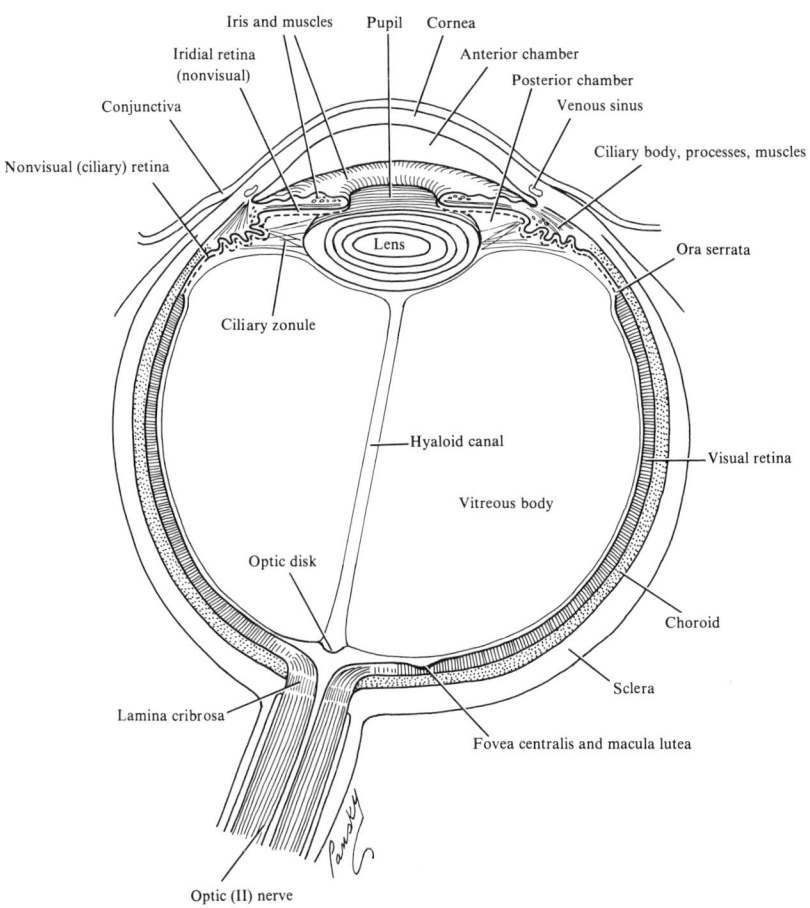

FIGURE 79. **Anatomy of the mature eye.**

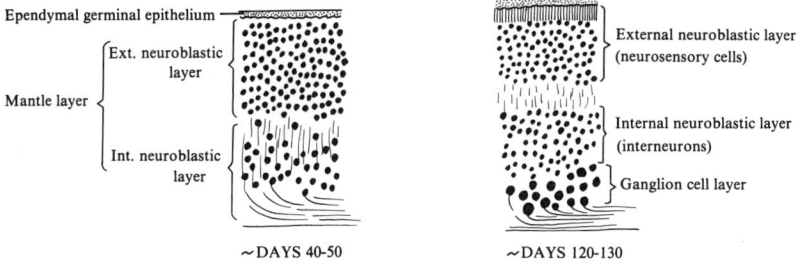
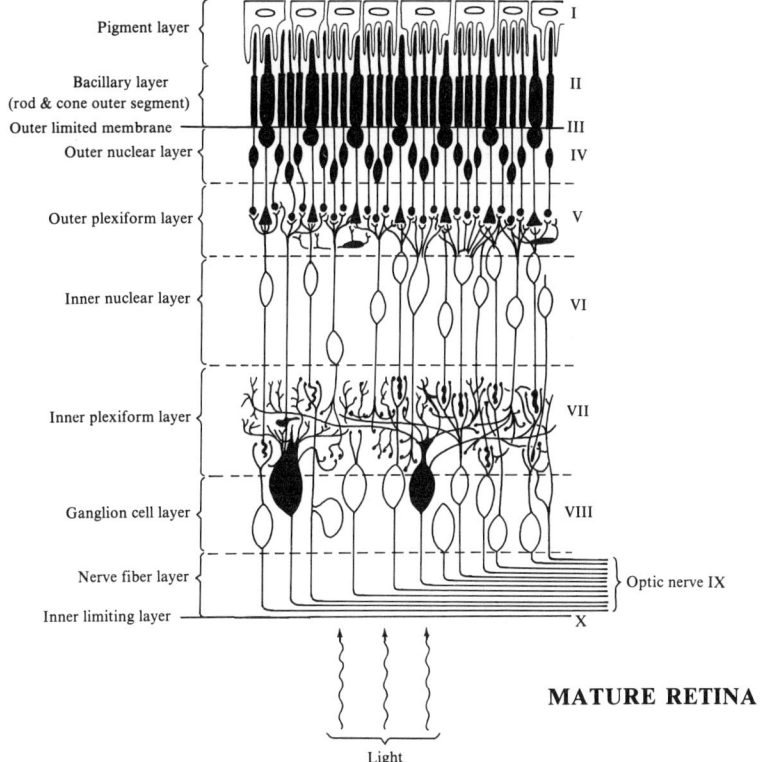

FIGURE 80. Retinal histogenesis (developmental and adult).

168. THE EYE: LENS, CHOROID, SCLERA, CORNEA, AND OPTIC NERVE

I. The lens
A. SHORTLY AFTER FORMATION OF THE LENS VESICLE, the cells of the posterior wall elongate anteriorly and form long fibers which gradually fill the vesicle lumen
B. THE PRIMARY LENS FIBERS reach the epithelium of the anterior wall of the vesicle by the end of week 7 and form the nucleus of the lens
   1. Lens growth is not finished at this stage, but new, secondary lens fibers are continuously being added to the central core from cells in the equatorial zone

II. The choroid, sclera, and cornea
A. BY THE END OF WEEK 5 and when the optic cup and lens vesicle have formed, the eye primordium is completely surrounded by loose mesenchyme
   1. The mesenchyme differentiates into an inner layer (comparable to the pia mater of the brain) and an outer layer (comparable to the dura mater)
      a. The inner layer forms the highly vascularized pigmented layer, the *choroid*
      b. The outer layer develops into the *sclera* and is continuous with the dura mater around the optic nerve
   2. Mesenchyme layers over the anterior aspect of the eye differentiate in various ways
      a. The cells arrange themselves so that a space, the *anterior chamber,* splits the mesenchyme into a thin inner layer just in front of the lens and iris (the *iridopupillary membrane*) and a thick outer layer continuous with the sclera
         i. The anterior chamber itself is lined by flattened mesenchymal cells which form the posterior lining of the cornea and the anterior covering of the iridopupillary membrane. The membrane in front of the lens normally disappears
B. THE CORNEA, from outside to inside, is formed by
   1. An epithelial layer derived from surface ectoderm
   2. A layer of dense connective tissue, the substantia propria or stroma
   3. An epithelial layer that borders the anterior chamber
C. THE MESENCHYME that surrounds the eye primordium also invades the inside of the optic cup via the choroid fissure
   1. It participates in formation of the hyaloid vessels (during intrauterine life, supplying the lens and helping to form the vascular layer seen on the inner retinal surface)
   2. Forms a delicate network of fibers between the lens and the retina, the interstitial spaces, which later fill with a transparent gelatinous substance, the *vitreous body*

III. The optic nerve
A. THE OPTIC CUP is initially connected to the brain by the optic stalk
   1. The choroid fissure is a groove on the stalk's ventral surface
   2. The nerve fibers of the retina returning to the brain are found among the cells of the inner wall of the stalk
B. DURING WEEK 7, the choroid fissure closes
C. WITH AN INCREASING NUMBER OF NERVE FIBERS growing toward the brain, the inner stalk walls increase in size and fuse with the outer walls
   1. The cells of the inner layer provide a network of neuroglia cells that support the optic nerve fibers. Thus, the optic stalk is transformed into the optic nerve, and in its center is the hyaloid artery, later called the *central artery of the retina*

IV. The extrinsic motor muscles of the eye are derived from the peripheral mesenchyme

V. The eyelids are simple cutaneous folds, closed at first, but separate at about month 7. Their morphogenesis is totally independent of that of the eye

VI. The lacrimal glands are derived from small epithelial cords that penetrate the mesenchyme from the superoexternal area of the conjunctival sac

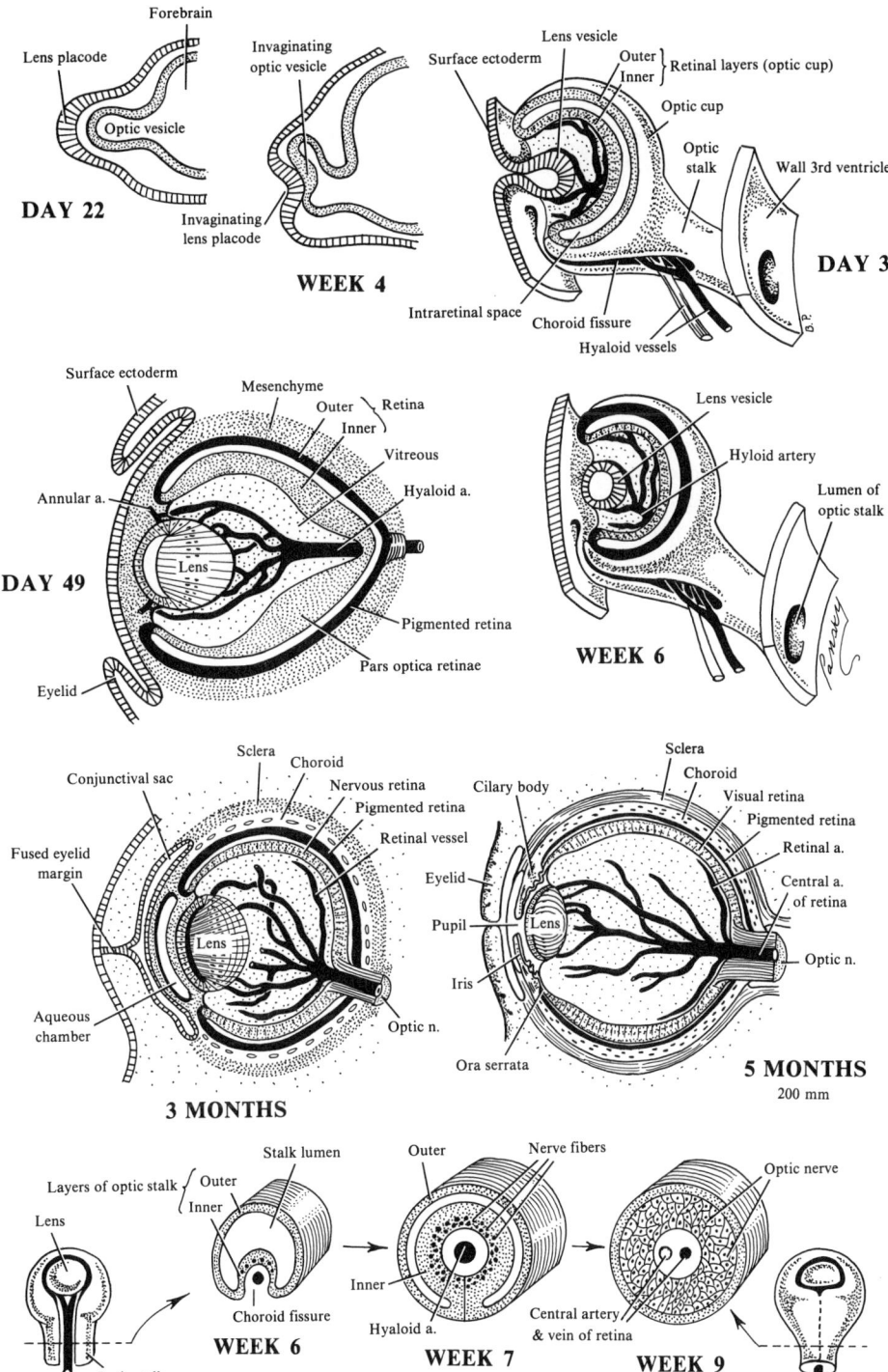

169. CONGENITAL MALFORMATIONS OF THE EYE

I. **Introduction:** gross malformations of the eye are formed during the period of organogenesis, approximately between days 20 and 60. However, the eye is vulnerable to teratogenic agents even after this period, especially at the time of histogenesis

II. **Disorders of basic organogenesis**
  A. CYCLOPIA: rare malformation; is the presence of a single median eye or 2 eyes, more or less fused on the midline
    1. This condition is never a single isolated malformation, but is usually a part of a complex syndrome associating ethmocephaly and arrhinencephaly, in varying degrees
  B. ANOPHTHALMIA refers to the uni- or bilateral absence of the eyes
    1. The eyelids and motor muscles usually are present, since the origin of these structures is independent of that of the eye
    2. This malformation generally is accompanied by other craniocerebral anomalies
    3. This malformation can be produced experimentally by pantothenic acid deficiency, hypervitaminosis A, hypoglycemia-producing sulfamides, as well as other agents
  C. MICROPHTHALMIA: the eyeball and the lens are small and more or less malformed
    1. This malformation frequently is associated with coloboma
    2. This condition can be produced experimentally by the use of the same substances listed under B.3
    3. The etiology of this malformation in humans is uncertain, but a genetic origin is probable, and irradiation during the first few weeks of pregnancy has been implicated
  D. RETINOCELE is a herniation of the retina into the sclera, caused by failure of the optic choroid fissue to close. If the hernia is severe, it can result in protrusion of the eyeball
  E. CONGENITAL CATARACT: in this condition, the structure of the lens is altered and becomes opaque during intrauterine life
    1. Cataract can be produced experimentally by the administration of chemicals or drugs (thyroxine, for example) to the mother
    2. In humans, rubella, certain other viral infections, and toxoplasmosis are frequent causes of this malformation when they are contracted by the mother during her first 2 months of pregnancy
      a. If the mother is infected after week 7 of pregnancy, the lens escapes damage, but the child may be deaf as a result of imperfect development of the cochlea
    3. Some cases of congenital cataract are thought to be of genetic origin
  F. COLOBOMA represents the persistence of the optic (choroid) fissure after week 7 of development. The coloboma may involve the retina and the iris
    1. Its cause is unknown, and visual problems vary with the severity of the malformation

III. **Secondary disorders**
  A. BUPHTHALMOS OR CONGENITAL GLAUCOMA is due to a problem of venous circulation (drainage of the aqueous humor), usually related to dysgenesis of the venous system circumscribing the iris
    1. It has been produced experimentally by giving glucagon to a pregnant rat
  B. INCOMPLETE REGRESSION OF THE HYALOID ARTERY causes problems of vision only if the remnants of the artery are considerable
  C. PERSISTENCE OF THE PUPILLARY MEMBRANE causes only minor problems
  D. ANOMALIES OF EYE DIMENSIONS may be the cause of *myopia* (the optic axis is too long) or hyperopia (the optic axis is too short)
    1. An irregularity of curvature of the lens or cornea results in *astigmatism*
  E. RETROLENTAL FIBROPLASIA is fibrosis of the vitreous body with folding of the retina

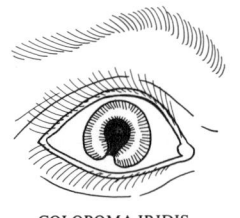
COLOBOMA IRIDIS

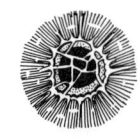
PARTIALLY PERSISTANT IRIDOPUPILLARY MEMBRANE

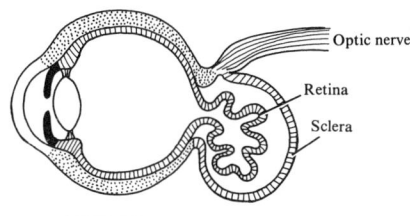
RETINOCELE

CYCLOPS

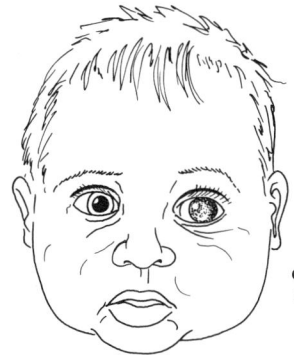
CONGENITAL GLAUCOMA
Buphthalmos

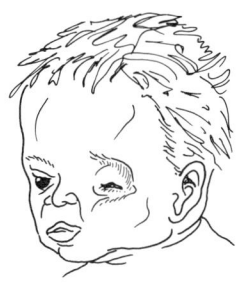
UNILATERAL ANOPHTHALMIA

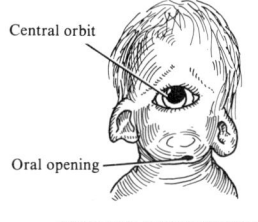
CYCLOPS HYPOGNATHUS
Nose absent, mouth rudimentary

170. THE VESTIBULOCOCHLEAR SYSTEM: THE EXTERNAL EAR AND THE EARDRUM (TYMPANIC MEMBRANE)

I. **Introduction:** development of the ear, the complex organ of hearing and balance, involves all 3 embryonic germ layers
   A. ECTODERM is the origin of the internal and external ears
   B. ENTODERM takes part in the formation of the middle ear
   C. MESODERM plays a role in the formation of all 3 parts of the ear

II. **Ontogenesis and phylogenesis:** the internal ear is the first to appear. Only the internal ear is seen in fish. The middle and external ears first appear in amphibians

III. **The external ear**
   A. THE EXTERNAL EAR IS DERIVED from the dorsal part of the first branchial (pharyngeal) groove and the external covering of arches I and II that border it
      1. The first branchial groove gives rise to a massive cellular cord or *meatal plug* by proliferation of the surface ectoderm. The cord reaches the tympanic cavity and in month 7, hollows out, to form the external auditory meatus
      2. Bone is developed in "membrane" around the inner part of the canal forming the *tympanic ring*, which expands after birth to form the bony external meatus
      3. Hairs and ceruminous glands are developed as ingrowths of the lining epithelium
      4. The nerve supply of the canal and adjacent parts of the eardrum is mainly from the mandibular nerve via its auriculotemporal branch. The posterior part of the canal and adjoining region of the drum are supplied by the vagus (X) nerve

IV. **The eardrum, or tympanic membrane**
   A. THE EARDRUM CONSISTS OF
      1. The ectodermal epithelial lining at the bottom of the external auditory meatus which forms its outer coat
      2. The entodermal epithelial lining of the tympanic cavity which forms its inner coat
      3. An intermediate layer of loose connective tissue containing the handle of the malleus and the chorda tympani (VII) nerve
   B. THE MAJOR PORTION OF THE EARDRUM is firmly attached to the handle of the malleus and is formed only after dissolution of the mesenchyme surrounding the ossicles
      1. The handle of the malleus and the chorda tympani nerve are trapped between the ectoderm of the meatal plug and the entoderm of the tympanic cavity and lie in the thin mesoderm between the 2
   C. THE REMAINDER OF THE EARDRUM forms the separation between the external auditory meatus and the original tubotympanic recess
      1. The meatal plug, arising as a solid ingrowth of surface ectoderm, comes into contact with the lateral wall and the adjacent part of the floor of the tubotympanic recess, accounting for the obliquity of the drum in the adult

V. **The auricle or pinna** is formed by the coalescence of a number of mesenchymal proliferations that form small tubercles or hillocks which appear around the upper portion of the branchial groove at about day 40 of gestation (in about a 13-mm embryo)
   1. The anterior tubercles are derived from the mandibular side of the first branchial groove, and the posterior tubercles from the hyoid side
   2. There are 3 tubercles seen on each side of the external auditory meatus
   3. The hillocks fuse and are gradually formed into the definitive auricle
      a. The first hillock forms the tragus; the second, the crus of the helix; the third, the helix; the fourth, the antihelix, the fifth, the antitragus; and the sixth, the lower part of the helix and lobule
   4. The development of the external ear is completed about month 4 (135 mm)
   5. Developmental abnormalities of the auricle are not uncommon

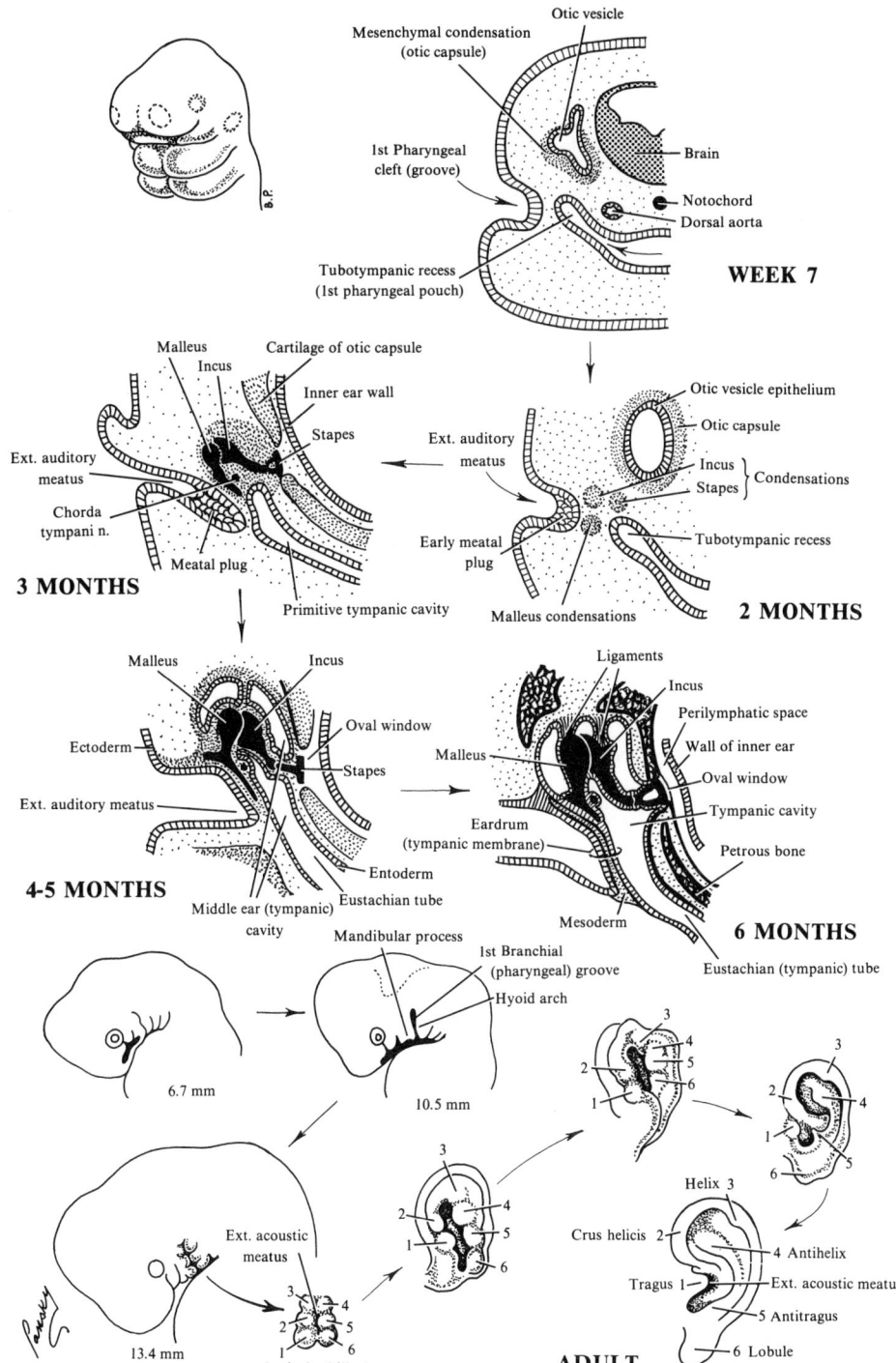

171. THE VESTIBULOCOCHLEAR SYSTEM: THE INTERNAL EAR—MEMBRANOUS LABYRINTH

I. **The auditory placode** appears toward week 3 in the human embryo, in the area of the rhombencephalon. Primary induction comes from the anterior chordomesoderm and secondarily from the rhombencephalon
   A. THE PLACODE gradually invaginates as a result of the fact that cellular proliferation is more intense internally than it is externally. As a result, the *auditory* or *otic vesicle* is formed, isolated from the superficial ectoderm by its active invagination and by the thrust of the lateral mesenchyme. The vesicle forms the *membranous labyrinth*
   B. THE SIMPLE OTIC VESICLE gives rise to the basic internal ear structure in about 50 days. The liquid it contains is found in all its derivatives and is called *endolymph*, which is supplied by specific vessels adjacent to the epithelium
   C. THE STATOACOUSTIC GANGLION CELLS arise from the inferior internal surface of the vesicle. Their dendrites are in contact with the sensory structures of the internal ear, and their axons conduct impulses toward the central nervous system

II. **The primary auditory vesicle** constricts to form the utricle, semicircular canals, and endolymphatic duct, dorsally, and the saccule and cochlear duct, ventrally
   A. THE ENDOLYMPHATIC DUCT OR SAC is an evagination, which is seen about day 30 on the internal surface of the vesicle and gradually elongates dorsally. During the second half of pregnancy, it elongates to reach the dura mater across the surrounding mesenchyme. It is important in the resorption of endolymphatic fluid
   B. THE UTRICLE
      1. The semicircular canals are derivatives of the utricle. The external half of the utricle gives rise to 2 flat evaginations. One is *vertical* and *sagittal* with respect to the embryo, and the other is *horizontal* (*lateral*), with its apex toward the exterior
         a. The vertical primordium gives rise to the *anterior* (*superior*) and *posterior semicircular canals*. The anterior (superior) canal differentiates first at about day 36. Its central portion is rapidly resorbed to form the canal. The posterior primordium undergoes a similar type of development after several hours. The common median portion of the 2 primordia persists to form the *common trunk* of the 2 canals
         b. The horizontal (lateral) primordium is perpendicular to the first 2 and undergoes a similar development after 3 or 4 days. Morphogenesis of all 3 canals is complete by about day 50
      2. One of the ends of each canal opens into the utricle via a swelling, the *ampulla*, wherein the sensory organs of balance develop
      3. The final transformation of this canal system consists of a 90° displacement of the anterior canal to the outside
         a. Thus, the 3 canals are placed in the 3 planes of space, and from here on they show only an increase in size until about day 70
   C. THE SACCULE
      1. The saccule shows a ventral evagination, the *cochlear duct,* at about day 36. The duct elongates progressively. It is straight at first, but soon spirals as a result of unequal growth of its internal and external surfaces
      2. By day 70 of gestation, the cochlear duct has 2½ spiral turns and its morphogenesis is finished
      3. The saccule, like the utricle, contains sensory organs for balance, whereas the cochlea develops cells especially adapted for hearing
   D. THE SEMICIRCULAR CANALS, the utricle and saccule, the endolymphatic duct, and the cochlear duct comprise the totally enclosed system called the *membranous labyrinth*. The latter is surrounded by mesenchyme which will give rise to the *osseous labyrinth*
      1. The membranous labyrinth is filled with *endolymph;* the bony labyrinth is filled with *perilymph*

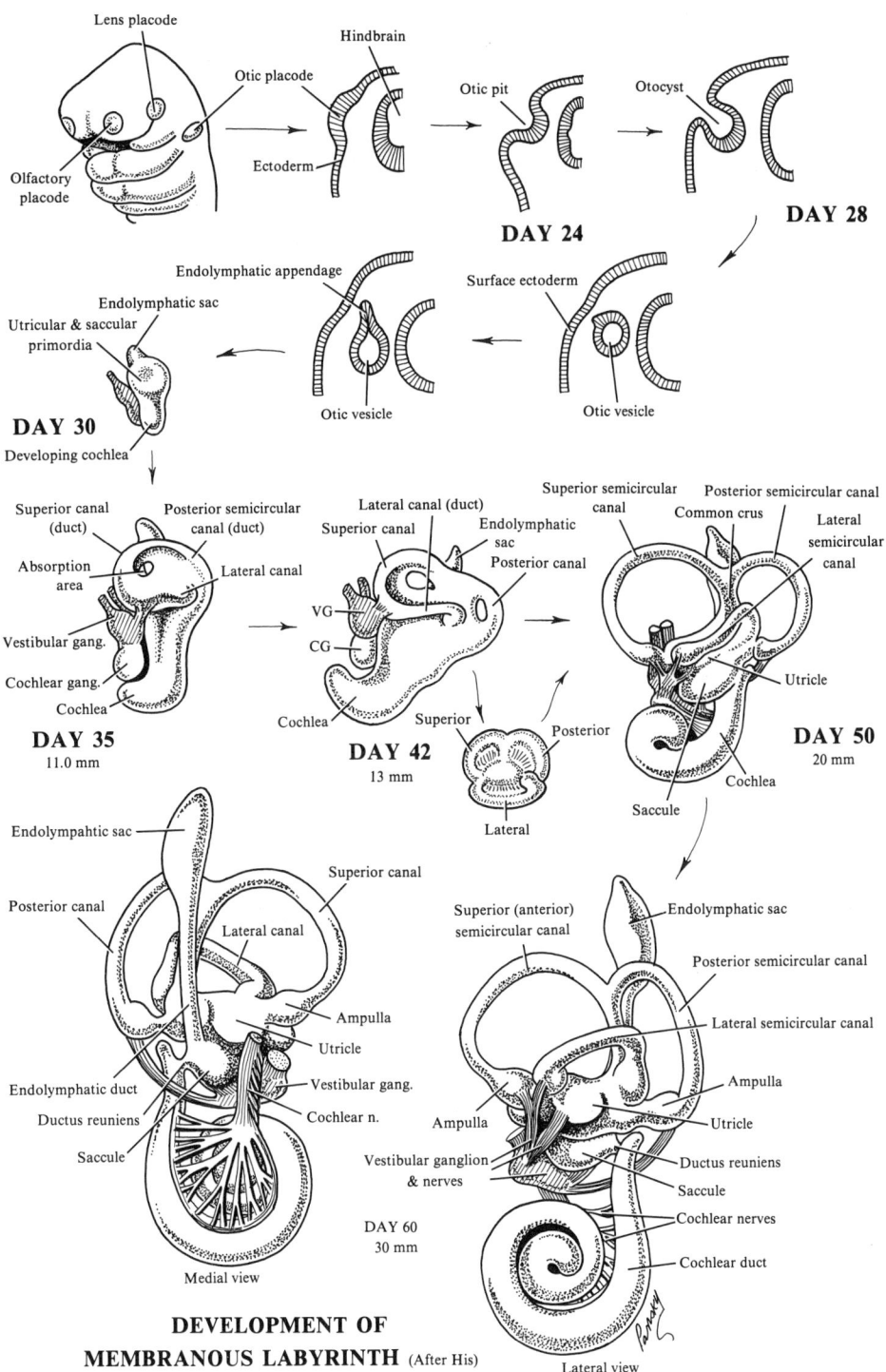

DEVELOPMENT OF MEMBRANOUS LABYRINTH (After His)

- 447 -

172. THE VESTIBULOCOCHLEAR SYSTEM: THE INTERNAL EAR—BONY LABYRINTH

I. **Development of the peripheral mesenchyme surrounding the membranous labyrinth**
A. THE OSSEOUS OR BONY LABYRINTH: the membranous labyrinth is embedded in mesenchyme, and at some distance from the membranous labyrinth, the peripheral mesenchyme becomes organized into cartilage at about week 5, and begins to form bone or the so-called otic capsule, after week 8. Gradually, the entire membranous labyrinth is encased in a bony shell.
B. PERILYMPHATIC SPACES are derived from the mesenchyme between the membranous labyrinth and the osseous labyrinth
   1. As the osseous labyrinth develops, the mesenchyme is transformed into a large-meshed reticulum which contains the *perilymphatic fluid*
      a. The perilymphatic spaces corresponding to the cochlea are divided into a *vestibular space* or *vestibule* within which lie the saccule and utricle, the *scala vestibuli* (continuous with the vestibule), and the *scala tympani*
      b. A perilymphatic space also exists around the semicircular canals
      c. The scala vestibuli and scala tympani, both related to the cochlear duct, are independent of each other as a result of incomplete mesenchymal resorption
         i. Thus, the mesenchyme gives rise outside to the *spiral ligament* which gives rise to the *basilar membrane,* and inside to a bony plate called the *spiral lamina* or *plate*
         ii. The 2 spaces (scala vestibuli and tympani) are united at the end of the bony cochlea by a small orifice, the *helicotrema,* which forms in the spiral plate at the end of month 3
      d. Thus, the modiolus and osseous spiral lamina (plate) of the cochlea are not preformed in cartilage, but are ossified directly from connective tissue
   2. The perilymphatic spaces are connected with the meningeal spaces (the subarachnoid space) by a fine duct, the *cochlear* or *perilymphatic duct,* which runs through the otic capsule opposite the saccule. Resorption of perilymph takes place through this pathway

II. **Formation of the neural sensory fibers**
A. THE GANGLIONIC CELLS derived from the auditory placode form 2 clusters
   1. One cluster, the *vestibular ganglion* or *ganglion of Scarpa,* is joined to the vestibular portion of the labyrinth
   2. The other, the *cochlear ganglion* or *ganglion of Corti,* is joined to the cochlear duct
   3. The dendrites of these cells reach the sensory epithelium of the internal ear, and the axons pass toward the metencephalon of the brainstem, bunching together to form the *statoacoustic (VIII) cranial nerve* which leaves the bony labyrinth through the *internal acoustic meatus* by way of bony canals that form a channel in the bone called the *modiolus*
   4. After metencephalic connections (nuclei of nerve VIII), the vestibular fibers reach the cerebellum to control subconscious balance, and the cochlear fibers reach the internal (medial) geniculate body and then the temporal cortex for conscious sound sensations

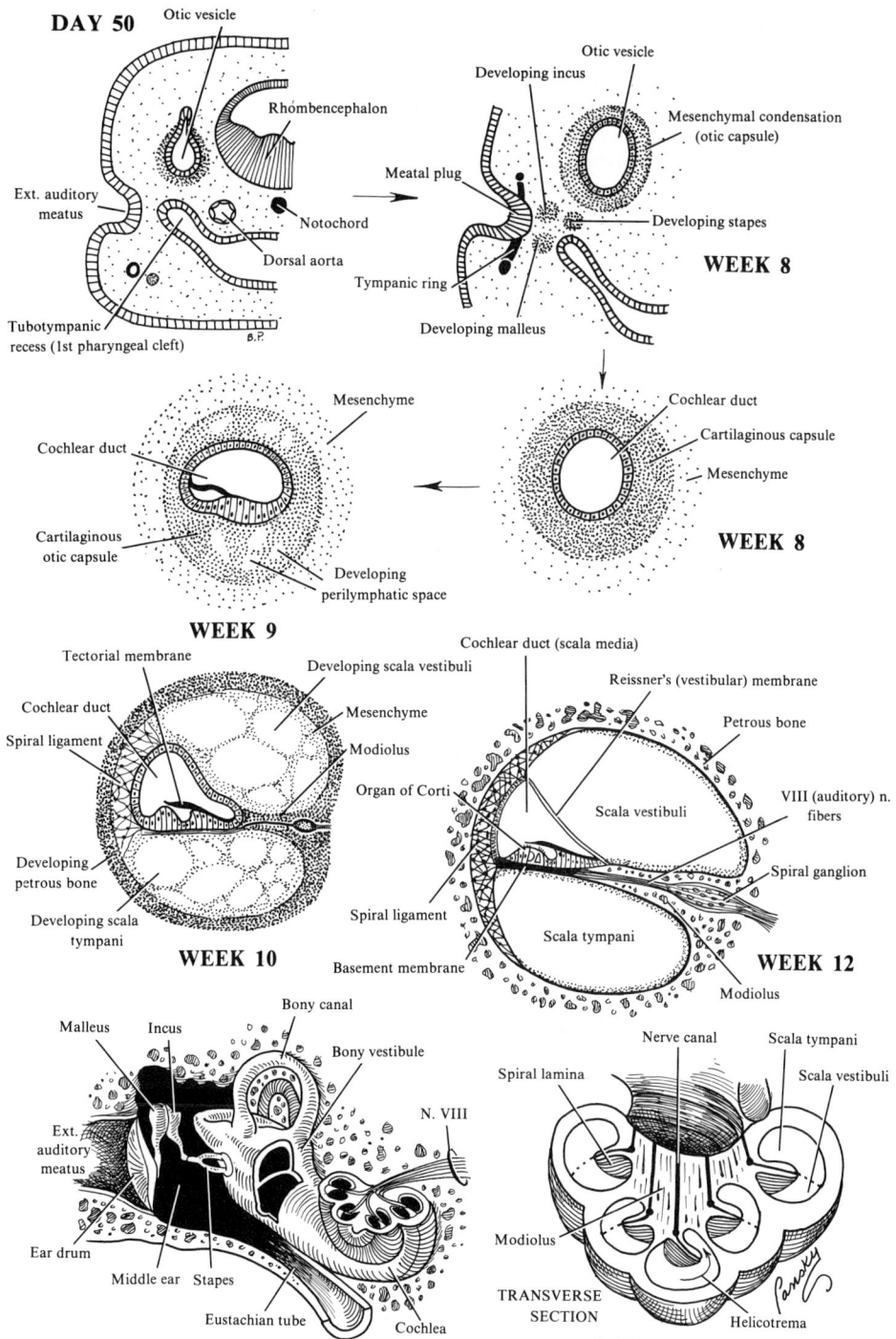

173. THE VESTIBULOCOCHLEAR SYSTEM: HISTOGENESIS OF THE INTERNAL EAR

I. Histogenesis of the semicircular canals
A. THE COLUMNAR EPITHELIUM lining the auditory vesicle (and its derivatives) flattens, except in the regions of the ampullae
   1. In the ampullae, toward day 50, the cells acquire sensory characteristics under the influence of the dendrites coming from the associated ganglion of Scarpa (vestibular). At their apical pole, they produce the *hair cells* and the *gelatinous mass of the cupula*, which rests on the hair cells
      a. The cupula is mobile. During rotation of the head there is displacement of the endolymph which, in turn, excites the sensory cells
   2. The cells, their hairs or flagella, and the cupula form the *crista of the ampulla.* They are fully differentiated about day 70 of gestation
   3. The *maculae,* which differentiate in the utricle and saccule, are analogous organs (like the cristae)
      a. Their sensory hairs are covered by a gelatinous mass, the *otolithic membrane.* On the membrane's free surface are found calcified structures called *otoliths*
      b. The maculae are highly sensitive to linear acceleration

II. Histogenesis of the cochlea
A. EPITHELIAL DERIVATIVES
   1. The ventral surface of the cochlear tube thickens at about day 70. Cellular proliferations take place involving the external and internal areas which are separated by a small depression, the *spiral sulcus*
      a. The *tectorial membrane,* fibrous and gelatinous, is derived from the internal portion of the so-called *spiral limbus*
      b. The *organ of Corti* is derived from the external portion
         i. Between months 3 and 5, some of the epithelial cells give rise to various categories of *supporting cells*
         ii. Here, too, sensory differentiation appears to need the presence of dendrites from the associated spiral ganglion (ganglion of Corti), and in month 5, fissures in these groups of cells make their appearance. One fissure forms the *spiral sulcus,* which clearly separates the organ of Corti from the limbus, and the other produces the *canal of Corti* which isolates the inner ciliated cells from the outer ciliated cells
   2. *Reissner's* or *vestibular membrane* is derived from the dorsal surface of the cochlear duct. It becomes very thin and remains unstratified
   3. The external surface of the tube has a thicker epithelium and many vessels and is called the *stria vascularis.* This area is said to produce the endolymphatic fluid
B. MEMBRANOUS DERIVATIVES
   1. The fibrillar *basement membrane* is derived from mesenchyme lining the ventral surface of the organ of Corti
      a. It is inserted on the ligament and the spiral lamina and is lined by the mesenchymal lamina bordering the scala tympani
   2. Cochlear histogenesis usually is completed by month 6 of gestation
   3. The basement membrane acts as a frequency analyzer. It is sensitive to high frequencies toward its base and to low frequencies near its apex. The sensitivity of the ear to low frequencies seems to correspond with progressive elongation of the cochlea

III. Function of the major derivatives:
vibrations reach the vestibule via the bony ossicles and set up vibrations in the perilymph of the scala vestibuli. From the latter, vibrations pass through the helicotrema to the scala tympani, resulting in distortion of the basement membrane. This produces movement in the sensory cells whose apical hairs stroke the tectorial membrane. This, in turn, excites the nerve endings

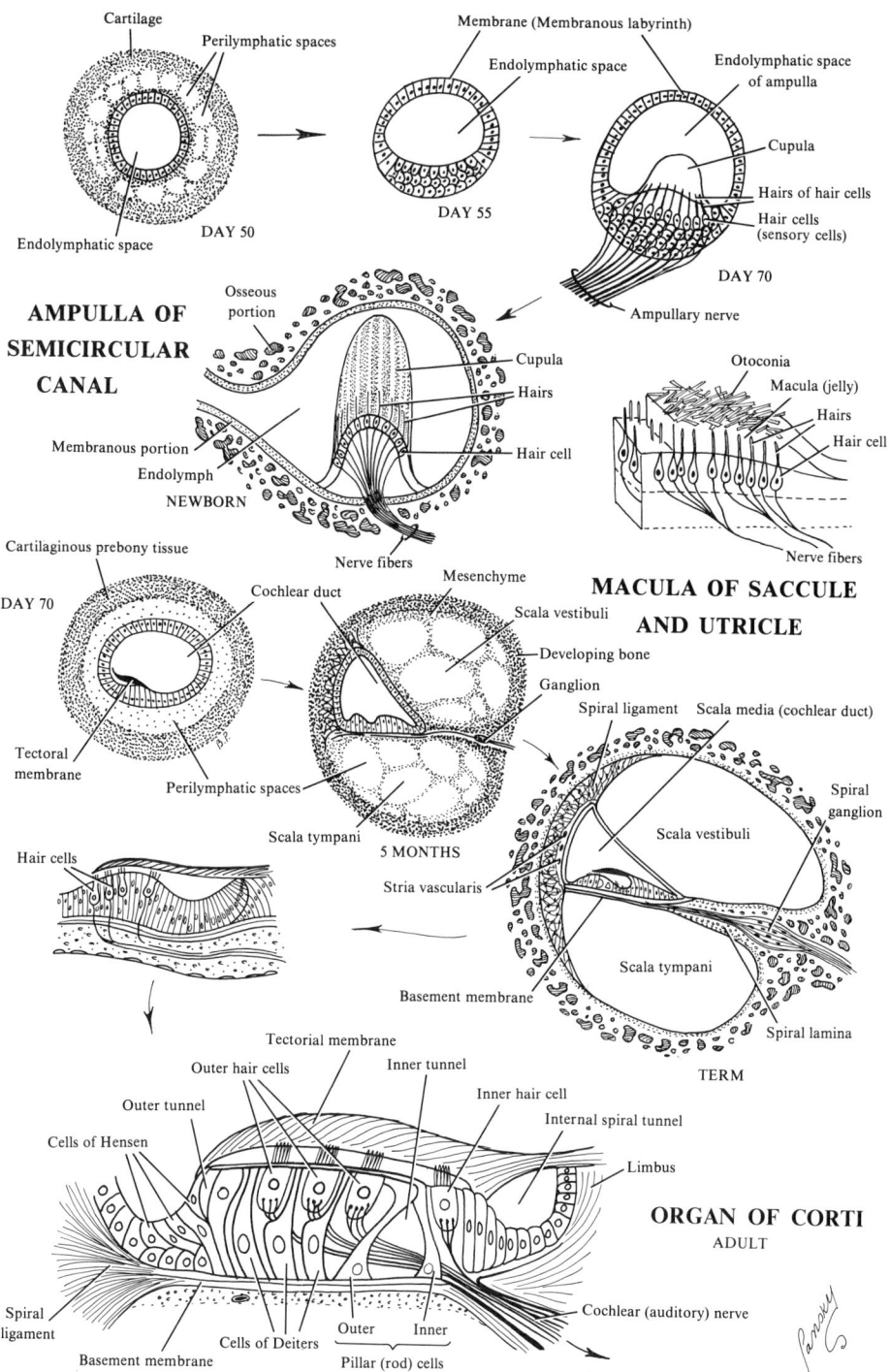

174. THE VESTIBULOCOCHLEAR SYSTEM: THE MIDDLE EAR

I. **The tympanic cavity of the middle ear** is of entodermal origin and is derived from the first pharyngeal pouch, an outpocketing of the pharynx, which grows laterally in the direction of the external acoustic meatus. The pouch, lined with epithelium of entodermal origin, appears in embryos at about week 4.
   A. THE POUCH GROWS RAPIDLY in a lateral direction and at about the end of month 6 of gestation, its external wall comes into contact wth the floor of the first ectodermal cleft (deep end of the external auditory meatus)
      1. A thin mesodermal plate persists between the ectodermal and entodermal epithelial structures, and the combination of all 3 layers forms the *tympanic membrane* or *eardrum*
   B. THE DISTAL PART of the pharyngeal pouch, the *tubotympanic recess,* widens and gives rise to the primitive tympanic cavity. The proximal portion remaining narrows and forms the *auditory* or *eustachian tube*
      a. The tube's pharyngeal orifice is surrounded by a considerable amount of lymphoid tissue, the *tubal* or *pharyngeal tonsil* (*adenoids*)
         i. Nasal inflammations associated with the tubal tonsillar swelling often result in occlusion of the tube and inflammation of the tympanic cavity resulting in *otitis media* (seen in young children quite frequently)
   C. DURING LATE FETAL LIFE, the tympanic cavity expands dorsally and posteriorly to form the *tympanic antrum*
      1. Its walls are covered with epithelium of entodermal origin
   D. AFTER BIRTH, the bone of the developing mastoid process is invaded by the epithelium of the tympanic cavity and epithelial-lined air sacs are created (pneumatization)
      1. Later, most of the mastoid air cells contact the antrum and the tympanic cavity

II. **The ossicles of the middle ear**
   A. BY THE END OF WEEK 7, the mesenchyme above the primitive tympanic cavity demonstrates a number of condensations caused by proliferation of the dorsal tips of pharyngeal arches I and II
      1. The condensations become the cartilaginous precursors of the auditory ossicles, the *malleus,* the *incus,* and the *stapes*
         a. The malleus and incus are derived from pharyngeal arch I (Meckel's cartilage) and the stapes is derived from arch II (Reichert's cartilage)
   B. THE OSSICLES appear during the first half of fetal life but remain embedded in mesenchyme until month 8, when the surrounding tissue dissolves
   C. THE ENTODERMAL LINING of the primitive tympanic cavity gradually extends along the wall of the newly developed space, and the tympanic cavity becomes almost twice as large as originally
   D. WHEN THE OSSICLES are entirely free from the surrounding mesenchyme, the entodermal epithelium connects them in a mesenterylike manner to the cavity wall. The supporting ligaments of the ossicles develop in these mesenteries
   E. SINCE THE MALLEUS IS DERIVED from pharyngeal arch I, its muscle, the *tensor tympani,* is innervated by the mandibular branch of the trigeminal (V) nerve
   F. SINCE THE STAPES is of pharyngeal arch II, its muscle, the *stapedius,* is innervated by the facial (VII) nerve

III. **Further development:** at birth, the cavities of the middle ear fill with air via the eustachian tube. With the ossicles, they form the system used for transmitting vibrations to the inner ear
   A. THE PART OF THE OSSEOUS LABYRINTH opposite the stapes remains thin and becomes the *oval window* (*fenestra vestibuli*) which opens into the vestibule; below this, another thinning of the bony labyrinth forms the *round window* (*fenestra cochleae*) which opens into the scala tympani. Both are closed by membranes

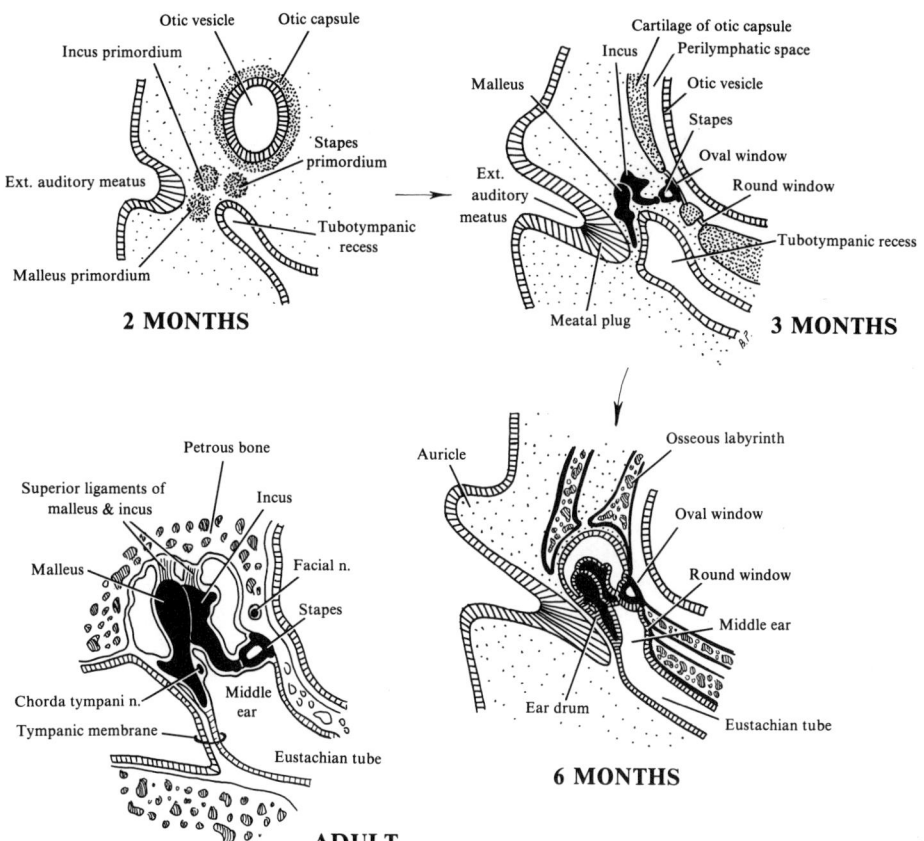

DERIVATIVES OF PHARYNGEAL ARCH CARTILAGES

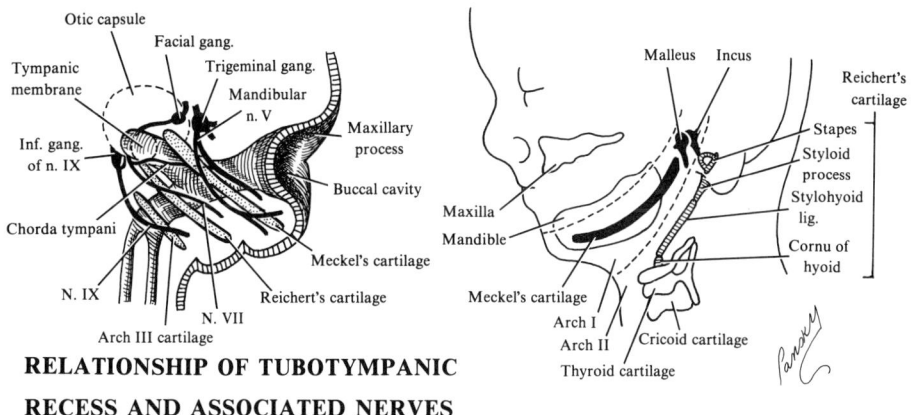

RELATIONSHIP OF TUBOTYMPANIC RECESS AND ASSOCIATED NERVES

HUMAN FETUS

175. CONGENITAL MALFORMATIONS OF THE VESTIBULOCOCHLEAR SYSTEM

I. **Malformations of the vestibulocochlear system** can be classified into 3 different groups, corresponding to the specific embryologic characteristics of each of the ear's 3 parts
   A. LESIONS OF THE INTERNAL EAR
      1. Cochlear
         a. Rubella in the second month of pregnancy is often a cause of lesions here since this is when the internal ear differentiates
            i. The epithelium of the cochlea and vestibule is altered, but the organ of Corti has some intact regions, so that children affected do perceive some deep frequencies
      2. The semicircular canals: few malformations are known, although some have been seen with the use of thalidomide
   B. LESIONS OF THE MIDDLE EAR: these usually involve the ossicles
      1. After an infection, the mesenchymal plate separating the ossicles may become sclerotic, impede movement, and may result in total deafness
      2. Congenital fixation of the stapes results in severe congenital conductive deafness, although the remainder of the ear is normal. The stapes is fixed to the bony labyrinth
         a. May be due to a failure of differentiation of the annular ligament of the stapes
      3. Defects of the malleus and stapes are often associated with abnormalities of branchial arch I, e.g., in hypoplasia of the mandible as seen in micrognathia
   C. MALFORMATIONS OF THE EXTERNAL EAR
      1. May result from the absence or nonunion of the primordial tubercles
      2. Major variations of the auricle have been associated with serious internal abnormalities, such as kidney malformations
      3. Abnormal position of the ear: usually associated with abnormal mandibular development, such as agnathia or micrognathia
         a. Instead of moving to the sides of the head, the ears develop at the site of the primordia namely, at the level of the first branchial groove. Thus, at birth, the ears are seen at the angle of the missing jaw (otocephalus)
      4. Auricular appendages or tags are common and due to accessory auricular hillocks
      5. Absence and hypoplasia of the auricles are rare and are associated with arch I syndrome where there is a failure of the auricular hillocks to develop (anotia) or they are suppressed in their development (microtia)
      6. Auricular sinus and fistulas: sinuses are usually preauricular and fistulas connect the exterior with the tympanic cavity
      7. Atresia of the external auditory meatus: failure of the meatal plug to canalize

II. **Congenital deafness** is usually associated with deaf-mutism
   A. MAY BE CAUSED BY
      1. An abnormal development of the membranous and bony labyrinths
      2. Malformations of the eardrum and ossicles
      3. In extreme cases, the tympanic cavity and the external auditory meatus are completely absent
   B. OTHER CAUSES OF DEAFNESS
      1. Hereditary
      2. Environmental and other factors affecting the mother early in pregnancy
         a. Rubella virus, affecting the embryo in weeks 7 and 8 of development, can cause severe damage to the organ of Corti; diabetes; erythroblastosis fetalis; hypothyroidism; toxoplasmosis; and x-radiation
   C. MAJOR VARIATIONS OF THE AURICLE have been associated with serious internal abnormalities, such as seen with kidney malformations

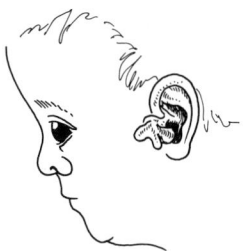

MULTIPLE AURICULAR
APPENDAGES

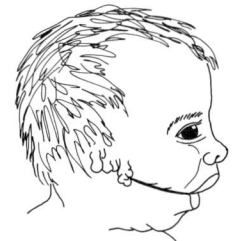

EXTERNAL EAR ANOMALY WITH
MAXILLOMANDIBULAR FISSURE

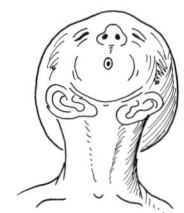

OTOCEPHALUS AND SYNOTIA WITH
MICROSTOMIA AND AGNATHIA

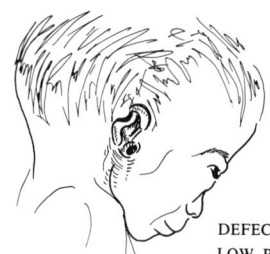

DEFECTIVE HELIX AND
LOW-PLACED AURICLE
Drug defect

MICROTIA

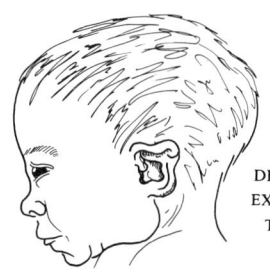

DEFORMED
EXTERNAL EAR
Trisomy 18

MALFORMED
AURICLE

HYPOPLASIA OF AURICLE
With absence of external
auditory meatus

FISTULA AURIS
Probed

SYNOTIA WITH
MACROSTOMIA AND
MANDIBULAR
HYPOSTOMIA

176. THE HYPOPHYSIS (PITUITARY GLAND): GLANDULAR PRIMORDIUM

I. **Introduction:** the hypophysis or pituitary gland is an unpaired gland located in the bony sella turcica of the sphenoid bone of the skull. It is found in all vertebrates and consists of 2 major parts of different embryonic origin
   A. THE GLANDULAR PART arises from an evagination (outpocketing) of the ectodermal epithelium covering the roof of the stomodeum, and during embryonic development, the glandular primordium becomes located anterior to the neural primordium
   B. THE NEURAL OR DIENCEPHALIC PART arises from an evagination of the floor of the third ventricle (from neuroectoderm)
   C. THE GLANDULAR PRIMORDIUM is induced first by the anterior end of the notochordal system (or the prochordal plate). This system next induces the neural primordium or infundibulum. From this time on, each primordium affects the development of the other by means of reciprocal induction

II. **Development of the glandular primordium**
   A. THE CELLS OF THE STOMODEAL SURFACE ECTODERM just ahead of the buccopharyngeal membrane become thicker than the rest, at about day 21 of gestation (7 somite stage). This *placodal primordium* is very near the wall of the diencephalon and just in front of the notochord
   B. THE FLAT PRIMORDIUM then invaginates and penetrates the mesenchyme in the direction of the diencephalon as the *diverticulum of Rathke* and forms *Rathke's pouch,* which grows toward the brain
      1. Just behind the buccopharyngeal membrane, the entodermal epithelium also forms another pouch, the *pouch of Seessel,* which is involved in the formation of the glandular hypophysis in lower vertebrates, but which usually is not present in humans; however, it may be seen and related to certain tumors, such as a craniopharyngioma
      2. By week 5, Rathke's pouch has elongated and become constricted at its attachment to the oral epithelium by the *pharyngohypophyseal stalk,* which regresses and eventually disappears during week 6. A remnant of this stalk may persist and give rise to a *pharyngeal hypophysis* in the pharyngeal roof
         a. Rarely, accessory masses of anterior lobe tissue may occur outside the capsule of the gland, but within the sella turcica, or in the substance of the bone
   C. DURING SUBSEQUENT DEVELOPMENT, cells of the anterior wall of Rathke's pouch proliferate actively and give rise to the *pars distalis* of the pituitary gland
      1. Later, a small extension of the pars distalis, the *pars tuberalis,* extends up and around the infundibular stem
   D. THE EXTENSIVE PROLIFERATION of the anterior wall of Rathke's pouch reduces the lumen to a narrow residual cleft
      1. This cleft generally is not recognizable in the human adult gland and usually is represented by a *zone of cysts*
   E. IN HUMANS, CELLS OF THE POSTERIOR WALL OF RATHKE'S POUCH do not proliferate but give rise to the thin, poorly defined *pars intermedia*
   F. THUS, THE ADENOHYPOPHYSIS or glandular portion of the gland consists of the pars distalis and tuberalis, which make up the anterior lobe, and the pars intermedia, all arising from oral ectoderm from the roof of the stomodeum

III. **Histogenesis** begins at about month 4
   A. THE SURROUNDING MESENCHYME appears to induce the glandular character of the cellular differentiation which takes place from the same cellular layer. The cells also reflect the bilateral symmetry seen during the gland's development
      1. The *basophil (PAS⁺) cells* are much more numerous in the anteromedial portion of the gland than in its posterolateral portions, whereas the *acidophils (PAS⁻)* show the opposite arrangement
      2. The *chromophobe cells* are spread almost uniformly through the gland

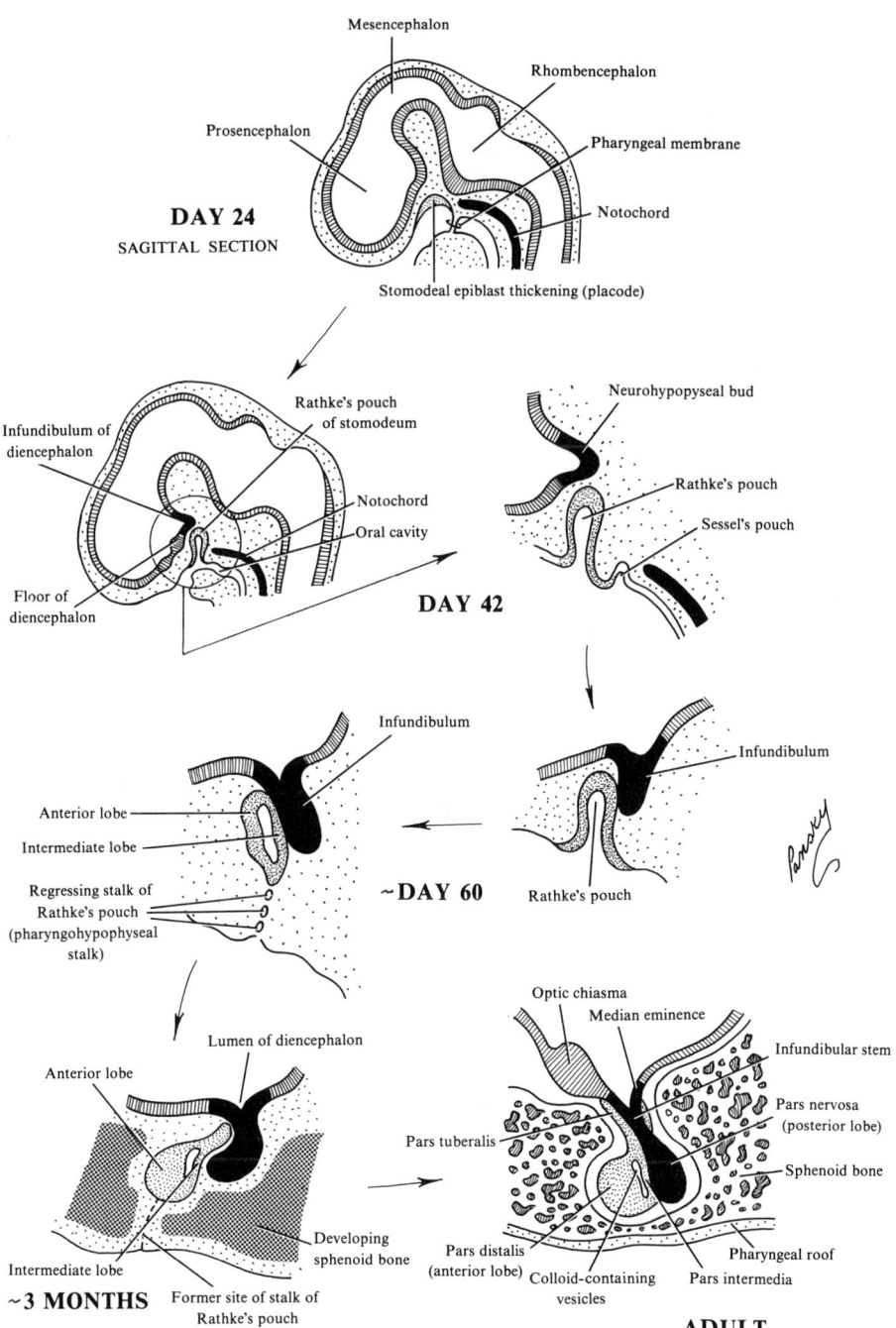

177. THE HYPOPHYSIS (PITUITARY GLAND): NEURAL PRIMORDIUM AND PORTAL SYSTEM

I. **The floor of the third ventricle, just behind the glandular primordium** becomes depressed and produces the *infundibulum,* at about day 40 of gestation
  A. THE INFUNDIBULUM will give rise to the *median eminence,* the *infundibular stem,* and the *pars nervosa,* all of which make up the *neurohypophysis*
  B. THE DEPRESSION extends progressively toward the glandular primordium, and about day 45, its ventral end forms a diverticulum, which thickens. Its lumen also gradually fills. Thus, the *pars nervosa* or *posterior lobe* is formed and becomes attached to the posterior wall of Rathke's pouch
    1. The pars nervosa is attached to the diencephalon by a stalk, the *infundibular stem*
    2. The median eminence is the slightly prominent segment of the infundibulum of the hypothalamus, proximal to the infundibular stem, just below the third ventricle.

II. **The neural lobe** differentiates during month 4, and one can see specific neuroglial cells appearing, the *pituicytes.* These may be glandular, but this as yet has not been proven

III. **The neurohypophysis** is then colonized by axons coming from the hypothalamus
  A. THESE AXONS form the *hypothalamic-hypophyseal tract,* the pathway for hypothalamic neurosecretions, which are rich in polypeptides and are elaborated by the hypothalamic nuclei from cells which appear to be both neural and glandular
    1. The paraventricular and supraoptic nuclei represent some of these nuclei
    2. The neurosecretory material is then conducted to the neurohypophysis by axons of these cells. This can be seen in the fetus at about month 4 of gestation

IV. **Relationship between the neural and glandular hypophysis**
  A. A PORTION OF THE NEUROSECRETION from the hypothalamus passes to the glandular hypophysis through the blood via a "portal" system. The latter involves
    1. The superior hypophyseal arteries originate from the internal carotid arteries and form a capillary bed in the proximal half of the neurohypophysis
       a. These capillaries collect the neurosecretory material
       b. Venules which follow them pass into the pars tuberalis and the anterior lobe of the gland, where they form a new capillary system which releases the neurosecretory material
       c. The above arrangement represents the portal system of the hypophysis, which constitutes the essential pathway of neuroadenohypophyseal relationships
    2. The inferior hypophyseal arteries also arise from the internal carotid arteries
       a. They irrigate the neural lobe and do not appear to be connected to the portal system
       b. The veins which follow carry away the hormones of the posterior lobe, which are modified neurosecretions, into the dural sinuses
  B. THERE ARE NO KNOWN NERVES to the anterior lobe of the pituitary gland except those following the blood vessels. The posterior lobe, however, has a direct connection, as previously mentioned, to the inferior surface of the brain, and this part of the gland is innervated by fibers from neuron cell bodies located in nuclei of the hypothalamus
    1. These neurons contain granules of neurosecretion in their cell bodies and axons, and it is thought that the hormones produced by the posterior lobe are actually produced in these neurons and stored in the posterior lobe

V. **Tumors of the neurohypophysis** are very uncommon
  A. INFUNDIBULOMA: a rare tumor that appears to be of neurohypophyseal derivation and stimulates the structural pattern of the infundibulum
    1. Seen in children, it grows slowly in the floor of the third ventricle, causing pressure effects on neighboring structures
    2. The cells resemble pituicytes, and the tumor has a distinctive vascular pattern

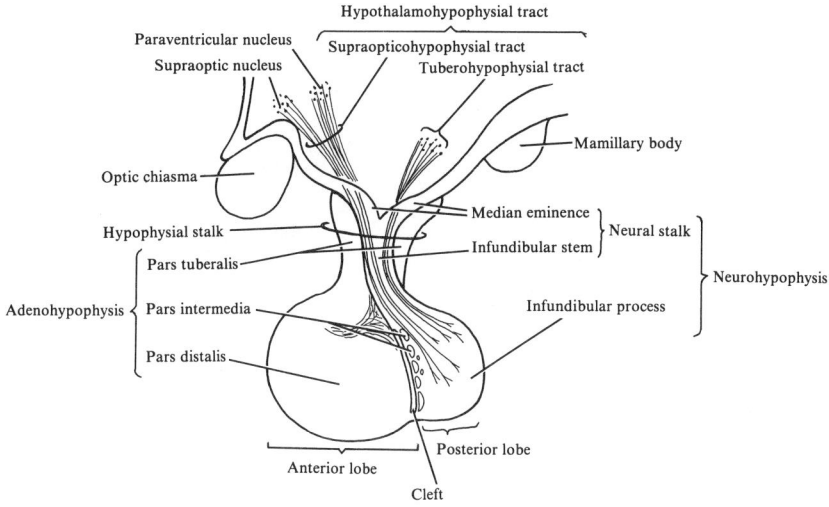
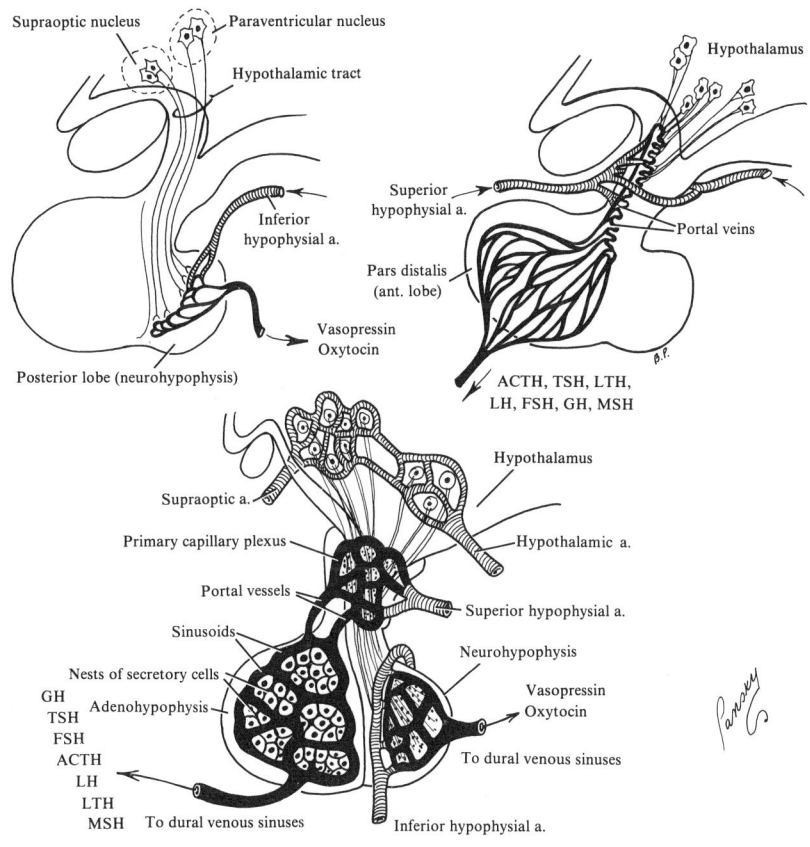

178. ROLE OF THE HYPOPHYSIS (PITUITARY GLAND): PHYSIOLOGY AND PATHOLOGY

I. **Introduction:** all endocrine receptors are dependent on stimulation by the hypophysis after birth. In the fetus, however, some endocrine glands partly or completely function without hypophyseal control
   A. HYPOPHYSECTOMY apparently does not affect fetal growth, which appears to be independent of growth hormone (in contrast to juvenile growth)
   B. ABLATION OF THE FETAL HYPOPHYSIS, experimentally, results in severe atrophy of the thyroid gland and adrenals, but only results in a slowing of gonadal function

II. **Neuroendocrine relations:** hypothalamohypophyseal control
   A. IN THE FETUS: hypothalamic control seems less essential than in the adult. Only ACTH secretion appears to require nervous stimulation
   B. IN THE ADULT: all the tropic hormones of the anterior pituitary seem to be generally under hypothalamic nuclear control
      1. The neurosecretion material includes inhibitory and releasing factors
      2. Neurosecretion also includes oxytocin and vasopressin (antidiuretic, ADH) hormones of the posterior pituitary which originate in hypothalamic nuclei

III. **Gonadotropic hormones**
   A. FOLLICLE-STIMULATING HORMONE (FSH): important in the development of the ovarian follicle as well as in spermatogenesis
   B. LUTEINIZING HORMONE (LH) influences testicular endocrine activity after the prenatal period, ovulation, and appearance of the ovarian corpus luteum. Synergistic with FSH
   C. PROLACTIN (LACTOGENIC) HORMONE: postpartum lactation in women. Important for the maintenance of progesterone secretion by the corpus luteum of pregnancy in the rat
   D. THYROTROPIC HORMONE (TSH) stimulates growth and function of the thyroid gland
   E. SOMATOTROPIC OR GROWTH HORMONE (STH) promotes body growth, fat metabolism, and inhibition of glucose utilization
   F. ADRENOCORTICOTROPIC HORMONE (ACTH): development of the fetal adrenal cortex and the primordium of the cortical zone of the definitive cortex

IV. **The intermediate lobe** produces the melanotropic hormone (MSH) or intermedin
   A. MSH STIMULATES PIGMENT MOVEMENT in the melanophores from cell body to cell processes. Seen in the retinal pigmented epithelium of the mammalian eye

V. **The pineal gland (epiphysis) and hypophyseal relations:** its function is antagonistic to the gonadotropic cells of the pituitary in the fetus and in prepubertal children
   A. IF EXPERIMENTALLY DESTROYED or there is a destructive pineal tumor, there is precocious development of the sex organs and secondary sex characteristics
   B. THE ENDOCRINE ROLE of the pineal apparently stops at puberty. Its secretion, melatonin, however, may be involved in the timing of human adolescence

VI. **Malformations of the hypophysis (pituitary gland)**
   A. AGENESIS: appears to be the result of a disturbance of notochordal induction
   B. DOUBLING OF THE GLAND: produced experimentally by riboflavin deficiency, hypervitaminosis A, and some tranquilizers
   C. CRANIOPHARYNGIOMAS (Rathke's pouch tumor) constitute about 3-5% of intracranial neoplasms, are usually benign with effects due to pressure, are operable, and especially affect children. Seen intracranially with symptoms like anterior pituitary tumors
      1. Onset of symptoms is usually before the age of 15 years
      2. Usually form in front of the anterior pituitary lobe resulting in visual problems
      3. It is a heterogenous, intrasellar tumor, usually partly solid and partly cystic with keratinized areas and foci of necrosis. Cholesterol crystals and calcification are common. The latter are seen easily on x-rays

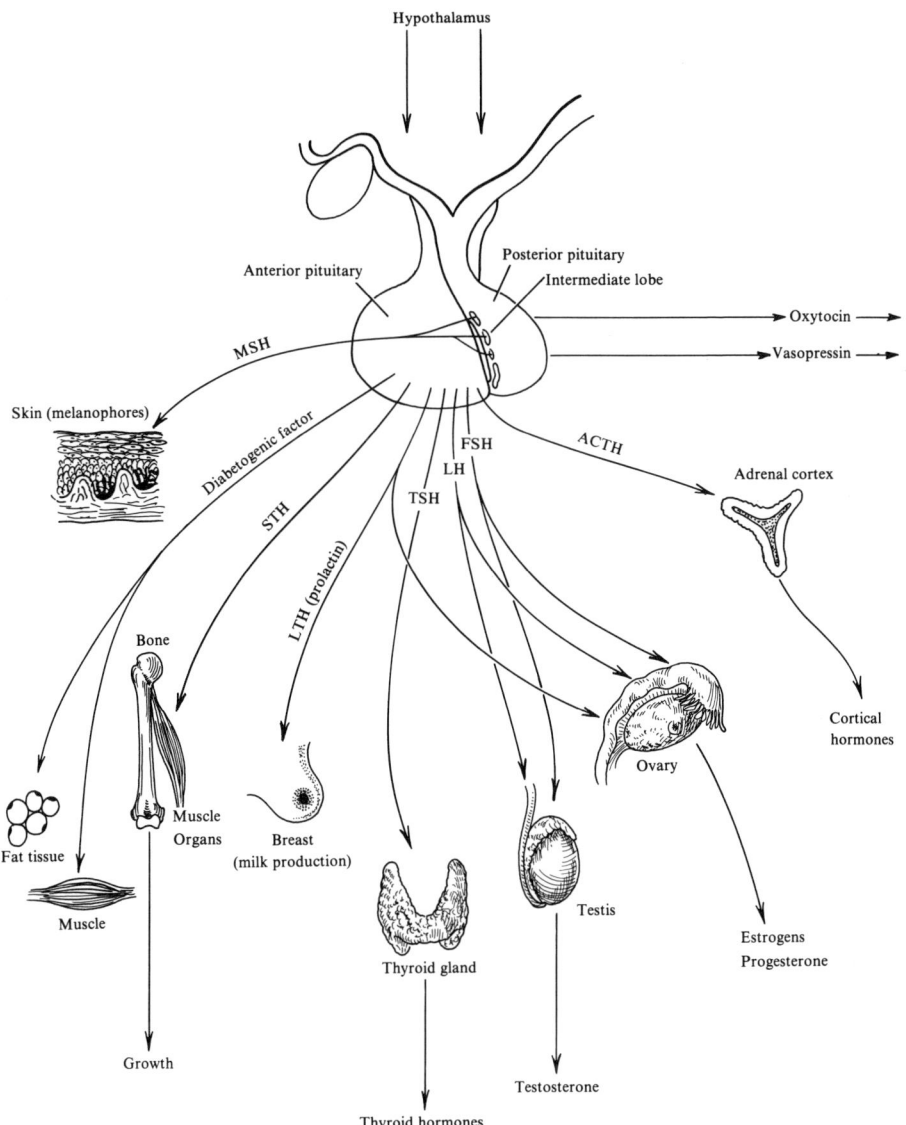

HORMONES OF PITUITARY GLAND

179. THE PARAGANGLIONIC SYSTEM: THE PARAGANGLIA

I. **Introduction:** the primordial neural crest cells of the sympathetic ganglia are not all transformed into nervous cells. Some of the sympathogonia migrate beyond the sympathetic chain and take on a glandular character. They form the paraganglionic or chromaffin body system, so named because the paraganglia are often associated with the sympathetic ganglia, and their cells are also derived from the same cells. The mechanism of regulation of their migration is unknown
   A. PARAGANGLIONIC TISSUE is found dispersed in relation to the sympathetic ganglia, the abdominal aorta (the organs of Zuckerkandl), and beginning in embryos of about 10 mm, similar cells migrate into the primordium of the suprarenal gland to form the suprarenal medulla
   B. THE CELLS OF THE SYMPATHOCHROMAFFIN SYSTEM combine some of the properties of secretory cells and of nerve cells and are homologous in the suprarenal gland with postganglionic neurons
      1. The other elements of this system undergo a slow atrophy after birth

II. **The paraganglia**
   A. INTRODUCTION: the sympathogonia detach themselves from the sympathetic primordium and differentiate into glandular cells during month 2 of gestation
      1. Small groups of cells are formed behind the peritoneum (retroperitoneal), in the connective tissue capsule of the suprarenal gland, in the thoracic and abdominal sympathetic chains, in the genital glands, in the epicardium, near and around the kidneys, and elsewhere
   B. MANY OF THESE PARAGANGLIA REGRESS when the adrenal medulla becomes functional after birth, e.g., the aortic chromaffin body called the body of Zuckerkandl, located near the origin of the inferior mesenteric artery
      1. Other large cell clusters persist but are usually fragmented, and the resulting paraganglia come in contact with sympathetic ganglionic cells or even some blood vessels
   C. FUNCTION OF THE PARAGANGLIA: they are probably involved in the production of epinephrine (adrenaline) or norepinephrine (noradrenaline)
      1. The above 2 hormones are secreted in the fetus near the end of month 4. They help maintain the fetal blood pressure. After birth, the adrenal medulla and the autonomic nervous system take over this function
      2. Some paraganglia cells are located near the vagus (X) nerve and are found to secrete acetylcholine. These cells, however, do not show the same staining reaction as those of the adrenaline-producing paraganglia, which selectively take up chromium salts and account for the chromaffin reaction
      3. The cells of the paraganglionic system are distinguished from the neuroblasts by their smaller size and by their characteristic reaction to staining with dichromate salts (yellow) and to ferric chloride (green)
         a. The staining is probably due to the presence of cytoplasmic droplets which have been tentatively identified with the precursors of epinephrine and norepinephrine
      4. Paraganglionic tissue liberates adrenaline and acetylcholine more abundantly than do the ordinary nerve endings, thus, can be considered to be in support of organs that function continuously, like the heart or the vasosensory system that controls circulation

III. **Pathology of the paraganglionic system**
   A. PHEOCHROMOCYTOMA AND PARAGANGLIOMA are tumors that involve this system
      1. The tumor may affect the adrenal medulla or the adrenal-producing paraganglia
      2. An abnormally persisting organ of Zuckerkandl may become tumorous
   B. THE MOST FREQUENT SYMPTOM of involvement of this system is *paroxysmal hypertension*

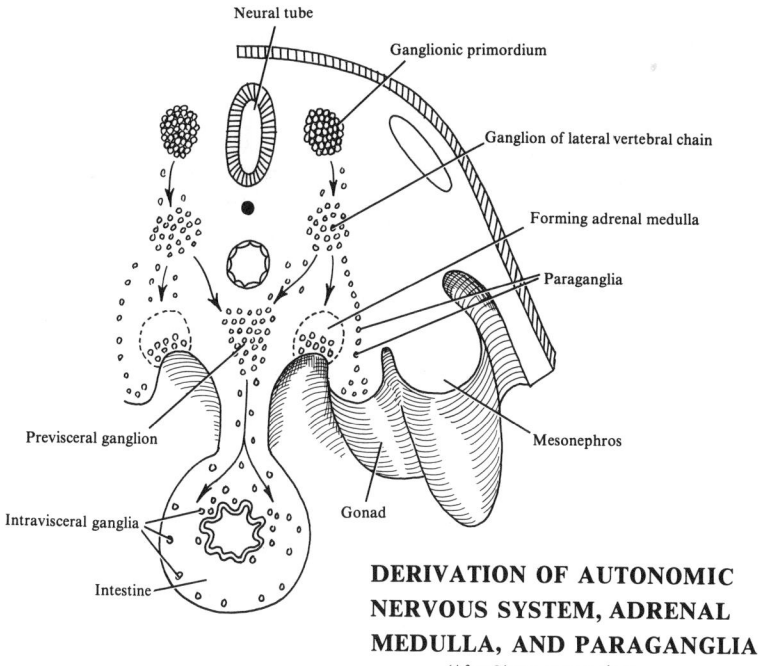

DERIVATION OF AUTONOMIC NERVOUS SYSTEM, ADRENAL MEDULLA, AND PARAGANGLIA
(After Giroud and LeLiévre)

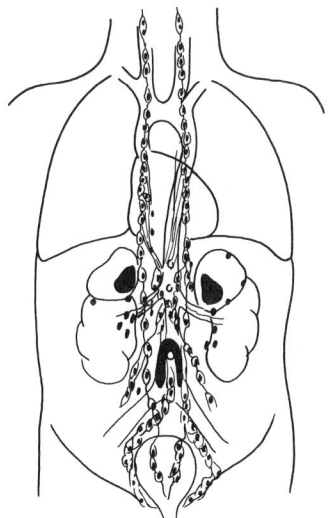

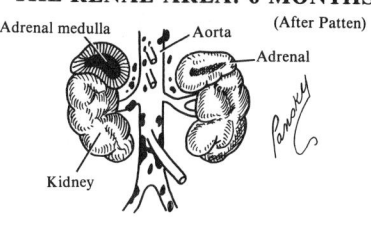

PARAGANGLIONIC (CHROMAFFIN) TISSUE OF THE RENAL AREA: 6 MONTHS
(After Patten)

DISTRIBUTION OF CHROMAFFIN TISSUE IN THE NEONATAL HUMAN BODY
(After R. E. Copeland)

180. DEVELOPMENT OF THE ADRENAL (SUPRARENAL) GLAND

I. **Introduction:** the adrenals are glandular masses found above the kidneys and consist of a *cortical area* of mesodermal origin and a *medullary area* of ectodermal origin

A. THE CORTICAL PRIMORDIUM comes from a wide plate of coelomic epithelium in the most internal region of the mesonephric blastema, between the mesenteric root and the gonadal primordium
   1. This origin accounts for the presence of accessory paratesticular and paraovarian accessory cortical masses which are sometimes encountered
   2. At the beginning of month 2 (8-mm fetus), under induction by the primary ureter (wolffian duct), mesothelial (coelomic epithelial) cells proliferate and penetrate the underlying mesenchyme. They multiply quickly and differentiate into large acidophilic cells which surround the medullary primordium and form the *fetal* or *primitive suprarenal cortex*
   3. Shortly after, toward the end of month 3, a second wave of cells from the coelomic epithelium (mesothelium) penetrates the mesenchyme and surrounds the original acidophilic cell mass. These smaller cells form the *definitive cortex of the gland*
      a. The small basophilic cells will form the future *glomerular* and *fascicular* zones of the definitive cortex
      b. After birth, the fetal cortex regresses rapidly, except for its outer layer which differentiates into the *reticular zone* of the cortex
   4. The adult structure of the cortex is not achieved until near puberty
   5. Prior to month 5, the cortex appears to develop autonomously. After this time, its development depends on hypophyseal corticotropic hormone (ACTH)

B. THE MEDULLARY PRIMORDIUM occurs at about day 45 of gestation and results from the assembling of sympathogonia from the sympathetic chain (of ectodermal origin) in the region near the developing mesodermal cortical primordium
   1. While the fetal cortex is forming, the invading sympathogonia cells become arranged in clusters and cords. The cells do not form nerve processes, but rather, stain yellow-brown with chrome salts and are called *chromaffin cells*. The staining is probably due to epinephrine and norepinephrine in the cells.
   2. At birth, the medulla is only slightly developed and is not yet functional. The definitive cortex is only a peripheral ring and makes up only 15–20% of the glandular parenchyma

II. **At birth:** the structure of the external glomerular zone is still not precise, but the zona fasciculata is seen readily and is directly continuous with the fetal zone
   1. The fetal zone begins to regress, but is not completely gone until the second year
   2. While the fetal zone is regressing, the zona glomerulosa and fasciculata develop, and the zona reticularis makes its appearance
   3. Development of the definitive cortex and its physiologic activity is regulated by ACTH, and it is not completely differentiated until 18 to 21 months after birth

III. **Vascularization of the adrenal gland**
   1. The adrenal arteries take their major origin from the abdominal aortic system
      a. On reaching the gland, the arteries ramify with some going directly to the medulla, but the majority nourish the extended capillary network of the glomerulosa and reticularis. From the latter, blood flows into vessels of the medulla and finally leaves the gland via a large central vein at the gland hilus
      b. Thus, there is an important vascular relationship between the cortex and medulla

IV. **Innervation of the adrenal gland:** preganglionic sympathetic fibers to the gland do not synapse in the sympathetic ganglia but go directly to the gland and synapse in ganglia in the cortex and medulla. From these, postganglionic fibers supply the blood vessels. However, the majority of preganglionic fibers go directly to the cells of the medulla

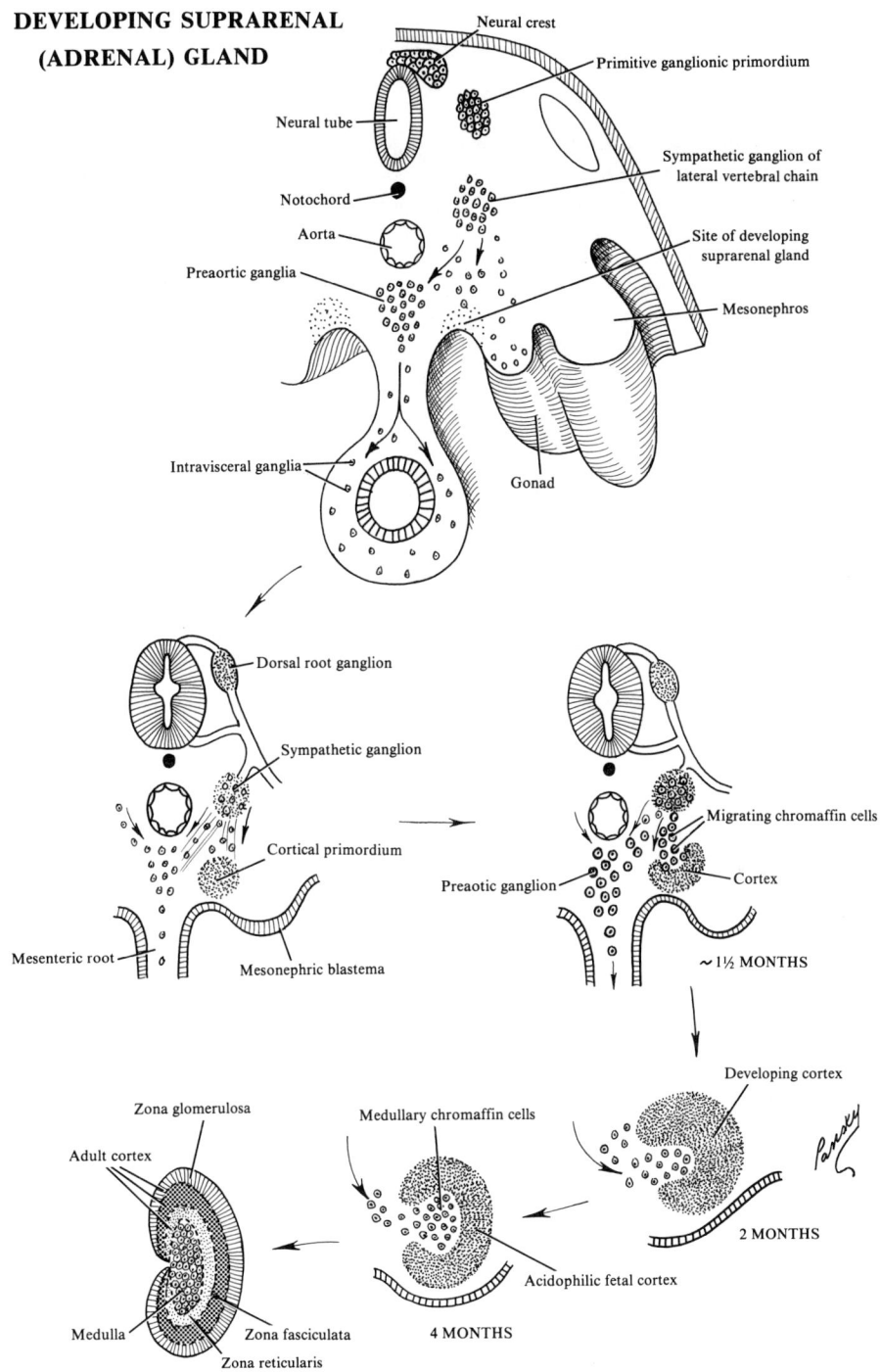

181. PATHOLOGY ASSOCIATED WITH THE ADRENAL GLAND

I. **Anencephaly:** the relationship of the adrenal and pituitary gonads
A. IN ANENCEPHALICS, the anterior pituitary is often only slightly affected. However, since the hypothalamus is missing, the neurohypophysis contains no neurosecretion, and without this, ACTH is not secreted (this is the case even in the presence of an intact anterior pituitary lobe)
  1. It should be noted that anencephaly has little effect before month 5 of fetal life since development of the adrenal up to this point appears to be autonomous
  2. After month 5, development of the fetal cortex cannot occur without ACTH, thus, in the anencephalic, there is an involution of the adrenal cortex leading to agenesis or hypoplasia
B. IN HYDROCEPHALUS, the hypothalamus is undamaged. The adrenals develop normally

II. **Female pseudohermaphroditism with precocious virilization**
A. THIS CONDITION PRODUCES late fetal masculinization of the female genital tract
B. THE MOST COMMON CAUSE of this condition is the *adrenogenital syndrome,* resulting from congenital virilizing adrenal hyperplasia
  1. There is no ovarian abnormality
  2. Excessive adrenal production of androgens causes masculinization of the external genitalia varying from clitoral enlargement to almost male genitalia
  3. Clitoral hypertrophy is common, as are partial fusion of the labia majora and a persistent urogenital sinus
  4. Rarely, fusion of the labioscrotal and urogenital folds is so complete that the urethra passes through the penis and the infants are raised as males
  5. Prompt recognition and treatment of the associated adrenal inbalance are essential
C. THIS ANOMALY occurs quite frequently and is of genetic origin
  1. Transmission of autosomal recessive, and the genotype of the affected person is actually female (XX)
D. THIS CONDITION IN THE MALE causes hypermasculinization with precocious virilization

III. **Ectopic adrenal glands:** the adrenal tissue is found beneath the kidney capsule

IV. **Accessory cortical tissue:** this accessory tissue is most often associated with the definitive adrenal gland itself, but it may be found retroperitoneally on the posterior abdominal wall or in the pelvis. Medullary tissue is not usually present

V. **Agenesis of the adrenal:** unilateral agenesis of the gland is almost always associated with agenesis of the kidney on the same side
A. BOUND UP WITH the failure of the entire nephrogenic ridge to form, early in week 4

VI. **Congenital adrenal hypoplasia** usually manifests itself shortly after birth with many of the symptoms of Addison's disease

VII. **Fusion of the suprarenal glands:** seen when kidneys are also fused across the midline

VIII. **Cushing's syndrome:** fully developed Cushing's syndrome has not been observed as yet, but hirsuitism and pubic hair at birth have been described with clitoral enlargement
A. CUSHING'S SYNDROME has been described at about 17 months of age along with adrenogenital syndromes
B. SOME CASES have been described with symptoms appearing within weeks after birth, suggesting a prenatal disturbance
C. IT IS CAUSED BY ADRENAL HYPERPLASIA or tumor which produces excessive amounts of hormones from the cortex. Clinical features vary, depending on the type of adrenal hormones being produced in excess

CHIEF CONGENITAL ANOMALIES
OF SUPRARENAL (ADRENAL) GLANDS

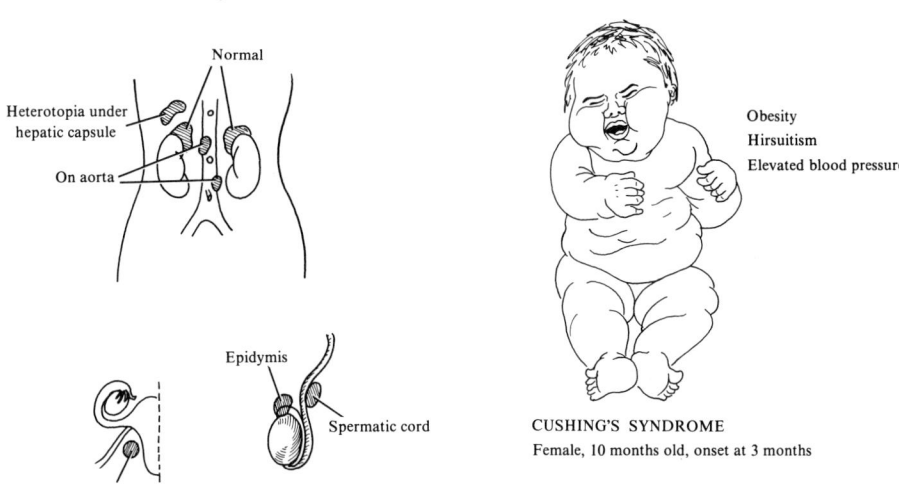

SUPRARENAL HETEROTOPIA

CUSHING'S SYNDROME
Female, 10 months old, onset at 3 months

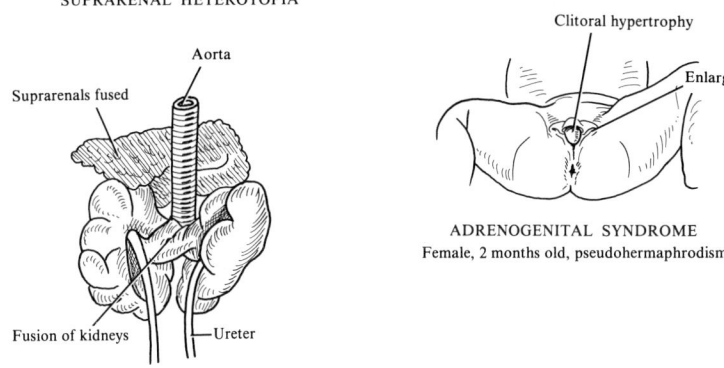

FUSION OF SUPRARENAL GLANDS BEHIND AORTA

ADRENOGENITAL SYNDROME
Female, 2 months old, pseudohermaphrodism

APPENDIXES

APPENDIX I. CORRELATED HUMAN DEVELOPMENT

I. **Age (weeks): 2.5**
A. SIZE (CR IN MM): 1.5
   1. Body form: embryonic disk flat. Primitive streak prominent. Neural groove indicated
   2. Mouth: not developed
   3. Pharynx and derivatives: not developed
   4. Digestive tube and glands: gut not as yet distinct from yolk sac
   5. Respiratory system: not developed
   6. Coelom and mesenteries: extraembryonic coelom present. Embryonic coelom almost ready to make its appearance
   7. Urogenital system: allantois present
   8. Vascular system: blood islands appear on chorion and yolk sac. Cardiogenic plate reversing
   9. Skeletal system: head process for notochordal plate is present
   10. Muscular system: not developed
   11. Integumentary system: ectoderm seen as a single layer
   12. Nervous system: neural groove is indicated
   13. Sense organs: not developed

II. **Age (weeks): 3.5**
A. SIZE (CR IN MM): 2.5
   1. Body form: neural groove deepens and closes except at ends. Somites 1–16± present. Cylindric body constricting from yolk sac. Branchial arches 1 and 2 seen
   2. Mouth: mandibular arch is prominent. Stomodeum a definitive pit. Oral membrane ruptures
   3. Pharynx and derivatives: pharynx broad and flat. Pharyngeal pouches forming. Thyroid gland makes its appearance
   4. Digestive tube and glands: foregut and hindgut present. Yolk sac broadly attached at midgut. Liver bud is present. Cloaca and cloacal membrane are seen
   5. Respiratory system: respiratory primordium appears as a groove on pharyngeal floor
   6. Coelom and mesenteries: embryonic coelom a U-shaped canal with a large pericardial cavity. Septum transversum seen. Mesenteries forming at this time. Mesocardium begins to atrophy
   7. Urogenital system: all pronephric tubules are formed. Pronephric duct growing caudally as a blind tube. Cloaca and cloacal membrane are present
   8. Vascular system: primitive blood cells and vessels are present. Embryonic blood vessels form a paired symmetric system. The heart tubes fuse, bend in an S-shape, and the *heartbeat begins*
   9. Skeletal system: mesodermal segments are appearing (1–16±). Older somites do not demonstrate sclerotomes. Notochord appears as a cellular rod
   10. Muscular system: mesodermal segments appearing (1–16±). Older somites begin to demonstrate myotome plates
   11. Integumentary system: no new developments
   12. Nervous system: neural groove is prominent but rapidly closing. Neural crest is a continuous band
   13. Sense organs: optic vesicle and auditory placode are seen. Acoustic ganglia appearing

III. **Age (weeks): 4.0**
A. SIZE (CR IN MM): 5.0
   1. Body form: branchial arches completed. Flexed heart is prominent. Yolk stalk is slender. All 40 somites are present. Limb buds appear. Otocyst and eye are evident. Body is now C-shaped and flexed

2. Mouth: maxillary and mandibular processes are prominent. Tongue primordia is present. Rathke's pouch becomes evident
3. Pharynx and derivatives: 5 pharyngeal pouches are present. Pouches 1-4 have closing plates. The primary tympanic cavity is indicated. The thyroid gland is now a stalked sac
4. Digestive tube and glands: esophagus is short. The stomach is spindle-shaped. Intestine is a simple tube. Liver cords, ducts, and gallbladder are forming. Both pancreatic buds appear. The cloaca is at its full development
5. Respiratory system: trachea and paired lung buds become prominent. Laryngeal opening is a simple slit
6. Coelom and mesenteries: coelom still a continuous system of cavities. Dorsal mesentery is a complete median structure. Omental bursa is now indicated
7. Urogenital system: pronephros has degenerated. Pronephric (mesonephric) duct reaches the cloaca. Mesonephric tubules are differentiating rapidly. Metanephric bud is seen pushing into secretory primordium
8. Vascular system: hematopoiesis is seen on the yolk sac. The paired aortae fuse. The aortic arches and cardinal veins are completed. The dilated heart shows a sinus, atrium, ventricle, and bulbus
9. Skeletal system: all 40 somites are present. Sclerotomes are massed as primitive vertebrae around the notochord
10. Muscular system: all 40 somites are present
11. Integumentary system: not remarkable
12. Nervous system: neural tube is closed. Three primary vesicles of brain are seen. Nerves and ganglia are forming. Ependymal, mantle, and marginal layers are seen
13. Sense organs: optic cup and lens pit are forming. Auditory pit becomes the closed detached otocyst. Olfactory placodes arise and differentiate nerve cells

IV. **Age (weeks): 5.0**
A. SIZE (CR IN MM): 8.0
1. Body form: nasal pits are present. Tail is prominent. Heart, liver, and mesonephros are prominent. Umbilical cord organizes
2. Mouth: jaws are outlined. Rathke's pouch is now a stalked sac
3. Pharynx and derivatives: pharyngeal pouches develop dorsal and ventral diverticulae. Thyroid becomes bilobed. Thyroglossal duct atrophies
4. Digestive tube and glands: tail-gut atrophies. Yolk stalk detaches. Intestine elongates into a loop. Cecum is now evident
5. Respiratory system: bronchial buds presage future lung lobes. Arytenoid cartilage swellings and epiglottis are indicated
6. Coelom and mesenteries: pleuropericardial and pleuroperitoneal membranes are forming. Ventral mesogastrium pulls away from the septum
7. Urogenital system: mesonephros reaches its caudal limit. Ureteric and pelvic primordia become distinct. Genital ridge bulges
8. Vascular system: primitive vessels extend into head and limbs. Vitelline and umbilical veins are transforming. Myocardium is condensing. Cardiac septa appearing. Spleen becomes evident
9. Skeletal system: condensations of mesenchyme presage many future bones
10. Muscular system: premuscle masses are seen in head, trunk, and limbs
11. Integumentary system: epidermis acquiring a second layer (periderm)
12. Nervous system: 5 brain vesicles are seen. Cerebral hemispheres bulging. Nerves and ganglia more clearly seen. Suprarenal cortex beginning to form
13. Sense organs: choroid fissure is prominent. Lens vesicle is free. Vitreous anlage is appearing. Otocyst elongates, and endolymphatic duct is budded off. Olfactory pit deepens

V. **Age (weeks): 6.0**
A. SIZE (CR IN MM): 12.0
   1. Body form: upper jaw components are prominent but still separate. Lower jaw halves fuse. Head becomes dominant in size. Cervical flexure is marked. External ear appears. Limbs become more clearly recognizable
   2. Mouth: lingual primordia fusing. Foramen cecum is established. Labiodental laminae begin to appear. Parotid and submandibular gland buds appear
   3. Pharynx and derivatives: thymic sacs, ultimobranchial sacs, and solid parathyroids are conspicuous and ready to detach. Thyroid gland becomes solid, converts into plates
   4. Digestive tube and glands: stomach rotating, intestinal loop under torsion. Hepatic lobes are identifiable. Cloaca is subdividing
   5. Respiratory system: definitive pulmonary lobes are indicated. Bronchi are subbranching. Laryngeal cavity temporarily obliterated
   6. Coelom and mesenteries: pleuropericardial communications close. Mesentery expands as the intestine forms a loop
   7. Urogenital system: cloaca subdividing. Pelvic anlage sprouts pole tubules. Sexless gonads and genital tubercle are prominent. Müllerian duct appearing
   8. Vascular system: hematopoiesis in the liver is seen. Aortic arches are transforming. Left umbilical vein and ductus venosus become important. Bulbus is absorbed into right ventricle. Heart acquires its general definitive shape
   9. Skeletal system: first appearance of chondrification centers. Desmocranium seen
   10. Muscular system: myotomes become fused into a continuous column and spread ventrally. Muscle segmentation generally lost
   11. Integumentary system: milk line is now present
   12. Nervous system: 3 primary flexures of brain are seen. Diencephalon is large. Nerve plexuses are present. Epiphysis recognizable. Sympathetic ganglia forming segmental masses. Meninges are beginning to appear
   13. Sense organs: optic cup shows nervous and pigment layers. Lens vesicle thickens. Eyes are set at 160°. Nasolacrimal duct seen. Modeling of external, middle, and internal ear is beginning. Vomeronasal organ seen

VI. **Age (weeks): 7.0**
A. SIZE (CR IN MM): 17.0
   1. Body form: branchial arches are lost. Cervical sinus is obliterated. Face and neck are forming. Digits are evident. Back straightening. Heart and liver determine shape of body ventrally. Tail is now regressing
   2. Mouth: lingual primordia merge into a single tongue. Separate labial and dental laminae are visible. Jaws are formed and begin to ossify. Palate folds are present and are separated by the tongue
   3. Pharynx and derivatives: thymus becomes elongated and loses its lumen. Parathyroids become trabeculate and associate with the thyroid. Ultimobranchial bodies fuse with the thyroid. The thyroid becomes crescentic
   4. Digestive tube and glands: stomach attaining final shape and position. Duodenum is temporarily occluded. Intestinal loops herniate into cord. Rectum separates from bladder-urethra. Anal membrane ruptures. Dorsal and ventral pancreatic primordia fuse
   5. Respiratory system: larynx and epiglottis are well outlined with a T-shaped orifice. Laryngeal and tracheal cartilages are foreshadowed. Conchae begin to appear. Primary choanae are rupturing
   6. Coelom and mesenteries: the pericardium is extended by splitting from body wall. Mesentery is expanding rapidly as the intestine coils. Ligaments of the liver become very prominent

7. Urogenital system: mesonephros is at the height of its differentiation. Metanephric collecting tubules begin branching. Earliest metanephric secretory tubules differentiating. Bladder-urethra separates from the rectum. The urethral membrane is beginning to rupture
8. Vascular system: cardinal veins are transforming. The inferior vena cava is visible. The atrium, ventricle, and bulbus are partitioned. Cardiac valves are present. Stem of the pulmonary vein is absorbed into the left atrium. The anlage of the spleen is prominent
9. Skeletal system: the chondrocranium is seen. Chondrification is now more general
10. Muscular system: muscles are differentiating rapidly throughout body and are assuming their final shapes and relationships
11. Integumentary system: there is mammary thickening
12. Nervous system: cerebral hemispheres are becoming large. Corpus striatum and thalamus are prominent. Infundibulum and Rathke's pouch are in contact. Choroid plexuses are appearing. Suprarenal medulla begins to invade the cortex
13. Sense organs: choroid fissure closes, enclosing the central artery. Nerve fibers invade the optic stalk. The lens loses its cavity by elongating lens fibers. Eyelids are forming. The fibrous and vascular coats of the eye are beginning to form. The olfactory sacs open into the mouth

VII. Age (weeks): 8.0
A. SIZE (CR IN MM): 23.0
1. Body form: nose is flat. Eyes are far apart. Digits are well formed. Growth of gut makes body evenly rotund. Head is elevating. The fetal state is now reached
2. Mouth: tongue muscles are well differentiated. The earliest taste buds are indicated. Rathke's pouch detaches from the mouth. The sublingual glands are now appearing
3. Pharynx and derivatives: the auditory tube and tympanic cavity are evident. The sites of the tonsils and their fossae are evident. The thymic gland halves unite and become solid. The thyroid gland follicles are forming
4. Digestive tube and glands: small intestine is coiling within the cord. The intestinal villi are developing. The liver is very large in relative size
5. Respiratory system: the lung is becoming glandlike by branching of the bronchioles
6. Coelom and mesenteries: pleuroperitoneal communications close. Pericardium is a very large sac. The diaphragm is completed, including its musculature. The diaphragm also completes its descent
7. Urogenital system: the testis and ovary are now distinguishable as such. Müllerian ducts are nearing the urogenital sinus and are about ready to unite with the uterovaginal primordium. The genital ligaments are indicated
8. Vascular system: the main blood vessels are assuming their final plan. The primitive lymph sacs are present. The sinus venosus is absorbed into the right atrium. The atrioventricular bundle is present
9. Skeletal system: the first indications of ossification are evident
10. Muscular system: definitive muscles of the trunk, limbs, and head are well represented, and the fetus is now capable of some movement
11. Integumentary system: mammary primordia are seen as globular thickenings
12. Nervous system: the cerebral cortex begins to acquire its typical cells. The olfactory lobes are visible. The dura and pia-arachnoid are distinct. Chromaffin bodies are seen
13. Sense organs: the eyes are converging rapidly. The external, middle, and internal ears are assuming their final form. The taste buds are appearing. The external nares are plugged.

VIII. Age (weeks): 10.0
A. Size (CR in mm): 40.0
1. Body form: head is erect. Limbs are well modeled. Nailfolds are indicated. The umbilical hernia is reduced
2. Mouth: fungiform and vallate papillae are differentiating. The lips are separate from the jaws. The enamel organs and dental papillae are forming. The palate folds are fusing
3. Pharynx and derivatives: the thymic epithelium is transforming into reticulum and thymic corpuscles. The ultimobranchial bodies disappear as such
4. Digestive tube and glands: the intestines withdraw from the umbilical cord and assume their characteristic position. The anal canal is formed. The pancreatic alveoli are present
5. Respiratory system: the nasal passages are partitioned by fusion of the septum and palate. The nose is cartilaginous. The laryngeal cavity is reopened and the vocal folds appear
6. Coelom and mesenteries: the processus (saccus) vaginales are forming. The intestine and its mesentery withdraw from the umbilical cord
7. Urogenital system: the kidneys are able to secrete. The bladder expands as a sac. The genital ducts of the opposite sex are degenerating. The bulbourethral and vestibular glands are appearing. The vagina sacs are forming
8. Vascular system: the thoracic duct and peripheral lymphatics are developed. Early lymph glands are appearing. Enucleated red blood cells predominate in the blood
9. Skeletal system: ossification centers are more common. The chondrocranium is at the height of its development
10. Muscular system: the perineal muscles are developing slowly
11. Integumentary system: intermediate cells are added to the epidermis. The periderm cells are prominent. The nail fields are indicated. The earliest hair follicles begin developing on the face
12. Nervous system: the spinal cord attains its definitive internal structure
13. Sense organs: the iris and ciliary bodies are organizing. The eyelids are fused. The lacrimal glands are budding. The spiral organ of Corti begins to differentiate

IX. Age (weeks): 12.0
A. Size (CR in mm): 56.0
1. Body form: the head is still a dominant feature. The nose acquires its bridge. Sex is readily determined by external inspection
2. Mouth: the filiform and foliate papillae are elevating. Tooth primordia form prominent cups. The cheeks are seen. Fusion of the palate is complete
3. Pharynx and derivatives: the tonsillar crypts begin to invaginate. The thymus forms its medulla and is becoming more lymphoid. The thyroid attains its typical form
4. Digestive tube and glands: muscle layers of the gut are present. Pancreatic islets are appearing. Bile is being secreted
5. Respiratory system: the nasal conchae are prominent. Glands of the nasal cavity are forming. The lungs are acquiring their definitive form
6. Coelom and mesenteries: the omentum has become an expansive apron which is partly fused to the dorsal body wall. The mesenteries are free and exhibit their usual relationships. Extension of the coelom into the umbilical cord is obliterated
7. Urogenital system: the uterine horns are absorbed. The external genitalia attain distinctive features. The mesonephric and rete tubules complete the male duct. The prostate and seminal vesicles begin to appear. The hollow viscera are beginning to form muscular walls
8. Vascular system: blood formation is beginning in the bone marrow. The blood vessels acquire accessory coats

9. Skeletal system: the notochord is degenerating very rapidly. Ossification is spreading rapidly. A number of bones are well defined
10. Muscular system: smooth muscle layers are becoming evident in the hollow viscera
11. Integumentary system: the epidermis is now 3-layered. The corium and subcutaneous tissue are now distinct
12. Nervous system: the brain attains its general structural features. The spinal cord demonstrates cervical and lumbar enlargements. The cauda equina and filum terminale make their appearance. Neuroglial types are beginning to differentiate
13. Sense organs: the characteristic organization of the eye is attained. The retina is now becoming layered. The nasal septum and plate fusions are completed

X. Age (weeks): 16.0
A. SIZE (CR IN MM): 112.0
1. Body form: face takes on a "human" appearance. Hair of the head is appearing. Muscles become spontaneously active. Body is beginning to outgrow head
2. Mouth: hard and soft palates are differentiating. The hypophysis is acquiring its definitive structure
3. Pharynx and derivatives: lymphocytes are beginning to accumulate in the tonsils. The pharyngeal tonsil (adenoids) is beginning to develop
4. Digestive tube and glands: gastric and intestinal glands are developing. The duodenum and colon become affixed to the posterior abdominal wall. Meconium is collecting
5. Respiratory system: the accessory nasal sinuses are developing. The tracheal glands appear. Mesoderm is still abundant between the pulmonary alveoli. Elastic fibers make their appearance in the lungs
6. Coelom and mesenteries: greater omentum is fusing with the transverse mesocolon and colon. The mesoduodenum, ascending mesocolon, and descending mesocolon are attaching to the posterior abdominal wall
7. Urogenital system: the kidneys attain their characteristic shape. The testis is in position for its descent into the scrotum. The uterus and vagina are recognizable as entities. The mesonephros is involuted
8. Vascular system: blood formation is now active in the spleen. The heart musculature is much condensed
9. Skeletal system: most bones are clearly indicated throughout the body. Joint cavities appear
10. Muscular system: cardiac muscle which appeared in earlier weeks is now more condensed. Muscular movements *in utero* can be detected
11. Integumentary system: the epidermis begins adding additional layers. The body hair begins to develop. The sweat glands appear. The first sebaceous glands begin to differentiate
12. Nervous system: the hemispheres now conceal much of the brain. The cerebral lobes are delimited. The corpora quadrigemina appear. The cerebellum attains some prominence
13. Sense organs: the eye, ear, and nose are nearing their typical appearance. The general sense organs are differentiating

XI. Age (weeks): 20.0-40.0
A. SIZE (CR IN MM): 160.0-350.0
1. Body form: lanugo hair appears in week 20. Vernix caseosa collects in week 20. Body becomes better proportioned but lean in week 24. Fetus is wrinkled, lean, and red, and eyelids reopen in week 28. Testes invade the scrotum in week 32. Body rounds out, fat collects, and wrinkling smooths out in weeks 32 to 40

2. Mouth: enamel and dentine deposited in week 20. Lingual tonsil forming in week 20. Permanent teeth primordia indicated in weeks 24 to 32. Milk teeth are unerupted at birth
3. Pharynx and derivatives: the tonsils are structurally typical in week 20
4. Digestive tube and glands: lymph nodules and muscularis mucosae of the gut are present in week 20. The ascending colon becomes recognizable in week 24. The appendix lags behind the cecum in growth at week 24. The deep esophageal glands are indicated in week 28. The plicae circulares are represented in week 32
5. Respiratory system: the nose begins ossifying in week 20. The nostrils reopen in week 24. The cuboidal epithelium of the lung alveoli is disappearing in week 24. Pulmonary branching is only two-thirds completed by week 40. The frontal and sphenoidal sinuses are still quite incomplete by week 40
6. Coelom and mesenteries: the mesenterial attachments are completed by week 20. The vaginal sacs are passing into the scrotal sacs between weeks 28 and 36
7. Urogenital system: the female urogenital sinus is becoming a shallow vestibule by week 20. The vagina regains its lumen by week 20. The uterine glands appear in week 28. The scrotum is solid until sacs and testes descend in weeks 28 to 36. The tubules of the kidney cease forming at birth
8. Vascular system: blood formation is increasing in the bone marrow but is decreasing in the liver between weeks 20 and 40. The spleen acquires its typical structure by week 28. A number of fetal blood vessels are discontinued by week 40
9. Skeletal system: the carpal, tarsal, and sternal bones ossify late, some after birth. Most epiphyseal centers appear after birth, many during adolescence
10. Muscular system: the perineal muscles finish their development by week 24
11. Integumentary system: the vernix caseosa is seen in week 20. The epidermis cornifies by week 20. The nail plates begin in week 20. Hairs emerge by week 24. The mammary primordia bud in week 20, and the buds hollow out and branch by week 32. The nail reaches the fingertip by week 36. Lanugo hair is prominent in week 28 and is shed in week 40
12. Nervous system: the commissures are completed by week 20. Myelinization of the cord begins in week 20. The cerebral cortex is typically layered by week 24. The cerebral fissures and convolutions appear rapidly in week 28. Myelinization of the brain begins in week 40
13. Sense organs: the nose and ear ossify in week 20. The vascular tunic of the lens is at its greatest by week 28. The retinal layers are completed and light perception is possible by week 28. Taste sense is present in week 32. The eyelids reopen between 28 and 32. The mastoid cells are still unformed by week 40. At birth, the ear is still deaf to sounds

APPENDIX II. GERM LAYER DERIVATIVES

I. Ectoderm
  A. Surface ectoderm
    1. Epidermis
    2. Hair and nails
    3. Cutaneous and mammary glands
    4. Anterior pituitary gland
    5. Enamel of teeth
    6. Inner ear
    7. Lens of eye
  B. Neuroectoderm
    1. Neural crest origin
       a. Cranial and sensory ganglia and nerves
       b. Medulla of adrenal gland
       c. Pigment cells
    2. Neural tube origin
       a. Central nervous system
       b. Retina of eye
       c. Pineal body (gland)
       d. Posterior pituitary gland

II. **Mesoderm**
  A. Head mesoderm
    1. Skull
    2. Muscles and connective tissue of head
    3. Dentine of teeth
  B. Paraxial mesoderm
    1. Muscles of trunk
    2. Skeleton, except for skull
    3. Dermis of skin
    4. Connective tissue
  C. Intermediate mesoderm
    1. Urogenital system
       a. Includes the gonads, ducts, and accessory glands
  D. Lateral plate mesoderm
    1. Connective tissue and muscles (smooth and cardiac) of the viscera
    2. Serous membranes of pleura, pericardium, and peritoneum
    3. Blood and lymph cells
    4. Cardiovascular and lymphatic systems
    5. The spleen
    6. The adrenal cortex
    7. Bone marrow

III. **Endoderm**
  A. Epithelial parts of
    1. The larynx and trachea
    2. The bronchi
    3. The lungs
    4. The pharynx
    5. The thyroid gland
    6. The tympanic cavity
    7. The pharyngotympanic (eustachian) tube

    8. The tonsils
    9. The parathyroid glands
B. Epithelium of
    1. The gastrointestinal tract and its glands
    2. The liver
    3. The pancreas
    4. The urinary bladder
    5. The urachus
    6. Vagina and vestibule
    7. Urethra and glands

APPENDIX III. CRITICAL PERIODS OF HUMAN DEVELOPMENT (SENSITIVITY TO TERATOGENS)

Major morphologic abnormalities occur during weeks 3 to 7. However, physiologic defects and minor morphologic abnormalities do occur from week 8 to term

I. Weeks 1 and 2: period of dividing zygote, implantation, and bilaminar embryo
  A. THE EMBRYO IS NOT NORMALLY SUSCEPTIBLE to teratogens during this period
   1. A substance will damage either all or most of the cells at this time, resulting in prenatal death, or the embryo will survive with few defects, if any

II. Week 3
  A. THIS IS A HIGHLY SENSITIVE PERIOD for the developing heart and central nervous system
   1. Highly sensitive for the heart from the middle of week 3 to week 6
   2. Highly sensitive period for the central nervous system from the beginning of week 3 through early in week 6

III. Week 4
  A. DURING THIS PERIOD, the eyes, ears, arms, and legs begin to develop
   1. Sensitive period for the *eyes* is the middle of week 4 to the middle of week 8
   2. Sensitive period for the *ears* is the middle of week 4 to the middle of week 9
   3. Sensitive period for the *arms* is the middle of week 4 to the end of week 7
   4. Sensitive period for the *legs* is the middle of week 4 to the end of week 7

IV. Week 6
  A. THE TEETH are most sensitive between the end of week 6 until the end of week 8
  B. THE PALATE is most sensitive between the end of week 6 until early in week 9

V. Week 7
  A. THE EXTERNAL GENITALIA are most sensitive from the middle of week 7 until the end of week 9

VI. Periods of lesser sensitivity to teratogens
  A. THE CENTRAL NERVOUS SYSTEM: from early in week 6 to term
  B. THE HEART: from late in week 6 to the end of week 8
  C. THE ARMS: during week 8
  D. THE EYES: middle of week 8 to term
  E. THE LEGS: during week 8
  F. THE TEETH: during weeks 9 and 10
  G. THE PALATE: during week 9
  H. THE EXTERNAL GENITALIA: late in week 9 to term
  I. THE EAR: middle of week 9 to the end of week 16

APPENDIX IV. TERATOGENS KNOWN TO CAUSE HUMAN MALFORMATIONS

I. Androgenic agents
   A. ETHISTERONE AND MORESTHISTERONE
      1. Implicated with varying degrees of masculinization of female fetuses, e.g., most have labial fusion and clitoral hypertrophy
   B. DIETHYLSTILBESTROL (DES)
      1. May not be teratogenic (?) to the embryo or fetus, but is carcinogenic to female offspring in later life (16 to 22 years)

II. Antitumor agents
   A. AMINOPTERIN
      1. Implicated in a wide range of skeletal defects and malformations of the CNS, particularly anencephaly
   B. BUSULFAN (MYLERAN) alternating with 6-mercaptopurine
      1. Implicated with stunted growth, skeletal abnormalities, corneal opacities, cleft palate, and hypoplasia of various organs
   C. METHOTREXATE
      1. Implicated with multiple malformations, especially skeletal

III. Sedative-hypnotics
   1. Thalidomide: its use results in meromelia and other limb malformations, as well as malformations of the external ears, heart, and digestive tract

IV. Infectious agents
   A. CYTOMEGALOVIRUS
      1. Implicated with microcephaly, hydrocephalus, microphthalmia, microgyria, and mental retardation
   B. RUBELLA VIRUS
      1. Implicated with cataracts, chorioretinitis, deafness, microphthalmia, and congenital heart defects
   C. TOXOPLASMA GONDII
      1. Implicated with microcephaly, microphthalmia, hydrocephalus, and chorioretinitis
   D. HERPES SIMPLEX VIRUS
      1. Implicated with microcephaly, microphthalmia, and retinal dysplasia
   E. VARICELLA ZOSTER VIRUS
      1. May cause congenital malformations similar to those from other viruses
   F. SYPHILIS MICROORGANISM (TREPONEMA PALLIDUM)
      1. Implicated with wasting of fetal tissues, malformations of the teeth, fetal meningitis, mental retardation, hydrocephalus, deafness, and central nervous system disease
      2. Pneumonia alba is a form of chronic pneumonia seen in stillborn and newborn infants dying of congenital syphilis

V. Therapeutic radiation
   1. Implicated with microcephaly and skeletal malformations
   2. Mutations in fetal germ cells have also been noted

VI. Alkaloids
   A. CAFFEINE
      1. It is questionable as to whether or not, depending on level of usage, it is implicated in congenital malformations of the human embryo
   B. NICOTINE
      1. Its implications with human congenital malformations is not certain, but it does have an effect on fetal growth, resulting in increased premature delivery, thus, lower weight infants

VII. **Alcohol (chronic)** has been implicated with prenatal and postnatal growth deficiency, mental retardation, microcephaly, short palpebral fissures, maxillary hypoplasia, abnormal palmar creases, joint abnormalities, and congenital heart disease

VIII. **Antibiotics**
   A. TETRACYCLINE
      1. In the second and third trimesters of pregnancy has been implicated with tooth enamel hypoplasia, yellow-brown discoloration of the deciduous teeth, distortion of bone growth, and possibly congenital cataract
   B. STREPTOMYCIN
      1. Implicated with deafness in infants
   C. PENICILLIN
      1. Appears to have no teratogenic effects

IX. **Anticoagulants**
   A. WARFARIN
      1. May result in fetal hemorrhage
      2. Implicated with hypoplasia of the nasal bones

X. **Anticonvulsants**
   A. TRIMETHADIONE (TRIDONE) OR PARAMETHADIONE (PARADIONE)
      1. Implicated with fetal dysmorphia, cardiac defects, cleft palate, intrauterine growth retardation, and digital hypoplasia
   B. DILANTIN, PHENYTOIN, AND PHENYTOIN WITH BARBITURATE
      1. Have been experimentally implicated with cleft palate
      2. In the human, there is some indication of involvement with hypoplasia of the terminal phalanges

XI. **Corticosteroids** are apparently weak teratogens in the human, but experimentally have been shown to cause cleft palate and cardiac defects

XII. **Insulin:** does not appear to be teratogenic in the human embryo

XIII. **Thyroid drugs**
   A. POTASSIUM IODIDE (KI), ^{131}IODINE, AND PROPYLTHIOURACIL have all been implicated in congenital goiter

XIV. **LSD and marihuana:** ideas about their implication in congenital defect formation have been conflicting and not fully proven. There is some indication that they may cause some limb malformations and some severe CNS abnormalities

XV. **Environmental chemicals (industrial pollutants and food additives)** have not been specifically shown to cause abnormal defects

XVI. **Mercury** has been implicated with fetal Minamata disease (caused by eating fish contaminated with mercury) with neurologic and behavioral disturbances such as cerebral palsy, brain damage, mental retardation, and blindness

XVII. **Mechanical factors on the uterus and fetus:** such factors are rarely involved in the formation of abnormal defects

APPENDIX V. PHYSIOLOGIC DEVELOPMENT OF THE CENTRAL NERVOUS SYSTEM

I. Prenatal stages

A. DEVELOPMENT OF CELLULAR FUNCTION: physiologic development of cells usually begins with and parallels their morphologic development, and the 2 processes continually interact with one another

B. TYPICAL ENZYME SYSTEMS appear in the spinal cord cells before they are seen in the brain. Examples of these are succinic dehydrogenase, ATPase, and cytochrome oxidase

C. THE APPEARANCE OF NISSL BODIES (cytoplasmic RNA) marks the beginning of increased protein synthesis, resulting in the formation of axon and dendritic processes

D. SIGNS OF FUNCTIONAL ACTIVITY, such as the onset of electrical activity and reactivity resulting in muscular contractions, as well as suppression of the related reflexes by cephalic structures, are superimposed on the morphologic and chemical changes with cerebral maturation

E. CEREBRAL MATURATION is slower and more gradual than that of the rest of the CNS and corresponds to the duration of cortical histogenesis. It may be physiologically evaluated by the spontaneous activity of the brain recorded from the skull by the electroencephalogram
   1. Electrical potentials recorded across the amniotic membrane suggest that this activity begins at about day 50 of intrauterine life
   2. Its maturation, which is a function of dendritic development of the neurons, as well as enzymatic development, is usually completed at about 11 years of age
   3. The fetus of 7 months shows anarchic activity with interhemispheric asymmetry, indicating an immature cortex and commissures
   4. At birth, there is slow, more coordinated activity of about 3 to 4 cycles per second with the onset of some symmetry
   5. From 2 to 3 years of age, more rapid activity with alpha waves of about 6 to 7 cycles per second are seen with symmetric activity of greater amplitude being well organized in the occipital regions but less well in the frontal regions
   6. At 13 to 14 years of age, there is still more rapid activity of from 8 to 12 cycles per second, and the alpha waves are well organized on the entire cerebral surface

F. DURING MATURATION OF THE BRAIN, there are special needs and requirements for oxygen and glycogen
   1. The oxygen consumption in the adult is about 25% of that used by the entire body, whereas in the newborn and young child, it can be as high as 60%. Thus, neonatal anoxia is very serious and may result in intracranial hemorrhage, epilepsy, or even psychomotor retardation

G. OVERALL DEVELOPMENT: physiologic development parallels histogenesis and begins in the spinal cord. It then follows in the derivatives of the rhombencephalon, the mesencephalon, and the prosencephalon to end with the development of the cerebral cortex. In a sense, it conforms to phylogenetic evolution
   1. Fetal stages of development
      a. Muscular reactions to external stimuli are first seen at about week 8
      b. Spontaneous movements, a sign of medullary maturation, are seen in week 9
      c. Osteotendinous reflexes are seen in month 6
      d. The respiratory centers of the medulla are functional at month 5 and, since maturation of pulmonary alveolar epithelium occurs at about 6 months of gestation, viability is theoretically possible at this age
      e. Archaic reflexes involving subcortical centers are possible
         i. Sucking at month 5 of gestation
         ii. Grasping at month 6 of gestation
         iii. The inexcitability of the cerebral cortex until this time appears to indicate

that these movements are independent of the cortex and may represent very rudimentary instinctive reactions, since they are also seen in anencephalics
   f. Cerebral maturation begins between the months 6 and 7 of gestation, when the basic structures are all completed, although some disease processes may slow down this development, and newborns may then show a psychomotor retardation of 1 to several months

II. Postnatal stages
A. THE POSTNATAL PERIOD is a continuation of the fetal state in terms of nervous system function. Behavior is predominantly reflex and purely subcortical. Movement is instinctual and rudimentary, consisting of flexion and extension or simple reflexes such as crying and coughing
   1. The neocortex becomes excitable about day 10, but in a very weak and diffuse manner, and for a long time, movement is generalized and awkward
   2. Gradually, autonomic movements come under cortical control and are more elaborate, and behavior becomes progressively imitative and expressive
   3. Structural developments in the cortex foretell these activities: neural development coincides with myelinization which proceeds in a cephalocaudal direction
      a. The first fibers to be myelinated are those coming from the motor, visual, and auditory cortex areas
      b. The last fibers to become myelinated at the end of gestation and just before birth are those coming from the association areas
B. THE MAJOR CLINICAL STAGES OF DEVELOPMENT postnatally are
   1. Regression of the archaic reflexes is seen between the first and third months
   2. Ability to completely right one's head, with stability (head control), from a prone position is seen at about 3 months
   3. Sitting and development of prehension (use of thumb and index finger) usually occurs at about 8 or 9 months
   4. Standing usually occurs at about 9 or 10 months
   5. Walking takes place at about 12 to 15 months
   6. The first words are usually spoken between the 18th month and 2 years
   7. Cerebral maturation usually ends at about 11 years of age
C. GENERAL BRAIN DEVELOPMENT
   1. The cerebral cortex has a surface of about 700 cm^2 at birth, 950 cm^2 at about 5 months, and about 1700 cm^2 at about 2 years of age, after which time, the surface no longer increases
   2. The brain weighs from 300 to 350 at birth or about one-tenth of its body weight, and its weight increases, as does its volume, mostly during the first 2 years of life. It weighs 800 g at 1 year and 1350 g at maturity
      a. The growth of the brain takes place essentially in the hemispheres, particularly in the frontal lobes
      b. Increase of brain weight continues until about the age of 14 years, but at a slower pace, and is due especially to the multiplication of the neuroglial cells and to the neuronal fiber growth. The central nervous system has most of its neurons at the time of birth
      c. In the adult, the brain represents only about 2.5% of its body weight as a result of general growth of the total body mass
      d. With growth, a complex pattern of sulci and gyri develops. These permit a considerable increase in the volume of the cerebral cortex without requiring an extensive increase in cranial volume or a reshaping of the cranial vault
   3. Myelinization starts at the fourth fetal month
   4. The cranial nerves are myelinized at birth, and the spinal nerves are completely myelinized by 3 years of age

D. NORMAL REFLEXES
1. Moro's reflex (embrace): when the infant is startled by a jarring of the table or crib or by a loud noise, he draws his legs up and brings his arms around, as in an embrace
2. Tonic neck reflex: in the resting state, the infant's posture is maintained by flexor tonicity of the arms and legs. Lateral rotation of the head to one side abolishes flexor tone on that side, causing extension of the arm and leg. This reflex is usually not developed fully until 1 month of age
3. Grasp reflexes: when the palm is stimulated by one's finger, the infant grasps and holds on; when the sole of the foot is stimulated from the heel forward, the toes turn downward
4. The deep tendon reflexes are present but tire easily
5. The abdominal reflexes are inconstant
6. Babinski's reflex is present, but there is no ankle clonus, and it disappears at about 10 to 16 months
7. Chvostek's sign is positive in 50% of newborn infants during the first week
8. The pupils react to light with contraction, but there may be secondary dilation
9. Swimming and walking reflexes are present during the first weeks
10. Rooting, sucking, and swallowing reflexes are important to feeding
    a. Rooting reflex: when the infant smells milk, he turns his head to find the source, and when the cheek is touched by a smooth object, the mouth turns toward the object and the lips open as if to grasp a nipple

E. DEVELOPMENT IS RELATED TO MYELINIZATION and is not a steady process but a pattern of sequences of rapid and slow growth. Motor and sensory controls develop from above and proceed downward so that eye control develops before hand and leg control. Development is related to 3 functioning levels of the CNS: brainstem, archipallium, and neopallium
1. The newborn functions at brainstem levels
2. Archipallium, which includes part of the temporal lobe, cingulate gyrus, and basal ganglia, supervenes on the brainstem and can be considered to be responsible for the basic emotions and some primitive motor and sensory control
3. Neopallium, which includes most of the cerebral hemisphere, has intellectual rather than emotional function and is responsible for skill, discrimination, and fine movements
4. Clinical application of the above developmental patterns is important
    a. Changes in physical signs in static brain lesions
        i. Upper limb paresis becomes apparent at 5 to 6 months
        ii. Lower limb paresis becomes apparent at 10 to 12 months
        iii. Abnormalities of coordination, namely, athetoid and involuntary movements, become apparent between 18 and 24 months

APPENDIX VI. PREMATURITY

I. Introduction: any newborn infant born alive who weighs 2500 g (5.50 lb) or less is considered a premature infant. This would include some infants of apparently full-term gestation

II. Signs of premature infancy (clinical and x-ray features)
A. WEIGHT AND LENGTH are less than 2500 g and 47 cm at birth. The gestation period is usually 28 to 36 weeks
B. THE CRY is more feeble than that of a full-term infant, sucking is weaker, and there is general weakness with sluggish movements
C. THE SKIN is thin and wrinkled, there are abundance of lanugo and minimal amount of subcutaneous tissue, and the nails are soft
D. THE HEAD appears large, but the circumference is less than 33 cm, with a characteristic wizened facies
E. THE TEMPERATURE is low and unstable
F. THERE IS A TENDENCY to cyanosis and irregular respirations
G. JAUNDICE may be prominent and continue longer than in full-term infants
H. FEEDING DIFFICULTIES are common with vomiting more frequent and a tendency to loose stools
I. INITIALLY THE WEIGHT LOSS is often greater and weight recovery not rapid, in contrast to a full-term infant
J. THE DISTAL FEMORAL AND PROXIMAL TIBIAL OSSIFICATION CENTERS may be absent. In the full-term infant, these are present including those of the calcaneus, cuboid, talus, and proximal tibia

III. Some complications of prematurity
A. A WEAK GAG AND COUGH REFLEX and an immature respiratory center that requires stronger afferent stimuli for response; incompletely developed alveoli; reduced vascularity of the pulmonary capillaries, sparse pulmonary elastic tissue; poor muscle tone and weak movements of the intercostal muscles and diaphragm plus softness and pliability of the bones of the thoracic cage which reduces intrathoracic pressure; and the presence of fetal hemoglobin which releases oxygen less readily to tissues
B. CYANOSIS OF THE EXTREMITIES and edema; paucity of vascular elastic tissue and low body reserves of vitamin C, resulting in capillary fragility and potential hemorrhage
C. SUCKING REFLEX is very weak and hepatic immaturity predisposes to the development of increased bilirubinemia with jaundice. The slow rate of secretion of digestive enzymes and gastric acid causes diminished tolerance, resulting in vomiting, diarrhea, and poor absorption of fats and minerals
D. DECREASED RENAL FUNCTION produces dehydration and acidosis
E. THE FAULTY CONTROL OF BODY TEMPERATURE is associated with hypothermia and hyperthermia due to inadequate function of the sweating mechanism, decreased body insulation (less fatty tissue), and low total heat production due to body inactivity and poor muscle development
F. THE POOR STORAGE OF MINERALS, vitamins, and immune materials makes the premature infant subject to rickets, anemia, and a variety of infections

IV. Prognosis
A. THE SURVIVAL RATE is directly in proportion to the weight of the premature infant at birth, and the prognosis improves as the period of gestation is lengthened
B. PREMATURITY is one of the leading primary causes of neonatal deaths, and most of these occur in the first 24 hours, fewer in the second 24 hours, and even fewer after that
C. THE MAJOR CONTRIBUTING FACTORS FOR DEATH include anoxia, intracranial hemorrhage, hyaline membrane with resorption atelectasis, pneumonia and other infections, congenital malformations, and blood dyscrasias

APPENDIX VII. REFERENCES

Allan, F. D.: *Essentials of Human Embryology,* 2nd ed. Oxford University Press, New York, 1969.
Arey, L. B.: *Development Anatomy,* 7th ed. W. B. Saunders Company, Philadelphia, 1974.
Barber, A. N.: *Embryology of the Human Eye.* The C. V. Mosby Company, St. Louis, 1955.
Barnes, A. C. (ed.): *Intra-Uterine Development.* Lea & Febiger, Philadelphia, 1968.
Becker, R. F.; Wilson, J. W.; and Gehweiler, J. A.: *The Anatomical Bases of Medical Practice.* The Williams & Wilkins Company, Baltimore, 1971.
Benirschke, K., and Driscoll, S. G.: *The Pathology of the Human Placenta.* Springer-Verlag, New York, 1967.
Benninghoff, A., and Goerttler, K.: *Lehrbuch der Anatomie des Menschen,* 3rd ed. Urban and Schwartzenberg, München and Berlin, 1960.
Blandau, R. J. (ed.): *The Biology of the Blastocyst.* University of Chicago Press, Chicago, 1971.
Blechschmidt, E.: *The Stages of Human Development Before Birth.* W. B. Saunders Company, Philadelphia, 1961.
Bloom, W., and Fawcett, D. W.: *A Textbook of Histology,* 10th ed W. B. Saunders Company, Philadelphia, 1975.
Boyd, J. D., and Hamilton, W. J.: *The Human Placenta.* W. Heffer and Sons, Ltd., Cambridge, 1970.
Bremer, J. L.: *Congenital Anomalies of the Viscera: Their Embryological Basis.* Harvard University Press, Cambridge, Mass., 1957.
Copenhaver, W. M.; Bunge, R. P.; and Bunge, M. B.: *Bailey's Textbook of Histology,* 16th ed. The Williams & Wilkins Company, Baltimore, 1971.
Corliss, C. E.: *Patten's Human Embryology: Elements of Clinical Development.* McGraw-Hill Book Company, New York, 1976.
Corning, H. K.: *Lehrbuch der Entwicklungsgeschicte der Menschen.* Verlag von J. F. Bergmann, München and Wiesbaden, 1921.
Davies, J.: *Human Developmental Anatomy.* The Ronald Press Company, New York, 1963.
DeHann, R., and Ursprung, H.: *Organogenesis.* Holt, Rinehart and Winston, Inc., New York, 1965.
Eastman, N. J.; Hellman, L. M.; Pritchard, J. A.; and Wynn, R. M.: *Williams Obstetrics,* 13th ed. Appleton-Century-Crofts, New York, 1966.
Emery, J.: *The Anatomy of the Developing Lung.* William Heinemann, Ltd., London, 1969.
Federman, D. D.: *Abnormal Sexual Development: A Genetic and Endocrine Approach to Differential Diagnosis.* W. B. Saunders Company, Philadelphia, 1967.
Fitzgerald, M. J. T.: *Human Embryology.* Harper and Row, Hagerstown, Md., 1978.
Fraser, F. C., and Nora, J. J.: *Genetics of Man.* Lea & Febiger, Philadelphia, 1975.
Gardner, E. J.: *Principles of Genetics,* 5th ed. John Wiley & Sons, New York, 1975.
Gardner, L. I.: *Endocrine and Genetic Diseases of Childhood and Adolescence,* 2nd ed. W. B. Saunders Company, Philadelphia, 1975.
Gorlin, R. J.; Pindborg, J. J.; and Cohen, M. M.: *Syndromes of the Head and Neck,* 2nd ed. McGraw-Hill Book Company, New York, 1975.
Gray, S. W., and Skandalakis, J. E.: *Embryology for Surgeons.* W. B. Saunders Company, Philadelphia, 1972.
Greenhill, J. P. (ed.): *Obstetrics,* 13th ed. W. B. Saunders Company, Philadelphia, 1965.
Haines, R. W., and Mohiuddin, A.: *Human Embryology,* 4th ed. E. and S. Livingstone, Ltd., Edinburgh and London, 1968.
Ham, A. W., and Cormack, D. H.: *Histology,* 8th ed. J. B. Lippincott Company, Philadelphia, 1979.
Hamerton, J.: *Human Cytogenetics. Clinical Cytogenetics.* Vol. II. Academic Press, Inc. New York, 1971.
Hamilton, W. J., and Mossman, H. W.: *Human Embryology,* 4th ed. The Williams & Wilkins Company, Baltimore, 1972.

Harrison, R. G.: *A Textbook of Human Embryology,* 2nd ed. Blackwell Scientific Publications, Oxford, 1963.

Haymaker, W.: *Bing's Local Diagnosis in Neurological Diseases,* 14th ed. The C. V. Mosby Company, St. Louis, 1956.

Hertig, A. T.: *Human Trophoblast.* Charles C Thomas, Springfield, Ill., 1968.

Holmes, R. L., and Sharp, J. A.: *The Human Nervous System: A Developmental Approach.* J. and A. Churchill, London, 1969.

Jirásek, J. E.: *Development of the Genital System and Male Pseudohermaphroditism.* Johns Hopkins Press, Baltimore, 1971.

Jones, H. W., Jr., and Scott, W. W.: *Hermaphroditism, Genital Anomalies and Related Endocrine Disorders.* The Williams & Wilkins Company, Baltimore, 1958.

Keith, J. D.; Rowe, R. D.; and Vlad, P. (eds.): *Heart Disease in Infancy and Childhood,* 3rd ed. Macmillan Publishing Co., Inc., New York, 1977.

Kollman, J.: *Handatlas der Entwicklungsgeschichte des Menschen,* 2nd ed. Jena, Verlag von Gustav Fischer, 1907.

Langebartel, D. A.: *The Anatomical Primer: An Embryological Explanation of Human Gross Morphology.* University Park Press, Baltimore, 1977.

Langman, J.: *Medical Embryology,* 4th ed. The Williams & Wilkins Company, Baltimore, 1981.

Last, R. J.: *Anatomy, Regional and Applied,* 5th ed. The Williams & Wilkins Company, 1972.

Lemire, R. J.; Loeser, J. D.; Leech, R. W.; and Alvord, E. C.: *Normal and Abnormal Development of the Human Nervous System.* Harper and Row, Hagerstown, Md., 1975.

Marshal, W. A.: *Development of the Brain.* Oliver and Boyd, Edinburgh, 1968.

McCrory, W. W.: *Developmental Nephrology.* Harvard University Press, Cambridge, Mass., 1972.

Moore, K. L. (ed.): *The Sex Chromatin.* W. B. Saunders Company, Philadelphia, 1966.

Moore, K. L.: *The Developing Human,* 2nd ed. W. B. Saunders Company, Philadelphia, 1977.

Moss, A. J., and Adams, F. H.: *Heart Disease in Infants, Children and Adolescents.* The Williams & Wilkins Company, Baltimore, 1968.

Nora, J. J., and Fraser, F. C.: *Medical Genetics: Principles and Practice.* Lea & Febiger, Philadelphia, 1974.

O'Rahilly, R.: *Developmental Stages in Human Embryos.* Part A: *Embryos of the First Three Weeks.* Carnegie Institute of Washington, D.C., 1973.

Page, E. W.; Villee, C. A.; and Villee, D. B.: *Human Reproduction,* 2nd ed. W. B. Saunders Company, Philadelphia, 1976.

Pansky, B., and Allen, D. J.: *Review of Neurosciences.* Macmillan Publishing Co., Inc., New York, 1980.

Patten, B. M.: *Human Embryology,* 3rd ed. McGraw-Hill Book Co., New York, 1968.

Philipp, E. E.; Barnes, J.; and Newton, U. (eds.): *Scientific Foundations of Obstetrics and Gynecology.* William Heinemann, Ltd., London, 1970.

Reid, D. E.; Ryan, K. J.; and Benirschke, K.: *Principles and Management of Human Reproduction.* W. B. Saunders Company, Philadelphia, 1972.

Remnick, H.: *Embryology of the Face and Oral Cavity.* Fairleigh Dickinson University Press, Rutherford, N.J., 1970.

Romanes, G. J. (ed.): *Cunningham's Textbook of Anatomy,* 10th ed. Oxford University Press, New York, 1964.

Ross, R. B., and Johnston, M. C.: *Cleft Lip and Palate.* The Williams & Wilkins Company, Baltimore, 1972.

Rubin, A.: *Handbook of Congenital Malformations.* W. B. Saunders Company, Philadelphia, 1967.

Saxén, L., and Rapola, J.: *Congenital Defects.* Holt, Rinehart and Winston, Inc., New York, 1969.

Schlegel, R. J., and Gardner, L. I.: Ambiguous and abnormal genitalia in infants: differential diagnosis and clinical management; in L. I. Gardner (ed.): *Endocrine and Genetic Diseases of Childhood and Adolescence.* 2nd ed. W. B. Saunders Company, Philadelphia, 1975.
Shettles, L. B.: *Ovum Humanum.* Hafner Publishing Company, New York, 1960.
Sicher, H., and Bhaskar, S. N.: *Orban's Oral Histology and Embryology,* 7th ed. The C. V. Mosby Company, St. Louis, 1972.
Smithels, R. W.: Drugs and Human Malformations. *Adv. Teratol.,* **1**:251, 1966.
Smithels, R. W.: Environmental Teratogens of Man, *Br. Med. J.,* **32**:27, 1976.
Snell, R. S.: *Clinical Embryology for Medical Students,* 2nd ed. Little, Brown and Company, Boston, 1975.
Tandler, J.: *Tratado de Anatomía Sistemática,* 4th vol. Salvat Editores, S. A., Barcelona, 1933.
Taussig, H. B.: Cardiac Abnormalities. In Fishbein, M. (ed.): *Birth Defects.* J. B. Lippincott Company, Philadelphia, 1963.
Testut L., and Latarjet, A.: *Traité D'Anatomie Humaine,* 9th ed. G. Doin et Cie, Paris, 1949.
Torpin, R.: *The Human Placenta.* Charles C Thomas, Springfield, Ill., 1969.
Tuchmann-Duplessis, H.; David, G.; and Haegel, P.: *Illustrated Human Embryology.* Springer-Verlag, New York, 1972.
Vaughan, V. C., and McKay, R. J. (eds.): *Nelson Textbook of Pediatrics.* W. B. Saunders Company, Philadelphia, 1975.
Villee, C. A. (ed.): *The Placenta and Fetal Membranes.* The Williams & Wilkins Company, Baltimore, 1960.
Villee, D. B.: *Human Endocrinology: A Developmental Approach.* W. B. Saunders Company, Philadelphia, 1975.
Waisman, H. A., and Kerr, G. R.: *Fetal Growth and Development.* McGraw-Hill Book Company, New York, 1970.
Warkany, J.: *Congenital Malformations.* Year Book Medical Publishers, Inc., Chicago, 1971.
Warwick, R., and Williams, P. L. (eds.): *Gray's Anatomy,* 35th Brit. ed. W. B. Saunders Company, Philadelphia, 1973.
Williams, P. L.; Wendell-Smith, C. P.; and Treadgold, S.: *Basic Human Embryology.* J. B. Lippincott Company, Philadelphia, 1966.

INDEX

Figures in **boldface** refer to pages on which illustrations appear.

Abdomen, 110, **120**, **121**
   organs of, herniation into thorax by, 122, **125**
Abdominal pregnancy, 30, **31**
Abdominal reflexes, 485
Abdominal wall, congenital absence of, 228, **230**
Acardia, 82
Accessory nerve, 388, **389**, **390**, **391**, 402, **405**, **406**
Acetylcholine, 462
Achondroplasia, 78, **83**, 172
Acini, of pancreas, 216, **217**
Acoustic meatus, external, 444, **445**
   atresia, 454
Acrania, 182
Acrocephaly, **183**
Acromegaly, 172
Acrosome reaction, of sperm, 32, **33**
Adenocarcinoma, diethylstilbestrol (DES) as cause of, 481
Adenohypophysis, 456, **457**
"Adenoids," 332, **335**, 452
Aditus, laryngeal, 150, **151**
Adrenal glands, 462, 464, **465**
   innervation, 464
   pathology of, 466, **467**
   vascularization, 464
Adrenogenital syndrome, 466, **467**
   as causing female pseudohermaphroditism, 466, **467**
Aeration, of lungs, 156, **157**
Aganglionic megacolon, congenital (Hirschsprung's disease), 232, **233**, 430, **431**
Age, bone, determination of, 180
   embryonic, estimation of, 60, **61**, 471-77
   fetal, estimation of, 180, 471-77
Age (weeks), system development, 471-77
Agenesis, of adrenal, 466
   anal, 234, **235**
   anorectal, 234, **235**
   bronchial, 158, **159**
   of corpus callosum, 418, 424
   of lungs, 158, **159**
   of midgut, 228
   of penis, 282, **283**
   renal, 250, **251**
   of uterus, 286, **287**
   of vagina, 286, **287**
Agnathia, 454
Air cells, 154, **155**, **157**
Ala, orbitalis, **181**
   temporalis, **181**

Alae, of nose, 140, **141**
Alar plate, of diencephalon, 406, **407**
   of mesencephalon, 400, **401**
   of metencephalon, 394, **395**
   of myelencephalon, 392, **393**
   of spinal cord, 374, **375**, 376, **377**
Albinism, 194
Alcohol, possible teratogenic effects of, 482
Alecithal egg, 36
Alkaloids, possible teratogenic effects of, 481
Allantoenteric diverticulum. *See* Diverticulum, allantoenteric
Allantois, 46, **47**, 84, **85**, 86, **87**
   effect of tailfold on, 84, **85**
   fate of, 84, **85**, 86
   formation of, 46, **47**, 84, **85**, 86, **87**
   significance of, 46, 84, **85**, 86, **87**
Alopecia, congenital, 198
Alveolar ducts, 154, **155**
Alveolar epithelial cells, type I, 154, **155**, 156, **157**
   type II, 154, **155**, 156, **157**
Alveolar period, of lung development, 154, **155**
Alveolar-capillary membrane, 154, 156, **157**
Alveolus(i), dental, 202, **203**, **205**
   immature, dental, 202, **203**
   pulmonary, 154, **155**, **157**
Amastia, 200
Ambiguous external genitalia, differential diagnosis of, **289**
Amelia, **83**, **189**
Ameloblast, 202, **203**
Amelogenesis imperfecta, 204
Amino acids, fetal growth and, 72
Aminopterin, teratogenic effects of, 481
Amnioblast, 38, **39**
Amniocentesis, 86
Amniochorionic membrane, 98, **99**
Amnioectodermal junction, 88, **89**
Amnion, 84, **85**, 86, **87**
Amniotic cavity, 38, **39**, 84, **85**
Amniotic fluid, 86
   composition of, 86
   exchange of, 86
   function of, 86
   origin of, 86
   sampling of, 86
   volume of, 86
Amniotic sac, 86, **87**
Amphibian (frog) brain, **369**
Amphioxus, nervous system, 366, **369**
Ampulla(e), of inner ear, 446, **447**, 450, **451**
   of uterine tube, 22, **23**
   of Vater, 216, **217**

Amygdaloid nucleus, 410, **411**
Anal canal, 222, **223**
  lymphatics of, 222
  nerves of, 222
  vessels of, 222, **223**
Anal agenesis, 234, **235**
Anal fold, 222, 278, **279**
  membrane, 222, **223**, 266, **267**, 278, **279**
Anal pit, 208, 222, **223**
Anastomosis, arteriovenous, and monozygotic twins, 74, **76**
  placental vascular, and multiple pregnancies, 74, **75**, **76**
Anatomic position, definition of, 2, **3**
  in embryos, 2, **3**
Anatomy, descriptive terms in, 2, **3**
  developmental, definition, 2, **3**
Anchoring villi, 56, **57**, 96, **97**, **99**
Androgen(s), 258
Androgenic agents, teratogenic effects of, 481
Anencephaly, 384, 422, **423**, 466
Angioblast, 294
Angiogenesis, 54, **55**, 294, **295**
Ankyloglossia, 138
Annelids, nervous system, 366, **368**
Annulus fibrosus, 174, **175**
Anoderm, 222, **223**
Anodontia, 204
Anogenital raphé, **279**
Anonychia, 198
Anophthalmia, 442, **443**
Anorectal canal, 222, **223**, 248, **249**
  agenesis of, 234, **235**
  congenital malformation of, 234, **235**
Anorectal line, 222, **223**
Anotia, 454
Anovulatory cycle, of menstrual cycle, 28, **29**
Anterior chamber, **437**, **438**, 440, **441**
Antigens, inherited, **93**
Antibiotics, possible teratogenic effects of, 482
Antibodies, maternal, placental transfer of, 92, **93**
Anticoagulants, possible teratogenic effects of, 482
Anticonvulsants, possible teratogenic effects of, 482
Anti-Rh antibodies, 92, **93**
Antitumor agents, teratogenic effects of, 481
Antrum, of mastoid, 132
  of ovarian follicle, 14, **15**
  tympanic, 252
Anus, 222, **223**
  agenesis of, 234, **235**
  congenital malformations of, 234, **235**
  ectopic, 234
  insufficient, 234
Aorta, 313, 317, 324, **325**
  arch of, 324, **325**
  atresia, 352, **354**
  coarctation of, 340, **341**
  dorsal, 294, **295**, 318, **319**
  stenosis of, 352, **354**
  transposition of, 340, **341**, **342**, 343
Aortic arches, 318, **319**, 320, **321**, 324, **325**
  congenital malformations of, 340, **341**, **342**
  derivations of, 320, **321**, 324, **325**
Aortic arch arteries. *See* Aortic arches
Aortic sac, 320
Aortic valve, 314, **315**, **316**
Aortic valvular atresia, 352, **354**
Aortic vestibule, 312
Aorticopulmonary septum, 312, **313**
  defects of, 348, **350**
Aplasia, congenital thymic, 134
  of intestine, 228
  of respiratory tract, 158, **159**
  of urinary system, 250, **251**
Apocrine glands, 196
Apodymy, 134
Appendage(s), auricular, 454, **455**
  cutaneous, 196–**203**
Appendicular skeleton, 178, **179**, **180**, 184, **185**
  malformations of, 188, **189**
Appendix, of epididymis, **259**, **262**, 264, **265**, **288**
  of testis, **259**, **262**, 264, **265**, **288**
  vermiform, 218, **219**, 220, **221**
    pelvic, 220
    retrocecal, 220
    retrocolic, 220
  vesiculosa, **288**
Appendixes, 471–86
Aqueduct, cerebral (of Sylvius), 362, **364**, **365**, 400, **401**
Aqueous chamber, **437**, **438**, 440, **441**
Arachnodactyly, 188, **189**
Arachnoid membrane, 382, **383**, 420, **421**
Arachnoid trabeculae, 420, **421**
Arachnoid villi, 420, **421**
Arch(es). *See also* Aortic arches
  branchial, 58, **59**, 128, **129**, **130**, **131**
    components of, 128, **129**, **130**, **131**
    derivatives of, 128, **129**, **130**, **131**
    fate of, 128, **129**, **130**, **131**
    innervation of, 128, **129**, **130**, **131**
  hyoid, 58, **59**, 128, **129**, **130**, **131**
    derivatives and innervation of, 128, **129**, **130**, **131**
  mandibular, 58, **59**, 128, **129**, **130**, **131**
    derivatives and innervation of, 128, **129**, **130**, **131**
  neural, 174, **175**
  pharyngeal, 128, **129**, **130**, **131**
  postoral, **129**
  thyrohyoid, 128, **129**, **130**, **131**
  vertebral, 174, **175**
Archeocortex, 428, **409**, 414, **415**, 416
Archeostriatum, 410, **411**
Archicerebellum, 396, **399**
Archiencephaly, 134

Archipallium, 408, **409**, 410, **411**, 414, **415**, 416, **417**
Areola, 200, **201**
Arm buds, 58, **59**
Arms. *See* Limbs
Arnold-Chiari malformation, 384, 426
Arrector pili muscle, 196, **197**
Arrhinencephaly, 424, 434
Arterial system, 320–25
Arterio-capillary-venous system, of fetal circulation, 318, **319**, 320
Arteriosus, ductus, 324, **325**, 336, **337**, 338
Artery(ies), allantoic, 318, **319**
   aorta, **313**, **317**, 320, **321**
   aortic arch. *See* Aortic arches
   axillary, **322**
   basal (basilar), 420, **421**
   brachial, **322**
   brachiocephalic, 324, **325**
   branchial arch, 320, **321**, 324, **325**
   carotid, common, 324, **325**
     external, **321**, 324, **325**
     internal, **321**, 324, **325**, 420, **421**
     primitive, 324, **325**
   celiac, 320, **321**
   central, of retina, 440, **441**
   cerebral, anterior, **421**
     middle, **421**
   communicans, anterior, 420, **421**
     posterior, 420, **421**
   femoral, **323**
   great, complete transposition of, **348**, 351
   hyaloid, 436, **437**, 440, **441**
   hyoid, 324
   hypophyseal, 458, **459**
   iliac, common, 320, **323**
     external, 320
     internal (hypogastric), 320, **323**
   intercostal, 320, **321**, 325
   intersegmental, fate of, 320, **321**
   intestinal, **321**
   of lower extremity, **323**
   maxillary, 324, **325**
   mesenteric, inferior, 320, **321**
     superior, 218, **219**, 320, **321**
   ophthalmic, **421**
   paired segmental, 318, **319**
   pulmonary, 152, **153**, **313**, **317**, 324, **325**
   radial, **322**
   renal, **321**
   retinal, 440, **441**
   segmental, 320, **321**
   spinal, **321**
   spiral endometrial (uterine), **103**
   stapedial, 324
   subclavian, **321**, **322**, 324, **325**
     right, abnormal origin of, 340, **341**
   Sylvian (middle cerebral), **421**
   tibial, **323**
   ulnar, **322**
   umbilical, 88, **89**, **91**, 318, **319**, 320, **321**, 338
     derivatives of, 318, **319**, 320, **321**, 338
     fate of, 320, **321**, 338
   of upper extremity, **322**
   ventral, primitive, 318, **319**
   vertebral, **321**, 420, **421**
   vesical, superior, 320
   vitelline, 318, **319**
Articular cartilage, 172, **173**
Articular system, 172, **173**
Arytenoid cartilage, 128, 150, **151**
   swellings, 150, **151**
Astroblast, **379**
Astrocyte, 378, **379**
   fibrillar, 378, **379**
   fibrous, **379**
   protoplasmic, 378, **379**
Atelectasis, 156
Athelia, 200
Atresia, aortic, 352, **354**
   biliary, extrahepatic, 224, **226**
   cervix, **285**, 286, **287**
   duodenal, 228, **229**
     amniotic fluid accumulation in, 224
   esophageal, 158, **159**, 224, **225**
     amniotic fluid accumulations in, 224
   of external acoustic meatus, 454
   of gallbladder, 224
   intestinal, **225**, 228, **229**, **231**
   membranous, **225**, 234
   mitral, **354**
   of penile urethra, **285**
   pulmonary, 352, **354**
   rectal, 234
   tracheal, 158, **159**, 225
   of tricuspid valve, 352, **354**
   of uterine canal, **275**
   vaginal, **287**
Atrial septal defects, 344, **346**, **347**
Atrichia congenita, 198
Atrioventricular canal, common, 300, **301**, 304, **305**, **307**
   cushions, 304, **305**, **307**
   partitioning (septation), 304, **305**, **307**
   persistent, **347**
Atrioventricular node, 314, **316**
Atrioventricular valves, 304, **305**, 314, **316**, **317**
Atrioventriculares communis, 344, **347**
Atrium(a), 300, **301**, **311**
   common, 300, **301**, **311**
   primitive, 300, **301**, **311**
     partitioning of, 300, **301**, **311**
   walls of, 310, **311**
Auditory meatus, 132
Auditory (otic) placode, 446, **447**
Auditory (otic) vesicle, 446, **447**
Auricle(s), of external ear, 444, **445**
   of heart, **307**, 310, **311**
     abnormalities of, 134, **135**
Auricular cysts, 134

Auricular hillocks, 444, **445**
Auricular pits, 134
Auricular sinus, 454, **455**
Auricular tags, 454, **455**
Autonomic nervous system, 428–33, **463**
Autosomes, abnormal, 18, **19**
   normal, **9, 15**
   trisomy of, 78, **81**
Axial skeleton, 174, **175**
   congenital malformations of, 172, 182, **183**
Axons, primitive, 378, **379**
Azygos lobe, of lung, 158
Azygos vein, 308, **309**

Babinski reflex, 485
Band(s) (bars), sternal, 174
Bare area, of liver, 214, **215**
Barriers, blood-brain, 420
   blood-gas, lung, 154
Bartholin, glands of, 272, **274**
Basal (basilar) artery, 420, **421**
Basal nuclei, 408, **409**
Basal plate(s), of mesencephalon, 400, **401**
   of metencephalon, 394, **395**
   of myelencephalon, 388–93
   of skull, 178, **179, 181**
   of spinal cord, 374, **375**, 376, **377**
Basilar (basement) membrane, 448, **449**, 450, **451**
Basis pendunculi, 400, **401**
Basket cell, 396, **398**
Bell stage, of tooth development, 202, **203**
Bicervical uterus, 286, **287**
Bicornuate uterus, 286, **287**
Bifid tongue, 138
Bilaminar, embryo, 38, **39**
   germ disk, 38, **39**
Bile duct, common, 214, **215**, 216, **217**
   atresia of, 224
Bile pigment, 214
Biliary apparatus, 214, **215**
   congenital anomalies and malformations, **226**
   intrahepatic portion of, 214, **215**
   variations in, **226**
Bilocular uterus, 286, **287**
Bird (chicken) brain, **369**
Birth, calculating time of, 72
"Birth control pills," 28
Birth weight, low, causes of, 72
Bladder, ectopia, 252
   exstrophy, 252
   gall, 214, **215**
   trigone, 248, **249**
   urinary, 248, **249, 288**
Blastema, metanephrogenic, 244, **245**
Blastocele, 36, **37**
Blastocyst, 17, 36, **37**, 38, **39**, 40, **41**
   definition of, 36, **37**
   implantation of, 30, **31**, 36, **37**, 38, **39**, 40, **41**

   early stages of, **17**, 36, **37**, 38, **39**, 40, **41**
   sites of, 30, **31**
   sections of, **17, 39**
Blastomere(s), 32, **33**, 36, **37**
   definition of, 36, **37**
Blood-brain barrier, 420
Blood
   circulation of, fetal and neonatal, 292, **293**
      placental, 294, **295**
      through primitive heart, 292, **293**, 302, 318, **319**
      uteroplacental, primitive, 292, **293**, 318, **319**
   formation of, 54, **55**, 294, **295**
      role of allantois in, 54, **55**
      role of yolk sac in, 54, **55**, 294, **295**
Blood cells, 294, **295**
Blood islands, 294, **295**, 318, **319**
   vessel, formation, 294, **295**, 318, **319**
Blood vessels. *See* Artery(ies); Vein(s)
Bochdalek, foramen of, 122
Body(ies), cavities, 110, **111**
   ciliary, of eye, 436, **437, 438**
   mamillary, 406, **407**
   mesonephric, 260, 276, **277**
   pineal, 406, **407**
   perineal, 278, **279**
   polar, 14, **15**
   stalk, 40, **41**, 48, 88, **89**
   of tongue, 132, **133**, 138, **139**
   ultimobranchial, 132, **133**
   vitreous, **437, 438**, 440, **441**
   wolffian. *See* Mesonephros
Body cavities. *See* Coelom
Bone(s). *See also* Skeletal system
   age of, determination of, **180**
   of branchial arch origin, 178, **180**
   cancellous, **171**
   compact, **171**
   histogenesis, 168, **169, 170, 171**, 172, **173**
   hyoid, 128, **130**
   irregular, 172
   limb, **170, 171**
   long, **171**
   marrow, 168, **170**
   of middle ear, 444, **445**, 452, **453**
   occipital, malformations of, 422
   of skull, 178, **179, 181**
   spongy, 168
   temporal, 178, **179, 181**
Bone cells, 168, **169, 170, 171**
Bone-forming cells, 168, **169, 170, 171**
Bony labyrinth, of inner ear, 448, **449**
Bowman's capsule, 240, **241**, 244, **245**
Brachiocephalic vein, 308, **309**
Brachydactyly, 188
Brain, **409**
   comparative, **409**
   congenital malformations of, 422–27
   effect of headfold on, 64, **65**

flexures of, 362, **363**, 364, **365**
vascularization of, 420, **421**
vesicles of, 392–**415**
Brainstem, 386–**421**
  nerves of, **390–401**
  nuclei of, **390–401**
  pathways of, 386
Branchial apparatus, 128–**45**
  arches of. *See* Branchial arches
  pharyngeal pouches of, 128, **129**, **130**, **131**, 132, **133**
    derivatives of, 128, **129**, **130**, **131**, 132, **133**
    structures resulting from, 128, **129**, **130**, **131**, 132, **133**
    congenital malformations of, 134, **135**
Branchial arches, 58, **59**, 128, **129**, **130**, **131**
  arteries of, 128, **129**
  cartilage of, 128, **129**, **130**
    derivatives of, 128, **129**, **130**
  components of, 128, **129**, **130**
    structures derived from, 128, **129**, **130**
  fate of, 128, **129**, **130**
  innervation of, 128, 129, **130**
Branchial cyst, 132, **135**
Branchial fistula, 134, **135**
Branchial grooves, 128, **129**
Branchial membranes, 132
Branchial sinus, congenital, 132, **133**
Branchial vestige, 134
Breasts, 200, **201**
  congenital malformations of, 200, **201**
Brevicollis, 182
Broad ligaments, 272, **273**, 275
Bronchi, 128, 150, **151**, 152, **153**, 157
  abnormalities of, 158, **159**
Bronchial buds, 150, **151**, 152, **153**
Bronchial cyst, congenital, 158
Bronchiole(s), respiratory, 152, **153**, 154, **155**, 157
Bronchopulmonary segments, 152, **153**
Buccopharyngeal membrane, 44, **45**, 48, 208, **209**, 358
  formation of, 44, **45**, 208, **209**, 358
Bud(s), arm, 58, **59**
  bronchial, 150, **151**, 152, **153**
  hepatic, 214, **215**
  leg, 58, **59**
  limb, 58, **59**
  lung, **133**, 150, **151**, 152, **153**
  mammary, 200, **201**
  neurohypophyseal, **457**
  pancreatic, 216, **217**
  taste, 138, **139**
  tooth, 202, **203**
  ureteric, 248, **249**
    division of, 248, **249**
Bud stage, of tooth development, 202, **203**
Bulb(s), hair, 196, **197**
  olfactory, 414, **415**, 434, **435**
  sinovaginal, 272, **274**

of vestibule, **288**
Bulbar ridges, 312, **313**
Bulbourethral glands, 4, **5**, 264, **265**, **288**
Bulboventricular flange, 300, **301**
  fold, 300
  loop, 300, **301**
  spur, **307**
  sulcus, 300, **301**
Bulbus cordis, 300, **301**, 304, **305**, 312, **313**
  anomalies of, 348
  fate of, 300, **301**, 304, **305**, 312, **313**
  partitioning of, 312, **313**
Bundle(s), of His, 314, **316**
Buphthalmos, 442, **443**
Bursa(e), infracardiac 116, 212
  omental, 116, **117**, **211**, 212, **213**
Busulfan (Myleran), teratogenic effects of, 481

Caecum. *See* Cecum
Calcitonin, 132
Calyx(ces), renal, 244, **245**, **288**
Canal(s), anal, 222, **223**
  anorectal, 222, **223**, 248, **249**
  atrioventricular, 304, **305**, 307, **313**
    partitioning of, 304, **305**, **307**, **313**
  central, of spinal cord, 376, **377**
  ependymal (central), 376, **377**
  inguinal, 268, **269**
  neurenteric, 46, **47**, **48**
  notochordal, 46, **47**, **48**
  pericardioperitoneal, 110, **111**, **112**, **113**
  pleural, 122, **123**
  pleuroperitoneal, **303**
  semicircular, 446, **447**
  spinal, **375**, **377**
  uterovaginal, 260
  vertebral. *See* Canal, spinal
  vesicourethral, 248
Canalicular period, of lung development, 154, **155**
Cap stage, of tooth development, 202, **203**
Capillary(ies), fetal, endothelium of, 294, **295**
Capsule(s), Bowman's, 240, **241**, 244, **245**
  external, 410, **411**
  glomerular, 240, **241**, 244, **245**
  internal, 410, **411**
  otic, 448, **449**
Cardiac development, major stages, **339**
Cardiac jelly, 296
Cardiac muscle, 162
Cardiac part of stomach, 210, **211**
Cardiac skeleton, 314, **316**
Cardiac valves, 314, **315**, **316**, **317**
Cardiogenic area, 296, **297**, **298**
  during embryonic development, 296, **297**, **298**
  formation of heart tubes in, 296, **297**, **299**
  plate, 296, **297**
Cardiogenic cords, 296, **297**

Cardiovascular system, 54, **55**, 292–355
  aortic arches of. *See* Aortic arches
  atria and ventricles of, 300, **301, 311**
  changes in, at birth, 336, **337**
  congenital malformations of, 340–**43**
  fetal and neonatal circulation of, 292, **293**
  heart tubes of, 296, **297**
  primitive, 294, **295**, 296, **297**
Carotid arteries, 324, **325**, 420, **421**
Cartilage(s), arytenoid, 128, 150, **151**
  branchial arch, 178, **180**
    derivatives of, 178, **180**
  bronchial, 128, 152
  corniculate, 150, **151**
  cricoid, 128
  cuneiform, 150, **151**
  histogenesis of, 168, **169, 170**
  hyaline, 168, **169**
  laryngeal, 128, 150, **151**
  Meckel's, 128, **130, 131**, 178, **180, 181**, 452, **453**
  model of bone, **170**
  Reichert's, 128, 178, **180**, 452, **453**
  thyroid, 128
  tracheal, 128, 150
  viscerocranial derivatives, 178, **180**
Cartilaginous joint, 172, **173**
Cataract, congenital, 442
Cauda equina, 374, 382, **383**
Caudate nucleus, 410, **411**
Cavity(ies), 110–**17**
  amniotic, 40, **41, 43**, 84, **85**
  body. *See* Coelom
  chorionic, 40, **41, 43**
  myencephalic, 386
  nasal, 140, **141**, 148, **149**
  pericardial, 110, 114, **115**, 298, **299**, 302, **303**
  peritoneal, 110, **303**
  pleural, 110, **303**
  tympanic, 132, 444, **445**, 452, **453**
  of yolk sac, 52
Cecal diverticulum, 220, **221**
Cecum, 218, **219**, 220, **221**
  congenital malformations of, 228, **230**, 232, **233**
  inverted, **233**
  mobile, 232
  retroperitoneal, **233**
  subhepatic, 232, **233**
Celiac artery, 320, **321**
Cell(s), acidophil, 456
  air, 154, **155**
  alveolar epithelial, type I, 154, **155**, 156, **157**
    type II, 154, **155**, 156, **157**
  astrocytes, 378, **379**
  basket, 396, **398**
  basophil, 456
  bipolar, **379, 381**
  blood, 294, **295**
  bone, 168, **169, 171**

  bone-forming, 168, **169, 171**
  C, of thyroid, 132
  chromaffin, **381**, 402
  chromophobe, 456
  Clara, **157**
  decidual, 98, **99**
  Deiter's, **451**
  endothelial, 294, **295**
  follicular, 256, **257**
  ganglion, 360
  germ. *See* Germ cells
  glial, 378, **379**
  Golgi, 396, **398**
  granule (granular), 396, **398**
  hair, 192, **193**, 196, **197**
  hemocytoblasts, 294
  of Hensen, **451**
  Hofbauer, 97
  inner mass of, 36, **37**, 38, **39**
  interstitial (of Leydig), **6, 11**, 258
  Kupffer, 214
  of Langerhans. *See* Cytotrophoblast
  Leydig, **6, 11**, 258
  megaloblasts, 294
  microglial, 378, **379**
  multipolar, 378, **379, 381**
  myoepithelial, 196
  nerve, 378, **379**
  neural, of cord, 374
  neural crest, 360, **361**, 380, **381**, 402, **403**, 428, **429**, 462, **463**
    derivatives of, 360, **361**, 380, **381**, 402, **403**, 428, **429**, 462, **463**
  neurilemmal, 380, **381**
  neuroepithelial, 360, 374, 376, **377**
  neurosensory, 378, **379**
  nucleated red, 294, **295**
  olfactory, 434
  oligodendroglia, 378, **379**
  parafollicular, 132
  pigment, 360
  pillar, of inner ear, **451**
  pituicytes, 458
  primordial germ, 256, **257**
  Purkinje, 396, **398**
  pyramidal, **417**
  reproductive, 256, **257**
  satellite, **381**, 402, **403**
  Schwann, 360, 380, **381**
  secretory, 196
  Sertoli, **11**, 256, 258
  sex, 8, 256, **257**
  spongioblasts, 378, **379**
  stellate, 396, **398**
  sustentacular (of Sertoli), 258
Cell division, nondisjunction during, 18, **19**
Cell mass, **37**, 38, **39**
  inner, **37**, 38, **39**
  outer, **37**, 38, **39**
Cell migration, 68, **69**

Cementoblast, 202, **203**
Cementoenamel junction, 202, **203**
Cementum, 202, **203**, **205**
Center, primary ossification, 168, **170**
Central artery of retina, 440, **441**
Central canal, 362, **363**, **364**, 376, **377**
Central nervous system. *See also* Brain; Spinal cord
   cells of, histogenesis of, 416, **417**
   origin of, 362, **363**, **364**, **365**
Centrum, of vertebra, 174, **175**
Cephalic flexure, 362, **363**, **364**, **365**
Cephalocaudal folding, 64, **65**, **66**, **67**
Cephalopagus, 74, **77**
Cerebellar agenesis, 424
Cerebellar cortex, 396–**99**
Cerebellar plate, 396, **397**
Cerebellar swellings, 396, **397**, **398**
Cerebellum, **365**, 386, 394–**97**
Cerebral aqueduct, 362, **364**, **365**, 400, **401**
Cerebral commissures, 406, **407**, 418, **419**
Cerebral cortex, **409**, 410–**13**, 416, **417**
   histogenesis of, 416, **417**
Cerebral hemispheres, 362, **363**, **364**, **365**, 410–**13**
Cerebral maturation, 483
Cerebral peduncles, **387**, 400, **401**
Cerebral veins, 327
Cerebromedullary pathways, 386
Cerebrospinal fluid, 420
Cervical cyst, 134, **135**
   sinus, 58, **59**, 132, **133**
Cervical flexure, 362, **363**, **364**, **365**
Cervical pregnancy, 30, **31**
Cervical rib, 182, **183**
Cervical sinus, 128
Cervical thyroid, 136, **137**
Cervix, **274**
   atresia of, **285**
Chamber(s), anterior, posterior, aqueous, **437**, **438**, 440, **441**
Chemicals, environmental, teratogenic effects of, 482
Chiasma, optic, 418, **419**
Childbirth, 90, **91**
Chimeras, 74
Choanae, definitive, 148, **149**
   primitive, 148, **149**
Chondrification, 168, **169**
Chondroblasts, 44, 168, **169**
Chondrocranium, 178, **181**
Chondrocytes, **169**
Chondrodystrophy, hypoplastic, 172
Chorda tympani nerve, 138, **431**, 444
Chordae tendinae, 314, **316**, **317**
Chordata, nervous system, 366, **369**
Chordee, of penis, 282
Chordomesoderm, 446
Chorioallantoic placenta, circulation, 96, **97**
Chorioepithelioma, 42, 104

Chorion, **52**, **85**, 94, **95**, **99**
   embryonic, **52**, **85**, 94, **95**
   fetal, **85**, 94, **95**, **97**, **99**
   smooth, **85**, 94, **95**
   villous, **85**, 94, **95**, **97**, **99**
Chorion frondosum, **85**, **87**, 94, **95**
Chorion laeve, **85**, 94, **95**
Chorionic cavity, 40, **41**, **87**
   gonadotropins, 104, **105**
   plate, 40, **41**, **97**, **99**, 102, **103**
   vessels, **99**
   villi, 56, **57**, 94, **95**, **96**, **97**
Choroid, 438, 440, **441**
Choroid fissure, 410, **411**, 436, **437**, 440, **441**
   plexus, 392, **393**, 406, **407**, 410, **411**, 420, **421**
Chromaffin body (aortic), 462
Chromaffin cells, **381**, 402, 462, **463**
Chromaffin system, 462, **463**
   distribution in body (neonatal), **463**
   of renal area, **463**
Chromatin pattern, sex, in differential diagnoses of ambiguous external genitalia, 289
Chromosome(s)
   abnormalities of, 18, **19**, 78, **80**, **81**, 82
     in limb malformations, **83**
     numerical, 78
     retarded fetal growth and, 78, **80**, **81**, 82, **83**
     structural, **79**
   effect of sex determination, 78
   normal composition of, 8, **9**
   sex, complex of, abnormal, 18, **19**, 78, **80**, **81**
     nondisjunction of, 78
     in normal oocyte and sperm, 8, **9**, 14, **15**
     trisomy of, 78, **81**
   somatic. *See* Autosome
   X, 260
   Y, 260
Ciliary body, 436, **437**, **438**
Ciliary muscle, 436, **437**, **438**
Ciliary zonule, 436, **437**, **438**
Circle, Willis, 420, **421**
Circulation, at birth, 336, **337**
   chorioallantoic, 96, **97**
   maternal-placental, 96, **97**, 101, **103**
   through primitive heart, 318, **319**, 336, **337**
   uteroplacental, primitive, 336, **337**
   vitelline, 336, **337**
Circulatory system, 292, **293**, 336, **337**
   cardiovascular system of, 292–**355**
   congenital malformations of, 340–**55**
   lymphatic system of, 332–**35**
   stages of development, 292, **293**
Cisterna chyli, 332, **333**
Claustrum, **407**, 410, **411**
Clavicle, **177**, 178, **179**
   abnormalities of, 182, **183**
Cleavage, definition of, 36, **37**
   stages of, 36, **37**

Cleft(s), facial, 144, **145**
  foot, 188, **189**
  hand, 188, **189**
  lip, 144, **145**
    and cleft palate, differentiation of, 144, **145**
  palate, 144, **145**
    and cleft lip, differentiation of, 144, **145**
  pharyngeal, 128, **129**, **130**, **131**, 132, **133**
  sternum, **176**, 182, **183**
  tongue, 138
  uvula, 144, **145**
Cleidodysostosis, 182
Clitoris, 278, **279**, **288**
  duplication of, **285**
Cloaca, 222, **223**
  exstrophy, 252
  partitioning of, 222, **223**
  persistent, **235**
Cloacal eminence, 266, **267**
Cloacal fold, 266, **267**, 278, **279**
  membrane, **45**, **47**, **48**, 68, **223**, 266, **267**, 274, 278, **279**, **359**
Closing plug, embryonic, 38, **41**
Clubfoot, 188, **189**
Clubhand, 188, **189**
Coagulum "plug," 38, **39**, **41**
Coarctation, of aorta, 340, **341**
Cochlea, **447**
Cochlear duct, 446, **447**, 448, **449**
  ganglion, 446, **447**
  nerve, 402, **404**, **405**
  nucleus, 390
Coelenterates, nervous system, 366, **367**
Coelom, 50, **51**, **52**, 84, **85**, 110, **111**, **112**, **113**
  division of, 110, **111**, **112**, **113**
    congenital malformations related to, 228
  extraembryonic, 40, **41**, **52**, 84, **85**
  intraembryonic, 50, **51**, **52**, 110, **111**
    effect of headfold on, 110, **111**
    pericardial, effect of headfold on, 110, **111**
    peritoneal, effect of transverse fold on, 110, **111**
  Coelomic cavity, 110, **111**
Coelosomy, 82
Collecting tubule, renal, 244, **245**, **288**
Colliculus(i), seminal, 264, **265**
  superior (anterior) and inferior (posterior), of midbrain, 400, **401**
Coloboma iridis, congenital, 442, **443**
Colon, 218, **219**, 220, **221**, 222, **223**
  congenital malformations, **230**
Colostrum, 200
Column(s), vertebral, 174, **175**, **176**
  abnormalities, 182, **183**
Commissure(s), anterior, 406, **407**, **415**, 418, **419**
  cerebral, 406, **407**, 418, **419**
  corpus callosum, 418, **419**
  fornix, 418, **419**

habenular, 406, **407**, 418, **419**
hippocampal, 418, **419**
labial, posterior, 278, **279**
posterior, 406, **407**
spinal cord, 378
Compact bone, 168, **171**
Conchae, nasal, 148, **149**
Conducting system, 314, **316**
Cones, of retina, 436, **439**
Congenital adrenal hypoplasia, 466
Congenital glaucoma, 442, **443**
Congenital hip dislocation, 188
Congenital malformations. *See* Malformations, congenital
Congenital myxedema, 424
Congenital thymic aplasia, 134
Conjoined twins, 74, **77**
Conjunctiva, **437**, **438**
Conjunctival sac, **437**, **441**
Connecting stalk, effect of tailfold on, **85**, 88, **89**
  during embryonic development, 40, **41**, **85**, 88, **89**
Connective tissue, embryonic, 62, **63**, 478
Connexus, interthalamic, 406, **407**
Constrictors of pharynx, 128, **130**
Contraceptives, oral, possible teratogenic effects of, 481
Control, epigenetic, 68
  genetic, 68
Conus cordis, 300, **301**, **307**
Conus medullaris, **383**
Convolutions, cerebral, 414, **415**
Copula, of tongue, 138, **139**
Cor trioculare biventriculare, 347
Cord(s), cardiogenic, 296, **297**
  cortical, 270, **271**
  fibrous, intestinal, **233**
  medullary ovarian, 270, **271**
  nephrogenic, 238, **239**
  Pflügers, 270, **271**
  seminiferous, 258, **259**
  sex, primary, 256, **257**, 258, **259**
    secondary, 270, **271**
  spinal, 374, **375**, 376, **377**
    congenital malformations of, 384, **385**
    fissures of, **383**
  testicular, 258, **259**
  umbilical, 56, **85**, **87**
    covering of, **85**
    knots in, 88
    looping of, 218, **219**, 220, **221**
  urogenital, 240, 260
  vitelline, **233**
  vocal, 150
Corium, 192, **193**
Cornea, **437**, **438**, 440, **441**
Corniculate cartilage, 150, **151**
Corona radiata, 14, **15**, 24, **25**
Coronary sinus, **307**, 308, **309**

Corpus, albicans, 26
  callosum, **407**, 414, **415**, 418, **419**
    agenesis of, 418, 424
  cavernosum, **5**, 266, **267**, **288**
  luteum, 24, **25**, 26
  spongiosum, **5**, 266, **267**, **288**
  striatum, **407**, **409**, **411**
Corpuscle(s), mesonephric, 240, **241**
  renal, 240, **241**, 244, **245**
Correlated human development, Appendix I, 471–77
Cortex, **409**, 410–**13**, 416, **417**
  adrenal, 464, **465**
  cerebellar, 396, **397**, **398**, **399**
  cerebral, **409**, 410–**13**, 416, **417**
    histogenesis of, 416, **417**
  fetal, 416, **417**
  hippocampal, 414, **415**
  of indifferent gonads, **288**
  of suprarenal (adrenal) glands, 464, **465**
Corti, organ of, 450, **451**
Cortical cords, 258, **259**, 270, **271**
Cortiocosteroids, possible teratogenic effect, 482
Costodiaphragmatic recess, 125
Cotyledons, placental, 94, **95**, 96, **97**, 102, **103**
Cowper's glands, 4, **5**, 264, **265**
Cranial mennigocele, 384, **385**, 422, **423**
Cranial nerves, 402, **404**, **405**
  sutures, 178, **181**
    premature closing of, 182
Craniopagus, 74, **77**
Craniopharyngioma, 456, 460
Craniorrhachischisis, 82, 422
Cranioschisis, 182
Craniostenosis, 182
Craniosynostosis, 182
Cranium, 178, **181**
Cremasteric fascia, **269**
Crest, neural, 46, 360, **361**
  cells of, 46, 360, **361**
Cretinism, 172
Cri-du-chat syndrome, **80**
Cricoid cartilage, 128
Cricothyroid muscle, 128, **130**
Crista(e) ampullaris, 450, **451**
Crista terminalis, **307**, 310, **311**
Critical periods of development, Appendix III, 480
Crown, of tooth, 202, **205**
Crura, of diaphragm, 122, **124**, **125**
Crus cerebri, 400, **401**
Cryptorchidism, 268, **269**, 282, **284**
Cumulus oophorus, 24, **25**
Cuneate nucleus, **390**, **391**
Cuneiform cartilages, 150, **151**
Cup(s), optic, 436, **437**
Cupula, 450, **451**
Curvature, greater, of stomach, 210, **211**
  lesser, of stomach, 210, **211**

Cushing's syndrome, 466, **467**
Cushion(s), endocardial, 304, **305**, **307**
  defects of, 344, **347**
Cusp(s), cardiac valve, 314, **315**, **316**, **317**
Cutaneous appendages, 196–**203**
Cutaneous innervation, 186, **187**
Cutaneous nerve distribution, 370, **373**
Cuticle, 198, **199**
Cycle(s), reproductive (sexual), 24, **29**
Cyclopia, 134, 424, **425**, 442, **443**
Cyst(s), auricular, congenital, of ear, 134
  of bile duct, **226**
  branchial, 132, 134, **135**
  bronchial, 158
  cervical, 134, **135**
  Gärtner's duct, **277**
  thyroglossal duct, 136
  urachal, 252
  vitelline, 232, **233**
Cystic duct, 214, **215**
Cystic lymphangioma (hygroma), 352, **355**
Cytomegalovirus, teratogenic effects of, 481
Cytotrophoblast, 36, **37**, 38, **39**, 40, **41**, 56, **57**, 94, **95**, 96, **97**
  shell of, 56, **57**

Dandy-Walker deformity, 426
Deafness, congenital, 454
Decidua, **85**, **95**, 98, **99**
  regions of, **85**, **95**, 98, **99**
Decidual cells, 98, **99**
Decidual formation, 98, **99**
Decidual plate, 96, **97**, 98, **99**
  septa, 98, **99**, 102, **103**
Deciduous teeth, 202, **203**
Defect(s), atrial septal, 344, **345**, 346
  diaphragmatic, posterolateral, 122, **125**
  membranous septal, 348, **349**, **350**, **351**
  muscular septal, 348, **349**, **350**, **351**
  pericardial, congenital, 114, **345**
  ventricular septal, 348, **349**, **350**, **351**
Dendrite(s), primitive, 378, **379**
Dental epithelium, 202, **203**
  inner, 202, **203**
  lamina, 202, **203**
  outer, 202, **203**
  papilla, 202, **203**
  process, 202, **203**
Dental nucleus, 202, **203**
Dental sac, 202, **203**
Dentate gyrus, 414, **415**
Dentate nucleus, 396, **398**
Dentate (pectinate) line, 222, **223**
Dentine, 202, **203**, **205**
Dentinogenesis imperfecta, 204
Derivatives, of embryonic urogenital structures, **288**
  germ layer, 62, **63**
Dermal papilla, 192, **193**, 196, **197**

Dermal root sheath, 196, **197**
Dermal sinus, 384, **385**
Dermatoglyphics, 192
Dermatome(s), 370, **372, 373**
  of limbs, **53,** 162, **163,** 370, **373**
Dermis, 192, **193,** 360
Dermomyotome, **53,** 162, **163**
Development. *See also* Embryo; Fetus
  abnormal, 42
  control of, 34, **35**
  critical period of, 72, **73, 83,** 471–77, 480
  definition of, 68, **69**
  mechanisms of, 68, **69**
  postnatal, 484–85
  prenatal, 471–77, 480, 483–84
    week 1, 36, **37**
    week 2, 38, **39,** 40, **41,** 42, **43,** 471
    week 2, review of, 42, **43**
    week 3, 44, **45,** 46, **47,** 50, **51,** 54, **55,** 56, **57,** 471
    weeks 4–6, 58, **59,** 471, 472, 473
    weeks 7 and 8, 60, **61,** 473, 474
    weeks 9 to birth, 70, **71,** 72, **73,** 475, 476, 477
Dextrocardia, 344, **345**
Diakinesis, 10
Diaphragm, 110, 122, **123, 124, 125**
  congenital malformations of, 122
  crura of, 122, **125**
  eventration of, 122
  positional changes and innervation, 122
  tendon, central, 122, **125**
Diaphragmatic hernia, 122, **125**
Diaphragmatic ligament, 260
Diaphyseoepiphyseal junction, 168, **170**
Diaphysis, 168, **170, 171,** 172, **173, 180**
Diencephalon, 362, **363, 364, 365,** 406, **407**
Diethylstilbestrol, as cause of adenocarcinoma, 481
Differentiation, cellular, sexual, **9, 11, 15**
Diffusion, in placental transfer, 100, **101,** 102, **103**
DiGeorge's syndrome, 134
Digestive system, 208–**35**
  congenital malformations of, 224, **225, 226, 227**
  foregut, derivatives of, 208, **209,** 210, **211**
  hindgut, derivatives of, 208, **209,** 222, **223**
  midgut, derivatives of, 208, **209,** 218, **219,** 220, **221**
Digits, supernumerary, 188, **189**
Dilator pupillae muscles, 436, **437**
Diploid number, restoration of, 32
Diplopodia, **189**
Diplotene stage, 10
Discoid kidney, 253
Disk(s), embryonic, bilaminar, 38, **39**
  intervertebral, 174, **175**
  trilaminar, 44, **45**
Dispermy, 32

Distal convoluted tubule, 244, **245**
Diverticulum(a), allantoenteric, **87**
  antimesenteric, **229**
  duodenal, 228, **229**
  false, intestinal, **229**
  hepatic, 214, **215**
  ileal, 228, 232, **233**
  jejunal, 228
  laryngotracheal, **151**
  Meckel's, 86, 218, **231,** 232, **233**
  metanephric, 244, **245**
  Rathke's, 456, **457**
  thyroid, 132, **133,** 136, **137**
  tracheal, 158
  urachal, 348, **349**
Dizygotic twins, 74, **75**
Dorsal mesentery, 210, **211,** 212, **213**
  in development of stomach, 210, **211,** 212, **213**
Dorsal root ganglion(a), 380, **381**
Double monsters, 82
Douglas's (uterorectal) pouch, 272, **273, 274, 275**
Down's syndrome, 78, **81**
Drugs, placental transfer, 100, **101**
  teratogenic effects, 481, 482
Duct(s), alveolar, 154, **155**
  bile, common, 214, **215,** 216, **217**
  cochlear, 446, **447,** 448, **449**
  collecting, renal, 244, **245**
  cystic, 214, **215**
  ejaculatory, 4, **5,** 248, 264, **265, 288**
  endolymphatic, 446, **447**
  Gärtner's, 272, **273, 277, 288**
  genital, 262, **263,** 264, **265**
  of Haller, 264
  lactiferous, 200, **201**
  lymphatic, 332, **333**
  mesonephric, 240, **241,** 258, **259,** 260, **261, 262, 263,** 264, **265, 288**
    remnants of, 240, **241,** 276, **277**
  Müllerian, 260, **261, 262, 263,** 264, **265,** 272, **273, 275, 288**
  nasolacrimal, 140
  omphalomesenteric (vitelline), **65,** 219
  pancreatic, 216, **217, 227**
  papillary, 244, **245**
  paramesonephric, 260, **261,** 264, **265,** 272, **273, 275, 288**
    remnants of, 264, **265**
  pronephric, 238, **239**
  Santorini, 216, **217**
  semicircular, 448, **449**
  Stensen's, 138
  thoracic, 332, **333**
  thyroglossal, 136, **137**
    remnants of, 136, **137**
  vitelline, 232
  Wharton's, 138
  Wirsung, 216, **217**

Wolffian, 240, **241, 262, 263,** 264, **265**
Ductule(s), efferent, 4, **5,** 258, **259, 288**
Ductus arteriosus, 324, **325,** 336, **337,** 338
   closure of, 324, 336, **337,** 338
   derivatives of, 324, **325,** 338
   in fetal circulation, 324, **325,** 336, **337,** 338
   patent, 340, **341**
Ductus deferens, 4, **5,** 258, **259, 262, 263,** 264, **265, 288**
Ductus epididymidis, 4, **5,** 258, **259,** 264, **265, 288**
Ductus reuniens, **447**
Ductus venosus, 308, 330, **331,** 338
   derivatives of, 330, **331,** 338
   in fetal circulation, 330, **331,** 338
Duodenohepatic ligament, 212, **213**
Duodenum, 212, **213**
   congenital malformations of, **225, 229, 230**
   fixation of, 212, **213,** 218, **219,** 220, **221**
Duplication, of clitoris, **285**
   intestinal, **229**
   of penis, **285**
   of upper urinary tract, 250, **251**
   of uterus, **285**
Dura mater, 382, **383,** 420, **421**
Dysgenesis, cerebral, 424

Ear, 446–53
   congenital malformations of, 454, **455**
   external, 444, **445**
   internal, 446–51
   middle, 452, **453**
Eardrum, 132, **133,** 444, **445,** 452, **453**
Eccrine glands, 196, **197**
Ectoderm, 38, **39,** 44, **45,** 62, **63,** 478
   derivatives of, 38, **39,** 62, **63,** 192, **193,** 358, 478
   neural, 46, **47,** 358, **359,** 360, **361**
   surface, 192, **193,** 358
Ectodermal dysplasia, 194
Ectomesenchyme, 360, 420
Ectopia, adrenal, 466
Ectopia cordis, 344, **345**
Ectopia vesicae, 252, **253**
Ectopic pregnancy, 30, **31**
Ectopic sinus, 234
Ectopic testis, 268, **269**
Ectopic ureteral orifices, 250
Ectrodactyly, **189**
Efferent ductules, 4, **5,** 258, **259**
Ejaculatory ducts, 4, **5,** 248, 264, **265, 288**
Elastic cartilage, 168
Electrolytes, placental transfer of, 100, **101**
Embryo. *See also* Embryonic period
   age of, estimation of, 60
   bilaminar, formation of, 38, **39**
   chorion of, **43, 52**
   connecting stalk of, 40, **41, 85,** 88, **89**
   critical period of development of, 58, **59,** 60, **61**
   definition of, 58
   vs. fetus, 70
   folding of, 64, **65, 66, 67**
   trilaminar, formation of, 44, **45**
   week 1, 36, **37**
   week 2, 38, **39,** 40, **41**
   week 3, 44, **45,** 46, **47, 48, 49,** 50, **51,** 52, **53,** 54, **55,** 56, **57**
   weeks 4–6, 58, **59**
   weeks 7 and 8, 60, **61**
Embryoblast, 36, **37,** 38, **39**
Embryogenesis, sections of, **48**
Embryonic disk, bilaminar, 38, **39**
   changes in, 38, **39,** 44, **45**
   trilaminar, 44, **45**
Embryonic folding and flexion, 64, **65**
Embryonic period, 58, **59**
   critical period of development of, 58, **59**
   definition of, 58, **59**
   estimating fertilization age during, 60, **61,** 471–77
Embryonic pole, **39**
Embryonic pharynx, **137**
Eminence(s), cloacal, 266, **267**
   hypobranchial, **131,** 138, **139,** 150, **151**
Enamel, in tooth development, 202, **203, 205**
   hypoplasia, 204
Enamel epithelium, 202, **203**
Enamel organ, 202, **203**
Enamel reticulum, 202, **203**
Encephalocele, 422, **423**
Endbrain, 362, **363, 364, 365**
Endocardial cushions, 304, **305,** 307
   defects of, 344, **347**
Endocardium, 296, **299**
Endochondral ossification, 168, **169,** 170
Endocrine(s), placental secretion of, 104, **105**
Endolymph, 446
Endolymphatic duct, 446, **447**
Endolymphatic sac, 446, **447**
Endometrial veins, 100, **101,** 102, **103**
Endometrium, 20, **21,** 272
   cyclic changes of, ovaries and, 24, **25,** 26, **27, 28, 29**
Enterocystoma, 232
Entoderm (endoderm), 38, **39,** 62, **63,** 478, 479
   derivatives of, 38, **39,** 62, **63,** 478, 479
Environment, congenital malformations caused by, 482
Epaxial division, of myotomes, 162, **163**
Ependymal layer, **377, 379,** 416, **417**
Epiblast, 38, **39**
Epicardium, 296, **299**
Epidermis, 62, **63,** 192, **193**
Epididymis, 4, **5,** 258, **259,** 264
   appendix of, **259, 262, 263,** 264, **265, 288**
Epiglottis, **131,** 150, **151**
Epimere, 162, **163**

Epinephrine, 462
Epiphyseal ossification center, **170**, 172, **173**
Epiphyseal plate, **170**
Epiphysis, 168, **170**, 172, **173**, **180**
Epiploic foramen of Winslow, 118, 212, **213**
Epispadias, 282, **283**
Epithalamus, 406, **407**
Epithelial cells, alveolar, type I, 154, **155**, 156, **157**
    type II, 154, **155**, 156, **157**
Epithelial root sheath, 202, **203**, **205**
Epithelium, enamel, 202, **203**, **205**
    laryngeal, 150
    olfactory, 148, **434**, **435**
    tubular, 150, **151**
Epitrichium, 192, **193**
Eponychium, 198, **199**
Epoöphoron, 270, **273**, 276, **277**, **288**
Erythroblastosis fetalis, 92, **93**
Esophageal atresia, 158, **159**, 224, **225**
    amniotic fluid accumulation in, 224
Esophagotracheal fistula, **225**
Esophagotracheal septum, 151
Esophagus, **131**, 150, **151**, 210, **211**
    congenital malformations of, 224, **225**
    hernia, 122
Estradiol, 27
Estrogen, in follicular development, 24, 26, **27**
    placental secretion of, 104, **105**
Estrous cycle, in mammals, 24
Ethmocephaly, 424, 434
Eustachian tube, 132, **133**, **445**
Eventration, of abdominal viscera, 228
Exchanges, fetal-maternal, 100, **101**
Excretory system, 238–**53**, **262**, **263**
    congenital malformations of, 250–**53**
Exocoelomic cavity, 38, **39**
    membrane, 38, **39**
Exomphalos, 228
Exstrophy, of cloaca, 252
    of urinary bladder, 252
External auditory meatus, **445**
External genitalia, female, 278, **279**
    male, 266, **267**
Extraembryonic blood vessels, 294, **295**
Extraembryonic coelom, 40, **41**, 84, **85**
Extraembryonic mesoderm, **39**, 40, **41**, 44, 84, **85**
    somatic (somatopleure), 40, **41**, 84, **85**
    splanchnic (splanchnopleure), 40, **41**, 84, **85**
Extralobular (lung) sequestrations, 158, **159**
Extrauterine sites, of pregnancy, 30, **31**
Eye, 58, **59**, 436–**41**
    congenital malformations of, 442, **443**
Eyeball, mature, **438**
Eyelids, **437**, 440, **441**

Face, 140, **141**
    congenital malformations of, 144, **145**
    cutaneous nerves of, **141**
Facial muscles, of expression, 128, **130**
Facial nerve, 128, **130**, **131**, 402, **404**, **405**, **431**, **433**
Facial swellings, 140, **141**
Factors, teratogenic, 481, 482
Falciform ligament, 118, **119**, **120**, 121
Fallopian tube. *See* Uterine tube
Fallot, tetralogy of, 352, **353**
False knots, 88
Falx cerebri, **421**
Fascia, cremasteric, **269**
    spermatic, **269**
Fat, brown, 70
    subcutaneous, 72
    white, 72
Female genital system, 270–**79**
Female pronucleus, **16**, 32, **33**
Feminization, testicular, 280, **281**
Fenestra cochleae, 452, **453**
Fenestra vestibuli, 452, **453**
Fertilization, 32, **33**
    anomalies of, 30
    consequences of, 32
    definition, 32
    morphologic changes in, 32
Fertilization age, of embryo, 60, **61**, 471–77
    of fetus, 180, 471–77
Fetal circulation, 102, **103**, 336, **337**
    structures associated with, adult derivatives of, 102, **103**, 336, **337**, 338
Fetal-maternal exchange, 100, **101**
Fetal-maternal incompatibility, 92, **93**
Fetal membranes, 84, **85**
Fetal period, 70, **71**
    definition of, 70, **71**
    estimating fertilization age during, 70
    growth during, 70, **71**
    highlights of, 70, **71**
Fetal placental hydrops, 92
Fetal red blood cells, 92, **93**
Fetus. *See also* Fetal period
    age of, estimation of, 70, 71
    definition of, 70
    vs. embryo, 70
    effect of drugs on, 480, 481, 482
    transition from embryo to, 70
    viability of, 70
    week 8, 474
    week 9, 70, **71**
    week 10, 70, **71**, 475
    week 11, 70, **71**
    week 12, 475
    week 13, 70, **71**
    week 16, 476
    week 17, 70, **71**
    week 25, 72, **73**, 476
    week 29, 72, **73**, 476
    week 36, 72, **73**, 476
Fiber(s), commissural, 418, **419**

lens, 440, **441**
postganglionic, 428, **429, 432, 433**
preganglionic, 428, **429, 432, 433**
Purkinje, 162, 396, **398**
Fibroblasts, 44
Fibrocartilage, 168
Fibroplasia, retrolental, 442
Fibrous joint, 172, **173**
Field(s), nail, 198, **199**
Filum terminale, 382, **383**
Fimbriae, of uterine tube, 276, **277**
Finger rays, 58, **59**
Fingerprints, 192
Fingers, 60, **61**
  congenital malformations of, 188, **189**
First arch syndrome, 134, **135**
First polar body, 14, **15**
Fish brain, **369**
Fissure(s), of cerebellum, 396, **397, 399**
  choroidal, 410, **411**
  optic (choroid), 436, **437,** 440, **441**
  of Rolando, 412, **413**
  sylvian (lateral), 412, **413**
Fistula(s), anoperineal, **235**
  auricular, 454, **455**
  branchial, 134, **135**
  bronchoesophageal, 158, **159**
  of neck, 134, **135**
  oronasal, **135**
  pharyngeal, 132
  pharyngocutaneous, 132
  rectal, 252, **253**
  rectourethral, 252, **253**
  rectourinary, 252
  rectovaginal, **235**
  rectovesical, 252, **253**
  tracheoesophageal, 158, **159**
  umbilical, 232
  umbilicoileal, **233**
  urachal, 252
  urorectal, **235**
  vitelline, 232
Fixation, of duodenum, 218, **219,** 220, **221**
  of hindgut, 222
  of midgut, 218, **219,** 220, **221**
Flexion, cephalocaudal, 64, **65, 66, 67**
Flexure(s), of brain, 362, **363, 364, 365**
  caudal, 58, **59,** 64, **65, 66, 67**
  cephalic, 362, **363, 364, 365**
  cervical, 362, **363, 364, 365**
  hepatic, **219**
  pontine, 362, **363, 364,** 465
Flocculonodular lobe, 396, **397, 399**
Flocculus, 396, **397, 399**
Floor plates, 376, **377, 379**
  of mesencephalon, 400, **401**
  of metencephalon, 394, **395**
  of myelencephalon, 388, **389, 391**
Fluid(s). *See also* Amniotic fluid
  cerebrospinal, 420

Fold(s), anal, 222, 278, **279**
  cloacal, 266, **267,** 278, **279**
  genital, 266, **267,** 278, **279**
  glossoepiglottic, **151**
  head, of embryo, 58, **59,** 64, **65**
  longitudinal, 64, **65, 66, 67**
  nail, 198, **199**
  neural, 46, **47, 49,** 64, **65,** 359, 360, **361**
  tail, of embryo, 58, **59,** 64, **65, 66, 67**
  transverse, of embryo, 58, **59,** 64, **65, 66, 67**
  urethral, 266, **267**
  urogenital, 266, **267, 288**
  vestibular, of larynx, 150
  vocal, 150
Folding, of embryo, 64, **65**
Foliate papillae, of tongue, 138, **139**
Follicle(s), 14, **15,** 24, **25**
  Graafian, 24, **25**
  growing, 14, **15,** 24, **25**
  hair, 196, **197**
  ovarian, 14, **15**
  primary, 14, **15,** 24, **25,** 270
  primordial, 14, **15,** 24, **25,** 270, **271**
  of thyroid gland, 136
Follicle-stimulating hormone, 24, 26, **27,** 34, **35**
Follicular antrum, 24, **25**
  cells, 24, **25**
  phase, of menstrual cycle, 28, **29, 35**
Fontanelles, 178, **181**
Foot, congenital abnormalities of, 82, **83,** 188, **189**
  plate of, **59,** 184, **185**
Foramen(ina), of Bochdalek, 122
  cecum, **131,** 136, **137**
  epiploic, 118, **120**
  incisive, 142, **143**
  interventricular, 312, **313,** 406, **407,** 410, **411**
    closure of, 312
  of Luschka, 362, 382, **383**
  of Magendie, 362, 382, **383**
  magnum, 178, **181**
  of Monro, 362, **364, 365**
  of Morgagni, 122
  of Winslow, 118, 212, **213**
Foramen ovale, 304, **306,** 336, **337,** 338
  congenital malformations of, 344, **345,** 346
  in fetal circulation, 336, **337,** 338
  patent, 344, **345,** 346
  valve of, 304, **306**
Foramen primum, 304, **305, 306**
  patent, 344, **347**
Foramen secundum, 304, **305, 306**
Forebrain, 362, **363, 364**
  embryonic, 362, **363, 364**
Foregut, 208, **209,** 210–17
  derivatives of, 208, **209,** 210, **211**
  embryonic, 208, **209,** 210–17
  malformations of, 224, **225, 226, 227**
Foreskin, penile, 266, **267**
Formation, reticular, 386

Fornices, of vagina, 272, **274**
Fornix (hippocampal) commissure, 414, 418, **419**
Fornix system, 414, **415**
Fossa(e), nasal, 148, **149**
Fossa ovalis, 304, **307**
  limbus of, 304, **307**
Fossa, tonsillar, 132
Fourth ventricle, 362, **364, 365**
Fovea centralis, **438**
Fraternal twins. See Twins, fraternal
Freemartins, 74
Frenulum, of lips, 140
Frontal prominence, **181**
Frontonasal prominence, 140, **141**
Fungiform papillae, of tongue, 138, **139**
Fusion, of heart tubes, 296, **297**, 302, **303**

Gallbladder, 214, **215**
  atresia of, 224
Gametes, 8, **9**
Gametogenesis, 8, **9**, 14, **15**
  abnormal, 18, **19**
Ganglion(a), 402, **404, 405**
  autonomic, 402, **403**
  cardiac, 402, **403**
  celiac, 402, **403**, 428, **429, 432**
  cell layer, 417
  cells, 360
  cerebral, 402, **404, 405**
  cervical, 428, **429, 432**
  ciliary, **431, 433**
  cochlear, 402, **404, 405, 447**, 448
  collateral, 402, **403**
  of Corti. See Ganglion, cochlear
  dorsal root, 380, **381, 403, 429**
  Meissner's, 402
  mesenteric, 402, 428, **432**
  otic, **431, 433**
  parasympathetic, 402, 428, **430, 431, 433**
  of peripheral nervous system, 360, **361**
  preaortic, 428, **429, 432**
  prevertebral, 402
  primordia, 360, **361**
  renal, **403**
  root, dorsal, 360, **361**
  Scarpa's. See Ganglion, vestibular
  sensory, 380, **381**
  sphenopalatine, **431, 433**
  spinal, 380, **381**
  spiral, **449,** 450
  statoacoustic, 402, **404, 406,** 446, **447**
  submandibular, **431, 433**
  sympathetic, 402, **403**, 428, **429, 432**
  terminal, 402
  thoracic, 428, **432**
  vestibular, 402, **404, 406, 447,** 448
  visceral, 402, **403,** 428, **429,** 430, **431, 432, 433**

  of Zuckerkandl, 462
Gärtner's duct, 272, **273, 277**
  cyst of, **277**
Gastrulation, 44, **45**
Gases, placental transfer of, 100, **101**
Gastric atresia, **225**
Gastroschisis, 228
Gastrulation, 44, **45**
Gene(s), malformations caused by, 78, **80, 81,** 82, **83**
  mutant, malformations caused by, 78, **80, 81,** 82, **83**
  retarded fetal growth and, 78, **80, 81,** 82, **83**
Genetic sex, determination of, 256
Genital, ducts, **262, 263,** 264, **265**
  folds, 266, **267,** 278, **279**
  ridges, 256, **257**
  swellings, 266, **267,** 278, **279**
  system, 256–**89**
  tracts, primitive, 260, **261, 262, 263,** 264, **265**
  tubercle, 266, **267,** 278, **279**
  vestigial structures derived from, 264, 276, **277**
Genital system, 256–**89**
  congenital malformations of, 282–**89**
  female, 270–**79**
  male, 256–**69**
Genitalia, external, 266, **267,** 278, **279**
  ambiguous, differential diagnosis of, **289**
Genitourinary system, definitive male, **262, 263**
Germ cells, 270
  haploid, **9, 10, 15**
  primitive, yolk sac and, 38, **39**
  primordial, migration of, 256, **257,** 270, **271**
  transport, 270
  viability of, 270
Germ disk, bilaminar, 38, **39**
  trilaminar, 44, **45**
Germ layers, of embryo, derivatives of, 38, **39,** 44, **45,** 62, **63,** 478, **479**
Gestation, period of, 70, **71,** 72, **73**
  division into trimesters during, 70, **71**
Gigantism, 172
Gingivae, 140
Girdle(s), pectoral (shoulder), 178, **179, 180**
  pelvic, 178, **179, 180**
Gland(s)
  adrenal, 462, 464, **465**
    fetal, 462, 464, **465**
  apocrine, 196
  Bartholin's, 272, **274**
  bronchial, 152
  bulbourethral, 4, **5,** 264, **265, 288**
  ceruminous, 444
  Cowper's, 4, **5**
  eccrine, 196, **197**
  independent, of epidermis, 196
  lacrimal, 440
  mammary, 200, **201**
    abnormalities of, 200, **201**

merocrine, 196, **197**
parathyroid. *See* Parathyroid glands
paraurethral, of Skene, 248, 272, **288**
parotid, 138
pineal, 406, **407**, 460
pituitary, 456–**61**
prostate, 4, **5**, 248, **249**, 264, 265
salivary, 138
sebaceous, 196, **197**
seminal, 264, **265**
sublingual, 138
submandibular, 138
suprarenal, 462, 464, **465**
  fetal, 462, 464, **465**
sweat, 196, **197**
thymus, 132, **133**
thyroid, 132, **133**
  abnormalities of, 172
  descent of, 132, **133**, 136, **137**
  diverticulum of, 132, **133**, 136, **137**
urethral, 272, **288**
vestibular, 272, **274**, **288**
Glandular plate, of penis, 266, **267**
Glandular urethra, 266, **267**
Glans clitoridis, **288**
Glans penis, 5, 266, **267**, **288**
Glaucoma, congenital, 442, **443**
Glia cells, 378, **379**
Glioblasts, 378, **379**
Globus pallidus, **407**, 410, **411**
Glomerular capsule, 240, **241**, 247
Glomerular chamber, 240, **241**, 247
Glomerulus(i), **239**, **241**, 247
Glossoepiglottic fold, lateral, **151**
Glossopharyngeal nerve, 128, **130**, **131**, 388, **389**, **390**, **391**, 402, **404**, **405**, 430, **431**, **433**
Glottis, primitive, **131**, 150, **151**
Gonad, differentiated, 4, **5**, **25**
  indifferent, 256, **257**
Gonadal dysgenesis, 256, 280
  ridges, 256, **257**
Gonadal primordium, 240, 256, **257**
Gonadal sex, determination of, 260
Gonadotropin(s), human chorionic, 104
  hypophyseal, 24, 34, **35**
  multiple pregnancies and, 74
  placental secretion of, 104
Graafian follicle, 24, **25**
Gracile nucleus, 390, **391**
Granular layer, 396, **398**
Granule(s), cells, 396, **398**
Grasp reflex, 485
Greater omentum, 116, **117**
Groove(s), branchial (pharyngeal), 128, **129**, 444, **445**
  labiogingival, 140
  laryngotracheal, **131**, 150, **151**
  nasolacrimal, **131**, 140, **141**
  neural, 46, **47**, **49**, 359, 360, **361**

primitive, 44, **45**
urethral, 266, **267**, **279**
urogenital, 266, **267**
Growth, of face, 58, **59**
  during fetal period, 70, **71**, 72, **73**
  of uterus, during pregnancy, 90, **91**
Gubernaculum testes, **262**, **263**, 64, **265**, 268, **269**, **288**
Gums, 140, **141**
Gut. *See also* Foregut; Hindgut; Midgut
  primitive, 208, **209**, 218, **219**, 222, **223**
Gynecomastia, male, **201**
Gyrus, cingulate, **415**
  dentate, 414, **415**
  parahippocampal, **415**
  paraterminal, **415**

Habenular commissure, 406, **407**
  nuclei, 406
Hair, abnormalities of, 194
  development of, 196, **197**
  epithelial root sheath, 196, **197**
  follicle, 196, **197**
  lanugo, 196
  papilla, 196, **197**
  shaft, 196, **197**
  terminal, 196
  vellus type, 196
Hair cells, of inner ear, 450, **451**
Hand development, **179**, **180**
Hand plates, 58, **59**
"Harelip," 144, **145**
Hartmann's pouch, **226**
Haversian canals, **171**
Haversian systems, 168, **171**
Head process. *See* Notochordal process
Headfold, of embryo, 64, **65**, 110, **111**
Heart, 296, **297**
  congenital abnormalities of, 344–54
  contraction of, 302
  effect of headfold on, 296, **297**, **298**, **299**, 302, **303**
  primitive, circulation through, 292, **293**, 294, **295**, **297**, 302
Heart loop, 300, **301**
Heart prominence, **59**, 363
Heart tubes, 296, **297**
  embryonic, 296, **297**
  fusion of, 296, **297**, 302, **303**
Heart wall, formation of, 296, **299**
Helicotrema, 448, **449**
Hemangioma, capillary, 194, **195**
Hematopoietic function, 214, 292, **293**
  islands, 294, **295**
Hemiazygos vein, 328, **329**
Hemichordates, nervous system, 366
Hemispheres, cerebral, 362, **363**, **364**, **365**, 410–13
Hemochorial placenta, 96, 100, **101**

Hemocytoblasts, 294
Hemolytic disease, fetal (perinatal), 92, **93**, 292
Hemophilia, 78
Hemopoietic tissue, 292
Hemorrhoidal vessels, 222, **223**
Henle, loop of, 244, **245**
Hensen's node, 44, **45**
Hepatic, diverticulum, 214, **215**
　epithelial cords, 292
　flexure, **219**
　hematopoiesis, 292
　segment, inferior vena cava, 328, **329**
　sinusoids, 292, 330, **331**
　　in fetal circulation, 292, 330, **331**
　veins, 330, **331**
Hermaphroditism, 280
　pseudo, 280, **281**
　true, 280
Hernia, diaphragmatic, congenital, 122, **125**
　esophageal, 122
　hiatal, congenital, 122
　inguinal, congenital, 268, **269**, 282, **284**
　of Morgagni, 122
　retrosternal, congenital, 122
　umbilical, congenital, 228
Herpes simplex virus, teratogenic effects of, 481
Herpes zoster, 370
Heterotopias, adrenal, **467**
　cerebral, 424
Heterotopic pancreatic tissue, 227
Heuser's membrane, 38, **39**
Hiatal hernia, congenital, 122
Hillock(s), auricular, 444, **445**
Hindbrain, 362, **363**, 364
Hindgut, derivatives of, 208, **209**, 222, **223**
　effect of tailfold on, 64, **65, 66, 67**
　fixation of, 222
　malformations of, 234, **235**
Hip, dislocation of, congenital, 188
Hippocampal (fornix) commissure, 418, **419**
Hippocampal cortex, **407**, 414, **415**
Hippocampus, **411**, 414, **415**, 416
Hirschsprung's disease (congenital aganglionic megacolon), 232, 430, **431**
His, bundle of, 314, **316**
Histogenesis, adrenal, 464, **465**
　of anterior hypophysis, 456
　cortex, cerebellar, 396, **398**
　　cerebral, 46, **417**
　of inner ear, 450, **451**
　retinal, **439**
Hofbauer cells, **97**
Hormonal aspects, of pregnancy, **27**, 28, **29**, 104, **105**
　of implantation, 26, **27**, 28, **29**, 34, 35
Hormone(s), 104, **105**
　androgenic, 258
　antidiuretic (ADH), 460
　chorionic-gonadotropic, 104, **105**

corticotrophic (ACTH), **459**, 460, **461**
estrogen, 24, 26, **27**
　placental secretion of, 104, **105**
follicle-stimulating (FSH), 24, 26, **27**, 34, **35**, **459**, 460, **461**
follicle-stimulating-hormone releasing, 35
gonadotropins, 34, **35**
　human chorionic, 104, **105**
　multiple pregnancies and, 74
　placental secretion of, 104, **105**
growth, **459**, 460, **461**
hypophyseal, 34, **35**, **459**, 460, **461**
lactogenic, 104, **459**, 460, **461**
luteinizing (LH), 24, 26, **27**, 34, **35**, **459**, 460, **461**
luteinizing-releasing, 35
melanotropic (MSH), **459**, 460, **461**
ovarian, 26, **27**
placental transfer of, 100, **101**
progesterone, in ovarian cycle, 26, **27**
　placental secretion of, 104, **105**
prolactin, **459**, 460, **461**
protein, placental secretion of, 104
relationships in female, **27**
somatomammotropin, human chorionic, 104
　placental secretion of, 104
somatotropic (STH), **459**, 460, **461**
steroid, placental secretion of, 104, **105**
thyrotropic (TSH), **459**, 460, **461**
Horn(s), anterior, 374, **375**
　lateral, 374, **375**
　intermediate, 376, **377**
　posterior, 374, **375**
　of sinus venosus, 300, **301**
Horseshoe kidney, 250, **251**
Humor, aqueous, 437, **438**, 440, **441**
Hyaline cartilage, 168, **169**
Hyaline membrane disease, 156
Hyaloid artery, 436, **437**, 440, **441**
　persistence of, 440, **441**
Hydatid of Morgagni, 277, **288**
Hydatidiform mole, 42, 104, **105**
Hydra, nervous system of, 366, **367**
Hydrocele, 268, **269**, 282, **284**
Hydrocephalus, 426, **427**, 466
Hydronephrosis, 252, **253**
Hydrops, 92
Hydroureter, **253**
Hygroma, 352, **355**
Hymen, 272, **274**, **288**
　imperforate, **285**
Hyoid arch, 128, **129**, **130**, **131**
　derivatives and innervation of, 128, **129**, **130**, **131**
Hyoid artery, 324
Hyoid bone, 128, **130**
Hypaxial division of myotomes, 162, **163**
Hyperopia, 442
Hyperpituitarism, 172
Hyperplasia, adrenal, congenital, 466

virilizing, 466
Hypertrichosis, 198
Hypoblast (entoderm), 38, **39, 41, 43**
Hypobranchial eminence, **131**, 138, **139**, 150, **151**
Hypodermis, **193**
Hypoglossal nerve, 388, **389, 390, 391**, 402, **404, 405**
Hypomere, 162, **163**
Hyponychium, 198, **199**
Hypophyseal portal system, 458, **459**
Hypophysis, 456–**61**
　anterior lobe, 456, **457**
　malformations of, 460
　pharyngeal, 456
　physiology and pathology, 460, **461**
　posterior lobe, 456, **457**
Hypoplasia, enamel, 204
　of respiratory tract, 158, **159**
　of urinary system, 250, **251**
Hypospadias, 282, **283**
Hypothalamic-hypophyseal tract, 458, **459**
Hypothalamic nuclei, 406
Hypothalamic sulcus, 406, **407**
Hypothalamohypophyseal control, 460
Hypothalamus, 406, **407, 411**
Hypothyroidism, 172

Ichthyosis, 192, 194, **195**
Icterus, 92, **93**
　nuclear, 92
Idiozome, 256
Ileum, 218, **219**, 220, **221**
　congenital malformations of, **230**
Iliac artery, common, 320, **323**
　internal, 320, **323**
Iliac lymph sacs, 332, **333**
Iliac vein, 328, **329**
Imperforate anus, 234, **235**
Implantation, of blastocyst, 30, 34, 36, **37**
　abnormal sites, 30, **31**
　changes in uterine mucosa, 34
　early stages of, 36, **37**
　hormonal aspects of, 34, **35**
　interstitial, 38, **39**
　sites of, 30
Incisive foramen, of hard palate, 142, **143**
Incisor, 202, **203**
Incompatibility, fetal-maternal, 92, **93**
　blood, 92, **93**
Incus, 128, **130**, 452, **453**
Induction, 68, **69**, 358
　during embryonic development, 68, **69**, 358
　secondary, 68, **69**, 358
Infection, fetal, placental transfer and, 82
Infectious agents, teratogenic effects of, 82, **83**, 481
Inferior vena cava, 308, **309**, 328, **329**

absence, 328, 340, **341**
double, 328, **341**
in fetal circulation, 308, **309**, 328, **329**
hepatic segment, 328, **329**
renal segment, 328, **329**
sacrocardinal segment, 328, **329**
suprahepatic segment, 328, **329**
Infracardiac bursa, 116, 212
Infundibular stem, 458, **459**
Infundibuloma, 458
Infundibulum, cardiac, 312
　hypophyseal, 406, **407**, 456, **457**, 458
　stem of, 456, **457**
　of uterine tube, 22, **23**
Inguinal canals, 268, **269**
Inguinal hernia, congenital, 268, **269**, 282, **284**
Inguinal rings, 268, **269**
Inheritance, autosomal, 14, **15**
　sex-linked, 14, **15**
Inheritance, congenital malformations caused by, 18, **19**
Inner cell mass, 36, **37**, 38, **39**
Insula, of cerebral cortex, **411**
Insulin, 216
　fetal growth and, 216
　possible teratogenic effects of, 482
Integumentary system, 192
　congenital malformations of, 194, **195**
Interatrial septation, 304, **305**, 306
Intercostal arteries, 174, **175**, 320, **321**
　vein, left superior, 326
Intermediate horns, 376, **377**
Intermediate mesoderm, 50, **51**, 238, **239**
Internal capsule, 410, **411**
　carotid artery, **321**, 324, **325**, 420, **421**
　ear, 446, **447**, 448, **449**
　　histogenesis of, 450, **451**
　glomeruli, 242, **241**, 245
　iliac artery, 320, **323**
Interrupted aortic arch, 340, **341**
Intersegmental arteries, 174, **175**
Interstitial cells, of Leydig, 258
Interthalamic connexus, 406, **407**
Interventricular foramen of Monro, 312, **313**, 406, **407**, 410, **411**
　closure of, 312
Interventricular groove, 310
Interventricular septum, 304, **305**, 307, 310, **311**
　anomalies, 348, **349, 350, 351**
　defect, 348, **349, 350, 351**
　membranous, 312
　muscular, 312
Intervertebral disk, 174, **175**
Intervetebal foramina, 174, **175**
Intervillous spaces, of placenta, 96, **97**
Intestinal loop, abnormal rotation, 228, **230**
　distal, 218, **219**
　primary, 218, **219**
　proximal, 218, **219**
　reversed rotation, 228, **230**

Intestines, 218, **219**, 220, **221**
   congenital malformations of, 228–**33**
Intracartilaginous ossification, 168, **169**, **170**
Intraembryonic blood vessels, 294, **295**
   coelom, 50, **51**
Intraembryonic mesoderm, 44, **45**
Intraembryonic vascular system, 294, **295**
Intrahepatic portion, of biliary apparatus, 214, **215**
Intramembranous ossification, 168, **169**
Intraretinal space, 436, **437**
Intrauterine sites, of pregnancy, 30, **31**
Invertebrate nervous system, 366, **367**, **368**
Inverted nipple, 200
Iridopupillary membrane, 440
Iris, 436, **437**, **438**, 440, **441**
   coloboma of, congenital, 442, **443**
Ischemic phase, of menstrual cycle, 28, **29**
Island(s), blood, 294, **295**
Islets of Langerhans, 216, **217**
Isthmus of rhombencephalon, **365**

Jacobson's organ, 434
Janus-type (janiceps) cephalothoracopagus, 82
Jejunum, 218, **219**, 220, **221**
   diverticulum of, 218, **219**
Jelly, cardiac, 296
   Wharton's, 88
Joint(s), cartilaginous, 172, **173**
   development of, 172, **173**
   fibrous, 172, **173**
   interphalangeal, **173**
   knee, **173**
   sutures, 172
   synovial, 172, **173**
Jugular, lymph sacs, 332, **333**
   veins, 326, **327**
Junction(s), cementoenamel, 205
   diaphyseoepiphyseal, 168, **170**, **171**, **173**
   duodenojejunal, 218, **219**, **221**
   ileocecal, 218, **219**, **221**

Karyotype, normal, 79
Kidney(s), 244, **245**, 246, **247**
   abnormal rotation, 252, **253**
   agenesis of, 250, **251**
   ascent of, 246, **247**
   congenital malformations of, 250, **251**, 252, **253**
   definitive, 246, **247**
   discoid, 252
   double, 250, **251**
   fetal, 246, **247**
   horseshoe, 250, **251**
   lobation, 246, **247**
   pelvic, 246, **247**
   permanent, 244, **245**, 246, **247**

   polycystic disease of, bilateral, congenital, 250, **251**
   positional changes of, 246, **247**
   supernumerary, 250
   vascular supply, 246, **247**
Klinefelter's syndrome, 78, **81**, 280, **281**
Klippel-Feil syndrome, 182
Knot(s), of umbilical cord, 88
   primitive, 88
Kupffer cells, 214

Labia, majora, 266, 278, **279**, **288**
   minora, 278, **279**, **288**
Labiogingival groove, 140
Labiogingival lamina, 140
Labioscrotal swelling, 266, **267**
Labor, 90, **91**
Labyrinth(s), of inner ear, 446, **447**, 448, **449**
Lacrimal glands, 440
Lacrimal sac, 140
Lactiferous ducts, 200, **201**
Lacunae, bone, **171**
   networks of, 40, **41**, 42
   trophoblastic, 40, **41**, 42
Lamina(e), cribrosa of ethmoid bone, 434, **435**
   dental, 202, **203**
   labiogingival, 140
   terminalis, **364**, **365**, 410, **411**, **413**, 418, **419**
Langerhans, islets of, 216, **217**
Langhans, cells or layers of. *See* Cytotrophoblast
Lanugo, 70, 196
Laryngeal aditus, 150, **151**
Laryngeal cartilages, 128
Laryngeal epithelium, 150
Laryngeal muscles, 150
Laryngeal nerves, relation to aortic arches, 150
Laryngeal ventricles, 150
Laryngeal web, 158
Laryngopharynx, **151**
Laryngotracheal diverticulum, **151**
Laryngotracheal groove, **131**, 150, **151**
Laryngotracheal tube, 150, **151**
Larynx, 128, 150, **151**
   vestibular fold of, 150
   vocal fold of, 150
Lateral cervical cyst, 134, **135**
   lingual swellings, 138, **139**
   plate mesoderm, 50, **51**
   sinus, 58, **59**, 132, **133**
   ventricles, brain, 362, **364**, **365**, 412, **413**
Layer(s), basal, of epidermis, 192, **193**
   compact (of placenta), 98, **99**
   ependymal, **377**, **379**, 416, **417**
   germ, of embryo, derivatives of, 62, **63**
   granular, 396, **398**
   mantle, of cerebral cortex, 416, **417**
      of spinal cord, 374, **375**, 376, **377**, **379**
   marginal, of cerebral cortex, 416, **417**

of spinal cord, 374, **375**, 376, **377**, 379
matrix cell, **377**
molecular, 416, **417**
neuroepithelial, of spinal cord, 374, 376, **377**
plexiform, **439**
retina, **439**
spongy (of placenta), 98, **99**
superficial cortical, of cerebellum, 396, **398**
of uterine wall, 20, **21**
"Left-sided colon," 228, **230**
Left superior vena cava, 340, **343**
Leg buds, **59**, 60, **61**, 184, **185**
Legs, 60, **61**, 184, **185**
Lens, 58, **59**, 436, **437**, **438**, 440, **441**
  fibers, primary, 440, **441**
Lens placode, 58, **59**, 436, **437**, **441**
Lens vesicle, 436, **437**, 440, **441**
Lentiform nucleus, 410, **411**
Leptomeninges, 382, **383**, 420, **421**
Leptotene stage, 10
Lesser curvature of stomach, 210, **211**
  omentum, 116, **117**, 118, **119**, **120**, **121**, 212, **213**
  peritoneal sac, 116, **117**
Leydig, interstitial cells of, **6**, **13**, 258
Lienorenal ligament, 116, **117**
Ligament(s), broad, 272, **273**, 275
  coronary, 118, **121**
  diaphragmatic, 260
  duodenohepatic, 212, **213**, 214, **215**
  falciform, 118, **119**, **120**, **121**, 213, 214, **215**
  gastrohepatic, 214, **215**
  gastrolienal, 116, **117**, **121**, 213
  gastrophrenic, **121**
  inguinal, 276, **277**
  lienorenal, 116, **117**, **121**, 213
  median umbilical, 86
  ovarian, **23**, 276, **277**, **288**
  periodontal, 202, **203**
  proper, of ovary, **275**, 276, **277**
  round, of liver, 118, **119**, **120**, **121**
    of uterus, **23**, 272, **273**, **275**, **288**
  sphenomandibular, 130
  spiral, 448, **449**, 450, **451**
  stylohyoid, 128, **453**
  suspensory, of ovary, **23**, 276, **277**
  umbilical, median, 86
Ligamentum arteriosum, 324, **325**
Ligamentum teres, hepatis, 118, **119**, **120**, **121**, 330, **331**
  uteri, 275
Ligamentum venosum, 330, **331**
Limb(s), 58, **59**, 60, **61**
  congenital malformations of, 188, **189**
  dermatomes of, 186, **187**
  fetal development of, 184, **185**
  innervation of, 184
  positional changes of, 184
Limb bones, **179**, **180**, 184, **185**
Limb buds, 58, **59**, 60, **61**, 184, **185**

Limb muscles, 162, **163**
Limbic system structures, **415**
Limbus fossa ovalis, 304, **307**
Line, intersphincteric, **223**
Line(s), pectinate (dentate), 222, **223**
Lingual nerve, **130**
Lingual swellings, of tongue, 138, **139**
Lingual thyroid, 136, **137**
Lingual tonsil, 332, **335**
Lip(s), 140, **141**
  cleft, 144, **145**
    cleft palate and, differences between, 144, **145**
  rhombic, 394, **395**, 396, **397**
Liquor folliculi, **15**
Lissencephaly, 424
Liver, 214, **215**
  bare area, 118, **120**, **121**
  fetal, visceral surfaces of, 214, **215**
  malformations of, 224, **226**
  variations in structure of, **226**
Lobation, of kidney, 246, **247**
Lobe(s), 412, **413**
  anterior, of hypophysis, 456, **457**
  azygos, 158
  of cerebellum, 396, **397**, **398**, 399
  frontal, 412, **413**
  of insula, **411**
  limbic, 414, **415**
  neural, 458, **459**
  occipital, 412, **413**
  parietal, 412, **413**
  posterior, of hypophysis, **457**, 458, **459**
  pyramidal, of thyroid gland, 136, **137**
  renal, 244
  temporal, 412, **413**, 414, **415**
  tracheal, 158
Lobster-claw deformity, 188, **189**
Long bone, growth of, 168, **170**
Loop(s), bulboventricular, 300, **301**
  of heart, 300, **301**
  of Henle, 244, **245**
  midgut, 218, **219**, 220, **221**
Looping, of umbilical cord, 218, **219**, 220, **221**
LSD, possible teratogenic effects of, 482
Lumbar puncture, 382
Lumbricus terrestris, nervous system, 366, **368**
Lunar months, fetal growth, 73
Lung(s), 150, **151**, 152, **153**, 154, **155**
  aeration, 152
  agenesis, 158, **159**
  congenital malformations of, 158, **159**
Lung buds, **133**, 150, **151**, 152, **153**
Lunula, of nail, 198, **199**
Luschka, foramina of, 362, 382, **383**
Luteal cells, 26
Luteal phase, of menstrual cycle, 28, **29**
Luteinizing hormone, 24, 26, **27**, 34, **35**
Lymph glands (nodes), 332, **333**, **334**
Lymph nodules, of palatine tonsil, 332

Lymph sacs, 332, **333**
Lymph sinus, 332
Lymphatic system, 332–35
　congenital malformations of, 352, **355**
Lymphatic vessels, 332, **333**

Macrodactyly, **189**
Macroglossia, 138
Macrogyria, 424, **425**
Macrostomia, **145**
Macula lutea, **438**
Maculae sacculi, 450, **451**
Maculae utriculi, 450, **451**
Magendie, foramen of, 362, 382, **383**
Male genitourinary system, definitive
　　　development of, **262, 263, 264, 265**
Male pronucleus, **16,** 32, **33**
Malformations, congenital, 78, **80, 81,** 82, **83**
　aganglionic dystony, 232
　of brain, 82, **83,** 422–27
　of branchial apparatus, 134, **135**
　of cardiovascular system, 340–**55**
　causes of, 78, 82
　　environment factors in, 78, 82, **83**
　　genetic factors in, 78, **80, 81,** 82, **83**
　coelom division and, 228
　of digestive system, 224–**35**
　of ear, 454, **455**
　embryology and, 78, **80, 81,** 82, **83**
　of excretory system, 250–**53**
　experimental, 82
　of eye, 442, **443**
　of genital system, 282–**89**
　　female, 286–**89**
　　male, 282, **283, 284, 285**
　of head and neck, 82, **83**
　of heart, 344–**54**
　of hypophysis, 460
　of limbs, 82, **83**
　of lymphatic system, 352, **355**
　of muscles, 162
　of nervous system, 82, **83,** 384, **385,** 422–27
　of olfactory system, 434
　of reproductive system, 282–**89**
　of respiratory system, 82, **83,** 158, **159**
　of skeletal system, 82, **83**
　of special sense organs, 434, 442, **443,** 454, **455**
　of urinary system, 250–**53**
　uterovaginal, 286, **287**
　of vestibulocochlear system, 454, **455**
Malleus, 128, **130,** 444, **445,** 452, **453**
Mamillary body, 406, **407**
Mammary buds, 200, **201**
Mammary glands, 200, **201**
　abnormalities of, 200, **201**
　line, 200, **201**
　pits, 200, **201**
　ridge, 200, **201**

Mandible, 128
Mandibular arch, 128, **129, 130, 131**
　derivatives and innervation of, 128, **129, 130**
　process, 58
　prominence, 140, **141**
Mandibulofacial dysostosis, 134
Mantle(s), myoepicardial, 296
Mantle layer, of cerebral cortex, 416, **417**
　of spinal cord, 374, **375, 379**
Marfan's syndrome, 188, **189**
Marginal layer, of cerebral cortex, 416, **417**
　of spinal cord, 374, **375, 379**
Marijuana, possible teratogenic effects of, 482
Massa intermedia, 406, **407**
Mastoid antrum, 132, 452
Mastoid process, 452
Maternal plasma cells, **93**
Matrix, cartilaginous, 168, **169**
Maturation division, 8, **9, 10,** 14, **15**
Maxillary artery, 324, **325**
　process, **129, 131**
　prominence, 140, **141**
　swelling, 140, **141**
Measurement(s), in estimating embryonic
　　　stages, 60, **61**
Meatal plug, 444, **445**
Meatus, acoustic, external, 132, **133,** 444, **445**
　atresia, 454
Mechanical factors, and limb malformations, 188
　possible teratogenic effects of, 482
Mechanisms of normal development, 68, **69**
Meckel's cartilage, 128, **130, 131,** 178, **180, 181,** 452, **453**
Meckel's diverticulum, 86, 218, **231,** 232, **233**
Meconium, 214
Medial lemniscus, **391**
Median cleft lip, 144, **145**
Median eminence, 458, **459**
Median septum, of tongue, 138, **139**
Medulla, of indifferent gonads, **288**
　of suprarenal adrenal glands, 402, **403,** 464, **465**
Medulla oblongata, 388, **389, 391**
Medullary velum, anterior, 396, **398, 399**
　posterior, 396, **398, 399**
Medullocerebral pathways, 386
Megacolon, aganglionic, congenital
　　　(Hirschsprung's disease), 232, **233,** 430, **431**
Megaesophagus, 225
Meiosis, 8, **9, 10**
　divisions in, **10**
Melanin, production of, 192, 196
Melanoblast, 192, 196, 402
Melanocyte, 192, **193,** 196
Membrane(s), alveolar-capillary, 154, 156, **157**
　amniochorionic, 98, **99**
　anal, 62, 222, **223,** 266, **267,** 278, **279**
　arachnoid, 382, **383,** 420, **421**

basement, of alveolar cells, 157
basilar (basement), 448, **449**, 450, **451**
branchial, 128
buccopharyngeal, 44, **45, 48**, 208, **209**, 358, **359**
   formation of, 44, **45**, 208, **209**, 358, **359**
cloacal, 44, **45, 223**, 266, **267**, 274, 278, **279**, **359**
exocoelomic, 38, **39**
fetal, 84, **85**
Heuser's, 38, **39**
iridopupillary, 440
oronasal, 148, **149**
oropharyngeal, 44, **45, 48**, 62, **133**, 142, 208, **209**, 358
   formation of, 44, **45, 48**, 208, **209**, 358
otolithic, 450, **451**
pellucida, 14, **15**, 17, **25**
placental, 100, **101**
pleuropericardial, 114, **115**, 302, **303**
pleuroperitoneal, 114, **115**, 122, **123, 124**, 303
pupillary, 440, **441**
   persistent, 442, **443**
Reissner's, **449**, 450
respiratory, 154, 156, **157**
tectorial, **449**, 450, **451**
tympanic, 132, **133**, 452, **453**
urogenital, 62, 222, **223**, 266, **267**, 278, **279**
vestibular, **449**, 450
Membranous atresia, **225**, 234
Membranous labyrinth, of inner ear, 446, **447**
Membranous neurocranium, 178, **181**
Membranous septal defect, 348, **349, 350**
Membranous viscerocranium, 178, **181**
Meninges, of brain, 420, **421**
   of spinal cord, 374, **375**, 382, **383**
Meningocele, 384, **385**, 422, **423**
   cranial, 422, **423**
   spina bifida and, 384, **385**
Meningoencephalocele, 422, **423**
Meningohydroencephalocele, 422, **423**
Meningomyelocele, 384, **385**
   spina bifida and, 384, **385**
Meninx, primitive, 420, **421**
Menses, 28
"Menstrual age," 24
Menstrual cycle, hormonal interrelations in, 24, 28, **29**
   phases of, 28, **29**
   pregnancy and, 28, **29**
Menstrual period, last, in estimating embryonic age, 60
Menstruation, cessation of, during pregnancy, 28, **29**
Mental retardation, congenital, 481, 482
Mercury, teratogenic effects of, 482
Merocrine glands, 196, **197**
Meromelia, **83**, 189
Mesencephalon, 362, **363, 364, 365**, 386, **387**, 400, **401**

Mesenchymal cells, 84, **85**
Mesenchyme, 84, **85**
   premandibular, **230**
Mesenteric arteries, superior and inferior, 320, **321**
   vein, superior, 330, **331**
Mesentery(ies), 110, **111, 112**, 118, **119, 120, 121, 213, 221**
   common, 118, **119, 120, 121, 221**
   dorsal, 110, **111, 112**, 118, **119, 120, 121, 213, 221**
     in development of stomach, 118–**21, 213, 221**
   embryonic, 110, **111, 112**, 118, **119, 120, 121, 221**
   gonadal, 240, **243**
   primitive, 118, **119, 120, 121, 221**
   proper, 110, **111, 112**, 118, **119, 120, 121, 221**
   urogenital, 240, **243**, 260, **261**
   ventral, 110, **111, 112**, 118, **119, 120, 121, 213, 214, 215**
   Wolffian body, 240, **241, 242, 243**
Mesocardium, 296, **299**, 302, **303**
Mesocolon, 118
Mesoderm, 62, **63**, 478
   derivatives of, 62, **63**, 478
   embryonic, 62, **63**
   extraembryonic, 40, **41**, 44, **45**, 84, **85**
     somatic (somatopleuric), 40, **41**, 50, **51**, 84, **85**, 110, **111**
     splanchnic (splanchnopleuric), 40, **41**, 50, **51**, 84, **85**, 110, **111**
   intermediate, 50, **51**, 238, **239**
   intraembryonic, 44, **45**, 50, **51**
   lateral, 50, **51**, 238
   metanephric mass, 244, **245**
   parachordal, 358, **359**
   paraxial, 50, **51**, 238
   somatic (parietal) layer of, 50
   splanchnic (visceral) layer of, 50
Mesodermal germ layer, 62, **63**, 478
Mesoduodenum, 118, **120, 213**
Mesoesophagus, 122
Mesogastrium, 116, **117**, 118, **119, 120, 121, 211, 212, 213**
Mesonephric body, 260
Mesonephric corpuscle, 240, **241, 243**
Mesonephric duct, 240, **241, 242, 243**, 258, **259**, 260, **261, 262, 263**, 264, **265**, 288
   remnants of, 240, **241, 242**, 276, **277**
Mesonephric nephrotomes, 240, **241**
Mesonephric tubule, 240, **241**, 252, **262**, 264, **288**
   blood vessels of, **241**
Mesonephric vesicles, 240, **241**
Mesonephros, 238, **239**, 240, **241, 242, 243**
   regression of, 240
   relationships of, **243**
Mesorchium, 258
Mesosalpinx, 272, **275**

Mesovarium, 270, **275**
Metamerization, 162, 370, **371**, 372, **373**
Metanephric blastema, 244, **245**
Metanephric diverticulum, 244, **245**
Metanephric mass, of mesoderm, 244, **245**
Metanephric tubule, 244, **245**
Metanephros, 244, **245**, 246, **247**
Metencephalon, 362, **364**, **365**, 386, **387**, 394, **395**, 396, **397**
Microcephaly, 182, 424, **425**
Microglia cells, 378, **379**
Microglossia, 138
Micrognathia, 138, 454
Microgyry, 424
Micropenis, 282, **283**
Microphthalmia, 442
Microstomia, congenital, **145**
Microtia, 454, **455**
Midbrain, 362, **363**, **364**, 365
  flexure of, 362, **363**, **364**, 365
Middle ear, 452, **453**
  cavity, 452, **453**
Midgut, 208, **209**, 218, **219**, 220, **221**
  congenital malformations of, 228–33
  derivatives of, 208, **209**
  fixation and rotation of, 218, **219**, 220, **221**
  formation of, 208, **209**, 218, **219**, 220, **221**
  volvulus, 232, **233**
Midgut loop, 218, **219**
"Midkidney," 240
Migration, ovarian, 272, **275**
  primordial germ cells, 86, 256, **257**, 270, **271**
  testicular, 268, **269**
"Milk lines," 200, **201**
Milk teeth, 202, **203**
Mitotic division, 9, **10**, **15**
Mitral valve, 314, **316**, **317**
Mittelschmerz, 34
Mixed gonadal dysgenesis, 280
Modiolus, 448, **449**
Mohr's syndrome, 144
Mole, hydatidiform, 42, 104, **105**
Mongolism. *See* Down's syndrome
Monosomy, definition of, 18
Monozygotic twins, 74, **76**
Mons pubis, 278, **279**
Monsters, 82, **83**
Morgagni, foramen of, 122
  hydatid of, **277**, **288**
Moro's reflex, 485
Morphogenesis, 58, **59**
Morula, **17**, 36, **37**
  definition of, 36, **37**
Motor neuron, **372**, 378, **379**
  root, anterior, 378, 380
    posterior, 378, 380
Mouth, primitive, 140, **141**
Mucoviscidosis, 232
Müllerian duct, 260, **261**, **262**, **263**, 272, **273**, **275**, **288**
  tubercle, 260, 272, **274**
Multiple pregnancies, 74, **75**
Muscular septal defect, 348, **349**, 350, **351**
Muscular system, 162–65
Muscle(s), arrector pili, 196, **197**
  body groups, **164**
  branchial arch, derivatives of, 162, **163**
  branchiomeric, **165**
  cardiac, 162
  ciliary, 436, **437**
  congenital malformations of, 162
  cricothyroid, 128, **130**
  dilator pupillae, 162, 436, **437**
  facial expression, 128
  of iris, 162
  limb, 162, **163**
  mastication, 128, **130**
  ocular, **130**, 162
  papillary, 314, **316**, **317**
  platysma, 128
  skeletal, 162, **163**, 164, **165**
  smooth, 162
  sphincter pupillae, 162, 436, **437**
  stapedius, 452, **453**
  stylopharyngeal, 128, **130**
  tensor tympani, 452, **453**
  tongue, 162
  trachealis, 150
  visceral, 162
Musculi pectinati, **307**
Myelencephalon, 362, **364**, **365**, 386, **387**, 388–93
  alar sensory and roof plates, 392, **393**
  basal motor plate of, 388–91
Myelin, formation of, 380, **381**
Myelin sheath, 380, **381**
Myelination, 380, **381**
  related to development, 485
Myelocele, 384, **385**
Myelomeningocele, 384
Myeloschisis, **385**
Myoblasts, 162
Myocardium, 296, **299**
Myocele, 50, 162, **163**
Myoepicardial mantle, 296
Myoepithelial cells, 196
Myofibrils, 162
Myometrium, 20, **21**, 272
Myopia, 442
Myotomes, 50, **53**, 162, **163**, **164**, **165**, 370, **372**
  cervical, 162, **163**, **164**
  occipital, 138, **164**
  preotic, **164**
  sacral, **164**
  thoracic, 162, **163**, **164**
Myotomic muscles, 162, **163**, **164**, **165**

Nail(s), 198, **199**
  abnormalities of, 198

Nail field, 198, **199**
Nail folds, 198, **199**
Nail matrix, 198, **199**
Nail plate, 198, **199**
Nasal cavities, 140, **141,** 148, **149, 435**
  conchae, **143,** 148, **149, 435**
  pits, **59, 131,** 140, **141,** 148, **149, 435**
  placodes, 140, **141, 435**
  prominences (swellings), **131,** 140, **141,** 148, **149, 435**
  sacs, **59, 131,** 140, **141,** 148, **149, 435**
  septum, **143**
    agenesis of, 134
Nasolacrimal duct, 140
  groove, **131,** 140, **141**
Nasopharynx, **435**
Natal teeth, 204
Neck, congenital malformations of, 134, **135**
Necturus, brain, **369**
Neocerebellum, 396, **399**
Neocortex, 408, **409,** 414, **415**
Neoencephalon, 408
Neopallium, 408, **409,** 410, **411,** 414, **415,** 416, **417**
Neostriatum, 410, **411**
Nephrocele, **239**
Nephrogenic cord, 238, **239**
Nephrogenic loop, 244, **245**
Nephron, 244, **245,** 246, **247**
Nephrostome, **239**
Nephrotome(s), 238, **239,** 240, **241**
Nerve(s), abducens, 388, **389, 390,** 394, **395,** 402, **404, 405**
  accessory, 388, **389, 390, 391,** 402, **404, 405**
  acoustic, **390,** 402, **404, 405,** 448
  brainstem, 388, **389, 390, 391**
  branchial arch, derivatives of, 128, **129, 130, 131**
  chorda tympani, 138, **405, 431,** 444
  cochlear, **390,** 392, **393**
  cranial, 388, **389, 390, 391,** 402, **404, 405**
  cutaneous, 186, **187,** 360, **363**
  facial, 128, **130, 131,** 138, 394, **395,** 402, **404, 405,** 430, **431, 433**
  glossopharyngeal, 128, **130, 131,** 138, 388, **389, 390, 391,** 402, **404, 405,** 430, **431, 433**
  hypoglossal, 138, 388, **389, 390, 391,** 402, **404, 405**
  inferior alveolar, **130**
  lacrimal, **390**
  laryngeal, relation to aortic arches, 128, **130, 131**
  lingual, **130,** 138
  oculomotor, 388, **389, 390,** 400, **401,** 402, **404, 405,** 430, **431, 433**
  olfactory, 402, **405,** 434, **435**
  optic, 402, **405,** 436–**41**
  pharyngeal arch, 128, **129, 130, 131**
  phrenic, **404**
  pudendal, **432**
  spinal, 402, **404**
  statoacoustic, **390,** 392, **393,** 448
  of tongue, 138, 388, **389, 390, 391**
  trigeminal, 128, **130, 131,** 394, **395,** 402, **404, 405**
    bulbospinal part, 392, **393**
    mandibular branch, 128, **130, 131,** 138, **405,** 444
    maxillary branch, **405**
  trochlear, 388, **389, 390,** 400, **401,** 402, **404, 405**
  vagus, 128, **130, 131,** 388, **389, 390, 391,** 394, **395,** 402, **404, 405,** 430, **431, 433,** 444
    dorsal motor nucleus, 388, **391**
    inferior laryngeal branch, 128, **405**
    recurrent laryngeal branch, 128, **405**
    superior laryngeal branch, 128, **130, 131,** 138, **405**
  vestibular, **390,** 392, **393,** 448
Nerve cells, differentiation of, 378, **379**
Nervus intermedius, **390**
Nervi erigentes, **431, 433**
Nervous system, 358–**467** *See also* Nerve(s); Brain; Spinal cord
  central, 362, **363, 364, 365**
  congenital malformations of, 82, **83,** 384, **385,** 422–27
  early development of, 358, **359,** 360, **361**
  histogenesis of, 416, **417**
  induction of, 358
  metameric organization, 370, **371,** 372, **373**
  parasympathetic, 428, 430, **431, 433**
  peripheral, 402, **403, 404**
  phylogenesis of, 366–**69**
  sympathetic, 428, **429, 432**
Neural arch, 174, **175**
Neural crest, 46, **49,** 360, **361,** 380, **381**
Neural crest cells, 46, 360, **361,** 380, **381,** 402, **403,** 428, **429,** 462, **463**
  derivatives of, 46, **49,** 360, **361,** 380, **381,** 402, **403,** 428, **429,** 462, **463**
  removal of, 380
Neural ectoderm, 358, **359,** 360, **361**
Neural folds, 46, **47, 49,** 64, **65,** 359, 360, **361**
Neural groove 46, **47, 49,** 359, 360, **361**
Neural plate, 46, **47, 49,** 359, 360, **361**
Neural tube, 46, **47, 49,** 360, **361,** 362, **363, 364, 365**
Neurenteric canal, 46, **47, 48**
Neurilemma cells, 380, **381**
  sheath, 380, **381**
Neuroblasts, 376, **377,** 378, **379,** 428, **429**
  organization of, 376, **377,** 378, **379**
Neurocranium, 178, **179, 181**
Neuroectoderm, 46, **47**
  abnormalities of, 194, **195**
Neuroepithelial cells, of spinal cord, 376, **377**
Neuroepithelium, 374, 376, **377**

Neuroglia cells, 378, **379**
Neurohypophyseal bud, **457**, 458, **459**
Neurohypophysis, 458, **459**
Neuron, 374, **375**, 378, **379**
　association, 374, **375**, 378
　motor, 372, 374, **375**, 378, **379**
　sensory, 372, 374, **375**, 378, **379**
Neuropore, anterior, 46, 58, **59, 129**, 360, **361**
　posterior, 46, 58, **59, 129**, 360, **361**
Neurulation, 46, **47**, 62, 358, **359**, 360, **361**
Nevi, pigmented, congenital, 194, **195**
Nidation. *See* Implantation
Nipple, 200, **201**
　abnormalities of, 200, **201**
　inverted, 200
Node(s), atrioventricular, 314, **316**
　lymph, 332, **333, 334**
　primitive, 44, **45**, 48
　sinoatrial, 314, **316**
Nodule(s), of cerebellum, 396, **397, 399**
　lymph, of palatine tonsil, 132, 332, **335**
Nondisjunction, 78
　as cause of trisomy, 78
Nonrotation of midgut, 228
Noradrenaline, 462
Nose, 140, **141**, 148, **149, 435**
　absence of, 148
　congenital malformations of, 148
Notochord, 46, **47, 48**, 358, **359**, 360, **361**
　dorsal, in vertebrates, 366, **369**
　in vertebral development, 68, **69**, 174, **175**
Notochordal canal, 46, **47, 48**
Notochordal plate, 46, **47**
Notochordal process, 44, **45, 359**
Nucleus(i), acoustic, **390**
　ambiguus, 388, **390, 391**
　amygdaloid, 410, **411**
　brainstem, 388, **390, 391**
　of Burdach, 386, **390**
　caudate, 410, **411**
　cerebellar, 386
　cuneatus, **390, 391**
　dentate, 396, **398**
　dorsal motor, of vagus, 388, **391**
　Edinger-Westphal, **390**, 400, **401**
　of Goll, 386, **390**
　gracilis, **390, 391**
　hypoglossal, 388, **390, 391**
　inferior salivatory, 388, **390, 391**
　lateral, of III, **390**
　lenticular, 410, **411**
　mesencephalic of V, **401**
　olivary, 386, **387, 391**
　paraventricelar, 396, **398**, 458, **459**
　pontine, 386, **387**, 394, **395**
　pulposus, 174, **175**
　ruber (red), 386, **387**, 400, **401**
　salivatory, **390**, 394, **395**
　solitary tract, **390, 391**
　of spinal V tract, **390, 391**
　superior salivatory, 394, **395**
　supraoptic hypothalamic, 406, 458, **459**
　thalamic, 406
　vestibular, **390**
Nutrients, placental transfer of, 100, **101**
　role of yolk sac in, 86

Oblique vein, of left atrium (of Marshall), 308, **309**
Occipital bone, malformations of, 422
　sclerotomes, 168, 178, **181**
　somites, **130**, 138
Ocular muscles, 162
Oculomotor nerve, 388, **389, 390**, 400, **401**, 430, **431, 433**
Odontoblast(s), 202, **203**
Odontoblastic process, 202, **203**
Olfaction, 434, **435**
Olfatory bulbs, 365, 414, **415**, 434, **435**
　cells, 434, **435**
　nerves, 402, **405**, 434, **435**
　region, 148, 434, **435**
　tract, **365**, 414, **415**
Oligodactyly, **189**
Oligodendroblast, 378, **379**
Oligodendrocyte, **379**
Oligodendroglia, 378, **379**
Oligohydramnios, 86, 244
Olivary nucleus, 386, **387**
Omental bursa, 116, **117, 211**, 212, **213**
Omentum, greater, 116, **117**, 118, **119, 120, 121**, 212, **213**
　lesser, 116, **117**, 118, **119, 120, 121**, 212, **213**, 214, **215**
Omphalocele, 228, **230**
Omphalomesenteric (vitelline) duct, 232
　system, 326, **327**
　vein, 308, **309**, 310, 326, **327**, 330, **331**
Oocyte(s), 14, **15**
　abnormal, 42
　comparison of sperm and, 14
　definition of, 14
　female pronucleus of, 16, 32, **33**
　fertilization of, 32, **33**
　maturation of, 14, **15**
　normal, 14, **15**
　parts of, 14, **15**
　primary stage, 14, **15**
　secondary, 14, **15**
　transport of, 30, **31**
　viability of, 14
Oogenesis, 14, **15**
　abnormal, 18, **19**
　comparison of spermatogenesis and, 14
Oogonium(a), 14, **15**, 270, **271**
Optic chiasma, 418, **419**
Optic cup, 436, **437, 441**
Optic disk, **438**

Optic fissures, 436, **437**, 440, **441**
Optic nerves, 402, **405**, 436, **437**, **438**, **439**, 440, **441**
Optic stalks, 436, **437**
Optic vesicles, 436, **437**
Ora serrata, **438**
Organ of Corti, 450, **451**
Organogenetic period, of embryonic development, 58, **59**
Organs, genital, external, 266, **267**
Organs, of Zuckerkandl, 462
Oronasal fistula, **135**
Oronasal membrane, 148, **149**
Oropharyngeal membrane, 44, **45**, **48**, 62, **133**, **142**, 208, **209**, 358
  formation of, 44, **45**, **48**, 208, **209**, 358
Osseous labyrinth, of inner ear, 448, **449**
Osteoblasts, 44, 168, **169**
Osteoclasts, 168
Osteocytes, 168
Osteon, **171**
Ostium
  primum, 304, **305**, 306
    defect, 344, **346**, **347**
  secundum, 304, **305**, 306
    defect, 344, **346**, **347**
Otic capsule, **445**, 448, **449**
Otic pit, 58, **59**, 447
Otic (auditory) placode, 446, **447**
Otic vesicle, **445**, 446, **447**
Otitis media, 452
Otocyst, **447**
Otoliths (otoconia), 450, **451**
Otolithic membrane, 450, **451**
Oval foramen, 304, **306**
Ovarian cortex, 270, **271**
  cycle, 14, **15**, 24, **25**, 26, **27**
  follicular development in, 14, **15**
  medulla, 270, **271**
  teratoma, 30
Ovarian differentiation, 270, **271**
Ovarian follicle, 14, **15**, 24, **25**, 270, **271**, 288
Ovarian ligament, **23**, 276, **277**, 288
Ovarian pregnancy, 30, **31**
Ovarian stroma, 270
Ovary(ies), **25**, **288**
  cyclic changes in, 24, **25**
  descent of, 272, **275**
  endocrine activity of, 34, **35**
  sections of, 270, **271**
  stigma of, 30
Oviduct, 22, **23**
Ovulation, 4, **25**, 26, **27**, 36, **37**
Ovum. *See* Oocyte
Oxycephaly, 182, **183**
Oxytocin, **459**, 460, **461**

Pacchionian bodies, 420, **421**
Pacemaker, primitive, 314, **316**

Pachytene stage, meiosis, 10
Palate, 142, **143**
  cleft, 144, **145**
    cleft lip and, differentiation of, 144, **145**
  congenital malformations of, 144, **145**
  hard, 142, **143**
  primary, 140, 142, **143**
  secondary, 142, **143**
  soft, 142, **143**
Palatine process, 142, **143**, 148
Palatine raphé, 142, **143**
Palatine shelves, 142, **143**
Palatine tonsils, 132, 332, **335**
Paleocerebellum, 396, **399**
Paleocortex, 408, **409**, 416
Paleopallium, 408, **409**, 410, **411**, 414, **415**, 416, **417**
Paleostriatum, 410, **411**
Pallidum, 406
Pallium, 408, **409**, 410, **411**, 416
Pancreas, 216, **217**
  accessory tissue, 224
  acini of, 216, **217**
  annular, 224, **227**
  congenital malformations of, 224
  histogenesis of, 216
Pancreatic bladder, 224
Pancreatic buds, 216, **217**, **227**
Pancreatic ducts, 216, **217**, **227**
Pancreatic tissue, heterotopic, **227**
Papilla(e), dental, 202, **203**
  dermal, 192, **193**, 196, **197**
  of hair, 196, **197**
  of tongue, 138, **139**
Papillary ducts, renal, 244, **245**
Papillary muscles, 314, **316**, **317**
Paradidymis, **262**, **263**, 264, **288**
Parafollicular cells, 132
Paraganglia, 462, **463**
Paraganglioma. *See* Pheochromocytoma
Paraganglionic system, 462, **463**
Parahippocampal gyrus, **415**
Paramecium, nervous system, 367
Paramesonephric ducts, 260, **261**, 264, **265**, **288**
  remnants of, 264, **265**
Parametrium, 272
Paranasal, air sinuses, 148
Paraphysis, 406
Parasympathetic nervous system, 428, 430, **431**, **433**
Parathyroid glands, 132, **133**
Paraurethral glands, of Skene, 248
Paraventricular nuclei, 396, **398**, 458, **459**
Paraxial mesoderm, 50, **51**
Parietal pleura, 154, **155**
Paroöphoron, 276, **277**, **288**
Parotid gland, 138
  innervation of, 388, **390**, **391**
Paroxysmal hypertension, 462
Pars cystica, liver bud, 214, **215**

Pars distalis, 456, **457**
Pars hepatica, liver bud, 214, **215**
Pars intermedia, 456, **457**
Pars nervosa, **457**, 458, **459**
Pars tuberalis, 456, **457**
Parthenogenesis, 30
Parturition, 90, **91**
Patent ductus arteriosus, 340, **341**
Patent foramen ovale, 344, **345**, **346**
Patent foramen primum, 344, **346**, **347**
Pattern(s), dermatomal, 370, **372**, **373**
Pecten, **223**
Pectinate (dentate) line, 222, **223**
Pectoral girdle, 178, **179**, **180**
Peduncle(s), cerebellar, 394, **395**
  cerebral, 387, 400, **401**
Pelvic girdle, 178, **179**, **180**
Pelvis, renal, 244, **245**
  duplication of, 250, **251**
Penicillin, effect on human embryo of, 482
Penile hypospadias, 282, **283**
Penile urethra, 4, **5**, 266, **267**
Penis, 4, **5**, 266, **267**
  agenesis of, 282, **283**
  congenital malformations of, 282, **283**, **285**
Penoscrotal hypospadias, 282, **283**
Pericardial cavity, 110, 114, **115**, **298**, **299**, 302, **303**
Pericardial defects, congenital, 114, **345**
Pericardioperitoneal canal, 110, **111**, **112**, **113**, 114, **115**
Pericardium, 296, **299**
  visceral, 296, **299**
Periderm, 192, **193**
Perilymph, 446, 448
Perilymphatic space, 448, **449**
Perimetrium, 20, **21**
Perineal body, 222, **223**, 278, **279**
Perineal hypospadias, 282
Perineum, 222, **223**, 278, **279**
Periodontal ligament, 202, **203**
Periosteal bone, 168, **170**
  ossification of, 168, **170**
Periosteum, 168, **170**
Periotic capsule, 445
Peripheral nervous system, 402, **403**, **404**
Peritoneal cavity, 110, **303**
Peritoneal sac, greater, 115, **119**, **120**, **121**
  lesser, 116, **117**
Peritoneum, visceral, of liver, 214, **215**
Persistent atrioventricular canal, 348
  iridopupillary membrane, 440
  truncus arteriosus, 348, **350**
Pflüger's tubes. See Cords, cortical
Phalanges, 178, **179**, **180**, 184, **185**
  congenital malformations of, 188, **189**
Phallus, 266, **267**, **288**
  derivatives of, 266, **267**, **288**
Pharyngeal arches, 128, **129**, **130**, **131**
  clefts, 128, **129**, **130**, **131**, 132, **133**

grooves, 128, **129**, **130**, **131**, 444, **445**
Pharyngeal hypophysis, 456
Pharyngeal pouches, 132, **133**, **151**, 452
  derivatives of, 132, **133**
Pharyngeal tonsil, 332, **335**, 452
Pharyngohypophyseal stalk, 456, **457**
Pharyngotympanic tonsil, 332, **335**
Pharyngotympanic tube, 132, **133**, **445**, 452, **453**
Pharynx, 210, **211**
  primitive, **137**, 210, **211**
Pheochromocytoma, 462
Philtrum, of upper lip, 140, **141**
Phocomelia, 82, **83**, **189**
Phrenic nerve, 122
Phrygian cap, **226**
Phylogenesis of nervous system, 366–**69**
  of telencephalon, 408, **409**
Physiologic development of CNS, Appendix V, 483–**85**
Pia-arachnoid layer of spinal cord, 374, **375**, 382, **383**
Pia mater, 382, **383**, 420, **421**
Pierre Robin syndrome, 134
Pigment, bile, 214
  retinal, 436
Pigment cells, 360
Pineal gland (body), 406, **407**, 460
Pinna, 444, **445**
Piriform area, **411**
Pit(s), anal, 208, 222, **223**
  auricular, congenital, 134
  nasal, **59**, **131**, 140, **141**, 148, **149**
  olfactory, 434, **435**
  otic, 58, **59**, **447**
  primitive, 44, **45**
Pituicytes, 458
Pituitary gland, 456–**61**
Placenta, 94, **95**
  activities of, 100, **101**
  appearance of, 94, **95**
  "barrier" of, 100, **101**
  chorioallantoic, 96
  circulation via, 96, **97**, 102, **103**
    fetal, 96, **97**, 98, **99**, 102, **103**
    maternal, 96, **97**, 98, **99**, 102, **103**
  decidual formation, 98, **99**
  endocrine secretion by, 104
  endotheliochorial, 100, **101**
  epitheliochorial, 100, **101**
  estrogen secretion by, 104
  fetal component of, **97**, **99**
  full-term, 94, **95**, 100, **101**
  hemochorial, 96, 100, **101**
  intervillous spaces of, 96, **97**, **101**, 102, **103**
  maternal component of, 98, **99**
  metabolism of, 100, **101**
  morphology, 94, **95**, **97**, **99**, **101**, **103**
  permeability, 100, **101**
  physiology, 100, **101**
  structure of, 96, **97**, 98, **99**

syndesmochorial, 100, **101**
tissues of, **99**
villi, 96, **97**
villous type, 96
Placenta previa, 30
Placental membrane (barrier), 100, **101**
Placental septum, 96, **97**
Placental stage, of circulatory system, 292, **293**
Placental surface, **95**, 100, **101**
Placental transfer, 100, **101**
 auditory. *See* Placode, otic
Placode(s), lens. *See* Placode, optic
 nasal, 140, **141**
 olfactory, 434, **435**
 optic, 58, **59**, 436, **437**, **441**
 otic (auditory), 446, **447**
Plagiocephaly, 182, **183**
Planaria, nervous system, **367**
Plane(s), anatomic, 2, **3**
Plate(s)
 alar, 374, **375**, 376, **377**
  of diencephalon, 406, **407**
  of mesencephalon, 400, **401**
  of metencephalon, 394, **395**
  of myelencephalon, 92, **93**, 392, **393**
 basal, of mesencephalon, 400, **401**
  of metencephalon, 394, **395**
  of myelencephalon, 388–**91**
  of neural tube, 374, **375**
  of placenta, 98, **99**
  of spinal cord, 376, **377**
 cerebellar, 396, **397**
 chorionic (placenta), 40, **41**
 decidual, 96, **97**, 98, **99**
 floor, 376, **377**, **379**
  of mesencephalon, 400, **401**
  of metencephalon, 394, **395**
  of myelencephalon, 388–**91**
 hand, 58, **59**
 intermediate, 238, **239**
 lateral, 50, **51**
 mesodermal, 238, **239**
 motor, 376, **377**
 nail, 198, **199**
 neural, 46, **47**, **48**, **49**, 359, 360, **361**, 374, **375**
 notochordal, 46, **47**, **359**
 preputial epithelial, 266, **267**
 prochordal, 40, **41**, 44, **45**
 roof, 376, **377**, **379**
  of diencephalon, 406, **407**
  of mesencephalon, 400, **401**
  of metencephalon, 396, **397**
  of myelencephalon, 392, **393**
 sensory, 376, **377**
 of spinal cord, 374, **375**, 376, **377**
 urethral, 266, **267**
 vaginal, 272, **274**
Pleura, parietal, 154, **155**
 visceral, 154, **155**
Pleural canals, 122, **123**

Pleural cavities, 110, 114, **115**, 152, 154, **155**, **303**
Pleuropericardial membranes, 110, 114, **115**, 302, **303**
 congenital defects of, 114
Pleuroperitoneal membranes, 114, **115**, 122, **123**, **124**, **303**
Plexus, Auerbach's, 428, **432**
 brachial, 382, **383**, **404**
 choroid, 392, **393**, 406, **407**, 410, **411**, 420, **421**
 lumbar, 382, **383**, **404**
 Meissner's, 428, **432**
 sacral, 382, **383**, **404**
 solar, 428, **432**
Plug(s), closing (coagulum), 38, **39**
 meatal, 444, **445**
Polar body(ies), 14, **15**
Polycystic kidney, bilateral, congenital, 250, **251**
Polydactyly, 188, **189**
Polygyny, 32
Polyhydramnios, 86, 224, 228
Polymastia, 200, **201**
Polythelia, 200, **201**
Pons, 394, **395**
Pontine flexure, 362, **363**, **364**, 365
 nuclei, 394, **395**
Pore of Kohn, **155**
Porencephaly, 424, **425**
Portal system, hypophyseal, 458, **459**
Portal vein, 308, 330, **331**
"Port-wine stain," 194, **195**
Postductal coarctations, of aorta, 340, **341**
Posterior cardinal vein, 308, **309**, 318, **319**, 326, **327**
 commissure, 418, **419**
 neuropore, 46, 58, **59**, 129, 360, **361**
 root neurons, 372, 374, **375**, 378, **379**
Postganglionic fibers, 428, **429**, **432**, **433**
Postnatal brain and nervous system development, 484
Postoral branchial arches, **129**
Potassium iodide, teratogenic effects of, 482
Pouch(es)
 branchial (pharyngeal), of Sessel, **129**, 452
  derivatives of, 132, **133**
 Hartmann's, **226**
 pharyngeal, 132, **133**, 151
 Rathke's, 456, **457**
 Sessel's, **129**, 456, **457**
 uterorectal (of Douglas), 272, **273**, **274**, **275**
 uterovesical, 272, **273**, **274**, **275**
Preaortic ganglia, 428, **429**, **432**
Precartilage, 168, **169**
Predentin, 202, **203**
Preductal coarctation, of aorta, 340, **341**
Preganglionic fibers, 428, **429**, **432**, **433**
Pregnancy(ies), cessation of menstrual period during, 28
 corpus luteum, 26, **27**, **29**

Pregnancy(ies) (Cont.)
  extrauterine (ectopic) sites of, 30, **31**
    abdominal, 30, **31**
    ovarian, 30, **31**
    tubal, 30, **31**
  growth of uterus during, 90, **91**
  intrauterine sites of, 30, **31**
    cervical, 30, **31**
  menstrual cycle and, 28, **29**
  multiple, 74, **75, 76**
    fetal growth and, 74, **75, 76**
  test, biologic, 104
    immunologic, 104, **105**
Prematurity, Appendix VI, 486
  complications of, 486
Pregnandiol, **105**
Premenstrual phase, of menstrual cycle, 28, **29**
Prepuce, penile, 266, **267**
Preputial epithelial plate, 266, **267**
Primary, follicle, 14, **15**, 270
  oocyte, 14, **15**, 270
  palate, 142, **143**
  sex cells, 256, **257**
  sex cords, 258, **259**, 270, **271**
  spermatocyte, 8, **9, 11**
Primary ossification centers, **179**, 180
Primary villi, 40, **41**
Primitive atrium, 300, **301**, 304, **305**
  partitioning of, 304, **305, 306**
Primitive cardiovascular system, 294, **295**
Primitive choanae, 148, **149**
Primitive circulatory network, 318, **319**
Primitive endothelial cardiac tubes, 54
Primitive genital system, 256, **257**, 260, **261**
Primitive groove, 44, **45, 48**
Primitive gut, 62, 86, **87**, 208, **209**
Primitive heart, circulation through, 294, 300, 302
Primitive knot or node, 44, **45**, 48, **359**
Primitive meninx, 420
Primitive mouth, 140, **141**
Primitive pit, 44, **45**
Primitive plasma, 54, **55**
Primitive primary bronchus, 152, **153**
Primitive streak, 44, **45, 359**
Primitive vascular system, 292, **293**, 294, **295, 297**, 318, **319**, 326, **327**
Primitive ventricle, partitioning of, 304, **305, 307**
Primitive yolk sac, 38, **39**
Primordial follicle, 14, **15**, 24, **25**, 270, **271**
  germ cells, migration of, 86, 256, **257**, 270, **271**
Primordium, gonadal, 256, **257**
  of liver, 292
  pancreatic, 216, **217**
  uterovaginal, 272–75
Primum type of atrial septal defect, 344, **346, 347**
Process(es), mastoid, 452
  maxillary, **129, 131**
  notochordal, 46, **47**
  odontoblastic, 202, **203**
  palatine, 142, **143**, 148
  spinous, 174, **176**
  styloid, 128, **130**
  transverse, 174, **176**
  uncinate, 216, **217**
Processus vaginalis, 268, **269**
Prochordal plate, 40, **41**, 44, **45**, 358
  embryonic, 40, **41**, 44, **45**, 358
Prochordates, nervous system, 366, **369**
Proctodeum, 208, **209**, 222
Progestational phase, of menstrual cycle, 28, **29**
Progesterone, in ovarian cycle, 26, **27**
  placental secretion of, 104, **105**
Prolactin, lactogenic, 104, **459**, 460, **461**. See also Hormone
Proliferative phase, of menstrual cycle, 28, **29**
Prominence(s), frontonasal, 140, **141**
  mandibular, 140, **141**
  maxillary, 140, **141**
Pronephric ducts, 238, **239**
Pronephros, 238, **239**
Pronucleus, female, of oocyte, **16**, 32, **33**
  male, of sperm, **16**, 32, **33**
Prosencephalon, 362, **363, 364**
Prostate gland, 4, **5**, 248, **249**, 264, **265**, 288
Prostatic urethra, **262**, 264, **265**
Prostatic utricle, **262**, 264, **265**, 288
Protein hormones, placental secretions of, 104, **105**
Proteus, amoeba, nervous system, 366, **367**
Proximal convoluted tubule, 244, **245**
Pseudoglandular period, of lung development, 154, **155**
Pseudohermaphroditism, 466
  female, 466
  male, 466
Pudendal nerve, **432**
Pulmonary arch, 324, **325**
  arteries, **313, 317**, 330, **331**
Pulmonary atresia, 158, **159**, 352, **353**
Pulmonary lining epithelium, 154, **155**, 156, **157**
Pulmonary stenosis, 158, **159**
Pulmonary surfactant, 154, 156, **157**
Pulmonary trunk, 312, **313**, 324, **325**
Pulmonary valve, 314, **315, 316**
  stenosis of, 348, **351**
Pulmonary veins, 152, 330, **331**
Pulmonary venous connections, anomalous, 340, **343**
Pulmonary vessels, 152, **153**
Pulp, dentin, 202, **203**, 204, **205**
Pupil, 436, **437, 438, 441**
Pupillary membrane, 440, **441**
  persistent, 442, **443**
Purkinje fibers, 162, 396, **398**
Putamen, **407**, 410, **411**
Pygopagus, 74, **77**

Pyloric stenosis, 224, **225**
Pyramidal lobe, of thyroid gland, 136, **137**
Pyramids, of medulla, 388, **389, 391**

Quadrigemina, corpora. *See* Colliculi
Quadruplets, 74
Quintuplets, 74

Rachischisis, 384, **385**
Radiation, teratogenic effects of, 481
Radioactive iodine (^{131}I), teratogenic effects of, 482
Ramus, postganglionic, 428, **429, 432, 433**
  preganglionic, 428, **429, 432, 433**
Ranvier node, **381**
Raphé, anogenital, **279**
  scrotal, 266, **267**
Rathke's pouch, 456, **457**
  tumor of, 460
Recess(es), costodiaphragmatic, 122, **125**
  inferior, of lesser peritoneal sac, 212, **213**
  superior, of lesser peritoneal sac, 212, **213**
  tubotympanic, 132, 444, **445**, 452, **453**
Rectal atresia, 234
  columns, **223**
  fistula, 252, **253**
  valves, **223**
Rectum, 222, **223**
  malformations of, 234
Recurrent laryngeal nerve, 128, **405**
Reduction division, 9, 10, **15**
References, Appendix VII, 487–89
Reflexes, normal, 485
Regression, embryonic development, 68
Regulation, embryonic development, 68
Reichert's cartilage, 128, 178, **180**, 452, **453**
Reissner's membrane, **449**, 450
Renal agenesis, 250, **251**
Renal aplasia, 250, **251**
Renal calyces, 244, **245**
Renal corpuscle, 244, **245**
Renal duplications, 250, **251**
Renal ectopia, 252, **253**
Renal hypoplasia, 250, **251**
Renal lobe, 244
Renal pelvis, 244, **245**
  duplications of, 250, **251**
Renal vein, 246, **247**, 328, **329**
Renal vesicle, 244, **245**
Renal vessels, multiple, 252, **253**
Reproductive cycles, 24, 26, **27**, 28, **29**
Reproductive system, 256–**89**
  congenital malformations of, 280–**89**
  female, 20, **21, 22, 23**, 285, 286, **287**, 289
  male, 4, **5**, 282–**85**, 289
Reptile brain, **369**

Respiration, initiation of, 154
Respiratory bronchiole, 154, **155**
Respiratory distress syndrome, 156
Respiratory diverticulum, 150, **151**, 152, **153**
Respiratory membrane, 154, 156, **157**
Respiratory movements, 156, **157**
Respiratory primordium, **153**, 154, **155**
Respiratory system, 148–**59**
  congenital malformations of, 158, **159**
Restiform body, **391**
Rete ovarii, 270, **271, 288**
Rete testis, 4, **5**, 258, **259, 262**
Reticular formation, 386
Reticulum, enamel, 202, **203**
Retina, 436, **437, 438, 439**
  arteries of, 440, **441**
  pars caeca, 436, **437, 438**
  pars ciliaris, 436, **437, 438**
  pars iridica, 436, **437, 438**
  pars optica, 436, **437, 438**
Retinal pigment, 436
Retinocele, 442, **443**
Retrolental fibroplasia, 442
Retroscrotal penis, 282, **283**
Reversed rotation, intestinal loop, 228, **230**
Rh antibody, 92, **93**
  antigen, 92, **93**
  immunization, 92, **93**
Rhinencephalon, **365**, 408, 414, **415**
Rhinodymy, 134
Rhombencephalic isthmus, **365**
Rhombencephalon, 362, **363, 364**
Rhombic lip, 394, **395**, 396, **397**
Ribs, 174, **177**
  congenital malformations of, 182, **183**
Ridge(s), gonadal (genital), **242**, 256, **257**
  mesonephric, **242**
  urogenital, 238, **239, 242**
Right aortic arch, **342**
Ring(s), inguinal, 268, **269**
  tympanic, 444
Rods, of retina, 436, **439**
Roof plate, 376, **377, 379**
  of diencephalon, 406, **407**
  of mesencephalon, 400, **401**
  of metencephalon, 396, **397**
  of myelencephalon, 392, **393**
Root(s), dental, 202, **203, 205**
  of spinal nerves, **375**, 378, 380, **381**
  of tongue, 138, **139**
Root sheath, 196, **197**
Rosenmüller's body, 270, 276, **277**
Rosette, paraependymal, 424
Rotation
  of midgut, 218, **219**, 220, **221**
    reversed (mixed), 228, **230**
  of stomach, 210, **211**
Round ligament, of liver, 118, **119, 120, 121**
  of uterus, **23**, 272, **273, 275**
Rubella virus, teratogenic effects of, 481

Sac(s), amniotic, 86, **87**
  conjunctival, 437, **438**
  dental, 202, **203**
  endolymphatic, 446, **447**
  lacrimal, 140
  lymph, 332, **333**
  nasal, 148, **149**, 435
  peritoneal, greater, **119, 120, 121**
    lesser, 116, **117**
  terminal, of respiratory bronchiole, 152, **153**, 154, **155**
  yolk, 40, **41, 45, 52**, 86, **87**
    fate of, 86, **87**
    primitive (primary), 38, **39**, 86, **87**
    role of, in blood formation, 86, **87**
    secondary, 40, **41, 45, 52**, 86, **87**
    significance of, 86
Saccular portion, of otocyst, 446, **447**
Saccule, of inner ear, 446, **447**
Salivary glands, 138
Salivatory nucleus, 388, **390, 391**
Santorini, duct of, 216, **217**
Satellite cells, 402, **403**
Scala(e), tympani and vestibuli, 448, **449**, 451
Scaphocephaly, 182, **183**
Schwann cells, 360, 380, **381**
Sclera, 438, 440, **441**
Sclerotome, 50, **53**, 162, **163**, 168, **169**
Scoliosis, **183**
Scrotal septum, 266, **267**
  swelling, 266, **267**
Scrotum, 266, **267**, 268, **269**, 288
Sebaceous glands, 196, **197**
Sebum, 196
Secretory phase, of menstrual cycle, 28, **29**
Sections, anatomic, of embryo, **66, 67**
Secundum type of atrial septal defect, **346, 347**
Segment(s), bronchopulmonary, 152, **153**
Segmentation. *See* Metamerization
Sella turcica, 456, **457**
Semen, 4
Semicircular canals, 446, **447**
Semicircular ducts, 448, **449**
Semilunar valves, 314, **315, 316**
Seminal colliculus, **5**, 264, **265**
Seminal gland (vesicle), 264, **265**
Seminal vesicle, 4, **5**
Seminiferous cord, 258, **259**
Seminiferous tubules, 4, **5, 6**, 258, **259**, 288
Sense organs, 434–53
  congenital malformations of, 434, 442, **443**, 454, **455**
Septation, atrioventricular, 304, **305, 307**
  interatrial, 304, **305, 306**
  ventricular, 312, **313**
Septum(a), aorticopulmonary, 312, **313**
  decidual, 98, **99**, 102, **103**
  esophagotracheal, **151**
  interventricular, 304, **305, 307**
    absence of, 348, **349, 350**
  median, of tongue, 138, **139**
  membranous, defect of, 348, **349**
  nasal, **143**
  pellucidum, 418, **419**
  placental, 98, **101, 103**
  primum, 304, **305, 306**
  secundum, 304, **305, 306**
  spurium, **305**, 310, **311**
  tracheoesophageal, 150, **151**
  transversum, 114, **115**, 122, **123, 124**
  urorectal, 222, **223**, 248
Septum primum, 304, **305, 306**
  defects of, 344, **346, 347**
Septum secundum, 304, **305, 306**
  defects of, 344, **346, 347**
Septum transversum, 122, **123, 124, 125**
  in diaphragmatic development, 122, **123, 124, 125**
Sertoli, sustentacular cells of, **13**, 258
Sessel's pouch, **129**, 456, **457**
Sex, determination of, 260
Sex cells, 8, 256, **257**
Sex chromatin pattern, in differential diagnosis of ambiguous external genitalia, **289**
Sex chromosome(s), 78, **79, 80, 81**
  complex of, abnormal, 78, **79, 80, 81**
  nondisjunction of, 78
  in normal oocyte and sperm, 9, 14, **15**
  trisomy of, 78, **81**
Sex cords, 256, **257**, 258, **259**, 270, **271**
Sexual anomalies, of genetic and hormonal origin, 280, **281**
Shaft(s), bone, **170**
Sheath(s), myelin, 380, **381**
  root, 196, **197**
Shelves, palatine, 142, **143**
Shoulder (pectoral) girdle, 178, **179**, 180
Shunt(s), venous, 330, **331**
Siamese twins. *See* Twins, siamese
Sigmoid colon, 222
Sinoatrial folds, 308
Sinoatrial node, 314, **316**
Sinoatrial orifices, **309, 311**
Sinoatrial valve(s), 302, 310, **311**
Sinovaginal bulbs, 272, **274**
Sinus(es), auricular, 454, **455**
  branchial, congenital, 132
  cervical, 58, **59**, 132, **133**
    lateral, congenital, 134
  coronary, **307**, 308, **309**
  dermal, 384, **385**
  ectopic, 234
  horns, of heart, 300, **301**
  Müllerian, 260
  paranasal, air, 148
  subcardinal, 328, **329**
  superior sagittal, **421**
  thyroglossal duct, 136
  transverse, of pericardial cavity, 300
  urachal, 252

urogenital, 222, **223,** 248, **249,** 264, **265**
   division of, 248, **249,** 264, **265**
   venosus, 300, **301, 311**
Sinus venarum, 310, **311**
Sinus venosus, 300, **301,** 304, 308, **309,** 326, **327**
   changes in, 308, **309**
   fate of, 302, **303**
   horns of, 300, **301,** 302, **303**
Sinus venosus type atrial septal defect, 344, **347**
Sinusoid(s), embryonic, 40, **41, 43**
   in fetal circulation, 330, **331**
   hepatic, 330, **331**
Sirenomelia, **83,** 188
Situs inversus, 344, **345**
Skeletal musculature, 162, **163**
Skeletal system, 162–65. *See also* Bone(s)
   congenital malformations of, 172, 182, **183**
Skeleton, appendicular, 178, **179,** 180
   axial, 174, **175**
   cardiac, 314, **316**
Skene, paraurethral glands of, 248, 272, **288**
Skin, 192, **193**
   abnormalities of, 192, **193**
   fetal, 192, **193**
   histogenesis of, **193**
Skull, base of, 178, **179,** 181
   congenital malformations of, 182, **183**
   fetal, 178, **181**
   newborn, 178, **181**
   postnatal growth of, 178, **181**
   size of, 178
Small intestines, 218, **219**
   anomalies of, 228–33
Smooth chorion, 84, **85, 87,** 94, **95**
Smooth muscle, 162
Solitary tract, nucleus of, **390**
Somatic mesoderm layer, 50
Somatomammotropin, human chorionic hormone, 104
   placental secretion of, 104
Somatopleure, extraembryonic, 84, **85**
   intraembryonic, 50
Somite(s), 50, **51, 53,** 362, **363,** 370, **371**
   differentiation of, 50, **51, 53,** 362, **363,** 370, **371**
   embryonic development of, 50, **51, 53,** 362, **363,** 370, **371**
   occipital, **130,** 138
Space(s), coelomic, 50, **51**
   intervillous, of placenta, 96, **97**
   intraretinal, 436, **437**
   perilymphatic, 448, **449**
   subarachnoid, 382, **383,** 420, **421**
   subdural, **421**
   vestibular, 448, **449**
Sperm(s), 7, 8, **9, 11**
   abnormal, 18, **19**
   acrosome reaction of, 32, **33**
   comparison of oocyte and, 18
   definition of, 8

electron micrographs of, **7, 12**
male pronucleus of, 32, **33**
mature, **7**
normal, **11**
number of, 8
penetration of oocyte by, 32, **33**
transport of, 30, **31**
viability of, 8, 30
Spermatic cord cysts, 268, **269**
Spermatic fascia, **269**
Spermatid, 8, **9, 13**
Spermatocytes, primary and secondary, 8, **9, 12**
Spermatogenesis, 8, **9**
   abnormal, 18, **19**
   comparison of oogenesis and, 14
Spermatogonia, **6,** 8, **9,** 258
Spermatozoon. *See* Sperm(s)
Spermiogenesis, 8, **9, 11**
Sphenoid, 456, **457**
Sphenomandibular ligament, **130**
Sphincter pupillae muscles, 436, **437**
Spina bifida, 384, **385**
   meningocele and, 384, **385**
   meningomyelocele and, 384, **385**
   occulta, 182, **183,** 384, **385**
Spinal cord, 374, **375,** 376, **377,** 382, **383**
   congenital malformations of, 384, **385**
   embryonic, 374, **375,** 376, **377**
   meninges of, 374, **375,** 382, **383**
   positional changes of, 382, **383**
Spinal dermal sinus, 384, **385**
Spinal nerves, **175,** 370, **372, 375, 381**
   roots, 370, **372, 375, 377, 381**
Spiral ganglion, 450
   lamina (plate), 448, **449**
   ligament, 448, **449,** 450, **451**
   limbus, 450, **451**
Splanchnopleure, extraembryonic, 84, **85**
   intraembryonic, 50
Spleen, 216, **217,** 332, **335**
Spongioblast, 378, **379**
Stage(s), of tooth development, 202, **203,** 204, **205**
Stalk(s), connecting, 40, **41, 48,** 88, **89**
   effect of tailfold on, 64, **65,** 66, **67**
   optic, 436, **437**
   pharyngohypophyseal, 456, **457**
   pituitary, 456, **457**
   yolk, 88, **89**
     remnant of, 88, **89**
Stapedial arteries, 324
Stapedial muscle, 452, **453**
Stapes, 128, **130, 445,** 452, **453**
   fixation of, congenital, 454
Statoacoustic ganglion, 446, **447**
   nerves, **390**
Stellate cell, 396, **398**
Stem, of hypophyseal infundibulum, 406, **407,** 456, **457,** 458
Stem(s) villi, 40, **41,** 56, **57,** 96, **97**

Stem(s) villi (*Cont.*)
   primary, 40, **41,** 56, **57,** 96, **97**
   secondary, 56, **57,** 96, **97**
   tertiary, 56, **57,** 96, **97**
Stenosis, anal, **235**
   aortic valvular, 352, **354**
   duodenal, 228, **229**
   esophageal, 224, **225**
   intestinal, 228, **229, 231**
   pulmonary, 348, **351**
   pyloric, 224, **225**
   subaortic, 352
   tracheal, 158, **159**
Sternal bands (bars), 174
Sternum, 174, **176**
   cleft, **176,** 182, **183**
Stigma, ovarian, 34
Stomach, 210, **211**
   rotation of, 210, **211**
Stomodeum, 140, **141,** 208, **209**
Stomodymy, 134
Stratum, corneum, 192, **193**
   germinativum, 192, **193**
   granulosum, 192, **193**
   spinosum, 192, **193**
Streak, primitive, 44, **45,** 359
   effect of tailfold on, 64, **65, 66, 67**
   fate of, 44, **45**
Streptomycin, possible teratogenic effects of, 482
Striated bodies, 410, **411**
Striatum, 408, **409,** 410, **411**
Stylohyoid ligament, 128, **453**
Styloid process, 128, **130, 453**
Stylopharyngeal muscle, 128, **130**
Subarachnoid space, 382, **383,** 420, **421**
Subclavian artery(ies), **321, 322,** 324, **325**
   right, abnormal origin of, 340, **341**
Subcorium, 192, **193**
Sublingual glands, 138
Submandibular glands, 138
Substantia nigra, 386, 400, **401**
Sulcus(i), hypothalamic, 406, **407**
   median, of tongue, 138, **139**
Sulcus limitans, 376, **377,** 388, **389, 391**
Sulcus terminalis, 138, **139,** 310
Superfecundation, 30, 74
Superfetation, 30, 74
Superior vena cava, 308, **309,** 326, **327**
   abnormalities of, 326, 340, **343**
Supernumerary, digits, 188, **189**
   kidneys, 250
   nipples, 200, **201**
Supraoptic nucleus, 458, **459**
Suprarenal glands. *See* Adrenal glands
Suprasegmental brain structures, 386, **387**
Supratonsillar fossa, **133**
Surface(s), fetal, of full-term placenta, 94, **95**
   maternal, of full-term placenta, 94, **95**
Surface ectoderm, 192, **193,** 358

Surface epithelium, gonadal, **25**
Surfactant, pulmonary, 154, 156, **157**
Suspensory ligament, **23,** 436, **437, 438**
Sustentacular cells, of Sertoli, 258
Suture(s), cranial, 178, **181**
   premature closing of, 182
Sweat glands, 196, **197**
Swelling(s), arytenoid, 150, **151**
   cerebellar, 396, **397, 398**
   facial, 140, **141**
   genital, 266, **267,** 278, **279**
   labioscrotal, 266, **267, 288**
   lingual, of tongue, 138, **139**
   maxillary, 140, **141**
Sylvian fissure, 412, **413**
   nasal, **131,** 148, **149**
   scrotal, 266, **267**
Sympathetic nervous system, 428, **429, 432**
   chains, 428, **429, 432**
   neuroblasts, 428
   organ plexus, 428, **432**
Sympathogonia, 462, **463**
Symphysis pubis, **179**
Sympodia, 188
Syncytiotrophoblast, 36, **37,** 38, **39,** 40, **41,** 94, **95,** 96, **97**
Syndactyly, 188, **189**
Syndrome(s),
   adrenogenital, 466
      as cause of female pseudohermaphroditism, 466
   Arnold-Chiari, 384, 426
   cri-du-chat, 80
   Cushing's, 466, **467**
   DiGeorge, 134
   Down's, 78, **81**
   Klinefelter's, 78, **81,** 280, **281**
   Klippel-Feil, 182
   Marfan's, 188, **189**
   Mohr's, 144
   Pierre Robin, 134
   respiratory distress, 156
   thalidomide, 481
   Treacher-Collins, 134
   triple-X, 78
   trisomy, 78, **81**
   Turner's, 78, **80,** 280, **281**
Synotia, **455**
Synovial joint, 172, **173**
Synovial membrane, 172
Syphilis, 481
System(s), arterio-capillary-venous, of fetus, 54, **55,** 320–31
   articular, 168–**89**
   autonomic, 428–33
   cardiovascular, 54, **55,** 292–355
   chromaffin, 462, **463**
   development of, by weeks, Appendix I, 471–77
   digestive, 208–**35**

excretory, 238–53
genital, 256–**89**
Haversian, 168, **171**
integumentary, 192
lymphatic, 332–**35**
muscular, 162–**65**
nervous, 358–**467**
  parasympathetic, 430, **431, 433**
  sympathetic, 428, **429, 432**
olfactory, 434, **435**
omphalomesenteric, 326, **327**
paraganglionic, 462, **463**
portal, of hypophysis, 458, **459**
primitive vascular, **297**
reproductive, 256–**89**
respiratory, 148–**59**
skeletal, 168–**89**
umbilicoallantoic, 326, **327**
urinary, 238–**53**
venous, 326–**31**
vestibulocochlear, 444–**53**
visual, 436–**41**

Tag(s), auricular, 454, **455**
Tail, horse's. *See* Cauda equina
Tail bud, disappearance of, 60, 473
Tailfold, of embryo, 64, **65**
Talipes equinovarus, 188, **189**
Taste buds, 138, **139**
Tectorial membrane, **449**, 450, **451**
Teeth, 202, **203**, 204, **205**
  abnormalities of, 204, **205**
  deciduous, 202, **203**
  eruption, 204, **205**
  incisor, 202, **203**
  permanent, 202, 204, **205**
Tegmentum, **401**
Tela choroidea, 392, **393**
Telencephalon, 362, **363, 364, 365,** 408, **409,** 410, **411**
  phylogenesis of, 408, **409**
Temporal bone, 178, **179, 181**
Tendon(s), central, of diaphragm, 122, **125**
Tendon reflex, 485
Tensor tympani, 452, **453**
Teratogens, 82, 480, 481, 482
  chemicals, environmental, as, 82, 422, 481, 482
  congenital malformations caused by, 82, **83,** 422, 481, 482
  critical periods, of embryonic sensitivity to, **83,** 480
    of fetal sensitivity to, 480
  definition of, 82
  drugs as, 82, 481, 482
  exposure of embryo to, 82, **83,** 422, 480
  infectious agents as, 82, **83,** 422, 481
  therapeutic radiation as, 82, 481

Terminal hairs, 196
Terminal sac, of respiratory bronchiole, 154, **155**
Terminal sulcus, of tongue, 138, **139**
Tertiary stem villi, 56, **57,** 96, **97**
Testicular cord, 268, **269**
Testicular dysgenesis of Klinefelter, 280, **281**
Testicular feminization, 280, **281**
Testis(es), 4, **5,** 258, **259,** 288
  appendix, **259, 262, 263,** 264, **265**
  cords, 258, **259**
  descent of, 268, **269**
  ectopic, 268, **269,** 282
  fetal, 258, **259**
  gubernaculum, **259**
  interstitial tissue, **259**
  microscopy of, **6**
  undescended, 268, **269,** 282
Tetracycline, possible teratogenic effects of, 482
Tetralogy of Fallot, 352, **353**
Thalamus, 406, **407, 411**
Thalidomide, congenital malformations caused by, 82, **83,** 454
  teratogenic effects of, 82
Theca, externa, **15**
  interna, **15**
Thoracic duct(s), 332, **333**
  variations of, **355**
Thoracopagus, 74, **77,** 82
Thymus, 132, **133,** 335
  abnormalities of, 134, **135**
Thyroglossal duct, 136, **137**
  cysts of, 136
  remnants of, 136
  sinuses of, 136
Thyrohyoid arch, 128, **129**
Thyroid cartilage, 128
Thyroid drugs, teratogenic effects of, 482
Thyroid gland, 132, **133,** 136, **137**
  agenesis of, 172
  congenital malformations of, 136, **137**
  descent of, 136, **137**
  diverticulum of, 136, **137**
Thyroid tissue, accessory, 136, **137**
Thyroxine, 136
  in surfactant production, 156
Tissue(s), connective, embryonic, 62, **63,** 478
  hemopoietic, 292
  induced, embryonic, 68, **69,** 358
Toes, **61,** 70, **71,** 178, **179, 180**
  congenital malformations of, 184, **185**
Tomes' dentinal fibers, 202, **203**
Tongue, 132, **133,** 138, **139**
  congenital malformations of, 138
  root, 138, **139**
Tongue buds, 138, **139**
Tongue muscles, 162
"Tongue-tie," 138
Tonic neck reflex, 485
Tonsil(s), 132, **133,** 332, **335**

Tonsil(s) (*Cont.*)
  lingual, 332, **335**
  palatine, 132, 332, **335**
  pharyngeal, 332, **335**
  pharyngotympanic, 332, **335**
  tubal, 332, **335**
Tonsillar fossa, 132
Tooth. *See* Teeth
Tooth buds, 202, **203**
Tower skull, 182, **183**
Toxoplasma gondii, teratogenic effect of, 481
Trabeculae, arachnoid, 420, **421**
  carneae, 314, **317**
Trachea, 128, 150, **151**
  congenital malformations of, 158, **159**
Tracheoesophageal fistula, 158, **159**
Tracheoesophageal groove, 158
Tracheoesophageal septum, 150, **151**, 210
Tract, genital, 260, **261**, 264, **265**
  hypothalamohypophyseal, 458, **459**
  olfactory, **365**, 434, **435**
  pyramidal, 386
  tuberohypophyseal, **459**
Transposition of the great vessels, 348, **351**
Transverse colon, distal, 219, 222
Transverse folding (flexion), 64, **65, 66, 67**
Transverse sinus, of pericardial cavity, 302
Treacher-Collins syndrome, 134
Tricuspid atresia, 352, **354**
Trigeminal nerve, 128, **130**, **131**, **390**, 392, **393**, 402, **404, 405**
Trigone, of urinary bladder, 248, **249**
Trilaminar embryo, 44, **45**
Trilaminar embryonic disk, 44, **45**
Triple-X syndrome, 78
Triplets, 74
Trisomy, 18, 78, **81**
  of autosomes, 18, 78, **81**
  definition of, 18, 78
  of sex chromosomes, 78
Trisomy syndromes, 78, **81**
Trochlear nerves, 388, **389, 390**, 400, **401**, 402, **404, 405**
Trophoblast, 36, **37**, 56, **57**, 84, **85**, 94, **95**
  expansion of, 56, **57**, 84, **85**, 94, **95**
  during implantation, 36, **37**, 56, **57**, 84, **85**, 94, **95**
Truncoconal ridges, 312, **313**
Truncus arteriosus, 300, **301**, 312, **313**
  fate of, 312, **313**
  partitioning of, 312, **313**
    abnormalities due to, 348, **350**
  septum, 312, **313**
  swellings, 312, **313**
  unequal division of, 348, **350**
Trunk(s), pulmonary, 312, **313**, 324, **325**
Tubal pregnancy, 30, **31**
Tubal tonsil, 332, **335**, 452
Tube(s), auditory, 132, **133**, **445**, 452, **453**
  digestive, 208–35

  endocardial, 296, **297**
  eustachian, 132, **133, 445**, 452, **453**
  fallopian, 272, **273, 274, 275**, 276, **277**
  heart, 296, **297**
    embryonic, 296, **297**
    fusion of, 296, **297**
  laryngotracheal, 150, **151**
  neural, 46, **47**, 360, **361**, 362, **363, 364**, 365
  Pflügers, 270, **271**
  pharyngotympanic, 132, **133, 445**, 452, **453**
  uterine, 272, **273, 274, 275**, 276, **277**
Tubercle(s), genital, 266, **267**, 277, 278, **279**
  Müllerian, 260, **262**, 272, 274
Tuberculum impar, **131**, 138, **139**
Tubotympanic recess, 132, 444, **445**, 452, **453**
Tubular lung bud, 150, **151**
Tubule(s), collecting, renal, 244, **245**, 288
  distal, convoluted, 244, **245**
  mesonephric, 240, **241, 262**, 264, **288**
  metanephric, 244, **245**
  Pflüger's, 270, **271**
  pronephric, **239**
  proximal convoluted, 244, **245**
  renal collecting, 244, **245**
  seminiferous, 4, **5**, 258, **259, 262**, 288
  straight, 4, **5**
  uriniferous, **247**
Tubuli recti, of testis, 258, **259**, 262
Tunica, albuginea, 258, **259**, 262, **263**, 270, **271**
  vaginalis, 268, **269**
Turbinates, 148, **149**
Turner's syndrome, 78, **80**, 280, **281**
Twins, 74, **75**
  conjoined, 74, **77**
  diovular, 74, **75**
  dizygotic, 74, **75**
  fraternal, 74
  heredity and, 74
  identical, 74, **75**, **76**
  monovular, 74, **76**
  monozygotic, 74, **76**
  occurrence of, 74
  Siamese, 74, **77**
  unequal, **77**
Tympanic antrum, 252
Tympanic cavity, 132, 444, **445**, 452, **453**
Tympanic membrane, 132, **133**, 444, **445**, 452, **453**
Tympanic ring, 444

Ultimobranchial body, 132, **133**
Umbilical arteries, **103**, 318, **319**, 320, **321**, 338
  derivatives of, 318, **319**, 320, **321**, 338
  fate of, 320, **321**, 338
Umbilical cord, 56, **85, 87**, 88, **89**
  covering of, **85**, 88, **89**, 94, **95**
  fistula, 232
  knots in, 88
  looping of, 88

primitive, **85**
vessels of, 88, **89, 103**
Umbilical hernia, congenital, 228
Umbilical ligament, median, 86
Umbilical veins, **103**, 308, **309**, 338
    derivatives of, 338
    fate of, 338
    transformation of, 338
Umbilicoallantoic vascular network, 318, **319**, 326, **327**
Uncinate process, 216, **217**
Uncus, **415**
Unicornate uterus, **287**
Urachal cyst, 252
    fistula, 252
    sinus, 252
Urachus, 86, 348, **349**
    congenital malformations of, 252, **253**
Ureter(s), 244, **288**
    congenital malformations of, 250, **251**
Ureteral orifices, ectopic, 250
Ureteric buds, 244, **245**, 248, **249**
Ureteric pelvis, **288**
Ureterohydronephrosis, 252
Urethra, 4, **5**, 248, **249**, 278, **279**, **288**
    female, 276, **277**, 278, **279**, **288**
    glandular, 266, **267**
    male, 4, **5**, **262**, **263**, 264, **265**, 266, **267**, **288**
    membranous, **5**, **262**, **263**, 264, **265**
    penile, 4, **51** **262**, **263**, 264, **265**, 266, **267**
    prostatic, **5**, **262**, **263**, 264, **265**
Urethral folds, 266, **267**
    glands, 264, **265**
    groove, 266, **267**
    meatus, definitive external, 266, **267**, 276, **277**
    plate, 266, **267**
Urinary bladder, 248, **249**, **288**
    extrophy of, 252
Urinary expulsion system, **433**
Urinary retention mechanism, **432**
Urinary system, 238–53
    congenital malformations of, 250–**53**
Urine, formation of, prenatal, 244
Uriniferous tubules, **247**
Urogenital connections, 256, **257**
Urogenital cord, 240, **243**, 260
Urogenital folds, 266, **267**, **288**
    groove, 266, **267**
Urogenital membrane, 62, 222, **223**, 276
Urogenital mesentery, 240, **243**
    ridge, 238, **239**
    sinus, definitive, 248, **249**, 264, **265**, 274, 276
        membrane, 248, **249**, 264, 266, **267**, 276, **277**
        pelvic part, 248, **249**, 264, **265**, 274, 276
        phallic part, 248, **249**, 264, **265**, 274, 276
        primitive, 248, **249**, 264, **265**, 274
Urogenital system, 238–**53**
    embryonic, adult derivatives and vestigial remains, 256, 258, 260, **262**, **263**

genital or reproductive system of, 256, **257**, 258, **259**, 260, **261**, **262**, **263**
    congenital malformations of, 282–**89**
urinary or excretory system of, 238–**53**, **262**, **263**
    congenital malformations of, 250–**53**
Urorectal septum, 222, **223**, 248
Uterine canal, 20, **21**, 272, **273**, 274
Uterine cycle. *See* Menstrual cycle
Uterine tube, 22, **23**, 272, **273**, 274, **275**, 276, **277**, **288**
Uteroplacental circulation, primitive, 40, **41**
Uterorectal pouch (of Douglas), **21**, 272, **273**, 274, **275**
Uterovaginal primordium (canal), 260, 272, **273**, 276
Uterovesical pouch, **21**, 272, **273**, 274, **275**
Uterus, 20, **21**, 23, 90, **91**, 272, **273**, 274, **275**, **288**
    agenesis, 286
    atresia, 286, **287**
    congenital malformations of, 286, **287**
    growth of, during pregnancy, 90, **91**
    structure of, 20, **21**, 272, **273**, 274
Uterus bicornis, 286, **287**
Uterus cordiformis, 286, **287**
Uterus didelphys, 286, **287**
Uterus parvicollis, 286
Utricle, of inner ear, 446, **447**
    prostatic, **262**, **288**
Utricular part, of otocyst, **447**
Uvula, 142, **143**

Vagina, **21**, 272, **273**, 274, 276, **277**, 278, **279**, **288**
    agenesis of, 286, **287**
    septate, **285**
Vaginal fornices, 272, **274**
    plate, 272, **274**
    process, 268, **269**
Vagus nerve, 128, **130**, **131**, 388, **389**, **390**, **391**, 402, **404**, **405**, 430, **431**, **433**
    inferior laryngeal branch, 128, **130**
    nucleus, dorsal, 388, **391**
    recurrent laryngeal branch, **130**
    superior laryngeal branch, 128, **130**
Vallate papillae, of tongue, 138, **139**
Vallecula, **139**
Valve(s), aortic, stenosis of, 352, **354**
    atrioventricular, 304, **305**, 314, **316**, **317**
    bicuspid, 314, **316**, **317**
    cardiac, 314, **315**, **316**, **317**
    coronary sinus, **307**, 310, **311**
    of foramen ovale, **307**
    of inferior vena cava, **307**, 310, **311**
    mitral, 314, **316**, **317**
    semilunar, 314, **315**, **316**
    sinoatrial, 302

Valve(s) (*Cont.*)
    venous, 310, **311**
Varicella zoster virus, possible teratogenic effects of, 481
Vas deferens, 4, **5**, 258, **259, 262, 263**, 264, **265**
Vasa efferentia, 4, **5**, 258, **259**, **262**
Vascular islets, villus, 96, **97**
Vascular networks, primitive, 318, **319**
Vascularization, extraembryonic, 294, **295**
    intraembryonic, 54, **55**, 294, **295**
Vasopressin, **459**, 460, **461**
Vater's ampulla, 216, **217**
Vein(s), azygos, 308, **309**, 328, **329**
    brachiocephalic, 308, **309**, 326, **327**
    cardiac development and, 308, **309**
    cardinal, anterior, 308, **309**, 318, **319**, 326, **327**
        common, 308, **309**, 318, 326, **327**
        middle, **311**
        posterior, 308, **309**, 318, **319**, 326, **327**
    cerebral, **327**
    congenital malformations of, 326
    endometrial, **101**
    gonadal, 328, **329**
    hemiazygos, 328, **329**
    hepatic, 308
    hypophyseal, 458, **459**
    iliac, common, 328, **329**
    inferior vena cava, 308, **309**, 328, **329**
    intercostal, left superior, 326, **327**
    internal, of Wolffian body, 328
    jugular, 326, **327**
    mesenteric, superior, 330, **331**
    oblique, of left atrium, 308, **309**
    omphalomesenteric, 326, **327**, 330, **331**
    ovarian, **329**
    portal, 308, 330, **331**
    pulmonary, 152, 330, **331**
    renal, left, 328, **329**
    spermatic, **329**
    subcardinal, 328, **329**
    subclavian, 326, **327**
    superior vena cava, 308, **309**, 326, **327**
    supracardinal, **327**, 328, **329**
    suprarenals, **329**
    umbilical, **89**, 308, **309**, 318, **319**, 338
        derivatives of, 318, **319**, 338
    umbilicoallantoic, 326, 330, **331**
        fate of, 330, **331**
        transformation of, 330, **331**
    vitelline, 308, **309**, 326, **327**, 330, **331**
Vena cava
    inferior, absence, 328, **343**
        double, 328, **343**
        in fetal circulation, 308, **309**, 328, **329**
    superior, 308, **309**, 326, **327**
        abnormalities of, 326, **343**
Venous return, congenital malformations of, 326, 328, **343**
Venous shunts, **331**
Venous valves, 310, **311**
Ventral mesentery, 118, **119, 120, 121, 213**, 214, **215**
Ventricle(s), fourth, 362, **364, 365**, 386, **387, 389, 390, 391**
    of heart, 300, **301**, 314, **316, 317**
        primitive, partitioning of, 300, **301**
    laryngeal, 150
    lateral, 362, **364, 365, 407**, 410, **411**, 412, **413**
    septation of, **307**, 312, **313**
    third, 362, **364, 365, 407, 411**
Ventricular, septal defects, 348, **349, 350**
    septum, 312, **313**
Ventricular wall, 314, **316, 317**
Vermis, of cerebellum, 396, **397, 398, 399**
Vernix caseosa, 70, 192
Vertebra(e), ossification of, 174, **175, 176**
Vertebral arch, 174, **175**
Vertebral artery, 420, **421**
Vertebral body, 174, **175, 176**
Vertebral column, 174, **175, 176**
    abnormalities of, 182, **183**
    fused, 182, **183**
Vertebrate nervous system, 366, **369**
Verumontanum, 264, **265**
Vesicle(s), auditory (otic), 446, **447**
    brain, 386, **387**
    ectodermal, **48**
    entodermal, **48**
    fifth, 386, **387**
    fourth, 386, **387**
    lens, 436, **437**, 440, **441**
    mesonephric, 240, **241**
    nephrotomal, 238, **239**
    optic, 436, **437**
    otic, 446, **447**
    seminal, 4, **5**, 264, **265**
    third, 386, **387**
    trigone, 248, **249, 262**
Vessel(s), blood, embryonic. *See also* Artery(ies); Vein(s)
    great, complete transposition of, 348, **351**
    congenital abnormalities of, 348, **351**
    lymphatic, 332–**35**
    placental, **55, 89, 95, 99, 101**, 102, **103**
    renal, multiple, 252, **253**
    umbilical, 88, **89**
    umbilicoallantoic, 318, **319**, 326, **327**
Vestibular fold, of larynx, 150
Vestibular ganglion, **447**
Vestibular glands, 272, **274**
Vestibular membrane, **449**, 450
Vestibule, aortic, 312
    of inner ear, 448, **449**
    vaginal, **274**, 276, **277**
Vestibulocochlear system, 444–53
    congenital malformations of, 454, **455**
Villi, 56, **57**, 94, **95**, 96, **97**, **99**
    anchoring, 56, **57**, 96, **97**, **99**
    arachnoid, 420, **421**

chorionic, 56, **57**, 94, **95**, 96, **97**, **99**
floating, 56, **57**, 96, **97**, **99**
primary, 40, **41**, 56, **57**, 96, **97**
secondary, 56, **57**, 96, **97**
stem, 40, **41**, 56, **57**, 96, **97**
tertiary, 56, **57**, 96, **97**
tree, 56, **57**, 96, **97**
Villus hairs, 196
Virilization, 280
   abnormal, 280
Virus(es), congenital malformations caused by, 82, **83**, 422
   placental transfer, 100, **101**
   teratogenic effects, 481
Viscera, abdominal, eventration of, 228
Visceral pericardium, 296, **299**
   pleura, 154, **155**
Viscerocranium, 178, **179**, **181**
Visual organs, 436–**41**
   congenital malformations of, 442, **443**
Vitelline cyst, 232, **233**
   duct, **65**, **219**
      remnants of, 232
   stage, of circulatory system, 292, **293**
   veins, 308, **309**, 310, 326, **327**
   vessels, 310, **311**
Vitreous body, **437**, **438**, 440, **441**
Vocal cords, 150
Vocal folds, 150
Volvulus, intestinal, 228, **229**, 232, **233**
   of midgut, 228, **229**, 232, **233**

Waste products, placental transfer of, 100, **101**
Web(s), laryngeal, 158
Webbed digits, 188
Weight, fetal, in estimating fetal age, 70, **71**, 72, **73**
Wharton's jelly, 88
White matter of spinal cord, 374, 376, **377**
Willis, circle of, 420, **421**
Window(s), of middle ear, oval, 452, **453**
   round, 452, **453**

Winslow, foramen of, 118, 212, **213**
Wirsung's duct, 216, **217**
"Witches' milk," 200
Wolffian body, 240, **241**, **242**, **243**
   duct, 240, **241**, 258, **259**, **262**, **263**, 264, **265**, 276, **277**, **288**
Worms, primitive, nervous system, 366, **367**

X chromosomes, 260
   effect on sex, determination of, 260
X-rays, diagnostic, possible teratogenic effects of, 481

Y chromosomes, 260
   effect on sex determination of, 260
Yolk sac, 40, **41**, **45**, **52**, 84, **85**, 86, **87**, 208, **209**
   capillaries, 292, 294, **295**
   fate of, 84, **85**, 86, **87**, 208, **209**
   primitive (primary), 38, **39**, 84, **85**, 86, **87**, 208, **209**
   role, in blood formation, 86
   secondary, 40, **41**, **52**, 84, **85**, 86, **87**
   significance of, 84, **85**, 86, **87**
   stalk of, 88, **89**

Zona fasciculata, 464, **465**
Zona glomerulosa, 464, **465**
Zona pellucida, 14, **15**, **17**, **25**
   definition of, 14
   during ovulation, 24, **25**
   penetration of, by sperm, 33, **34**
Zona reticularis, 464, **465**
Zone(s), of urogenital sinus, 248, **249**
Zonular fibers, 436, **437**, **438**
Zuckerkandl, organs of, 462
Zygote, cleavage of, 36, **37**
   definition of, 36, **37**
   formation of, 36, **37**
Zygotene stage, 10